Output parameters for various open collector gates

Family	Example	V_{OL} (V)	I_{OL} (A)
Standard TTL	7405	0.4	0.016
Schottky TTL	74S05	0.5	0.02
Low power Schottky TTL	74LS05	0.5	0.008
High speed CMOS	74HC05	0.33	0.004
High voltage output TTL	7406	0.7	0.040
High voltage output TTL	7407	0.7	0.04
Silicon monolithic IC	75492	0.9	0.25
Silicon monolithic IC	75451–75454	0.5	0.3
Darlington switch	ULN-2074	1.4	1.25
MOSFET	IRF-540	varies	28

Output parameters for various open emitter gates

Family	Example	V_{CE} (V)	I_{CE} (A)
Silicon monolithic IC	75491	0.9	0.05
Darlington switch	ULN-2074	1.4	1.25
MOSFET	IRF-540	varies	28

MC9S12C32 interrupt vectors and CodeWarrior numbers

Address	Number	Interrupt Source
$FFFE	0	Reset
$FFFC	1	COP Clock Monitor Fail Reset
$FFFA	2	COP Failure Reset
$FFF8	3	Unimplemented Instruction Trap
$FFF6	4	SWI
$FFF4	5	XIRQ
$FFF2	6	IRQ
$FFF0	7	Real Time Interrupt, RTIF
$FFEE	8	Timer Channel 0, C0F
$FFEC	9	Timer Channel 1, C1F
$FFEA	10	Timer Channel 2, C2F
$FFE8	11	Timer Channel 3, C3F
$FFE6	12	Timer Channel 4, C4F
$FFE4	13	Timer Channel 5, C5F
$FFE2	14	Timer Channel 6, C6F
$FFE0	15	Timer Channel 7, C7F
$FFDE	16	Timer Overflow, TOF
$FFDC	17	Pulse Acc. Overflow, PAOVF
$FFDA	18	Pulse Acc. Input Edge, PAIF
$FFD8	19	SPI, SPIF or SPTEF
$FFD6	20	SCI, TDRE TC RDRF IDLE
$FFD2	22	ATD Sequence Complete, ASCIF
$FFCE	24	Key Wakeup J, PIFJ [7:6]
$FFB6	36	CAN wakeup
$FFB4	37	CAN errors
$FFB2	38	CAN receive
$FFB0	39	CAN transmit
$FF8E	56	Key Wakeup P, PIFP [7:0]

Parameters of typical transistors used by microcomputer to source or sink current

Type	NPN	PNP	Package	$V_{be(SAT)}$	$V_{ce(SAT)}$	h_{fe} min/max	I_c
General purpose	2N3904	2N3906	TO-92	0.85 V	0.2 V	100	10 mA
General purpose	PN2222	PN2907	TO-92	1.2 V	0.3 V	100	150 mA
General purpose	2N2222	2N2907	TO-18	1.2 V	0.3 V	100/300	150 mA
Power transistor	TIP29A	TIP30A	TO-220	1.3 V	0.7 V	15/75	1 A
Power transistor	TIP31A	TIP32A	TO-220	1.8 V	1.2 V	25/50	3 A
Power transistor	TIP41A	TIP42A	TO-220	2.0 V	1.5 V	15/75	3 A
Power Darlington	TIP120	TIP125	TO-220	2.4 V	2.0 V	1000 min	3 A

General specification of various types of resistor components[1]

Type	Range	Tolerance	Temperature coef	Max power
Carbon composition	1 Ω to 22 MΩ	5 to 20 %	0.1 %/°C	2 W
Wire-wound	1 Ω to 100 kΩ	>0.0005 %	0.0005 %/°C	200 W
Metal film	0.1 Ω to 10^{10} Ω	>0.005 %	0.0001 %/°C	1 W
Carbon film	10 Ω to 100 MΩ	>0.5 %	0.05 %/°C	2 W

[1] Wolf and Smith, *Student Reference Manual*, Prentice Hall, pg. 272, 1990.

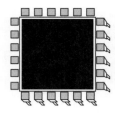

Embedded Microcomputer Systems

Real Time Interfacing
Second Edition

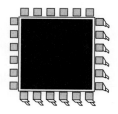

Embedded Microcomputer Systems

Real Time Interfacing
Second Edition

Jonathan W. Valvano
University of Texas at Austin

Australia • Canada • Mexico • Singapore • Spain • United Kingdom • United States

Embedded Microcomputer Systems: Real Time Interfacing, 2nd Edition
by Jonathan W. Valvano

Associate Vice-President and Editorial Director:
Evelyn Veitch

Publisher:
Chris Carson

Developmental Editor:
Kamilah Reid Burrell/Hilda Gowans

Permissions Coordinator:
Vicki Gould

Production Services:
RPK Editorial Services

Copy Editor:
Harlan James

Proofreader:
Erin Wagner

Indexer:
RPK Editorial Services

Production Manager:
Renate McCloy

Creative Director:
Angela Cluer

Interior Design:
John Edeen

Cover Design:
Andrew Adams

Compositor:
International Typesetting and Composition

Printer:
Transcontinental Printing

COPYRIGHT © 2007 by Nelson, a division of Thomson Canada Limited.

Printed and bound in Canada
2 3 4 07

For more information contact Nelson, 1120 Birchmount Road, Toronto, Ontario, Canada, M1K 5G4. Or you can visit our Internet site at http://www.nelson.com

Library of Congress Control Number: 2005938475
ISBN-13: 978-0-534-55162-9
ISBN-10: 0-534-55162-9

North America
Nelson
1120 Birchmount Road
Toronto, Ontario M1K 5G4
Canada

Asia
Thomson Learning
5 Shenton Way #01-01
UIC Building
Singapore 068808

Australia/New Zealand
Thomson Learning
102 Dodds Street
Southbank, Victoria
Australia 3006

Europe/Middle East/Africa
Thomson Learning
High Holborn House
50/51 Bedford Row
London WC1R 4LR
United Kingdom

Latin America
Thomson Learning
Seneca, 53
Colonia Polanco
11560 Mexico D.F.
Mexico

Spain
Paraninfo
Calle/Magallanes, 25
28015 Madrid, Spain

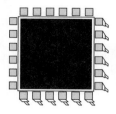

Preface

Embedded computer systems, which are electronic systems that include a microcomputer to perform a specific dedicated application, are ubiquitous. Every week millions of tiny computer chips come pouring out of factories like Freescale and Mitsubishi and find their way into our everyday products. Our global economy, food production, transportation system, military defense, communication systems, and even quality of life depend on the efficiency and effectiveness of these embedded systems. As electrical and computer engineers we play a major role in all phases of this effort: planning, design, analysis, manufacturing, and marketing.

This book is unique in several ways. Like any good textbook, it strives to expose underlying concepts that can be learned today and applied later in practice. The difference lies in the details. You will find that this book is rich with detailed case studies that illustrate the basic concepts. After all, engineers do not simply develop theories but rather continue all the way to an actual device. During my years of teaching I have found that the combination of concepts and examples is an effective method of educating student engineers. Even as a practicing engineer, I continue to study actual working examples whenever I am faced with learning new concepts.

Also unique to this book is its simulator, called *Test Execute and Simulate* (TExaS). This simulator, like all good applications, has an easy learning curve. It provides a self-contained approach to writing and testing microcomputer hardware and software. It differs from other simulators in two aspects. If enabled, the simulator shows you activity internal to the chip, like the address/data bus, the instruction register, and the effective address register. In this way the application is designed for the educational objectives of understanding how a computer works. On the other end of the spectrum, you have the ability to connect such external hardware devices as switches, keyboards, LEDs, LCDs, serial port devices, motors and analog circuits. Logic probes, voltmeters, oscilloscopes, and logic analyzers are used to observe the external hardware. The external devices together with the microcomputer allow us to learn about embedded systems. The simulator supports many of the I/O port functions of the microcomputers like interrupts, serial port, output compare, input capture, key wake up, timer overflow, and the ADC. You will find the simulator on the CD that accompanies this book. Get the CD out now, run the *Readme.exe,* install TExaS, and follow the tutorial example. In particular, double-click the **tut.uc** file in the MC9S12 subdirectory and run the tutorial on the simulator. If you are still having fun, then run the other four tutorials: **tut2.*** shows the simple serial I/O functions, **tut3.*** is an ADC data acquisition example, **tut4.*** shows interrupting serial I/O functions, and **tut5.*** is an interrupting square-wave generator. Although many programs are included in the application, these five give a broad overview of the capabilities of the simulator. The screen shot in Figure P.1 was obtained when running the **tut2.rtf** example. Notice these features in the figure: (1) address/data bus activity, (2) embedded figures in the source code, (3) external hardware, (4) voltmeters and logic probes, and (5) an oscilloscope.

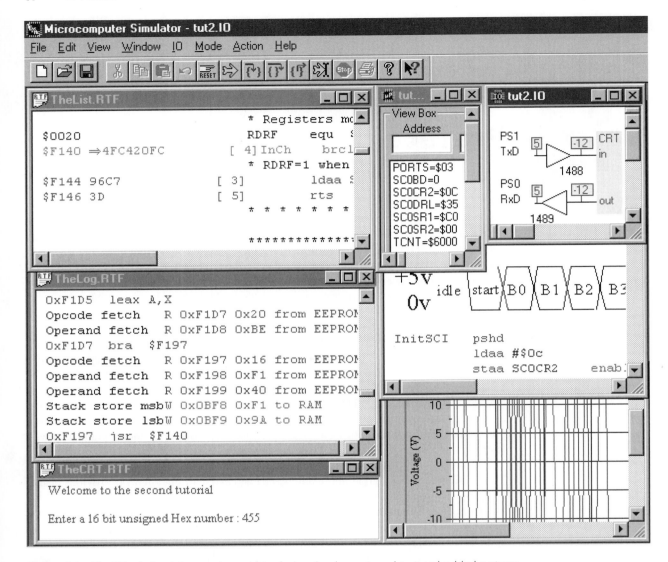

Figure P.1 The TExaS simulator can be used to design, implement, and test embedded systems.

P.1 General Objectives

This book constitutes an in-depth treatment of the design of embedded microcomputer systems. It includes the hardware aspects of interfacing, advanced software topics like interrupts, and a systems approach to typical embedded applications. The book is unique to other microcomputer books because of its in-depth treatment of both the software and hardware issues important for real-time embedded applications. The specific objectives of the book include the understanding of:

1. Advanced architecture
   - Timing diagrams
   - Memory and I/O device interfacing to the address/data bus
2. Interfacing external devices to the computer
   - Switches and keyboards
   - LED and LCD displays

■ DC and stepper motors
■ Amplifiers, analog filters, DAC and ADC
■ Synchronous and asynchronous serial ports
3. Advanced programming
■ Debugging of real time embedded systems
■ Interrupts and real time events
■ Signal generation and timing measurements
■ Threads and semaphores
4. Embedded applications
■ Data acquisition systems with digital filters
■ Linear and fuzzy logic control systems
■ Simple communication networks

P.2 Prerequisites

This book is intended for a junior- or senior-level course in microcomputer programming and interfacing. We assume the student has knowledge of:

■ Microcomputer programming
■ Digital logic (multiplexers, Karnaugh maps, tristate logic)
■ Data structures in C (queues, stacks, linked lists)
■ Test equipment like multimeters and oscilloscopes
■ Discrete analog electric circuits (resistors, capacitors, inductors, and transistors)

Although this book focuses on how to design embedded systems, extensive tutorials on the CD help students with the important issues of how to program in assembly language and how to program in C. A section later in this preface contains more information about what is on the CD.

P.3 Structure of the Book

We will use `this font` whenever displaying either assembly language or high-level code written in C. Usually, we present assembly language followed by C software for each case study. For the most part, software that is encapsulated in a box has equivalent examples for the 6811 and the 6812. The program number in the caption will assist you in finding the software on the CD. All programs in an entire chapter are grouped together into one file. For example, the assembly and C programs in Chapter 1 can be found on the CD as files Chap1.asm and Chap1.c in the corresponding MC6811 or MC6812 subdirectory. For each chapter there are four files, grouped by microcomputer (MC6811 or MC6812) and by language (assembly and C).

When presenting C code, sometimes the implementation for the different computers is so identical that only one version is given, but usually there are important differences. Usually C code is given for the MC68HC11 and MC68HC12. The C code in this book was written compiled and tested using the ImageCraft compilers (ICC11 for the 6811) and Metrowerks CodeWarrior for the 6812. We have been very successful transporting these software examples to other compilers with only minor syntax modifications required. The major differences between the compilers are the syntax for embedded assembly and defining interrupt handlers; we have tried using various free SmallC compilers, but the software in this book is too complex to be compiled using SmallC. The biggest limitation of SmallC is its lack of support for data structures. In other words, substantial editing would be required to use SmallC in a course based on this book. Gnu C compilers are available for the 6811 and

6812, but the object code generated by these compilers is not as efficient as code generated with the ImageCraft or Metrowerks compilers

We assume you have access to the Freescale programming reference guides for your particular microcomputer that show the details of each assembly instruction. You should also have the microcomputer technical reference manuals, which detail the I/O ports you will be using. For example, if you are using input capture to measure periods on the Freescale 9S12C32, you will find basic principles and examples in this book, but you need to access the 9S12C32 Technical Summary for a complete list of all the details. In other words, you should to use this book along with the manuals from Freescale. Although these reference manuals are available as pdf documents on the CD, it might be better to order physical documents from Freescale's literature center, or to download the latest version from the Freescale web site.

Although software development is a critical aspect of embedded system design, this book is not intended to serve as an introduction to C programming. Consequently, you will find it convenient to have a C programming reference available. Fortunately, located on the CD that accompanies this book, there is a small reference called *Developing Embedded Software in C using ICC11/ICC12/*Metrowerks written as an HTML document. Although this reference is not as complete a programming guide as some books on C, it is specific for writing embedded software for the 6811 and/or 6812.

In this book we will discuss programming style, and develop debugging strategies specific for embedded real time systems from both an assembly language and a C perspective. Due to the nature of single-chip computers (they are very slow and have very very little RAM when compared to today's desktop microcomputers) most of the available C compilers for the single-chip microcomputers used in this book do not support objects, or floating-point. Floating-point should be used only in situations where the range of values spans many orders of magnitude or where the range of values is unknown at software design time. Our experience is that numbers used in embedded systems usually have a narrow and known range, so integer math is sufficient. On the other hand, there is some interest for applying object-oriented approaches of C++ to embedded system design. A few illustrations of object-oriented design can be found in this book.

Chapters 1 and 2 can serve as an introduction to assembly language programming. Most embedded systems engineers agree that a working knowledge of assembly is necessary even when virtually all of our software is written in C. In specific, we believe we must know enough assembly language to be able to follow the assembly listing files generated by our compiler. This understanding of assembly language is vital when debugging, writing interrupt handlers, calculating real-time events, and considering reentrancy. Consequently, detailed information on assembly language programming is included on the CD. In particular you will find microcomputer data sheets, interactive on-line help as part of the TExaS simulator, and a short introduction to assembly language programming as an HTML document.

The electrical components used in this book span a wide range. Examples include the 2764 PROM, 1N914 diode, 2N2222 transistor, 7406 open collector TTL driver, 74LS74 LSTTL flip flop, 74HC573 high speed CMOS transparent latch, Max494 op amp, L293 interface driver, Hitachi 44780 LCD driver, IRF540 MOSFET, 6N139 optocoupler, Max232 driver, Dallas Semiconductor DS1620 temperature controller, and Analog Devices DAC8043 digital-to-analog converters. It is unrealistic for each student to have a personal library that contains the data books for all these devices. On the other hand it is appropriate for the company or university to establish a reference library that can be accessed by all the students. The circuit diagrams in this book usually include chip numbers and component values, but not pin numbers or circuit board layout information. Consequently, with the appropriate data sheets available, most circuits can be readily built.

P.4   How to Teach a Course Based on This Book

The first step in the design of any course is to create a list of educational objectives, along with an understanding of the topics taught in previous classes and the topics needed as prerequisites to subsequent classes. Some important topics like computer architecture and modular software design may be included in multiple courses. Armed with this list, the instructor (or department committee) searches for a book that covers most of the topics in an effective manner. It is unrealistic to teach the entire contents of this book in a single one-semester class, but the rationale in writing this book was to cover a wide range of possible classes. There are two approaches to selecting a subset of material from this book to cover. The first approach is to pick a microcomputer and programming language. For example, one could teach just assembly language on the 6811 or just C programming on the 6812. In this situation, one simply skips the other cases.

The other approach to selecting the appropriate subset is to pick and choose topics. For example, a junior-level laboratory class might introduce the student to microcomputer interfacing. This class might focus on interfacing techniques and may cover Chapters 1 to 4, 6 to 8 and a little bit of Chapters 11 to 13. For these students, the remaining parts of the book become a resource to them for projects later in school or on the job. Another possibility is a senior-level project laboratory class. The objectives of this class might focus on the systems aspect of real time embedded systems. In this situation, some microcomputer programming has been previously taught, so this course might cover the advanced interfacing techniques in Chapters 5, 9, 10, and 15 and the applications in Chapters 12 to 14. For these students, the first half of the book becomes a review, allowing them to properly integrate previous learned concepts to solve complex embedded system applications.

In most departments analog circuit design (for example, op amps and analog filters) is taught in separate classes, so Chapter 11 will be a review chapter. Specific and detailed information about analog circuit design was included in this book to emphasize the system integration issues when designing embedded systems. In other words, developing embedded systems does not rely solely on the tools of computer and software engineering but rather involves all of electrical, computer, and software engineering.

The next important decision to make is the organization of the student laboratory. As engineering educators we appreciate the importance of practical "hands on" experiences in the educational process. On the other hand, space, staff, and money constraints force all of us to compromise, doing the best we can. Consequently, we present three laboratory possibilities that range considerably in cost. Indeed, you may wish to mix two or more approaches into a hybrid simulated/physical laboratory configuration. We do believe that the role of simulation is becoming increasingly important as the race for technological superiority is run with shorter and shorter design cycle times. On the other hand, we should expose our students to all the phases of engineering design, including problem specification, conceptualization, simulation, construction, and analysis.

In the first laboratory configuration, we use the traditional approach to an interfacing laboratory. A physical microcomputer development board is made available for each laboratory group of two students. There are numerous possibilities here. Companies such as Technological Arts, Axiom, Wytec, and Shuan Shizu produce development systems. In addition to the microcomputer board, each group will need a power supply, a prototyping area to build external circuits, and the external I/O devices themselves. A number of shared development/debugging stations will also have to be configured. It is on these dedicated PC-compatible computers that the assembler or compiler is installed. If you develop software in assembly, then the TExaS simulator can be used to edit and assemble software. TExaS will create the standard S19 object code records for downloading. An early version of ICC11 can be used to create 6811 EVB code. In most cases, when programming in C, a current version of a C cross-compiler is greatly preferable. As mentioned earlier, the specific

C examples in this book were developed with ImageCraft's ICC11 and ICC12 (ImageCraft, 706 Colorado Ave. Suite 10-88, Palo Alto, CA 94303 *http://www.imagecraft.com*). The Metrowerks CodeWarrior with educational license is an excellent choice for developing 6812 software. Test equipment like an oscilloscope, a digital multimeter, and a signal generator are required at each station. Expensive equipment like logic analyzers and printers can be shared. Some mail-order companies sell used or surplus electronics that can be configured into laboratory experiments (for possibilities, see my web site *http://www.ece. utexas.edu/~valvano/book.HTML*). Many laboratory assignments are available using this traditional configuration. For universities that adopt this book, you will be allowed to download these assignments in Microsoft Word format, then rewrite, print out, and distribute to your students laboratory assignments based on these example laboratory assignments. Because of the detailed and specific nature of the laboratory setup, rewriting will certainly be necessary. The 9S12C32 Metrowerks CodeWarrior board from Technological Arts and the Metrowerks cross-compiler is the specific configuration presented in the example laboratory assignments, but the assignments are appropriate for most microcomputer development boards based on the 6811 or 6812.

The second laboratory configuration is based entirely on the TExaS simulator. Each book comes with a CD that allows the student to install the application on a single computer. Students, for the most part, work off campus and come to a teaching assistant station for help or laboratory grading. In this configuration you can either develop software in assembly using the TExaS assembler or develop C programs using the demonstration version of ICC11 or CodeWarrior. The simulator itself becomes the platform on which the laboratory assignments are developed and tested. As mentioned earlier, this freeware version of ICC11 only can be used to create 6811 EVB code. Fortunately, the simulator supports the 6811 EVB architecture. For examples of this ICC11/TExaS combination, run some of the demonstration examples in the ICC11 subdirectory of the TExaS application. The educational license of Metrowerks CodeWarrior supports code up to 12 K. Laboratory assignments are also available using the simulator. Again, for universities that adopt this book, you will be allowed to download these assignments in Microsoft Word format, then rewrite, print out, and distribute to your students laboratory assignments based on these example laboratory assignments.

The third configuration performs simulation of 6812 C programs. In this laboratory setup, a standard PC-compatible laboratory room can be used. TExaS and Metrowerks CodeWarrior are installed, and the students perform laboratory assignments in this central facility. Other than these two pieces of software, no additional setup costs are required. The available laboratory assignments mentioned in both the preceding paragraphs could be adapted for this configuration. The advantage of this approach is that C programming is performed on the newest microcomputer technology (for example, 6812) with only a modest expense.

The exercises at the end of each chapter can be used to supplement the laboratory assignments. In actuality, these exercises, for the most part, were collected from old quizzes and final examinations. Consequently, these exercises address the fundamental educational objectives of the chapter, without the overwhelming complexity of a regular laboratory assignment.

P.5 What's on the CD?

The Readme.exe is a 15-minute introductory tutorial about TExaS. This document does not need to be copied to your hard drive; you can simply watch the movie from the CD itself.

TExaS is a complete editor, assembler, and simulator for the 6811, and 6812 microcomputers. It simulates external hardware, I/O ports, interrupts, memory, and program execution. It is intended as a learning tool for embedded systems. This software is not freeware, but the purchase of the book entitles the owner to install one copy of the program. The **Texas** directory contains the installer, which you must do before using the application. Performing a typical installation, TExaS creates the following 12 subdirectories:

- MC6805   containing 6805 assembly examples
- MC6808   containing 6808 assembly examples
- MC6811   containing 6811 assembly examples
- MC6812   containing MC68HC812A4 assembly examples
- MC9S12   containing 9S12C32 assembly examples
- 9S12C32   containing 9S12C32 CodeWarrior C examples
- ICC11   containing 6811 C examples using the freeware ICC11 compiler
- ICC11a   containing 6811 ICC11 C examples
- ICC12   containing MC68HC812A4 ICC12 C examples
- GCC12   containing MC68HC812A4 C examples using the GCC12 compiler
- 6811 Manuals   containing 6811 datasheets
- 6812 Manuals   containing 6812 datasheets

The Custom installation allows you to pick and choose among these 12 choices. The subdirectory ICC11 also contains the ImageCraft Freeware compiler. You can run the existing ICC12 examples, but to edit and recompile you will need the commercial C compiler. The Metrowerks CodeWarrior or the Gnu GCC12 require separate installations.

The **PDF** directory contains many data sheets in Adobe pdf format. This information does not need to be copied to your hard drive; you can simply read the data sheets from the CD itself. In particular there are data sheets for microcomputers, digital logic, memory chips, op amps, ADCs, DACs, timer chips, and interface chips. If you wish to download more current data sheets, or search for other devices, we have a page of web links on the site *http://www.ece.utexas.edu/~valvano/book.HTML.*

The **example** directory contains software from the book. For example, all the assembly language programs from Chapter 1 can be found in the file "Chap.1.asm." Similarly, all the C language programs from Chapter 1 can be found in the file "Chap1.c." The versions for the various microcomputers are located in the corresponding subdirectories MC6811, and MC6812. Not all programs have versions for each microcomputer.

The **assembly** directory contains an HTML document describing how to program in assembly for embedded systems using the TExaS application. This document does not need to be copied to your hard drive; you can simply read the HTML document from the CD itself. (Note also that the TExaS application itself contains a lot of information about assembly language development as part of its on-line help).

The **embed** directory contains an HTML document describing how to program in C for embedded systems. This document does not need to be copied to your hard drive; you can simply read the HTML document from the CD itself.

The **lab** directory contains software that could be used in a laboratory setting.

The **Metrowerks** directory contains an educational version of the 6812 C compiler. This limited version can be used to develop small programs that are less than 12 K bytes of object code. This application must be installed before it can be used.

P.6   Acknowledgments

Many shared experiences contributed to the development of this book. First, I would like to acknowledge the many excellent teaching assistants I have had the pleasure of working with. Some of these hard-working, underpaid warriors include Pankaj Bishnoi, Rajeev Sethia, Adson da Rocha, Bao Hua, Raj Randeri, Santosh Jodh, Naresh Bhavaraju, Ashutosh Kulkarni, Bryan Stiles, V. Krishnamurthy, Paul Johnson, Craig Kochis, Sean Askew, George Panayi, Jeehyun Kim, Vikram Godbole, Andres Zambrano, Ann Meyer, Hyunjin

Shin, Anand Rajan, Anil Kottam, Chia-ling Wei, Jignesh Shah, Icaro Santos, David Altman, Nachiket Kharalkar, Robin Tsang, Byung Geun Jun, John Porterfield, Daniel Fernandez, and James Fu. I dreamed of writing this book the first time I taught microcomputer interfacing on the old Motorola MC6809 in 1981. Over the intervening years my teaching assistants have contributed greatly to the contents of this book, particularly to its laboratory assignments. In a similar manner, my students have recharged my energy each semester with their enthusiasm, dedication, and quest for knowledge.

I would also like to thank the reviewers who provided such excellent feedback including N. Alexandridis, George Washington University; David W. Capson, McMaster University; Lee D. Coraor, The Pennsylvania State University; Subra Ganesan, Oakland University; Voicu Groza, University of Ottawa; and William R. Murray, California Polytechnic State University, San Luis Obispo.

Second, I appreciate the patience and expertise of my fellow faculty members at the University of Texas at Austin. From a personal perspective, Dr. John Pearce has provided much needed encouragement and support throughout my career. Also, Dr. John Cogdell and Dr. Francis Bostick helped me with analog circuit design, and Dr. Baxter Womack and Dr. Robert Flake provided good information about control systems. The book and accompanying software include many finite-state machines derived from the digital logic examples explained to me by Dr. Charles Roth. I continue to appreciate the encouragement and support of Dr. G. Jack Lipovski. An outside observer might conclude that Dr. Jack and I enjoy taking opposite sides of every issue. In actuality, this friendly competition makes us organize our otherwise erratic thoughts, and in the process everything we do is the better for it.

Last, I appreciate the valuable lessons of character and commitment taught to me by my grandparents and parents. Most significantly, I acknowledge the love, patience, and support of my entire family, especially my wife Barbara and my children, Ben, Dan, and Liz.

Good luck! JONATHAN W. VALVANO

Contents

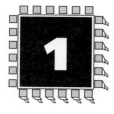

1 Microcomputer-Based Systems

Chapter 1 objectives are to introduce:

❏ Embedded systems
❏ Practical aspects of digital logic
❏ Architecture of the Freescale MC68HC711E9
❏ Architecture of the Freescale MC9S12C32
❏ Parallel port input/output operations

The overall objective of this book is to teach the design of embedded systems. It is an effective approach to learn new techniques by doing them. But, the dilemma in teaching a laboratory-based topic such as embedded systems is that there is a tremendous volume of details that first must be learned before microcomputer hardware and software systems can be designed. The approach taken in this book is to learn by doing, starting with very simple problems and building up to more complex systems later in the book.

We will begin with a short section introducing some terminology and the basic components of a computer system. In order to understand the context of our designs, we will overview the general characteristics of embedded systems. It is in these discussions that we develop a feel for the range of possible embedded applications. Because courses taught using this book typically have a lab component, we will review some practical aspects of digital logic. We then introduce the specific architectures of the Freescale MC68HC711E9 and MC9S12C32. For more detailed information concerning your specific microcomputer, refer to the respective Freescale manual. Data sheets for the devices we will use can be found in PDF format on the CD that accompanies this book. At the end of the chapter, we will discuss prototyping methods to build embedded systems, and present a simple example with binary inputs and outputs.

Even though we will design systems based specifically on the MC68HC711E9 and MC9S12C32 devices, these solutions can, with little effort, be implemented on other versions of the 6811 and 6812. If you are studying just one of these microcontrollers, you can skip the sections that apply to the other device. If your overall goal is to develop assembly language software, then the C code can serve to clarify the software algorithms. If your overall goal is to develop C code, then I strongly advise you to learn a little assembly language so that you can understand the machine code that your compiler generates. From this understanding, you can evaluate, debug, and optimize your system.

1.1   Computer Architecture

A *computer* system combines a processor, random access memory (RAM), read only memory (ROM), and input/output (I/O) ports, as shown in Figure 1.1. *Software* is an ordered sequence of very specific instructions that are stored in memory, defining exactly what and when certain tasks are to be performed. The *processor* executes the software by retrieving and interpreting these instructions one at a time. A *microprocessor* is a small processor, where small refers to size (i.e., it fits in your hand) and not computational ability. For example, the Intel Pentium and the PowerPC are microprocessors. A *microcomputer* is a small computer, where again small refers to size (i.e., you can carry it) and not computational ability. For example, a desktop PC is a microcomputer. A very small microcomputer, called a *microcontroller*, contains all the components of a computer (processor, memory, I/O) on a single chip. The Freescale MC68HC711E9 and MC9S12C32 used in this book are microcontrollers. Because a microcomputer is a small computer, this term can be confusing because it is used to describe a wide range of systems from an 8-bit 6811 running at 2 MHz with 512 bytes of memory to a personal computer with a state-of-the-art 64-bit processor running at multi-GHz speeds having terabytes of storage.

Figure 1.1
The basic components of a computer system include processor, memory, and I/O.

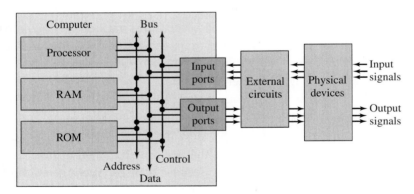

The computer can store information in *RAM* by writing to it, or it can retrieve previously stored data by reading from it. Most RAMs are *volatile*, meaning if power is interrupted and restored, the information in the RAM is lost. Information is programmed into *ROM* using techniques more complicated than writing to RAM. On the other hand, retrieving data from a ROM is identical to retrieving data from RAM. ROMs are *nonvolatile*, meaning if power is interrupted and restored, the information in the ROM is retained. Some ROMs are programmed at the factory and can never be changed. A Programmable ROM (*PROM*) can be erased and reprogrammed by the user, but the erase/program sequence is typically 10000 times slower than the time to write data into a RAM. Some PROMs are erased with ultraviolet light and programmed with voltages, whereas electrically erasable PROMs (*EEPROM*) are both erased and programmed with voltages. For most of the systems in this book, we will store instructions and constants in ROM and place variables and temporary data in RAM.

Checkpoint 1.1: What are the differences between a microcomputer, a microprocessor, and a microcontroller?

The external devices attached to the computer provide functionality for the system. An *input port* is hardware on the computer that allows information about the external world to be entered into the computer. The computer also has hardware called an *output port* to send information out to the external world. An *interface* is defined as the collection of the I/O port, external electronics, physical devices, and the software, which combine to allow the computer to communicate with the external world. An example of an input interface is a switch, where the operator moves the switch, and the software can recognize the switch position. An example of an output interface is a light-emitting diode (LED), where the software can turn the light on and off, and the operator can see whether or not the light is shining. There is a wide range of possible inputs and outputs, which can exist in either digital or analog form. In general, we can classify I/O interfaces into four categories

> parallel—binary data is available simultaneously on a group of lines
> serial—binary data is available one bit at a time on a single line
> analog—data is encoded as an electrical voltage, current or power
> time—data is encoded as a period, frequency, pulse width or phase shift

Checkpoint 1.2: What are the differences between an input port and an input interface?

Checkpoint 1.3: List three input interfaces available on a personal computer.

Checkpoint 1.4: List three output interfaces available on a personal computer.

In this book, numbers that start with $ (e.g., $64) are specified in hexadecimal, which is base 16. In C, we start hexadecimal numbers with 0x (e.g., 0x64). Intel assembly language adds an "H" at the end to specify hexadecimal (e.g., 64H). Texas Instruments uses "h" (e.g., 64h).

In a system with *memory-mapped I/O*, as shown in Figure 1.1, the I/O ports are connected to the processor in a manner similar to memory. I/O ports are assigned addresses, and the software accesses I/O using reads and writes to the specific I/O addresses. The software inputs from an input port using the same instructions as it would if it were reading from memory. Similarly, the software outputs from an output port using the same instructions as it would if it were writing to memory. The processor, memory, and I/O are connected together by an address bus, a data bus, and a control bus. Together, these buses direct the data transfer between the various modules in the computer. A *bus* is defined as a collection of signals, which are grouped for a common purpose. For example, the *address bus* on the 6811/6812 is 16 signals (A15-A0), which together specify the memory address ($0000 to $FFFF) that is currently being accessed. The address specifies both which module (input, output, RAM or ROM) as well as which cell within the module will communicate with the processor. The *data bus* contains the information that is being transferred, which on the 6811 is 8 signals (D7-D0). On the 6812, the data bus is 16 bits (D15-D0), but it can transfer either 8-bit or 16-bit data. The *control bus* specifies the timing and the direction of the transfer. We call a complete data transfer a *bus cycle*. In a simple computer system, like the 6811/6812, only two types of transfers are allowed, as shown in Table 1.1.

Table 1.1
Simple computers generate two types of cycles.

Type	Address Driven by	Data Driven by	Transfer
Memory Read Cycle	Processor	RAM, ROM or Input	Data copied to processor
Memory Write Cycle	Processor	Processor	Data copied to Output or RAM

In this simple system, the processor always controls the address (where to access), the direction (read or write), and the control (when to access.) The MC9S12C32 has 2048 bytes of RAM located at addresses $3800 to $3FFF. Figure 1.2 illustrates how the processor fetches the 8-bit contents of location $3800 using a read cycle. Assume memory at address $3800 has the value $98. The processor first places the RAM address $3800 on the address bus, then the processor issues a read command on the control bus. The memory will respond by placing its $98 information on the data bus, and lastly the processor will accept the data and terminate the read command.

Figure 1.2
A memory read cycle copies data from RAM, ROM, or an input device into the processor.

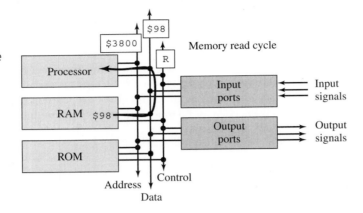

Figure 1.3 illustrates how the processor stores the 8-bit value $25 into RAM location $3800 using a write cycle. The processor first places the RAM address $3800 on the address bus. Next, the processor places the $25 information on the data bus, and then the processor issues a write command on the control bus. The memory will respond by storing the $25 information into the proper place, and after the processor is sure the memory has captured the data, it will terminate the write command.

Figure 1.3
A memory write cycle copies data from the processor into RAM or an output device.

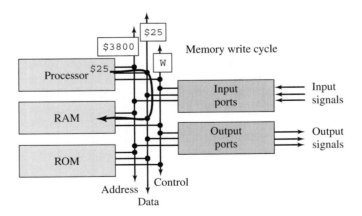

You see that if we wish to transfer data from an input device into RAM, we must first transfer it from input to the processor, then from the processor into RAM. In some microcontrollers, such as the Freescale MC68HC708XL36 and in all desktop PCs, we can transfer data directly from input to RAM or from RAM to output using *direct memory access* (DMA). The *bandwidth* of an I/O device is the number of information bytes/sec that can be transferred. Because DMA is faster, we will use this method to interface high bandwidth devices such as disks and networks. During a read DMA cycle (Figure 1.4), data flows directly from the memory to the output device. Many systems support DMA, which transfers data from memory to memory (See Chapter 10).

Figure 1.4
A DMA read cycle copies data from RAM, ROM, or an input device into an output device.

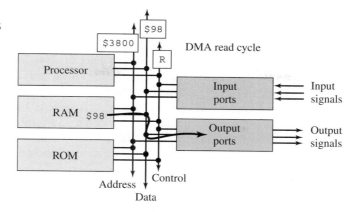

Figure 1.5
A DMA write cycle copies data from the input device into RAM or an output device.

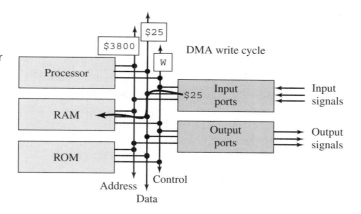

During a write DMA cycle (Figure 1.5) data flows directly from the input device to memory.

There is an alternative to memory-mapped I/O. In a system with *isolated I/O*, the processor uses a separate bus, and the software uses separate instructions to perform input and output. The Intel x86 family of processors utilizes isolated I/O. Because input/output is performed with special I/O instructions, errant software is less likely to cause inadvertent I/O functions. In contrast, systems with memory-mapped I/O are easier to design, and the software is easier to write.

The processor within the 6811 and 6812 microcontrollers has four major components, as illustrated in Figure 1.6. The *bus interface unit* (BIU) reads data from the bus during a read cycle, and writes data onto the bus during a write cycle. The 6811 and 6812 have a single

Figure 1.6
The four basic components of 6811 and 6812 processors.

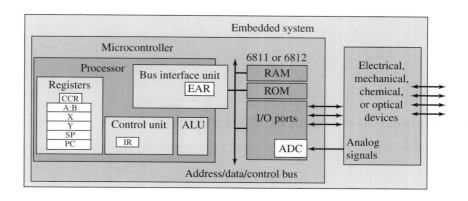

processor and do not support DMA. Therefore, the BIU always drives the address bus and the control signals of the bus. The *effective address register* (EAR) contains the memory address used to fetch the data needed for the current instruction.

The *control unit* (CU) orchestrates the sequence of operations in the processor. The CU issues commands to the other three components. The *instruction register* (IR) contains the *operation code* (or *opcode*) for the current instruction. Most 6811/6812 opcodes are 8 bits wide, but some are 16 bits. Most instructions have two parts, the opcode that defines the function to perform, and an *operand* that specifies the data to be used. In an embedded system the software is converted to machine code, which is a list of instructions, and stored in nonvolatile memory. When the system is running, instructions one at a time are fetched from memory and executed.

The *registers* are high-speed storage devices located in the processor. Registers do not have addresses like regular memory, but rather they have specific functions explicitly defined by the instruction. *Accumulators* are registers that contain data. *Index registers* contain addresses. The *program counter* (PC) points to the memory containing the instruction to execute next. In an embedded system, the PC usually points into nonvolatile memory (e.g., ROM, EPROM, or EEPROM). The information stored in nonvolatile memory (e.g., the instructions) is not lost when power is removed. The *stack pointer* (SP) points to the RAM and defines the stack. The stack is an extremely important component of software development and can be used to pass parameters, save temporary information, and implement local variables. The internal RAM of the 6811/6812 is volatile memory, meaning its information is lost when power is removed. On some systems, such as calculators and PDAs, a separate battery powers the RAM, creating nonvolatile RAM. The *condition code register* (CCR) contains the status of the previous operation, as well as some operating mode flags such as the interrupt enable bit. This register is called the *flag register* on the Intel computers.

The *arithmetic logic unit* (ALU) performs arithmetic and logic operations. Addition, subtraction, multiplication, and division are examples of arithmetic operations. And, or, exclusive or, and shift are examples of logical operations.

Checkpoint 1.5: For what do the acronyms CU DMA BIU ALU stand?

In general, the execution of an instruction goes through four phases. First, the computer fetches the machine code for the instruction by reading the value in memory pointed to by the program counter (PC). Some instructions are only one byte long, while others are two or more bytes. After each byte of the instruction is fetched, the PC is incremented. At this time, the instruction is decoded, and the effective address is determined (EAR). Many instructions require additional data, and during phase 2, the data is retrieved from memory at the effective address. Next, the actual function for this instruction is performed. Often, the computer bus is idle at this time, because no additional data is required. During the last phase, the results are written back to memory. All instructions have a phase 1, but the other three phases may or may not occur for any specific instruction. Each of the phases may require one or more bus cycles to complete. Each bus cycle reads or writes one piece of data. All 6811 bus cycles transfer 8-bit data, whereas the 6812 bus cycle can transfer 8-bit or 16-bit data.

Phase	Function	R/W	Address	Comment
1	Instruction fetch	read	PC++	Put into IR,
2	Data read	read	EAR	Data passes through ALU,
3	Operation	none		ALU operations, set CCR
4	Data store	write	EAR	Results stored in memory

1.2 Embedded Computer Systems

An *embedded computer system* is an electronic system that includes a microcomputer such as the Freescale 6811 or 6812 that is configured to perform a specific dedicated application, drawn previously as Figure 1.6. To better understand the expression *embedded microcomputer system*, consider each word separately. In this context, the word embedded means "hidden inside so one can't see it." The software that controls the system is programmed or fixed into ROM and is not accessible to the user of the device. Even so, *software maintenance*, which is verification of proper operation, updates, fixing bugs, adding features, extending to new applications, updating end user configurations, is still extremely important. In this book, we will develop techniques that facilitate this important aspect of system design. Embedded systems have these four characteristics.

First, embedded systems typically perform a single function. Consequently, they solve only a limited range of problems. For example, the embedded system in a microwave oven may be reconfigured to control different versions of the oven within a similar product line. Still, a microwave oven will always be a microwave oven, and you can't reprogram it to be a dishwasher. What makes each embedded system unique are the I/O ports of the micro-controller and the external devices interfaced to them.

Second, embedded systems are tightly constrained. There are typically very specific per-formance parameters within which the system must operate. For example, a cell-phone car-rier typically gets 832 radio frequencies to use in a city, a hand-held video game must cost less than $50, an automotive cruise control system must operate the vehicle within 3 mph of the set-point speed, and a portable MP3 player must operate for 12 hours on one battery charge.

Third, many embedded systems must operate in *real time*. In a real-time computer system, we can put an upper bound on the time required to perform the input-calculation-output sequence. A real-time system can guarantee a worst-case upper bound on the response time between when the new input information becomes available and when that information is processed. Another real-time requirement that exists in many embedded systems is the exe-cution of periodic tasks. A periodic task is one that must be performed at equal time intervals. A real-time system can put a small and bounded limit on the interval between when a task should be run and when it is actually run. Because of the real-time nature of these systems, microcomputers like the 6811 and 6812 have a rich set of features to handle all aspects of time.

The fourth characteristic of embedded systems is their small memory requirements. There are exceptions to this rule, such as those that process video or audio, but most have memory requirements measured in thousands of bytes. As of 2005, the median program size of an embedded system application has grown to 32K bytes. As the size and complexity of the embedded system software continues to grow, the extended memory features found in the 6812 family will become critical.

Checkpoint 1.6: What is an embedded system?

The computer engineer has many design choices to make when building a real-time embedded system. Often, defining the problem, specifying the objectives, and identifying the constraints are harder than actual implementations. In this book, we will develop com-puter engineering design processes, introducing fundamental methodologies for problem specification, prototyping, testing, and performance evaluation. We will illustrate this design methodology by presenting detailed solutions for the two following, Freescale microcontrollers:

- Freescale MC68HC711E9
- Freescale MC9S12C32

In this book we will refer to devices such as the Freescale MC68HC711E9 simply as 6811. The different versions of the microcomputers contain varying amounts of memory and input/output (I/O) devices. A typical automobile now contains an average of ten microcon-trollers. In fact, upscale homes may contain as many as 150 microcontrollers, and the average consumer now interacts with microcontrollers up to 300 times a day. As shown in Figure 1.7, the general areas that employ embedded microcomputers encompass every field of engineering:

- Communications
- Automotive

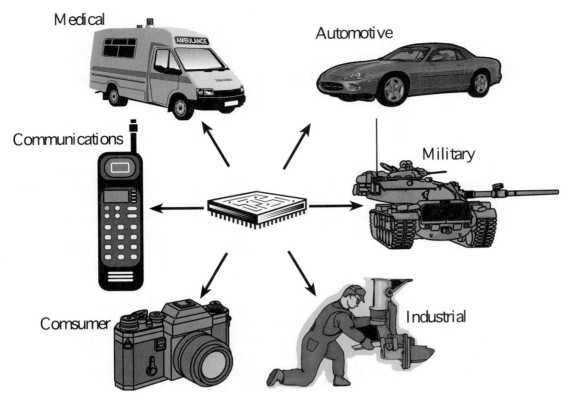

Figure 1.7
Example embedded computer systems.

- Military
- Medical
- Consumer
- Machine control

Table 1.2 presents typical embedded microcomputer applications and the function performed by the embedded microcomputer. Each microcomputer accepts inputs, performs calculations, and generates outputs. We must also learn how to interface a wide range of inputs and outputs that can exist in either digital or analog form.

Checkpoint 1.7: There is a microcomputer embedded in an alarm clock. List three operations the software must perform.

In contrast, a *general-purpose computer system* typically has a keyboard, disk, and graphics display and can be programmed for a wide variety of purposes. Typical general-purpose applications include word processing, electronic mail, business accounting, scientific computing, and data base systems. General-purpose computers have the opposite of the four characteristics previously listed. First, they can perform a wide and dynamic range of functions. Because the general-purpose computer has a removable disk or network interface, new programs can easily be added to the system. The user of a general-purpose computer does have access to the software that controls the machine. In other words, the user decides which operating system to run and which applications to launch. Second, they are loosely constrained. For example, the Java machine used by a web browser will operate on a extremely wide range of computer platforms. Third, general-purpose machines do not run in real time. Yes, we would like the time to print a page on the printer to be fast, and we would like web page to load quickly, but there are no guaranteed

Table 1.2
Embedded system
applications.

	Function Performed by the Microcomputer
Consumer:	
Washing machine	Controls the water and spin cycles
Exercise equipment	Measures speed, distance, calories, heart rate
Remote controls	Accepts key touches, sends infrared pulses
Clocks and watches	Maintains the time, alarm, and display
Games and toys	Entertains the user, joystick input, video output
Audio/video electronics	Interacts with the operator, enhances performance
Communication:	
Phone answering machines	Plays outgoing and saves incoming messages
Telephone system	Switches signals and retrieves information
Cellular phones, pagers	Interacts with key pad, microphone, and speaker
Automotive:	
Automatic breaking	Optimizes stopping on slippery surfaces
Noise cancellation	Improves sound quality, removing noise
Theft deterrent devices	Allows keyless entry, controls alarm
Electronic ignition	Controls sparks and fuel injectors
Power windows and seats	Remembers preferred settings for each driver
Instrumentation	Collects and provides necessary information
Military:	
Smart weapons	Recognizes friendly targets
Missile guidance systems	Directs ordnance at the desired target
Global positioning systems	Determines where you are on the planet
Surveillance	Collects information about enemy activities
Industrial:	
Point-of-sale systems	Accepts inputs and manages money
Set-back thermostats	Adjusts day/night thresholds saving energy
Traffic control systems	Senses car positions, controls traffic lights
Robot systems	Inputs from sensors, controls the motors
Inventory systems	Inputs from bar code readers, prints labels
Automatic sprinklers	Controls the wetness of the soil
Medical:	
Infant apnea monitors	Detects breathing, alarms if stopped
Glucose monitors	Measures blood sugar levels in diabetics
Cardiac monitors	Measures heart function, alarms if problem
Drug delivery	Administers proper doses
Cancer treatments	Controls doses of radiation, drugs, or heat
Pacemakers	Sends pulses to the heart to make it beat
Prosthetic devices	Increases mobility for the handicapped

response times for these types of activities. In fact, the real-time tasks that do exist (such as sound recording, burning CDs, and graphics) are actually performed by embedded systems built into the computer. Fourth, general-purpose computers employ billions, if not trillions, of memory cells.

The most common type of general-purpose computer is the personal computer (e.g., the Pentium-based IBM-PC compatible and the PowerPC-based Macintosh). Computers more powerful than the personal computer can be grouped in the workstation ($10,000 to $50,000 range) or the supercomputer categories (above $50,000). See the web site **www.top500.org** for a list of the fastest computers on the planet. These computers often employ multiple processors and have much more memory than the typical personal computer. The workstations and supercomputers are used for handling large amounts of information (business applications), running large simulations (weather forecasting), searching (**www.google.com**), or performing large calculations (scientific research). This book will not cover the general-purpose computer, although many of the basic principles of embedded computers do apply to all types of systems.

The I/O interfaces are a crucial part of an embedded system because they provide necessary functionality. Most personal computers have the same basic I/O devices (mouse, keyboard, display, CD, USB, etc.) In contrast, there is no common set of I/O that all embedded system have. The software together with the I/O ports and associated interface circuits give an embedded computer system its distinctive characteristics. A *device driver* is a set of software functions that facilitate the use of an I/O port. Another name for device driver is *application programmer interface* (API). In this book we will study a wide range of I/O ports supported by the Freescale microcomputers. Parallel ports provide for digital input and/or outputs. Serial ports include the synchronous *Serial Peripheral Interface* (SPI) and asynchronous *Serial Communications Interface* (SCI), which have a wide range of software selectable baud rates and are optimized to minimize CPU overhead. The SPI also enables synchronous communication between the microcontroller and peripheral devices such as

> Shift registers
> Liquid Crystal Display (LCD) drivers
> Analog to digital converters
> Other microprocessors

Analog to digital converters (ADC) convert analog voltages to digital numbers, and they are available with 8-bit and 10-bit resolution. The timer features on the Freescale microcomputers include

> Fixed periodic rate interrupts
> Computer Operating Properly (COP) protection against software failures
> Pulse accumulator for external event counting or gated time accumulation
> Pulse Width Modulated outputs (PWM)
> Event counter system for advanced timing operations
> Input capture used for period and pulse width measurement
> Output compare used for generating signals and frequency measurement

1.3 The Design Process

1.3.1 Top-Down Design

In this section, we introduce the design process. The process is called *top-down*, because we start with the high-level designs and work down to low-level implementations. The basic approach is introduced here, and the details of these concepts are presented throughout the remaining chapters of the book. As we learn software/hardware development tools and techniques, we can place them into the framework presented in this section. As illustrated in Figure 1.8, the development of a product follows an analysis-design-implementation-testing cycle. For complex systems with long life-spans, we traverse multiple times around the development cycle. For simple systems, a one-time pass may suffice.

During the **analysis phase**, we discover the requirements and constraints for our proposed system. We can hire consultants and interview potential customers in order to gather this critical information. A *requirement* is a general parameter that the system must satisfy. We begin by rewriting the system requirements, which are usually written in general form, into a list of detailed *specifications*. In general, specifications are detailed parameters describing how the system should work. For example, a requirement may state that the system should fit into a pocket, whereas a specification would give the exact size and weight of the device. For example, suppose we wish to build a motor controller. During the analysis phase, we would determine obvious specifications such as range, stability, accuracy, and response time. There may be less obvious requirements to satisfy, such as weight, size,

Figure 1.8
System development cycle.

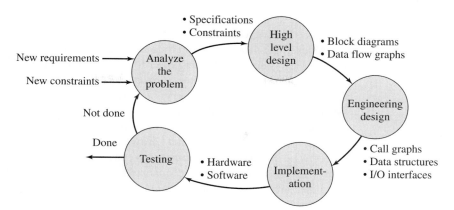

battery life, product life, ease of operation, display readability, and reliability. Often, improving the performance of one parameter can be achieved only by decreasing the performance of another. This art of compromise defines the tradeoffs an engineer must make when designing a product. A *constraint* is a limitation, within which the system must operate. The system may be constrained as to such factors as cost, safety, compatibility with other products, use of specific electronic and mechanical parts as employed in other devices, interfaces with other instruments and test equipment, and development schedule. The following measures are often considered during the analysis phase of a project:

Safety: The risk to humans or the environment
Accuracy: The difference between desired and actual parameter performance
Precision: The number of distinguishable measurements
Resolution: The smallest change that can be reliably detected
Response time: The time difference between triggering event and resulting action
Bandwidth: The amount of information processed per time unit
Maintainability: The flexibility with which the device can be modified
Testability: The ease with which proper operation of the device can be verified
Compatibility: The conformity of the device to existing standards
Mean time between failure: The reliability of the device
Size and weight: The physical space required by the system
Power: The amount of energy it takes to operate the system
Nonrecurring engineering cost (NRE cost): The one-time cost to design and test the product
Unit cost: The cost required to manufacture one additional product
Time-to-prototype: The time required to design, build, and test an example system
Time-to-market: The time required to deliver the product to the customer
Human factors: The degree to which our customers enjoy or appreciate the product

Checkpoint 1.8: What's the difference between a requirement and a specification?

During the **high-level design** phase, we build a conceptual model of the hardware and software system. It is in this model that we employ as much abstraction as appropriate. The project is broken in modules or subcomponents. Modular design will be presented in Chapter 2. During this phase, we estimate the cost, schedule, and expected performance of the system. At this point we can decide whether the project has a high enough potential for profit. A *data flow graph* is a block diagram of the system, showing the flow of information. Arrows point from source to destination. The rectangles represent hardware components and the ovals are software modules. We use data flow graphs in the high-level design, because they describe the overall operation of the system while hiding the details of how it works. Issues such as safety (e.g., Isaac Asimov's first Law of Robotics: "*A robot may not harm a human being, or, through inaction, allow a human being to come to harm*") and testing (e.g., we need to verify that our system is operational) should be addressed during the high-level design.

An example data flow graph for a motor controller is shown in Figure 1.9. The requirement of the system is to deliver power to a motor so that the speed of the motor equals the desired value set by the operator using a keypad. In order to make the system easier to use and to assist in testing, a *liquid crystal display* (LCD) is added. The sensor converts motor speed into an electrical voltage. The amplifier converts this signal into the 0 to +5V voltage range required by the ADC. The ADC converts analog voltage into a digital sample. The ADC routines, using the ADC and timer hardware, collect samples and calculate voltages. Next, this software uses a table data structure to convert voltage into measured speed. The user will be able to select the desired speed using the Keypad interface. The desired and measured speed data are passed to the Controller software, which will adjust the power output in such a manner as to minimize the difference between the measured speed and the desired speed. Finally, the power commands are output to the *actuator* module. The actuator interface converts the digital control signals to power delivered to the motor. The measured speed and speed error will be sent to the LCD module. The solution to this problem will be presented in Chapter 13.

Figure 1.9
A data flow graph showing how signals pass through a motor controller.

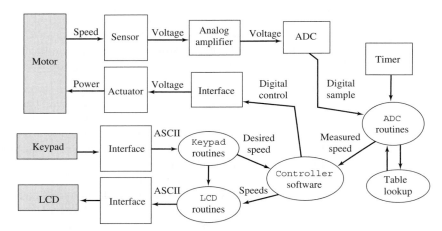

The next phase is **engineering design**. We begin by constructing a preliminary design. This system includes the overall top-down hierarchical structure, the basic I/O signals, shared data structures, and the overall software scheme. At this stage there should be a simple and direct correlation between the hardware/software systems and the conceptual model developed in the high-level design. Next, we finish the top-down hierarchical structure, and build mock-ups of the mechanical parts (connectors, chassis, cables, etc.) and user software interface. Sophisticated 3-D CAD systems can create realistic images of our system. Detailed hardware designs must include mechanical drawings. It is a good idea to have a second source, which is an alternative supplier that can sell us our parts if the first source can't deliver on time. *Call-graphs* are a graphical way to define how the software/hardware modules interconnect. A hierarchical system will have a tree-structured call graph. *Data structures* include both the organization of information and mechanisms to access the data. Again safety and testing should be addressed during this low-level design.

A call-graph for this motor controller is shown in Figure 1.10. Again, rectangles represent hardware components and ovals show software modules. An arrow points from the calling routine to the module it calls. The I/O ports are organized into groups and placed at the bottom of the graph. A high-level call-graph, like the one shown in Figure 1.11, shows only the high-level hardware/software modules. A detailed call-graph would include each software function and I/O port. Normally, hardware is passive and the software initiates hardware/software communication, but as we will learn in Chapter 4, it is possible for the hardware to interrupt the software and cause certain software modules to be run. In this

Figure 1.10
A call flow graph for a motor controller.

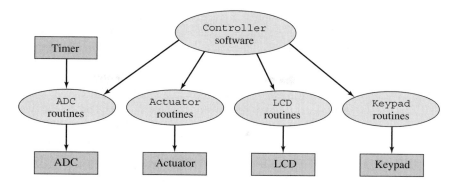

system, the timer hardware will cause the ADC software to collect a sample at a regular rate. The Controller software calls the Keypad routines to get the desired speed, calls the ADC software to get the motor speed at that point, determines what power to deliver to the motor, and updates the actuator by sending the power value to the Actuator interface. The Controller software calls the LCD routines to display the status of the system. As we will see in Chapters 11–15, acquiring data, calculating parameters, and outputting results at a regular rate is strategic when performing digital processing in embedded systems.

> **Checkpoint 1.9:** What confusion could arise if two software modules were allowed to access the same I/O port? This situation would be evident on a call-graph if the two software modules had arrows pointing to the same I/O port.

> **Observation:** If module A calls module B, and B returns data, then a data flow graph will show an arrow from B to A, but a call-graph will show an arrow from A to B.

The next phase is **implementation**. An advantage of a top-down design is that implementation of subcomponents can occur concurrently. During the initial iterations of the development cycle, it is quite efficient to implement the hardware/software using simulation. One major advantage of simulation is that it is usually quicker to implement an initial product on a simulator than to construct a physical device out of actual components. Rapid prototyping is important in the early stages of product development. This allows for more loops around the analysis-design-implementation-testing cycle, which in turn leads to a more sophisticated product.

Recent software and hardware technological developments have made significant impacts on the software development process for embedded microcomputers. The simplest approach is to use a cross-assembler or cross-compiler to convert source code into the machine code for the target system. The machine code can then be loaded into the target machine. Debugging embedded systems with this simple approach is very difficult for two reasons. First, the embedded system lacks the usual keyboard and display that assist us when we debug regular software. Second, the nature of embedded systems involves the complex and real-time interaction between the hardware and software. These real-time interactions make it impossible to test software with the usual single-stepping and print statements.

The next technological advancement that has greatly affected the manner in which embedded systems are developed is simulation. Because of the high cost and long times required to create hardware prototypes, many preliminary feasibility designs are now performed using hardware/software simulations. A simulator is a software application that models the behavior of the hardware/software system. If both the external hardware and software program are simulated together, even through the simulated time is slower than the clock on the wall, the real-time hardware/software interactions can be studied.

During the **testing** phase, we evaluate the performance of our system. First, we debug the system and validate basic functions. Next, we use careful measurements to optimize performance, such as static efficiency (memory requirements), dynamic efficiency (execution speed), accuracy (difference between truth and measured), and stability (consistent operation). Debugging techniques are presented in Chapter 2.

Maintenance is the process of correcting mistakes, adding new features, optimizing for execution speed or program size, porting to new computers or operating systems, and reconfiguring the system to solve a similar problem. No system is static. Customers may change or add requirements or constraints. To be profitable, we probably will wish to tailor each system to the individual needs of each customer. Maintenance is not really a separate phase, but rather involves additional loops around the development cycle.

Figure 1.11
System development process illustrating bottom-up design.

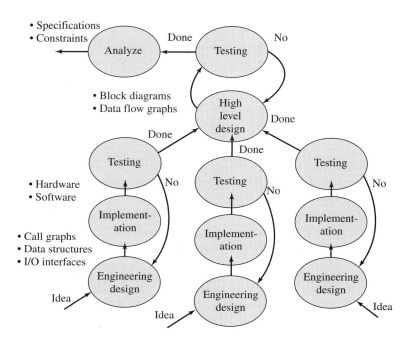

1.3.2 Bottom-Up Design

Figure 1.8 describes top-down design as a cyclic process, beginning with a problem statement and ending up with a solution. With a *bottom-up* design we begin with solutions and build up to a problem statement. Many innovations begin with an idea, "what if . . .?" In a bottom-up design, one begins with designing, building, and testing low-level components. Figure 1.11 illustrates a two-level process, combining three subcomponents to create the overall product. This hierarchical process could have more levels and/or more components at each level. The low-level designs can occur in parallel. The design of each component is cyclic, iterating through the design-build-test cycle until the performance is acceptable. Bottom-up design may be inefficient because some subsystems may be designed, built, and tested but never used. As the design progresses the components are fit together to make the system more and more complex. Only after the system is completely built and tested does one define the overall system specifications. The bottom-up design process allows creative ideas to drive the products a company develops. It also allows one to quickly test the feasibility of an idea. If one fully understands a problem area and the scope of potential solutions, then a top-down design will arrive at an effective solution most quickly. On the other hand, if one doesn't really understand the problem or the scope of its solutions, a bottom-up approach allows one to start off by learning about the problem.

> ***Observation:*** A good engineer knows both bottom-up and top-down design methods, choosing the approach most appropriate for the situation at hand.

1.4 Digital Logic and Open Collector

Normal digital logic has two states: high and low. There are four currents of interest, as shown in Figure 1.12, when analyzing whether the inputs of the next stage are loading the output. I_{IH} and I_{IL} are the currents required of an input when high and low, respectively. Similarly, I_{OH} and I_{OL} are the maximum currents available at the output when high and low. In order for the output to properly drive all the inputs of the next stage, the maximum available output current must be larger than the sum of all the required input currents for both the high and low conditions.

Figure 1.12
Sometimes one output must drive multiple inputs.

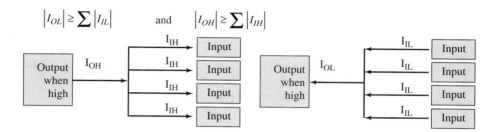

$$|I_{OL}| \geq \sum |I_{IL}| \quad \text{and} \quad |I_{OH}| \geq \sum |I_{IH}|$$

When we design circuits using devices all from a single logic family, we can define *fan out* as the maximum number of inputs, one output can drive. For *transistor-transistor logic* (TTL) logic we can calculate *fan out* from the input and output currents:

$$\text{fan out} = \text{minimum} \, ((I_{OH}/I_{IH}), (I_{OL}/I_{IL}))$$

The fan out of high speed *complementary metal-oxide semiconductor* (CMOS) devices (e.g., the 6811 and 6812), is determined by capacitive loading and not by the currents. The *slew rate* of a signal is the slope of the voltage versus time during the time when the logic level switches between low and high. A similar parameter is the *transition time*, which is the time it takes for an output to switch between high and low. There is a capacitive load for each CMOS input connected to a CMOS output. As this capacitance increases the slew rate decreases, which will increase the transition time. For circuits that mix devices from one family with those from another, we must look individually at the input and output currents, voltages, and capacitive loads. Table 1.3 shows typical current values for the various digital logic families. The MC9S12C32 allows full drive (10 mA) and reduced drive (2 mA) modes.

Table 1.3
The input and output currents of various digital logic families and microcomputers.

Family	Example	I_{OH}	I_{OL}	I_{IH}	I_{IL}	Fan Out
Standard TTL	7404	0.4 mA	16 mA	40 μA	1.6 mA	10
Schottky TTL	74S04	1 mA	20 mA	50 μA	2 mA	10
Low-power Schottky TTL	74LS04	0.4 mA	4 mA	20 μA	0.4 mA	10
High-speed CMOS	74HC04	4 mA	4 mA	1 μA	1 μA	
Freescale microcomputer	MC68HC11E	0.8 mA	1.6 mA	1 μA	1 μA	
Freescale microcomputer	MC9S12C32	10 mA	10 mA	1 μA	1 μA	
Intel microcomputer	87C51 P0	7 mA	3.2 mA	10 μA	10 μA	
	87C51 P1,P2,P3	60 μA	1.6 mA		50 μA	

Observation: For TTL devices the logic low currents are much larger than the logic high currents.

Figure 1.13 compares the input and output voltages for many of the digital logic families. V_{IL} is the voltage below which an input is considered a logic low. Similarly, V_{IH} is the voltage

Figure 1.13
Voltage thresholds for
various digital logic
families.

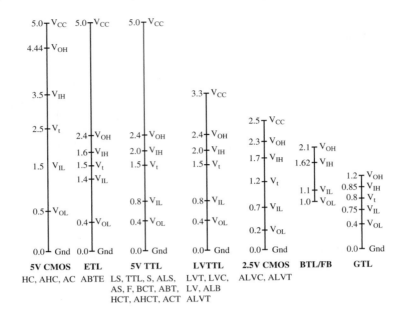

above which an input is considered a logic high. V_{OH} is the output voltage when the signal
is high. In particular, if the output is a logic high, and the current is less than I_{OH}, then the
voltage will be greater than V_{OH}. Similarly, V_{OL} is the output voltage when the signal is low.
In particular, if the output is a logic low and the current is less than I_{OL}, then the voltage
will be less than V_{OL}. The maximum output current specification on the 6811 and 6812 is
25 mA, which is the current above which it will cause damage. Normally, we design the
system so the output currents are less than I_{OH} and I_{OL}. V_t is the typical threshold voltage,
which is the voltage at which the input usually switches between logic low and high. For-
mally however, an input is considered as indeterminate for voltages between V_{IL} and V_{IH}.
The five parameters that affect our choice of logic families are

- Power supply voltage (e.g., +5V, 3.3V etc.)
- Power supply current (e.g., will the system need to run on batteries?)
- Speed (e.g., clock frequency and propagation delays)
- Output drive, I_{OL}, I_{OH} (e.g., does it need to drive motors or lights?)
- Noise immunity (e.g., electromagnetic field interference)

Common error: If the voltage applied to a high speed CMOS input pin exists
between V_{IL} and V_{IH} for extended periods of time, permanent damage may occur.

Checkpoint 1.10: The 6811 and 6812 are HC devices. How will the 6811 and 6812
interpret an input pin as the input voltage changes from 0, 1, 2, 3, 4, to 5V? That is, for
each voltage, will it be considered as a logic low, as a logic high, or as indeterminate?

Checkpoint 1.11: Considering both voltage and current, can the output of a
74HC04 drive the input of a 74LS04?

Checkpoint 1.12: Considering both voltage and current, can the output of a 74LS04
drive the input of a 74HC04?

A very important concept used in computer technology is *tristate logic*, which has three
output states: high, low, and off. Other names for the off state are HiZ, floating, and tristate.
As shown in Figure 1.14, the 74HC245 chip has eight bidirectional tristate drivers. The triangle shape, drawn with a signal on the top or the bottom of the triangle, is used to specify tristate control output in logic diagrams. The 74HC245 chip is active when the output enable,
OE, input is low, and the direction is controlled by the **DIR** input. For example, if **OE**=0 and

DIR=0, then the **B1-B8** are inputs and **A1-A8** are outputs, and each output A_n equals the corresponding B_n input. Conversely, if **OE**=0 and **DIR**=1, then the **A1-A8** are inputs and **B1-B8** are outputs, and each output B_n equals the corresponding A_n input. When **OE** is high, all the outputs will be off (floating). Devices like the 74HC245 are used in microcomputer systems because of the large output current and bidirectional tristate outputs. Tables 1.4 and 1.5 illustrate the wide range of technologies available for digital logic design. Not all logic families have the same choice of logic functions. t_{pd} is the propagation delay from input to output.

Figure 1.14
Block diagram of a
74HC245 tristate driver.

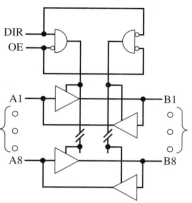

Family Technology	V_{IL} V_{IH}	V_{OL} V_{OH}	I_{OL}	I_{OH}	I_{CC}	t_{pd}
LVT—Low-Voltage BiCMOS	LVTTL	LVTTL	64	−32	190	3.5
ALVC—Advanced Low-Voltage CMOS	LVTTL	LVTTL	24	−24	40	3.0
LVC—Low-Voltage CMOS	LVTTL	LVTTL	24	−24	10	4.0
ALB—Advanced Low-Voltage BiCMOS	LVTTL	LVTTL	25	−25	800	2.0
AC—Advanced CMOS	CMOS	CMOS	12	−12	20	8.5
AHC—Advanced high-Speed CMOS	CMOS	CMOS	4	−4	20	11.9
LV—Low-Voltage CMOS	LVTTL	LVTTL	8	−8	20	14
Units			mA	mA	μA	ns

Table 1.4
Comparison of the output drive, power supply current, and speed of various 3.3V logic '245 gates.

Family Technology	V_{IL} V_{IH}	V_{OL} V_{OH}	I_{OL}	I_{OH}	I_{CC}	t_{pd}
AHC—Advanced High-Speed CMOS	CMOS	CMOS	8	−8	0.04	7.5
AHCT—Advanced High-Speed CMOS	TTL	CMOS	8	−8	0.04	7.7
AC—Advanced CMOS	CMOS	CMOS	24	−24	0.04	6.5
ACT—Advanced CMOS	TTL	CMOS	24	−24	0.04	8.0
HC—High-Speed CMOS Logic	CMOS	CMOS	6	−6	0.08	21
HCT—High-Speed CMOS Logic	TTL	CMOS	6	−6	0.08	30
ABT—Advanced BiCMOS	TTL	TTL	64	−32	0.25	3.5
74F—Fast Logic	TTL	TTL	64	−15	120	6.0
BCT—BiCMOS	TTL	TTL	64	−15	90	6.6
AS—Advanced Schottky Logic	TTL	TTL	64	−15	143	7.5
ALS—Advanced Low-Power Schottky	TTL	TTL	24	−15	58	10
LS—Low Power Schottky logic	TTL	TTL	24	−15	95	12
S—Schottky Logic	TTL	TTL	64	−15	180	9
TTL—Transistor-Transistor Logic	TTL	TTL	16	−0.4	22	22
Units			mA	mA	mA	ns

Table 1.5
Comparison of the output drive, power supply current, and speed of various 5V logic '245 gates.

The 74LS04 is a low-power Schottky NOT gate, as shown in Figure 1.15. It is called Schottky logic because the devices are made from Schottky transistors. The output is *high* when the transistor Q4 is active, driving the output to +5V. The output is *low* when the transistor Q5 is active, driving the output to 0.

Figure 1.15
Transistor implementation of a low-power Schottky NOT gate.

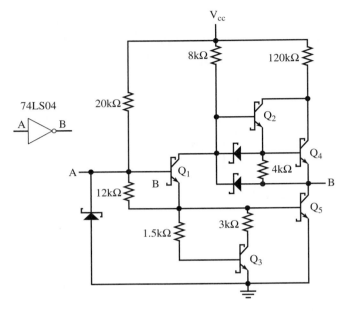

The 74HC04 is a high-speed CMOS NOT gate, as shown in Figure 1.16. The output is *high* when the transistor Q1 is active, driving the output to +5V. The output is *low* when the transistor Q2 is active, driving the output to 0. Since the 6811 and 6812 are made with high-speed CMOS logic, their outputs behave like the Q1/Q2 "push/pull" transistor pair. The 6811 and 6812 output ports are not inverting. That is, when you write a "1" to an output port, then the output voltage goes high. Similarly, when you write a "0" to an output port, then the output voltage goes low. Analyses of the circuit in Figure 1.16 reveal some of the basic properties of high-speed CMOS logic. First, because of the complementary nature of the P-channel (the one on the top) and the N-channel (the one on the bottom) transistors, when the input is constant (continuously high or continuously low), the supply current, I_{cc}, is very low. Second, the gate will require supply current only when the output switches from low to high or from high to low. This observation leads to the design rule that the power required to run a high-speed CMOS system is linearly related to the frequency of its clock, because the frequency of the clock determines the number of transitions per second. Along the same lines, we see that if the voltage on input A exists between V_{IL} and V_{IH} for extended periods of time, then both Q1 and Q2 are partially active, causing a short from V_{cc} to ground. This condition can cause permanent damage to the transistors. Third, since the input A is connected to the gate of the two MOS transistors, the input currents will be very small (1 μA). In other words, the input impedance (input voltage divided by input current) of the gate is very high. Normally, a high input impedance is a good thing, except when the input is not connected. If the input is not connected, then it takes very little input currents to cause the logic level to switch.

Figure 1.16
Transistor implementation of a high-speed CMOS NOT gate.

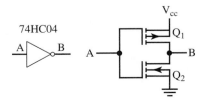

Common error: If unused input pins on a CMOS microcontroller are left unconnected, then the input signal may oscillate at high frequencies depending on the EM fields in the environment, wasting power unnecessarily.

Maintenance tip: It is a good design practice to connect unused CMOS inputs to ground or connect them to +5V.

Open collector logic has outputs with two states: low and off. The 7405 is a TTL open collector NOT gate, as shown in Figure 1.17. When drawing logic diagrams, we add the "x" on the output to specify open collector logic. It is called open collector because the collector pin of Q3 is not connected, or left open. The output is *off* when there is no active transistor driving the output. In other words, when the input is low, the output floats. This "not driven" condition is called the open collector state. The output is low when the transistor Q3 is active, driving the output to 0.

Figure 1.17
Transistor implementation of a regular TTL open collector NOT gate.

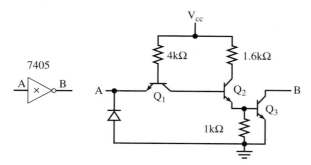

The 74HC05 is a high-speed CMOS open collector NOT gate, as shown in Figure 1.18. The output is *off* when there is no active transistor driving the output. In other words, when the input is low, the output floats. The output is low when the transistor Q2 is active driving the output to 0. Technically, the 74HC05 implements *open drain* rather than open collector, because it is the drain pin of Q2 that is left open. In this book, we will use the terms open collector and open drain interchangeably to refer to digital logic with two states (low and off). The data sheets of the 6811 and 6812 refer to open collector logic as *wire or mode* (WOM).

Figure 1.18
Transistor implementation of a high-speed CMOS open collector NOT gate.

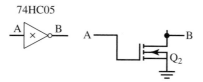

Because of the multiple uses of open collector, many microcomputers can implement open collector logic. All the ports of the Intel 8051 are inherently open collector. In the 6811, when the CWOM bit in the PIOC is set to 1, the outputs of Port C have two output states: low and off. Also in the 6811, the DWOM bit in the SPCR specifies whether Port D output pins are regular or open drain. The 6812 PORTS also supports open collector outputs by setting the SWOM bit.

Observation: The 7405 and 74HC05 are inverting open collector examples, but most microcomputer output ports are not inverting. That is, when you write a "1" to an open-collector output port, then the output floats, and when you write a "0" to an open-collector output port, then the output voltage goes low.

In general, we can use an **open collector NOT** gate to control the current to a device, such as a relay, a *light emitting diode* (LED), a solenoid, a small motor, or a small light. We used the open collector NOT gate in the LED interface shown in Figure 1.19 to control the current to our diode. When input to the 7405 is high (p=1, which means +5V), the output is low (q=0, which means 0V). In this state, a 10 mA current is applied to the diode, and it

Figure 1.19
Open collector used to
interface a light emitting
diode.

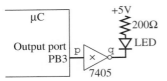

lights up. But, when the input is low (p=0, which means +0V), the output floats (q=HiZ, which is neither high or low). This floating output state causes the LED current to be zero, and the diode is dark.

When needed for digital logic, we can convert an open collector output to a digital signal using a pull-up resistor from the output to +5V. In this way, when the open collector output floats, the signal will be a digital high. How do we select the value of the pullup resistor? In general the smaller the resistor, the larger the I_{OH} it will be able to supply when the output is high. On the other hand, a larger resistor does not waste as much I_{OL} current when the output is low. One way to calculate the value of this pull-up resistor is to first determine the required output high voltage, V_{out}, and output high current, I_{out}. To supply a current of at least I_{out} at a voltage above V_{out}, the resistor must be less than:

$$R \le (+5 - V_{out})/I_{out}$$

As an example, we will calculate the resistor value for the situation where the circuit needs to drive five regular TTL loads. We see from Figure 1.13 that V_{out} must be above V_{IH} (2V) in order for the TTL inputs to sense a high logic level. We can add a safety factor and set V_{out} at 3V. In order for the high output to drive all five TTL inputs, I_{out} must be more than five I_{IH}. From Table 1.3, we see that I_{IH} is 40 μA, so I_{out} should be larger than 5•40 μA or 0.2 mA. For this situation the resistor must be less than 10 kΩ.

Another example of open collector logic occurs when interfacing switches to the microcontroller. The circuit in the left of Figure 1.20 shows a mechanical switch with one terminal connected to ground. In this circuit, when the switch is pressed, the voltage r is zero. When the switch is not processed, the signal r floats. The circuit in the middle of Figure 1.20 shows the mechanical switch with a 10k pull-up resistor attached the other side. When the switch is pressed, the voltage at s still goes to zero, because the resistance of the switch (less than 0.1Ω) is much less than the pull-up resistor. But now, when the switch is not pressed, the pull-up resistor creates a +5V at s. This circuit is shown connected to an input pin of the microcontroller. The software, by reading the input port, can determine whether or not the switch is processed. If the switch is pressed, the software will read zero, and if the switch is not pressed, the software will read one. The circuit on the right of Figure 1.20 also interfaces a mechanical switch to the microcontroller, but it implements positive logic using a pull-down resistor. The signal t will be high if the switch is pressed and low if it is released.

Figure 1.20
Switch interface.

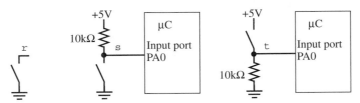

Observation: Some of the ports on the 6812 implement pull-up or pull-down resistors, so the interfaces shown in Figure 1.20 can be made without the resistor.

1.5   Digital Representation of Numbers

1.5.1
Fundamentals

Numbers are stored on the computer in binary form. In other words, information is encoded as a sequence of 1's and 0's. On most computers, the memory is organized into 8-bit bytes. This means each 8-bit byte stored in memory will have a separate address. *Precision* is the number of distinct or different values. We express precision in alternatives, decimal digits, bytes, or binary bits. *Alternatives* are defined as the total number of possibilities. For example, an 8-bit number scheme can represent 256 different numbers. An 8-bit *digital to analog converter* (DAC) can generate 256 different analog outputs. An 8-bit *analog to digital converter* (ADC) can measure 256 different analog inputs. We use the expression $4\frac{1}{2}$ decimal digits to mean 20,000 alternatives and the expression $4\frac{3}{4}$ decimal digits to mean 40,000 alternatives. The 1/2 decimal digit means twice the number of alternatives or one additional binary bit. The 3/4 decimal digit means four times as many alternatives or two additional binary bits. For example, a voltmeter with a range of 0.00 to 9.99V has a three decimal digit precision. Let the operation [[x]] be the greatest integer of x. E.g., [[2.1]] is rounded up to 3. Tables 1.6 and 1.7 illustrate various representations of precision.

Table 1.6
Relationship between bits, bytes, and alternatives as units of precision.

Binary Bits	Bytes	Alternatives
8	1	256
10		1024
12		4096
16	2	65536
20		1,048,576
24	3	16,777,216
30		1,073,741,824
32	4	4,294,967,296
n	[[n/8]]	2^n

Table 1.7
Definition of decimal digits as a unit of precision.

Decimal Digits	Alternatives
3	1000
$3\frac{1}{2}$	2000
$3\frac{3}{4}$	4000
4	10000
$4\frac{1}{2}$	20000
$4\frac{3}{4}$	40000
5	100000
n	10^n

Observation: A good rule of thumb to remember is $2^{10 \bullet n} \approx 10^{3 \bullet n}$.

Checkpoint 1.13: How many binary bits correspond to $2\frac{1}{2}$ decimal digits?

Checkpoint 1.14: About how many decimal digits can be presented in a 64-bit 8-byte number? You can answer this without a calculator, just using the "rule of thumb."

The *hexadecimal* number system uses base 16 as opposed to our regular decimal number system, which uses base 10. Hexadecimal is a convenient mechanism for humans to represent binary information, because it is extremely simple for us to convert back and forth between binary and hexadecimal. Hexadecimal number system is often abbreviated as "hex." A *nibble* is defined as 4 binary bits, which will be one hexadecimal digit. In mathematics, a subscript of 2 means binary, but in assembly language we will use the prefix % to signify binary numbers. The hexadecimal digits are 0, 1, 2, 3, 4, 5, 6, 7, 8, 9, A, B, C, D, E, and F. In assembly language we use the prefix $ to signify hexadecimal, and in C we use the prefix 0x. To convert from binary to hexadecimal, you simply separate the binary number into groups of four binary bits (starting on the right), then convert each group of four bits into one hexadecimal digit. For example, if you wished to convert %10100111, first you would group it into nibbles 1010 0111, then you would convert each group 1010=A 0111=7, yielding the result of $A7. To convert hexadecimal to binary, you simply substitute the 4-bit binary for each hexadecimal digit. For example, if you wished to convert $B5D1, you substitute B=1011 5=0101 D=1101 1=0001, yielding the result of %1011010111010001.

Checkpoint 1.15: Convert the binary number %111011101011 to hexadecimal.

Checkpoint 1.16: Convert the hex number $3800 to binary.

Checkpoint 1.17: How many binary bits does it take to represent $12345?

1.5.2
8-Bit Numbers

A byte contains 8 bits as shown in Figure 1.21, where each bit $b_7, \ldots, b_0$ is binary and has the value 1 or 0. We specify b_7 as the *most significant bit* or MSB, and b_0 as the least significant bit or LSB.

Figure 1.21
8-bit binary format.

b7	b6	b5	b4	b3	b2	b1	b0

If a byte is used to represent an unsigned number, then the value of the number is

$$N = 128 \cdot b_7 + 64 \cdot b_6 + 32 \cdot b_5 + 16 \cdot b_4 + 8 \cdot b_3 + 4 \cdot b_2 + 2 \cdot b_1 + b_0$$

Notice that the significance of bit n is 2^n. There are 256 different unsigned 8-bit numbers. The smallest unsigned 8-bit number is 0, and the largest is 255. For example, %00001010 is 8+2 or 10.

Checkpoint 1.18: Convert the binary number %01101010 to unsigned decimal.

Checkpoint 1.19: Convert the hex number $32 to unsigned decimal.

The *basis* of a number system is a subset from which linear combinations of the basis elements can be used to construct the entire set. The basis represents the "places" in a "place-value" system. For positive integers, the basis is the infinite set $\{1, 10, 100, ...\}$, and the "values" can range from 0 to 9. Each positive integer has a unique set of values such that the dot-product of the **value-vector** times the **basis-vector** yields that number. For example, 2345 is (. . ., 2,3,4,5)•(. . ., 1000,100,10,1), which is 2*1000+3*100+4*10+5. For the unsigned 8-bit number system, the basis is

$$\{1, 2, 4, 8, 16, 32, 64, 128\}$$

The values of a binary number system can only be 0 or 1. Even so, each 8-bit unsigned integer has a unique set of values such that the dot-product of the values times the basis yields that number. For example, 69 is (0,1,0,0,0,1,0,1)•(128,64,32,16,8,4,2,1), which equals 0*128+1*64+0*32+0*16+0*8+1*4+0*2+1*1.

Checkpoint 1.20: Give the representations of decimal 35 in 8-bit binary and hexadecimal.

Checkpoint 1.21: Give the representations of decimal 200 in 8-bit binary and hexadecimal.

One of the first schemes to represent signed numbers was called *one's complement*. It was called one's complement because to negate a number, you complement (logical not) each bit. For example, if 25 equals 00011001 in binary, then −25 is 11100110. An 8-bit one's complement number can vary from −127 to +127. The most significant bit is a sign bit, which is 1 if and only if the number is negative. The difficulty with this format is that there are two zeros +0 is 00000000, and −0 is 11111111. Another problem is that ones complement numbers do not have basis elements. These limitations led to the use of two's complement.

The *two's complement* number system is the most common approach used to define signed numbers. It was called two's complement because to negate a number, you complement each bit (like one's complement), then add 1. For example, if 25 equals 00011001 in binary, then −25 is 11100111. If a byte is used to represent a signed two's complement number, then the value of the number is

$$N = -128 \cdot b_7 + 64 \cdot b_6 + 32 \cdot b_5 + 16 \cdot b_4 + 8 \cdot b_3 + 4 \cdot b_2 + 2 \cdot b_1 + b_0$$

Observation: One usually means two's complement when one refers to signed integers.

There are 256 different signed 8-bit numbers. The smallest signed 8-bit number is −128, and the largest is 127. For example, %10000010 equals −128+2 or −126.

Checkpoint 1.22: Are the signed and unsigned decimal representations of the 8-bit hex number $35 the same or different?

For the signed 8-bit number system the basis is

$$\{1, 2, 4, 8, 16, 32, 64, -128\}$$

Observation: The most significant bit in a two's complement signed number will specify the sign.

Notice that the same binary pattern of %11111111 could represent either 255 or −1. It is very important for the software developer to keep track of the number format. The computer can not determine whether the 8-bit number is signed or unsigned. You, as the programmer, will determine whether the number is signed or unsigned by the specific assembly instructions you select to operate on the number. Some operations like addition, subtraction, and shift left (multiply by 2) use the same hardware (instructions) for both unsigned and signed operations. On the other hand, multiply, divide, and shift right (divide by 2) require separate hardware (instruction) for unsigned and signed operations. For example, the 6811/6812 multiply instruction, `mul`, operates only on unsigned values. So if you use the `mul` instruction, you are implementing unsigned arithmetic. The Freescale 6812 has both unsigned, `emul`, and signed, `emuls`, multiply instructions. So if you use the `emuls` instruction, you are implementing signed arithmetic.

Observation: To take the negative of a two's complement signed number we first complement (flip) all the bits, then add 1.

Checkpoint 1.23: Give the representations of −35 in 8-bit binary and hexadecimal.

Checkpoint 1.24: Why can't you represent the number 200 using 8-bit signed binary?

Common error: An error will occur if you use signed operations on unsigned numbers, or use unsigned operations on signed numbers.

Maintenance tip: To improve the clarity of your software, always specify the format of your data (signed versus unsigned) when defining or accessing the data.

1.5.3 Character Information

We can use bytes to represent characters with the **American Standard Code for Information Interchange** (ASCII) code. Standard ASCII is actually only 7 bits, but is stored using 8-bit bytes with the most significant byte equal to 0. Some computer systems use the 8th bit of the ASCII code to define additional characters such as graphics and letters in other alphabets. The 7-bit ASCII code definitions are given in the Table 1.8. For example, the letter "V" is in the $50 row and the 6 column. Putting the two together yields hexadecimal $56.

Checkpoint 1.25: How is the character 0 represented in ASCII?

One way to encode a character string is to use null-termination. In this way, the characters of the string are stored one right after the other, and the end of the string is signified by the NUL character (0). For example, the string "Valvano" is encoded as the following eight bytes: $56,$61,$6C,$76,$61,$6E,$6F,$00.

Checkpoint 1.26: How is "Hello World" encoded as a null-terminated ASCII string?

1.5.4 16-Bit Numbers

A *word* or *double byte* contains 16 bits, where each bit b15, . . . , b0 is binary and has the value 1 or 0, as shown in Figure 1.22.

Table 1.8
Standard 7-bit ASCII.

		BITS 4 to 6							
		0	1	2	3	4	5	6	7
	0	NUL	DLE	SP	0	@	P	`	p
B	1	SOH	DC1	:	1	A	Q	a	q
I	2	STX	DC2	!	2	B	R	b	r
T	3	ETX	DC3	#	3	C	S	c	s
S	4	EOT	DC4	$	4	D	T	d	t
	5	ENQ	NAK	%	5	E	U	e	u
0	6	ACK	SYN	&	6	F	V	f	v
	7	BEL	ETB	`	7	G	W	g	w
T	8	BS	CAN	(	8	H	X	h	x
O	9	HT	EM	)	9	I	Y	i	y
	A	LF	SUB	*	:	J	Z	j	z
3	B	VT	ESC	+	;	K	[	k	{
	C	FF	FS	,	<	L	\	l	|
	D	CR	GS	–	=	M	]	m	}
	E	SO	RS	.	>	N	^	n	~
	F	S1	US	/	?	O	_	o	DEL

Figure 1.22
16-bit binary format.

b15	b14	b13	b12	b11	b10	b9	b8	b7	b6	b5	b4	b3	b2	b1	b0

If a word is used to represent an unsigned number, then the value of the number is

$$N = 32768 \cdot b_{15} + 16384 \cdot b_{14} + 8192 \cdot b_{13} + 4096 \cdot b_{12}$$
$$+ 2048 \cdot b_{11} + 1024 \cdot b_{10} + 512 \cdot b_9 + 256 \cdot b_8$$
$$+ 128 \cdot b_7 + 64 \cdot b_6 + 32 \cdot b_5 + 16 \cdot b_4 + 8 \cdot b_3 + 4 \cdot b_2 + 2 \cdot b_1 + b_0$$

There are 65536 different unsigned 16-bit numbers. The smallest unsigned 16-bit number is 0, and the largest is 65535. For example, %0010000110000100 or $2184 is 8192+256+ 128+4 or 8580. For the unsigned 16-bit number system the basis is

{1, 2, 4, 8, 16, 32, 64, 128, 256, 512, 1024, 2048, 4096, 8192, 16384, 32768}

There are also 65536 different signed 16-bit numbers. The smallest two's complement signed 16-bit number is −32768 and the largest is 32767. For example, %1101000000000100 or $D004 is −32768+16384+4096+4 or −12284. If a word is used to represent a signed two's complement number, then the value of the number is

$$N = -32768 \cdot b_{15} + 16384 \cdot b_{14} + 8192 \cdot b_{13} + 4096 \cdot b_{12}$$
$$+ 2048 \cdot b_{11} + 1024 \cdot b_{10} + 512 \cdot b_9 + 256 \cdot b_8$$
$$+ 128 \cdot b_7 + 64 \cdot b_6 + 32 \cdot b_5 + 16 \cdot b_4 + 8 \cdot b_3 + 4 \cdot b_2 + 2 \cdot b_1 + b_0$$

For the signed 16-bit number system the basis is

{1, 2, 4, 8, 16, 32, 64, 128, 256, 512, 1024, 2048, 4096, 8192, 16384, −32768}

Common error: An error will occur if you use 16-bit operations on 8-bit numbers, or use 8-bit operations on 16-bit numbers.

Maintenance tip: To improve the clarity of your software, always specify the precision of your data when defining or accessing the data.

When we store 16-bit data into memory, it requires two bytes. Since the memory systems on most computers are byte addressable (a unique address for each byte), there are two possible ways to store in memory the two bytes that constitute the 16-bit data. Freescale microcomputers implement the *big endian* approach, which stores the most significant part first. Intel microcomputers implement the *little endian* approach, which stores the least

Figure 1.23
Example of big and little endian formats of a 16-bit number.

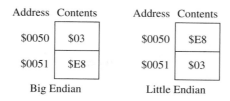

In the foregoing two examples, we normally would not pick out individual bytes (e.g., the $12) but rather capture the entire multiple byte data as one nondivisible piece of information. On the other hand, if each byte in a multiple byte data structure is individually addressable, then both the big and little endian schemes store the data in first-to-last sequence. For example, if we wish to store the four ASCII characters '6811' as a string, which is $3638313100 at locations $50-$54, then the ASCII '6' =$36 comes first in both big and little endian schemes.

The terms "big and little endian" comes from Jonathan Swift's satire *Gulliver's Travels*. In Swift's book, a Big Endian refers to people who crack their egg on the big end. The Lilliputians were Little Endians, because they insisted that the only proper way is to break an egg on the little end. The Lilliputians considered the Big Endians as inferiors. The Big and Little Endians fought a long and senseless war over the best way to crack an egg.

significant part first. The PowerPC is *biendian*, because it can be configured to efficiently handle both big and little endian. Figure 1.23 shows two ways to store the 16-bit number 1000 ($03E8) at locations $50-$51.

We also can use either the big endian or the little endian approach when storing 32-bit numbers into memory that is byte (8-bit) addressable. Figure 1.24 shows the big and little endian formats that could be used to store the 32-bit number $12345678 at locations $50-$53.

Figure 1.24
Example of big and little endian formats of a 32-bit number.

Address	Contents		Address	Contents
$0050	$12		$0050	$78
$0051	$34		$0051	$56
$0052	$56		$0052	$34
$0053	$78		$0053	$12
Big Endian			Little Endian	

1.5.5 Fixed-Point Numbers

We will use fixed-point numbers when we wish to express values in our software that have noninteger values. A **fixed-point number** contains two parts. The first part is a **variable integer**, called **I**. This integer may be signed or unsigned. An unsigned fixed-point number is one that has an unsigned variable integer. A signed fixed-point number is one that has a signed variable integer. The **precision** of a number is the total number of distinguishable values that can be represented. The precision of a fixed-point number is determined by the number of bits used to store the variable integer. On the 6811 or 6812, we typically use 8 bits or 16 bits. Extended precision can be implemented, but the execution speed will be slower because the calculations will have to be performed using software algorithms rather than hardware instructions. This integer part is saved in memory and is manipulated by software. These manipulations include but are not limited to add, subtract, multiply, divide, convert to BCD and convert from BCD. The second part of a fixed-point number is a **fixed constant**, called Δ. This value is fixed, and cannot be changed during execution of the program. The fixed constant is not stored in memory. Usually we specify the value of this fixed content using software comments to explain our fixed-point algorithm. The value of the fixed-point number is defined as the product of the two parts:

$$\text{fixed-point number} \equiv \mathbf{I} \cdot \Delta$$

The **resolution** of a number is the smallest difference that can be represented. In the case of fixed-point numbers, the resolution is equal to the fixed constant (Δ). Sometimes we express the resolution of the number as its units. For example, a decimal fixed-point number with a resolution of 0.001V is really the same thing as an integer with units of mV. When interacting with a human operator, it is usually convenient to use **decimal fixed-point**. With decimal fixed-point the fixed constant is a power of 10.

$$\text{decimal fixed-point number} = I \bullet 10^m \text{ for some constant integer m}$$

Again, the integer **m** is fixed and is not stored in memory. Decimal fixed-point is easy to display, whereas **binary fixed-point** is easier to use when performing mathematical calculations. With binary fixed-point the fixed constant is a power of 2.

$$\text{binary fixed-point number} = I \bullet 2^n \text{ for some constant integer n}$$

Observation: If the range of numbers is known and small, then the numbers can be represented in a fixed-point format.

Checkpoint 1.27: Give an approximation of π, using the decimal fixed-point ($\Delta = 0.001$) format.

Checkpoint 1.28: Give an approximation of π, using the binary fixed-point ($\Delta = 2^{-8}$) format.

For the 6811, which has an 8-bit ADC and a range of 0 to +5V, its resolution is about 5V/256 or 0.02V. It would be appropriate to store voltages on the 6811 as 16-bit unsigned fixed-point numbers with a resolution of 0.01V. For the 9S12C32, which has a 10-bit ADC and a range of 0 to +5V, its resolution is about 5V/1024 or 0.005V. It would be appropriate to store voltages on the 9S12C32 as 16-bit unsigned fixed-point numbers with a resolution of 0.001V. The resolution is choosen so that no information is lost.

It is very important to carefully consider the order of operations when performing multiple integer calculations. For example, assume we wished to calculate M=(53*N)/100, where M and N are integers. There are two mistakes that can happen. The first error is **overflow**, and it is easy to detect. Overflow occurs when the result of a calculation exceeds the range of the number system. In this example, if N is an 8-bit unsigned number, then 53*N can overflow the 0 to 255 range. One solution of the overflow problem is promotion. **Promotion** is the action of increasing the inputs to a higher precision, performing the calculation at the higher precision, checking for overflow, then demoting the result back to the lower precision. In this example, the 53, N, and 100 are all converted to 16-bit unsigned numbers. (53*N)/100 is calculated in 16-bit precision. The result can be verified to be in the 0 to 255 range, then converted back to 8-bit precision. The other error is called **drop-out**. Drop-out occurs during a right shift or a divide, and the consequence is that an intermediate result loses its ability to represent all of the values. To avoid drop-out, it is very important to divide last when performing multiple integer calculations. If we divided first, e.g., M=53*(N/100), then the values of M would be only 0, 53, or 106. We could have calculated M=(53*N+50)/100 to implement rounding to the closest integer. The value 50 is selected because it is about one half of the divisor.

When adding or subtracting two fixed-point numbers with the same Δ, we simply add or subtract their integer parts. First, let x,y,z be three fixed-point numbers with the same Δ. Let $x = I \bullet \Delta$, $y = J \bullet \Delta$, and $z = K \bullet \Delta$. To perform z=x+y, we simply calculate K=I+J. Similarly, to perform z=x−y, we simply calculate K=I−J. When adding or subtracting fixed-point numbers with different fixed parts, we must first convert two the inputs to the format of the result before adding or subtracting. This is where binary fixed-point is more convenient, because the conversion process involves shifting rather than multiplication/division.

For multiplication, we have z=x•y. Again, we substitute the definitions of each fixed-point parameter and solve for the integer part of the result

$$K = (I \bullet J)/\Delta$$

For division, we have z=x/y. Again, we substitute the definitions of each fixed-point parameter and solve for the integer part of the result

$$K=(I \bullet \Delta)/J$$

Again, it is very important to carefully consider the order of operations when performing multiple integer calculations. We must worry about overflow and drop-out.

We can use these fixed-point algorithms to perform complex operations using the integer functions of our 6811/6812. For example, consider the following digital filter calculation.

$$y=x-0.0532672 \bullet x_1 + x_2 + 0.0506038 \bullet y_1 - 0.9025 \bullet y_2$$

In this case, the variables y, y_1, y_2, x, x_1, and x_2 are all integers, but the constants will be expressed in binary fixed-point format. The value -0.0532672 will be approximated by $-14 \bullet 2^{-8}$. The value 0.0506038 will be approximated by $13 \bullet 2^{-8}$. Lastly, the value -0.9025 will be approximated by $-231 \bullet 2^{-8}$. The fixed-point implementation of this digital filter is

$$y=(256 \bullet x - 14 \bullet x_1 + 256 \bullet x_2 + 13 \bullet y_1 - 231 \bullet y_2) >> 8$$

Common error: Lazy or incompetent programmers use floating-point in many situations where fixed-point would be preferable.

Observation: As the fixed constant is made smaller, the accuracy of the fixed-point representation is improved, but the variable integer part also increases. Unfortunately, larger integers require more bits for storage and calculations.

Checkpoint 1.29: Using a fixed constant of 10^{-3}, rewrite the digital filter $y=x-0.0532672 \bullet x_1 + x_2 + 0.0506038 \bullet y_1 - 0.9025 \bullet y_2$ in decimal fixed-point format.

1.6   Common Architecture of the 6811 and the 6812

In 1968, two unhappy engineers named Bob Noyce and Gordon Moore left the Fairchild Semiconductor Company and created their own company, which they called Integrated Electronics (Intel). Working for Intel in 1971, Federico Faggin, Ted Hoff, and Stan Mazor invented the first single-chip microprocessor, the Intel 4004 (Figure 1.25). It was a four-bit processor designed to solve a very specific application for a Japanese company called Busicon. Busicon backed out of the purchase, so Intel decided to market it as a "general

Figure 1.25
The first microprocessors (http://www.cpu-world.com).

1971
Intel 4004

1974
Intel 8080

1974
Motorola 6800

1978
Motorola 6809

purpose" microprocessing system. The product was a success, which lead to two more powerful microprocessors: the Intel 8008 in 1974, and the Intel 8080 also in 1974. Both the Intel 8008 and the Intel 8080 were 8-bit microprocessors using N-channel metal-oxide semiconductor (NMOS) technology. Seeing the long-term potential for this technology, Motorola released its MC6800 in 1974, which was also an 8-bit processor with about the same capabilities as the 8080. Although similar in computing power, the 8080 and 6800 had very different architectures. The 8080 used isolated I/O and handled addresses in a fundamentally different way from data. Isolated I/O defines special hardware signals and special instructions for input/output. On the 8080, certain registers had capabilities designed for addressing, whereas other registers had capabilities for specific for data manipulation. In contrast, the 6800 used memory-mapped I/O and handled addresses and data in a similar way. As defined previously, input/output on a system with memory-mapped I/O is performed in a manner similar to accessing memory.

During the 1980s and 1990s, Motorola (von Neumann architecture) and Intel (Harvard architecture) traveled down similar paths. The microprocessor families from both companies developed bigger and faster products: Intel 8085, 8088, 80x86, . . . and the Motorola 6809, 68000, 680x0. . . . During the early 1980's another technology emerged—the microcontroller. In sharp contrast to the microprocessor family, which optimized computational speed and memory size at the expense of power and physical size, the microcontroller devices minimized power consumption and physical size, striving for only modest increases in computational speed and memory size. Out of the Intel architecture came the 8051 family (www.semiconductors.philips.com), and out of the Motorola architecture came the 6805, 6811, and 6812 microcontroller family (**www. freescale.com**). Many of the same fundamental differences that existed between the original 8-bit Intel 8080 and Motorola 6800 have persisted during thirty years of microprocessor and microcontroller developments. In 1999, Motorola shipped its 2 billionth MC68HC05 microcontroller. In 2004, Motorola spun off its microcontroller products as Freescale Semiconductor, and remained No. 2 in market share in microcontrollers overall and No. 1 in market share in microcontrollers for automotive applications (Gartner Dataquest). Microchip is the No. 1 supplier of 8-bit microcontrollers.

In this section, common features of the 6811 and 6812 are presented. In subsequent sections, information specific to each processor is presented. From an assembly language perspective, the 6812 is a superset of the 6811. In other words, assembly language programs written for the 6811 can be reassembled and run on the 6812. The stack pointer operates slightly differently on the two computers, but this difference will not prevent most 6811 programs from running without modification on the 6812. Software written for embedded systems such as the 6811 and 6812 are tightly coupled to their I/O devices. Although the 6812 has more I/O ports than the 6811, the 6812 I/O structure is not a superset of the 6811. Consequently, a significant effort, translating I/O interface software, would be required to convert an embedded system based on the 6811 system to run on a 6812. The machine code for the two processors is different, so software developed on one must be reassembled to run on the other.

Observation: Freescale should have developed versions of the 6812 with I/O ports and pin outs identical to its most popular versions of the 6811.

1.6.1 Registers

The 6811 and 6812 registers are depicted in Figure 1.26. Registers A and B concatenated together form a 16-bit accumulator, Register D, with Register A containing the most significant byte. Typically Registers A and B contain data (numbers) whereas Registers X and Y contain addresses (pointers.) The stack pointer operates slightly differently on the 6811 and the 6812. On the 6811, SP+1 points to the top element of the stack. On the 6812, SP points to the top element of the stack. Register PC (program counter) points to the current instruction.

Figure 1.26
The 6811 and 6812
have six registers.

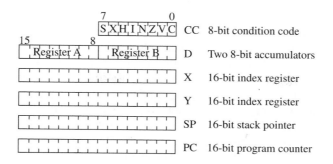

The condition code bits are shown in Figure 1.27. The N, Z, V, and C bits signify the status of the previous ALU operation. Many instructions set these bits to signify the result of the operation. When S=1, the stop instruction is disabled. When X=0, XIRQ interrupts are allowed. Once X is set to zero, the software cannot set it back to 1. The H bit is used for binary coded decimal (BCD) addition. When I=0, IRQ interrupts are enabled. Interrupts are discussed in Chapter 4.

Figure 1.27
The 6811 and 6812
condition code bits.

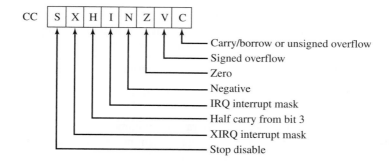

Checkpoint 1.30: To prevent (disable) interrupts, what value should the I bit be?

Checkpoint 1.31: List all the registers that can hold 16-bit addresses?

**1.6.2
Terminology**

This chapter focuses on 6811 and 6812 architecture, but now we introduce a few simple instructions in order to understand how the microcomputer works. When describing the action of an assembly instruction we will use the following notation.

```
w is a signed 8-bit -128 to +127 or unsigned 8-bit 0 to 255
n is a signed 8-bit -128 to +127
u is a unsigned 8-bit 0 to 255
W is a signed 16-bit -32787 to +32767 or unsigned 16-bit 0 to 65535
N is a signed 16-bit -32787 to +32767
U is a unsigned 16-bit 0 to 65535
=[addr] specifies an 8-bit read from addr
={addr} specifies a 16-bit read from addr using "big endian"
=<addr> specifies a 32-bit read from addr using "big endian"
[addr]= specifies an 8-bit write to addr
{addr}= specifies a 16-bit write to addr using "big endian"
<addr>= specifies a 32-bit write to addr using "big endian"
```

Assembly language instructions have four fields. The *label field* is optional and starts in the first column, it is used to identify the position in memory of the current instruction. You must choose a unique name for each label. The *opcode field* specifies the microcomputer command to execute. The 6812 includes all the 6811 instructions. The *operand field* specifies where to find the data to execute the instruction. We will see that opcodes have 0, 1, 2, or 3 operands. The *comment field* is also optional and is ignored by the computer, but it allows you to describe the software making it easier to understand. Good programmers but add comments to explain the software. The `ldaa` instruction reads 8 bits of data from memory and places them in register A. The `staa` instruction stores the 8-bit value from register A into memory. The `ldx` instruction reads 16 bits of data from memory and places them in register X. The `stx` instruction stores the 16-bit value from register X into memory. The first two assembly instructions copy the contents of Port A (location 0 on the 6812) to memory location $3800. The next two assembly instructions move a 16-bit value from locations $3802–$3803 into locations $3804–$3805.

label	opcode	operand	comment
here	ldaa	$0000	RegA=[$0000]
	staa	$3800	[$3800]=RegA
	ldx	$3802	RegX={3802}
	stx	$3804	{3804}=RegX

As we will learn in further along in the book, it is much better to add comments to explain how—or even better, why—we do the action. But for now we are learning what the instruction is doing, so in this chapter, comments will describe what the instruction does. The assembly language instructions (like the forgoing example) are translated into machine instructions. The `ldaa $0000` instruction is translated into 2 bytes of machine code:

Object code	instruction	comment
$96 $00	ldaa $0000	RegA=[$0000]

1.6.3 Addressing Modes

A fundamental issue in program development is the differentiation between data and address. It is in assembly language programming in general and addressing modes in specific that this differentiation becomes clear. When we put the number 1000 into register X, whether this is data or address depends on how the 1000 is used. Most instructions access memory to fetch parameters or save results. The addressing mode is the format the instruction uses to specify the memory location to read or write data. All instructions begin by fetching the machine instruction (opcode and operand) pointed to by the PC. Some instructions operate completely within the processor and require no memory data fetches. These instructions have no operand and are classified as *inherent*. For example, the `clra` instruction places a zero into register A. If the data is found in the instruction itself, the instruction uses *immediate* addressing mode. If the instruction uses the absolute address to specify the memory data location, the instruction uses either *direct* or *extended* addressing mode. Notice in Table 1.9 that the `ldaa` instruction can be used with the immediate, direct, and extended addressing modes. In particular, the `ldaa` instruction means "load into register A" and the addressing mode specifies the source of the data to be used. Many computers, including the 6811 and the 6812, use *PC-relative* addressing mode to encode branch instructions. PC-relative addressing makes the object code smaller and relocatable. Normally, the computer executes one instruction after another as they are listed in memory, except the branch instructions, which cause the program to jump to another place. For

Table 1.9
Simple addressing modes.

6811 code	6812 code	opcode	operand	comment
$4F	$87	clra		Reg A = 0 (inherent)
$86 24	$86 24	ldaa	#36	Reg A = $24 (immediate)
$96 24	$96 24	ldaa	36	Reg A = [$0024] (direct)
$B6 08 01	$B6 08 01	ldaa	$0801	Reg A = [$0801] (extended)
$20 40	$20 $40	bra	$F042	Jump from $F002 to $F042

example, the `bra $F000` instruction is an unconditional branch, causing the program to jump to location $F000. The addressing mode is called PC-relative because the machine code contains the address difference between where the program is now and the address to which the program will jump. There are many more addressing modes, but for now, these five addressing modes, as illustrated in Table 1.9, are enough to get us started. Notice that the assembly code for the 6811 and the 6812 are the same, but the machine code is slightly different. Therefore, when converting a 6811 program to run on a 6812, the original software must be reassembled using a 6812 assembler.

Checkpoint 1.32: What is the addressing mode used for?

In this section, these five addressing modes are introduced. These simple addressing modes are sufficient to understand most of the software presented in this book. The more complicated addressing modes are presented in Chapter 2.

1. Inherent addressing mode has no operand field. Sometimes there is no data for the instruction at all. For example, the `stop` instruction halts execution. Sometimes the data for the instruction is implied. For example, the `clrb` instruction sets register B to zero. In this case, the data value of zero is implied. On the other hand, sometimes the data must be fetched from memory, but the address of the data is implied. For example, the `pula` instruction will pop an 8-bit data from the stack and store it in register A. In particular, the data value pointed to by the SP is read from memory and stored into register A.

2. Immediate addressing mode uses a fixed data constant. The data itself is included in the machine code. For example, the `ldaa #36` instruction will store a data value of 36 into register A (Figure 1.28). Notice that the "36" itself is encoded in the machine code for the `ldaa #36` instruction. In assembly code, this mode is signified by the # sign.

Figure 1.28
Example of the immediate addressing mode.

Observation: With immediate mode addressing, the information is stored in the machine code.

Common error: It is illegal to use the immediate addressing mode with instructions that store data into memory (e.g., staa).

Checkpoint 1.33: What is the difference between `ldaa #36` and `ldaa #$24`?

3. Direct-page addressing mode uses an 8-bit address to access from addresses 0 to $00FF. In many computer systems outside the Freescale family this addressing mode is called *zero-page*. These addresses include the 6811 single chip RAM, but on the 6812 they reference the I/O ports. Some 6811 instructions can be used with both direct and extended addressing whereas others can be used with extended, but not direct mode. In assembly language, the < operator forces direct addressing. Figure 1.29 illustrates the execution of the `ldaa 36` instruction.

Figure 1.29
Example of the direct-page addressing mode.

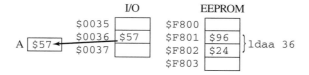

Observation: With direct and extended mode addressing, a fixed pointer to the information is stored in the machine code. The data itself may change dynamically, but its location is fixed.

Checkpoint 1.34: What is the difference between `ldaa #36` and `ldaa 36`?

4. Extended addressing mode uses a 16-bit address to access all memory and I/O devices. In many computer systems outside the Freescale family this addressing mode is called *direct*, because it can directly access all of memory. In this book, we will adhere to the Freescale terminology (direct means 8-bit addressing and extended means 16-bit addressing). Some 6811 instructions, which can be used with direct and indexed addressing, cannot be used with extended addressing. The > operator forces extended addressing. In general, when the address happens to fall in the 0 to $00FF range, the assembler will automatically use direct addressing, otherwise it uses extended addressing. Figure 1.30 illustrates the execution of the `ldaa $0801` instruction.

Figure 1.30
Example of the extended addressing mode.

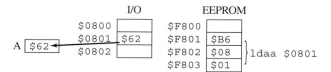

Common error: It is wrong to assume the < and > operators affect the amount of data that is transferred. The < and > operators will affect the addressing mode; that is how the address is represented.

Observation: Accesses to Registers A, B, and CC transfer 8 bits, whereas accesses to Registers D, X, Y, SP, and PC transfer 16 bits regardless of the addressing mode.

Checkpoint 1.35: What is the difference between `ldx #$0801` and `ldx $0801`?

Checkpoint 1.36: What is the difference between the instruction `ldaa $12` and the instruction `ldx $12`?

5. PC-relative addressing mode is used for the branch and branch to subroutine instructions. Stored in the machine code is not the absolute address of where to branch, but the 8-bit signed offset relative distance from the current PC value. When the branch address is being calculated, the PC already points to the next instruction. Calculating relative offsets gives first-time programs a lot of trouble, but lucky for us the assembler calculates it for us. It is explained here in order to better understand how the computer works, rather than being necessary for us to do while programming. The address of the next instruction is (location of instruction)+(number of bytes in the machine code). In the following example, assume the branch instruction is located at address $F880. The destination address is before the current instruction, which is called a backward jump.

```
bra $F840
```

The operand field for PC relative addressing is an 8-bit value called **rr,** which is calculated using the equation (destination address)−(location of instruction)−(size of the instruction). Since the bra op code is one byte ($20) and the operand is one byte, this instruction requires two bytes and the **rr** field is

```
$F840 - $F880 - 2 = -$42 = $BE
```

and the object code for this instruction will be $20BE. Again assume the branch instruction is located at address $F880. This time the destination address is after the current instruction, which is called a forward jump.

```
bra $F8C8
```

The **rr** field is

```
$F8C8 - $F880 - 2 = $46
```

and the object code for this instruction will be $2046.

> **_Common error:_** Since not every instruction supports every addressing mode, it would be a mistake to use an addressing mode not available for that instruction.

> **_Observation:_** The number of cycles required to execute instructions on the 6811 is fixed for each instruction. The time for each 6811 cycle is also fixed.

> **_Observation:_** Some of the conditional branch instructions on the 6812 require a different number of cycles to execute depending on whether or not the branch is taken. The cycle time when accessing external memory on a 6812 depends on the speed of the external memory. It also depends on whether the address is an even number or an odd number. These facts complicate the task of predetermining how long a 6812 program will take to execute.

> **_Observation:_** Relative addressing within a program block is essential for implementing relocatable code.

**1.6.4
Numbering
Scheme Used by
Freescale for the
6811 and the 6812**

The numbering scheme used by Freescale allows you to quickly determine the type of ROM used as the main program memory in the device, as shown in Table 1.10. In most cases the value at the end of the part number specifies the size of the program memory (e.g., MC68HC711E20 is 20K and MC9S12C32 is 32K), but sometimes for the 6811 it does not (e.g., MC68HC11D3 is 4K and MC68HC711E9 is 12K). The letter (or letters) placed after the 11 or 12 and before the final number specifies the series. For example, the MC68HC711E9 belongs to the "E" series. Please note that most Freescale microcontrollers have two or three memory modules, and the numbering scheme refers only to the main program memory. For example, according to the numbering scheme the MC9S12DP512 has 512K bytes flash EEPROM, but it also has 4K bytes of regular EEPROM and 12K bytes of RAM.

Table 1.10
Freescale uses a
numbering scheme to
specify the type of main
memory.

Number	Main Memory	Examples
None	ROM	MC68HC11E9 MC68HC12D60
7	EPROM	MC68HC711E9 MC68HC711D3
8	EEPROM	MC68HC811E2 MC68HC812A4
9	Flash EEPROM	MC68HC912B32 MC9S12C32

> **_Checkpoint 1.37:_** The MC9S12A64 has how much and what type of main memory?

1.7 6811 Architecture

**1.7.1
6811 Family**

In 1984, Motorola introduced the first 6811 microcontroller. This device continues to be popular in the automotive and educational markets because of its simple yet elegant architecture and its effective balance of low cost and moderate computational performance. The 6811 uses either two separate 8-bit accumulators (A,B) or one combined 16-bit accumulator (D). There are two 16-bit index registers (X and Y). Like all the Freescale microcomputers, the 6811 has powerful bit-manipulation instructions. The 6811 supports 16-bit add/subtract, 16×16 integer divide, 16×16 fractional divide, and 8×8 unsigned multiply. Although not as convenient as the 6812, the 6811 instruction set does lend itself to C compiler implementations, primarily because it handles 16-bit manipulations well.

In many applications, the 6811 provides a single-chip solution with mask programmed ROM or user-programmable EPROM. All family derivatives are also expandable for the incorporation of external memory in the design. A 4-channel Direct Memory Access (DMA) unit on some devices permits fast data transfer between two blocks of memory (including externally mapped memory in expanded mode), between registers, or between registers and memory. Within the 6811 family, there are five major series of microcontroller units. The following are examples of the features the 6811 can offer as shown through specific devices within each series:

D series—The 68HC11D3 chip with 4K bytes ROM offers an economical alternative for applications when advanced 8-bit performance is required with fewer peripherals and less memory.

E-series—The 68HC11E9 family comes in a wide range of I/O capabilities. It combines EEPROM and EPROM on a single chip. The E-series offers multiple memory sizes in a pin compatible package.

F-series—The MC68HC11Fl runs in expanded mode. This particular chip series stands out with its extra I/O ports, an increase in static RAM (1K bytes), chip selects, and a 5 MHz non-multiplexed bus.

K-series—The MC68HC11K4 offers high speed, large memories, a memory management unit, pulse width modulation along with standard I/O ports.

P-series—The 68HC11P2 offers a power saving programmable phase lock loop (PLL) based clock circuit along with many I/O pins, large memory and 3 SCI ports.

The 6811 can operate in one of four modes, as specified by the values on signals MODA and MODB that exist when the chip starts up after a reset. In *expanded mode*, some of the pins are used for the address and data bus so that external devices can be attached to the system. We will focus on expanded mode, which we can use during development, and single-chip mode, which we embed into our final product. The other two modes are bootstrap mode, which can be used to load programs into RAM and test mode, which is used by Freescale to verify that the chip is operational. In *single-chip mode*, the 6811 implements a complete microcomputer, where all I/O ports are available. This mode is used for the final product with the application software programmed into the ROM or EPROM. There are two approaches to develop hardware/software systems for 6811. The most popular approach is to operate the 6811 in expanded mode on a development system, so that user can download software into external RAM. Once the system is debugged, the program could be burned into ROM or EPROM, and the 6811 single chip computer can be embedded into the system. A second approach can be used when the program fits into EEPROM. In this case, we download programs directly into EEPROM without the need to have external RAM. The one-time-programmable (OTP) devices such as the MC68HC711E9 can be used in single-chip mode in applications that require a small number of devices to be built. The memory maps shown in this section are the default values and can be relocated by writing to the INIT and CONFIG registers. Figure 1.31 shows a 6811 in single-chip mode, and the interface circuits are built on a protoboard, as one would do during product development.

Figure 1.31
ADAPT-11 single-chip 6811 module from Technological Arts, used in a motor controller.

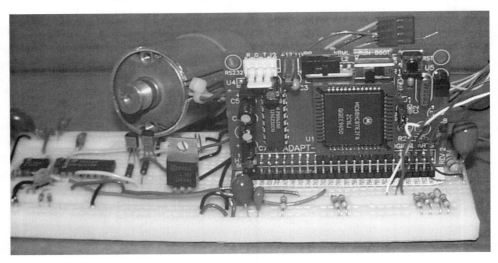

1.7.2
MC68HC711E9

In the single-chip mode, the MC68HC711E9 is a microcontroller with 12K bytes of internal EPROM, 512 bytes of EEPROM, and 512 bytes of RAM. There are five I/O ports on this 6811, called Port A, Port B, Port C, Port D, and Port E. All the ports are 8 bits wide except for Port D, which is only 6 bits. Observing the 38 I/O pins on the 6811, as shown at the bottom of Figure 1.32, we see that each signal has at least two definitions, meaning it can be used for more than one purpose. For example, the pin PE0 can be used as a digital input or as an analog input. Other versions of the 6811 have differing amounts of memory and I/O ports. On the other hand, notice that ports A, B, and E are not completely general. For example, pins PA6, PA5, PA4, and PB7-PB0 cannot be used as inputs. Pins PA2, PA1, PA0, and PE7-PE0 cannot be used as outputs.

Figure 1.32
Block diagram of the Freescale MC68HC711E9.

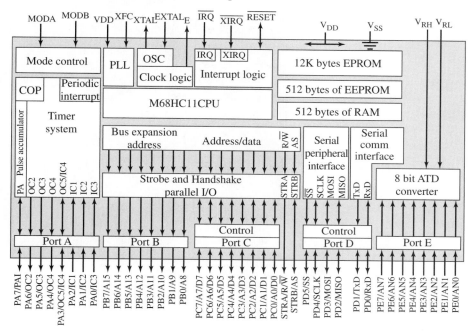

The address map of a single-chip MC68HC711E9 is shown in Table 1.11. Table 1.12 shows the configuration of the I/O pins on the E series 6811.

Address	Size	Device	Device	Contents
$0000 to $01FF	512	RAM	Random Access Memory	Variables and stack
$1000 to $103F	64	I/O	Input/output devices	
$B600 to $B7FF	512	EEPROM	Electrically erasable	Constants unique to each system
$D000 to $FFFF	12K	EPROM	OTP	Programs and fixed constants

Table 1.11
The MC68HC711E9 has 12K of EPROM and 512 bytes of RAM.

Port	Input Pins	Output Pins	Bidirectional Pins	Shared Functions
Port A	PA2-PA0	PA6-PA4	PA7, PA3	Timer
Port B	—	PB7-PB0	—	High Order Address
Port C	—	—	PC7-PC0	Low Order Address and Data Bus
Port D	—	—	PD5-PD0	SCI and SPI
Port E	PE7-PE0	—	—	Analog to Digital Converter

Table 1.12
The 6811 E series has five external I/O ports.

During the hardware and software development, we can use the 6811 in the expanded or microprocessor mode. This mode allows you to add external RAM and ROM. Running the software out of RAM shortens the edit/compile/download time, because it is faster to write to RAM then to program PROM. In expanded mode, 6811 ports B and C are used for the address and data bus. The Peripheral Recovery Unit (PRU, 6824) can be added to reconstruct I/O ports B and C. Many of the first commercial 6811 development systems utilized the 6811 in expanded mode, for example, the Technological Arts system shown in Figure 1.33.

Figure 1.33
ADAPT-11C24DX expanded mode 6811 module from Technological Arts.

Figure 1.34
Block diagram of the MC68HC11D3.

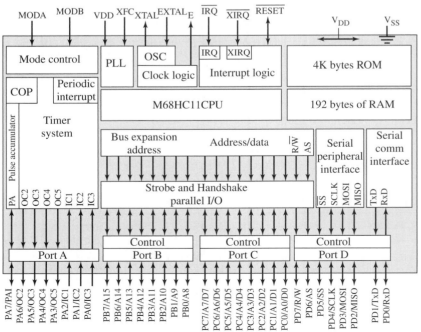

1.7.3
MC68HC11D3

Compare the block diagram of the MC68HC11D3 computer shown in Figure 1.34 with the MC68HC711E9 and notice that it has fewer I/O devices but is available in a smaller, less-expensive package. The on-board 192-byte RAM is initially located at $0040 after reset but can be placed at any other 4K boundary ($x040) by writing an appropriate value to the INIT register. The 64-byte register block originates at $0000 after reset but can be placed at any other 4K boundary ($x000) after reset by writing an appropriate value to the INIT register. There is no EEPROM or ATD on this version of the 6811. The 4K bytes of ROM are located at $C000 through $FFFF.

The MC68HC11D3 has four 8-bit I/O ports: A, B, C, and D. In single-chip and boot-strap modes, all ports are parallel I/O data ports. In expanded-multiplexed and test modes, Ports B and C and lines AS and R/W are a memory expansion bus. The address map of a single chip MC68HC11D3 is presented in Table 1.13.

Address (hex)	Size	Device	Device	Contents
$0000 to $003F	64	I/O	Input/output devices	
$0040 to $00FF	192	RAM	Random Access Memory	Variables and stack
$F000 to $FFFF	4K	ROM	Read only memory	Programs and fixed constants

Table 1.13
The MC68HC711D3 has 4K of EPROM and 192 bytes of RAM.

1.8 ■ 6812 Architecture

1.8.1
6812 Family

The 6812 is a highly integrated, general-purpose family of microcomputers with a 16-bit microcontroller architecture specifically designed for low-power consumption, as listed in Table 1.14. It is source code compatible with the popular 6811 8-bit microcontroller family, but some of the I/O ports operate slightly differently. Features include:

Low power consumption and low voltage operation at full bus speed
Single wire Background Debug™ module for minimally intrusive in-circuit
    programming and debugging
High-level language optimization

Series	ROM	RAM	I/O pins	Features
MC68HC812A4	4	1	91	EEPROM
MC68HC912B32	32	1	63	J1850 Byte Data Link
MC912D60A	60	2	84	
MC912DG128A	128	8	85	
MC9S12A	32 to 512	2 to 14	59 to 91	General Purpose with I^{2}C
MC9S12B	64 to 128	2 to 4	91	Automotive/Industrial with CAN
MC9S12C	32 to 128	2 to 4	60	Low Pin Count, Low Cost CAN
MC9S12D	32 to 512	2 to 12	59 to 91	Automotive/Industrial with CAN
MC9S12E	64 to 128	4 to 8	90	General Purpose, 3 V with D/A
MC9S12GC	16 to 128	2 to 4	60	Low Cost, Low Pin Count
MC9S12H	128 to 256	12	117	LCD/H-Bridge Drivers with CAN
MC9S12U	32	3.5	75	USB2.0
MC9S12XDP	512	32	119	12C, CAN, Peripheral Coprocessor
MC9S12NE	64	8	80	Ethernet, 12C

Table 1.14
The 6812 family includes many series, where ROM and RAM sizes are given in K bytes.

Flash EEPROM and byte-erasable EEPROM integrated on a single device
Fuzzy logic instructions
Enhanced arithmetic instructions over the 68HC11

The first 6812 products were the MC68HC812A4 (1K bytes of RAM and 4K bytes of EEPROM) and the MC68HC912B32 (1K bytes of RAM, 32K bytes of flash EEPROM). In fact, the first edition of this book featured these two devices. Currently, the 6812 family consists of a large number of devices with a wide range of memory and I/O configurations, as shown in Table 1.14. This book will focus on the MC9S12C32 because it is a low-cost full-featured device with minimal memory, making it ideally suitable for educational use. The fundamental concepts learned in this book can easily be adapted to other members of the Freescale microcontroller family.

1.8.2 MC9S12C32

The MC9S12C32 microcontroller unit (MCU) is a 16-bit device composed of many on-chip peripheral modules connected by an intermodule bus, as shown in Figure 1.35. Modules include a 16-bit central processing unit (HCS12), a system integration module (SIM), an asynchronous serial communications interface (SCI) module, a serial peripheral interface (SPI) module, an 8-channel 16-bit timer module, a pulse width modulation (PWM) module, an 8-channel 10-bit analog-to-digital converter (ATD), a 1 Mbps controller area network (CAN 2.0) module, 2K bytes of RAM, 32K bytes of Flash EEPROM, memory expansion logic, and a phase-locked loop (PLL). The PLL allows the software to increase or decrease the execution speed. The PWM module can utilize either Port P or Port T, which can be configured as either a 6-channel 8-bit device or a 3-channel 16-bit device. Ports J and P also support keypad wakeup interrupts.

Figure 1.35
Block diagram of a Freescale MC9S12C32.

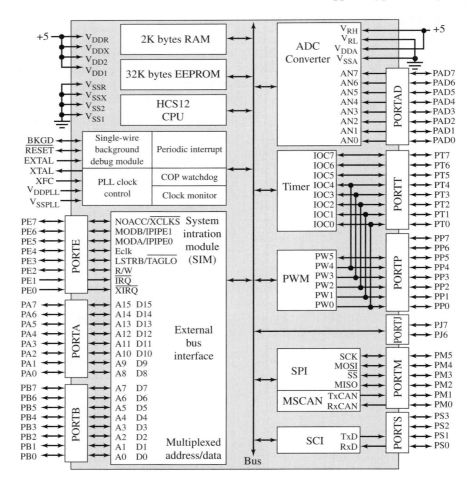

The MC9S12C32 can operate in one of eight modes, the mode being selected by the values of the three signals BKGD, MODA and MODB that exist when the device starts up after a reset. In this book, we will study only the three modes shown in Table 1.15. In *single-chip mode*, the 6812 contains the four major building blocks required to make a complete computer system: processor, I/O, RAM, and EEPROM. For most of the book, we will be using single-chip mode, where all ports are available for input/output. Because the 9S12 family is available with flash EEPROM ranging from 32K bytes to 512K bytes, most embedded projects can be developed directly in single-chip mode. On the other hand, the 9S12C32 family RAM sizes range from 2K to 12K bytes. In Chapter 9, we will use the two expanded modes to interface external devices to Ports A, B, and E. For embedded systems that require a large amount of read/write storage, we will use expanded modes to interface external RAM to the system. *Expanded narrow mode* creates a 16-bit address bus and an 8-bit data bus, whereas *expanded wide mode* implements both the address bus and the data bus as 16 bits.

BKGD	MODB	MODA	Mode	Port A	Port B
1	0	0	Normal Single Chip	In/Out	In/Out
1	0	1	Normal Expanded Narrow	A15-8/D15-8/D7-0	A7-A0
1	1	1	Normal Expanded Wide	A15-8/D15-8	A7-0/D7-0

Table 1.15
The MC9S12C32 has eight operating modes, but we will study these three modes.

We use flash EEPROM during development because it takes only minutes to perform an edit/assemble/download cycle. For projects requiring a large number of manufactured products, we can burn our program code into factory-programmed ROM and embed the microcomputer into our final product. For smaller projects, we simply embed the flash EEPROM device into our final product. The MC9S12C32 has 32K bytes of flash EEPROM (electrically erasable programmable read only memory), and 2048 bytes of RAM. Other versions of the 6812 have differing amounts of these types of memory. In single-chip mode, the 6812 implements a complete microcomputer, where all its I/O ports are available. This mode is used for the final product, with the application software programmed into the ROM. The address space of the input/output devices, and the RAM can be mapped on any 2K-boundary by software. In this book, we will use the address map shown in Table 1.16 for the single-chip MC9S12C32.

Address	Size	Device	Device	Contents
$0000 to $03FF	1K	I/O	Input/output devices	
$3800 to $3FFF	2K	RAM	Random Access Memory	Variables and stack
$4000 to $7FFF	16K	EEPROM	Electrically erasable PROM	Programs and fixed constants
$C000 to $FFFF	16K	EEPROM	Electrically erasable PROM	Programs and fixed constants

Table 1.16
The MC9S12C32 has 32K of EEPROM and 2K bytes of RAM.

The MC9S12C32 is available in three quad flat package (QFP) sizes. The larger chip packages have more pins, as shown in Table 1.17. The Technological Arts development system, shown in Figure 1.36, utilizes the 48-pin QFP package.

Port	48-pin	52-pin	80-pin	Shared Functions
Port A	PA0	PA2–PA0	PA7–PA0	Address/Data Bus
Port B	PB4	PB4	PB7–PB0	Address/Data Bus
Port E	PE7, PE4, PE1, PE0	PE7, PE4, PE1, PE0	PE7–PE0	System Integration Module
Port J	—	—	PJ7, PJ6	Key wakeup
Port M	PM5–PM0	PM5–PM0	PM5–PM0	SPI, CAN
Port P	PP5	PP5–PP3	PP7–PP0	Key wakeup, PWM
Port S	PS1–PS0	PS1–PS0	PS3–PS0	SCI
Port T	PT7–PT0	PT7–PT0	PT7–PT0	Timer, PWM
Port AD	PAD7–PAD0	PAD7–PAD0	PAD7–PAD0	Analog-to-Digital Converter

Table 1.17
The MC9S12C32 has nine external I/O ports.

Figure 1.36
MC9S12C32
development system
from Technological Arts
(#NC12C32).

Normally, the execution speed of a microcontroller is determined by an external crystal. The MC9S12C32, shown in Figure 1.36, has an 8 MHz crystal creating a 4 MHz E clock. This 6812 also has a phase-locked loop (PLL) that allows the software to adjust the execution speed of the computer. Program 1.1 will increase the E clock from 4 MHz to 24 MHz. The OSCCLK is the frequency of the crystal. Typically, the choice of frequency involves software execution speed versus electrical power.

Program 1.1
MC9S12C32 program
that increases the 4 MHz
E clock to 24 MHz.

```
void PLL_Init(void){
  SYNR = 0x02;
  REFDV = 0x00; // PLLCLK = 2*OSCCLK*(SYNR+1)/(REFDV+1)
  CLKSEL = 0x00;
  PLLCTL = 0xD1;
  while((CRGFLG&0x08) == 0){          // Wait for PLLCLK to stabilize.
  }
  CLKSEL_PLLSEL = 1; // Switch to PLL clock
}
```

1.8.3
MC68HC812A4

The MC68HC812A4 microcontroller unit (MCU) is a 16-bit device composed of on-chip modules connected by an intermodule bus, as shown in Figure 1.37. Modules include a 16-bit central processing unit (CPU12), a Lite integration module (LIM), two asynchronous serial communications interfaces (SCI0 and SCI1), a serial peripheral interface (SPI), a timer and pulse accumulation module, an 8-bit analog-to-digital converter (ATD), 1K bytes of RAM, 4K bytes of EEPROM, memory expansion logic with chip selects, key wakeup ports, and a phase-locked loop (PLL).

Figure 1.37
Block diagram of a Freescale MC68HC812A4.

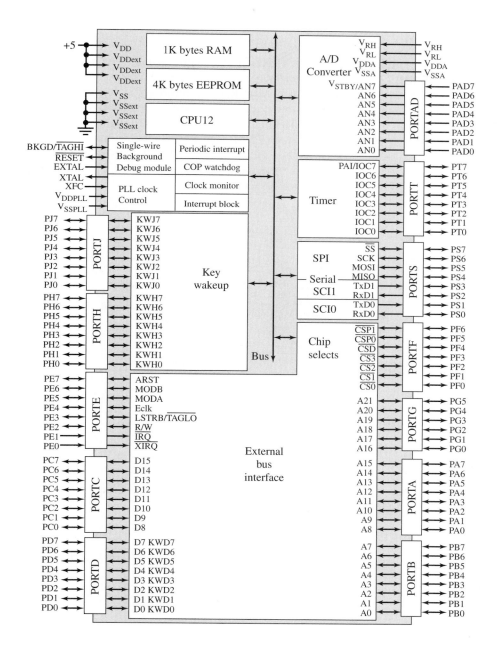

The address space of the Input/Output Devices and the RAM are mappable to any 2K space. The EEPROM is mappable to any 4K space. The address map of a single-chip MC68HC812A4 is shown in Table 1.18.

Address	Size	Device	Device	Contents
$0000 to $01FF	512	I/O	Input/output devices	
$0800 to $0BFF	1028	RAM	Random access memory	Variables and stack
$F000 to $FFFF	4096	EEPROM	Electrically erasable PROM	Programs and fixed constants

Table 1.18
The MC68HC812A4 has 4K of EEPROM and 1K bytes of RAM.

1.8.4 MC68HC912B32

The MC68HC912B32 microcontroller unit (MCU) is a 16-bit device composed of on-chip peripherals including a 16-bit central processing unit (CPU12), 32K bytes of flash EEPROM, 1K bytes of RAM, 768 bytes of regular EEPROM, an asynchronous serial communications interface (SCI), a serial peripheral interface (SPI), an 8-channel timer and 16-bit pulse accumulator, an 8-bit analog-to-digital converter (ADC), a four-channel pulse-width modulator (PWM), and a J1850-compatible *byte data link communications* module (BDLC), as shown in Figure 1.38. The MC68HC912B32 has full 16-bit data paths throughout; however, the multiplexed external bus can operate in an 8-bit narrow mode so single 8-bit wide memory can be interfaced for lower-cost systems.

Figure 1.38
Block diagram of a Freescale MC68HC912B32.

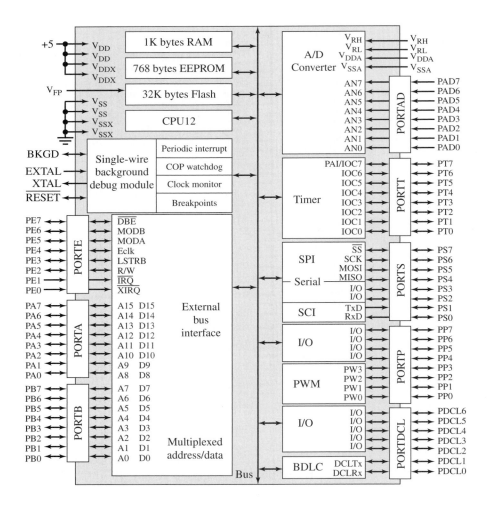

The address map of a single-chip MC68HC912B32 is shown in Table 1.19.

Address	Size	Device	Device	Contents
$0000 to $01FF	512	I/O	Input/output devices	
$0800 to $0BFF	1028	RAM	Random Access Memory	Variables and stack
$0D00 to $0FFF	768	EEPROM	Regular EEPROM	Fixed constants
$8000 to $FFFF	32768	EEPROM	Flash EEPROM	Programs

Table 1.19
The MC68HC912B32 has 32K bytes of flash EEPROM and 1K bytes of RAM.

1.9   Parallel I/O Ports

1.9.1
Basic Concepts of Input and Output Ports

A *parallel I/O port* is a simple mechanism that allows the software to interact with external devices. It is called parallel because multiple signals can be accessed all at once. An input port, which allows the software to read external digital signals, is usually read only. That means a read cycle access from the port address returns the values existing on the inputs at that time. In particular, the tristate driver (triangle-shaped circuit in Figure 1.39) will drive the input signals onto the data bus during a read cycle from the port address. A write cycle access to an input port usually produces no effect. A simple fixed input port, such as pin PA0 on the MC68HC711E9 or pin PE0 on the MC9S12C32, behaves similarly to the circuit shown in Figure 1.39. The digital values existing on the input pins are copied into the microcomputer when the software executes a read from the port address.

Figure 1.39
A read-only input port allows the software to read external digital signals.

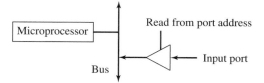

A latched input port behaves similarly to the circuit shown in Figure 1.40. The digital values existing on the input pins are copied into an internal latch on an appropriate edge of the external control signal. At a later time, the data is transferred to the microcomputer when the software executes a read from the latch address. Notice that this latched input port also supports the regular input function. In other words, the software has the option of reading the port address to get information directly from the input port pins or from the latch address to get information that existed at the time of the previous active edge of the external control signal. On the MC68HC711E9, the STRA signal can be configured to latch the Port C inputs into the PORTCL latch register, as shown in to Figure 1.40.

Figure 1.40
A latched input port allows the software to read external digital signals that are captured via the external control signal.

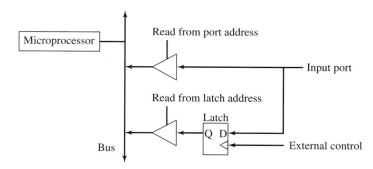

Checkpoint 1.38: What happens if the software writes to an input port?

Often the latched input port allows the software to select the strategic edge to affect the latch function. In other words, the software specifies whether the rise or fall of an external control signal latches the input data. It is important to remember that when the software reads from the latch address, it obtains the values of the input signals occurring at the time of the active edge of the external control signal. In this way, the external device can provide the data at the input and issue an edge on the control signal latching the data into the computer; the software can process the data at a later time without requiring the external device to maintain the valid data at the input. One of the limitations of this particular configuration is that it does not include a way for the software to signal back (acknowledge) to the external hardware that the previous data has been read by the software. In Chapter 3 we will develop more sophisticated handshaking protocols that provide for two-way synchronization.

Although an input device usually involves just the software reading the port, an output port can participate in both the read and write cycles very much like a regular memory. Figure 1.41 describes a "readable output port." For an 8-bit output port there will be 8 D flip flops to hold the values on the output pins. A write cycle to the port address will affect the values on the output pins. In particular, the microcomputer places information on the data bus and that information is clocked into the D flip flops. Since it is a readable output, a read cycle access from the port address returns the current values existing on the port pins. A "write-only output port" does not allow software to read the current values. There are no output-only ports on the 6812, but Port B on the MC68HC711E9 operates as shown in to Figure 1.41.

Figure 1.41
A readable output port allows the software to generate external digital signals.

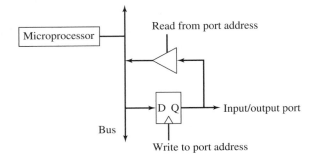

Checkpoint 1.39: What happens if the software reads from an output port as shown in Figure 1.41?

Some ports are fixed as inputs (e.g., the 6811 Port E) or outputs (e.g., the 6811 Port B). To make the microcontroller more marketable, most ports can be software-specified to be either inputs or outputs. Freescale uses the concept of a direction register to determine whether a pin is an input (direction register bit is 0) or an output (direction register bit is 1), as shown in Figure 1.42. We define a *ritual* as a program executed during startup that initializes hardware and software. If the ritual software makes direction bit zero, the port pin behaves like a simple input, and if it makes the direction bit one, the pin becomes a readable output. Ports C and D on the 6811 as well as most of the 6812 ports operate as shown in to Figure 1.42.

Intel uses open collector outputs to determine whether a pin is an input or an output, as shown in Figure 1.43. Port P0 on the Intel 8051 is open collector. If the software writes a "1" to the port, it behaves like a simple input port. If Port P0 is to be used as an output, you will need to add external pull-up resistors.

On the Intel 8051, Ports P1, P2, and P3 are open collector with internal pull-up resistors, as shown in Figure 1.44. If the software writes a "1" to the port, it behaves like a simple input port with the pullup to +5V. If the software writes a "0" to the port, the output will be pulled low, and if it writes a "1", the output will be pulled up by the internal resistor.

Figure 1.42
A bidirectional port can be configured as a read-only input port or a readable output port.

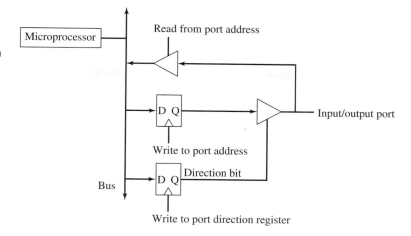

Figure 1.43
A bidirectional port is sometimes implemented with an open collector readable output port.

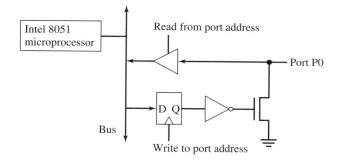

Figure 1.44
Some of the bidirectional ports on the Intel 8051 have internal pull-up resistors.

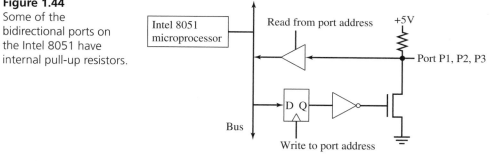

Common error: An output port that is created with open collector outputs with pull-up may not have enough I_{OH} to drive its external circuits.

Common error: Many program errors can be traced to confusion between I/O ports and regular memory. For example, you should not write to an input port, and sometimes you cannot read from an output port.

Observation: If a port pin is configured as a readable output but external loading causes the pin voltage to be different from the value written by the software, then a read from the port will return the voltage level at the pin and not the value last written by the software. This fact can be used by the software to detect excess loading by the external circuits.

1.9.2
Introduction to
I/O Programming
and the Direction
Register

On most embedded microcomputers, the I/O ports are memory mapped. This means the software accesses an input/output port simply by reading from or writing to the appropriate address. To make our software more readable we usually include symbolic definitions for the I/O ports. The assembly code in Program 1.2 does just that, defining an I/O port for the 6811 and 6812 microcomputers. A *pseudo-operation code* is a command to the assembler and usually does not create machine-executable code. Other names for pseudo-operation code are *pseudo-op* and *assembly directive*. The equ is a pseudo-op that creates a mapping from a symbol and a value. Given the code in Program 1.2, the instructions staa $1003 and staa DDRC produce the exact same machine code; therefore they perform the exact same function when executed. We prefer staa DDRC because it is easier to understand. Notice that these 8-bit ports have two I/O addresses. We use the direction register to specify which pins are input and which are output. We use the data register to perform input/output on the port. For a complete list of I/O ports and their addresses refer to the respective data sheets.

Program 1.2
Assembly definitions of an I/O port.

```
; MC68HC711E9                 ; MC9S12C32
PORTC equ $1003               PTT  equ $0240
DDRC  equ $1007               PTIT equ $0241
                              DDRT equ $0242
```

The C code in Program 1.3 defines these same I/O ports. The volatile qualifier means the contents at that address can change by means other than explicit software action of the current module. It is used to prevent the compiler from optimizing a C program that accesses I/O ports.

Program 1.3
C definitions of an I/O port.

```
// MC68HC711E9
#define PORTC  *(unsigned char volatile *)(0x1003)
#define DDRC   *(unsigned char volatile *)(0x1007)

// MC9S12C32
#define PTT    *(unsigned char volatile *)(0x0240)
#define PTIT   *(unsigned char volatile *)(0x0241)
#define DDRT   *(unsigned char volatile *)(0x0242)
```

The first step to utilize a bidirectional I/O port is to set the direction register. The direction register specifies bit for bit whether the corresponding pins are input (0) or output (1). To make pins 7-4 input, and pins 3-0 output you should set the direction register to $0F, as shown in Programs 1.4 and 1.5.

Program 1.4
Assembly software that initializes pins 7-4 to input and pins 3-0 to output.

```
; MC68HC711E9                 ; MC9S12C32
   ldaa #$0F                     ldaa #$0F
   staa DDRC                     staa DDRT
```

Program 1.5
C software that initializes pins 7-4 to input and pins 3-0 to output.

```
// MC68HC711E9                 // MC9S12C32
   DDRC = 0x0F;                   DDRT = 0x0F;
```

If a pin is programmed as an input, a write to that port has no effect on that pin. If a pin is programmed as an output, a write to that port will set or clear that pin. On the 6811, a read from a pin that is configured as an output will return the value present at that pin. On the 6812, we have two options when reading pins that are configured as outputs. If a pin on the 6812 is configured as an output, a read from the regular port address (e.g., PTT) will return the value that was previously written. In addition to the regular port addresses, the 6812 has separate input addresses (e.g., PTIT), which are read-only. If a pin on the 6812 is configured as an output, a read from the input address returns the value that exists on the pin that exists at that time. We could compare PTT to PTIT to determine whether any output pins are damaged or to detect excess loading.

> **Checkpoint 1.40:** Does the entire port need to be defined as input or output, or can some pins be input while others are output?

> **Observation:** We can create open collector outputs (zero, hiZ) by setting the data port to zero at the beginning of our software, then setting the direction register to the complement of the desired output whenever we want to change the output.

1.9.3 Our First Design Problem

In this example, we will design an embedded system that flashes LEDs in a 0101, 0110, 1010, 1001 binary repeating pattern.

Some problems are so unusual that they require the engineer to invent completely original solutions. Most of the time, however, the engineer can solve even complex problems by building the system from components that already exist. Creativity will still be required in selecting the proper components, making small changes in their behavior (tweaking), arranging them in an effective and efficient manner, then verifying the system satisfies both the requirements and constraints. When young engineers begin their first job, they are sometimes surprised to see that education does not stop with college graduation, but rather is a life-long activity. In fact, it is the educational goal of all engineers to continue to learn both processes (rules about how to solve problems) and products (hardware software components). As the engineer becomes more experienced, he or she has a larger toolbox from which processes and components can be selected.

The hardest step for most new engineers is the first one, where to begin? We begin by analyzing the problem to create a set of specifications and constraints. This system will need four LEDs, and the computer must be able to activate and deactivate them. Since the problem didn't specify power source, speed, color, or brightness, we could either put off these decisions until the engineering design stage in order to simplify the design or minimize cost, or we could go back to the customer and ask for additional requirements. In this case, the customer didn't care about power, speed, color or brightness, but did think minimizing cost was a good idea. Due to the nature of this book, we will constrain all our designs to include the 6811 or 6812. Because we have +5V microcomputer systems, we will specify the system to run on +5V power. We have in our stockroom very-lost-cost red LEDs with 2.2V 10 mA specification, so we will use them. Table 1.20 summarizes the specifications and constraints.

Table 1.20
Specifications and constraints of the LED output system.

Specifications	Constraints
Repeating pattern of 5, 6, 10, 9	6811/6812 based
Four 2.2V 10 mA red LEDs	Minimize cost
+5V power supply	Standard 5% resistors

It is often difficult to distinguish whether a parameter is a specification or a constraint. In actuality, when designing a system it often doesn't matter into which category a parameter falls, because the system must satisfy all specifications and constraints. Nevertheless, when documenting the device it is better to categorize parameters properly. Specifications generally define in a quantitative manner the overall system objectives as given to us by our customers. Constraints, on the other hand, generally define the boundary space within which we must search for a solution to the problem. If we must use a particular component, it is often considered a constraint. In this case we will be using the 6811 or the 6812. Constraints also are often defined as an inequality, such as the cost must be less than $50, or the battery must last for at least one week. Specifications on the other hand are often defined as a quantitative number, and the system satisfies the requirement if the system operates within a specified *tolerance* of that parameter. Tolerance can be defined as a percentage error or as a range with minimum and maximum values. For example, our LED output system is acceptable if it has four LEDs, but unacceptable if it has three or five of them. Similarly, it will be OK as long as the LED current is between 8 and 12 mA. If the current drops below 8 mA, we won't be able to see the LED, and if it goes above 12 mA, it might damage the LED.

> *Observation:* Defining realistic tolerances on our specifications will have a profound effect on system cost.

> *Checkpoint 1.41:* What are the effects of specifying a tighter tolerance (e.g., 1% when the problem asked for 5%)?

> *Checkpoint 1.42:* What are the effects of specifying a looser tolerance (e.g., 10% when the problem asked for 5%)?

The next step is the high-level design. The data flow graph in Figure 1.45 shows information as it flows from the `controller` software to the four LEDs. The data flow graph will be important during the subsequent design phases because the hardware blocks can be considered as a preliminary hardware block diagram of the system. The call graph, also shown in Figure 1.45, illustrates this is master/slave configuration in which the `controller` software will manipulate the four LEDs.

Figure 1.45
Data flow graph and call graph of the LED output system.

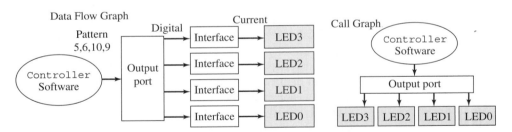

The hardware design of this system simply involves using four copies of the LED interface presented earlier in Figure 1.19. We can use the 7405 because its I_{OL} (16mA) is enough to drive the 10 mA LED. Notice the similarity of the data flow graph in Figure 1.45 with the hardware circuit in Figure 1.46. The data flow graph and call graph in this example look similar because this system just performs output. We will need to tweak the circuit adjusting the value of the resistor to produce the desired 2.2V 10 mA specification. If the V_{OL} of the 7405 is 0.4V, and the voltage across the LED is 2.2V, then the voltage across the resistor should be 5-2.2-0.4V or 2.4V. We calculate the resistor value using Ohm's Law—R is 2.4V/10 mA or 240Ω. Using standard resistor values with a 5% tolerance will make the product both cheaper and easier to build. One good way to tell whether a resistor value is standard is to go online and see which resistor values are in stock (e.g., www.digikey.com, www.jameco.com, www.mouser.com). In particular, 220Ω and 270Ω are two standard resistor values near 240Ω. If we were to use 220Ω, then the LED current would be (5-2.2-0.4V)/220Ω or 10.9 mA. Similarly, if we were to use 270Ω, then the LED current would be (5-2.2-0.4V)/270Ω or 8.9 mA. Both would have been acceptable, but we will use the 220Ω

Figure 1.46
Hardware circuit for the
LED output system.

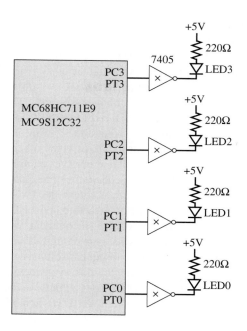

resistor because it is closer. It would have been more expensive to design this system with
240Ω resistors.

The software design of this system also involves using examples presented earlier with
some minor tweaking. The only data required in this problem is the 5,6,10,9 sequence. In
Chapter 3 we will consider solutions to this type of problem using data structures, but in this
first example, we will take a simple approach, not using a data structure. Figure 1.47 illustrates a software design process using *flowcharts*. We start with a general approach on the
left. Flowchart 1 shows that the software will initialize the output port and perform the output
sequence. As we design the software system, we fill in the details. This design process is

Figure 1.47
Software design for the
LED output system using
flowcharts.

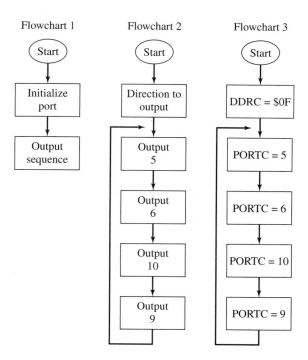

called *successive refinement*. It is also classified as top-down, because we begin with high-level issues and end with the low-level. In Flowchart 2, we set the direction register, then output the 5,6,10,9. It is at this stage that we figured out how to create the repeating sequence. Flowchart 3 fills in the remaining details, showing the details for the 6811 solution. The 6812 solution uses Port T; otherwise it is identical to the 6811 solution. To output the pattern 0101 to the LEDs, we will output a 5 to PORTC on the 6811 and PTT on the 6812. *Pseudo-code* is similar to high-level languages but lacks a rigid syntax. This means we can utilize whatever syntax we like. Flowcharts are good when the software involves complex algorithms with many decisions points causing control paths. On the other hand, pseudo-code may be better when the software is more sequential and involves complex calculations.

Many software developers use pseudo-code rather than flowcharts, because the pseudo-code itself can be embedded into the software as comments. Program 1.6 shows the assembly code implementation. Notice the similarity in structure between Flowchart 3 and this assembly code. Information following the semicolon is a comment, which allows programmers to document their software, but is ignored when converting to machine code. org is a pseudo-op instructing the assembler to place the subsequent software at the specified address. The first org specifies that the machine code for this system will be loaded into EPROM/EEPROM. Making the bottom four bits of the port outputs was previously presented as Program 1.4. In order to set the LEDs to the pattern 0101, we first bring the value of 5 into Register A, then we output the 5 by storing register A to the port. We create the repeating pattern by using an unconditional branch at the end, so the 5,6,10,9 output pattern occurs over and over. The *reset vector* on the 6811 and the 6812 is a 16-bit value stored at $FFFE and $FFFF. This 16-bit value is loaded into the program counter when the system starts up after a reset. In particular, this value specifies where our software will start execution. The last two lines of this program define the reset vector so our program will start at Main. Program 1.7 presents the C code implementation, which also is similar in structure to Flowchart 3.

Program 1.6
Assembly software for the LED output system.

```
; MC68HC711E9                      ; MC9S12C32
      org  $D000  ;ROM                   org  $4000  ;ROM
Main ldaa #$0F   ;make PC3-0        Main ldaa #$0F   ;make PT3-0
      staa DDRC   ;outputs                staa DDRT   ;outputs
Controller                         Controller
      ldaa #5                            ldaa #5
      staa PORTC ;set 0101               staa PTT   ;set 0101
      ldaa #6                            ldaa #6
      staa PORTC ;set 0110               staa PTT   ;set 0110
      ldaa #10                           ldaa #10
      staa PORTC ;set 1010               staa PTT   ;set 1010
      ldaa #9                            ldaa #9
      staa PORTC ;set 1001               staa PTT   ;set 1001
      bra  Controller                    bra  Controller
      org  $FFFE                         org  $FFFE
      fdb  Main  ;Reset vector           fdb  Main  ;Reset vector
```

Program 1.7
C software for the LED output system.

```
// MC68HC711E9                      // MC9S12C32
void main(void){  // make PC3-0     void main(void){// make PT3-0
   DDRC = 0x0F;  // outputs            DDRT = 0x0F; // outputs
   while(1){                           while(1){
    PORTC = 5;    // 0101               PTT = 5;     // 0101
    PORTC = 6;    // 0110               PTT = 6;     // 0110
    PORTC = 10;   // 1010               PTT = 10;    // 1010
    PORTC = 9;    // 1001               PTT = 9;     // 1001
   }                                   }
}                                   }
```

In order to test the system we need to build a prototype. One option is simulation. If you install the application Test EXecute And Simulate (TExaS) from the CD, you will find an implementation of this system as program Prog1_6. A screen shot of this implementation can be seen as Figure 1.48. A second option is to use a development system like the one shown in Figure 1.49. In this approach, you build the external circuits on a protoboard and use the debugger to download and test the software. A third approach, as illustrated in Figure 1.50, is typically used after a successful evaluation with one of the previous methods. In this approach, we design a printed circuit board (PCB) including both the external circuits and the microcontroller itself. Both the 6811 (using bootstrap mode) and the 6812 (using the background debug module) have facilities to download software onto the microcontroller. Notice the many test points (loops of wire) included in this design, in order to facilitate debugging.

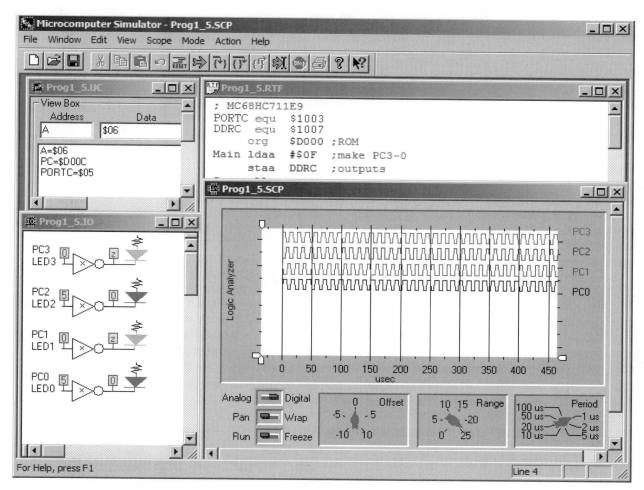

Figure 1.48
TExaS simulation of the LED output system.

During the testing phase of the project we observe that all four of the LEDs are continuously on. We use the software debugger to single-step our program, which correctly outputs the 0101, 0110, 1010, 1001 binary repeating pattern. During this single stepping the LEDs do come on and off in the proper pattern. Using a voltmeter on the circuit we observe the 0.4V signal on the output of the 7405 whenever the software wishes to turn the LED on.

Figure 1.49
MC9S12C32
development system
from Technological Arts
(#NC12C32SP).

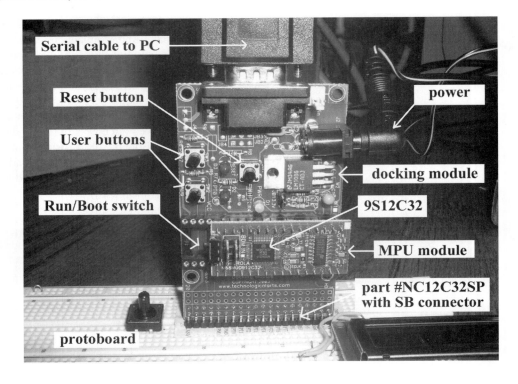

Serial cable to PC

power

Reset button

User buttons

docking module

Run/Boot switch

9S12C32

MPU module

part #NC12C32SP
with SB connector

protoboard

Figure 1.50
PCB layout of an
embedded 6811 system
with switches and LEDs.

We run the system at full speed again and observe two 7405 outputs on the oscilloscope, collecting data presented as Figure 1.51. Simulation, shown in Figure 1.48, also seems to be completely functional. All the electronic tests show the system is running properly, but our eyes still see all four LEDs continuously on.

Figure 1.51
Oscilloscope waveforms collected during the testing of the LED output system. The voltage sensitivity is 5V/division, and the time base is 1μs/division.

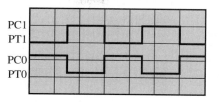

Checkpoint 1.43: What is the error in this design, and how do we fix it?

Portability is a measure of how easy it is to convert software that runs on one machine to run on another machine.

Observation: It is fairly easy to port 6811 assembly language to the 6812.

Observation: In general, C code is more portable than assembly language.

1.10   Choosing a Microcontroller

I chose to focus this book on the 6811 and the 6812, because year after year Freescale continues to be the leading supplier of microcontrollers. Sometimes, the computer engineer is faced with the task of selecting the microcontroller for the project. When faced with this decision, some engineers consider only those devices for which they have hardware and software experience. Fortunately, this blind approach often yields an effective and efficient product, because many microcontrollers overlap in their cost and performance. In other words, if a familiar microcontroller can implement the desired functions for the project, then it is often efficient to bypass that more perfect piece of hardware in favor of a faster development time. On the other hand, sometimes we wish to evaluate all potential candidates. It may be cost-effective to hire or train the engineering personnel so that they are proficient in a wide spectrum of potential microcontroller devices. There are many factors to consider when selecting a microcontroller including the following:

- Labor costs include training, development, and testing
- Material costs include parts and supplies
- Manufacturing costs depend on the number and complexity of the components
- Maintenance costs involve revisions to fix bugs and perform upgrades
- ROM size must be big enough to hold instructions and fixed data for the software
- RAM size must be big enough to hold locals, parameters, and global variables
- EEPROM must hold nonvolatile fixed constants that are field-configurable
- Speed must be fast enough to execute the software in real time
- I/O bandwidth determines the input/output rate
- 8-, 16-, or 32-bit data size should match most of the data to be processed
- Numerical operations may be required such as multiply, divide, signed, floating point
- Special functions may be required such as multiply/accumulate, fuzzy logic, complex numbers
- Parallel ports are needed for the input/output digital signals
- Serial ports are used to interface with other computers or I/O devices

- Timer functions are used to generate signals, measure frequency, measure period
- Pulse-width modulation is used for the output signals in many control applications
- ADC is used to convert analog inputs to digital numbers
- Package size and environmental issues affect many embedded systems
- Second source availability increases the chance parts can be purchased
- Availability of high-level language cross-compilers, simulators, emulators is important
- Power requirements, are important because many systems will be battery operated

When considering speed, it is best to compare time to execute a benchmark program similar to your specific application, rather than just comparing bus frequency. One of the difficulties is that the microcomputer selection depends on the speed and size of the software, but the software cannot be written without the computer. Given this uncertainty, it is best to select a family of devices with a range of execution speeds and memory configurations. In this way a prototype system with large amounts of memory and peripherals can be purchased for software and hardware development, and once the design is in its final stages, the specific version of the computer can be selected now that the memory and speed requirements for the project are known. In conclusion, while this book focuses on the 6811 and 6812 microcontrollers, it is expected that once the study of this book is completed, the reader will be equipped with the knowledge needed to select the proper microcontroller and complete the software design.

1.11 Exercises

For these questions, substitute your specific 6811 or 6812 for *YourComputer* as appropriate.

1.1 Is RAM volatile or nonvolatile?

1.2 Assuming *YourComputer* is running in expanded mode, what are the names of its bus signals?

1.3 Consider *YourComputer* with memory-mapped I/O. Assume there is no direct memory access (DMA) on your microcomputer. For this problem, we specify four classes of devices: *processor*, *RAM*, *ROM*, and *I/O*.
   a) List the devices that can drive the address bus during a CPU read cycle.
   b) List the devices that can drive the address bus during a CPU write cycle.
   c) List the devices that can drive the data bus during a CPU read cycle.
   d) List the devices that can drive the data bus during a CPU write cycle.
   e) List the devices that can drive the R/W line during a CPU read cycle.
   f) List the devices that can drive the R/W line during a CPU write cycle.
   g) List the devices that can receive the information from data bus during a CPU read cycle.
   h) List the devices that can receive the information from data bus during a CPU write cycle.

1.4 How many 74LS low-power Schottky inputs can an output of *YourComputer* drive?

1.5 Consider the current it takes to power a +5V 74xx245 (I_{CC}) as shown in Table 1.5. What is the qualitative difference between the CMOS devices and the non-CMOS devices? What is the explanation for the difference?

1.6 Design the circuit that interfaces a 3V 5 mA LED to *YourComputer*.

1.7 Design the circuit that interfaces a 2.5V 1 mA LED to *YourComputer*.

1.8 Redesign the switch interface shown in the middle of Figure 1.20 assuming the signal s is to be connected to a 74LS04 instead of the input port of the microcontroller. In particular, change the 10kΩ resistor value to the appropriate value for low-power Schottky logic.

1.9 How many alternatives does a 14-bit ADC have?

1.10 If a system uses a 12-bit ADC, about how many decimal digits will it have?

1.11 If a system requires 3½ decimal digits of precision, what is the smallest number of bits the ADC needs to have?

1.12 If a system requires $4^3/_4$ decimal digits of precision, what is the smallest number of bits the ADC needs to have?

1.13 Convert the following decimal numbers to 8-bit unsigned binary: 25, 63, 125, and 200.

1.14 Convert the following decimal numbers to 8-bit signed binary: 25, 63, -125, and -2.

1.15 Convert the following hex numbers to unsigned decimal: $25, $63, $A3, and $FE.

1.16 Convert the 16-bit binary number %0010000001101010 to unsigned decimal.

1.17 Convert the 16-bit hex number $1234 to unsigned decimal.

1.18 Convert the unsigned decimal number 1234 to 16-bit hexadecimal.

1.19 Convert the unsigned decimal number 10000 to 16-bit binary.

1.20 Convert the 16-bit hex number $1234 to signed decimal.

1.21 Convert the 16-bit hex number $ABCD to signed decimal.

1.22 Convert the signed decimal number 1234 to 16-bit hexadecimal.

1.23 Convert the signed decimal number -10000 to 16-bit binary.

1.24 List the registers in *YourComputer*?

1.25 How much RAM and ROM are in *YourComputer*? What are the specific address ranges of these memory components?

1.26 What are the bits in the CCR of *YourComputer*?

1.27 What are the bidirectional ports on *YourComputer*?

1.28 What are the unidirectional ports on *YourComputer* (if any)?

1.29 What is a direction register?

1.30 Why does the microcomputer have direction registers?

1.31 Write software that initializes 6811 Port C (or 6812 Port T), so pins 7,5,3,1 are output and the rest are input.

1.32 Write software that initializes 6811 Port C (or 6812 Port T), so pins 5,4 are output and the rest are input.

1.33 You can make the 6811 Port C generate open collector outputs by setting appropriate bit in the direction register, DDRC, to 1 and setting the CWOM bit in the PIOC register to 1. This open collector mode can be found in some of the ports of the Intel 8051 microcontroller family, but it is missing in the 6812.

Freescale claims you can easily upgrade 6811 systems to the more powerful 6812. You have just graduated from the Prestigious University, and the first task at your new high-paying job at *We Are Nerds, INC,* is to upgrade an existing *WAN INC* 6811 system to the 6812. It seems easy, so you begin by reading all about the 6812. The 6812 has more instructions and addressing modes, but luckily all existing 6811 assembly instructions and addressing modes are still available on the 6812. The 6812 has more parallel ports, more serial ports, more input captures, and more output compares. In particular, the existing system uses Port C, and the MC68HC812A4 has a parallel port C, PORTC, and a direction register, DDRC. So far so good, but all of a sudden BAM it hits you—looking at you square in the face is a big problem. The existing system uses open collector output mode with the 6811 bit CWOM set, but the 6812 has no open collector modes on any of its parallel port outputs. So now you, the young engineering genius from PU, must solve your first engineering problem.

Lucky for you, the engineers that worked on the problem previously were also from PU. They implemented the low-level access to PORTC as a device driver, and organized the software with a layered approach. This layered system will allow you to come in and replace the low (hardware access) level, without having to modify the upper levels. In particular, here are the existing low-level routines in both assembly and C.

```
;Initialize Parallel Port
;Input: Reg B specifies which port bits will be input(0) or
output(1)
;Outputs: none
Pinit  ldaa PIOC
       oraa #$20 ; set CWOM so outputs are open collector
       staa PIOC
       stab DDRC ; 1 means open collector output, 0 means input
       ldaa #$FF
       stab PORTC ; any output pins are initialized to HiZ
       rts
;Set output (only output pins are effected)
;Input: Reg B specifies new output values
;Outputs: none
Pout   stab PORTC ; modifies output pins only
       rts
; Get input
; Input: none
; Outputs: Reg B returned with current values of both inputs and
outputs
Pin    ldab PORTC
       rts
void Pinit(unsigned char direction){
  PIOC |= 0x20; // set CWOM so outputs are open collector
  DDRC = direction; // 1 means open collector output, 0 means input
  Pout(0xFF);} // any output pins are initialized to HiZ
void Pout(unsigned char data){
  PORTC = data;} // modifies output pins only
unsigned char Pin(void){
  return PORTC;}
```

Your specific task: Rewrite the three device driver routines to run on the 6812, either in assembly or in C. You may use any 6811 assembly language instructions and addressing modes. Refer to the 6812 parallel port using the symbol **PORTC** and to the direction register using the symbol **DDRC**. A global variable will be required.

Test your answer with this example: Assume the four input signals (C7–C4) are set by external hardware to C7=0,C6=1,C5=0, C4=0. Assume the four output signals (C3–C0) have +5V pull-up resistors.

```
void main(void){ unsigned char info;
    Pinit(0x0F);   // C7-C4 inputs(HiZ), C3-C0 outputs are HiZ
    info=Pin();    // info set to 0x4F (because of pullups)
    Pout(0x00);    // C7-C4 still inputs(HiZ), but C3-C0 are low
    info=Pin();    // info set to 0x40
    Pout(0x05);    // C7-C4,C2,C0 are HiZ, but C3,C1 are low
    info=Pin(); }  // info set to 0x45
```

1.12   Lab Assignments

The labs in this book involve the following steps:

Part a) During the analysis phase of the project, determine additional specifications and constraints. In particular, discover which microcontroller you are to use, whether you are to develop in assembly language or in C, and whether the project is to be simulated then built, just built or just simulated. For example, inputs can be created with switches, and outputs can be generated with LEDs. The SCI can be interfaced to a PC, and a communication program such as HyperTerminal can be used to interact with the system.

Part b) Design, build, and test the hardware interfaces. Use a computer-aided-drawing (CAD) program to draw the hardware circuits. Label all pins, chips, and resistor values. In this chapter, there will be one switch for each input and one LED for each output. Connect the switch interfaces to 6811/6812 input pins, and connect the LED interfaces to 6811/6812 output pins. Pressing the switch will signify a high input logic value. You should activate the LED to signify a high output logic value.

Part c) Design, implement, and test the software that initializes the I/O ports and performs the specified function. Often a main program is used to demonstrate the system.

Lab 1.1. The overall objective is to create a `not` gate. The system has one digital input and one digital output, such that the output is the logical complement of the input. Implement the design such that the complement function occurs in the software of the 6811/6812. If you are writing in assembly, you may wish to investigate the `coma lsra` and `lsla` instructions.

Lab 1.2. The overall objective is to create a 3-input `and` gate. The system has three digital inputs and one digital output, such that the output is the logical `and` of the three inputs. Implement the design such that the `and` function occurs in the software of the 6811/6812. If you are writing in assembly, you may wish to investigate the `anda lsra` and `lsla` instructions.

Lab 1.3. The overall objective is to create a 3-input `or` gate. The system has three digital inputs and one digital output, such that the output is the logical `or` of the three inputs. Implement the design such that the `or` function occurs in the software of the 6811/6812. If you are writing in assembly, you may wish to investigate the `oraa lsra` and `lsla` instructions.

Lab 1.4. The overall objective is to create a 2-input `exclusive or` gate. The system has two digital inputs and one digital output, such that the output is the logical `exclusive or` of the two inputs. Implement the design such that the exclusive or function occurs in the software of the 6811/6812. If you are writing in assembly, you may wish to investigate the `eora lsra` and `lsla` instructions.

Lab 1.5. The overall objective is to create a 3-input `voting` logic. The system has three digital inputs and one digital output, such that the output is high if and only if two or more inputs are high. This means the output will be low if two for more inputs are low. Implement the design such that the `voting` function occurs in the software of the 6811/6812. If you are writing in assembly, you may wish to investigate the `anda oraa lsra` and `lsla` instructions.

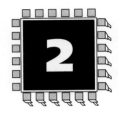

2 Design of Software Systems

Chapter 2 objectives are to:

❑ Create an overview of assembly language programming
❑ Develop techniques for writing modular or structured software
❑ Introduce layered software organization
❑ Present a software model for device drivers
❑ Define the concept of threads
❑ Describe effective techniques for software debugging

The ultimate success of an embedded system project depends on both its software and its hardware. Computer scientists pride themselves in their ability to develop quality software. Similarly, electrical engineers are well-trained in the processes to design both digital and analog electronics. Manufacturers, in an attempt to get designers to use their products, provide application notes for their hardware devices. The main objective of this book is to combine effective design processes with practical software techniques to develop quality embedded systems. As the size and especially the complexity of the software increase, the software development changes from simple "coding" to "software engineering," and the required skills also vary along this spectrum. These software skills include modular design, layered architecture, abstraction, and verification. Real-time embedded systems are usually on the small end of the size scale, but nevertheless these systems can be quite complex. Therefore the above-mentioned skills are essential for developing embedded systems. This chapter on software development is placed early in the book because writing good software is an art that must be developed and cannot be added on at the end of a project. Good software combined with average hardware will always outperform average software on good hardware. In this chapter we will outline various techniques for developing quality software, then apply these techniques throughout the remainder of the book.

2.1 Quality Programming

Software development is similar to other engineering tasks. We can choose to follow well-defined procedures during the development and evaluation phases, or we can meander in a haphazard way and produce code that is hard to test and harder to change. The ultimate goal of the system is to satisfy the stated objectives such as accuracy, stability, and I/O relationships. Nevertheless, it is appropriate to separately evaluate the individual components of the system. Therefore in this section we will evaluate the quality of our software. There are two categories of performance criteria with which we evaluate the "goodness" of our software. Quantitative criteria include dynamic efficiency (speed of execution), static efficiency (ROM and RAM program size), and accuracy of the results. Qualitative criteria center around ease of software maintenance. Another qualitative way to evaluate software is ease of understanding. If your software is easy to understand, then it will be:

Easy to debug (fix mistakes)
Easy to verify (prove correctness)
Easy to maintain (add features)

Common error: Programmers who sacrifice clarity in favor of execution speed often develop software that runs fast but doesn't work and can't be changed.

Golden Rule of Software Development
Write software for others as you wish they would write for you.

**2.1.1
Quantitative
Performance
Measurements**

To evaluate our software quality, we need performance measures. The simplest approaches to this issue are quantitative measurements. *Dynamic efficiency* is a measure of how fast the program executes. It is measured in seconds or CPU cycles. *Static efficiency* is the number of memory bytes required. Since most embedded computer systems have both RAM and ROM, we specify memory requirement in global variables, stack space, fixed constants, and program object code. The global variables plus maximum stack size must be less than the available RAM. Similarly, the fixed constants plus program size must be less than the ROM or PROM size. We can also judge our software system according to whether or not it satisfies given constraints, like software development costs, memory available, and timetable.

**2.1.2
Qualitative
Performance
Measurements**

Qualitative performance measurements include those parameters to which we cannot assign a direct numerical value. Often in life the most important questions are the easiest to ask but the hardest to answer. Such is the case with software quality. So we ask the following qualitative questions: Can we prove our software works? Is our software easy to understand? Is our software easy to change? Since there is no single approach to writing the best software, we can only hope to present some techniques that you may wish to integrate into your own software style. In fact, we will devote most of this chapter to the important issue of developing quality software. In particular, we will study self-documented code, abstraction, modularity, and layered software. These parameters indeed play a profound effect on the bottom-line financial success of our projects. Although quite real, because there often is not an immediate and direct relationship between a software's quality and profit, we may be tempted to dismiss its importance.

To get a benchmark on how good a programmer you are, we challenge you to two tests. In the first test, find a major piece of software that you have written over 12 months ago, then see if you can still understand it enough to make minor changes in its behavior. The second test is to exchange with a peer a major piece of software that you have both recently written (but not written together), then in the same manner see if you can make minor changes to each other's software.

Observation: You can tell that you are a good programmer if (1) you can understand your own code 12 months later and (2) others can make changes to your code.

2.2 Assembly Language Programming

**2.2.1
Introduction**

In this section, we will present assembly language syntax for the **TExaS** assembler. There are minor syntactical differences between the assembler in **CodeWarrior** and the one in **TExaS**. However, most example programs in this book will be syntactically correct for both assemblers. Figure 2.1 outlines the assembly language development process. To develop assembly language software, we first use an **editor** to create our **source code**. Source code contains specific commands in human-readable form. Next, we use an **assembler** to translate our source code into **object code**. Object code contains the specific commands in machine-readable form. When developing software for a real microcontroller, a **loader** is used to place the object code into the microcontroller's memory. We test our system with the aid of a debugger. If the loader burns the object code into EPROM, the program will exist in nonvolatile storage. Consequently, the power-on-reset will start our software after power is supplied to the microcontroller. The last two lines of the source code shown in Figure 2.1 define the reset vector, which is the starting location after a reset. Both the 6811 (bootstrap mode) and 6812 (background debug module) contain built-in features that assist in programming its EPROM, whereas other microcontrollers require a separate apparatus to program. Some hardware development systems use RAM to hold the program. In these systems, the loader downloads the object code into RAM. Since RAM is volatile, the programs must be loaded each time power is removed.

Figure 2.1
Assembly language
development process.

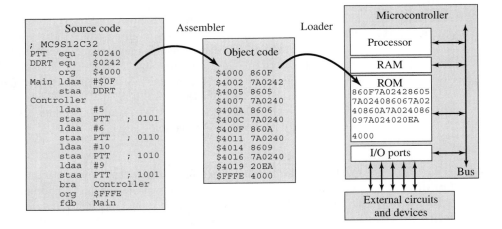

Assemblers are programs that process assembly language source program statements and translate them into executable machine language object files. Cross assemblers allow source programs written and edited on one computer (the host) generating executable code for another computer (the target). The executable code can either be simulated (using an application such as **TExaS** or **CodeWarrior**), or downloaded onto a real computer for execution.

The symbolic language used to code source programs to be processed by the assembler is called assembly language. The language is a collection of mnemonic symbols representing operations (i.e., machine instruction mnemonics or directives to the assembler), symbolic names, operators, and special symbols. The assembly language provides mnemonic operation codes for all machine instructions in the instruction set. The instructions are defined and explained in the Programming Reference Manual for the 6811/6812, which can be found in the **pdf** directory of the CD. These documents are also available for downloading from the **freescale.com** web site. A brief overview of each instruction and example usage can be found by executing **Help→OpCodes** with the **TExaS** application.

The assembly language also contains mnemonic directives that specify auxiliary actions to be performed by the assembler. These directives or pseudo-ops are not always translated into machine language. For more information execute **Help→PseudoOp**.

Most assemblers have two-passes, meaning it will scan through the source code twice. During the first pass, the source program is analyzed in order to develop the symbol table. A **symbol table** is a mapping between symbolic names (e.g., PTT) and their numeric values (e.g., $0240). During the second pass, the object file is created (assembled) using the symbol table that was developed in pass one. It is during the second pass that a **listing file** is also produced (see Program 2.1). The symbol table is recreated in the second pass. A phasing error occurs if any symbol table values calculated during the two passes are different.

Program 2.1

Assembly language listing of Program 1.5.

```
Copyright 2005-2006 Test EXecute And Simulate
                                     ; MC9S12C32
$0240                                PTT   equ   $0240
$0242                                DDRT  equ   $0242
$4000                                      org   $4000 ;ROM
$4000 860F          [ 1](  0){P      }Main ldaa  #$0F   ;make PT3-0
$4002 7A0242        [ 3](  1){wOP    }     staa  DDRT   ;outputs
$4005                                 Controller
$4005 8605          [ 1](  4){P      }     ldaa  #5
$4007 7A0240        [ 3](  5){wOP    }     staa  PTT    ;set 0101
$400A 8606          [ 1](  8){P      }     ldaa  #6
$400C 7A0240        [ 3](  9){wOP    }     staa  PTT    ;set 0110
$400F 860A          [ 1]( 12){P      }     ldaa  #10
$4011 7A0240        [ 3]( 13){wOP    }     staa  PTT    ;set 1010
$4014 8609          [ 1]( 16){P      }     ldaa  #9
$4016 7A0240        [ 3]( 17){wOP    }     staa  PTT    ;set 1001
$4019 20EA          [ 3]( 20){PPP    }     bra   Controller
$FFFE                                      org   $FFFE
$FFFE 4000                                 fdb   Main   ;Reset vector
**************Symbol Table********************
Controller     $4005
DDRT           $0242
Main           $4000
PTT            $0240
Assembly successful
```

Errors that occur during the assembly process (e.g., undefined symbol, illegal opcode, branch destination too far, etc.) are explained in the listing file. Program 2.1 contains the listing file for the program presented as Program 1.6 and shown in Figure 2.1. The listing file created by **TExaS** contains the source code, the object code, and the symbol table.

The **source code** is a file of ASCII characters usually created with an editor. Each source statement is processed completely before the next source statement is read. As each statement is processed, the assembler examines the label, operation code, and operand fields. The operation code table is scanned for a match with a known opcode. During the processing of a standard operation code mnemonic, the standard machine code is inserted into the object file. If an assembler directive is being processed, the corresponding action is taken.

Any errors that are detected by the assembler are displayed. **Object code** is the binary values (instructions and data) that, when executed by the computer, perform the intended function. The listing file contains the address, object code, and a copy of the source code. There will also be a symbol table describing where in memory the program and data will be loaded. The symbol table is a list of all the names used in the program along with the values. A symbol is created when you put a label starting in column 1. Examples of this type are `Main`, and `Controller`. The symbol table value for this type is the absolute memory address where the instruction, variable or constant will reside in memory. The second type of label is created by the `equ` pseudo-op, e.g., `PTT`. The value for this type of symbol is simply the number specified in the operand field. When the assembler processes an instruction with a symbol in it, it simply substitutes the fixed value in place of the symbol. It is good programming practice to use symbols that clarify (make it easier to understand) our programs. The symbol table for this example is given at the end of the listing file.

A compiler converts high-level language source code into object code. A cross-compiler also converts source code into object code and creates a listing file except that the object code is created for a target machine that is different from the machine running the cross-compiler. **TExaS** is a cross-assembler because it runs on an Intel computer but creates 6811/6812 object code. **ImageCraft ICC11/ICC12** and **Metrowerks CodeWarrior** include both a cross-assembler and a cross-compiler because they run on the PC and create 6811/6812 object code.

Checkpoint 2.1: What does the assembler do in pass 1?

Checkpoint 2.2: What does the assembler do in pass 2?

2.2.2
Assembly
Language Syntax

Programs written in assembly language consist of a sequence of source statements. Each source statement consists of a sequence of ASCII characters ending with a carriage return. Each source statement may include up to four fields: a label, an operation (instruction mnemonic or assembler directive), an operand, and a comment. We use pseudo-op codes in our source code to give instructions to the assembler itself. The `equ` is an assembly directive, and the `ldaa` is a regular machine instruction.

```
PORTA   equ    $0000   ; Assembly time constant
Inp     ldaa   PORTA   ; Read data from fixed address I/O data port
```

An assembly language statement contains the following fields.

Label Field can be used to define a symbol
Operation Field defines the operation code or pseudo-op
Operand Field specifies either the address or the data.
Comment Field allows the programmer to document the software.

Instructions with inherent mode addressing do not have an operand field. For example,

```
label   clra        comment
        deca        comment
        cli         comment
        inca        comment
```

The **label field** begins in the first column of a source statement. The label field can take one of the following three forms:

A. An asterisk (*) or semicolon (;) as the first character in the label field indicates that the rest of the source statement is a comment. Comments are ignored by the assembler, and are printed on the source listing only for the programmer's information. Examples:

```
* This line is a comment
; This line is also a comment
```

B. A white-space character (blank or tab) as the first character indicates that the label field is empty. The line has no label and is not a comment. These assembly lines have no labels:

```
ldaa 0
rmb  10
```

C. A symbol character as the first character indicates that the line has a label. Symbol characters are the upper-case letters A–Z, lower-case letters a–z, digits 0–9, and the special characters, period (.), dollar sign ($), and underscore (_). Symbols consist of at least one and at most 99 characters, the first of which must be alphabetic or the special characters period (.) or underscore (_). All characters are significant and upper and lower case letters are distinct.

Each label may be defined only once in your program. The exception to this rule is the set pseudo-op that allows you to define and redefine the same symbol. We typically use set to define the stack offsets for the local variables in a subroutine. The set pseudo-op allows two separate subroutines to re-use the same name for their local variables.

With the exception of the equ = and set directives, a label is assigned the value of the program counter of the first byte of the instruction or data being assembled. The value assigned to the label is absolute. Labels may optionally be ended with a colon (:). If the colon is used, it is not part of the label but merely acts to set the label off from the rest of the source line. Thus the following code fragments are equivalent:

```
here: deca
      bne   here

here  deca
      bne   here
```

A label may appear on a line by itself. The assembler interprets this as set the value of the label equal to the current value of the program counter. A label may also occur on a line with a pseudo-op.

The **operation field** occurs after the label field and must be preceded by at least one white-space character. The operation field must contain a legal opcode mnemonic or an assembler directive. Upper case characters in this field are converted to lower case before being checked as a legal mnemonic. Thus nop, NOP, and NoP are recognized as the same mnemonic. Entries in the operation field may be opcodes or directives.

Opcodes correspond directly to the machine instructions. The operation code includes any register name associated with the instruction. These register names must not be separated from the opcode with any white-space characters. Thus clra means clear accumulator A, but clr a means clear memory location identified by the label a.

Directives or *pseudo-ops* are special operation codes known to the assembler that control the assembly process rather than being translated into machine instructions. The directives that **TExaS** supports are described in detail later in this section.

The interpretation of the **operand field** is dependent on the contents of the operation field. The operand field, if required, must follow the operation field, and must be preceded by at least one white-space character. The operand field may contain a symbol, an expression, or a combination of symbols and expressions separated by commas. There can be no white-spaces in the operand field. For example, the following two lines produce identical object code in TExaS because of the space between data and + in the first line:

```
ldaa  data + 1
ldaa  data
```

Observation: The CodeWarrior assembler allows spaces within the operand field, and then requires a semicolon (;) be placed before each comment.

The operand field of machine instructions is used to specify the addressing mode of the instruction, as well as the operand of the instruction. Table 2.1 summarizes the operand field

Table 2.1
Example operands for
the 6811 and 6812.

Operand	Format	Example
no operand	inherent	`clra`
<expression>	direct, extended, or relative	`ldaa 4`
#<expression>	immediate	`ldaa #4`
<expression>,idx	indexed with address register	`ldaa 4,x`
<expr>,#<expr>	bit set or clear	`bset 4,#$01`
<expr>,#<expr>,<expr>	bit test and branch	`brset 4,#$01,there`
<expr>,idx,#<expr>,<expr>	bit test and branch	`brset 4,x,#$01,there`

formats on the 6811 and 6812. On the 6811 the index register, `idx`, is X or Y, but on the 6812 it can be X, Y, SP, or PC.

The 6812 assembly language includes some additional operand formats, as shown in Table 2.2. The accumulator offset, `acc`, is A, B or D, and the index register, `idx`, is X, Y, SP, or PC. The PC is not allowed with any of the predecrement, postdecrement, preincrement, or postincrement addressing modes.

Table 2.2
Additional example
operands for the 6812.

Operand	Format	Example
<expression>,idx+	indexed, post increment	`ldd  2,SP+`
<expression>,idx-	indexed, post decrement	`ldaa 4,Y-`
<expression>,+idx	indexed, pre increment	`ldaa 4,+X`
<expression>,-idx	indexed, pre decrement	`staa 1,-SP`
acc,idx	accumulator offset indexed	`ldaa A,X`
[<expression>,idx]	indexed indirect	`ldaa [4,X]`
[D,idx]	RegD indexed indirect	`ldaa [D,Y]`

2.2.3 Memory and Register Transfer Operations

The `w W U` terminology used in this section was presented earlier in Section 1.6.2. The 8-bit load instructions transfer data from memory into a register. In real life, when we *move* a box, *push* a broom, *load* a rifle, or *transfer* to a new job, there is a single physical object and the action changes the location of that object. Assembly language uses these same verbs, but the action will be different. It creates a copy of the data and places it at the new location. In other words, since the original data still exists, there are now two copies of the information. If the address U is between 0 and $00FF, then direct addressing mode will be used; otherwise, it will use extended addressing mode.

`ldaa`	#w	RegA=w	Load an 8-bit constant into RegA
`ldaa`	U	RegA=[U]	Load an 8-bit memory value into RegA
`ldab`	#w	RegB=w	Load an 8-bit constant into RegB
`ldab`	U	RegB=[U]	Load an 8-bit memory value into RegB

Condition code bits are set with R equal to the 8-bit memory contents loaded into the register.

N: result is negative $N=R_7$

Z: result is zero $Z = \overline{R_7} \cdot \overline{R_6} \cdot \overline{R_5} \cdot \overline{R_4} \cdot \overline{R_3} \cdot \overline{R_2} \cdot \overline{R_1} \cdot \overline{R_0}$

V: signed overflow $V=0$

The 16-bit load instructions also transfer information from memory into a register. Although D usually contains data and X,Y usually contain addresses, it is acceptable programming practice to place address or data information in any of these three registers. The stack pointer (called either S or SP) will always contain an address specifying the top of the stack. The program counter (PC) will always contain an address specifying the next instruction to execute.

ldd	#W	RegD=W	Load a 16-bit constant into RegD
ldd	U	RegD={U}	Load a 16-bit memory value into RegD
lds	#W	RegS=W	Load a 16-bit constant into RegS
lds	U	RegS={U}	Load a 16-bit memory value into RegS
ldx	#W	RegX=W	Load a 16-bit constant into RegX
ldx	U	RegX={U}	Load a 16-bit memory value into RegX
ldy	#W	RegY=W	Load a 16-bit constant into RegY
ldy	U	RegY={U}	Load a 16-bit memory value into RegY

Condition code bits are set with R equal to the 16-bit memory contents loaded into the register.

N: result is negative $N=R_{15}$

Z: result is zero

$$Z = \overline{R_{15}} \cdot \overline{R_{14}} \cdot \overline{R_{13}} \cdot \overline{R_{12}} \cdot \overline{R_{11}} \cdot \overline{R_{10}} \cdot \overline{R_9} \cdot \overline{R_8} \cdot \overline{R_7} \cdot \overline{R_6} \cdot \overline{R_5} \cdot \overline{R_4} \cdot \overline{R_3} \cdot \overline{R_2} \cdot \overline{R_1} \cdot \overline{R_0}$$

V: signed overflow $V=0$

The 6812 has two very convenient memory to memory move instructions, which set no flags.

movb	#w,addr	[addr]=w	Move an 8-bit constant into memory
movb	addr1,addr2	[addr2]=[addr1]	Move an 8-bit value memory to memory
movw	#W,addr	{addr}=W	Move a 16-bit constant into memory
movw	addr1,addr2	{addr2}={addr1}	Move a 16-bit value memory to memory

The 8-bit store instructions move data from a register to memory. The data in the register remains intact, so after executing one of these instructions there are two copies of the data.

| staa | U | [U]=RegA | Store RegA into memory |
| stab | U | [U]=RegB | Store RegB into memory |

Condition code bits are set with R equal to the 8-bit register contents stored into memory.

N: result is negative $N=R_7$

Z: result is zero $Z = \overline{R_7} \cdot \overline{R_6} \cdot \overline{R_5} \cdot \overline{R_4} \cdot \overline{R_3} \cdot \overline{R_2} \cdot \overline{R_1} \cdot \overline{R_0}$

V: signed overflow $V=0$

The 16-bit store instructions move from a register to memory.

std	U	{U}=RegD	Store RegD into memory
sts	U	{U}=RegS	Store RegS into memory
stx	U	{U}=RegX	Store RegX into memory
sty	U	{U}=RegY	Store RegY into memory

Condition code bits are set with R equal to the 16-bit register contents stored into memory.

N: result is negative $N=R_{15}$

Z: result is zero

$$Z = \overline{R_{15}} \cdot \overline{R_{14}} \cdot \overline{R_{13}} \cdot \overline{R_{12}} \cdot \overline{R_{11}} \cdot \overline{R_{10}} \cdot \overline{R_9} \cdot \overline{R_8} \cdot \overline{R_7} \cdot \overline{R_6} \cdot \overline{R_5} \cdot \overline{R_4} \cdot \overline{R_3} \cdot \overline{R_2} \cdot \overline{R_1} \cdot \overline{R_0}$$

V: signed overflow $V=0$

The following transfer operations use inherent addressing:.

xgdx	Swap RegD and RegX
xgdy	Swap RegD and RegY
clc	Clear carry bit, C=0
cli	Clear interrupt mask bit, enable interrupts, I=0
clv	Clear overflow bit, V=0
sec	Set carry bit, C=1
sei	Set interrupt mask bit, disable interrupts, I=1
sev	Set overflow bit, V=1
tap	Transfer A to CC, (can not change X bit from 0 to 1)
tpa	Transfer CC to A

2.2.4
Indexed
Addressing Mode

2.2.4.1
Indexed Addressing
on the 6811

6811 indexed addressing mode uses an 8-bit unsigned offset with either RegX or RegY. The 8-bit unsigned offset is included even if the offset is zero. Notice, that instructions that use RegX take less object code and execute faster than those using RegY do. Indexed mode is useful when addressing the 6811 I/O ports located from $1000 to $103F. Indexed mode is also useful when addressing the data structures and information on the stack. In each case, the 16-bit register is used as a pointer (index) and is unmodified by the instruction. The instruction staa 4,x has an object code $A704. The $A7 specifies the staa instruction with indexed mode addressing. The $04 is the index. The effective address will be X+4.

```
staa    4,x      ; [X+4]=RegA
```

Assuming Register X=$0023, the instruction staa 4,X will store a copy of the value in Register A at $0027 leaving Register X unchanged, as shown in Figure 2.2. The effective address (EA) is $0023+4=$0027.

Figure 2.2
Example of the 6811
indexed addressing
mode.

More indexed mode examples are shown as follows:

machine code	opcode	operand	comment
$A664	ldaa	100,x	;RegA = [X+100]
$18A420	anda	$20,y	;RegA = RegA&[Y+32]
$1C0301	bset	$03,x,#1	;[X+3] = [X+3]|1
$18ED05	std	5,y	;{Y+5} = RegD

> ***Observation:*** With indexed mode addressing the pointer to the information is calculated at run time, so the data and its location may change dynamically.

> ***Observation:*** 6811 instructions involving Register Y require an extra byte of machine code.

> ***Observation:*** The 6811 instructions bset bclr brset brclr do not operate with extended addressing. You have to use indexed addressing mode when using these instructions with the I/O ports.

2.2.4.2
Indexed Addressing on
the 6812

The 6812 instruction set has 15 addressing modes. Five of the modes were presented earlier, and the remaining modes are presented next.

1. Indexed addressing mode uses a fixed offset with the 16-bit registers: X, Y, SP, or PC. On the 6811, instructions that use register Y take more memory and run slower than the equivalent instruction using register X. The 6812 eliminates this speed and memory cost of using Reg Y. In addition, the 6812 indexing modes can be used with the stack pointer (SP) and the program counter (PC). The offset can be 5-bit (-16 to $+15$), 9-bit (-256 to $+127$), or 16-bit. Five-bit (-16 to $+15$) index mode requires one machine byte to encode the operand. In the first example that uses 5-bit indexed mode, $6A is the staa instruction and $5C is the index mode operand. Tables A.3 and A.4 of the Freescale CPU12 Reference Manual show the machine codes for the indexed instructions.

machine code	opcode	operand	comment
$6A5C	staa	-4,Y	; [Y-4] = RegA

Assuming Register Y=$0823, the instruction staa -4,Y will store a copy of the value in Register A at $081F leaving Register Y unchanged, as shown in Figure 2.3. The effective address (EA) is $0823-4=$081F.

Figure 2.3
Example of the 6812 indexed addressing mode.

Nine-bit (-256 to $+255$) indexed mode requires two machine bytes to encode the operand.

machine code	opcode	operand	comment
$6AE840	staa	$40,Y	; [Y+$40] = RegA

Assuming Register Y=$0823, the instruction staa $40,Y will store a copy of the value in Register A at $0863 leaving Register Y unchanged, as shown in Figure 2.4. The effective address (EA) is $0823+$40=$0863.

Figure 2.4
Another example of the 6812 indexed addressing mode.

Sixteen-bit indexed mode requires three machine bytes to encode the operand.

machine code	opcode	Operand	comment
$6AEA0200	staa	$200,Y	; [Y+$200] = RegA

Assuming Register Y=$0823, the instruction staa $200,Y will store a copy of the value in Register A at $0A23 leaving Register Y unchanged, as shown in Figure 2.5. The effective address (EA) is $0823+$200=$0A23.

Figure 2.5
A third example of the 6812 indexed addressing mode.

Due to the properties of 16-bit addition, the 16-bit offset can be interpreted either as unsigned (0 to 65535) or signed (-32768 to $+32767$.) Indexed mode is useful when addressing the data structures and information on the stack. In each case, the 16-bit register used as a pointer (index) is not modified by the instruction.

Common error: SP relative indexed addressing with a negative constant is usually defined as an illegal stack access.

2. Auto Pre/Post Decrement/Increment Indexed Addressing modes use the 16-bit registers: X, Y, or SP. The PC cannot be used with these index modes that modify the index register. In each case, the 16-bit register used as a pointer (index) is modified either before (pre) or after (post) the memory access. These modes are useful when addressing the data structures. The 6812 allows the programmer to specify the amount added to (subtracted from) the index register from 1 to 8. In each case assume RegY is initially 2345.

Post-increment examples are as follows:

```
staa 1,Y+   ;Store the value in RegA at 2345, then RegY=2346
staa 4,Y+   ;Store the value in RegA at 2345, then RegY=2349
```

Pre-increment examples are as follows:

```
staa 1,+Y   ;RegY=2346, then store the value in RegA at 2346
staa 4,+Y   ;RegY=2349, then store the value in RegA at 2349
```

Post-decrement examples are as follows:

```
staa 1,Y-   ;Store a copy of the value in RegA at 2345, then RegY=2344
staa 4,Y-   ;Store a copy of the value in RegA at 2345, then RegY=2341
```

Pre-decrement examples are as follows:

```
staa 1,-Y   ;RegY=2344, then store the value in RegA at 2344
staa 4,-Y   ;RegY=2341, then store the value in RegA at 2341
```

> **Observation:** Usually we would add/subtract one when accessing an 8-bit value and add/subtract two when accessing a 16-bit value.

> **Common error:** The improper use of these index modes with the SP can result in an illegal stack access or unbalanced stack.

3. Accumulator Offset Indexed Addressing mode uses two registers. The offset is located in one of the accumulators A, B or D, and the index (memory address) uses the 16-bit registers: X, Y, SP, or PC. In each case, the accumulator used for the offset and the index register used as a pointer (index) are not modified by the instruction. Examples:

```
ldab #4
ldy  #2345
staa B,Y    ;Store the value in RegA at 2349 (B & Y unchanged)
```

4. Indexed Indirect Addressing mode uses a fixed offset with the 16-bit registers: X, Y, SP, or PC. The fixed offset is always 16 bits. The fixed 16-bit value is added to the index register (X, Y, SP, or PC), and used to fetch a second 16-bit big endian address from memory. The load or store is performed at this second address, as shown in Figure 2.6. Indexed indirect mode is useful when data structures contain pointers. In each case, the 16-bit index register and the memory pointer are not modified by the instruction. For example,

```
ldy  #$2345
staa [-4,Y]   ;fetch 16-bit address from $2341, store $56 at $1234
```

Figure 2.6
Example of the 6812 indexed-indirect addressing mode.

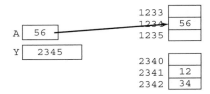

5. Accumulator D Offset Indexed Indirect Addressing mode uses two registers. The offset is located in accumulator D, and the index (memory address) is in one of the 16-bit registers: X, Y, SP, or PC. The value in D is added to the index register (X, Y, SP, or PC), and is used to fetch a second 16-bit big endian address from memory, as shown in Figure 2.7. The load or store is performed at this second address. This mode is also useful when data structures contain pointers. In each case, accumulator D and the index register used as a pointer (index) are not modified by the instruction. For example,

```
ldd  #4
ldy  #$2341
stx  [D,Y]  ;Store the value in RegX at $1234
```

Figure 2.7
Example of the 6812 accumulator-offset indexed-indirect addressing mode.

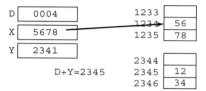

The 6812 load effective address instructions can only be used with indexed addressing mode. These instructions are very useful for manipulating the 16-bit registers. Let idx represent one of the preceding index addressing modes. They do not affect any condition code bits.

```
leax idx     ;RegX=EA
leay idx     ;RegY=EA
leas idx     ;RegS=EA
```

The basic idea is that the effective address is calculated in the usual manner. But rather than fetching the memory contents at that address as a regular load instruction (ldaa ldab ldd ldx ldy lds) would, this instruction puts the effective address itself into the register. In each of the following cases, the effective address, EA, is loaded into Register X.

```
leax m,r     ;IDX 5-bit index, EA=r+m (-16 to 15)
leax v,+r    ;IDX pre-increment r=r+v, EA=r   (1 to 8)
leax v,-r    ;IDX pre-decrement r=r-v, EA=r   (1 to 8)
leax v,r+    ;IDX post-increment, EA=r, r=r+v  (1 to 8)
leax v,r-    ;IDX post-decrement, EA=r, r=r-v  (1 to 8)
leax A,r     ;IDX Reg A offset     EA=r+A, zero padded
leax B,r     ;IDX Reg B offset     EA=r+B, zero padded
leax D,r     ;IDX Reg D offset     EA=r+D
leax q,r     ;IDX1 9-bit index     EA=r+q (-256 to 255)
leax W,r     ;IDX2 16-bit index EA=r+W (-32768 to 65535)
```

where r is Reg X, Y, SP, or PC, and the fixed constants are
  m is any signed 5-bit −16 to +15
  q is any signed 9-bit −256 to +255
  v is any unsigned 3-bit 1 to 8
  W is any signed 16-bit −32768 to +32767 or any unsigned 16-bit 0 to 65535

Observation: The leas -4,sp instruction subtracts four from the stack pointer which causes 4 bytes to be allocated on the stack.

Checkpoint 2.3: Write 6812 assembly code that sets Register Y equal to X+10.

2.2.5 Arithmetic Operations

It is important to remember that arithmetic operations (addition, subtraction, multiplication, and division) have constraints when performed with finite precision on a microcomputer. An overflow error occurs when the result of an arithmetic operation cannot fit into the finite precision of the result. For example, when two 8-bit numbers are added, the sum is 9 bits, and thus it may not fit into the 8-bit result. The same digital hardware (instructions) will be used to add and subtract unsigned and signed numbers. On the other hand, we will need separate overflow detection for signed and unsigned addition and subtraction (C bit and V bit).

It is common for computers to perform arithmetic operations using a register such as Register A. As we have seen, a *register* is a high-speed storage inside the processor. An *accumulator*, such as Register A, is a register with which arithmetic and logic operations can be performed. The following instructions are a few of the arithmetic functions available on the 6811/6812, which fetch data from memory and add/subtract it from the register. With immediate mode (#w) the 8-bit constant is located in the instruction itself. With direct mode and extended mode (U) the 8-bit data is fetched from memory location U. Recall that direct/extended mode affects the size of the address, not the size of the data. The size of the data will be determined by the size of the register into which the operation will be performed.

All microcomputers have a *condition code register* (CC or CCR) that specifies the status of the most recent operation. In this section, we will introduce the four condition code bits common to most microcomputers, shown in Table 2.3. If the two inputs to an addition or subtraction operation are considered as unsigned, then the C bit (carry) will be set if the result does not fit. In other words, after an unsigned operation, the C bit is set if the answer is wrong. If the two inputs to an addition or subtraction operation are considered as signed, then the V bit (overflow) will be set if the result does not fit. In other words, after a signed operation, the V bit is set if the answer is wrong.

Table 2.3
Condition code bits contain the status of the previous arithmetic or logical operation.

Bit	Name	Meaning after Addition or Subtraction
N	negative	result is negative
Z	zero	result is zero
V	overflow	signed overflow
C	carry	unsigned overflow

The adda and addb instructions add an 8-bit value from memory to the corresponding register.

These instructions work for both signed and unsigned data.

adda	#w	RegA=RegA+w	Add 8-bit constant to RegA
adda	U	RegA=RegA+[U]	Add 8-bit memory value to RegA
addb	#w	RegB=RegB+w	Add 8-bit constant to RegB
addb	U	RegB=RegB+[U]	Add 8-bit memory value to RegB

Condition code bits are set after $R=X+M$, where X is initial register value, R is the final register value.

N: result is negative $N = R_7$

Z: result is zero $Z = \overline{R_7} \cdot \overline{R_6} \cdot \overline{R_5} \cdot \overline{R_4} \cdot \overline{R_3} \cdot \overline{R_2} \cdot \overline{R_1} \cdot \overline{R_0}$

V: signed overflow $V = X_7 \cdot M_7 \cdot \overline{R_7} + \overline{X_7} \cdot \overline{M_7} \cdot R_7$

C: unsigned overflow $C = X_7 \cdot M_7 + M_7 \cdot \overline{R_7} + \overline{R_7} \cdot X_7$

Let N and M be 8-bit unsigned locations. The following assembly code implements $M=N+25$:

```
ldaa N
adda #25 ;RegA=N+25, error if C is set
staa M
```

The addd instruction adds a 16-bit value from memory to Register D. This instruction works for both signed and unsigned data.

addd	#W	RegD=RegD+W	Add 16-bit constant to RegD
addd	U	RegD=RegD+{U}	Add 16-bit memory value to RegD

Condition code bits are set after R=D+M, where D is initial register value, R is the final register value.

N: result is negative $N = R_{15}$

Z: result is zero $Z = \overline{R_{15}} \cdot \overline{R_{14}} \cdot \overline{R_{13}} \cdot \overline{R_{12}} \cdot \overline{R_{11}} \cdot \overline{R_{10}} \cdot \overline{R_9} \cdot \overline{R_8} \cdot \overline{R_7} \cdot \overline{R_6} \cdot \overline{R_5} \cdot \overline{R_4} \cdot \overline{R_3} \cdot$
$$\overline{R_2} \cdot \overline{R_1} \cdot \overline{R_0}$$

V: signed overflow $V = D_{15} \cdot M_{15} \cdot \overline{R_{15}} + \overline{D_{15}} \cdot \overline{M_{15}} \cdot R_{15}$

C: unsigned overflow $C = D_{15} \cdot M_{15} + M_{15} \cdot \overline{R_{15}} + \overline{R_{15}} \cdot D_{15}$

Let N and M be 16-bit unsigned locations. The following assembly code implements M=N+1000.

```
ldd  N
addd #1000 ;RegD=N+1000, error if C is set
std  M
```

Checkpoint 2.4: Write assembly code that adds a constant 2000 to Register Y.

These instructions subtract an 8-bit memory value from a register. The operation works for both signed and unsigned values. The compare and test instructions do not change the register value. The condition code bits can be used by a conditional branch instruction to compare the two values. If the numbers represent unsigned values, then follow a subtraction/compare with an unsigned conditional branch: beq bne bhi bhs blo bls. If the numbers represent signed values, then follow a subtraction/compare with a signed conditional branch: beq bne bgt bge blt ble.

cmpa	#w	RegA-w	Compare RegA to 8-bit constant
cmpa	U	RegA-[U]	Compare RegA to 8-bit memory value
cmpb	#w	RegB-w	Compare RegB to 8-bit constant
cmpb	U	RegB-[U]	Compare RegB to 8-bit memory value
suba	#w	RegA=RegA-w	Subtract 8-bit constant from RegA
suba	U	RegA=RegA-[U]	Subtract 8-bit memory value from RegA
subb	#w	RegB=RegB-w	Subtract 8-bit constant from RegB
subb	U	RegB=RegB-[U]	Subtract 8-bit memory value from RegB
tsta		RegA-0	Test RegA
tstb		RegB-0	Test RegB

Condition code bits are set after R=X−M, X is initial register value, and R is the final register value.

N: result is negative $N = R_7$

Z: result is zero $Z = \overline{R_7} \cdot \overline{R_6} \cdot \overline{R_5} \cdot \overline{R_4} \cdot \overline{R_3} \cdot \overline{R_2} \cdot \overline{R_1} \cdot \overline{R_0}$

V: signed overflow $V = X_7 \cdot \overline{M_7} \cdot \overline{R_7} + \overline{X_7} \cdot M_7 \cdot R_7$

C: unsigned overflow $C = \overline{X_7} \cdot M_7 + M_7 \cdot R_7 + R_7 \cdot \overline{X_7}$

Let N and M be 8-bit unsigned locations. The following assembly code implements M=N−10.

```
ldaa N
suba #10    ;RegA=N-10, error if C is set
staa M
```

These instructions subtract a 16-bit memory value from a register. Just like the 8-bit subtraction operators, these operators work for both signed and unsigned values. Again, the condition code bits can be used by a conditional branch instruction to compare the two values.

cpd	#W	RegD-W	Compare RegD to 16-bit constant
cpd	U	RegD-{U}	Compare RegD to 16-bit memory value
cpx	#W	RegX-W	Compare RegX to 16-bit constant
cpx	U	RegX-{U}	Ccompare RegX to 16-bit memory value
cpy	#W	RegY-W	Compare RegY to 16-bit constant
cpy	U	RegY-{U}	Compare RegY to 16-bit memory value
subd	#W	RegD=RegD-W	Subtract 16-bit constant from RegD
subd	U	RegD=RegD-{U}	Subtract 16-bit memory value from RegD

Condition code bits are set after R=X−M, X is initial register value, and R is the final register value.

N: result is negative $N = R_{15}$

Z: result is zero $Z = \overline{R_{15}} \cdot \overline{R_{14}} \cdot \overline{R_{13}} \cdot \overline{R_{12}} \cdot \overline{R_{11}} \cdot \overline{R_{10}} \cdot \overline{R_9} \cdot \overline{R_8} \cdot \overline{R_7} \cdot \overline{R_6} \cdot \overline{R_5} \cdot \overline{R_4} \cdot \overline{R_3} \cdot \overline{R_2} \cdot \overline{R_1} \cdot \overline{R_0}$

V: signed overflow $V = X_{15} \cdot \overline{M_{15}} \cdot \overline{R_{15}} + \overline{X_{15}} \cdot M_{15} \cdot R_{15}$

C: unsigned overflow $C = \overline{X_{15}} \cdot M_{15} + M_{15} \cdot R_{15} + R_{15} \cdot \overline{X_{15}}$

Let N and M be 16-bit unsigned locations. The following assembly code implements M=N−1000.

```
ldd  N
subd #1000 ;RegD = N-1000, error if C is set
std  M
```

There are increment and decrement instructions, which operate properly on either signed or unsigned values. These instructions use inherent addressing. The Z bit is set if the result is zero.

deca	RegA=RegA−1	Decrement RegA
decb	RegA=RegA−1	Decrement RegB
dex	RegX=RegX−1	Decrement RegX
dey	RegY=RegY−1	Decrement RegY
inca	RegA=RegA+1	Increment RegA
incb	RegB=RegB+1	Increment RegB
inx	RegX=RegX+1	Increment RegX
iny	RegY=RegY+1	Increment RegY

The `mul` instruction performs an 8-bit by 8-bit into 16-bit unsigned multiply, giving RegD equal to RegA times RegB, as shown in Figure 2.8. No overflow is possible.

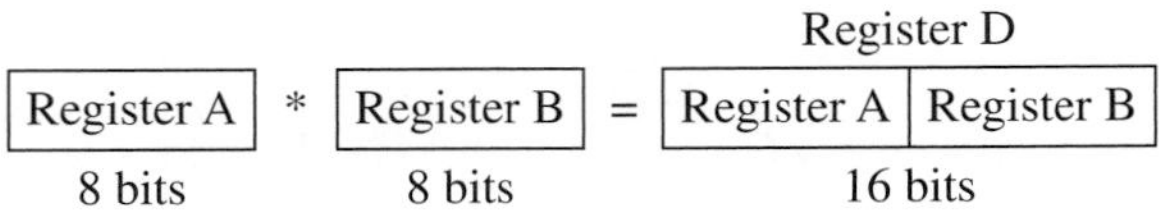

Figure 2.8
The `mul` instruction takes two 8-bit inputs and generates a 16-bit product.

Condition code bits are set after R=A*B.

C: R_7, set if bit 7 of the 16-bit result is one

Checkpoint 2.5: Prove the `mul` instruction can't overflow when multiplying two 8-bit unsigned numbers yielding a 16-bit product.

Let N and M be 8-bit unsigned locations. The following assembly code implements M=3*N.

```
ldaa N
ldab #3
mul      ;RegD=*N, error if RegA is not zero
stab M
```

The `idiv` instruction performs a 16-bit by 16-bit unsigned divide with remainder, giving RegX=RegD/RegX, as shown in Figure 2.9. Register D is the remainder.

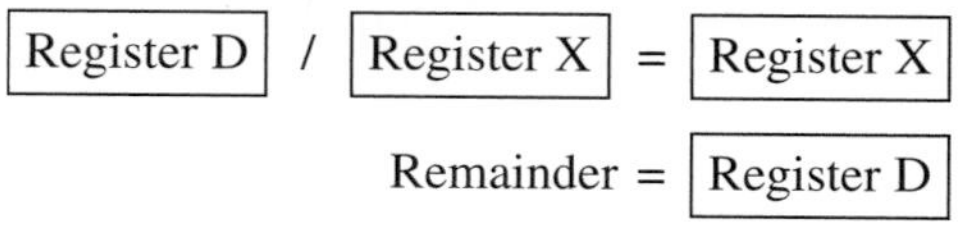

Figure 2.9
The `idiv` instruction takes two 16-bit inputs and generates a 16-bit quotient and a 16-bit remainder.

$$\boxed{\text{Register D}} \; / \; \boxed{\text{Register X}} \; = \; \boxed{\text{Register X}}$$

$$\text{Remainder} = \boxed{\text{Register D}}$$

Condition code bits are set after quotient=dividend/divisor or Q=D/X.

Z: result is zero, $Z = \overline{Q_{15}} \cdot \overline{Q_{14}} \cdot \overline{Q_{13}} \cdot \overline{Q_{12}} \cdot \overline{Q_{11}} \cdot \overline{Q_{10}} \cdot \overline{Q_9} \cdot \overline{Q_8} \cdot \overline{Q_7} \cdot \overline{Q_6} \cdot \overline{Q_5} \cdot \overline{Q_4}$
$\cdot \overline{Q_3} \cdot \overline{Q_2} \cdot \overline{Q_1} \cdot \overline{Q_0}$

V: 0

C: divide by zero, $C = \overline{X_{15}} \cdot \overline{X_{14}} \cdot \overline{X_{13}} \cdot \overline{X_{12}} \cdot \overline{X_{11}} \cdot \overline{X_{10}} \cdot \overline{X_9} \cdot \overline{X_8} \cdot \overline{X_7} \cdot \overline{X_6} \cdot \overline{X_5} \cdot \overline{X_4}$
$\cdot \overline{X_3} \cdot \overline{X_2} \cdot \overline{X_1} \cdot \overline{X_0}$

Checkpoint 2.6: Give a single mathematical equation relating the dividend, divisor, quotient, and remainder. This equation gives a unique solution as long as you assume the remainder is strictly less than the divisor.

Let N and M be 8-bit unsigned locations. The following assembly code implements M=(53*N+50)/100, using promotion. Notice that this overall operation can not overflow because the result will be less than or equal to (53*255+50)/100=135.

```
ldaa N     ;RegA=N   (between 0 and 255)
ldab #53
mul        ;RegD=53*N (between 0 and 13515)
addd #50   ;RegD=53*N+50 (between 0 and 13565)
ldx  #100
idiv       ;RegX=(53*N+50)/100 (between 0 and 135)
xgdx       ;RegB
stab M
```

Checkpoint 2.7: Let N and M be 8-bit unsigned locations. Write assembly code to implement M=(10*N)/51.

The `fdiv` instruction also performs a 16-bit by 16-bit unsigned divide with remainder. In contrast, this instruction calculates RegX=(2^{16}*RegD)/RegX, as shown in Figure 2.10. RegD is the remainder.

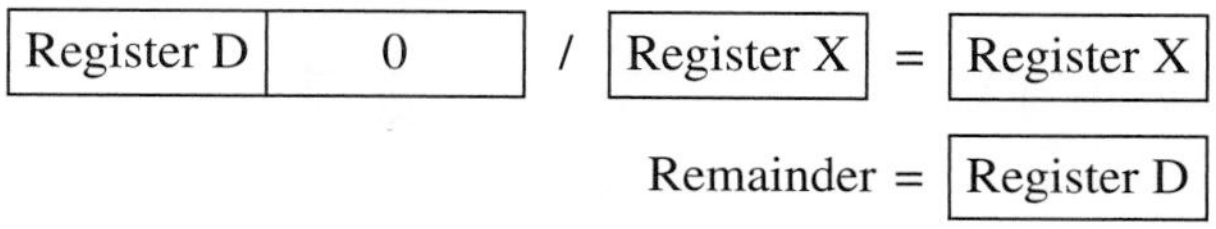

Figure 2.10
The `fdiv` instruction takes two 16-bit inputs and generates a 16-bit quotient and a 16-bit remainder.

$$\boxed{\text{Register D} \quad 0} \; / \; \boxed{\text{Register X}} \; = \; \boxed{\text{Register X}}$$

$$\text{Remainder} = \boxed{\text{Register D}}$$

Condition code bits are set after R=(65536*D)/X.

Z: result is zero, $Z = \overline{R_{15}} \cdot \overline{R_{14}} \cdot \overline{R_{13}} \cdot \overline{R_{12}} \cdot \overline{R_{11}} \cdot \overline{R_{10}} \cdot \overline{R_9} \cdot \overline{R_8} \cdot \overline{R_7} \cdot \overline{R_6} \cdot \overline{R_5} \cdot \overline{R_4}$
$\cdot \overline{R_3} \cdot \overline{R_2} \cdot \overline{R_1} \cdot \overline{R_0}$

V: overflow if RegX is less than or equal to RegD, result >$FFFF

C: divide by zero, $C = \overline{X_{15}} \cdot \overline{X_{14}} \cdot \overline{X_{13}} \cdot \overline{X_{12}} \cdot \overline{X_{11}} \cdot \overline{X_{10}} \cdot \overline{X_9} \cdot \overline{X_8} \cdot \overline{X_7} \cdot \overline{X_6} \cdot \overline{X_5} \cdot \overline{X_4}$
$\cdot \overline{X_3} \cdot \overline{X_2} \cdot \overline{X_1} \cdot \overline{X_0}$

Let N and M be 16-bit unsigned locations. The following assembly code implements M=12.34*N. We approximate 12.34 by 65536/5311.

```
ldd  N
ldx  #5311
fdiv      ;RegX=(65536*N)/5311
stx  M
```

Checkpoint 2.8: Let N and M be 16-bit unsigned locations. Write assembly code using `fdiv` to implement M=2.5*N.

2.2.6 Extended Precision Arithmetic Instructions on the 6812

When designing the 6812, Freescale added a few instructions not available on the 6811. These instructions are quite useful when implementing mathematical calculations. For 6811 systems, these operations are so convenient that we will need subroutines for the same calculations. Many of these subroutines can be found in the `math.rtf` file in the MC6811 folder as part of the **TExaS** simulator.

The `emul` instruction performs a 16-bit by 16-bit unsigned multiply RegY:D= RegY*RegD, as shown in Figure 2.11. The `emuls` instruction is a 16-bit by 16-bit signed multiply, using the same registers and generating the same condition code bits.

Figure 2.11
The `emul` and `emuls` instructions take two 16-bit inputs and generate a 32-bit product.

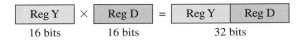

Condition code bits after R=Y*D

N: result is negative, $N = R_{31}$

Z: result is zero, $Z = \overline{R_{31}} \cdot \overline{R_{30}} \cdot \ldots \cdot \overline{R_1} \cdot \overline{R_0}$

C: R_{15}, bit 15 of the result

The `ediv` instruction performs a 32-bit by 16-bit unsigned divide RegY=(Y:D)/RegX; RegD is remainder, as shown in Figure 2.12. The `edivs` instruction is a 32-bit by 16-bit signed divide, using the same registers. The overflow bit calculation is different, but the other three condition code bits are the same.

Figure 2.12
The `ediv` and `edivs` instructions perform extended precision division.

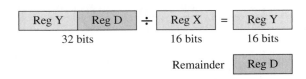

Condition code bits after R=(Y:D)/X

N: result is negative (undefined after an overflow or a divide by zero), $N=R_{15}$

Z: result is zero (undefined after an overflow or a divide by zero)

$$Z = \overline{R_{15}} \bullet \overline{R_{14}} \bullet \overline{R_{13}} \bullet \overline{R_{12}} \bullet \overline{R_{11}} \bullet \overline{R_{10}} \bullet \overline{R_9} \bullet \overline{R_8} \bullet \overline{R_7} \bullet \overline{R_6} \bullet \overline{R_5} \bullet \overline{R_4}$$
$$\bullet \overline{R_3} \bullet \overline{R_2} \bullet \overline{R_1} \bullet \overline{R_0}$$

V: overflow (undefined after a divide by zero),

`ediv` result >$FFFF

`edivs` result >$7FFF or less than −$8000

C: divide by zero, $C = \overline{X_{15}} \bullet \overline{X_{14}} \bullet \overline{X_{13}} \bullet \overline{X_{12}} \bullet \overline{X_{11}} \bullet \overline{X_{10}} \bullet \overline{X_9} \bullet \overline{X_8} \bullet \overline{X_7} \bullet \overline{X_6} \bullet \overline{X_5} \bullet \overline{X_4}$
$$\bullet \overline{X_3} \bullet \overline{X_2} \bullet \overline{X_1} \bullet \overline{X_0}$$

Let N and M be 16-bit unsigned locations. The following assembly code implements M=(53*N+50)/100, using promotion. Notice that this overall operation cannot overflow because the result will be less than or equal to (53*65535+50)/100=34734.

```
        ldd   N      ;RegA=N   (between 0 and 65535)
        ldy   #53
        emul         ;RegY:D=53*N (between 0 and 3473355)
        addd  #50     ;RegD=53*N+50 (between 0 and 3473405)
        bcc   skip
        iny
skip    ldx   #100
        ediv         ;RegY=(53*N+50)/100 (between 0 and 34734)
        sty   M
```

The `emacs` instruction performs a 16-bit by 16-bit signed multiply, followed by a 32-bit signed addition. It uses indexed addressing to access the two 16-bit inputs and extended addressing to access the 32-bit sum. Recall that {X} and {Y} represent the 16-bit contents pointed to by Registers X and Y, respectively. If we define <U> as the 32-bit contents of memory location U, then `emacs U` calculates

$$<U>=<U>+\{X\}*\{Y\}$$

Condition code bits, first P=X*Y, then R=M+P (M,P,R are 32 bits)

N: result is negative, $N=R_{31}$

Z: result is zero, $Z = \overline{R_{31}} \bullet \overline{R_{30}} \bullet \ldots \bullet \overline{R_1} \bullet \overline{R_0}$

V: signed overflow (after addition), $V = P_{31} \bullet M_{31} \bullet \overline{R_{31}} + \overline{P_{31}} \bullet \overline{M_{31}} \bullet R_{31}$

C: unsigned overflow (after addition), $C = P_{31} \bullet M_{31} + M_{31} \bullet \overline{R_{31}} + \overline{R_{31}} \bullet P_{31}$

This instruction is quite useful for calculating fixed-point equations, as illustrated in Program 2.2. We place the input variables xx,yy,zz consecutively in RAM, and the constants 902,−1810,45 consecutively in ROM. Registers X and Y are not automatically incremented, so the program performs that task explicitly.

**2.2.7
Shift Operations**

The shift instructions use inherent addressing. The N bit is set if the result is negative. The Z bit is set if the result is zero. The V bit is set on a signed overflow, and is detected by a change in the sign bit. The C bit is the carry out after the shift.

`asla`	RegA=RegA*2	Signed shift left, same as `lsla`
`aslb`	RegB=RegB*2	Signed shift left, same as `lslb`
`asld`	RegD=RegD*2	Signed shift left, same as `lsld`
`lsla`	RegA=RegA*2	Unsigned shift left, same as `asla`
`lslb`	RegB=RegB*2	Unsigned shift left, same as `aslb`

lsld	RegD=RegD*2	Unsigned shift left, same as asld
asra	RegA=RegA/2	Signed shift right
asrb	RegB=RegB/2	Signed shift right
asrd	RegD=RegD/2	Signed shift right
lsra	RegA=RegA/2	Unsigned shift right
lsrb	RegB=RegB/2	Unsigned shift right
lsrd	RegD=RegD/2	Unsigned shift right
rola		Rotate RegA (C←A7← . . . A0←C)
rolb		Rotate RegB (C←B7← . . . ←B0←C)
rora		Rotate RegA (C→A7→ . . . →A0→C)
rorb		Rotate RegB (C→B7→ . . . →B0→C)

Program 2.2
Fixed-point calculation using the 6812 emac instruction.

```
      org   $3800
; rr=0.902*xx-1.81*yy+0.045*zz
; first we can convert this equation to fixed point
; without introducing any error:
; rr=(902*xx-1810*yy+45*zz)/1000
xx    ds    2
yy    ds    2
zz    ds    2
rr    ds    2      ; result
acc   ds    4      ; temporary 32-bit result
      org   $4000
cc    dc.w  902,-1810,45
Calc  ldx   #xx    ; pointer to data
      ldy   #cc    ; pointer to coefficients
      movw  #0,acc ; initially clear temporary result
      movw  #0,acc+2
      ldaa  #3     ; number of terms
loop  emacs acc    ;acc=acc+{X}*{Y}
      leax  2,x
      leay  2,y
      dbne  A,loop
      ldy   acc
      ldd   acc+2  ;Y:D=902*xx-1810*yy+45*zz
      ldx   #1000
      edivs
      sty   rr
      rts
```

When programming in C, the shift is a binary operation. In other words, the << and >> operators take two inputs and yield one output, e.g., r=m>>n. But at the machine level (i.e., assembly programming), the shift operators are actually unary operations, e.g., r=m>>1. The assembly instructions used for shifting will shift one bit at a time. If you want to shift multiple times, you will have to execute the instruction multiple times. The logical shift right (LSR) is the equivalent to an unsigned divide by 2, as shown in Figure 2.13. A zero is shifted into the most significant position, and the carry flag will hold the bit shifted out.

Figure 2.13
8-bit logical shift right.

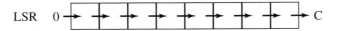

Figure 2.14
8-bit arithmetic shift right.

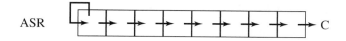

ASR

The arithmetic shift right (ASR) is the equivalent to a signed divide by 2, as shown in Figure 2.14. Notice that the sign bit is preserved and the carry flag will hold the bit shifted out.

The same shift left operation works for both unsigned and signed multiply by 2, as shown in Figure 2.15. In other words, the arithmetic shift left (ASL) is identical to the logical shift left (LSL). A zero is shifted into the least significant position, and the carry bit will contain the bit that was shifted out.

Figure 2.15
8-bit shift left.

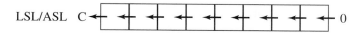

LSL/ASL

The **roll** operations can be used to create multiple-byte shift functions. Roll right and roll left are shown in Figure 2.16. In each case, the carry is shifted into the 8-bit byte, and the carry bit will contain the bit that was shifted out.

Figure 2.16
8-bit roll right and 8-bit roll left.

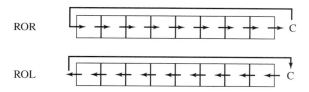

ROR

ROL

2.2.8 Logical Operations

Most 8-bit logical instructions take two inputs, one from a register and the other from memory. All but the `bita bitb` instructions put the result back in the register. The N bit will be set if the result is negative. The Z bit will be set if the result is zero. These logical instructions will clear the V bit and leave the C bit unchanged.

anda	#w	RegA=RegA&w	Logical and RegA with a constant
anda	U	RegA=RegA&[U]	Logical and RegA with a memory value
andb	#w	RegB=RegB&w	Logical and RegB with a constant
andb	U	RegB=RegB&[U]	Logical and RegB with a memory value
bita	#w	RegA&w	Logical and RegA with a constant
bita	U	RegA&[U]	Logical and RegA with a memory value
bitb	#w	RegB&w	Logical and RegB with a constant
bitb	U	RegB&[U]	Logical and RegB with a memory value
coma		RegA=$FF−RegA, RegA=~RegA Complement RegA	
comb		RegB=$FF−RegB, RegB=~RegB Complement RegB	
eora	#w	RegA=RegA ^ w	Exclusive or RegA with a constant
eora	U	RegA=RegA ^ [U]	Exclusive or RegA with a memory value
eorb	#w	RegB=RegB ^ w	Exclusive or RegB with a constant
eorb	U	RegB=RegB ^ [U]	Exclusive or RegB with a memory value
oraa	#w	RegA=RegA | w	Logical or RegA with a constant
oraa	U	RegA=RegA | [U]	Logical or RegA with a memory value
orab	#w	RegB=RegB | w	Logical or RegB with a constant
orab	U	RegB=RegB | [U]	Logical or RegB with a memory value

Condition code bits are set, where R is the result of the operation.

N: result is negative $N=R_7$

Z: result is zero $Z= \overline{R_7} \cdot \overline{R_6} \cdot \overline{R_5} \cdot \overline{R_4} \cdot \overline{R_3} \cdot \overline{R_2} \cdot \overline{R_1} \cdot \overline{R_0}$

V: signed overflow $V=0$

The following C code uses the shift and **or** operations to combine two parts into one number. High and Low are unsigned 4-bit components, which will be combined into a single unsigned 8-bit Result. We will assume both High and Low are bounded within the range of 0 to 15. The expression High<<4 will perform four logical shift lefts.

```
Result = (High<<4)|Low;
```

The assembly code for this operation is

```
ldaa High    ;read value of High
lsla          ;shift into position
lsla
lsla
lsla
oraa Low     ;combine the two parts together
staa Result ;save answer
```

To illustrate how the foregoing program works, let $0\ 0\ 0\ 0\ h_3\ h_2\ h_1\ h_0$ be the value of High, and let $0\ 0\ 0\ 0\ l_3\ l_2\ l_1\ l_0$ be the value of Low. The ldaa instruction brings High into register A. The four lsla instructions move the High into bit positions 4-7, the oraa instruction combines High and Low, and the staa instruction stores the combination into Result.

0	0	0	0	h_3	h_2	h_1	h_0	value of High
0	0	0	h_3	h_2	h_1	h_0	0	after first lsla
0	0	h_3	h_2	h_1	h_0	0	0	after second lsla
0	h_3	h_2	h_1	h_0	0	0	0	after third lsla
h_3	h_2	h_1	h_0	0	0	0	0	after last lsla
0	0	0	0	l_3	l_2	l_1	l_0	value of Low
h_3	h_2	h_1	h_0	l_3	l_2	l_1	l_0	result of the oraa instruction

Checkpoint 2.9: Assume PORTB is an output port. Write assembly code that just clears bit 4, leaving the other 7 bits unchanged.

Checkpoint 2.10: Assume PORTB is an output port. Write assembly code that just sets bit 3, leaving the other 7 bits unchanged.

2.2.9
Subroutines and the Stack

We begin this section with a general description of the stack, and introduce the basic concepts common to both the 6811 and 6812. In general, we initialize the stack pointer (RegSP) into RAM using the lds instruction, which is usually done once at the beginning of the program. In the classical definition of the stack, there are just two operations one can perform: **push** and **pull**. Some computers define the two stack operations as push and pop. The push function saves data on the top of the stack, and the pull function removes data from the top of the stack. For example, the psha instruction will push the value in RegA onto the stack, leaving RegA unchanged. The pula instruction will pull (or pop) a value off the stack bringing it into RegA. The pull operation does modify the stack such that the pulled data is no longer on the stack. The stack implements last in first out (LIFO) behavior. The following code pushes the numbers 1, 2, and 3 in that order.

```
ldaa #1
psha                ; push 1 on the stack
ldaa #2
```

```
psha                ; push 2 on the stack
ldaa #3
psha                ; push 3 on the stack
```

After these three push operations, the stack would contain these numbers with the 3 on the top, as shown in Figure 2.17. The top entry of the stack contains the newest data (i.e., the data pushed last). On the 6811, RegSP points to the entry just above the top element. On the 6812, RegSP points to the top element.

Figure 2.17
The stack holding three elements, with the 3 on top.

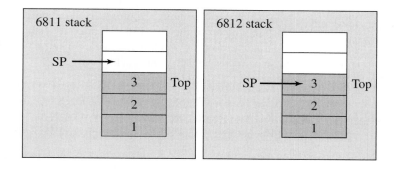

At this point if one were to pull from the stack (e.g., execute `pula`), the 3 would be returned, and 2 would now be on the top of the stack, as shown in Figure 2.18.

Figure 2.18
The stack holding two elements, with the 2 on top.

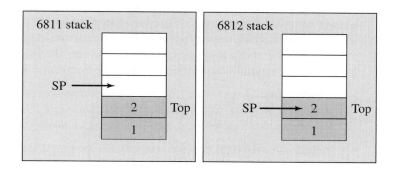

The push and pull instructions use inherent addressing and do not modify the condition code. The push instructions produce two copies of the data, one on the stack and the other still in the register. The pull instructions remove the data from the stack, so there will be only one copy of the data left, which is in the register.

psha	Push RegA on the stack
pshb	Push RegB on the stack
pshx	Push RegX on the stack
pshy	Push RegY on the stack
des	RegSP=RegSP−1 (reserve space on the stack)
pula	Pull value from stack, put in RegA
pulb	Pull value from stack, put in RegB
pulx	Pull value from stack, put in RegX
puly	Pull value from stack, put in RegY
ins	RegSP=RegSP+1 (discard top of stack)

The stack is used for many purposes. A common use is temporary storage. If a piece of information is important, we can push it on the stack. Later, when we wish to retrieve the data, we pull it off the stack.

Checkpoint 2.11: Assume you have two 8-bit global variables M and N. Write assembly code that switches the values in M and N using just the `ldaa staa psha` and `pula` instructions.

There are two additional stack operations called **stack read** and **stack write**. The stack read operation allows you to retrieve data previously pushed on the stack without modifying the data or the stack pointer. The stack write operation allows you to change a value previously pushed on the stack without changing the stack pointer. Even though these two stack operations are not part of the classical definition of a stack, they will be essential for implementing parameter passing and local variables. The following are important instructions that greatly facilitate the use of the stack.

tsx Transfer RegSP to RegX, 6811 does RegX=RegSP+1, 6812 does RegX=RegSP
tsy Transfer RegSP to RegY, 6811 does RegY=RegSP+1, 6812 does RegY=RegSP
txs Transfer RegX to RegSP, 6811 does RegSP=RegX−1, 6812 does RegSP=RegX
tys Transfer RegY to RegSP, 6811 does RegSP=RegY−1, 6812 does RegSP=RegY

The instruction `tsx` will move a copy of the stack pointer into Register X. Although `tsx` and `tsy` work a little differently on the 6811 versus the 6812, in both cases the register points to the top element of the stack.

Procedures and functions are programs that can be called to perform specific tasks. Some programming environments differentiate functions (return a value) from procedures (do not return a value). In assembly language, we use the term **subroutine** for all subprograms whether or not they return a value. Subroutines allow us to develop modular software. In assembly language, we will use either the `bsr` or `jsr` instruction to call a subroutine, and we will use the `rts` instruction to return from the subroutine. The `bsr` and `jsr` instructions will push the return address on the stack. The return address is the address of the instruction immediately after the branch to subroutine instruction. The `rts` will pull the return address from the stack, returning the program to the place from which the subroutine was called.

Observation: Since the `bsr` instruction uses relative addressing, it can only be used to call a subroutine near the current instruction. Since the `jsr` instruction allows extended addressing, it can be used to call a subroutine anywhere in memory.

We will study the concept of subroutines using the simple example shown in Program 2.3. The input parameter to the subroutine is passed in using RegA. The subroutine adds one and returns the result also in RegA. The main program calls the subroutine using the `bsr` instruction. The subroutine returns back to the main program using the `rts` instruction. The numbers on the right specify the sequence of execution for the first three times through the loop.

Program 2.3
Simple program showing how to use the `bsr` and `rts` instructions to implement a subroutine.

```
        org   $F800
main  lds   #$00FF .........................1
        clra ................................2
loop  bsr   Add1 ; branch to subroutine .....3   7   11
        bra   loop ..........................6  10   14
;*******Add1*********
;Purpose: Subtract one from RegA
; Input: RegA
;Output: RegA=Input+1
Add1  inca        ; adds one to RegA .......4   8   12
        rts ...............................5   9   13
        org $FFFE
        fdb main
```

Program 2.4
6811 assembly listing of
Program 2.3.

```
$F800                              org  $F800
$F800 8E00FF  {ppp    }main lds   #$00FF
$F803 4F      {pf     }     clra
$F804 8D02    {ppnfss }loop bsr   Add1    ;branch to subroutine
$F806 20FC    {pfn    }     bra   loop
$F808 4C      {pf     }Add1 inca       ;adds one to RegA
$F809 39      {pfxuu  }     rts
$FFFE                         org $FFFE
$FFFE F800                    fdb main
```

We begin the study by looking at the listing file generated by the assembler, shown as Program 2.4.

Figure 2.19 shows the stack before and after the `bsr` instruction is executed. We can also understand the execution of `bsr` by looking at the cycles it generates ppnfss. The 6812 runs faster and more efficiently, but performs basically the same sequence of operations, so we'll study the 6811. The 6811 needs 6 cycles to execute `bsr`. During the first two cycles it fetches the opcode and operand. At this point the PC is $F806, which will be the return location. During the next two cycles, the effective address ($F808) is calculated. The read $FFFF and read $F808 cycles perform no useful work, except to use up time while the processor is performing internal calculations. The last two cycles push the return address on the stack.

```
Opcode fetch   R 0xF804 0x8D from ROM
Operand fetch  R 0xF805 0x02 from ROM
Null Cycle     R 0xFFFF 0x00 from ROM
Dummy PC fetch R 0xF808 0x4C from ROM
Stack store lsbW 0x00FF 0x06 to RAM
Stack store msbW 0x00FE 0xF8 to RAM
```

Figure 2.19
The stack before and after execution of the `bsr` instruction.

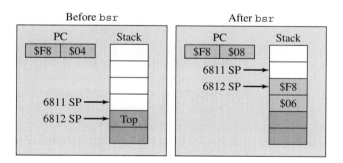

The `rts` instruction will return to the program that called the subroutine. Figure 2.20 shows the stack before and after the `rts` instruction is executed. From the assembly listing, we see that the `rts` generates the 5 cycles pfxuu. Again, the 6812 runs faster and more efficiently, but performs basically the same sequence of operations. During the first cycle it fetches the opcode. The read $F80A and read $00FD cycles perform no useful work, except to use up time while the processor is performing internal calculations. The last two cycles pull the return address from the stack.

```
Opcode fetch   R 0xF809 0x39 from ROM
Dummy PC fetch R 0xF80A 0xFF from ROM
Dummy SP read  R 0x00FD 0x00 from RAM
Stack read msb R 0x00FE 0xF8 from RAM
Stack read lsb R 0x00FF 0x06 from RAM
```

Figure 2.20
The stack before and after execution of the rts instruction.

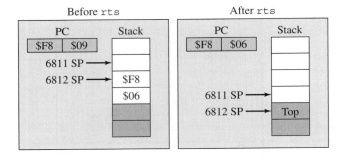

2.2.10
Branch Operations

Normally the computer executes one instruction after another in a linear fashion. In particular, the next instruction to execute is found immediately following the current instruction. We use branch instructions to deviate from this straight line path.

bcc target	Branch to target if C=0
bcs target	Branch to target if C=1
beq target	Branch to target if Z=1
bne target	Branch to target if Z=0
bmi target	Branch to target if N=1
bpl target	Branch to target if N=0
bra target	Branch to target always
brn target	Branch to target never
bvc target	Branch to target if V=0
bvs target	Branch to target if V=1
jmp target	Branch to target always, extended addressing

The following branch instructions must follow a subtract compare or test instruction, such as suba subb sbca sbcb subd cba cmpa cmpb cpd cpx cpy tsta tstb tst.

bge target	Branch if signed greater than or equal to, if $(N \wedge V)=0$, or $(\sim N \cdot V + N \cdot \sim V)=0$
bgt target	Branch if signed greater than, if $(Z + N \wedge V)=0$, or $(Z + \sim N \cdot V + N \cdot \sim V)=0$
ble target	Branch if signed less than or equal to, if $(Z + N \wedge V)=1$, or $(Z + \sim N \cdot V + N \cdot \sim V)=1$
blt target	Branch if signed less than, if $(N \wedge V)=1$, or $(\sim N \cdot V + N \cdot \sim V)=1$
bhs target	Branch if unsigned greater than or equal to, if C=0, same as bcc
bhi target	Branch if unsigned greater than, if C+Z=0
blo target	Branch if unsigned less than, if C=1, same as bcs
bls target	Branch if unsigned less than or equal to, if C+Z=1

Conditional execution is an important aspect of software programming. Two values are compared, and certain blocks of program are executed or skipped depending on the results of the comparison. In assembly language it is important to know the precision (e.g., 8-bit, 16-bit) and the format of the two values (e.g., unsigned, signed). It takes three steps to perform a comparison. You begin by reading the first value into a register. For 8-bit values you can use either Register A or Register B. 16-bit values can be loaded into Register D, Register X, or Register Y. The second step is to compare the first value with the second value.

You can use either a subtract instruction (suba subb subd) or a compare instruction (cmpa cmpb cpd cpx cpy). These instructions set the condition code bits. The last step is a conditional branch. Table 2.4 lists some simple comparisons. When testing for equal, or not equal, it doesn't matter whether the numbers are signed or unsigned. In the following examples, we assume G1, G2 are 8-bit variables.

Table 2.4
Conditional structures that test for equality.

C Code	Assembly Code

`if(G2 == G1){`
`  isEqual();`
`}`
	```ldaa G2```
	```cmpa G1```
	```bne next ;skip if not equal```
	```jsr isEqual ;G2==G1```
	```next```

`if(G2 != G1){`
`  isNotEqual();`
`}`
	```ldaa G2```
	```cmpa G1```
	```beq next ;skip if equal```
	```jsr isNotEqual ;G2!=G1```
	```next```

Common error: It is an error to use an 8-bit comparison to test two 16-bit values.

When testing for greater than or less than, it does matter whether the numbers are signed or unsigned. Table 2.5 lists some 8-bit unsigned comparisons. When comparing unsigned values, the instructions bhi blo bhs and bls should follow the subtraction or comparison instruction. To convert these examples to 16 bits, change the ldaa G2 to ldd H2 and the cmpa G1 to cpd H1.

Table 2.5
Unsigned 8-bit conditional structures.

C Code	Assembly Code

`if(G2 > G1){`
`  isGreater();`
`}`
	```ldaa G2```
	```cmpa G1```
	```bls next ;skip if G2<=G1```
	```jsr isGreater ;G2>G1```
	```next```

`if(G2 >= G1){`
`  isGreaterEq();`
`}`
	```ldaa G2```
	```cmpa G1```
	```blo next ;skip if G2<G1```
	```jsr isGreaterEq ;G2>=G1```
	```next```

`if(G2 < G1){`
`  isLess();`
`}`
	```ldaa G2```
	```cmpa G1```
	```bhs next ;skip if G2>=G1```
	```jsr isLess ;G2<G1```
	```next```

`if(G2 <= G1){`
`  isLessEq();`
`}`
	```ldaa G2```
	```cmpa G1```
	```bhi next ;skip if G2>G1```
	```jsr isLessEq ;G2<=G1```
	```next```

When comparing signed values, the instructions `bgt` `bls` `bge` and `ble` should follow the subtraction or comparison instruction.

Checkpoint 2.12: When implementing `if(N>25)isGreater();` why is it important to know whether N is signed or unsigned?

Common error: It is an error to use an unsigned conditional branch when comparing two signed values. Similarly, it is a mistake to use a signed conditional branch when comparing two unsigned values.

Quite often the microcomputer is asked to wait for events or to search for objects. Both of these operations are solved using the `while` or `do-while` structure. The while loop, illustrated in the Figure 2.21, will wait until Port A bit 0 equals 1. The operation is defined by the C code

```
while((PORTA&0x01)==0){}
```

Figure 2.21
Flowchart of a `while`
structure.

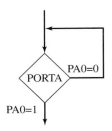

The program begins with reading Port A. Bit 0 is selected using the logical *and* function. Selecting certain bits with the *and* function is called **masking**. The `anda` instruction sets the Z bit, which is utilized by the `beq` instruction. The software will loop over and over while PA0 equals zero. The operation in assembly is

```
loop ldaa PORTA
     anda #$01  ;look at just PA0
     beq  loop  ;continue when PA0 is 1
```

2.2.11 Assembler Pseudo-ops

Pseudo-ops are specific commands to the assembler that are interpreted during the assembly process. An alternative name for pseudo-op is **assembly directive**. A few of them create object code, but most do not. There are many assemblers available developing 6811/6812 assembly code. Although they all use the standard Freescale opcodes, the spelling of the pseudo-op codes varies. The **TExaS** assembler supports many of the various dialects. The pseudo-op codes supported by this assembler are shown in Table 2.6. If you plan to export

Table 2.6
Assembly directives
supported by **TExaS**.

Group A	Group B	Group C	Meaning
org	org	.org	Specific absolute address to put subsequent object code
=	equ		Define a constant symbol
	set		Define or redefine a constant symbol
dc.b db	fcb	.byte	Allocate byte(s) of storage with initialized values
	fcc		Create an ASCII string (no termination character)
dc.w dw	fdb	.word	Allocate word(s) of storage with initialized values
dc.l dl		.long	Allocate 32-bit long word(s) of storage with initialized values
ds ds.b	rmb	.blkb	Allocate bytes of storage without initialization
ds.w		.blkw	Allocate bytes of storage without initialization
ds.l		.blkl	Allocate 32-bit words of storage without initialization
end	end	.end	Signifies the end of the source code (**TExaS** ignores these)

software developed with **TExaS** to another application, then you should limit your use only the pseudo-ops compatible with that application. Group A is supported by Freescale's MCUez, and HiWare (now Metrowerks). Group B is supported by Freescale's DOS level AS11 and AS12. Group C are are used by ImageCraft's ICC11 and ICC12.

Equate symbol to a value

```
<label> equ <expression> (<comment>)
<label> = <expression> (<comment>)
```

The equ (or =) directive assigns the value of the expression in the operand field to the label; see Program 2.5. The label cannot be redefined anywhere else in the program. The expression cannot contain any forward references or undefined symbols. Equates with forward references are flagged as a phasing error.

Program 2.5
A constant implemented with equ might make the program easier to change.

```
; MC68HC711E9                    ; MC9S12C32
      org  0                           org  $3800
size  equ  5                     size  equ  5
data  rmb  size                  data  rmb  size
      org  $D000                        org  $4000
sum   ldaa #size                 sum   ldaa #size
      ldx  #data                       ldx  #data
      clrb                             clrb
loop  addb 0,x                   loop  addb 1,x+
      inx                              dbne A,loop
      deca                             rts
      bne  loop
      rts
```

The equ pseudo-op is used to define the I/O ports and to access the elements of a data structure.

> ***Programming tip:*** Use equ definitions only if it makes the program easier to understand, to debug, or to change.

Redefinable equate symbol to a value

```
<label> set <expression> (<comment>)
```

The set directive assigns the value of the expression in the operand field to the label. The set directive assigns a value other than the program counter to the label. Unlike the equ pseudo-op, the label can be redefined within the program. Although allowed, it is probably a mistake to use forward references. The use of this pseudo-op with forward references will not be flagged with a phasing error. Local variable names created with the set directive could be reused in another subroutine. More information about local variables can be found in Section 2.5.1.

Form constant byte

```
(<label>) fcb <expr>(,<expr>,...,<expr>) (<comment>)
(<label>) dc.b <expr>(,<expr>,...,<expr>) (<comment>)
(<label>) db <expr>(,<expr>,...,<expr>) (<comment>)
(<label>) .byte <expr>(,<expr>,...,<expr>) (<comment>)
```

The fcb directive may have one or more operands separated by commas. The value of each operand is truncated to eight bits and is stored in a single byte of the object program. Multiple operands are stored in successive bytes. The operand may be a numeric constant, a character constant, a symbol, or an expression. If multiple operands are present, one or more of them can be null (two adjacent commas), in which case a single byte of zero will be assigned for that operand. If an operand is larger than the range of an 8-bit number (-128 to $+255$), the result is truncated without a warning, and the least significant 8 bits are used.

A string can be included, which is stored as a sequence of ASCII characters. The delimiters supported by TExaS are " ' and \. The string does not include a null-termination, so if desired, the programmer must explicitly terminate it. The following three examples produce identical null-terminated strings.

```
str1 fcb "Hello World",0
str2 fcb 'Hello World',0
str3 fcb \Hello World\,0
```

The stepper motor controller shown in Program 2.6 uses the fcb definitions to store the four stepper motor output values.

Program 2.6
A stepper motor controller using fcb.

```
; MC68HC711E9
size  equ  4
PORTB equ $1004  ;PB3-0 to stepper
      org  $D000
main  ldaa #size
      ldx  #steps
step  ldab 0,x
      inx
      stab PORTB    ;step motor
      deca
      bne  step
      bra  main
steps fcb  5,6,10,9 ;out sequence
      org  $FFFE
      fdb  main
```

```
; MC9S12C32
size  equ  4
PTT   equ  $0240
DDRT  equ  $0242
      org  $4000
main  movb #$FF,DDRT ;PT3-0 outputs
run   ldaa #size
      ldx  #steps
step  movb 1,x+,PTT  ;step motor
      dbne A,step
      bra  run
steps fcb  5,6,10,9  ;out sequence
      org  $FFFE
      fdb  main
```

Form constant character string

```
(<label>) fcc <delimiter><string><delimiter> (<comment>)
```

The fcc directive is used to store ASCII strings into consecutive bytes of memory. The byte storage begins at the current program counter. The label is assigned to the address of the first byte in the string. Any of the printable ASCII characters can be contained in the string. The string is specified between two identical delimiters. The first non-blank character after the fcc directive is used as the delimiter. The delimiters supported by **TExaS** are " ' and \. Examples:

```
LABEL1  FCC   'ABC'
LABEL2  fcc   "Jon Valvano"
LABEL4  fcc   /Welcome to FunCity!/
```

The first line creates the ASCII characters **ABC** at location LABEL1. Be careful to position the fcc code away from executable instructions. The assembler will produce object code as it would for regular instructions, one line at a time. For example, the following would crash because after executing the ldx instruction, the microcontroller would try to execute the ASCII characters "Trouble" as instructions.

```
     ldaa 100
     ldx  #Strg
Strg fcc  "Trouble"
```

Typically we collect all the `fcc`, `fcb`, `fdb` together and place them at the end of our program, so that the microcomputer does not try to execute the constant data. The ASCII string generated by `fcc` is not null-terminated, so if a termination is needed, you must add it explicitly using either

```
Strg1 fcc   "happy"
      fcb   0
```

or

```
Strg2 fcb   "happy",0
```

Form double byte

```
(<label>) fdb <expr>(,<expr>,...,<expr>) (<comment>)
(<label>) dc.w <expr>(,<expr>,...,<expr>) (<comment>)
(<label>) dw <expr>(,<expr>,...,<expr>) (<comment>)
(<label>) .word <expr>(,<expr>,...,<expr>) (<comment>)
```

The `fdb` directive may have one or more operands separated by commas. The 16-bit value corresponding to each operand is stored into two consecutive bytes of the object program (big endian). The storage begins at the current program counter. The label is assigned to the address of the first 16-bit value. Multiple operands are stored in successive 16-bit words. The operand may be a numeric constant, a character constant, a symbol, or an expression. If multiple operands are present, one or more of them can be null (two adjacent commas), in which case two bytes of zeros will be assigned for that operand. The `fdb` has been used many times so far in the book to define the reset vector.

Define 32-bit constant

```
(<label>) dc.l <expr>(,<expr>,...,<expr>) (<comment>)
(<label>) dl <expr>(,<expr>,...,<expr>) (<comment>)
(<label>) .long <expr>(,<expr>,...,<expr>) (<comment>)
```

The `dl` directive may have one or more operands separated by commas. The 32-bit value corresponding to each operand is stored into four consecutive bytes of the object program (big endian). The storage begins at the current program counter. The label is assigned to the address of the first 32-bit value. Multiple operands are stored in successive bytes. The operand may be a numeric constant, a character constant, a symbol, or an expression. If multiple operands are present, one or more of them can be null (two adjacent commas), in which case four bytes of zeros will be assigned for that operand. In the following finite state machine, the `dl` definitions are used to define 32-bit constants.

```
S1  dl     100000,$12345678
S2  .long  1,10,100,1000,10000,100000,1000000,10000000
S3  dc.l   -1,0,1
```

Set program counter origin

```
org<expression> (<comment>)
.org<expression> (<comment>)
```

The `org` directive changes the program counter to the value specified by the expression in the operand field. Subsequent statements are assembled into memory locations starting with the new program counter value. If no `org` directive is encountered in a source program, the program counter is initialized to zero. Expressions cannot contain forward references or undefined symbols. The `org` statements in Programs 2.5 and 2.6 place the variables in RAM and the programs in EEPROM. The `org` statement is also used to set the reset vector.

Reserve multiple bytes

```
(<label>)  rmb    <expression>  (<comment>)
(<label>)  ds     <expression>  (<comment>)
(<label>)  ds.b   <expression>  (<comment>)
(<label>)  .blkb  <expression>  (<comment>)
```

The rmb directive causes the location counter to be advanced by the value of the expression in the operand field. This directive reserves a block of memory the length of which in bytes is equal to the value of the expression. The block of memory reserved is not initialized to any given value. The expression cannot contain any forward references or undefined symbols. This directive is commonly used to reserve a scratchpad or table area for later use.

Checkpoint 2.13: Why can't you use a forward reference in an rmb directive?

Reserve multiple words

```
(<label>)  ds.w   <expression>  (<comment>)
(<label>)  .blkw  <expression>  (<comment>)
```

The ds.w directive causes the location counter to be advanced by 2 times the value of the expression in the operand field. This directive reserves a block of memory the length of which in words (16-bit) is equal to the value of the expression. The block of memory reserved is not initialized to any given value. The expression cannot contain any forward references or undefined symbols. This directive is commonly used to reserve a scratchpad or table area for later use.

ds.l Reserve multiple 32-bit words

```
(<label>)  ds.l   <expression>  (<comment>)
(<label>)  .blkl  <expression>  (<comment>)
```

The ds.l directive causes the location counter to be advanced by 4 times the value of the expression in the operand field. This directive reserves a block of memory the length of which in words (32-bit) is equal to the value of the expression. The block of memory reserved is not initialized to any given value. The expression cannot contain any forward references or undefined symbols. This directive is commonly used to reserve a scratchpad or table area for later use.

2.2.12 Memory Allocation

Memory allocation is the decision of where in memory we put the various pieces of our software. The memory on a PC-compatible computer is physically configured as a simple linear array. In other words, if you have 256 Mbytes of RAM, then this memory exists as a continuous linear object with no fundamental difference in the behavior of one memory cell to the next. Although the memory itself forces no structure in the way it is used, the Intel x86 processors have implemented a memory access scheme that requires the programmer to separate in memory segments (e.g., machine codes, global variables, and local variables). The term **x86** refers to any Intel processor from the 8086 through the current Pentiums. There can be more than three segments, but three are enough to illustrate the point. The mechanism to access these segments is called **segmentation**. Figure 2.22 shows a simple view of the memory allocation on the Intel x86 family.

In particular, when the Pentium fetches a machine code, it uses two registers. The code segment selector (CS) points to the beginning of the code segment, and the instruction pointer (IP) contains the offset within this segment of the opcode to fetch. Similarly, when the Pentium accesses a global variable, it uses two different registers. The data segment selector (DS) points to the beginning of the data segment, and a data pointer (e.g., DI) contains the offset within this segment of global variable. Lastly, when the Pentium accesses a local variable, it uses a stack segment selector (SS) and either the stack pointer (SP) or the base pointer (BP). The stack segment selector (SS) points to the beginning of the stack

Figure 2.22
The Intel x86 uses segmented memory allocation.

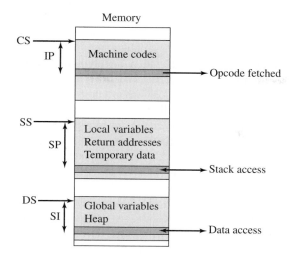

segment, and a stack pointer (SP or BP) contains the offset within this segment of local variable. Segmentation forces the programmer to allocate in memory information that has similar properties. In other words, all the machine codes are placed in one group, the global variables are in another group, and the stack is in a third group. This allocation scheme provides for protection so that the errors or stack overflow, stack underflow, accessing an illegal pointer do not modify machine codes.

We will allocate memory on our embedded system in a fashion similar to segmentation but for a different reason. Because different types of memory on an embedded computer behave in different fashions, it makes sense to group together in memory information that has similar properties or usage. Typical examples of this grouping include global variables, the heap, local variables, fixed constants, and machine instructions. Figure 2.23 shows a typical memory allocation scheme for an embedded system.

Figure 2.23
We place variables in RAM and programs in ROM on an embedded system.

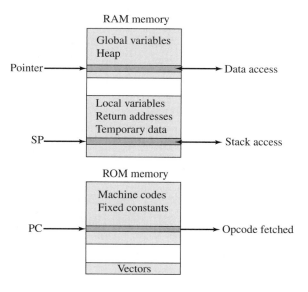

Global variables are permanently allocated and are usually accessible by more than one program. We must use global variables for information that must be permanently available, or for information that is to be shared by more than one module. We will see many applications later in this book of the first-in-first-out (FIFO) queue, a global data structure that is shared by more than one module. Some software systems use a heap to dynamically allocate and release memory (e.g., **malloc**). This information can be shared or not shared depending on which modules have pointers to the data. The heap is efficient in situations where storage is needed for only a limited amount of time. Local variables are usually allocated on the stack at the beginning of the subroutine/function, used within the subroutine/function, and deallocated at the end of the subroutine/function. Local variables are not shared with other modules. Fixed constants do not change and include information such as numbers, strings, sounds, and pictures. Just as with the heap, the fixed constants can be shared, or not shared, depending on which modules have pointers to the data. As we saw in the previous chapter, the assembler or compiler translates our software into machine instruction (opcodes and operands) that when executed perform the intended operations. For single-chip microcomputers, there are three types of memory. The RAM contains temporary information that is lost when the power is shunt off (i.e., volatile.) This means that all variables allocated in RAM must be explicitly initialized at run time by the software. Most C compilers initialize all RAM-based global variables to zero, but others do not. It is good software development practice to set globals to the desired initial value explicitly. As we saw in the previous chapter, many Freescale microcomputers have a little bit of EEPROM. The ROM is a low-cost nonvolatile storage that can be programmed only once.

In an embedded application, we must put global variables, the heap, and local variables in RAM because these types of information can change during execution. When software is to be executed on a regular computer, the machine instructions are usually read from a mass storage device (e.g., a disk) and loaded into memory. Because the embedded system usually has no mass storage device, the machine instructions and fixed constants must be stored in nonvolatile memory. If there is both EEPROM and ROM on our microcomputer, we put some fixed constants in EEPROM and some in ROM. If it is information that we may wish to change in the future, we could put it in EEPROM. Examples include language-specific strings, calibration constants, finite-state machines, and system ID numbers. This allows us to make minor modifications to the system by reprogramming the EEPROM without throwing the chip away. If our project involves producing a small number of devices, then the machine instructions can be placed in EPROM or EEPROM. For a project with a large volume it will be cost effective to place the machine instructions in ROM.

Program 2.7 is a simple illustration of how we allocate various sections of our software using the `org` pseudo-op. The program outputs to Port C (Port T on the 6812) the sequence

Program 2.7
Memory allocation places variables in RAM and programs in ROM.

```
; MC68HC711E9                    ; MC9S12C32
      org   $0000 ;RAM                 org   $3800 ;RAM
cnt   rmb   1     ;global        cnt   rmb   1     ;global

      org   $B600 ;EEPROM              org   $4000 ;EEPROM
const fcb   5     ;amount to add const fcb   5     ;amount to add

      org   $D000 ;ROM           init  movb #$FF,DDRT ;outputs
init  ldaa  #$FF                       clr   cnt
      staa  DDRC  ;outputs             rts
      clr   cnt                  main  lds   #$4000 ;sp=>RAM
      rts                              bsr   init
main  lds   #$00FF ;sp=>RAM      loop  ldaa  cnt
      bsr   init                       staa  PTT   ;output
loop  ldaa  cnt                        adda  const
      staa  PORTC ;output              staa  cnt
      adda  const                      bra   loop
      staa  cnt
      bra   loop                       org   $FFFE ;EEPROM
                                       fdb   main  ;reset vector
      org   $FFFE ;ROM
      fdb   main ;reset vector
```

0,5,10,15, ... The global variable is placed at the start of the RAM, the stack is initialized to the top of RAM (and grows down), and the program is placed in ROM. The constants are placed in the 6811 EEPROM.

2.3 Self-Documenting Code

2.3.1 Comments

The goal of this section is to present ideas concerning software documentation in general and writing comments in particular. Maintaining software is the process of fixing bugs, adding new features, optimizing for speed or memory size, porting to new computer hardware, and configuring the software system for new situations. Maintenance is the *most important* phase of software development. Documentation should therefore assist software maintenance. In many situations the software is not static but is continuously undergoing changes. Because of this liquidity, we believe that flowchart and software manuals are not good mechanisms for documenting programs because it is difficult to keep these types of documentation up to date when modifications are made. Therefore, the term *documentation* in this book refers almost exclusively to comments that are included in the software itself. There are two types of readers of our comments. Our client is someone who will use our software, incorporating it into a larger system. Client comments focus on the *policies* of the software. What are the possible valid inputs? What are the resulting outputs? What does the software do? How does one call the software? What are the error conditions? The other reader of our comments is a colleague, who is someone charged with software maintenance. Colleague comments focus on the *mechanisms* of the software. How does it work? What algorithms are used? How was the software tested? How does one change the software?

> **Observation:** The simulator, TExaS, that accompanies this book is one of the few regular software development applications that allows you to add color drawings into the comment fields.

As software developers, our goal is to produce code that not only solves our current problem but can serve as the basis of our future problems. To reuse software we must leave our code in a condition such that future programmers (including ourselves) can easily understand its purpose, constraints, and implementation. Documentation is not something tacked onto software after it is done, but rather it is a discipline built into it at each stage of the development. Writing comments as we develop the software forces us to think about what the software is doing and more importantly why we are doing it. Therefore, we should carefully develop a programming style that provides appropriate comments. A comment that tells us why we perform certain functions is more informative than comments that tell us what the functions are. We should assume the reader of our comment already knows the syntax of the language. The following are examples of bad comments because they provide no additional information:

```
X=X+4;        /* add 4 to X */
Flag=0;       /* set Flag=0 */
```

> **Common error:** A comment that simply restates the operation does not add to the overall understanding.

> **Common error:** Putting a comment on every line of software often hides the important information.

Good comments assist us now while we are debugging and will assist us later when we are modifying the software, adding new features, or using the code in a different context.

The following are examples of good comments because they explain why the function is being executed:

```
X=X+4;    /* 4 is added to correct for the offset (mV) in the transducer */
Flag=0;   /* means no key has been typed */
```

When a variable is defined, we should add comments to explain how the variable is used. If the variable has units, then it is appropriate to include them in the comments. It may be relevant to specify the minimum and maximum values. A typical value and what it means often will clarify the usage of the variable. For example:

```
short SetPoint;
/* The desired temperature for the temperature control system
   16-bit signed temperature with a resolution of 0.5C, a range of -55C to +125C
   a value of 25 means 12.5C, a value of -25 means -12.5C */
```

When a constant is used, we could add comments to explain what the constant means. If the number has units, then it is appropriate to include them in the comments. For example:

```
V=999;    /* 999mV is the maximum possible voltage */
err=1;    /* error code of 1 means out of range */
```

When a subroutine or function is defined, there will be two types of comments. The first type is directed to the user of the routine (client). These comments explain how the function is to be used, how to pass parameters, what sort of errors might happen, and how the results are returned. If we are writing in C language, then these comments should be included in the *.h file along with the function prototypes. If we are writing in assembly language, then these comments should be included at the beginning of the subroutine. If the parameters have units, then it is appropriate to include them in the comments. Just like a variable, it may be relevant to specify the minimum and maximum values for the I/O parameters. Typical I/O values and what they mean often will clarify the usage of the function. Frequently we give entire software examples showing how the functions could be used. The second type of comments is directed to the programmer responsible for debugging and software maintenance (colleague). These comments explain how the function works. Generally we separate these comments from the ones intended for the user of the function. This separation is the first of many examples in this book of the concept "separation of policy from mechanism." The policy is what the function does, and the mechanism is how it works. Specifically, we place this second type of comments within the body of the function. If we are writing in C, then these comments should be included in the *.c file along with the function implementation.

Self-documenting code is software written in a simple and obvious way such that its purpose and function are self-apparent. Descriptive names for variables, constants, and functions will go a long way to clarify their usage. To write wonderful code like this, we first must formulate the problem, organizing it into clear, well-defined subproblems. How we break a complex problem into small parts goes a long way toward making the software self-documenting. The concepts of abstraction, modularity, and layered software, all presented later in this chapter, address this important issue of software organization.

Observation: The purpose of a comment is to assist in debugging and maintenance.

We should use careful indenting, and descriptive names for variables, functions, labels, and I/O ports. Liberal use of C language #define and assembly language equ provide explanation of software function without cost of execution speed or memory requirements. A disciplined approach to programming is to develop patterns of writing that you consistently follow. Software developers are not like short story writers. When writing software it is OK to use the same *function outline* over and over again. In the programs of this chapter, notice the following assembly language style issues:

1. Begins and ends with a line of *s
2. States the purpose of the function
3. Gives the I/O parameters, what they mean, and how they are passed
4. Different phases (submodules) of the code delineated by a line of -'s

Observation: It is better to write clear and simple software that is easy to understand without comments than to write complex software that requires a lot of extra explanation to understand.

#define statements, if used properly, can clarify our software and make our software easy to change. Notice in Program 2.8, that if one changes the value of size, then a bug would occur in initialize.

Program 2.8
An inappropriate use of #define.

```
#define size 10
short data[size];
void initialize(void){ short j
   for(j=0;j<10;j++)
      data[j]=0;
};
```

It is proper to use size in all places that refer to the size of the data array (Program 2.9).

Program 2.9
An appropriate use of #define.

```
#define size 10
short data[size];
void initialize(void){ short j
   for(j=0;j<size;j++)
      data[j]=0;
};
```

Common error: A programmer may employ a #define or equ for the sole purpose of making the software easier to read, and a software bug may occur if you change the value of the constant.

Software documentation is an important communication tool between software developers. It also provides invaluable information for software that will be modified in the future. The approach to good documentation is to provide information that enhances understanding, use, and modification. It is good practice to tailor documentation specifically to the intended reader. For example, we might give different information to a user of our module (programmer writing code that calls our module) than to a developer (programmer responsible for testing and upgrading). Clearly state in the comments:

Purpose of the module
Input parameters
   How passed (call by value, call by reference)
   Appropriate range (does the module assume the input is within range?)
   Format (8 bit/16 bit, signed/unsigned, etc.)
Output parameters
   how passed (return by value, return by reference)
   format (8 bit/16 bit, signed/unsigned, etc.)
Example inputs and outputs if appropriate
Error conditions
Example calling sequence
Local variables and their significance

**2.3.2
Naming
Convention**

Choosing names for variables and functions involves creative thought, and it is intimately connected to how we feel about ourselves as programmers. Of the policies presented in this section, our naming conventions may be the hardest habit for us to break. The difficulty is that there are many conventions that satisfy the "easy to understand" objective. Good names reduce the need for documentation. Poor names promote confusion, ambiguity, and mistakes. Poor names can occur because code has been copied from a different situation and inserted into our system without proper integration (i.e., changing the names to be consistent with the new situation). They can also occur in the cluttered mind of a second-rate programmer, who hurries to deliver software before it is finished.

Names should have meaning. If we observe a name out of the context of the program in which it exists, the meaning of the object should be obvious. The object `TxFifo` is clearly the transmit first-in-first-out circular queue. The function `LCD_OutString` will output a string to the LCD display.

Avoid ambiguities. Don't use variable names in our system that are vague or have more than one meaning. For example, it is vague to use `temp`, because there are many possibilities for temporary data—in fact it might even mean temperature. Don't use two names that look similar but have different meanings.

Give hints about the type. We can further clarify the meaning of a variable by including phrases in the variable name that specify its type. For example, `dataPt timePt putPt` are pointers. Similarly, `voltageBuf timeBuf pressureBuf` are data buffers. Other good phrases include `Flag Mode U L Index Cnt`, which refer to Boolean flag, system state, unsigned 16-bit, signed 32-bit, index into an array, and a counter, respectively.

Use the same name to refer to the same type of object. For example, everywhere we need a local variable to store an ASCII character we could use the name `letter`. Another common example is to use the names `i j k` for indices into arrays. The names `V1 R1` might refer to a voltage and a resistance. The exact correspondence is not part of the policies presented in this section, just that a correspondence should exist. Once another programmer learns which names we use for which types of object, understanding our code becomes easier.

Use a prefix to identify public objects. An underline character will separate the module name from the function name. As an exception to this rule, we can use the underline to delimit words in all upper-case name (e.g., `#define MIN_PRESSURE 10`). Functions that can be accessed outside the scope of a module will begin with a prefix specifying the module to which it belongs. It is poor style to create public variables, but if they need to exist, they too would begin with the module prefix. The prefix matches the file name containing the object. For example, if we see a function call, `LCD_OutString("Hello world");` we know the public function belongs to the LCD module, where the policies are defined in `LCD.h` and the implementation in `LCD.c`. Notice the similarity between this syntax (e.g., `LCD_init()`) and the corresponding syntax we would use if programming the module as a class in C++ (e.g., `LCD.init()`). Using this convention, we can easily distinguish public and private objects.

Use upper and lower case to specify the scope of an object. We will define I/O ports and constants using no lower-case letters, as if typing with caps-lock on. In other words, names without lower-case letters refer to objects with fixed values. `TRUE FALSE` and `NULL` are good examples of fixed-valued objects. As mentioned earlier, constant names formed from multiple words will use an underline character to delimit the individual words—for example `MAX_VOLTAGE UPPER_BOUND FIFO_SIZE`. Global objects will begin with a capital letter, but may also include some lower-case letters. Local variables will begin with a lower-case letter, and may or may not include upper-case letters. Since all functions are global, we can start function names with either an upper-case or lower-case letter. Using this convention, we can distinguish constants, globals, and locals.

Observation: An object's properties (public/private, local/global, constant/variable) are always perfectly clear at the place where the object is defined. The importance of the naming policy is to extend that clarity also to the places where the object is used.

Use capitalization to delimit words. Names that contain multiple words should be defined using a capital letter to signify the first letter of the word. Recall that the case of the first letter specifies whether it is local or global. Some programmers use the underline as a word-delimiter, but except for constants, we will reserve underline to separate the module name from the variable name. Table 2.7 presents examples of the naming convention used in this book.

Table 2.7
Examples of names.

Type	Examples
constants	`CR SAFE_TO_RUN PORTA STACK_SIZE START_OF_RAM`
local variables	`maxTemperature lastCharTyped errorCnt`
private global variable	`MaxTemperature LastCharTyped ErrorCnt`
public global variable	`DAC_MaxVoltage Key_LastCharTyped Network_ErrorCnt`
private function	`ClearTime wrapPointer InChar`
public function	`Timer_ClearTime RxFifo_Put Key_InChar`

Checkpoint 2.14: How can you tell whether a function is private or pubic?

Checkpoint 2.15: How can you tell whether a variable is local or global?

2.4   Abstraction

2.4.1 Definitions

Software abstraction is when we can define a complex problem with a set of basic abstract principles. If we can construct our software system using these building blocks, then we have a better understanding of the problem, because we can separate what we are doing from the details of how we are getting it done. This separation also makes it easier to optimize. It provides for a proof of correct function and simplifies both extensions and customization. A good example of abstraction is the *finite-state machine* (FSM) implementation. The abstract principles of FSM development are the inputs, outputs, states, and state transitions. If we can take a complex problem and map it into a FSM model, then we can solve it with simple FSM software tools. Our FSM software implementation will be easy to understand, debug, and modify. Other examples of software abstraction include *proportional integral derivative* (PID) digital controllers, fuzzy logic digital controllers, neural networks, and linear systems of differential equations. In each case, the problem is mapped into a well-defined model with a set of abstract yet powerful rules. Then, the software solution is a matter of implementing the rules of the model.

Linked lists are lists or nodes where one or more of the entries is a pointer (link) to other nodes of similar structure. We can have statically allocated fixed-size linked lists that are defined at assemble or compile time and exist throughout the life of the software. On the other hand, we implement dynamically allocated variable-size linked lists that are constructed at run time and can grow and shrink in size. We will use a data structure similar to a linked list, called a *linked structure,* to build a FSM controller. Linked structures are very flexible and provide a mechanism to implement abstraction.

The FSM controller is a good example of the concept of program abstraction. A well-defined model or framework is used to solve our problem (implemented with a linked structure). The three advantages of abstraction are (1) it can be faster to develop because a lot of the building blocks preexist, (2) it is easier to debug (prove correct) because it separates conceptual issues from implementation, and (3) it is easier to change. An important factor when implementing FSMs using linked structures is that there should be a clear and one-to-one mapping between the FSM and the linked structure; that is, there should be one structure for each state.

We will present two implementations of finite state machines. The **Moore FSM** has an output that depends on state, and the next state depends on input and current state. On the other hand, the **Mealy FSM** has an output that depends on both the input and the state, and the next state depends on input and current state. We will use a Moore implementation if there is an association between a state and an output. There can be multiple states with the same output, but the output defines in part what it means to be in that state. For example, in a traffic light controller, the state of green light on the North road (red light on the East road) is caused by outputting a specific pattern to the traffic light. Conversely, we will use a Mealy implementation if the output causes the state to change. In this situation, we do not need a specific output to be in that state; rather, the outputs are required to cause the state transition. For example, to make a robot stand up, we perform a series of outputs causing the state to change from sitting to standing. Although we can rewrite any Mealy machine as a Moore machine and vice versa, it is better to implement the format that is more natural for the particular problem. In this way the state graph will be easier to understand.

Checkpoint 2.16: What are the differences between a Mealy and Moore finite-state machine?

One of the common features in many finite-state machines is a time delay. We will learn very elaborate mechanisms to handle time in Chapters 4–6, but in this section we will implement a simple time delay. Both the 6811 and 6812 have a 16-bit timer register, called TCNT, which increments at a regular rate. This counter is incremented at a fixed rate, and the software can read its value to know the current time.

2.4.2
6811 Timer Details

On the 6811, TCNT is located at address \$100E (see Table 2.8). The rate at which TCNT is incremented is determined by two prescale bits (PR1 and PR0) in the TMSK2 register (\$1024) as shown in Table 2.9. Every time the TCNT register overflows from \$FFFF to 0, the TOF flag in the TFLG2 register is set. The TOF condition will cause an interrupt if the arm bit TOI equals 1.

Address	msb															lsb	Name
\$100E	15	14	13	12	11	10	9	8	7	6	5	4	3	2	1	0	TCNT

Address	Bit 7	6	5	4	3	2	1	Bit 0	Name
\$1024	TOI	RTII	PAOVI	PAII	0	0	PR1	PR0	TMSK2
\$1025	TOF	RTIF	PAOVF	PAIF	0	0	0	0	TFLG2

Table 2.8
6811 timer ports.

Table 2.9
Assuming a 2 MHz
E clock on the 6811,
PR1 and PR0 define the
TCNT rate.

PR1	PR0	Divide by	TCNT Period	TCNT Frequency
0	0	1	500 ns	2 MHz
0	1	4	2 μs	500 kHz
1	0	8	4 μs	250 kHz
1	1	16	8 μs	125 kHz

The flags in the TFLG2 register are cleared by writing a 1 into the specific flag bit we wish to clear. For example, writing an \$80 into TFLG2 will clear the TOF flag.

2.4.3
6812 Timer Details

Table 2.10 shows some of the MC9S12C32 timer registers. On the 6812, TCNT is a 16-bit unsigned counter that is incremented at a rate determined by three bits (PR2, PR1, and PR0) in the TSCR2 register as shown in Table 2.11. Every time the TCNT register overflows from

Address	msb														lsb	Name
$0044	15	14	13	12	11	10	9	8	7	6	5	4	3	2	1	0	TCNT

Address	Bit 7	6	5	4	3	2	1	Bit 0	Name
$0046	**TEN**	TSWAI	TSFRZ	TFFCA	0	0	0	0	TSCR1
$004D	TOI	0	0	0	TCRE	**PR2**	**PR1**	**PR0**	TSCR2
$004F	TOF	0	0	0	0	0	0	0	TFLG2

Table 2.10
MC9S12C32 timer ports.

Table 2.11
Assuming a 4 MHz
E clock on the
MC9S12C32, PR2, PR1,
and PR0 define the
TCNT rate.

PR2	PR1	PR0	Divide by	TCNT Period	TCNT Frequency
0	0	0	1	250 ns	4 MHz
0	0	1	2	500 ns	2 MHz
0	1	0	4	1 µs	1 MHz
0	1	1	8	2 µs	500 kHz
1	0	0	16	4 µs	250 kHz
1	0	1	32	8 µs	125 kHz
1	1	0	64	16 µs	62.5 kHz
1	1	1	128	32 µs	31.25 kHz

$FFFF to 0, the TOF flag in the TFLG2 register is set. The TOF condition will cause an interrupt if the arm bit TOI equals 1. Chapter 4 discusses **TOF interrupts**.

2.4.4
Time Delay
Software Using
the Built-in Timer

In order to use TCNT on the 6812, you must first set bit 7 of the TSCR1 register. Program 5.16 shows how to use TCNT to create a time delay on the MC9S12C32. The Timer_Init function is not necessary for the 6811, but the 10000 constant should be increased to 20000, because TCNT counts every 500 ns. The delay parameter to the assembly Timer_Wait subroutine can be any number from 1 to 32767.

```
Timer_Init  ; Enable TCNT at 1us
    movb #$80,TSCR1
    movb #$04,TSCR2    ;prescale
    rts
; Reg D is the time to wait in cycles
Timer_Wait
    addd TCNT   ;end of wait time
wloop cpd  TCNT   ;stop when RegD<TCNT
    bpl   wloop
    rts
; RegY is the time to wait in 10ms
Timer_Wait10ms
    ldd   #10000      ;10000us=10ms
    bsr   Timer_Wait  ;wait 10ms
    dey
    bne   Timer_Wait10ms
    rts
```

```
void Timer_Init(void){
    TSCR1 = 0x80; // enable TCNT
    TSCR2 = 0x04; // 1us TCNT
}
void Timer_Wait(unsigned short cycles){
unsigned short startTime = TCNT;
    while((TCNT-startTime) <= cycles){}
}
// 10000us equals 10ms
void Timer_Wait10ms(unsigned short delay){
unsigned short i;
    for(i=0; i<delay; i++){
        Timer_Wait(10000);  // wait 10ms
    }
}
```

Program 2.10
Timer functions that implement a time delay.

Checkpoint 2.17: Explain how the timer prescale affects the range and resolution of the `Timer_Wait` function.

2.4.5
Moore Finite-State Machine Traffic Light Controller

In this section we will design a traffic light controller using a Moore FSM. The goal is to maximize traffic flow, minimize waiting time at a red light, and avoid accidents. The outputs of a Moore FSM are only a function of the current state. The intersection has two one-way roads: North and East, as shown in Figure 2.24. It will have two inputs (car sensors on North and East roads) and six outputs (one for each light in the traffic signal). The six traffic lights are interfaced to Port B. The two sensors are connected to Port A, such that

00 means no cars exist on either road
01 means there are cars on the East road
10 means there are cars on the North road
11 means there are cars on both roads

Figure 2.24
Traffic light interface.

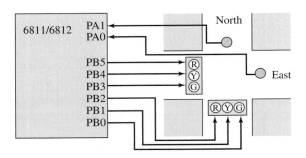

The first step in designing the FSM is to create some states. A Moore implementation was chosen because the output pattern (which lights are on) defines which state we are in. Each state is given a symbolic name:

goN,   100001 makes it green on North and red on East
waitN,   100010 makes it yellow on North and red on East
goE,   001100 makes it red on North and green on East
waitE,   010100 makes it red on North and yellow on East

The output pattern for each state is drawn inside the state circle. The time to wait for each state is also included. How the machine operates will be dictated by the input-dependent state transitions. We create decision rules defining what to do for each possible input and for each state. For this design we can list heuristics describing how the traffic light is to operate:

If no cars are coming, we will stay in a green state.
To change from green to red, we will implement a yellow light of exactly 5 seconds.
Green lights will last at least 30 seconds.
If cars are only coming in one direction, we will move to and stay green in that direction.
If cars are coming in both directions, we will cycle through all four states.

Finally, we implement the heuristics by defining the state transitions, as illustrated in Figure 2.25. Instead of using a graph to define the finite state machine, we could have used a table, as shown in Table 2.12.

Figure 2.25
Graphical form of a
Moore FSM that
implements a traffic
light.

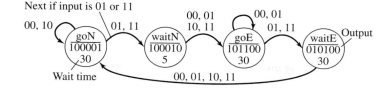

Table 2.12
Tabular form of a Moore
FSM that implements a
traffic light.

State \ Input	00	01	10	11
goN, 100001,30	goN	waitN	goN	waitN
waitN, 100010,5	goE	goE	goE	goE
goE, 001100,30	goE	goE	waitE	waitE
waitE, 010100,5	goN	goN	goN	goN

The first step in designing the software is to decide on the sequence of operations.

1. Initialize timer and directions registers
2. Specify initial state
3. Perform FSM controller
    a) Output to traffic lights, which depends on the state
    b) Delay, which depends on the state
    c) Input from sensors
    d) Change states, which depends on the state and the input

The second step is to define the FSM graph using a data structure. Program 2.11 shows two possible implementations of the Moore FSM. The implementation on the left uses a table

Program 2.11
Two 6812
C implementations of a
Moore FSM.

```
// Table implementation
const struct State {
  unsigned char Out;
  unsigned short Time;
  unsigned char Next[4];};
typedef const struct State STyp;
#define goN   0
#define waitN 1
#define goE   2
#define waitE 3
STyp FSM[4]={
  {0x21,3000,{goN,waitN,goN,waitN}},
  {0x22, 500,{goE,goE,goE,goE}},
  {0x0C,3000,{goE,goE,waitE,waitE}},
  {0x14, 500,{goN,goN,goN,goN}}};
unsigned char Input;
void main(void){
unsigned char n; // state number
  Timer_Init();
  DDRB = 0xFF;
  DDRA &= ~0x03;
  n = goN;
  while(1){
    PORTB = FSM[n].Out;
    Timer_Wait10ms(FSM[n].Time);
    Input = PORTA&0x03;
    n = FSM[n].Next[Input];
  }
}
```

```
// Pointer implementation
const struct State {
  unsigned char Out;
  unsigned short Time;
  const struct State *Next[4];};
typedef const struct State STyp;
#define goN   &FSM[0]
#define waitN &FSM[1]
#define goE   &FSM[2]
#define waitE &FSM[3]
STyp FSM[4]={
  {0x21,3000,{goN,waitN,goN,waitN}},
  {0x22, 500,{goE,goE,goE,goE}},
  {0x0C,3000,{goE,goE,waitE,waitE}},
  {0x14, 500,{goN,goN,goN,goN}}};
STyp *Pt;  // state pointer
unsigned char Input;
void main(void){
  Timer_Init();
  DDRB = 0xFF;
  DDRA &= ~0x03;
  Pt = goN;
  while(1){
    PORTB = Pt->Out;
    Timer_Wait10ms(Pt->Time);
    Input = PORTA&0x03;
    Pt = Pt->Next[Input];
  }
}
```

data structure, where each state is an entry in the table and state transitions are defined as indices into this table. The one on the right uses a linked structure, where each state is a node and state transitions are defined as pointers to other nodes. The four `Next` parameters define the input-dependent state transitions. The wait times are defined in the software as fixed-point decimal numbers with units of 0.01 s, giving a range of 10 ms to 655.35 s. Using good labels makes the program easier to understand; in other words `goN` is more descriptive than `&fsm[0]`.

A 6811 implementation results simply by removing `Timer_Init();`   `DDRB=0xFF;` `DDRA&=~0x03;` The MC68HC711E9 runs at 2 MHz, so the 10000 constant in `Wait10ms` will have to be changed to 20000.

Observation: The table implementation requires less memory space for the FSM data structure, but the pointer implementation will run faster.

Program 2.12 shows 6811 and 6812 assembly language implementations of the Moore FSM. On microcontrollers that have both ROM and EEPROM we can place the FSM data structure in EEPROM and the assembly language program in ROM, as illustrated in the MC68HC11E9 version in Program 2.12. This allows us to make minor modifications to the

```
;MC68HC11E9 (ROM version)
      org $B600 ; Put FSB in EEPROM
OUT   equ 0    ;offset for output
WAIT  equ 1    ;offset for time
NEXT  equ 3    ;offset for next
goN   fcb $21  ;North green, East red
      fdb 3000 ;30sec
      fdb goN,waitN,goN,waitN
waitN fcb $22  ;North yellow, East red
      fdb 500  ;5sec
      fdb goE,goE,goE,goE
goE   fcb $0C  ;North red, East green
      fdb 3000 ;30 sec
      fdb goE,goE,waitE,waitE
waitE fcb $14  ;North red, East yellow
      fdb 500  ;5sec
      fdb goN,goN,goN,goN


      org  $D000    ;code in ROM
Main  lds  #$01FF   ;stack init
      ldx  #goN     ;State pointer
FSM   ldab OUT,x
      stab PORTB     ;Output
      ldy  WAIT,x   ;Time delay
      bsr  Timer_Wait10ms
      ldab PORTA     ;Read input
      andb #$03      ;just bits 1,0
      lslb           ;2 bytes/address
      abx            ;add 0,2,4,6
      ldx  NEXT,x   ;Next state
      bra  FSM

      org  $FFFE
      fdb  Main      ;reset vector
```

```
;MC9S12C32
      org $4000 ; Put in ROM
OUT   equ 0    ;offset for output
WAIT  equ 1    ;offset for time
NEXT  equ 3    ;offset for next state
goN   fcb $21  ;North green, East red
      fdb 3000 ;30sec
      fdb goN,waitN,goN,waitN
waitN fcb $22  ;North yellow, East red
      fdb 500  ;5sec
      fdb goE,goE,goE,goE
goE   fcb $0C  ;North red, East green
      fdb 3000 ;30 sec
      fdb goE,goE,waitE,waitE
waitE fcb $14  ;North red, East yellow
      fdb 500  ;5sec
      fdb goN,goN,goN,goN
Main  lds  #$4000    ;stack init
      bsr  Timer_Init ;enable TCNT
      movb #$FF,DDRB ;PB5-0 are lights
      movb #$00,DDRA ;PA1-0 are sensors
      ldx  #goN     ;State pointer
FSM   ldab OUT,x
      stab PORTB     ;Output
      ldy  WAIT,x   ;Time delay
      bsr  Timer_Wait10ms
      ldab PORTA     ;Read input
      andb #$03      ;just bits 1,0
      lslb           ;2 bytes/address
      abx            ;add 0,2,4,6
      ldx  NEXT,x   ;Next state
      bra  FSM
      org  $FFFE
      fdb  Main      ;reset vector
```

Program 2.12
Assembly programs for a Moore FSM traffic light controller.

finite-state machine (add/delete states, change input/output values) by changing the linked structure without modifying the assembly language controller. In this way small modifications/upgrades/options to the finite-state machine can be made by reprogramming the EEPROM without throwing the chip away.

The FSM approach makes it easy to change. To change the wait time for a state, we simply change the value in the data structure. To add more states (e.g., put a red/red state after each yellow state, which will reduce accidents caused by bad drivers running the yellow light), we simply increase the size of the fsm[] structure and define the Out, Time, and Next fields for these new states.

To add more output signals (e.g., walk and left-turn lights), we simply increase the precision of the Out field. To add two more input lines (e.g., wait button, left-turn car sensor), we increase the size of the next field to Next[16]. Because now there are four input lines, there are 16 possible combinations, where each input possibility requires a Next value specifying where to go if this combination occurs. In this simple scheme, the size of the Next[] field will be 2 raised to the power of the number of input signals.

2.4.6 Mealy Finite-State Machine Robot Controller

The goal of this section is to design a finite-state machine robot controller, as illustrated in Figure 2.26. Because the outputs cause the robot to change states, we will use a Mealy implementation. The outputs of a Mealy FSM depend on both the input and the current state. This robot has mood sensors, that are interfaced to Port A. The robot has four possible mutually exclusive conditions

00 OK, the robot is feeling fine
01 Tired, the robot energy levels are low
10 Curious, the robot senses activity around it
11 Anxious, the robot senses danger

Figure 2.26
Robot interface.

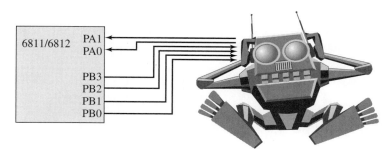

There are four actions this robot can perform, which are triggered by pulsing (make high, then make low) one of the four signals interfaced to Port B.

PB3 SitDown, assuming the robot is standing, it will perform a sequence of moves to sit down

PB2 StandUp, assuming the robot is sitting, it will perform a sequence of moves to stand up

PB1 LieDown, assuming the robot is sitting, it will perform a sequence of moves to lie down

PB0 SitUp, assuming the robot is sleeping, it will perform a sequence of moves to sit up

For this design we can list heuristics describing how the robot is to operate:

If the robot is OK, it will stay in the state it is currently in.
If the robot's energy levels are low, it will go to sleep.
If the robot senses activity around it, it will awaken from sleep.
If the robot senses danger, it will stand up.

These rules are converted into a finite-state machine graph, as shown in Figure 2.27. Each arrow specifies both an input and an output. For example, the "Tired/SitDown" arrow from Standing to Sitting states means if we are in the Standing state and the input is Tired, then we will output the SitDown command and go to the Sitting state. Mealy machines can have time delays, but this example just didn't have time delays.

Figure 2.27
Mealy FSM for a robot controller.

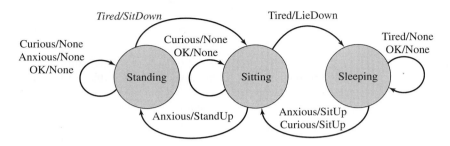

The first step in designing the software is to decide on the sequence of operations.

1. Initialize directions registers
2. Specify initial state
3. Perform FSM controller
   a) Input from sensors
   b) Output to the robot, which depends on the state and the input
   c) Change states, which depends on the state and the input

The second step is to define the FSM graph using a linked data structure. Program 2.13 shows two possible implementations of the Mealy FSM. The implementation on the left defines the outputs as simple numbers, where each pulse is defined as the bit mask required to cause that action. The one on the right uses functions to affect the output. The four Next parameters define the input-dependent state transitions. A 6811 implementation results simply by removing DDRB=0xFF; DDRA&=~0x03;

```
// outputs defined as numbers
const struct State{
  unsigned char Out[4];     // outputs
  const struct State *Next[4]; // Next
};
typedef const struct State StateType;
#define Standing &fsm[0]
#define Sitting  &fsm[1]
#define Sleeping &fsm[2]

#define None 0x00
#define SitDown 0x08  // pulse on PB3
#define StandUp 0x04  // pulse on PB2
#define LieDown 0x02  // pulse on PB1
#define SitUp   0x01  // pulse on PB0

StateType FSM[3]={
{{None,SitDown,None,None},     //Standing
  {Standing,Sitting,Standing,Standing}},
{{None,LieDown,None,StandUp}, //Sitting
  {Sitting,Sleeping,Sitting,Standing }},
{{None,None,SitUp,SitUp},     //Sleeping
  {Sleeping,Sleeping,Sitting,Sitting}}
};
```

```
// outputs defined as functions
const struct State{
  void (*CmdPt)[4](void);     // outputs
  const struct State *Next[4]; // Next
};
typedef const struct State StateType;
#define Standing &fsm[0]
#define Sitting  &fsm[1]
#define Sleeping &fsm[2]
void None(void){};
void SitDown(void){
  PORTB=0x08; PORTB=0;} // pulse on PB3
void StandUp(void){
  PORTB=0x04; PORTB=0;} // pulse on PB2
void LieDown(void){
  PORTB=0x02; PORTB=0;} // pulse on PB1
void SitUp(void) {
  PORTB=0x01; PORTB=0;} // pulse on PB0
StateType FSM[3]={
{{&None,&SitDown,&None,&None}, //Standing
  {Standing,Sitting,Standing,Standing}},
{{&None,&LieDown,&None,&StandUp},//Sitting
  {Sitting,Sleeping,Sitting,Standing }},
{{&None,&None,&SitUp,&SitUp},   //Sleeping
```

```
StatePtr *Pt;  // Current State                {Sleeping,Sleeping,Sitting,Sitting}}
unsigned char Input;                         };
void main(void){                             StatePtr *Pt;  // Current State
  DDRB = 0xFF;      // Output to robot        unsigned char Input;
  DDRA &= ~0x03;    // Input from sensor      void main(void){
  Pt = Standing;    // Initial State           DDRB = 0xFF;      // Output to robot
  while(1){                                     DDRA &= ~0x03;    // Input from sensor
    Input = PORTA&0x03;    // Input=0-3         Pt = Standing;    // Initial State
    PORTB =Pt->Out[Input]; // Pulse             while(1){
    PORTB = 0;                                    Input = PORTA&0x03;    // Input=0-3
    Pt = Pt->Next[Input];  // next state          (*Pt->CmdPt[Input])(); // function
  }                                               Pt = Pt->Next[Input];  // next state
}                                               }
                                             }
```

Program 2.13
Two C implementations of a Mealy Finite-State Machine.

Again proper memory allocation is required if we wish to implement a standalone or embedded system. The org pseudoops are used to place the FSM data structure in EEPROM and the assembly language program in ROM of a single chip 6811, as shown in Program 2.14.

```
;MC68HC11E9                                   ;MC9S12C32
         org  $B600 ;EEPROM                            org  $4000 ;EEPROM
Out      equ  0        ;Index for output     Out      equ  0        ;Index for output
Next     equ  4        ;Index for next state Next     equ  4        ;Index for next state
None     equ  0        ;no pulse             None     equ  0        ;no pulse
SitDown  equ  $08   ;pulse on PB3            SitDown  equ  $08   ;pulse on PB3
StandUp  equ  $04   ;pulse on PB2            StandUp  equ  $04   ;pulse on PB2
LieDown  equ  $02   ;pulse on PB1            LieDown  equ  $02   ;pulse on PB1
SitUp    equ  $01   ;pulse on PB0            SitUp    equ  $01   ;pulse on PB0
Standing fcb None,SitDown,None,None          Standing fcb None,SitDown,None,None
     fdb Standing,Sitting,Standing,Standing      fdb Standing,Sitting,Standing,Standing
Sitting  fcb None,LieDown,None,StandUp       Sitting  fcb None,LieDown,None,StandUp
     fdb Sitting,Sleeping,Sitting,Standing       fdb Sitting,Sleeping,Sitting,Standing
Sleeping fcb None,None,SitUp,SitUp           Sleeping fcb None,None,SitUp,SitUp
     fdb Sleeping,Sleeping,Sitting,Sitting       fdb Sleeping,Sleeping,Sitting,Sitting
                                             Main  movb #$FF,DDRB ;PB3-0 output
         org  $D000      ;ROM                       movb #$00,DDRA ;PA1-0 inputs
Main   ldx  #Standing ;current state                ldx  #Standing ;current state
LL     ldab PORTA      ;Read Sensors         LL     ldab PORTA      ;Read Sensors
       andb #$03        ;Just bits1-0               andb #$03        ;Just bits1-0
       abx              ;Base+input                 abx              ;Base+input
       ldaa Out,x       ;Fetch output               ldaa Out,x       ;Fetch output
       staa PORTB       ;Pulse function             staa PORTB       ;Pulse function
       clr  PORTB                                   clr  PORTB
       abx              ;Base+2*input               abx              ;Base+2*input
       ldx  Next,x                                  ldx  Next,x
       bra  LL          ;Infinite loop             bra  LL          ;Infinite loop
       org  $FFFE                                   org  $FFFE
       fdb  Main        ;reset vector               fdb  Main        ;reset vector
```

Program 2.14
Two assembly implementations of a Mealy Finite-State Machine.

Observation: In order to make the FSM respond quicker, we could implement a time delay function that returns immediately if an alarm condition occurs. If no alarm exists, it waits the specified delay.

Checkpoint 2.18: What happens if the robot is sleeping and then becomes anxious?

2.5 Modular Software Development

In this section we introduce the concept of modular programming and demonstrate that it is an effective way to organize our software projects. There are three reasons for forming modules. Functional abstraction allows us to reuse a software module from multiple locations. Complexity abstraction allows us to divide a highly complex system into smaller, less complicated components. The third reason is portability. If we create modules for the I/O devices, then we can isolate the rest of the system from the hardware details. This approach will be presented later in Section 2.6 on layered software.

**2.5.1
Local Variables
in Assembly
Language**

Because their contents are allowed to change, all variables must be allocated in RAM and not ROM. A **local variable** is temporary information used only by one software module. Local variables are typically allocated, used, then deallocated. The information stored in a local variable is not permanent. This means if we store a value into a local variable during one execution of the module, the next time that module is executed the previous value is not available. Examples include loop counters, temporary sums. We use a local variable to store data that is temporary in nature. We can implement a local variable using the stack or registers. Reasons why we place local variables on the stack include

- dynamic allocation/release allows for reuse of memory
- limited scope of access provides for data protection
- only the program that created the local variable can access it
- since an interrupt will save registers and create its own stack frame, the code is reentrant.
- since absolute addressing is not used, the code is relocatable
- the number of variables is only limited by the size of the stack allocation, more than registers

Checkpoint 2.19: How do you create a local variable in C?

A **global variable** is information shared by more than one program module. For example, we use globals to pass data between the main (or foreground) process and an interrupt (or background) process. Global variables are not deallocated. The information they store is permanent. Examples include time of day, date, user name, temperature, pointers to shared data. On the 6811 and 6812, we use absolute addressing (direct or extended) to access their information.

Observation: Sometimes we store temporary information in global variables out of laziness. This practice is to be discouraged because it wastes memory and may cause the module not to be reentrant.

Checkpoint 2.20: How do you create a global variable in C?

Checkpoint 2.21: How does the `static` modifier affect locals, globals, and functions in C?

Checkpoint 2.22: How does the `const` modifier affect a global variable in C?

A LIFO stack is implemented in hardware by most computers. It can be used for local variables (temporary storage), saving return addresses during subroutine calls, passing parameters to subroutines, and to save registers during the processing of an interrupt. The first advantage of placing local variables on the stack is that the storage can be dynamically allocated before usage and deallocated after usage. The second advantage is the facilitation of reentrant software. The 6811 stack operates differently from most other computers. However, the 6812 stack operates similar to most computers.

On the 6811, the stack pointer (SP) points to the free space that the next PUSH will store into, as shown in Figure 2.28. To PUSH a byte on the stack, first the stack pointer (SP) is decremented, then the byte is stored at the location pointed to by SP+1. To PULL a byte from the stack, first the byte is read from memory pointed to by SP+1, then SP is incremented.

Figure 2.28
6811 and 6812 stack.

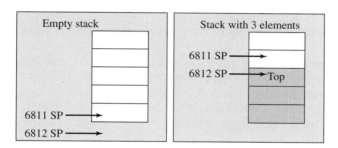

On the 6812, the stack pointer (SP) points to the top entry of the stack, also shown in Figure 2.28. To PUSH a byte on the stack, first the stack pointer (SP) is decremented, then the byte is stored at the location pointed to by SP. To PULL a byte from the stack, first the byte is read from memory pointed to by SP, then SP is incremented.

To access local variables on the 6811 stack, the stack pointer must first be transferred into RegX (or RegY). The instruction tsx will move a copy of the stack pointer into Register X. Although the tsx and tsy instructions work a little differently on the 6811 versus the 6812, in both cases the instruction causes the index register to point to the top element of the stack, as shown in Figure 2.29. The tsx and tsy instructions do not modify the stack pointer.

Figure 2.29
The tsx instruction creates a stack frame pointer.

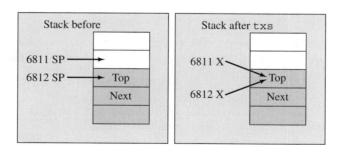

Then, index mode reads and writes are possible. For example to read the next to the top byte on both the 6811 and 6812,

```
tsx         ;RegX points to the top byte of the stack
ldaa 1,x    ;RegA = the next to the top byte
```

The LIFO stack has a few rules:

1. Program segments should have an equal number of pushes and pulls;
2. Stack accesses (PUSH or PULL) should not be performed outside the allocated area;

3. Stack reads and writes should not be performed within the *free area*,
4. Stack PUSH should first decrement SP, then store the data,
5. Stack PULL should first read the data, then increment SP.

Programs that violate rule number 1 will probably crash when an `rts` instruction pulls an illegal address of the stack at the end of a subroutine. The **TExaS** simulator will usually recognize this error as an illegal memory access when the processor tries to fetch an opcode at this incorrect address. The backdump command will be useful to retrace the steps leading up to the crash.

Violations of rule number 2 can be caused by a stack underflow or overflow. Stack underflow is caused when there are more pulls than pushes, and is always the result of a software bug. The **TExaS** simulator will recognize this error as an illegal memory access when the processor tries to pull data from an address that doesn't exist. A stack overflow can be caused by two reasons. If the software mistakenly pushes more than it pulls, then the stack pointer will eventually overflow its bounds. Even when there is exactly one pull for each push, a stack overflow can occur if the stack is not allocated large enough. Stack overflow is a very difficult bug to recognize, because the first consequence occurs when the computer pushes data onto the stack and overwrites data stored in a global variable. At this point the local variables and global variables exist at overlapping addresses. Setting a breakpoint at the first address of the allocated stack area allows you to detect a stack overflow situation.

Checkpoint 2.23: How do you specify the size of the stack?

The following 6812 assembly code violates rule 3, and will not work if interrupts are active. The objective is to save register A onto the stack. When an interrupt occurs, registers will automatically be pushed on the stack, destroying the data.

```
staa -1,sp   ;Store zero onto the stack (***illegal***)
```

To use the stack, one first allocates, then saves. The following assembly code also violates rule 3, because it first stores it on the stack, then allocates space. The objective is to push a zero onto the stack.

```
tsx        ;RegX points to the top of the stack
dex
clr 0,x    ;Store zero onto the stack (***illegal***)
des        ;Make space for the zero
```

If an interrupt were to occur between the `clr` and `des` instructions, the zero will be destroyed when all the registers are pushed on the stack by the interrupt process. The proper technique is to allocate first,

```
des      ;Allocate stack space first
tsx      ;RegX points to the top of the stack
clr ,x   ;Store zero onto the stack
```

The 6811 push instructions (e.g., `psha pshx`) do store data on the stack first, then decrement the stack pointer. These 6811 operations do not violate rule 4, because the store and decrement operations are atomic. An atomic operation is a sequence that once started will always finish and cannot be interrupted. Most instructions on the 6811 and 6812 are atomic. The exceptions are `wai rev` and `revw`, which can be suspended to process an interrupt.

Stack implementation of local variables has four stages: binding, allocation, access, and deallocation.

1. Binding is the assignment of the address (not value) to a symbolic name. This address will be the actual memory location to store the local variable. The assembler binds the symbolic name to a stack index. The computer calculates the actual location during execution. For example:

```
sum  set  0   ;16-bit local variable, stored on the stack
```

Checkpoint 2.24: Why is set better than equ for binding?

2. Allocation is the generation of memory storage for the local variable. The computer allocates space during execution by decrementing the SP. In this first example, the software allocates the local variable by pushing a register on the stack. An 8-bit push (e.g., psha) creates an unitialized 8-byte local variable, and a 16-bit push (e.g., pshx) creates an unitialized 16-byte local variable The value in the register is irrelevant, the instruction is used because it decrements the SP.

```
    pshx    ;allocate sum
```

In this next example, the software allocates the local variable by decrementing the stack pointer. This local variable is also uninitialized.

```
    des     ;allocate sum
    des
```

If you wished to allocate the 16-bit local and initialize it to zero, you could execute.

; 6811 or 6812	; 6812 only
`    ldx  #0`	`        movw #0,2,-sp ;allocate sum=0`
`    pshx     ;allocate sum=0`	

In this last example, the technique provides a mechanism for allocating large amounts of uninitialized stack space. This example allocates 20 bytes for the structure big[20]. Local variables are so important that the 6812 has special instructions to simplify the implementation of local variables.

; 6811 or 6812	; 6812 only
`    tsx       ;allocate big[20]`	`    leas  -20,sp ;allocate big[20]`
`    xgdx`	
`    subd  #20`	
`    xgdx`	
`    txs`	

3. The **access** to a local variable is a read or write operation that occurs during execution. In the next code fragment, the local variable sum is set to 0.

; 6811 or 6812	; 6812 only
`    tsx    ;X points to locals`	`    movw #0,sum,sp    ;sum=0`
`    ldd  #0`	
`    std  sum,x  ;sum=0`	

In the next code fragment, the local variable sum is incremented.

; 6811 or 6812	; 6812 only
`    tsx`	`    ldd  sum,sp`
`    ldd  sum,x`	`    addd #1`
`    addd #1`	`    std  sum,sp   ;sum=sum+1`
`    std  sum,x  ;sum=sum+1`	

4. Deallocation is the release of memory storage for the location variable. The computer deallocates space during execution by incrementing SP. In this first example, the software deallocates the local variable by pulling a register from the stack.

```
pulx    deallocate sum
```

> *Observation:* When the software uses the "push-register" technique to allocate and the "pull-register" technique to deallocate, it looks as though it were saving and restoring the register. Because most applications of local variables involve storing into the local, the value pulled will NOT match the value pushed.

In this next example, the software deallocates the local variable by incrementing the stack pointer.

```
ins
ins     deallocate sum
```

In this last example, the technique provides a mechanism for allocating large amounts of stack space.

```
; 6811 or 6812                        ; 6812 only
   tsx        deallocate big[20]         leas 20,sp deallocate big[20]
   ldab   #20
   abx
   txs
```

> *Checkpoint 2.25:* Write a 6811/6812 subroutine that allocates then deallocates three 8-bit locals.

The 6812 provides a negative offset index addressing mode, whereas on the 6811 only positive index values can be used. With either computer, it is possible to establish a stack frame pointer using either register X or Y. It is important in this implementation that once the stack frame pointer is established (e.g., the tsx instruction), that the stack frame register (X) not be modified. The term **frame** refers to the fact that the value of the stack frame pointer is fixed. Because the stack frame pointer should not be modified, every subroutine will save the old stack frame pointer of the function that called the subroutine (e.g., pshx at the top) and restore it before returning (e.g., pulx at the bottom.) The stack frame will allow you to use the txs to deallocate the local variables. Local variable access uses indexed addressing mode. The difference between the two versions is the position of the stack frame pointer. As always, the 6811 example will function properly on the 6812. This example will be extended to include subroutine parameters later in the chapter. Notice the subroutine deallocates the locals by moving the stack frame pointer back into SP with the txs instruction.

> *Observation:* One advantage of using a stack frame is that the tsx instruction needs to be executed only once at the beginning of the function.

> *Observation:* Another advantage of using a stack frame is that you can push and pull within the body of the function, and still be able to access local variables using their symbolic name.

> *Observation:* One disadvantage of using a stack frame is that a register is dedicated as the frame pointer, and thus it is unavailable for general use.

This example calculates the sum of the first 100 numbers. Program 2.15 shows the assembly code that implements this C function. The result will be returned by value in Register D.

```
unsigned short calc(void){ unsigned short sum,n;
   sum = 0;
   for(n=100;n>0;n--){
     sum=sum+n;
   }
   return sum;
}
```

```
; 6811 or 6812                        ; 6812 only
; *****binding phase***********       ; *****binding phase************
sum set  0  16-bit number             sum  set  -4  16-bit number
n   set  2  16-bit number             n    set  -2  16-bit number
; *******allocation phase *****       ; *******allocation phase ******
calc pshx   ;save old Reg X           calc pshx        ;save old Reg X
     pshx   ;allocate n                    tsx         ;stack frame pointer
     pshx   ;allocate sum                  leas -4,sp  ;allocate n,sum
     tsx    ;stack frame pointer
; ********access phase ********       ; ********access phase ********
     ldd  #0                               movw #0,sum,x   ;sum=0
     std  sum,x  ;sum=0                     movb #100,n,x  ;n=100
     ldd  #100                        loop ldd  n,x     ;RegD=I
     std  n,x    ;n=100                     addd sum,x   ;RegD=sum+n
loop ldd  n,x    ;RegD=n                    std  sum,x   ;sum=sum+n
     addd sum,x  ;RegD=sum+n                ldd  n,x     ;n=n-1
     std  sum,x  ;sum=sum+n                 subd #1
     ldd  n,x    ;n=n-1                     std  n,x
     subd #1                                bne  loop
     std  n,x
     bne  loop
; ******deallocation phase ***       ; *****deallocation phase *****
     txs         ;deallocation             txs         ;deallocation
     pulx        ;restore old X             pulx        ;restore old X
     rts         ;RegD=sum                  rts         ;RegD=sum
```

Program 2.15
Assembly language implementation of a function with two local 16-bit variables.

2.5.2 Modules

The key to completing any complex task is to break it down into manageable subtasks. Modular programming is a style of software development that divides the software problem into distinct and independent modules. The parts are as small as possible, yet relatively independent. Complex systems designed in a modular fashion are easier to debug because each module can be tested separately. Industry experts estimate that 50 to 90% of software development cost is spent in maintenance. All five aspects of software maintenance:

■ Correcting mistakes
■ Adding new features
■ Optimizing for execution speed or program size
■ Porting to new computers or operating systems
■ Reconfiguring the software to solve a similar related program

are simplified by organizing the software system into modules. The approach is particularly useful when a task is large enough to require several programmers (Figure 2.30).

Figure 2.30
Block diagram of a
software module.

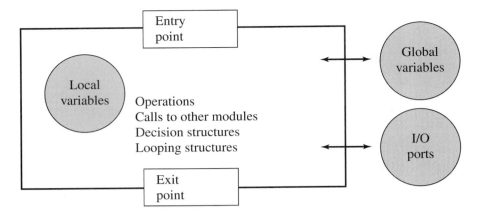

A *program module* is a self-contained software task with clear entry and exit points. We make the distinction between a module and the assembly language subroutine or C language function. A module can be a collection of subroutines or functions that in its entirety performs a well-defined set of tasks. The collection of serial port I/O functions presented in Section 2.7.2 can be considered one module. A collection of 32-bit math operations is another example of a module. Modular programming involves both the specification of the individual modules and the connection scheme whereby the modules are connected together to form the software system. While the module may be called from many locations throughout the software, there should be a common interface.

In assembly language, the entry point is used to call the subroutine. In this example, the input parameter (e.g., global variable ss) is first pushed on the stack by the instruction

```
movb ss,1,-sp.
```

Next, the instruction jsr sqrt calls the subroutine. After the subroutine returns, the output parameter is in Register B. The input parameter is no longer needed, so ins is executed to discard it. Finally, the output parameter is stored in the global variable tt by the instruction stab tt.

```
movb ss,1,-sp  ;push parameter on the stack (binary fixed point)
jsr  sqrt      ;subroutine call to the module "sqrt"
ins
stab tt        ;save result
```

The stack contents are shown in Figure 2.31 as the calling routine pushes the input parameter ss on the stack, calls the subroutine (Figure 2.32) and then the subroutine allocates three local variables.

> **Common error:** In many situations the input parameters have a restricted range. It would be inefficient for both the module and the calling routine to check for valid input. On the other hand, an error may occur if neither one checks for valid input.

The entry point for a C module (Figure 2.33) is also the name of the function, and it is also used to call the function [e.g., tt=sqrt(ss);].

The main module is the entry point of the entire program. The Freescale microcomputers employ a reset vector to specify where to begin execution after a reset. For Freescale microcomputers, the reset vector location is usually the last 2 bytes of the program ROM (EEPROM) space. The following assembly language software defines the reset vector:

```
org $FFFE
fdb main
```

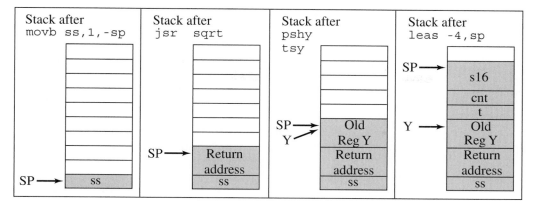

Figure 2.31
6812 stack contents before and after the function call.

Figure 2.32
Example of a 6812 assembly language module.

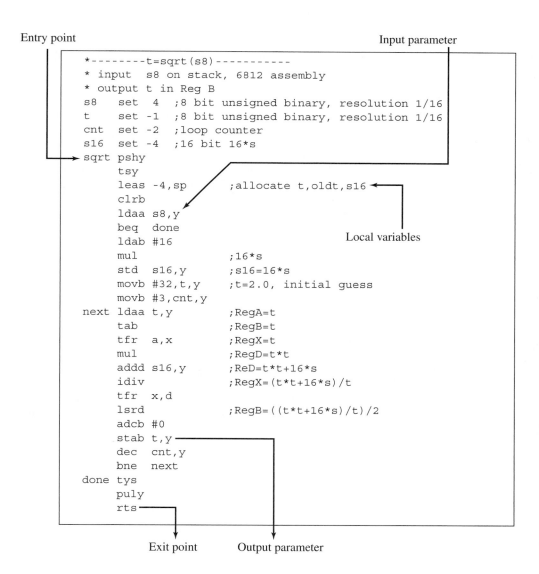

```
*--------t=sqrt(s8)-----------
* input  s8 on stack, 6812 assembly
* output t in Reg B
s8   set  4  ;8 bit unsigned binary, resolution 1/16
t    set -1  ;8 bit unsigned binary, resolution 1/16
cnt  set -2  ;loop counter
s16  set -4  ;16 bit 16*s
sqrt pshy
     tsy
     leas -4,sp        ;allocate t,oldt,s16
     clrb
     ldaa s8,y
     beq  done
     ldab #16
     mul               ;16*s
     std  s16,y        ;s16=16*s
     movb #32,t,y      ;t=2.0, initial guess
     movb #3,cnt,y
next ldaa t,y          ;RegA=t
     tab               ;RegB=t
     tfr  a,x          ;RegX=t
     mul               ;RegD=t*t
     addd s16,y        ;ReD=t*t+16*s
     idiv              ;RegX=(t*t+16*s)/t
     tfr  x,d
     lsrd              ;RegB=((t*t+16*s)/t)/2
     adcb #0
     stab t,y
     dec  cnt,y
     bne  next
done tys
     puly
     rts
```

Entry point   Input parameter

Local variables

Exit point   Output parameter

Figure 2.33
Example of a
C language module.

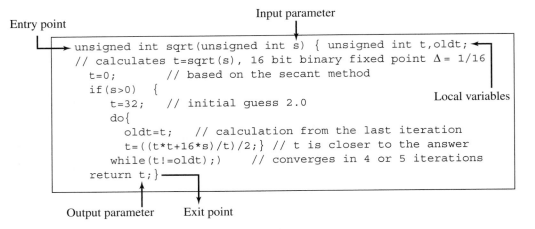

An *exit point* is the ending point of a program module. The exit point of a subroutine is used to return to the calling routine (e.g., rts). We need to be careful about exit points. It is important that the stack be properly balanced at all exit points. Similarly if the subroutine returns parameters, then all exit points should return parameters in an acceptable format.

Common error: It is an error if all the exit points of an assembly subroutine do not balance the stack and return parameters in the same way.

If the main program has an exit point, it either stops the program or returns to the debugger. The *input parameters* (or arguments) are pieces of data passed from the calling routine into the module during execution. The *output parameter* (or argument) is information returned from the module back to the calling routine after the module has completed its task.

It is easy to return multiple parameters in assembly language. If just a few parameters need to be returned, we can use the registers. In this simple example, the numbers 1, 2, 3, 4 are to be returned:

```
module: ldaa #1
        ldab #2
        ldx  #3
        ldy  #4
        rts        ; returns four parameters in 4 registers
********calling sequence******
        jsr  module
* Reg A,B,X,Y have four results
```

If many parameters are needed, then the stack can be used. Space for the output parameters is allocated by the calling routine, and the module stores the results into those stack locations.

```
data1 equ 2
data2 equ 3
data3 equ 4
data4 equ 5
module movb #1,data1,sp ;first parameter onto stack
       movb #2,data2,sp ;second parameter onto stack
       movb #3,data3,sp ;third parameter onto stack
       movb #4,data4,sp ;fourth parameter onto stack
       rts
```

```
********calling sequence******
        leas  -4,sp   ;allocate space for results
        jsr   module
        pula           ;first parameter from stack
        staa  first
        pula           ;second parameter from stack
        staa  second
        pula           ;third parameter from stack
        staa  third
        pula           ;fourth parameter from stack
        staa  fourth
```

There are two approaches for returning multiple parameters in C. In the first approach, we pass a pointer to the module and return the parameters through the pointer.

```
void module(short *pt){
   (*pt)=1;    *(pt+1)=2;   *(pt+2)=3;   *(pt+3)=4;}
void main(void){ short data[4];
     module(data);}
```

In the second approach we create a structure, and return the parameters within the structure:

```
struct data{ short first,second,third,fourth;};
typedef struct data dataType;
dataType module(void){ dataType myData;
  myData.first=1;  myData.second=2;  myData.third=3;  myData.fourth=4;
 return myData;}
void main(void){ dataType theData;
     theData=module();}
```

2.5.3 Dividing a Software Task into Modules

The overall goal of modular programming is to enhance clarity. The smaller the task, the easier it will be to understand. *Coupling* is defined as the influence one module's behavior has on another module. To make modules more independent we strive to minimize coupling. Obvious and appropriate examples of coupling are the I/O parameters explicitly passed from one module to another. On the other hand, information stored in shared global variables can be quite difficult to track. In a similar way shared accesses to I/O ports can also introduce unnecessary complexity. Global variables cause coupling between modules that complicates the debugging process, because now the modules may not be able to be separately tested. On the other hand, we must use global variables to pass information into and out of an interrupt service routine and from one call to an interrupt service routine to the next call. Another problem specific to embedded systems is the need for fast execution, coupled with the limited support for local variables. On the 6811 it is possible, but inefficient, to implement local variables on the stack. Consequently, many 6811 programmers opt for the less elegant, yet faster approach of global variables. When passing information through global variables is required, it is better to use a well-defined abstract technique like a FIFO queue. We assign a logically complete task to each module. The module is logically complete when it can be separated from the rest of the system and placed into another application. The interfaces are extremely important. The interfaces determine the policies of our modules. In other words, the interfaces define the operations of our software system. The interfaces also represent the coupling between modules. In general we wish to minimize the amount of information passing (i.e., bandwidth) between the modules yet maximize the number of modules. Of the following three objectives when dividing a software project into subtasks, only the first one really matters:

- Make the software project easier to understand
- Increase the number of modules
- Decrease the interdependency (minimize coupling)

We can develop and connect modules in a hierarchical manner: Construct new modules by combining existing modules. In Figure 2.34 the modules of the software project are organized into a call graph. An arrow points from the calling routine to the module it calls. The I/O ports are organized into groups (e.g., all the serial port I/O registers are in one group). This graph allows us to see the organization of the project. Figure 2.34 shows a call graph for the simple example of outputting decimal numbers via the serial port. To make simpler call graphs on large projects we can combine multiple related subroutines into a single module. The main program is at the top, and the I/O ports are at the bottom. In a hierarchical system the modules are organized both in a horizontal fashion (grouped together by function) and in a vertical fashion (decisions on overall policies at the top and implementation details at the bottom). Since one of the advantages of breaking a large software project into subtasks is concurrent development, it makes sense to consider concurrency when dividing the tasks. In other words, the modules should be partitioned in such a way that multiple programmers can develop the subtasks as independently as possible. On the other hand, careful and constant supervision is required as interfaces are designed and modules are connected together.

Figure 2.34
A call graph showing how the main program outputs numbers.

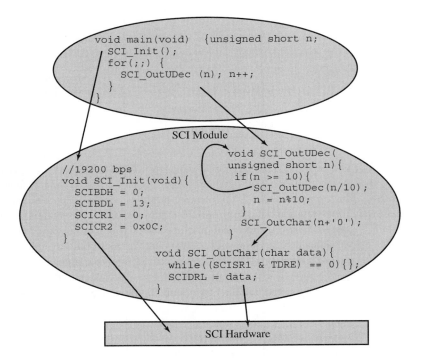

```
void main(void)   {unsigned short n;
  SCI_Init();
  for(;;) {
    SCI_OutUDec (n); n++;
  }
}
```

SCI Module

```
//19200 bps
void SCI_Init(void){
  SCIBDH = 0;
  SCIBDL = 13;
  SCICR1 = 0;
  SCICR2 = 0x0C;
}
```

```
void SCI_OutUDec(
  unsigned short n){
  if(n >= 10){
    SCI_OutUDec(n/10);
    n = n%10;
  }
  SCI_OutChar(n+'0');
}
```

```
void SCI_OutChar(char data){
  while((SCISR1 & TDRE) == 0){};
  SCIDRL = data;
}
```

SCI Hardware

Observation: Recursive functions do not fall into this hierarchical calling structure.

Observation: If module A calls module B and module B calls module A, then you have created a special situation that must account for these mutual calls.

In Figure 2.34 SCI_Init will initialize the serial port interface, establishing the baud rate and protocol format. SCI_OutChar transmits a single 16-bit byte using the serial port. SCI_OutDec accepts an 8-bit unsigned byte and calls SCI_OutChar multiple times to transmit the three ASCII characters. For example, if n=145, then the three ASCII characters will be $31, $34, and $35. The details of this example will be presented later in Chapters 3 and 7.

There are two approaches to hierarchical programming. The top-down approach starts with a general overview, like an outline of a paper, and builds refinement into subsequent layers. A top-down programmer was once quoted as saying:

"Write no software until every detail is specified"

It provides a better global approach to the problem. Managers like top-down because it gives them tighter control over their workers. The top-down approach works well when an existing operational system is being upgraded or rewritten. On the other hand, the bottom-up approach starts with the smallest detail, building up the system "one brick at a time." The bottom-up approach provides a realistic appreciation of the problem because we often cannot appreciate the difficulty or the simplicity of a problem until we have tried it. It allows programmers to start coding immediately and gives programmers more input into the design. For example, a low-level programmer may be able to point out features that are not possible and suggest other features that are even better. Some software projects are flawed from their conception. With bottom-up design, the obvious flaws surface early in the development cycle.

We believe bottom-up is better when designing a complex system and specifications are open-ended. On the other hand, top-down is better when you have a very clear understanding of the problem specifications and the constraints of your computer system. The **TExaS** simulator was actually written twice. The first pass was programmed bottom-up and served only to provide a clear understanding of the problem and the features and limitations of our hardware. We literally threw all the source code in the trash and programmed the second version in a top-down manner.

One of the biggest mistakes beginning programmers make is the inappropriate usage of I/O calls (e.g., screen output and keyboard input). Our explanation for their foolish behavior is that they haven't had the experience yet of trying to reuse software they have written for one project in another project. For example, assume you wrote and tested a function that found the median of three numbers. The goal of this first program was to display the result, so you solved it as shown in Program 2.16.

```
unsigned int Median (unsigned int u1, unsigned int u2, unsigned int u3) { unsigned int result;
  printf("The inputs are %d, %d, %d.\n",u1,u2,u3);
  if(u1>u2)
    if(u2>u3)    result=u2;     // u1>u2,u2>u3        u1>u2>u3
      else
        if(u1>u3) result=u3;    // u1>u2,u3>u2,u1>u3 u1>u3>u2
        else      result=u1;    // u1>u2,u3>u2,u3>u1 u3>u1>u2
  else
    if(u3>u2)    result=u2;     // u2>u1,u3>u2        u3>u2>u1
      else
        if(u1>u3) result=u1;    // u2>u1,u2>u3,u1>u3 u2>u1>u3
        else      result=u3;    // u2>u1,u2>u3,u3>u1 u2>u3>u1
  printf("The median is %d.\n",result);
  return(result):}
```

Program 2.16
A median function that is not very portable.

Software portability of this function is diminished because it is littered with printf's. To use this function in another situation, you will almost certainly have to remove the printf statements. In general we avoid interactive I/O at the lowest levels of the hierarchy; rather we return data and flags and let the higher-level program do the interactive I/O. Often we add keyboard input and screen output calls when testing our software. It is important to remove the I/O that is not directly necessary as part of the module function. This allows you to reuse these functions in situations where screen output is not available or appropriate. Obviously screen output is allowed if that is the purpose of the routine.

Common error: Performing unnecessary I/O in a subroutine makes it harder to reuse at a later time.

From a formal perspective, I/O devices are considered as global because I/O devices reside permanently at fixed addresses. From a syntactic viewpoint any module has access to any I/O device. To reduce the complexity of the system we will restrict the number of modules that actually do access the I/O device. It will be important to clarify which modules have access to I/O devices and when they are allowed to access it. When more than one module accesses an I/O device, it is important to develop ways to arbitrate (which module goes first if two or more want to access simultaneously) or synchronize (make a second module wait until the first is finished). These arbitration issues will be presented in detail in both Chapters 4 and 5.

Information hiding is similar to minimizing coupling. It is better to separate the mechanisms of software from its policies. We should separate what the function does (the relationship between its inputs and outputs) from how it does it. It is good to hide certain inner workings of a module, and simply interface with the other modules through the well-defined I/O parameters. For example, we could implement a FIFO by maintaining the current byte count in a global variable, CNT. A good module will hide how CNT is implemented from its users. If the user wants to know how many bytes are in the FIFO, it calls one of the FIFO routines that returns the count. A badly written module will not hide CNT from its users. The user simply accesses the global variable CNT. If we update the FIFO routines, making them faster or better, we might have to update all the programs that access CNT, too. The object-oriented programming environments provide well-defined mechanisms to support information hiding. This separation of policies from mechanisms is discussed further in Section 2.6 on layered software.

The *keep it simple stupid* approach tries to generalize the problem so that it fits an abstract model. Unfortunately, the person who defines the software specifications may not understand the implications and alternatives. As software developers, we always ask ourselves these questions:

"How important is this feature?"
"What if it worked this different way?"

Sometimes we can restate the problem to allow for a simpler (and possibly more powerful) solution.

**2.5.4
Rules for
Developing
Modular Software
in Assembly
Language**

The objective of this section is to present modular design rules and illustrate these rules with assembly language examples. This set of rules is meant to guide, not control. In other words, the rules serve as general guidelines rather than as fundamental law.

The single entry point is at the top. In assembly language, we place a single entry point at the first line of the code. This guarantees that registers will be saved and local variables will be properly allocated on the stack. By default, C functions have a single entry point. Placing the entry point at the top provides a visual marker for the beginning of the subroutine.

The single exit point is at the bottom. We prefer to use a single exit point as the last line of the subroutine. Many good programmers employ multiple exit points for efficiency reasons. In general, we must guarantee that the registers, stack, and return parameters are at a similar and consistent state for each exit point. In particular we must deallocate local variables properly. If you do employ multiple exit points, then we suggest you develop a means to visually delineate where one subroutine ends and the next one starts. You could use one line of comments to signify the start of a subroutine and a different line of comments to show the end of it, as in Program 2.17.

Observation: Having the first and last lines of a module be the entry and exit points makes it easier to debug, because it will be easy to place debugging instruments (like breakpoints).

Program 2.17
An assembly subroutine that uses comments to delineate its beginning and end.

```
;***************Abs***********************
; Input: RegA is signed 8 bit
; Output: Reg A is absolute value 0 to 127
Abs: tsta      ; already positive?
     bpl  ok
     nega
ok   rts
* ------------end of Abs ----------------
```

Common error: If you place a debugging breakpoint on the last `rts` of a subroutine with multiple exit points, then sometimes the subroutine will return without generating the break.

Write structured programs. A structured program is one that adheres to a strict list of program structures (e.g. sequence, if-then, and do-while). When we program in C (with the exception of `goto,` which, by the way, you should never use), we are forced to write structured programs because of the syntax of the language. One technique for writing structured assembly language is to adhere to the same strict list of program structures available in C. In other words, restrict the assembly language branching to configurations that mimic the software behavior of `if,` `if-else,` `do-while,` `while,` `for,` and `switch.` Assembly language examples of these control structures are included with the **TExaS** simulator in the files UIF.RTF, SIF.RTF, WHILE.RTF, and FOR.RTF. Structured programs are much easier to debug, because execution proceeds only through a limited number of well-defined pathways. When we reuse existing assembly branching structures, then our debugging can focus more on the overall function and less on how the details are implemented.

The registers must be saved. When working on a software team, it is important to establish a rule whether or not subroutines will save/restore registers. Establishing this convention is especially important when a mixture of assembly and C is being used or if the software project remains active for long periods of time. It is safest to save and restore registers that are modified (most programmers do not save/restore the CCR). Exceptions to this rule can be made for those portions of the code where speed is most critical.

Common error: If the calling routine expects a subroutine to save/restore registers and it doesn't, then information will be lost.

Observation: If the calling routine does not expect a subroutine to save/restore registers and it does, then the system executes a little slower and the object code is a little bigger than it could be.

Common error: When a mixture of C and assembly language programs is integrated, then an error may occur when the compiler is upgraded because there may be a change if registers are saved/restored or in how parameters are passed.

Use high-level languages whenever possible. It may seem odd to have a rule about high-level languages in a section about assembly language programming. In general we should use high-level languages when memory space and execution speed are less important than portability and maintenance. When execution speed is important, you could write the first version in a high-level language, run a profiler (which will tell you which parts of your program are executed the most), then optimize the critical sections by writing them in assembly language. If a C language implementation just doesn't run fast enough, you could consider a more powerful compiler or a faster microcomputer.

Minimize conditional branching. Every time software makes a conditional branch, there are two possible outcomes that must be tested (branch or not branch). For example, assume we wish to add two 16-bit numbers u3 = u2 + u1. A conditional branch could be avoided by solving the problem in another way, as shown in Program 2.18.

Program 2.18
Sometimes we can remove a conditional branch and simplify the program.

```
; no conditional branch
add16b ldaa u1+1   ;lsb
       adda u2+1
       staa u3+1
       ldaa u1     ;msb
       adca u2
       staa u3
       rts
```

```
; uses conditional branch
add16a ldaa u1+1   ;lsb
       adda u2+1
       staa u3+1
       ldaa u1     ;msb
       bcc  noc
       inca         ;carry
noc    adda u2
       staa u3
       rts
```

Observation: Software can be made easier to understand by reworking the approach to reduce the number of conditional branches.

2.6 Layered Software Systems

As the size and complexity of our software systems increase, we learn to anticipate the changes that our software must undergo in the future. In particular, we can expect to redesign our system to run on newer and more powerful hardware platforms. A similar expectation is that better algorithms may become available. The objective of this section is to use a layered software approach to facilitate these types of changes.

We can use the call graph defined in Section 2.5 to visualize software layers. A module in a layer can call a module within the same layer or a module in a layer below it. Some layered systems restrict the calls only to modules within the same layer or to a module in the most adjacent layer below it. If we place all the modules that access the I/O hardware in the bottommost layer, we can call this layer a *hardware abstraction layer* (HAL). Each layer of modules only calls modules of the same or lower levels, but not modules of higher level. Usually the top layer consists of the main program. In a multithreaded environment (e.g., Unix, Windows) there can be multiple main programs at the topmost level, but for now assume there is only one main program. The arrows in Figure 2.35 point from the calling module to the module it calls.

To develop a layered software system we begin with a modular system. The main advantage of layered software is the ability to separate the modules into groups or layers such that one layer may be replaced without affecting the other layers. For example, you could change which ports the printer is connected to by modifying the low level without any changes to the middle or high levels. Figure 2.35 depicts a layered implementation of a printer interface. In a similar way, you could replace the IEEE488 printer with a serial printer by replacing the bottom two layers. If we were to employ buffering and/or data compression to enhance communication bandwidth, then these algorithms would be added to the middle level. A layered system should allow you to change the implementation of one layer without requiring redesign of the other layers.

A *gate* is used to call from a higher- to a lower-level routine. Another name for this gate is *application program interface* (API). The gates provide a mechanism to link between the layers. Because the size of the software on an embedded system is small, it is possible and

Figure 2.35
A layered approach to interfacing an IEEE488 parallel port printer.

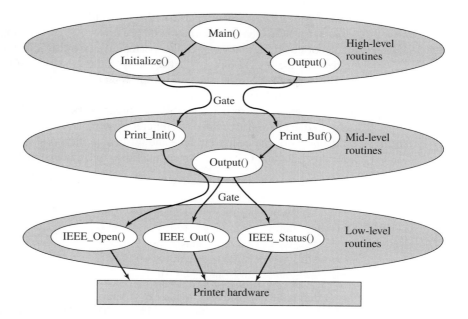

appropriate to implement a layered system using standard function calls by simply assembling/compiling all software together. We will see in the next section that the gate can be implemented by creating a header file with prototypes to public functions.

The following rules apply to layered software systems:

1. A module may make a simple call to other modules in the same layer.
2. A module may make a call to a lower-level module only by using the gate.
3. A module may not directly access any function or variable in another layer (without going through the gate).
4. A module may not call a higher-level routine.
5. A module may not modify the vector address of another level's handler(s). For example:
   The software interrupt handler address, memory contents at $FFF6, is specified by the middle level. The low-handler address, memory contents at $FF80, is specified by the low level.
6. (Optional) A module may not call farther down than the immediately adjacent lower level.
7. (Optional) All I/O hardware access is grouped in the lowest level.
8. (Optional) All user interface I/O (InString, OutString, OutDec, etc.) is grouped in the highest level unless it is the purpose of the module itself to do such I/O.

The purpose of rule 6 is to allow modifications at the low layer to not affect operation at the highest layer. On the other hand, for efficiency reasons you may wish to allow module calls farther down than the immediately adjacent lower layer. To get the full advantage of layered software, it is critical to design functionally complete interfaces between the layers. The interface should support all current functions as well as provide for future expansions.

2.7 Device Drivers

2.7.1
Basic Concept of Device Drivers

A device driver consists of the software routines that provide the functionality of an I/O device. The driver consists of the interface routines that the operating system or software developer's program calls to perform I/O operations as well as the low-level routines that configure the I/O device and perform the actual I/O. The issue of the separation of policy from mechanism is very important in device driver design. The policies of a driver include the list of functions and the overall expected results. In particular, the policies can be summarized by the interface routines that the operating system or software developer can call to access the device. The mechanisms of the device driver includes the specific hardware and low-level software that actually performs the I/O. As an example, consider the variety of mass storage devices that are available. Floppy disk, RAM disks, integrated device electronics (IDE) hard drive, Small Computer Systems Interface (SCSI) hard drive, tape, and even a network can be used to save and recall data files. A simple serial port system might have the following C level interface functions, as explained in the following prototypes:

```
void SCI_Init(unsigned short baudRate);   // Enable serial port
char SCI_InChar(void);                     // Input an ASCII character
void SCI_InString(char *buffer, unsigned short maxSize); // Input a String
unsigned short SCI_InUDec(void);           // Input a 16-bit number
void SCI_OutChar(char letter);             // Output an ASCII character
void SCI_OutUDec(unsigned short number);   // Output a 16-bit number
void SCI_OutString(char *buffer);          // Output String
```

Building a HAL is the same idea as separation of policy from mechanism. A diagram of this layered concept was shown in Figure 2.35. In the above file example, a HAL would treat all the potential mass storage devices through the same software interface. Another example of this abstraction is the way some computers treat pictures on the video screen and pictures printed on the printer. With the abstraction layer, the software developer's program draws lines and colors by passing the data in a standard format to the device driver, and the operating system redirects the information to the video graphics board or color laserwriter as appropriate. This layered approach allows one to mix and match hardware and software components but does suffer some overhead and inefficiency.

Low-level *device drivers* normally exist in the basic I/O system (BIOS) ROM and have direct access to the hardware. They provide the interface between the hardware and the rest of the software. Good low-level device drivers allow:

1. New hardware to be installed
2. New algorithms to be implemented
   a. Synchronization with gadfly, interrupts, or DMA
   b. Error detection and recovery methods
   c. Enhancements like automatic data compression
3. Higher-level features to be built on top of the low level
   a. Operating system features like blocking semaphores
   b. Additional features like function keys

and still maintain the same software interface. In larger systems like the Workstation and IBM PC, where the low-level I/O software is compiled and burned in ROM separate from the code that will call it, it makes sense to implement the device drivers as software interrupts (SWIs) and specify the calling sequence in assembly language. We define the "client programmer" as the software developer who will use the device driver. In embedded systems like we use, it is okay to provide device.H and device.C files that the client programmers can compile with their application. In a commercial setting, you may be able to deliver to the client only the device.H together with the object file. *Linking* is the process of resolving addresses to code and programs that have been compiled separately. In this way, the routines can be called from any program without requiring complicated linking. In other words, when the device driver is implemented with a TRAP, the linking is simple. In our embedded system, the compiler will perform the linking. The device driver software is grouped into four categories. Protected items can only be directly accessed by the device driver itself, and public items can be accessed by the client.

The concept of a device driver can be illustrated with the following design of a serial port device driver. There are typically four components of a device driver. In this section, the contents of the header file (SCI.h) will be presented, and the implementations will be developed in the next chapter.

1. Data structures: global (private) The first component of a device driver is private global data structures. To be private means only programs within the driver itself may directly access these variables. If the user of the device driver (e.g., a client) needs to read or write to these variables, then the driver will include public functions that allow appropriate read/write functions. One example of a private global variable might be an OpenFlag, which is true if the serial port has been properly initialized. The implementation developed in Chapter 3 will have no private global variables, but the SCI implementation developed in Chapter 7 will include a first-in-first-out (FIFO) queue, including the functions Fifo_Init, Fifo_Put, and Fifo_Get.

2. Initialization routines (public, called by the client once in the beginning) The second component of a device driver includes the public functions used to initialize the device. To be public means the user of this driver can call these functions directly. A prototype to public functions will be included in the header file (SCI.H). The names of public functions will begin with SCI_. The purpose of this function is to initialize the SCI hardware.

```
//------------SCI_Init------------
// Initialize Serial port SCI
// Input: baudRate is the baud rate in bits/sec
// Output: none
void SCI_Init(unsigned short baudRate);
```

3. Regular I/O calls (public, called by client to perform I/O) The third component of a device driver consists of the public functions used to perform input/output with the device. Because these functions are public, prototypes will be included in the header file (SCI.H). The input functions are grouped, followed by the output functions.

```
//------------SCI_InChar------------
// Wait for new serial port input
// Input: none
// Output: ASCII code for key typed
char SCI_InChar(void);

//------------SCI_InString------------
// Wait for a sequence of serial port input
// Input: maxSize is the maximum number of characters to look for
// Output: Null-terminated string in buffer
void SCI_InString(char *buffer, unsigned short maxSize);

//------------SCI_InUDec------------
// InUDec accepts ASCII input in unsigned decimal format
//      and converts to a 16-bit unsigned number (0 to 65535)
// Input: none
// Output: 16-bit unsigned number
unsigned short SCI_InUDec(void);

//------------SCI_OutChar------------
// Output 8-bit to serial port
// Input: letter is an 8-bit ASCII character to be transferred
// Output: none
void SCI_OutChar(char letter);
```

```
//------------SCI_OutString------------
// Output String (NULL termination)
// Input: pointer to a NULL-terminated string to be transferred
// Output: none
void SCI_OutString(char *buffer);

//------------SCI_OutUDec------------
// Output a 16-bit number in unsigned decimal format
// Input: 16-bit number to be transferred
// Output: none
// Variable format 1-5 digits with no space before or after
void SCI_OutUDec(unsigned short number);
```

4. Support software (private). The last component of a device driver is private functions. Because these functions are private, prototypes will not be included in the header file (SCI.H). We place helper functions and interrupt service routines in the category.

Notice that this SCI example implements a layered approach, as illustrated in Figure 2.34. The low-level functions provide the mechanisms and are protected (hidden) from the client programmer. The high-level functions provide the policies and are accessible (public) to the client. When the device driver software is separated into SCI.H and SCI.C files, you need to pay careful attention as to how many details you place in the SCI.H file. A good device driver separates the policy (overall operation, how it is called, what it returns, what it does, etc.) from the implementation (access to hardware, how it works, etc.) In general, you place the policies in the SCI.H file (to be read by the client) and the implementations in the SCI.C file (to be read by you and your coworkers). Think of it this way: if you were to write commercial software that you wished to sell for profit and you delivered the SCI.H file and its compiled object file, how little information could you place in the SCI.H file and still have the software system be fully functional. In object-oriented terms the policies will be public, and the implementations will be private.

> *Observation:* A layered approach to I/O programming makes it easier for you to upgrade to newer technology.

> *Observation:* A layered approach to I/O programming allows you to do concurrent development.

2.8 Object-Oriented Interfacing

2.8.1 Encapsulated Objects Using Standard C

Object-oriented software development in C++ involves three fundamental issues: encapsulation, polymorphism, and inheritance. *Encapsulation* is the grouping of functions and variables into a single class. C++ provides the mechanisms to implement modular software as described in this section. *Polymorphism* is the ability to reuse function names so that the exact operation depends on which class is being operated. *Inheritance* allows you to derive one class upon a previous class, reusing code and extending its functionality. For embedded systems, encapsulation is much more important than polymorphism and inheritance.

The example in this subsection will show you how to write C code that incorporates most of the important issues of encapsulation. This example also illustrates the top-down approach and includes three modules: the LCD interface, and some timer routines (Program 2.19). Notice that function names are chosen to reflect the module in which they are defined. If you are a C++ programmer, consider the similarities between this C function call LCD_clear() and a C++ LCD class and a call to a member function LCD.clear(). The *.H files contain function declarations and the *.C files contain the implementations.

For every function definition, the compiler generates an assembler directive declaring the function's name to be public. This means that every C function is a potential entry point

Program 2.19
Main program with three modules.

```
#include "HC12.H"
#include "LCD12.H"
#include "Timer.H"
void main(void){ char letter; int n=0;
    LCD_Init();
    Timer_Init();
    LCD_String("LCD");
    Timer_MsWait(1000);
    LCD_clear();
    letter='a'-1;
    while(1){
        if (letter=='z')
            letter='a';
        else
            letter++;
        LCD_putchar(letter);
        Timer_MsWait(250);
        if(++n==16){
            n=0;
            LCD_clear();
}}}
```

and so can be accessed externally. One way to create private/public functions is to control which functions have declarations. Now let us look inside the Timer.H and Timer.C files. To implement private and public functions we place the function declarations of the public functions in the Timer.H file (Program 2.20).

Program 2.20
Timer.H header file has public functions.

```
void Timer_Init(void);
void Timer_Wait10ms(unsigned short delay);
```

The implementations of all functions are included in the Timer.C file shown earlier as Program 2.10. We can apply this same approach to private and public global variables. Notice that in this case the global variable, CyclesPerMs, is private and cannot be accessed by software outside the Timer.C file (Program 2.21).

Program 2.21
Timer.C implementation file defines all functions.

```
unsigned short CyclesPerMs; // private global
void Timer_Init(void){ // public function
  TSCR1 |=0x80;        // TEN(enable)
  CyclesPerMs = 4000;  // 4000 counts per ms
}
void wait(unsigned short cycles){ // private function
unsigned short startTime = TCNT;
  while((TCNT-startTime) <= cycles){};  // wait 10ms
}
void Timer_MsWait(unsigned short time){ // public function
  for(;time>0;time--){
    wait(CyclesPerMs); // 1.00ms wait
  }
}
```

2.8.2
Object-Oriented
Interfacing Using
C++

The three characteristics of object-oriented programming are encapsulation, polymorphism, and inheritance. We defined these terms in Section 2.8.1 and discussed how to encapsulate modules using standard C. In this section we will introduce the concept of object-oriented interfacing using C++ by discussing the software environment on the IBM

PC compatible. Then, we will show an example of how C++ might provide support to improve the portability of our embedded system software.

Since its inception C++ has steadily replaced C as the preferred programming environment for developing both system programs and user applications for the *WinTel* platform (Windows operating system on an Intel microprocessor). The reasons for this software evolution (revolution?) stem from the inherent advantages of C++. Some of the fundamental difficulties for developing software for the *WinTel* platform are:

1. We need a common user interface on top of a multitude of similar but not identical computers.
2. The hardware platform makes a fundamental advancement every 6 months.
3. Many hardware and software companies act in concert to produce a product.
4. The newer software must run on the older computers.
5. The older software must run on the newer computers.
6. The hardware/software configurations may change at run time.

Because of these constraints, a layered software model was adopted so that changes in one aspect of the system could be made without having to reengineer the entire system. Objects in C++ allow the programmer to use hardware and software modules without complete knowledge of how they work. The member functions provide a clean yet powerful mechanism to implement the interface between the software modules. Especially at the hardware interface level, classes provide a mechanism for abstraction. In other words, the HAL is a set of C++ objects that define basic input/output operations.

There are some similarities but many differences between the *WinTel* and embedded platforms. If we examine the same six constraints for an embedded microcomputer, we see that only the first constraint is similar. In other words, we are interested in making embedded system software run on multiple microcomputers (code reuse). Software is portable if it is easy to convert it to run on another platform. The other five constraints for the most part do not exist in the embedded system development environment:

2. Embedded microcomputers have a much longer lifetime than a x86 microprocessor.
3. Usually a single company develops the hardware and software.
4., 5. Hardware and software are upgraded together.
6. Configurations are usually well-defined at compile time.

**2.8.3
Portability Using
Standard C and
C++**

Even though assembly and C are currently the primary software development approaches for embedded systems, it is appropriate to consider software development C++. As a case study, we will address the issue of portability using C and C++. First, we will show an enhanced C software implementation of the Moore FSM first presented in Section 2.4.5 and Program 2.11. To make this program more portable using C, we create #define macros for those parameters that are likely to change.

Programs 2.22 and 2.23 use the same Traffic Light FSM as Program 2.11.

Program 2.22
Enhanced C
implementation of the
Traffic Light FSM.

```
/* 6812 Port M bits 1,0 are input, Port T bits 1,0 are output */
#define   OutPort (*(unsigned char  volatile *)(0x0240))
#define   OutDDR  (*(unsigned char  volatile *)(0x0242))
#define   InPort  (*(unsigned char  volatile *)(0x0250))
#define   InDDR   (*(unsigned char  volatile *)(0x0252))
STyp *Pt;  // state pointer
unsigned char Input;
void main(void){
  Timer_Init(); // enable TCNT
  OutDDR  = 0xFF;
  InDDR   &= ~0x03;
```

```
        Pt = goN;
        while(1){
          OutPort = Pt->Out;
          Timer_Wait10ms(Pt->Time);
          Input = InPort&0x03;
          Pt = Pt->Next[Input];
        }
      }
```

```cpp
// a DDRAddress of 1 means fixed output port with no DDR
// a DDRAddress of 0 means fixed input port with no DDR
template <class T> class port{
  protected :
    unsigned char *PortAddress;   // pointer to data
    unsigned char *DDRAddress;     // pointer to data direction register
  public : port(unsigned short ThePortAddress, unsigned short TheDDRAddress){
    PortAddress = (unsigned char *)ThePortAddress; // pointer to I/O port
    DDRAddress  = (unsigned char *)TheDDRAddress;  // pointer to DDR
}
virtual short Initialize(unsigned char  data){
  if((int)DDRAddress==1){   // fixed output port
    return(data==0xFF);    // OK if initializing all bits to output
  }
  if(DDRAddress==0){       // fixed input port
    return(data==0);       // OK if initializing all bits to input
  }
  (*DDRAddress) = data;    // configure direction register
   return 1;               // successful
}
virtual void put(unsigned char  data){
  if((int)DDRAddress==0) return;  // fixed input
  if((*DDRAddress)==0) return;    // all input
  (*PortAddress) = data;          // output data to port
}
virtual unsigned char get(void){
  return (*PortAddress);  // input data from port
}
/* 6812 Port M bits 1,0 are input, Port T bits 1,0 are output */
port<unsigned char> OutPort(0x0240,0x0242);
port<unsigned char> InPort(0x0250,0x0252);

void main(void){ StateType *Pt;  unsigned char Input;
  Pt=SA;                         // Initial State
  OutPort.Initialize(0xFF);   // Make Output port outputs
  InPort.Initialize(0x00);    // Make Input port inputs
  Timer_Init();                  // Enable TCNT
  while(1){
    OutPort.put(Pt->Out);
    Timer_Wait10ms(Pt->Time); // Time to wait in this state
    Input=InPort.get()&0x03;  // Input=0,1,2,or 3
    Pt=Pt->Next[Input];
  }
}
```

Program 2.23
C++ implementation of the Traffic Light FSM.

In C++ we will define a class to describe a generic I/O port. The attributes of the port include its address, the existence and address of its data direction register, and its type T. "Program 2.23 uses the same FSM and Wait function as Program 2.22."

To configure this software for the 6811, we change the way the I/O ports are blessed.

```
// 6811 PortC bits 1,0 are input, Port B bits 1,0 are output
port<unsigned char> OutPort(0x1004,1);      // fixed output port
port<unsigned char> InPort(0x0003,0x0007);  // bidirectional port
```

To make this system truly portable we would have to create an object for the timer functions as well. You can derive classes from this class and override (enhance) the I/O functions. Additional member functions can also be added to enhance this C++ object, as in Program 2.24.

Program 2.24
Additional member functions for the I/O port class.

```
T operator = (T data){
        put(data);                  // output to port
        return data;}               // returns data itself
operator T () (T data){
        return get();}              // returns port data
virtual T operator |= (T data){
        put(data |= get());     // read modify write port access
        return data;}               // returns new data
virtual T operator &= (T data){
        put(data &= get());     // read modify write port access
        return data;}               // returns new data
virtual T operator ^= (T data){
        put(data ^= get());     // read modify write port access
        return data;}               // returns new data
```

Observation: The issues of software clarity, portability, and modularity are important no matter which programming language you are using.

2.9   Threads

2.9.1 Single-Threaded Execution

Software (e.g., program, code, module, procedure, function, subroutine) is a list of instructions for the computer to execute. A thread, on the other hand, is defined as the path of action of software as it executes. The expression "thread" comes from the analogy shown in Figure 2.36. This simple program prints the 8-bit numbers 000 001 002. . . . If we connect the statements of our executing program with a continuous line (the thread), we can visualize the dynamic behavior of our software.

The execution of the main program is called the *foreground thread.* In most embedded applications, the foreground thread executes a loop that never ends. We will learn later that this thread can be broken (execution suspended, then restarted) by interrupts and DMA.

2.9.2 Multithreading and Reentrancy

With interrupts we can create multiple threads. Some threads will be created statically, meaning they exist throughout the life of the software, while others will be created and destroyed dynamically. There will usually be one foreground thread running the main program, as in Figure 2.36. In addition to this foreground thread, each interrupt source has its own background thread, which is started whenever the interrupt is requested. Figure 2.37 shows a software system with one foreground thread and two background threads. The "key" thread is invoked whenever a key is touched on the keyboard, and the "time" thread is invoked every 1 ms in a periodic fashion. Because there is but one processor, the currently running thread must be suspended to execute another thread. In Figure 2.37 the suspension

Figure 2.36
Illustration of the
definition of a thread.

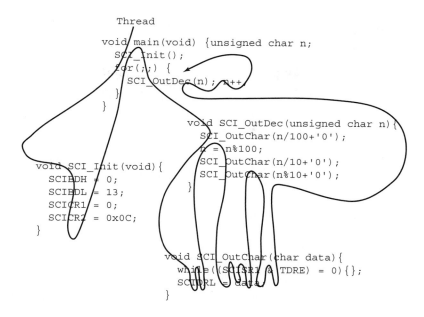

```
                              Thread
              void main(void) {unsigned char n;
                SCI_Init();
                for(;;) {
                  SCI_OutDec(n); n++;
                }
              }

                              void SCI_OutDec(unsigned char n){
                                SCI_OutChar(n/100+'0');
                                n = n%100;
                                SCI_OutChar(n/10+'0');
                                SCI_OutChar(n%10+'0');
              void SCI_Init(void){    }
                SCIBDH = 0;
                SCIBDL = 13;
                SCICR1 = 0;
                SCICR2 = 0x0C;
              }

                              void SCI_OutChar(char data){
                                while((SCISR1 & TDRE) = 0){};
                                SCIDRL = data;
                              }
```

Figure 2.37
Interrupts allow us to
have multiple
background threads.

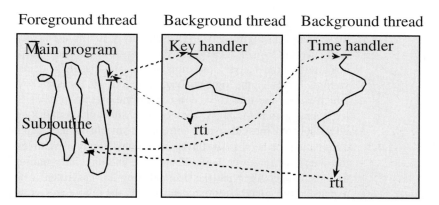

of the main program is illustrated by the two breaks in the foreground thread. When a key is touched, the main program is suspended, and a `Keyhandler` thread is created with an "empty" stack and uninitialized registers. When the `Keyhandler` is done, it executes RTI to relinquish control back to the main program. The original stack and registers of the main program will be restored to the state before the interrupt. In a similar way, when the 1-ms timer occurs, the main program is suspended again, and a `Timehandler` thread is created with its own "empty" stack and uninitialized registers. We can think of each thread as having its own registers and its own stack area. In Chapter 4, we will discuss in detail this approach to multithreaded programming. In Chapter 5, we will implement a preemptive thread scheduler that will allow our software to have multiple foreground threads.

A program segment is reentrant if it can be concurrently executed by two (or more) threads. In Figure 2.37 we can conceive of the situation where the main program starts executing a subroutine, is interrupted, and the background thread calls that same subroutine. For two threads to share a subroutine, the subroutine must be reentrant. To implement reentrant software, place local variables on the stack and avoid storing into I/O devices and global memory variables. The issue of reentrancy will be covered in detail later in Chapter 4.

2.10   Recursion

A recursive program is one that calls itself. When we draw a calling graph like the one in Figure 2.34, a circle is formed. Although many algorithms can be defined using recursion, it does require special care when implementing. In particular, recursive subroutines must be reentrant. For some sorting, computational, and database functions, recursion affords a more elegant solution. Recursive algorithms are often easy to prove correct and use less permanent memory but require more temporary stack space and execute slower than nonrecursive algorithms. For example, SCI_UDec the recursive implementation of the function is shown in Figure 2.34. Each call to the function creates a new stack frame.

2.11   Debugging Strategies

All programmers are faced with the need to debug and verify the correctness of their software. In this section we will study hardware-level probes like the logic analyzer and in-circuit emulator (ICE); software-level tools like simulators, monitors, and profilers; and manual tools like inspection and print statements.

2.11.1   Debugging Tools

Microcomputer-related problems often require the use of specialized equipment to debug the system hardware and software. Two very useful tools are the logic analyzer and the ICE. A logic analyzer (Figure 2.38) is essentially a multiple-channel digital storage scope with many ways to trigger. As a troubleshooting aid, it allows the experimenter to observe numerous digital signals at various points in time and thus make decisions based upon such observations. Typically, the logic analyzer is attached to the address/data bus in a passive manner, which allows the user to view the real-time execution of the software. One problem with logic analyzers is the massive amount of information that it generates. To use an analyzer effectively one must learn proper triggering mechanisms to capture data at appropriate times, eliminating the need to sift through volumes of output. With less expensive logic analyzers, the user must interpret the R/W, address, and data information by hand. This can be quite tedious, especially when the original program is written in a high-level language. More expensive logic analyzers can disassemble the output and show the assembly level instructions. This type of output is also difficult to interpret because the comments, labels, and program structure are not integrated into the assembly instructions. With the advent of today's microprocessor technology (in particular the cache, multiple instruction queues, branch predictions, and internal buses) it is very difficult to interpret software activity simply by observing read/write cycles to external memory. Even though the Freescale 8-bit microcomputer architectures are quite simple (having none of those above fancy features, they are therefore a good candidate for a logic analyzer), this discussion will focus on debugging techniques that can be applied to most computer architectures. The 6812 does have an instruction queue, so the address/data bus activity precedes program execution.

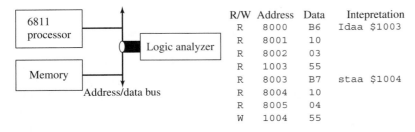

R/W	Address	Data	Intepretation
R	8000	B6	ldaa $1003
R	8001	10	
R	8002	03	
R	1003	55	
R	8003	B7	staa $1004
R	8004	10	
R	8005	04	
W	1004	55	

Figure 2.38
A logic analyzer and example output.

An ICE is a hardware debugging tool that recreates the I/O signals of the processor chip (Figure 2.39). To use an emulator, we remove the processor chip and insert the emulator cable into the chip socket. In most cases, the emulator/computer system operates at full speed. The ICE allows the programmer to observe and modify internal registers of the processor. An ICE is often integrated into a personal computer so that its editor, hard drive, and printer are available for the debugging process.

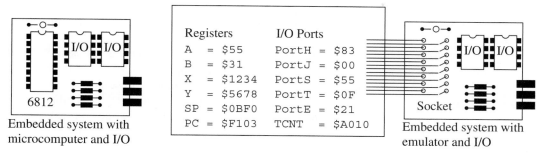

Figure 2.39
In-circuit emulator and example output.

Most software-based debuggers implement a breakpoint by replacing the existing instruction with a software trap (6811/6812 instruction `swi`). This procedure cannot be performed when the software is programmed in ROM. To debug this type of system we can use another class of emulator called the ROM emulator (Figure 2.40). This debugging tool replaces the ROM with cable connects to a dual-port RAM within the emulator. While the software is running, it fetches information from the emulator RAM just as if it were the ROM. While the software is halted, you can modify its contents.

Figure 2.40
In-circuit ROM emulator and example output.

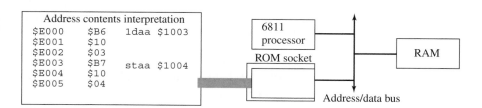

The only disadvantage of the ICE is its cost. To provide some of the benefits of this high-priced debugging equipment, the 6812 has a *background debug module* (BDM). The BDM hardware exists on the microcomputer chip itself and communicates with the debugging computer via a dedicated two- or three-wire serial interface. Although not as flexible as an ICE, the BDM can provide the ability to observe software execution in real time, the ability to set breakpoints, the ability to stop the computer, and the ability to read and write registers, I/O ports, and memory.

**2.11.2
Debugging
Theory**
Research in the area of program monitoring and debugging has not kept pace with developments in other areas of computer programming. This area is comparatively deficient in the structure and unity now common to programming languages. The area is compartmentalized because of specialized tools. For example, run-time profile generators and execution monitors are tools that have many commands and functions in common with "debuggers," but in practice, because of tool specialization, a user needs

to use two tools: a monitor to examine time behavior and a debugger for functional behavior. This area is also fragmented because of a multiplicity of terms. Terms such as program testing, diagnostics, performance debugging, functional debugging, tracing, profiling, instrumentation, visualization, optimization, verification, performance measurement, and execution measurement have specialized meanings, but they are also used interchangeably, and they often describe overlapping functions. For example, the terms profiling, tracing, performance measurement, or execution measurement may be used to describe the process of examining a program from a time viewpoint. But, tracing is also a term that may be used to describe the process of monitoring a program state or history for functional errors or to describe the process of stepping through a program with a debugger. Usage of these terms among researchers and users vary.

Furthermore, the meaning and scope of the term debugging itself is not clear. We hold the view that the goal of debugging is to maintain and improve software, and the role of a debugger is to support this endeavor. We define the debugging process as testing, stabilizing, localizing, and correcting errors. And in our opinion, although testing, stabilizing, and localizing errors are important and essential to debugging, they are auxiliary processes: The primary goal of debugging is to remedy faults or to correct errors in a program.

Although a variety of program monitoring and debugging tools are available today, in practice it is found that an overwhelming majority of users either still prefer or rely mainly upon "rough and ready" manual methods for locating and correcting program errors. These methods include desk checking, dumps, and print statements, with print statements being one of the most popular manual methods. Manual methods are useful because they are readily available, and they are relatively simple to use. But the usefulness of manual methods is limited: They tend to be highly intrusive, and they do not provide adequate control over repeatability, event selection, or event isolation. A real-time system, where software execution timing is critical, usually cannot be debugged with simple print statements, because the print statement itself will require too much time to execute.

We define a debugging instrument as software code that is added to the program for the purpose of debugging. A print statement is a common example of an instrument. Using the editor, we add print statements to our code that either verify proper operation or illustrate the programming errors. A key to writing good debugging instruments is to provide for a mechanism to reliably and efficiently remove them all when the debugging is done. Consider the following mechanisms as you develop your own unique debugging style.

- Place all print statements in a unique column (e.g., first column) so that the only code that exists in this column must be a debugging instrument.
- Define all debugging instruments as functions that all have a specific pattern in their names. In this way, the find/replace mechanism of the editor can be used to find all the calls to the instruments.
- Define the instruments so that they test a run-time global flag. When this flag is turned off, the instruments perform no function. Notice that this method leaves a permanent copy of the debugging code in the final system, causing it to suffer a run-time overhead, but the debugging code can be activated dynamically without recompiling. Many commercial software applications utilize this method because it simplifies "on-site" customer support.
- Use conditional compilation (or conditional assembly) to turn on and off the instruments when the software is compiled. When the compiler supports this feature, it can provide both performance and effectiveness.

The emergence of concurrent languages and the increasing use of embedded real-time systems place further demands on debuggers. The complexities introduced by the interaction of multiple events or time-dependent processes are much more difficult to debug than errors associated with sequential programs. The behavior of non-real-time sequential programs is reproducible: For a given set of inputs their outputs remain the same. In the case of concurrent or real-time programs this does not hold true. Control over repeatability, event selection, and event isolation is even more important for concurrent or real-time environments. **Intrusiveness** is the extent to which the debugging process itself affects the software under test. **Nonintrusive** debugging has no effect, and **minimally intrusive** debugging has a small but irrelevant effect.

The first step of debugging is to *stabilize* the system with the bug. In the debugging context, we stabilize the problem by creating a test routine that fixes (or stabilizes) all the inputs. In this way, we can reproduce the exact inputs over and over again. Once stabilized, if we modify the program, we are sure that the change in our outputs is a function of the modification we made in our software and not due to a change in the input parameters.

2.11.3 Functional Debugging

Functional debugging involves the verification of I/O parameters. It is a static process where inputs are supplied, the system is run, and the outputs are compared against the expected results. We will present seven methods of functional debugging.

2.11.3.1 Single Stepping or Trace

Many debuggers allow you to set the program counter to a specific address then execute one instruction at a time. The Metrowerhs debugger allows you to execute single assembly instructions or single C level instructions. The **TExaS** simulator provides Step, Few, StepOver, and StepOut commands. *Step* is the usual execute one assembly instruction. *Few* will execute some instructions and stop (you can set how many "some" is). *StepOver* will execute one assembly instruction, unless that instruction is a subroutine call, in which case the simulator will execute the entire subroutine and stop at the instruction following the subroutine call. *StepOut* assumes the execution has already entered a subroutine and will finish execution of the subroutine and stop at the instruction following the subroutine call.

2.11.3.2 Breakpoints without Filtering

A **breakpoint** is a mechanism to tag places in our software, which when executed will cause the software to stop.

2.11.3.3 Conditional Breakpoints

One of the problems with breakpoints is that sometimes we have to observe many breakpoints before the error occurs. One way to deal with this problem is the conditional breakpoint. Add a global variable called count and initialize it to zero in the ritual. Add the following conditional breakpoint to the appropriate location. And run the system again (you can change the 32 to match the situation that causes the error).

```
if(count==32)
        bkpt
```

Notice that the breakpoint occurs only on the 32nd time the break is encountered. Any appropriate condition can be substituted.

2.11.3.4 Instrumentation: Print Statements

The use of print statements is a popular and effective means for functional debugging. The difficulty with print statements in embedded systems is that a standard "printer" may not be available. Another problem with printing is that most embedded systems involve time-dependent interactions with their external environment. The print statement itself may be so slow that the debugging process itself causes the system to fail. The print statement is very *intrusive*. Therefore this section will focus on debugging methods that do not rely on the availability of a printer.

2.11.3.5 Instrumentation: Dump into Array without Filtering

One of the difficulties with print statements are that they can significantly slow down the execution speed in real-time systems. Many times the bandwidth of the print functions cannot keep pace with the existing system. For example, our system may wish to call a function 1000 times a second (or every 1 ms). If we add print statements to it that require 50 ms to perform, the presence of the print statements will significantly affect the system operation. In this situation, the print statements would be considered extremely intrusive. Another problem with print statements occurs when the system is using the same output hardware for its normal operation, as is required to perform the print function. In this situation, debugger output and normal system output are intertwined.

To solve both these situations, we can add a debugger instrument that dumps strategic information into an array at run time. We can then observe the contents of the array at a later time. One of the advantages of dumping is that the 6812 BDM module allows you to visualize memory even when the program is running. So this technique will be quite useful in systems with a BDM.

Assume `happy` and `sad` are strategic 8-bit variables. The first step when instrumenting a dump is to define a buffer in RAM to save the debugging measurements. The `cnt` will be used to index into the buffers. `cnt` must be initialized to zero, before the debugging begins. The debugging instrument, shown in Program 2.25, saves the strategic variables into the buffer.

Program 2.25
Instrumentation dump without filtering.

```
; global variables in RAM
size    equ 20
buffer rmb size*2
cnt     rmb 1
; programs in EEPROM
Save pshb
     pshx     ;save
     ldab cnt
     cmpb #size*2 ; full?
     beq  done
     ldx  #buffer
     abx      ; place to put next
     ldab happy
     stab 0,x ; save happy
     ldab sad
     stab 1,x ; save sad
     inc  cnt
     inc  cnt
done pulx
     pulb
     rts
```

```
// global variables in RAM
#define size 20
unsigned char buffer[size][2];
unsigned int cnt=0;

// dump happy and sad
void Save(void){
  if(cnt<size){
    buffer[cnt][0] = happy;
    buffer[cnt][1] = sad;
    cnt++;
  }
}
```

Next, you add `jsr Save` statements at strategic places within the system. You can either use the debugger to display the results or add software that prints the results after the program has run and stopped.

> **Observation:** You should save registers at the beginning and restore them back at the end so that the debugging instrument itself won't cause the software to crash.

2.11.3.6
Instrumentation: Dump into Array with Filtering

One problem with dumps is that they can generate a tremendous amount of information. If you suspect a certain situation is causing the error, you can add a filter to the instrument. A filter is a software/hardware condition that must be true to place data into the array. In this situation, if we suspect the error occurs when the pointer nears the end of the buffer, we could add a filter that saves in the array only when the pointer is above a certain value.

In the example shown in Program 2.26, the instrument saves the strategic variables into the buffer only when `sad` is greater than 100.

2.11.3.7
Monitor Using Fast Displays

Another tool that works well for real-time applications is the monitor. A monitor is an independent output process, somewhat similar to the print statement, but one that executes much faster and thus is much less intrusive. The LCD display can be an effective monitor for small amounts of information. Small LCDs can display up to 16 characters, while the larger ones can hold four lines by 40 characters. The hardware/software interface for such a display will be presented in Chapter 8. You can place one or more LEDs on individual

Program 2.26
Instrumentation dump
with filter.

```
Save pshb
     pshx    ;save
     ldab sad
     cmpb #100
     bls  done ; only when sad >100
     ldab cnt
     cmpb #size*2 ; full?
     beq  done
     ldx  #buffer
     abx      ; place to put next
     ldab happy
     stab 0,x ; save happy
     ldab sad
     stab 1,x ; save sad
     inc  cnt
     inc  cnt
done pulx
     pulb
     rts
```

```
// dump happy and sad
void Save(void){
  if(sad>100){
    if(cnt<size){
      buffer[cnt][0] = happy;
      buffer[cnt][1] = sad;
      cnt++;
    }
  }
}
```

otherwise unused output bits. Software toggles these LEDs to let you know what parts of the program are running. A LED is an example of a Boolean monitor.

Assume an LED is attached to Port B bit 6. Program 2.27 will toggle the LED.

Program 2.27
An LED monitor.

```
Toggle psha
       ldaa PORTB
       eora #$40
       staa PORTB
       pula
       rts
```

```
void Toggle(void){
  PORTB ^= 0x40; // flip LED
}
```

Next, you add `jsr Toggle` statements at strategic places within the system. On the 6812, the DDRB must be initialized so that bit 6 is an output before the debugging begins. You can either observe the LED directly or look at the LED control signals with a high-speed oscilloscope.

> **Observation:** When using LED monitors it is better to modify just the one bit, leaving the other seven as is. In this way, you can implement additional LED monitors on one port.

> **Checkpoint 2.26:** Write a debugging instrument that toggles Port A bit 3.

2.11.4
Performance
Debugging

Performance debugging involves the verification of timing behavior of our system. It is a dynamic process where the system is run, and the dynamic behaviors of the I/Os are compared against the expected results. We will present two methods of performance debugging, then apply the techniques to measure execution speed.

2.11.4.1
Instrumentation
Measuring with an
Independent Counter,
TCNT

There is a 16-bit counter, called TCNT, which is incremented every E clock. There is a prescaler that can be placed between the E clock and the TCNT counter. It automatically rolls over when it gets to $FFFF. If we are sure the execution speed of our function is less than 65,535 counts, we can use this timer to collect timing information with only a modest amount of intrusiveness.

2.11.4.2
Instrumentation Output
Port

Another method to measure real-time execution involves an output port and an oscilloscope. Assume an oscilloscope is attached to Port B bit 6. Program 2.28 can be used to set and clear the bit.

Program 2.28
Instrumentation output
port.

```
;6811                              ;6812
Set psha                           Set bset PORTB,#$40
    ldaa PORTB                         rts
    oraa #$40                      Clr bclr PORTB,#$40
    staa PORTB                         rts
    pula
    rts
Clr psha
    ldaa PORTB
    anda #$BF
    staa PORTB
    pula
    rts
```

Next, you add `jsr Set` and `jsr Clr` statements at strategic places within the system. On the 6812, the DDRB must be initialized so that bit 6 is an output before the debugging begins. You can observe the signal with a high-speed oscilloscope. For example, to measure the execution time of a subroutine called `Calculate`, we stabilize the system by calling it over and over. Using the scope, we can measure the width of the pulse on PB6, which will be the execution time of the subroutine `Calculate`.

```
loop jsr  Set
     jsr  Calculate    ; function under test
     jsr  Clr
     bra  loop
```

**2.11.4.3
Measurement of
Dynamic Efficiency**

There are three ways to measure dynamic efficiency of our software. To illustrate these three methods, we will consider measuring the execution time of the sqrt function presented earlier as Figures 2.32 and 2.33. The first method is to count bus cycles using the assembly listing (Program 2.28). This approach is appropriate only for very short programs and becomes difficult for long programs with many conditional branch instructions. Often this is a very tedious process, but luckily the **TExaS** assembler will look up and keep a running count of the number of cycles. The assembly pseudo-operation `org *` will reset the cycle counter, shown between the parentheses. A portion of the assembly output is presented in Program 2.29. Notice that the total cycle count for a 6812 implementation is 70 cycles. At 4 MHz, 70 cycles is 17.5 μs. Because the loop (between `next` and `bne next`) is executed exactly three times, the actual time will be 140 cycles, or 35 μs. For most programs it is actually very difficult to get an accurate time measurement using this technique.

The second method uses an internal timer called TCNT. Most Freescale microcomputers have this 16-bit internal register that is incremented at the bus frequency. If we are sure the function will complete in a time less than 65,535 bus cycles, then the internal timer can be used to measure execution speed empirically. The assembly language call to the function is modified so that TCNT is read before and after the subroutine call. The elapsed time is the difference. Since the execution speed may be dependent on the input data, it is often wise to measure the execution speed for a wide range of input parameters. There is a slight overhead in the measurement process itself. To be more accurate you could measure this overhead and subtract it from your measurements. Notice that in Program 2.30 the total time including parameter passing is measured.

This same technique can also be used in C language programs (Program 2.31).

The third technique can be used in situations where TCNT is unavailable or where the execution time might be larger than 65,535 counts. In this empirical technique we attach an unused output pin to an oscilloscope or to a logic analyzer. We will set the pin high before

```
$F019                                 org   *   ;reset cycle counter
$F019 35           [ 2](   0)sqrt pshy
$F01A B776         [ 1](   2)     tsy
$F01C 1B9C         [ 2](   3)     leas -4,sp        ;allocate t,oldt,s16
$F01E C7           [ 1](   5)     clrb
$F01F A644         [ 3](   6)     ldaa s8,y
$F021 2723         [ 3](   9)     beq  done
$F023 C610         [ 1](  12)     ldab #16
$F025 12           [ 3](  13)     mul               ;16*s
$F026 6C5C         [ 2](  16)     std  s16,y         ;s16=16*s
$F028 18085F20     [ 4](  18)     movb #32,t,y       ;t=2.0, initial guess
$F02C 18085E03     [ 4](  22)     movb #3,cnt,y
$F030 A65F         [ 3](  26)next ldaa t,y           ;RegA=t
$F032 180E         [ 2](  29)     tab                ;RegB=t
$F034 B705         [ 1](  31)     tfr  a,x           ;RegX=t
$F036 12           [ 3](  32)     mul                ;RegD=t*t
$F037 E35C         [ 3](  35)     addd s16,y          ;RegD=t*t+16*s
$F039 1810         [12](  38)     idiv               ;RegX=(t*t+16*s)/t
$F03B B754         [ 1](  50)     tfr  x,d
$F03D 49           [ 1](  51)     lsrd               ;RegB=((t*t+16*s)/t)/2
$F03E C900         [ 1](  52)     adcb #0
$F040 6B5F         [ 2](  53)     stab t,y
$F042 635E         [ 3](  55)     dec  cnt,y
$F044 26EA         [ 3](  58)     bne  next
$F046 B767         [ 1](  61)done tys
$F048 31           [ 3](  62)     puly
$F049 3D           [ 5](  65)     rts
$F04A 183E         [16](  70)     stop
```

Program 2.29
Assembly listing from TExaS of the sqrt subroutine.

Program 2.30
Empirical measurement of dynamic efficiency in assembly language.

```
before   rmb 2        ; TCNT value before the call
elasped  rmb 2        ; number of cycles required to execute sqrt
     movw TCNT,before
     movb ss,1,-sp ; push parameter on the stack (binary fixed point)
     jsr  sqrt      ; subroutine call to the module "sqrt"
     ins
     stab tt         ; save result
     ldd  TCNT       ; TCNT value after the call
     subd before
     std  elasped    ; execute time in cycles
```

Program 2.31
Empirical measurement of dynamic efficiency in C language.

```
unsigned short before,elasped;
void main(void){
    ss=100;
    before=TCNT;
    tt=sqrt(ss);
    elasped=TCNT-before;
}
```

the call to the function and set the pin low after the function call. In this way a pulse is created on the digital output with a duration equal to the execution time of the function. We assume Port B is available and that bit 7 is connected to the scope. By placing the function call in a loop, the scope can be triggered. With a storage scope or logic analyzer, the function need be called only once. Program 2.32 shows the assembly language measurement. Program 2.33 shows the same technique in C language.

Program 2.32
Another empirical measurement of dynamic efficiency in assembly language.

```
       movb #$FF,DDRB  ; make Port B an output
loop bset PORTB,#$80 ; set PB7 high
       ldaa #100       ; typical input
       jsr  sqrt       ; subroutine call to the module "sqrt"
       bclr PORTB,#$80 ; clear PB7 low
       bra  loop
```

Program 2.33
Another empirical measurement of dynamic efficiency in C language.

```
void main(void){unsigned char ss,tt;
   DDRB = 0xFF;   // PB7 is connected to a scope
   ss = 100;
   while(1){
     PORTB |= 0x80;  // set PB7 high
     tt = sqrt(ss);
     PORTB &= ~0x80; // clear PB7 low
   }
}
```

2.11.5
Profiling

Profiling is similar to performance debugging because both involve dynamic behavior. Profiling is a debugging process that collects the time history of strategic variables. For example, if we could collect the time-dependent behavior of the program counter, then we could see the execute patterns of our software. We can profile the execution of a multiple-thread software system to detect reentrant activity.

2.11.5.1
Profiling Using a Software Dump to Study Execution Pattern

In this section, we will discuss software instruments that study the execution pattern of our software. To collect information concerning execution we will add a debugging instrument that saves the time and location in an array (like a dump). By observing this data we can determine both a time profile (when) and an execution profile (where) of the software execution (Program 2.34).

Program 2.34
A time/ position profile dumping into a data array.

```
unsigned short time[100];
unsigned short place[100];
unsigned short n;
void profile(unsigned short p){
  time[n]=TCNT; // record current time
  place[n]=p;
  n++;
}
unsigned short sqrt(unsigned short s){ unsigned short t,oldt;
profile(0);
  t=0;        // based on the secant method
  if(s>0) {
profile(1);
     t=32;   // initial guess 2.0
     do{
profile(2);
        oldt=t;  // calculation from the last iteration
        t=((t*t+16*s)/t)/2;} // t is closer to the answer
     while(t!=oldt);}    // converges in 4 or 5 iterations
profile(3);
  return t;}
```

Since the debugging instrument is implemented as a function, we could read the return address off the stack (the place from which it was called). This is a good approach when we are adding/subtracting many debugging instruments.

2.11.5.2
Profiling Using an
Output Port

In this section, we will discuss a hardware/software combination to visualize program activity. Our debugging instrument will set output port bits. We will place these instruments at strategic places in the software. If we are using a regular oscilloscope, then we must stabilize the system so that the function is called over and over. We connect the output pins to a scope or logic analyzer and observe the program activity (Program 2.35).

Program 2.35
A time/position profile
using two output bits.

```
unsigned int sqrt(unsigned int s){ unsigned int t,oldt;
PTT=0;
  t=0;         // based on the secant method
  if(s>0) {
PTT=1;
    t=32;      // initial guess 2.0
    do{
PTT=2;
      oldt=t;  // calculation from the last iteration
      t=((t*t+16*s)/t)/2;} // t is closer to the answer
    while(t!=oldt);}    // converges in 4 or 5 iterations
PTT=3;
  return t;}
```

2.11.5.3
Thread Profile

When more than one program (multiple threads) is running, you could use the previous technique to visualize the thread that is currently active (the one running). For each thread, we assign an output pin. The debugging instrument would set the corresponding bit high when the thread starts and clear the bit when the thread stops. We would then connect the output pins to a multiple-channel scope to visualize in real time the thread that is currently running. For an example of this type of profile, run one of the thread.* examples included with the **TExaS** simulator and observe the logic analyzer.

2.12   Exercises

2.1 Write a subroutine, called FUZZY, that performs the I/O function shown in Figure 2.41. In and Th are inputs and Out is the result. All parameters are 8-bit unsigned integers. A typical calling shows that the two inputs, In and Th, are passed on the stack and the return parameter, Out, is returned in RegB.

Figure 2.41
Fuzzy logic membership
function.

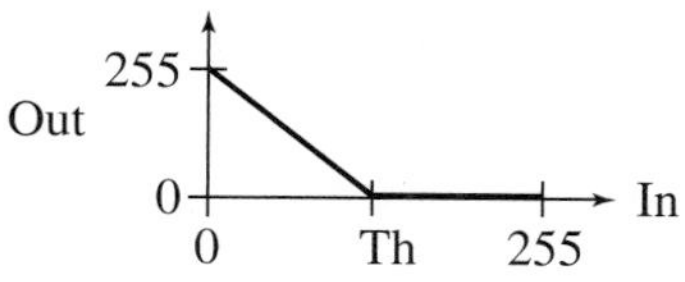

If In $\geq$ Th then Out = 0

If In < Th then Out = $\dfrac{255 \cdot (Th\text{-}In)}{Th}$

```
ldaa #150 value for Th
psha      Th pushed on the stack
ldaa #90  value for In
psha      In on the stack
jsr FUZZY your function
ins
ins       pop off Th and In
* Reg B = (255*(150-90))/150 = 102
```

You **must** use at least one local variable. Comments will be graded.

2.2 Consider the reasons why one chooses which technique to create a variable.
  a) List three reasons why one would implement a variable using a register.
  b) List three reasons why one would implement a variable on the stack and access it using RegX indexed mode addressing.
  c) List three reasons why one would implement a variable in RAM and access it using direct or extended mode addressing.

2.3 Write a subroutine that performs the following C function explicitly. Draw lines in your assembly code that delineate the implementation for each line of C explicitly. The input parameter, `channel`, is passed by value on the stack. You must implement the local variable, `result`, on the stack. The output parameter is returned in RegD. Notice that `Adr1` and `Adr2` are 8-bit unsigned, while `result` is 16-bit unsigned.

```
AtoD(channel) unsigned char channel; { unsigned int result;
    Adctrl=channel;                              /* Start ADC              */
    while ((Adstat& 0x80) == 0){};               /* Wait for ADC to finish */
    result=Adr1;                                 /* Read first ADC result  */
    result=result+Adr2;                          /* Combine two ADC results */
    return(result); }
```

An example calling sequence might be

```
        ldab #5        ; ADC channel 5
        pshb           ; parameter on stack (not in Reg B)
        jsr AtoD
        ins            ; discard input parameter
    * Reg D has result
```

2.4 List three factors that we can use to evaluate the "goodness" of a program.

2.5 Consider the following simple C program. To solve this problem you can read Chapters 6 and 10 of the HTML document in the embed folder on the accompanying CD or simply type it in to your compiler and observe the assembly listing file it creates.

```
int x1;
const int x3=1000;
int add3(int z1, int z2, int z3){ int y1;
    y1=z1+z2+z3;
/* ***** answer the questions when the program is here ***** */
    return(y);}
void main(void){ int y2; static int x2;
    x1=1000;
    x2=1000;
    y2=add3(x1,x2,x3);
```

  a) Where in memory are `x1`, `x2`, `x3`, `y1`, `y2`, `z1`, `z2`, `z3`, `add3`, and `main` allocated? For each object simply specify global RAM, stack RAM, or EEPROM. Answer the questions when the scope of execution is just after the addition in `add3`.
  b) Draw a stack picture at the point just after the addition in `add3`. Show in your picture the three parameters `z1`, `z2`, `z3` and the two local variables `y1`, `y2`. Show where Registers X and SP are pointing.

2.6 Let N,M,P be three 8-bit unsigned locations. Write assembly code to implement P=4*N+M.

2.7 Let N,M,P be three 8-bit unsigned locations. Write assembly code to implement P=(11*N+M)/19.

2.8 Let N,M,P be three 8-bit locations. Write assembly code to implement P=($03|N)&M.

2.9 Let PORTB be an 8-bit output port. Write assembly code to set bit 7 and clear bit 0.

2.10 Let PORTB be an 8-bit output port. Write assembly code to toggle bit 7 and set bit 0.

2.11 Assume RegA contains an ASCII character. Write assembly code that converts any lower-case letters (a–z) to upper case (A–Z). For example, if RegA is initially 'g', convert it to 'G'. Leave all other characters unchanged.

2.12 Assume RegA contains an ASCII character. Write assembly code that converts any upper-case letters (A–Z) to lower case (a–z). For example, if RegA is initially 'G', convert it to 'g'. Leave all other characters unchanged.

2.13 Two instructions were executed on a 6811. The following data was collected using a logic analyzer.

R/W	Address	Data
1	$E100	$49
1	$E101	$A7
1	$E101	$A7
1	$E102	$DF
1	$FFFF	$00
0	$00F0	$9B

a) Assuming the first cycle represents the start of a new instruction, list the instructions that were executed.

b) What is the value of Reg X?

2.14 Figure 2.42 depicts the initial state of the 6811 system.

Figure 2.42
6811 architecture diagram used in Exercise 2.14.

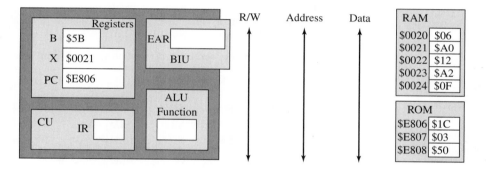

a) Which instruction will be executed next? Show both the opcode and the operand(s).

b) List below all the changes in the 6811 registers except the CCR, and/or 6811 memory that occur due to the execution of this instruction. Show only the changes. That is, if a register or memory is not modified by this instruction, do not include it.

2.15 You will describe in detail the execution of the branch to subroutine instruction.

Initially on the 6811, we have

```
SP = $00FD                $F000 $8D $30      bsr function
PC = $F000
```

Initially on the 6812, we have

```
SP = $0BFD                $F800 $07 $30      bsr function
PC = $F800
```

a) What is the absolute address of `function`? That is, where in memory is function located?

b) What is the return address? That is, where will the computer resume execution after the subroutine `function` finishes?

c) How many cycles does it take to execute?

d) For the 6811, show the R/W, address, and data bus for each cycle. For the 6812, simply list the operations that must occur to execute the instruction.

2.16 You will describe in detail the execution of this one 6811 instruction. Initially

```
B = $5B
X = $0001                 $F000 $E8 02       eorb 2,x
PC = $F000
```

At the end of each of its four cycles, fill in all boxes. In particular, determine the values of the R/W (1 is read and 0 is write), Address, Data, IR, EAR, and ALU operation. If there is no ALU function, write *none*. If a register (B,X,PC) or memory location changes, cross out the old value and put in the new value. Do **not** worry about the condition code register (CC). Answer the question using the starting point drawn in Figure 2.43.

Figure 2.43
6811 architecture diagram used in Exercise 2.16.

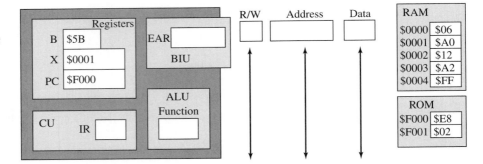

a) Update after first cycle.
b) Update after second cycle.
c) Update after third cycle.
d) Update after fourth and last cycle.

2.17 The original values of some registers/memory locations are specified for the 6811 or 6812 as follows:

C=1	RegA=$79	RegB=$AC	RegX=$2000	RegY=$4000
[$0030]=$A5	[$0031]=$B6	[$3000]=$C7	[$3001]=$D8	[$2000]=$00
[$2001]=$11	[$2002]=$22	[$2003]=$33	[$2004]=$44	[$2005]=$55
[$2006]=$66	[$4000]=$40	[$4001]=$41	[$4002]=$42	[$4003]=$43

a) Specify the addressing mode, the effective address, the opcode, the operand, and the number of cycles it takes to execute the following instructions.
b) Assume each instruction is at memory location 0. Start over with PC=0 each time and hand execute each instruction one at a time. That is, it is not a program that you execute one instruction after another. After each instruction is executed, indicate the registers and/or memory locations that are affected by the instructions. Give the new contents of the relevant registers and/or memory locations.

```
ldy   #$40
ldx   $30
ldy   $2002
ldy   2,X
ldy   2,Y
adca  #$30
adca  $30
```

2.18 Hand execute the following 6811/6812 program. After each instruction is executed, illustrate the condition of the stack and the values of the registers after each of the following instructions. The original condition is RegA=$55, RegB=$66, RegX=$A5B6, and RegY=$A1B2. Assume the stack is initially empty with the RegSP=$00FF (for the 6811) or $4000 (for the 6812).

```
psha
pshx
pshy
pulb
pula
pulx
pula
```

2.19 Choose from the following possibilities the most accurate description of the V bit in the CCR that occurs after an addition or subtraction. Choose from the following possibilities the most accurate description of the C bit in the CCR that occurs after an addition or subtraction. (Recall that the 6805 does not have a V bit.)

> The bit is set on a signed overflow,
> The bit is set on an unsigned overflow,
> The bit is set on a signed underflow,
> The bit is set on an unsigned underflow,
> The bit is set on either a signed overflow or a signed underflow,
> The bit is set on either an unsigned overflow or an unsigned underflow.

2.20 Rewrite the `Timer_Wait` function in Program 2.10 so that it continuously checks an alarm input on PA7. As long as PA7 is low (normal), it will wait the prescribed time. But if PA7 goes high (alarm), the wait function returns.

2.13 Lab Assignments

Lab 2.1. The overall objective is to create a **4-key digital lock**. The system has four digital inputs and one digital output. The LED will be initially on, signifying the door is locked. Define two separate key codes, one to lock and one to unlock the door. For example, if the keys are numbers 1, 2, 3, and 4, one possible key code is 23. This means if you push both the 2 and 3 keys (not pushing the 1, 4 keys) the door will unlock. Implement the design such that the unlock function occurs in the software of the 6811/6812.

Lab 2.2. The overall objective is to create a **line tracking robot**. The system has two digital inputs and two digital outputs. You can simulate the system with two switches and two LEDs, or build a robot with two DC motors and two optical reflectance sensors. Both sensor inputs will be on if the machine is completely on the line. One sensor input will be on and the other off if the machine is just going off the track. If the machine is totally off the line, then both sensor inputs will be off. Implement the controller using a finite state machine. Choose a Moore or Mealy format as appropriate.

Lab 2.3. The overall objective is to create an **enhanced traffic light controller**. The system has three digital inputs and seven digital outputs. You can simulate the system with three switches and seven LEDs. The inputs are North, East, and Walk. The outputs are six for the traffic light and one for a walk signal. Implement the controller using a finite state machine. Choose a Moore or Mealy format as appropriate.

Lab 2.4. The overall objective is to create an **8-key digital lock**. The system has eight digital inputs and one digital output. The LED will be initially off, signifying the door is locked. Define a key sequence to unlock the door. For example, if the keys are numbers 1, 2, . . . and 8, one possible key code is 556. This means if you push the 5, release the 5, push the 5, release the 5 and push the 6, then the door will unlock. The unlock operation will be a two-second pulse on the LED.

Lab 2.5. The overall objective is to design a **vending machine controller**. The system has five digital inputs and three digital outputs. You can simulate the system with five switches and three LEDs. The inputs are `quarter`, `dime`, `nickel`, `soda`, and `diet`. The `quarter` input will go high, then go low when a 25¢ coin is added to the machine. The `dime` and `nickel` inputs work in a similar manner for the 10¢ and 5¢ coins. The sodas cost 35¢ each. The user presses the `soda` button to select a regular soda and the `diet` button to select a diet soda. The `GiveSoda` output will release a regular soda if pulsed high, then low. Similarly, the `GiveDiet` output will release a diet soda if pulsed high, then low. The `Change` output will release a 5¢ coin if pulsed high, then low. Implement the controller using a finite state machine. Choose a Moore or Mealy format as appropriate. Because there are so many inputs and at most one is active at a time, you may wish to implement an FSM with a different format from the examples in the book.

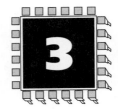

3 Interfacing Methods

Chapter 3 objectives are to:

❑ Introduce basic performance measures for I/O interfacing
❑ Outline various interfacing approaches
❑ Interface simple I/O devices using blind cycle synchronization
❑ Discuss the basic concepts of gadfly synchronization
❑ Describe general approach to I/O interface design
❑ Present the basic hardware/software for parallel port interfaces
❑ Introduce the general concept of a handshake interface, then present many examples
❑ Implement a serial port device driver

One factor that makes an embedded system different from a regular computer is the special I/O devices we attach to our embedded system. While the entire book addresses the design and analysis of embedded systems, this chapter serves as an introduction to the critical task of I/O interfacing. Interfacing includes both the physical connections of the hardware devices and the software routines that affect information exchange. The chapter begins with performance measures to evaluate the effectiveness of our system (latency, bandwidth, priority). As engineers we are not asked simply to design and build devices, but we also are required to evaluate our products. Latency and bandwidth are two quantitative performance parameters we can measure on our real-time embedded system. Next, the basic approaches to I/O interfacing are presented (blind cycle, gadfly, interrupts, periodic polling, and DMA). Although a complete understanding of interrupts and DMA won't come until you complete Chapters 4, 6, 7, and 10, the discussion in this chapter will point to situations that require these more powerful interfacing methods. The rest of the chapter presents simple examples to illustrate the blind cycle and gadfly approaches to interfacing.

3.1 Introduction

3.1.1 Performance Measures

Latency is the time between when the I/O device needs service and when service is initiated. Latency includes hardware delays in the digital gates plus computer hardware delays. Latency also includes software delays. For an input device, software latency (or software response time) is the time between new input data ready and the software reading the data. For an output device, latency is the delay from output device idle and the software giving the device new data to output. In this book, we will also have periodic events. For example, in our data acquisition systems, we wish to invoke the ADC at a fixed time interval. In this way we can collect a sequence of digital values that represent the continuous analog signal. Software latency in this case is the time between when the ADC is supposed to be started and when it is actually started. The microcomputer-based control system also employs periodic software processing. Similar to the data acquisition system, the latency in a control system is the time between when the control software is supposed to be run and when it is actually run. A *real-time* system is one that can guarantee a worst-case latency. In other words, there is an upper bound on the software response time. *Throughput* or *bandwidth* is the maximum data flow (bytes per second) that can be processed by the system. Sometimes the bandwidth is limited by the I/O device, while other times it is limited by computer software. Bandwidth can be reported as an overall average or a short-term maximum. *Priority* determines the order of service when two or more requests are made simultaneously. Priority also determines if a high-priority request should be allowed to suspend a low-priority request that is currently being processed. We may also wish to implement equal priority so that no one device can monopolize the computer. In some computer literature, the term *soft real time* is used to describe a system that supports priority.

3.1.2 Synchronizing the Software with the State of the I/O

One can think of the hardware as being in one of three states. The *idle* state occurs when the device is disabled or inactive. No I/O occurs in the idle state. When active (not idle), the hardware toggles between the *busy* and *done* states. For an input device, a status flag is set when new input data are available (Figure 3.1). The busy-to-done state transition will cause a gadfly loop (polling loop) to complete. Once the software recognizes that the input device has new data, it will read the data and ask the input device to create more data. These hardware state transitions are illustrated in Figures 3.1 and 3.3. It is the *busy-to-done* state transition that signals to the computer that service is required. When the hardware is in the done state, the I/O transaction is complete. Often the simple process of reading the data will clear the flag and request another input. Later in this chapter we will present examples of this type of interface.

Figure 3.1
The input device sets a flag when it has new data.

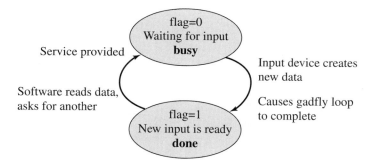

The problem with I/O devices is that they are usually much slower than software execution. Therefore, we need synchronization, which is the process of the hardware and software waiting for each other in a manner such that data are properly transmitted. A way to visualize this synchronization is to draw a state versus time plot of the activities of the hardware and software (Figure 3.2). For an input device, the software begins by waiting for new input. When the input device is busy, it is in the process of creating new input. When the input device is done, new data are available. When the input device makes the transition from busy to done, it releases the software to go forward. In a similar way, when the software accepts the input, it can release the input device hardware. The arrows from one graph to the other represent the synchronizing events. In this example, the time for the software to read and process the data is less than the time for the input device to create new input. This situation is called *I/O bound.* If the input device were faster than the software, a situation called *CPU bound,* then the software waiting time would be zero. From Figure 3.2 we can see that the bandwidth depends on both the hardware and the software.

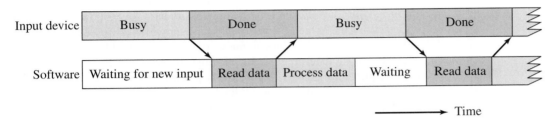

Figure 3.2
The software must wait for the input device to be ready.

This configuration is also labeled as *unbuffered* because the hardware and software must wait for each other during the transmission of each piece of data. A buffered system allows the input device to run continuously, filling a buffer as fast as it can. In the same way, the software can empty the buffer whenever it is ready and whenever data are in the buffer. We will implement a buffered interface in Chapter 4 using interrupts.

For an output device, a status flag is set when the output is idle and ready to accept more data (Figure 3.3). The busy-to-done state transition causes a gadfly loop (polling loop) to complete. Once the software recognizes that the output is idle, it gives the output device another piece of data to output. It will be important to make sure the software clears the flag each time new output is started.

Figure 3.3
The output device sets a flag when it has finished outputting the last data.

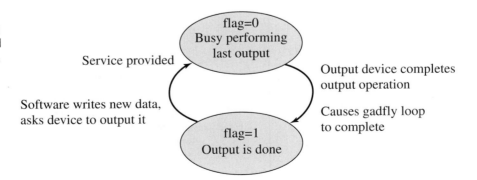

Figure 3.4 contains a state versus time plot of the activities of the output device hardware and software. For an output device, the software begins by generating data, then sending them to the output device. When the output device is busy, it is processing the data. Normally when the software writes data to an output port, that only starts the output process. The time it takes an output device to process data is usually longer than the software

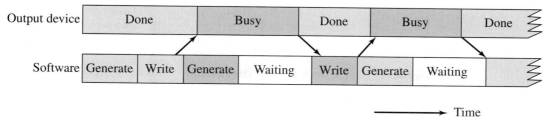

Figure 3.4
The software must wait for the output device to finish the previous operation.

execution time. When the output device is done, it is ready for new data. When the output device makes the transition from busy to done, it releases the software to go forward. In a similar way, when the software writes data to the output, it releases the output device hardware. The output interface illustrated in Figure 3.4 is also I/O bound because the time for the output device to process data is longer than the time for the software to generate and write it.

This output interface is also unbuffered, because when the hardware is done, it will wait for the software, and after the software generates data, it waits for the hardware. A buffered system would allow the software to run continuously, filling a buffer as fast as it wishes. In the same way, the hardware can empty the buffer whenever it is ready and whenever there are data in the buffer. We will implement a buffered interface in Chapter 4 using interrupts.

The purpose of our interface is to allow the microprocessor to interact with its external I/O device. There are five mechanisms to synchronize the microprocessor with the I/O device. Each mechanism synchronizes the I/O data transfer to the busy-to-done transition. The methods are discussed in the following paragraphs.

Blind cycle is a method whereby the software simply waits a fixed amount of time and assumes the I/O will complete after that fixed delay. For an input device, the software triggers (starts) the external input hardware, waits a specified time, then reads data from device. For an output device, the software writes data to the output device, triggers (starts) the device, then waits a specified time. We call this method *blind* because there is no status information about the I/O device reported to the computer software. This method will be used in situations where the I/O speed is short and predictable.

Gadfly or *busy waiting* is a software loop that checks the I/O status waiting for the done state. For an input device, the software waits until the input device has new data, then reads them from the input device. For an output device, the software writes data, triggers the output device, then waits until the device is finished. Another approach to output device interfacing is for the software to wait until the output device has finished the previous output, write data, then trigger the device. We will discuss these two approaches to output device interfacing later in the chapter. Gadfly synchronization will be used in situations where the software system is relatively simple and real-time response is not important.

An *interrupt* uses hardware to cause special software execution. With an input device, the hardware will request an interrupt when input device has new data. The software interrupt service will read the data from the input device and save them in a global structure. With an output device, the hardware will request an interrupt when the output device is idle. The software interrupt service will get data from a global structure, then write them to the device. Sometimes we configure the hardware timer to request interrupts on a periodic basis. The software interrupt service will perform a special function. A data acquisition system needs to read the ADC at a regular rate. Details of data acquisition systems can be found in Chapters 11 and 12. The Freescale microcomputers will execute special software when it tries to execute an illegal instruction. Other computers can be configured to request an interrupt on an access to an illegal address or a divide by zero.[1] Interrupt synchronization will be used in situations where the software system is fairly complicated or when real-time response is important.

Periodic polling uses a clock interrupt to periodically check the I/O status. With an input device, a ready flag is set when the input device has new data. At the next periodic interrupt, the software will read the data and save in global structure. With an output device, a ready flag is set when the output device is idle. At the next periodic interrupt, the software will get data from a global structure and write them. Periodic polling will be used in situations that require interrupts, but the I/O device does not support interrupt requests.

Direct memory access is an interfacing approach that transfers data directly to/from memory. With an input device, the hardware will request a DMA transfer when input device has new data. Without the software's knowledge or permission, the DMA controller will read from the input device and save in memory. With an output device, the hardware will request a DMA transfer when the output device is idle. The DMA controller will get data from memory, then write to the device. Sometimes we configure the hardware timer to request DMA transfers on a periodic basis. DMA can be used to implement a high-speed data acquisition system. Details of DMA can be found in Chapter 10. DMA synchronization will be used in situations where bandwidth and latency are important.

3.1.3 Variety of Available I/O Ports

Microcomputers perform digital I/O using their ports. In this chapter we will focus on the input and output of digital signals. Freescale microcontrollers have a wide variety of configurations, only few of which are illustrated in Table 3.1. Many of the port pins can be used for alternative functions other than parallel I/O. All 6812 microcontrollers include a background debug module. There are nine different 6805 families (B, C, F, J, K, L, P, SR, and SU). The 6808 has eighteen families (AB, AP, AS/AZ, BD, EY, G, GZ, JB/JT, JK/JL, KX, LB, LD, LJ/LK, MR, QL, QB/QT/QY, RF, and SR), the 68S08 has three families (AW, G, R.) The 6811 has nine families (A, D, E, F, G, K, L, M and P), the 6812 has four families (A, B, DG, DT), and the 68S12 as ten families (A, B, C, D, E, GC, H, NE, T, and UF). Basically, we first choose the processor type (CPU05, CPU08, CPU11, or CPU12), depending on our software processing needs. Next, we choose the family, depending on our I/O requirements. Lastly, we choose the particular part, depending on our RAM and ROM memory requirements.

Table 3.1
The number of I/O ports and alternative function.

	Port Pins	Alternative Functions
MC68HC708XL36	54	Serial, timer, signal generation,
MC68HC908JB16	21	Serial, timer, USB
MC68HC908LB8	18	PWM for Half Bridge, timer
MC68HC908RF2	12	Integrated RF Transmitter, timer
MC68HC11D3	32	Serial, timer, address/data bus
MC68HC711E9	40	Serial, timer, ADC, address/data bus
MC68HC812A4	93	Serial, timer, ADC, chip select, address/data bus
MC68HC912B32	64	Serial, timer, ADC, address/data bus, J1850,
MC9S12C32	60	Serial, timer, ADC, CAN
MC9S12H256B	99	Serial, timer, ADC, LCD/H-Bridge, CAN
MC9S12NE64	70	Serial, timer, ADC, 10/100 Base-T
MC9S12UF32	75	Serial, timer, USB 2.0

It is good practice to use the same technology of the microprocessor for the design of the I/O interface. When faced with the problem of designing an I/O interface to our microcomputer, we have the choice of using sophisticated devices made with large-scale integrated circuits [e.g.,

---

[1]The Freescale microcomputers do not provide for a divide-by-zero trap, but most computers do.

metal-oxide semiconductor (MOS), large-scale integration (LSI), very large scale integration (VLSI)] or simple devices made from small-scale integrated devices like standard TTL (e.g., 7400), low-power Schottky TTL (e.g., 74LS00) and high-speed CMOS (e.g., 74HC00).

The interfacing issues (speed, voltage levels, complexity) are matched when both the processor and I/O are designed from similar technologies. To produce a marketable LSI I/O device, one must:

- Increase the market by making the device flexible, able to perform many functions, and use only a fraction of its power
- Increase flexibility by making the device programmable and by decreasing pins while increasing function
- Increase yield by including redundancy, using a modular design, including on-chip diagnostics, making the device a standard size, and decreasing pins

Table 3.2 clearly shows us that LSI technology is appropriate for designing I/O devices.

Table 3.2
Advantages and disadvantages of using LSI technology to design I/O devices.

Advantages	Disadvantages
Shorter design time	Increased software complexity
Increased performance	Need to write a software ritual
Reduced size	Added LSI design costs
Fewer bugs	
Easier to maintain	
Easier to modify	
Increased flexibility	
Lower power	

To be successful each computer family needs a set of high-performance I/O devices. In single-chip microcomputer systems like the 6805, 6808, 6811, and 6812, most of these I/O devices are built-in. Similarly, Cyrix has an integrated microprocessor chip, Gx86, that implements the x86 microprocessor and the associated I/O ports (video controller, sound blaster, peripheral component interconnect (PCI) controller, etc.). We call the Gx86 an integrated microprocessor instead of a single-chip microcomputer because external memory is required to complete the system. Freescale has a line of integrated microprocessors built around the CPU32 (68000-like) processor. Examples include the 68332, 68333, and 68340.

In the early days of microcomputer interfacing before single-chip microcomputers, design engineers would first evaluate the needs of their project. They would select a basic microprocessor (like the 6800, 6809, or 68000) that could handle the software tasks. Then they added external RAM, PROM, and I/O devices to build the microcomputer system. Adding devices is a very expensive and complex process but provides for a wide range of possibilities.

The current trend in the microcomputer industry is customer-specific integrated circuits (CSICs). A similar term for this development process is application-specific integrated circuits (ASICs). With these approaches, the design engineers (customer) first evaluate the needs of their project. In many ways this new development process is similar to the "older way" of design, but now the design engineers work more closely with the microcomputer manufacturer. The design engineers together with the microcomputer manufacturer make a list of features the microcomputer requires. For example:

CPU type	CPU05, CPU08, CPU11, CPU12, CPU16, CPU32[2]
Memory	RAM, EEPROM, EPROM, Flash, ROM
ADC	8 or 10 bits
Timer	PWM, input capture, output compare, etc.
DMA	Number of channels

---

[2]The CPU within the 6805, 6808, 6811, 6812, 6816, 683xx microcontrollers.

Parallel ports	Key wakeup, pull-up, pull-down, reduced drive for low-power applications
Communication	SCI, SPI, CAN, USB, ethernet

The manufacturer then designs and manufactures a microcomputer specific for that project.

3.2 Handshake Protocols

3.2.1 6811 Handshake Protocol

The 6811 can create handshake protocols using pins STRA and STRB, associated with Ports B and C (Table 3.3). The STRA signal is always an input, and STRB is always an output. In this section, we will overview the available features, then we will present examples later in the chapter. Full handshake mode (HNDS=1) uses both STRA and STRB associated with Port C. On the other hand, simple strobe mode (HNDS=0) associates input STRA with Port C and output STRB with Port B. STRA can be configured to be active on either its rising or falling edge, and this edge can either simply set the STAF flag (STAI=0), or set the STAF flag and request an interrupt (STAI=1). If the EGA bit is 0, then a falling edge sets STAF. Conversely, if the EGA bit is 1, then a rising edge sets STAF. An interrupt will be generated if the flag bit (STAF) is set, the arm bit (STAI) is set, and the interrupts are enabled (I=0).

Table 3.3
6811 handshake ports.

Address	Bit 7	6	5	4	3	2	1	Bit 0	Name
$1002	STAF	STAI	CWOM	HNDS	OIN	PLS	EGA	INVB	PIOC
$1003	PC7	PC6	PC5	PC4	PC3	PC2	PC1	PC0	PORTC
$1004	PB7	PB6	PB5	PB4	PB3	PB2	PB1	PB0	PORTB
$1005	PCL7	PCL6	PCL5	PCL4	PCL3	PCL2	PCL1	PCL0	PORTCL
$1007	Bit 7	6	5	4	3	2	1	Bit 0	DDRC

Table 3.4 illustrates the five handshake protocols. The STAF bit is set on the active edge of STRA. For input modes, this edge will latch the Port C pins into the PORTCL register. The STAF flag is cleared by the software when it first reads PIOC with the flag set, followed by a read or write to PORTCL. The INVB bit specifics the active level of STRB output. In generating a pulse out, the STRB is normally in its inactive state; then the software step makes STRB active for two E clock cycles. In generating an interlocked handshake (HNDS=1, PLS=0), the active edge of STRA input makes the STRB output inactive, and the software step makes STRB active. If the CWOM bit is high, then the output bits on Port C are wire-or mode (open collector).

Mode	HNDS	OIN	Active Edge of STRA	Software Step to Clear STAF	PLS	Software Step to Make STRB Active
Simple strobe	0	-	Latch PC7-0 into PORTCL	Read PIOC, read PORTCL	-	Pulse out on write PORTB
Input handshake	1	0	Latch PC7-0 into PORTCL	Read PIOC, read PORTCL	0 1	Interlocked on read PORTCL Pulse out on read PORTCL
Output handshake	1	1	Makes PC7-0 outputs	Read PIOC, write PORTCL	0 1	Interlocked on write PORTCL Pulse out on write PORTCL

Table 3.4
6811 handshake modes.

Checkpoint 3.1: How do we initialize PIOC to make STRB be a positive logic pulse when we write to PORTB?

Checkpoint 3.2: What PIOC bit do we initialize to make a falling edge of STRA latch Port C inputs into the PORTCL register?

Checkpoint 3.3: What two steps do we perform to make Port C open collector outputs?

Common error: We cannot set or clear the STAF bit in the normal way by writing directly to the PIOC register.

3.2.2 MC68HC812A4 Key Wakeup Interrupts

The key wakeup mechanism on the 6812 allows an active edge (rise or fall) on an input signal to set a flag or generate an interrupt. On the MC68HC812A4, key wakeups are available on Ports D, H, and J (see Table 3.5). Any or all of these 24 pins can be configured as a key wakeup. Each of the 24 wakeup lines has a separate I/O pin (PORTD PORTH PORTJ) with a corresponding bit in the direction register bit (DDRD, DDRH, DDRJ). The essence of key wakeup is to make one or more these 24 lines an input, which will set the corresponding flag bit (KWIFD, KWIFH, KWIFJ) on the active edge. For Ports D and H, a falling edge on the input pin will set the corresponding flag bit. Wakeup interrupts on Port J can be configured on either the rising or falling edge. If the corresponding bit in the KPOLJ is 0, then a falling edge on Port J is active. Conversely, if the bit in the KPOLJ register is 1, then a rising edge on Port J will set the flag. Each of the 24 lines has a separate interrupt arm bit (KWIED, KWIEH, KWIEJ). A key wakeup interrupt will be generated if the flag bit is set, the arm bit is set, and the interrupts are enabled (I=0).

Address	Bit 7	6	5	4	3	2	1	Bit 0	Name
$0005	PD7	PD6	PD5	PD4	PD3	PD2	PD1	PD0	PORTD
$0007	Bit 7	6	5	4	3	2	1	Bit 0	DDRD
$0020	Bit 7	6	5	4	3	2	1	Bit 0	KWIED
$0021	Bit 7	6	5	4	3	2	1	Bit 0	KWIFD
$0024	PH7	PH6	PH5	PH4	PH3	PH2	PH1	PH0	PORTH
$0025	Bit 7	6	5	4	3	2	1	Bit 0	DDRH
$0026	Bit 7	6	5	4	3	2	1	Bit 0	KWIEH
$0027	Bit 7	6	5	4	3	2	1	Bit 0	KWIFH
$0028	PJ7	PJ6	PJ5	PJ4	PJ3	PJ2	PJ1	PJ0	PORTJ
$0029	Bit 7	6	5	4	3	2	1	Bit 0	DDRJ
$002A	Bit 7	6	5	4	3	2	1	Bit 0	KWIFJ
$002B	Bit 7	6	5	4	3	2	1	Bit 0	KWIFJ
$002C	Bit 7	6	5	4	3	2	1	Bit 0	KPOLJ
$002D	Bit 7	6	5	4	3	2	1	Bit 0	PUPSJ
$002E	Bit 7	6	5	4	3	2	1	Bit 0	PULEJ

Table 3.5
MC68HC812A4 key wakeup ports.

Another convenience of Port J is the available pull up or pull down resistors via the PUPSJ and PULEJ configuration registers, as illustrated in Table 3.6. Each of the eight pins of port J can be configured separately.

Table 3.6
Pull up modes of Port J.

DDRJ	PUPSJ Bit	PULEJ Bit	Port J Mode
1	0/1	0/1	regular output
0	0/1	0	regular input
0	0	1	input with passive pull down
0	1	1	input with passive pull up

A typical application of pull up is the interface of simple switches. Using pull up mode eliminates the need for an external resistor when interfacing a switch to Port J (see Figure 1.20). Three conditions must be simultaneously true for a key wakeup interrupt to be requested:

- the flag bit is set
- the arm bit is set, and
- the I bit in the 6812 CCR is 0

Even though there are 24 key wakeup lines, there are only three interrupt vectors. So, if two or more wakeup interrupts are used on the same port, it will be necessary to poll. Interrupt polling is the software function to look and see which of the potential sources requested the interrupt. A flag bit is cleared by writing a one to it. For example,

```
KWIFJ = 0x03;  // clears flag bits 1,0
```

Checkpoint 3.4: How do we clear a KWIFH bit 7?

Common error: We cannot set or clear a flag bit in the normal way by writing directly to the KWIFD, KWIFH, KWIFJ registers.

Checkpoint 3.5: Which 6812 pins could we use if we needed to recognize the occurrence of a falling edge on an input signal?

**3.2.3
MC9S12C32 Key
Wakeup
Interrupts**

The MC9S12C32 also has key wakeup interrupts, which are available on Ports J, and P (see Table 3.7). There are two pins on Port J (PJ7, PJ6), and eight pins on Port P (PP7, PP6, . . . PP0). Any or all of these ten pins can be configured as a key wakeup interrupt. Each of the ten wakeup lines has a separate I/O pin (PTJ, PTP), a direction register bit (DDRJ, DDRP), a flag bit (PIFJ, PIFP), an arm bit (PIEJ, PIEP), and a polarity bit (PPSJ, PPSP). The essence of key wakeup is to make one or more of these ten lines an input, which will set the corresponding flag bit on the active edge. Wakeup interrupts can be configured to be active on either the rising or the falling edge. If the corresponding bit in the PPSJ/PPSP is 0, then a falling edge will set the flag. Conversely, if the bit in the PPSJ/PPSP register is 1, then a rising edge will set the flag. A key wakeup interrupt will be generated if the flag bit is set, the arm bit is set, and the interrupts are enabled (I=0). If a pin is configured as an input,

Address	Bit 7	6	5	4	3	2	1	Bit 0	Name
$0268	PJ7	PJ6	-	-	-	-	-	-	PTJ
$0269	PJ7	PJ6	-	-	-	-	-	-	PTIJ
$026A	Bit 7	6	-	-	-	-	-	-	DDRJ
$026B	Bit 7	6	-	-	-	-	-	-	RDRJ
$026C	Bit 7	6	-	-	-	-	-	-	PERJ
$026D	Bit 7	6	-	-	-	-	-	-	PPSJ
$026E	Bit 7	6	-	-	-	-	-	-	PIEJ
$026F	Bit 7	6	-	-	-	-	-	-	PIFJ
$0258	PP7	PP6	PP5	PP4	PP3	PP2	PP1	PP0	PTP
$0259	PP7	PP6	PP5	PP4	PP3	PP2	PP1	PP0	PTIP
$025A	Bit 7	6	5	4	3	2	1	Bit 0	DDRP
$025B	Bit 7	6	5	4	3	2	1	Bit 0	RDRP
$025C	Bit 7	6	5	4	3	2	1	Bit 0	PERP
$025D	Bit 7	6	5	4	3	2	1	Bit 0	PPSP
$025E	Bit 7	6	5	4	3	2	1	Bit 0	PIEP
$025F	Bit 7	6	5	4	3	2	1	Bit 0	PIFP

Table 3.7
MC9S12C32 key wakeup ports.

then reads to PTJ/PTP return the same value as reads to PTIJ/PTIP, which will be the digital value at the input. Conversely, if a pin is configured as an output, then reads to PTJ/PTP return the most recent value written to the output port, while reads to PTIJ/PTIP will return the digital value at the input. The RDRJ/RDRP register determines the drive strength of an output signal. If the bit is 1, then the corresponding output will have 1/3 drive current. This mode is used to reduce supply current to the 6812.

Another convenience of Ports J and P is the available pull up or pull down resistors, as shown in Table 3.8. Each of the ten pins of Ports J and P can be configured separately.

DDRJ/DDRP	PPSJ/PPSP Bit	PERJ/PERP Bit	Port Mode
1	-	-	Regular output
0	-	0	Regular input
0	0	1	Input with passive pull up, rising edge
0	1	1	Input with passive pull down, falling edge

Table 3.8
Pull up/down modes of Ports J and P.

A typical application of pull up is the interface of simple switches. Using pull up mode eliminates the need for an external resistor when interfacing a switch (see Figure 1.20). Three conditions must be simultaneously true for a key wakeup interrupt to be requested:

- the flag bit is set
- the arm bit is set, and
- the I bit in the 6812 CCR is 0

Even though there are ten key wakeup lines, there are only two interrupt vectors, one for Port J and the other for Port P. So, if two or more wakeup interrupts are used on the same port, it will be necessary to poll. Interrupt polling is the software function to look and see which of the potential sources requested the interrupt. The flag bits are cleared by writing a one to it. For example

```
PIFP = 0x05;  // clears flag bits 2,0 of Port P
```

Checkpoint 3.6: How could you use PTIP and PTP registers to detect whether a Port P output signal is broken or overloaded?

Checkpoint 3.7: How do we clear a PIFJ bit 7?

Checkpoint 3.8: What bad thing happens if we use this code to clear bit 7 `PIFP|=0x80;`?

3.3 Blind Cycle Counting Synchronization

Blind cycle counting is appropriate when the I/O delay is fixed and known. This type of synchronization is blind because it provides no feedback from the I/O back to the computer.

3.3.1
Blind Cycle Printer Interface

For example, consider a printer that can print 10 characters every second. With blind cycle counting synchronization, there is no printer status signal from the printer telling the computer when the last character output is complete. A simple software interface would be to output the character, then wait 100 ms for it to finish (Figure 3.5).

Figure 3.5
A simple printer interface

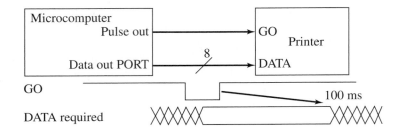

The subroutine that outputs one character follows these steps: (1) The software places the character to be printed on the output part, (2) the software issues a GO pulse (set GO to high, clear GO to low), and (3) the software waits the 100 ms for the character to be printed (Program 3.1).

Program 3.1
A software function that outputs to a simple printer.

```
void Output (unsigned char LETTER) {
    PORT=LETTER;              /* sets Port outputs */
    Pulse();                  /* pulses GO */
    Timer_MsWait(100);        /* Wait for 100 ms */
}
```

The advantage of blind cycle counting is that it is simple and predictable. It does not have the chance of hanging up (i.e., never returning). Unfortunately, there are several disadvantages of the blind cycle counting technique. If the output rate is variable (like a "carriage return," "tab," "graphics," or "formfeed"), then this technique is awkward. If the input rate is unknown (like a keyboard), this technique is inappropriate. The time delay is wasted. If the delay time is long (as it is in the above example), then this technique is dynamically inefficient. This wait time could be used to perform other useful functions. It does not allow for error checking or special conditions.

3.3.2 Blind Cycle ADC Interface

Nevertheless, blind cycle counting can be appropriate for simple high-speed interfaces. An ADC has an analog input (e.g., $0 \leq \text{In} \leq +5\text{V}$) and converts this analog signal into digital form (e.g., $0 \leq \text{data} \leq 255$). Consider the following example of a high-speed 8-bit ADC interface (Figure 3.6). A positive logic pulse, GO, starts the ADC conversion. The result, DATA, is available 5 μs later. There are no error conditions to consider in this problem.

Figure 3.6
A simple A/D interface.

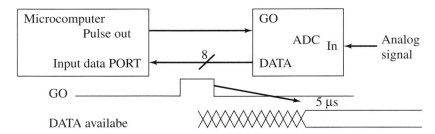

To perform one ADC conversion the subroutine does the following steps (Program 3.2). First, the software starts the ADC conversion by sending a GO pulse (set GO to high, clear GO to low). Second, the software waits for the ADC to convert the analog input signal into digital form (only takes 5 μs). Last, the software inputs the 8-bit result. It is a "blind" interface because there is no ADC status signal telling the software when the conversion is complete.

Program 3.2
A software function that
inputs from an ADC.

```
unsigned char Input(void); {
    Pulse();                      /* pulses GO      */
    Timer_Wait(5);                /* Wait for 5us   */
    return(PORT);                 /* Read ADC result */
}
```

3.4 Gadfly or Busy Waiting Synchronization

To synchronize the software with the I/O device, the microcomputer usually must be able to recognize the busy-to-done transition. With *gadfly* or *busy waiting* synchronization, the software checks a status bit in the I/O device and loops back until the device is ready. The gadfly loop must precede the data transfer for an input device (Figure 3.7).

Figure 3.7
A software flowchart for
gadfly input.

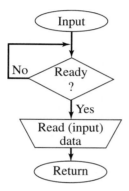

Two steps are involved when the software interfaces with hardware to perform an output function. One step is for the software to output the new data to an output port. This step usually executes in a short amount of time because it involves just a few instructions, with no backward jumps. The other software step is a gadfly loop that executes until the output device is ready. The time in this step is usually long compared to the other operations (I/O-bound situation). These two steps can be performed in either order as long as that order is consistently maintained, and we assume the device is initially ready. Polling before the output allows the computer to perform additional tasks while the output is occurring. Therefore, polling before the output will have a higher bandwidth than polling after the output. On the other hand, polling after the output allows the computer to know exactly when the output has been completed (Figure 3.8).

To illustrate the differences between polling before and after the write-data operation, consider a system with three printers. Each printer can print a character in 1 ms. In other words, a printer will be ready 1 ms after the write-data operation. We will also assume all three printers are initially ready. Since the execution speed of the microcomputer is fast compared to the 1 ms it takes to print a character, we will neglect the software execution time (I/O bound). In the gadfly-before-output system, all three outputs are started together and will operate concurrently. In the gadfly-after-output system, the software waits for the output on printer 1 to finish before starting the output on printer 2. In this system, the three outputs are performed sequentially—that is, about three times slower than the first case (Figure 3.9).

Figure 3.8
Two software flowcharts
for gadfly output.

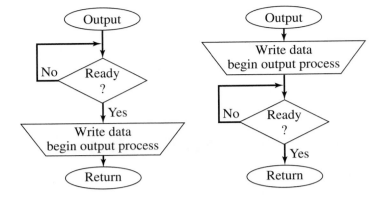

Figure 3.9
Two software flowcharts
for multiple gadfly
outputs.

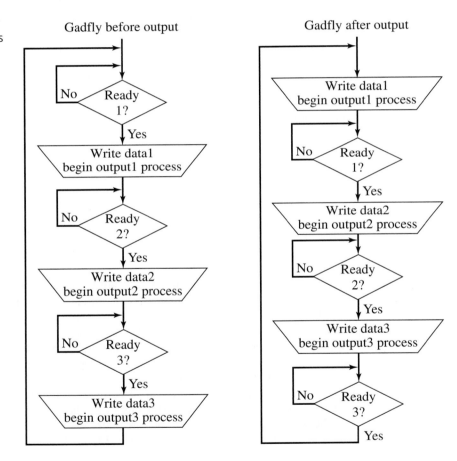

Time(ms)	Gadfly before output	Gadfly after output
0	Start 1,2,3	Start1
From 0 to 1	Wait for 1	Wait for 1
1	Start 1,2,3	Start 2
From 1 to 2	Wait for 1	Wait for 2
2	Start 1,2,3	Start 3
From 2 to 3	Wait for 1	Wait for 3

3	Start 1,2,3	Start1
From 3 to 4	Wait for 1	Wait for 1
4	Start 1,2,3	Start 2
From 4 to 5	Wait for 1	Wait for 2

Performance tip: Whenever we can establish concurrent I/O operations, we can expect an improvement in the overall system bandwidth.

To implement gadfly synchronization with multiple I/O devices, simply poll them in sequence and perform service as required. Figure 3.10 implements a fixed priority and does not allow high-priority devices to suspend the service of lower-priority devices. Therefore the software response time (latency) to high-priority devices will be poor.

Figure 3.10
A software flowchart for multiple gadfly inputs and outputs.

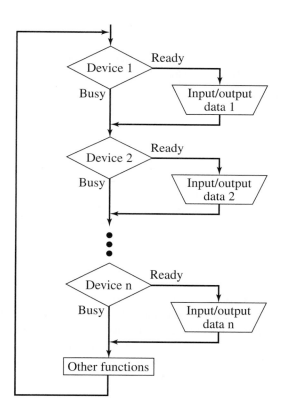

Interrupts, introduced later in the book, provide an efficient mechanism to reduce the software latency of high-priority devices (like power failure or temperature overflow).

3.5 Parallel I/O Interface Examples

A parallel interface encodes the information as separate binary bits on individual port pins, and the information is available simultaneously. For example, if we are transmitting ASCII characters, then the ASCII code is represented by digital voltages on seven (or eight) digital signals. We begin with parallel I/O devices because they are widely used and simple to understand.

The examples in this section can be run on the 6811 and the 6812. The differences involve the port names and port addresses. Some of the 6811 ports are fixed direction, and therefore they provide less flexibility but don't require setting the direction register (e.g., 6811 Port B is a fixed output port). The 6811 also has built-in handshaking hardware that uses STRA and STRB.

3.5.1 Blind Cycle Printer Interface

Consider the printer that can print 100 characters every second. We use the pulse output on GO to start the printer. It is a *blind cycle* interface because no printer status signal reports to the software when the character has been printed. This simple software interface outputs the character, then waits 10 ms for it to finish (Figure 3.11). The timing diagram of Figure 3.12 says the input data must be stable at the time of the rising edge of GO and must remain stable for at least 10 ms. After 10 ms, another output can be performed.

Figure 3.11
Hardware interface between a printer and the microcomputer.

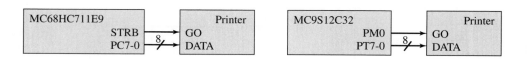

Figure 3.12
Timing diagram for simple parallel printer.

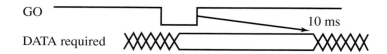

The 6812 uses simple output operations to implement the software interface. The 6811 ritual software sets Port B,C mode to no handshake, and STRB is a negative logic pulse output whenever the software writes to Port B. The subroutine outputs one character by sending a GO (STRB) pulse, then waits for the character to be printed.

The assembly language subroutines pass the data in and out using Register A (Program 3.3), or in C you could execute Program 3.4.

Program 3.3
Assembly language routines to initialize and output to a printer.

```
; MC68HC711E9
; PortB is DATA, STRB=GO
Init clr  PIOC
     rts




;Timer module from Program 2.6
Out  staa PORTB
     ldd  #20000
     bsr  Timer_Wait ;wait 10ms
     rts
```

```
; MC9S12C32
; PortT is DATA, PM0=GO
Init movb #$FF,DDRT  ;outputs
     movb #$01,DDRM
     bset PTM,#$01  ;GO=1
     bsr  Timer_Init ;Program 2.6
     rts
Out  staa PTT        ;set data
     bclr PTM,#$01   ;GO=0
     bset PTM,#$01   ;GO=1
     ldd  #10000     ;10000us=10ms
     bsr  Timer_Wait ;wait 10ms
     rts
```

Program 3.4
C language routines to initialize and output to a printer.

```
// MC68HC711E9
void Init(void){
  PIOC = 0x00;
}



void Out(unsigned char value){
  PORTB = value;
  Timer_Wait(20000);    // 10ms
}
```

```
// MC9S12C32
void Init(void){
  DDRT = 0xFF;    // outputs
  DDRM|= 0x01;
  PTM |= 1;        // GO=1
  Timer_Init();} // Program 2.6
void Out(unsigned char value){
  PTT = value;
  PTM&=~0x01;       // GO=0
  PTM|=0x01;        // GO=1
  Timer_Wait(10000);}  // 10ms
```

**3.5.2
Blind Cycle ADC
Interface**

Consider again the example of the high-speed 8-bit ADC. A positive logic pulse, GO, starts the ADC conversion. The result, DATA, is available 5 μs later. The interface is "blind" because there is no feedback from the ADC back to the software (Figure 3.13).

To perform an ADC conversion, the software issues a GO pulse, waits 5 μs, and reads the result (Figure 3.14).

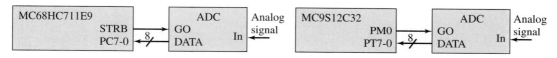

Figure 3.13
Hardware interface between an ADC and the microcomputer.

Figure 3.14
Timing diagram for an ADC interface.

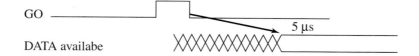

The ritual software sets the DATA port to inputs and the GO pin to an output. On the 6811 the STRB is configured as a positive logic pulse output on write Port B. The subroutine, In, starts the ADC conversion by sending a GO pulse, waits for the ADC to convert the analog input signal into digital form, then inputs the 8-bit result into Register A. (Program 3.5), or in C you could execute Program 3.6. The software delay is implemented using the timer system shown as Program 2.10.

Program 3.5
Assembly language routines to initialize and read from an ADC.

```
; MC68HC711E9
; PortC is DATA, STRB=GO
Init ldaa #$01  ;INVB=1
     staa PIOC
     clr  DDRC
     rts

In   staa PORTB ;GO pulse
     ldd  #20
     bsr  Timer_Wait ;10us
     ldaa PORTC
     rts
```

```
; MC9S12C32
; PortT is DATA, PM0=GO
Init  clr   DDRT      ;data input
      bset  DDRM,#$01
      bclr  PTM,#$01   ;GO=0
      bsr   Timer_Init ;Prog 2.6
      rts

In    bset  PTM,#$01 ;GO=1
      bclr  PTM,#$01 ;GO=0
      ldd   #10
      bsr   Timer_Wait ;10us
      ldaa  PTT
      rts
```

Program 3.6
C language routines to initialize and read from an ADC.

```
// MC68HC711E9
void Init(void){
  PIOC=0x01;   // GO=STRB
  DDRC=0;      // PORTC is data
}



unsigned char In(void){
  PORTB=value;      // GO pulse
  Timer_Wait(20);   // 10us
  return(PORTC);}
```

```
// MC9S12C32
void Init(void){
  DDRT = 0x00;    // input DATA
  DDRM|= 0x01;    // PM0 GO
  PTM &=~0x01;    // GO=0
  Timer_Init();}  // Program 2.6
unsigned char In(void){
  PTM |= 0x01;    // GO=1
  PTM &=~0x01;    // GO=0
  Timer_Wait(10); // 10us
  return(PTT);}
```

3.5.3 Gadfly Keyboard Interface Using Latched Input

The 6811 has true latched inputs. The 6812 implementation differ in that the data read is the value that exists when the read data port instruction is executed. On the 6811, the data read from PORTCL is the value that exists at the time of the rising edge of STROBE. On the 6812, the rise of STROBE triggers the software wait to complete, and the data are read (Figure 3.15).

Figure 3.15
Hardware interface between a simple keyboard and the microcomputer.

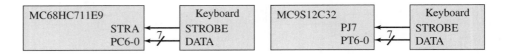

When the user types a key on this keyboard, the 7-bit ASCII code becomes available on the DATA, followed by a rise in the signal STROBE (Figure 3.16). The data remain available until the next key is typed. On the 6811, the simple latched input mode of Port C (HNDS bit in PIOC=0) is used to capture the input data on the rise of STROBE (STRA) and to set the flag STAF. On the 6811, latched input means the edge of a control signal (e.g., rise of STRA) clocks the input data into a register (e.g., Port CL). We use gadfly synchronization that is a software loop to wait for STAF to be set. In this interface STAF is set when new data are available. The two-step sequence, read parallel I/O control (PIOC) register with STAF set followed by read Port CL, will clear the STAF flag (Program 3.7).

Figure 3.16
Timing diagram for simple parallel keyboard.

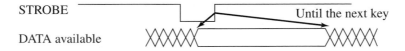

In the 6811, the STRA mode is used to wait for the rise of STROBE. The 6812 implementation uses the key wakeup feature. The PPSJ register determines if the rise (1) or fall (0) edge on the Port J inputs will set the corresponding bit in the PIFJ register. The 6812 Init routine sets PPSJ bit 7 to 1, signifying that the rise of STROBE will set bit 7 in the PIFJ register. The 6812 In routine first waits for PIFJ bit 7 to be set, then clears the PIFJ bit 7 (writing a 1 to this register clears the bit). Because of the gadfly loop, no software delays need to be calculated. Program 3.8 shows the same algorithms implemented in C.

Common error: CMOS inputs have a very large input impedance and a very small input current. If a CMOS input port is left unconnected, then it may begin to toggle, causing a significant power drain.

```
; MC68HC711E9
; PortC=DATA, STRA=STROBE
Init ldaa #$02  ;EGA=1
     staa PIOC
     ldaa #$80  ;PC7=output
     staa DDRC
     ldaa PIOC  ;STAF=0
     ldaa PORTCL
     clr  PORTC ;PC7=0
     rts
In   ldaa PIOC  ; wait for STAF
     bpl  In
     ldaa PORTCL
     rts
```

```
; MC9S12C32
; PT6-0 is DATA, PJ7=STROBE
Init movb #$80,DDRT   ;PT7 unused output
     bclr DDRJ,#$80
     bset PPSJ,#$80   ;rise on PJ7
     movb #$80,PIFJ   ;clear flag7
     clr  PTT         ;PT7=0
     rts

In   brclr PIFJ,#$80,In
     movb #$80,PIFJ ;clear flag7
     ldaa PTT
     rts
```

Program 3.7
Assembly language routines to initialize and read from a keyboard.

Program 3.8
C language routines to initialize and read from a keyboard.

```
// MC68HC711E9
void Init(void){
// PC6-0 is DATA, STRA input
unsigned char dummy;
  PIOC = 0x02;    // EGA=1
  DDRC = 0x80;    // STRA=STROBE
  PORTC = 0x00;   // PC7=0
  dummy = PIOC;   dummy = PORTCL;}
unsigned char In(void){
  while((PIOC & STAF)==0);
  return(PORTCL); }
```

```
// MC9S12C32
void Init(void){ // PJ7=STROBE
  DDRJ = 0x00;    // PT6-0 DATA
  DDRT = 0x80;    // PT7 unused output
  PPSJ = 0x80;    // rise on PJ7
  PIFJ= 0x80;}    // clear flag7
unsigned char In(void){
  while((PIFJ&0x80)==0); // wait
  PIFJ = 0x80;    // clear flag7
  return(PTT);
}
```

Observation: A gadfly loop like the one implemented in Programs 3.7 and 3.8 will hang (crash) the computer if the input device is broken and no STROBE pulse ever comes.

Performance tip: Unused CMOS inputs should be connected to +5 or to ground to prevent power loss.

Performance tip: Unused CMOS port pins could be programmed as outputs to prevent power loss.

Observation: Many pins on the 6812 can be configured with pull-ups to +5 or pull-downs to ground.

3.5.4
Gadfly ADC Interface Using Simple Input

To perform an ADC conversion, the software first issues a GO pulse. For the 6812 implementation, the ADC GO pulse is generated by writing a 1, then a 0 to the port where GO is connected. A gadfly loop waits for the rise of DONE. A simple input of the data (not latched) is used to capture the ADC input data (Figure 3.17).

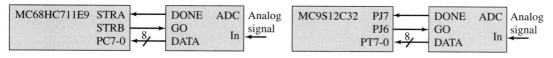

Figure 3.17
Hardware interface between an ADC and the microcomputer.

For the 6811 interface, the ADC GO pulse is generated by writing to Port B (HNDS=0 in the PIOC), and the simple latched input mode of Port C is used to capture the ADC input data on the rise of DONE (STRA) and to set the flag STAF. The rising edge of DONE (STRA) clocks the input data into Port CL. We use gadfly synchronization to wait for STAF to be set, meaning ADC conversion done. The two-step sequence, read PIOC with STAF set followed by read Port CL, will clear the STAF flag.

The timing diagram of Figure 3.18 says that the rise of GO will initialize the ADC, making DONE go low. The fall of GO will start the ADC conversion process. After 25 μs, the 8-bit digital result first becomes available on the DATA lines, followed by the rise of DONE. DONE will remain high, and the DATA will remain valid until the next ADC conversion operation.

Figure 3.18
Timing diagram for an ADC.

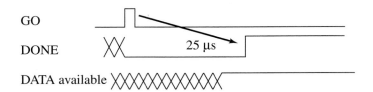

In the 6812 implementation, we explicitly make GO high, then low, and first wait for DONE = 1. In the 6811, the handshake mode automatically performs the interlocked communication. The 6812 implementation uses the key wakeup feature. The PPSJ register determines if the rise (1) or fall (0) edge on the Port J inputs will set the corresponding bit in the PIFJ register. The 6812 Init routine sets PPSJ bit 1 to 1, signifying that the rise of DONE will set bit 1 in the PIFJ register. The 6812 In routine first clears the PIFJ bit 1 (writing a 1 to this register clears the bit), issues the pulse on GO, then waits for PIFJ bit 1 to be set (Program 3.9).

Program 3.9
Assembly language routines to initialize and read from an ADC.

```
; MC68HC711E9                          ; MC9S12C32
; PortC=DATA STRA=DONE STRB=GO         ; PortT=DATA PJ7=DONE PJ6=GO
Init  ldaa #$03   ;EGA=1, INVB=1       Init movb #$40,DDRJ ;
      staa PIOC                        ; PJ7 input,PJ6 output
      ldaa #$00   ;PC=input                 movb #$80,PPSJ ;rise on PJ7
      staa DDRC                              clr   DDRT       ;PT7-0 inputs
      ldaa PIOC   ;STAF=0                    bclr PTJ,#$40   ;GO=0
      ldaa PORTCL                            rts
      rts                              In    movb #$80,PIFJ  ;clear flag7
In    staa PORTB ;GO pulse                  bset PTJ,#$40    ;GO=1
loop  ldaa PIOC  ;wait for STAF             bclr PTJ,#$40    ;GO=0
      bita #$80                        ;wait for rise of DONE
      beq  loop                        loop brclr PIFJ,#$02,loop
      ldaa PORTCL                           ldaa PTT
      rts                                   rts
```

The same algorithms in C are presented in Program 3.10.

Program 3.10
C language routines to initialize and read from an ADC.

```
// MC68HC711E9                          // MC9S12C32
void Init(void){                        void Init(void){ // PJ7=DONE in
// PortC=DATA STRA=DONE STRB=GO           DDRJ = 0x40;    // PJ6=GO out
unsigned char dummy;                      PPSJ = 0x80;    // rise on PJ7
  PIOC = 0x03;    // EGA=1 INVB=1         DDRT = 0x00;    // PT7-0 DATA in
  DDRC = 0x00;    // PC inputs            PTJ &=~0x40;}   // GO=0
  dummy = PIOC;                         unsigned char In(void){
  dummy = PORTCL;} // clear STAF          PIFJ=0x80;      // clear flag7
unsigned char In(void){                   PTJ |= 0x40;    // GO=1
  PORTB = 0;     // GO pulse              PTJ &=~0x40;    // GO=0
  while((PIOC & STAF)==0);                while((PIFJ&0x80)==0);
  return(PORTCL); }                       return(PTT);}
```

3.5.5
Gadfly External Sensor Interface Using Input Handshake

To input an 8-bit sensor reading, the software first waits for the next sensor reading to be ready. The new data available condition is signified by the rise of READY. After that, it sets the ACK low, signifying that the computer is processing the new input. Next, the computer reads the DATA. Then, the software sets the ACK high, signifying that the computer is done processing the current input and is ready to accept another. The 6812 implementation uses the key wakeup feature on Port J to wait for the rising edge of READY (Figure 3.19).

Figure 3.19
Handshaked hardware interface between a sensor and the microcomputer.

On the 6811, with the input handshake mode of Port C, the input data from the sensor is latched into Port CL on the active edge of STRA, the rising edge in this case. This rising edge of READY (STRA) also sets the flag STAF. The 6811 subroutine uses gadfly synchronization to wait for STAF to be set. The two-step sequence, read PIOC with STAF set followed by read Port CL, will clear the STAF flag. With PLS=0 and INVB=1, the rising edge of STRA automatically (without explicit software action) sets ACK (STRB) to 0 (opposite of INVB). The ACK (STRB) output goes to 1 (INVB) when the software reads Port CL. In this mode, the output ACK (STRB) is the complement of STAF flag. The falling edge of ACK is a signal from the 6811 to the sensor signifying that the 6811 has begun to process the current input (e.g., the data have been latched by the hardware but not yet read by the software). The rising edge of ACK signifies that the 6811 has finished processing the current input, and is ready to accept another.

This interface is called *handshaked* or *interlocked* because each event (1, 2, 3, 4, 5, 6) follows in sequence, one event after the other. The arrows in Figure 3.20 represent causal events. In a handshaked interface, one event causes the next to occur, and the arrows create a "head to tail" sequence. There is no specific minimum or maximum time delays for these causal events, except they must occur in sequence.

Figure 3.20
Handshaked timing diagram for a sensor.

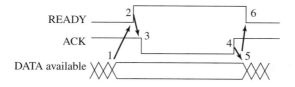

1. New DATA available from sensor
2. Rising edge of READY signifies new DATA available (software waits for this event)
3. Falling edge of ACK signifies computer is starting to process the data; computer reads the DATA
4. Computer makes ACK=1, signifying computer is done
5. Sensor no longer needs to maintain DATA on its outputs
6. Sensor makes READY fall, meaning DATA is not valid

One of the issues involved in handshaked interfaces is whether or not to wait for the falling edge (step 6.) In general, it is good engineering practice to make the system robust and perform a lot of testing. Because the 6811 and 6812 solutions utilize edge-sensitive triggers (STRA and key wakeup), they explicitly wait for both the READY=0 then READY=1 (Program 3.11). If we did not wait for READY=0, then a subsequent call to In might find READY still high from the last call and return the same DATA a second time.

Program 3.11
Handshaking assembly language routines to initialize and read from a sensor.

```
; MC68HC711E9                         ; MC9S12C32
; PortC=DATA STRA=READY STRB=ACK      ; PortT=DATA PJ7=READY PJ6=ACK
Init ldaa #$13   ;Input hndshake      Init movb #$40,DDRJ   ;PJ7 in,
     staa PIOC                        ; PJ7 in, PJ6 out
     ldaa #$00   ;PC=input                movb #$80,PPSJ   ;rise on PJ7
     staa DDRC                             movb #$80,PIFJ   ;clear flag7
     ldaa PIOC   ;STAF=0                   clr  DDRT        ;PT7-0 inputs
     ldaa PORTCL                           bset PTJ,#$40    ;ACK=1
     rts                                   rts
                                      In   brclr PIFJ,#$80,In
In   ldaa PIOC   ;wait for STAF            bclr PTJ,#$40    ;ACK=0
     bita #$80                             ldaa PTT         ;read DATA
     beq  loop                             movb #$80,PIFJ   ;clear flag7
     ldaa PORTCL ;read DATA                bset PTJ,#$40    ;ACK=1
     rts                                   rts
```

When initializing the bits in a control register like the PIOC, it is good programming practice to define the value for each bit and reason why that value was chosen. For example, after the 6811 program in Program 3.11 sets the PIOC to $13, a good comment would be as shown in Program 3.12.

Program 3.12
Example comment detailing what and why a value is used to initialize a control register.

```
; PortC ritual, Set PC7-PC0 inputs = sensor DATA
; PIOC ($1002)
; 7 STAF      Read Only Set on rise of STRA
; 6 STAI 0  Gadfly, no interrupts
; 5 CWOM 0  Normal outputs
; 4 HNDS 1  Input handshake
; 3 OIN  0
; 2 PLS  0  ACK=STRB goes to 0 on rise of READY
; 1 EGA  1  STAF set on rise of READY=STRA
; 0 INVB 1  ACK=STRB goes to 1 on a ReadCL
; STRB=ACK signifies 6811 status 0 means busy,1 means done
; STRA=READY rising edge when new sensor data is ready
```

Program 3.13
Handshaking C language routines to initialize and read from a sensor.

```
// MC68HC711E9
void Init(void){
// PortC=DATA STRA=READY STRB=ACK
unsigned char dummy;
  PIOC = 0x13;    // EGA=1 INVB=1
  DDRC = 0x00;    // PC inputs
  dummy = PIOC;
  dummy = PORTCL;} // clear STAF

unsigned char In(void){
  while((PIOC&STAF)==0);
  return(PORTCL); }
```

```
// MC9S12C32
void Init(void){// PJ7=READY in
  DDRJ = 0x40;   // PJ6=ACK out
  PPSJ = 0x80;   // rise on PJ7
  DDRT = 0x00;   // PT7-0 DATA in
  PIFJ= 0x80;    // clear flag7
  PTJ |= 0x40;}  // ACK=1
unsigned char In(void){
unsigned char data;
  while((PIFJ&0x80)==0);
  PTJ &=~0x40; // ACK=0
  data = PTT;  // read data
  PIFJ=0x80;   // clear flag7
  PTJ |=0x40;  // ACK=1
  return(data);}
```

These same algorithms can be implemented in C as shown in Program 3.13.

Observation: Programs written for embedded computers are tightly coupled (depend highly) on the hardware; therefore, it is good programming practice to document the hardware configuration in the software comments.

Handshaking is a very reliable synchronization method when connecting devices from different manufacturers and at different speeds. It also allows you to upgrade one device (e.g., get a newer and faster sensor) without redesigning both sides of the interface. Handshaking is used for the SCSI and the IEEE488 instrumentation bus.

3.5.6 Gadfly Printer Interface Using Output Handshake

To output a character on this printer, the user first outputs the 7-bit ASCII code to the DATA, followed by a pulse on the signal START. The completion of the output operation is signified by the rise of READY. The 6812 uses the key wakeup feature to wait for the rise of READY, and the 6811 uses the handshake mode to perform many of the functions directly in hardware (Figure 3.21).

Figure 3.21
Handshaking hardware interface between a printer and the microcomputer.

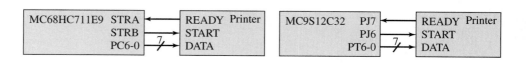

The printer will latch the new DATA on the rising edge of START. The setup time is the time before the edge (100 ns before ↑START, in this case) at which time the DATA must be valid. The hold time is the time after that same edge (20 ns after ↑START, in this case) at which time the DATA must continue to be valid. The printer is finished and ready to accept another character on the rise of READY (Figure 3.22).

Figure 3.22
Handshaked timing diagram for a printer.

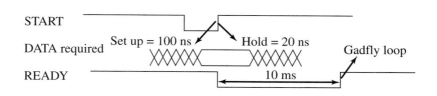

In the 6812 implementation, it is important to follow the sequence (1) set START=0, (2) output new DATA, and (3) set START=1. This sequence guarantees the setup and hold times for the printer. The 6811 solution uses output handshake that automatically satisfies the setup and hold (i.e., the Port C data first becomes valid, then the rising edge of STRB occurs). (Program 3.14). These same algorithms can be implemented in C as shown in Program 3.15).

```
; MC68HC711E9                          ; MC9S12C32
; PC=DATA STRA=READY STRB=START        ; PortT=DATA PJ7=READY PJ6=ACK
; 7 STAF    Set on rise of STRA        Init movb #$40,DDRJ  ;PJ7 in, PJ6 out
; 6 STAI 0  Gadfly, no interrpt             movb #$80,PPSJ  ;rise on PJ7
; 5 CWOM 0  Normal outputs                  movb #$80,PIFJ  ;clear flag7
; 4 HNDS 1  Output handshake                movb #$FF,DDRT  ;PT7-0 outputs
; 3 OIN  1                                  bset PTJ,#$40   ;START=1
; 2 PLS  1  Pulse on Write CL               rts
; 1 EGA  1  STAF set on rise READY     Out  movb #$80,PIFJ  ;clear flag7
; 0 INVB 0  Negative logic pulse            bclr PTJ,#$40   ;START=0
Init  ldaa #$1E  ;Output hndshke            staa PTT        ;write DATA
      staa PIOC                             bset PTJ,#$40   ;ACK=1
      ldaa #$FF  ;PC=output            wait brclr PIFJ,#$80,wait
      staa DDRC                             rts
      rts
Out   staa PORTCL ;out, pulse
wait  ldab PIOC   ;wait for STAF
      bitb #$80
      beq  wait
      rts
```

Program 3.14
Handshaking assembly language routines to initialize and write to a printer.

```
// MC68HC711E9                          // MC9S12C32
void Init(void){                        void Init(void){// PJ7=READY in
// PortC=DATA STRA=READY STRB=START       DDRJ = 0x40;  // PJ6=START out
  PIOC = 0x1E;   // output handshake      PPSJ = 0x80;  // rise on PJ7
  DDRC = 0xFF;   // PC outputs            DDRT = 0xFF;  // PT7-0 DATA out
}                                         PTJ |= 0x40;} // START=1
void Out(unsigned char data){           void Out(unsigned char data){
  PORTCL = data;                          PIFJ= 0x80;   // clear flag
  while((PIOC&STAF)==0);                  PTJ &=~0x40;  // START=0
}                                         PTT  = data;  // write data
                                          PTJ |= 0x40;  // START=1
                                          while((PIFJ&0x80)==0);}
```

Program 3.15
Handshaking C language routines to initialize and write to a printer.

3.5.7 Gadfly Synchronous Serial Interface to a Temperature Sensor

Many external modules use a synchronous serial interface because of its flexible design and low cost. Only four wires are required to connect an external sensor to the microcomputer. We introduce the Dallas Semiconductor DS1620 digital thermometer and thermostat as an example of a bit-banging serial interface. The purpose of presenting a serial interface is to compare and contrast it with the parallel interfacing in the rest of the chapter. The data sheet for this device is located on the accompanying CD (search for DS1620.pdf.) This device

Figure 3.23
Hardware interface between a temperature sensor/controller and the microcomputer.

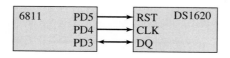 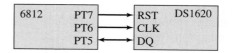

can be used as either a temperature sensor or a thermostat. We will develop another interface later in Chapter 7 that uses the SPI module. There are five types of communication between the computer and the DS1620. In all cases the RST and CLK lines are outputs from the computer and inputs to the DS1620 (Figure 3.23).

The ritual to initialize the DS1620 interface is presented first in Program 3.16. These same algorithms can be implemented in C as shown in Program 3.17.

Program 3.16
Assembly language initialization of the DS1620.

```
; 6811                                    ; 6812
; PD5=RST PD4=CLK PD3=DQ                  ; PT7=RST PT6=CLK PT5=DQ
DS_Init ldaa #$38   ;PD5-3 output         DS_Init ldaa #$E0   ;PT7-5 output
        staa DDRD                                 staa DDRT
        ldaa #$18   ;RST=0,CLK=1                  ldaa #$60   ;RST=0,CLK=1
        staa PORTD  ;DQ=1                          staa PTT    ;DQ=1
        rts                                       rts
```

Program 3.17
C language initialization of the DS1620.

```
// 6811                                   // 6812
void DS_Init(void){ // PD5=RST=0          void DS_Init(void){ // PT7=RST=0
  DDRD = 0x38;      // PD4=CLK=1            DDRT = 0xE0;      // PT6=CLK=1
  PORTD = 0x18;}    // PD3=DQ=1            PTT = 0x60;}       // PT5=DQ=1
```

The DS1620 is capable of sensing the current temperature, with a resolution of 0.5°C, a range of −55 to +125°C, and a conversion time of 1 s. The temperature data is encoded as 9-bit 2's complement signed binary fraction, with a ΔT of 0.5°C. The nine basis elements are −128, 64, 32, 16, 8, 4, 2, 1, and 0.5°C. In particular, some examples of temperature data are presented below in Table 3.9.

Table 3.9
Binary fixed point allows microcomputer to manipulate fractional values without a floating point.

Temperature, °C	Digital Value (Binary)	Digital Value (Hex)
+125.0°	011111010	$0FA
+64.0°	010000000	$080
+1.0°	000000010	$002
+0.5°	000000001	$001
0°	000000000	$000
−0.5°	111111111	$1FF
−16.0°	111100000	$1E0
−55.0°	110010010	$192

The chip also has two internal threshold EEPROM registers, TH and TL. When the temperature is above TH, the T_{HIGH} output goes to +5. When the temperature is below TL, the T_{LOW} output goes to +5. The third output, T_{COM}, implements a thermostat with hysteresis (Figure 3.24).

Figure 3.24
Thermostat control
response with hysteresis.

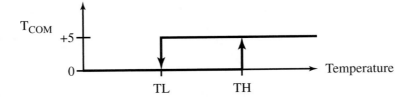

To send an 8-bit command to the DS1620, the microcomputer activates RST and clocks the 8-bit command code so that the data line DQ is stable during the 0 to 1 edge of the CLK. In this transmission, all three signals are outputs of the computer and inputs to the DS1620. The bits c0, c1, c2, c3, c4, c5, c6, and c7 form the 8-bit command. Examples include Start Convert Temperature ($EE) and Stop Convert Temperature ($22). It takes 1 s to convert temperature (Figure 3.25).

Figure 3.25
Synchronous serial
timing diagram for the
DS1620 send command.

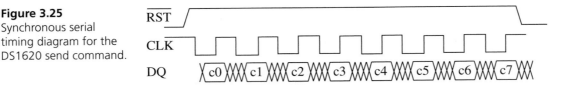

During the transmit phase the software will (1) set the CLK=0, (2) set the DQ output to the desired value, and (3) set the CLK to 1. In this way the DQ data are stable on the rising edge of the CLK. The functions to issue these simple commands are presented next. The command information is passed in RegA (Program 3.18). These same algorithms can be implemented in C as shown in Program 3.19.

Program 3.18
Assembly language
helper functions for the
DS1620.

```
; 6811                              ; 6812
;output 8 bits, Reg A=data          out8   ldab  #8          ;8 bits
out8   ldab  #8          ;8 bits    clop   bclr  PTT,#$40 ;CLK=0
clop   bclr  0,x,#$10 ;CLK=0               lsra              ;lsb first
       lsra              ;lsb first         bcc   set0
       bcc   set0                           bset  PTT,#$20 ;DQ=1
       bset  0,x,#$08 ;DQ=1                 bra   next
       bra   next                   set0   bclr  PTT,#$20 ;DQ=0
set0   bclr  0,x,#$08 ;DQ=0         next   bset  PTT,#$40 ;CLK=1
next   bset  0,x,#$10 ;CLK=1               decb
       decb                                bne   clop
       bne   clop                          rts
       rts                          DS_Start bset PTT,#$80 ;RST=1
DS_Start ldx   #PORTD                       ldaa  #$EE
       bset  0,x,#$20 ;RST=1                bsr   out8
       ldaa  #$EE                           bclr  PTT,#$80 ;RST=0
       bsr   out8                           rts
       bclr  0,x,#$20 ;RST=0        DS_Stop  bset PTT,#$80 ;RST=1
       rts                                  ldaa  #$22
DS_Stop  ldx   #PORTD                       bsr   out8
       bset  0,x,#$20 ;RST=1                bclr  PTT,#$80 ;RST=0
       ldaa  #$22                           rts
       bsr   out8
       bclr  0,x,#$20 ;RST=0
       rts
```

Program 3.19
C language helper functions for the DS1620.

```
// 6811
void out8(char code){ int n;
  for(n=0;n<8;n++){
     PORTD &= 0xEF;     // PD4=CLK=0
     if(code&0x01)
        PORTD |= 0x08;  // PD3=DQ=1
     else
        PORTD &= 0xF7;  // PD3=DQ=0
     PORTD |= 0x10;     // PD4=CLK=1
     code = code>>1;}}
void DS_Start(void){
  PORTD |= 0x20;     // PD5=RST=1
  out8(0xEE);
  PORTD &= 0xDF;}    // PD5=RST=0
void DS_Stop(void){
  PORTD |= 0x20;     // PD5=RST=1
  out8(0x22);
  PORTD &= 0xDF;}    // PD5=RST=0
```

```
// 6812
void out8(char code){ int n;
  for(n=0;n<8;n++){
     PTT &= 0xBF;     // PT6=CLK=0
     if(code&0x01)
        PTT |= 0x20;  // PT5=DQ=1
     else
        PTT &= 0xDF;  // PT5=DQ=0
     PTT |= 0x40;     // PT6=CLK=1
     code = code>>1;}}
void DS_Start(void){
  PTT |= 0x80;     // PT7=RST=1
  out8(0xEE);
  PTT &= 0x7F;}    // PT7=RST=0
void DS_Stop(void){
  PTT |= 0x80;     // PT7=RST=1
  out8(0x22);
  PTT &= 0x7F;}    // PT7=RST=0
```

The second type of communication involves writing the 8-bit configuration register. In this case, bits c0, c1, c2, c3, c4, c5, c6, and c7 form the 8-bit `Write Config` command ($0C), and the bits d0, d1, d2, d3, d4, d5, d6, and d7 are the new data written to the configuration register. For both the command and data transmission, the data line DQ is stable during the 0 to 1 edge of the CLK. Again, all three signals are outputs of the computer and inputs to the DS1620. It takes 5 ms to save a new configuration value in its EEPROM (Figure 3.26).

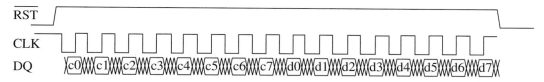

Figure 3.26
Synchronous serial timing diagram for the DS1620 send configuration.

Bits 7, 6, 5 of the configuration/status register are status bits (Table 3.10). The THF flag bit is set if the temperature ever goes above TH and cleared by writing a zero to the bit. Similarly, the TLF flag bit is set if the temperature ever goes below TL and is also cleared by writing a zero to the bit. The status bits are volatile, so the information is lost if the power is removed.

Table 3.10
The DS1620 has three status flags.

Bit	Mode	Configuration/status register meaning
7	DONE	1=Conversion done, 0=conversion in progress
6	THF	1=temperature above TH, 0=temperature below TH
5	TLF	1=temperature below TL, 0=temperature above TL

Bits 1, 0 of the configuration/status register are mode control bits (implemented in nonvolatile EEPROM) (Table 3.11).

Table 3.11
The DS1620 has two
control bits.

Bit	Status	Configuration/status register meaning
1	CPU	1=CPU control, 0=stand alone operation
0	1SHOT	1=one conversion and stop, 0=continuous conversions

The functions to set the configuration register are presented next in Program 3.20. The information is passed in RegA. These same algorithms can be implemented in C as shown in Program 3.21.

Program 3.20
C language functions to
set the configuration
register on the DS1620.

```
// 6811
void DS_Config(char data){
    PORTD |= 0x20;    // PD5=RST=1
    out8(0x0C);
    out8(data);
    PORTD &= 0xDF;}   // PD5=RST=0
```

```
// 6812
void DS_Config(char data){
    PTT |= 0x80;    // PT7=RST=1
    out8(0x0C);
    out8(data);
    PTT &= 0x7F;}   // PT7=RST=0
```

Program 3.21
Assembly language
functions to set the
configuration register on
the DS1620.

```
; 6811
DS_Config ldx #PORTD
          psha
          bset 0,x,#$20 ;RST=1
          ldaa #$0C
          bsr  out8
          pula
          bsr  out8
          bclr 0,x,#$20 ;RST=0
          rts
```

```
; 6812
DS_Config psha
          bset PTT,#$80 ;RST=1
          ldaa #$0C
          bsr  out8
          pula
          bsr  out8
          bclr PTT,#$80 ;RST=0
          rts
```

The third type of communication involves writing a 9-bit temperature threshold register. In this case, bits c0, c1, c2, c3, c4, c5, c6, and c7 form either the `Write TH` command ($01) or the `Write TL` command ($02), and the bits d0, d1, d2, d3, d4, d5, d6, d7, and d8 are the new 9-bit data written to the corresponding threshold register. For this mode too, all three signals are outputs of the computer and inputs to the DS1620. It takes 5 ms to program a new threshold value in its EEPROM (Figure 3.27).

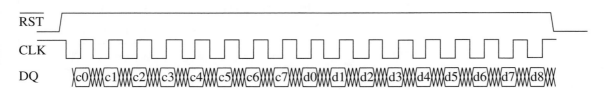

Figure 3.27
Synchronous serial timing diagram for the DS1620 write threshold.

The functions to issue these commands are presented next. The helper function is presented in Program 3.22.

Program 3.22
Assembly language 9-bit output helper function for the DS1620.

```
; 6811
;output 9 bits, Reg D=data
out9    ldx   #9        ;9 bits
olop    bclr  0,x,#$10  ;CLK=0
        lsrd            ;lsb first
        bcc   oset0
        bset  0,x,#$08  ;DQ=1
        bra   onext
oset0   bclr  0,x,#$08  ;DQ=0
onext   bset  0,x,#$10  ;CLK=1
        dex
        bne   olop
        rts
```

```
; 6812
out9    ldx   #9          ;9 bits
olop    bclr  PTT,#$40  ;CLK=0
        lsrd              ;lsb
        bcc   oset0
        bset  PTT,#$20  ;DQ=1
        bra   onext
oset0   bclr  PTT,#$20  ;DQ=0
onext   bset  PTT,#$40  ;CLK=1
        dex
        bne   olop
        rts
```

The routines shown in Program 3.23 access the threshold registers.

Program 3.23
Assembly language functions to set the threshold registers on the DS1620.

```
; 6811
;Reg D is temperature value
DS_WriteTH psha
        pshb
        ldx   #PORTD
        bset  0,x,#$20 ;RST=1
        ldaa  #$01
        bsr   out8
        pulb
        pula
        bsr   out9
        bclr  0,x,#$20 ;RST=0
        rts
;Reg D is temperature value
DS_WriteTL psha
        pshb
        ldx   #PORTD
        bset  0,x,#$20 ;RST=1
        ldaa  #$02
        bsr   out8
        pulb
        pula
        bsr   out9
        bclr  0,x,#$20 ;RST=0
        rts
```

```
; 6812
;Reg D is temperature value
DS_WriteTH
        pshd
        bset  PTT,#$80 ;RST=1
        ldaa  #$01
        bsr   out8
        puld
        bsr   out9
        bclr  PTT,#$80 ;RST=0
        rts
;Reg D is temperature value
DS_WriteTL
        pshd
        bset  PTT,#$80 ;RST=1
        ldaa  #$02
        bsr   out8
        puld
        bsr   out9
        bclr  PTT,#$80 ;RST=0
        rts
```

These same algorithms can be implemented in C as shown in Program 3.24.

The last two types of communication involve first sending an 8-bit command to the DS1620, followed by receiving information back from the sensor. In particular, this fourth type of communication receives an 8-bit status byte. In this case, bits c0, c1, c2, c3, c4, c5, c6, and c7 form the Read Config command ($AC), and the bits d0, d1, d2, d3, d4, d5, d6, and d7 are the current value of the configuration/status register. During the first 8 bits, the

Program 3.24
C language functions to
set the threshold
registers on the DS1620.

```
// 6811                                    // 6812
void out9(short code){ short n;            void out9(short code){ short n;
  for(n=0;n<9;n++){                          for(n=0;n<9;n++){
    PORTD &= 0xEF;    // PD4=CLK=0             PTT &= 0xBF;    // PT6=CLK=0
    if(code&0x01)                             if(code&0x01)
      PORTD |= 0x08; // PD3=DQ=1               PTT |= 0x20; // PT5=DQ=1
    else                                      else
      PORTD &= 0xF7; // PD3=DQ=0               PTT &= 0xDF; // PT5=DQ=0
    PORTD |= 0x10;    // PD4=CLK=1             PTT |= 0x40;    // PT6=CLK=1
    code = code>>1;}}                         code = code>>1;}}
void DS_WriteTH(short data){               void DS_WriteTH(short data){
  PORTD |= 0x20;    // PD5=RST=1             PTT |= 0x80;    // PT7=RST=1
  out8(0x01);                               out8(0x01);
  out9(data);                               out9(data);
  PORTD &= 0xDF;}   // PD5=RST=0            PTT &= 0x7F;}   // PT7=RST=0
void DS_WriteTL(short data){               void DS_WriteTL(short data){
  PORTD |= 0x20;    // PD5=RST=1             PTT |= 0x80;    // PT7=RST=1
  out8(0x02);                               out8(0x02);
  out9(data);                               out9(data);
  PORTD &= 0xDF;}   // PD5=RST=0            PTT &= 0x7F;}   // PT7=RST=0
```

DQ line is output from the computer, while during the last 8 bits it is an input to the computer. Once again the DQ line changes on the 1 to 0 edge of the CLK and is stable during the 0 to 1 edge. We will have to change the direction register halfway through to implement this protocol (Figure 3.28).

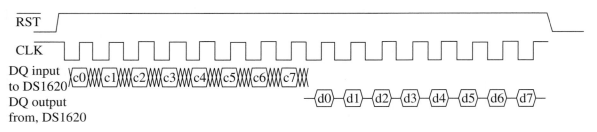

Figure 3.28
Synchronous serial timing diagram for the DS1620 read status.

During the receive phase of the communication, the software will (1) set the CLK=0, (2) read the DQ input, and (3) set the CLK to 1. The functions to issue these commands are presented next. The result is returned in RegA (Program 3.25). These same algorithms can be implemented in C as shown in Program 3.26.

The last type of communication involves reading temperature values from the DS1620. In this case, bits c0, c1, c2, c3, c4, c5, c6, and c7 form the Read TH command ($A1), the Read TL command ($A2), or the Read Temperature command ($AA). The bits d0, d1, d2, d3, d4, d5, d6, d7, and d8 are the current value of the corresponding temperature register (Figure 3.29).

The helper subroutine, in9, receives a 9-bit data frame from the DS1620 (Program 3.25).

The last three assembly subroutines allow you to read 9-bit temperature values from the DS1620 (Program 3.26).

Program 3.25
Assembly language functions to read the configuration register on the DS1620.

```
; 6811
;input 8 bits, Reg A=data
in8     ldy  #8         ;8 bits
        ldaa DDRD
        anda #$F7       ;DQ input
        staa DDRD
ilop    bclr 0,x,#$10 ;CLK=0
        lsra            ;lsb first
        brclr 0,x,#$08,inext
        oraa #$80  ;DQ=1
inext   bset 0,x,#$10  ;CLK=1
        dey
        bne  ilop
        ldab DDRD
        orab #$08       ;DQ output
        stab DDRD
        rts
;Reg A is config value
DS_ReadConfig
        ldx  #PORTD
        bset 0,x,#$20  ;RST=1
        ldaa #$AC
        bsr  out8
        bsr  in8
        bclr 0,x,#$20  ;RST=0
        rts
```

```
; 6812
;input 8 bits, Reg A=data
in8     ldx  #8         ;8 bits
        ldaa DDRD
        anda #$DF       ;DQ input
        staa DDRD
ilop    bclr PTT,#$40 ;CLK=0
        lsra            ;lsb
        brclr PTT,#$20,inext
        oraa #$80       ;DQ=1
inext   bset PTT,#$40  ;CLK=1
        dex
        bne  ilop
        ldab DDRT
        orab #$20       ;DQ output
        stab DDRT
        rts
;Reg A is config value
DS_ReadConfig
        bset PTT,#$80 ;RST=1
        ldaa #$AC
        bsr  out8
        bsr  in8
        bclr PTT,#$80 ;RST=0
        rts
```

Program 3.26
C language functions to read the configuration register on the DS1620.

```
// 6811
unsigned char in8(void){ short n;
unsigned char result;
  DDRD &= 0xF7; // PD3=DQ input
  for(n=0;n<8;n++){
    PORTD &= 0xEF;   // PD4=CLK=0
    result = result>>1;
    if(PORTD&0x08)
      result |= 0x80; // PD3=DQ=1
    PORTD |= 0x10;}  // PD4=CLK=1
  DDRD |= 0x08; // PD3=DQ output
  return result;}
unsigned char DS_ReadConfig(void){
unsigned char value;
  PORTD |= 0x20;   // PD5=RST=1
  out8(0xAC);
  value = in8();
  PORTD &= 0xDF;   // PD5=RST=0
  return value;}
```

```
// 6812
unsigned char in8(void){ short n;
unsigned char result;
  DDRT &= 0xDF; // PT5=DQ input
  for(n=0;n<8;n++){
    PTT &= 0xBF;    // PT6=CLK=0
    result = result>>1;
    if(PTT&0x20)
      result |= 0x80; // PT5=DQ=1
    PTT |= 0x40;}   // PT6=CLK=1
  DDRT |= 0x20; // PT5=DQ output
  return result;}
unsigned char DS_ReadConfig(void){
unsigned char value;
  PTT |= 0x80;    // PT7=RST=1
  out8(0xAC);
  value = in8();
  PTT &= 0x7F;    // PT7=RST=0
  return value;}
```

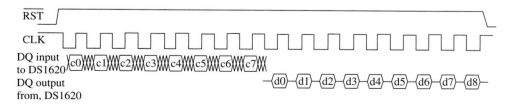

Figure 3.29
Synchronous serial timing diagram for the DS1620 read data.

Program 3.27
Assembly language 9-bit read helper function for the DS1620.

```
; 6811                                    ; 6812
;input 9 bits, Reg D=data                 ;input 9 bits, Reg D=data
in9    ldy  #9        ;9 bits            in9    ldx  #9        ;9 bits
       ldaa DDRD                                ldaa DDRT
       anda #$F7       ;DQ input                anda #$DF      ;DQ input
       staa DDRD                                staa DDRT
       clra                                     clra
jlop   bclr 0,x,#$10 ;CLK=0              jlop   bclr PTT,#$40 ;CLK=0
       lsrd           ;lsb first                lsrd          ;lsb
       brclr 0,x,#$08,jnext                     brclr PTT,#$20,jnext
       oraa #$01 ;DQ=1                          oraa #$01      ;DQ=1
jnext  bset 0,x,#$10 ;CLK=1             jnext  bset PTT,#$40 ;CLK=1
       dey                                      dex
       bne  jlop                                bne  jlop
       psha                                     ldab DDRT
       ldaa DDRD                                orab #$20       ;DQ output
       oraa #$08      ;DQ output                stab DDRT
       staa DDRD                                rts
       pula
       rts
```

Program 3.28
Assembly language functions to read the temperatures from the DS1620.

```
; 6811                                    ; 6812
;Reg D returned as TH value               ;Reg D returned as TH value
DS_ReadTH ldx  #PORTD                     DS_ReadTH bset PTT,#$80 ;RST=1
       bset 0,x,#$20 ;RST=1                     ldaa #$A1
       ldaa #$A1                                bsr  out8
       bsr  out8                                bsr  in9
       bsr  in9                                 bclr PTT,#$80 ;RST=0
       bclr 0,x,#$20 ;RST=0                     rts
       rts                                ;Reg D returned as TL value
;Reg D returned as TL value               DS_ReadTL bset PTT,#$80 ;RST=1
DS_ReadTL ldx  #PORTD                           ldaa #$A2
       bset 0,x,#$20 ;RST=1                     bsr  out8
       ldaa #$A2                                bsr  in9
       bsr  out8                                bclr PTT,#$80 ;RST=0
       bsr  in9                                 rts
       bclr 0,x,#$20 ;RST=0               ;Reg D returned as temperature
       rts                                DS_ReadT bset PTT,#$80 ;RST=1
;Reg D returned as Temperature                  ldaa #$AA
DS_ReadT ldx  #PORTD                             bsr  out8
       bset 0,x,#$20 ;RST=1                     bsr  in9
       ldaa #$AA                                bclr PTT,#$80 ;RST=0
       bsr  out8                                rts
       bsr  in9
       bclr 0,x,#$20 ;RST=0
       rts
```

The helper C function, in9, receives a 9-bit data frame from the DS1620 (Program 3.29).

Program 3.29
C language 9-bit read helper function for the DS1620.

```
// 6811
unsigned short in9(void){ short n;
unsigned short result=0;
  DDRD &= 0xF7; // PD3=DQ input
  for(n=0;n<9;n++){
    PORTD &= 0xEF;   // PD4=CLK=0
    result = result>>1;
    if(PORTD&0x08)
      result|=0x0100; // PD3=DQ=1
    PORTD |= 0x10;}   // PD4=CLK=1
  DDRD |= 0x08; // PD3=DQ output
  return result;}
```

```
// 6812
unsigned short in9(void){ short n;
unsigned short result=0;
  DDRT &= 0xDF;   // PT5=DQ input
  for(n=0;n<9;n++){
    PTT &= 0xBF;    // PT6=CLK=0
    result = result>>1;
    if(PTT&0x20)
      result |= 0x0100; // PT5=DQ=1
    PTT |= 0x40;}   // PT6=CLK=1
  DDRT |= 0x20;     // PT5=DQ output
  return result;}
```

This last interface can be implemented in C as shown in Program 3.30.

Program 3.30
C language functions to read the temperatures from the DS1620.

```
// 6811
unsigned short DS_ReadTH(void){
unsigned short value;
  PORTD |= 0x20;   // PD5=RST=1
  out8(0xA1);
  value = in9();
  PORTD &= 0xDF;   // PD5=RST=0
  return value;}
unsigned short DS_ReadTL(void){
unsigned short value;
  PORTD |= 0x20;   // PD5=RST=1
  out8(0xA2);
  value = in9();
  PORTD &= 0xDF;   // PD5=RST=0
  return value;}
unsigned short DS_ReadT(void){
unsigned short value;
  PORTD |= 0x20;   // PD5=RST=1
  out8(0xAA);
  value = in9();
  PORTD &= 0xDF;   // PD5=RST=0
  return value;}
```

```
// 6812
unsigned short DS_ReadTH(void){
unsigned short value;
  PTT |= 0x80;     // PT7=RST=1
  out8(0xA1);
  value = in9();
  PTT &= 0x7F;     // PT7=RST=0
  return value;}
unsigned short DS_ReadTL(void){
unsigned short value;
  PTT |= 0x80;     // PT7=RST=1
  out8(0xA2);
  value = in9();
  PTT &= 0x7F;     // PT7=RST=0
  return value;}
unsigned short DS_ReadT(void){
unsigned short value;
  PTT |= 0x80;     // PT7=RST=1
  out8(0xAA);
  value = in9();
  PTT &= 0x7F;     // PT7=RST=0
  return value;}
```

3.6 Serial Communications Interface (SCI) Device Driver

In this section we will develop a simple device driver using the Serial Communications Interface (SCI). This serial port allows the microcomputer to communicate with devices such as other computers, printers, input sensors, and LCD displays. Serial transmission involves sending one bit of a time, where the data is spread out over time. The total number of bits transmitted per second is called the *baud rate*. Most of the Freescale embedded microcomputers support at least one *Serial Communications Interface* or SCI. Before discussing the detailed operation of particular devices, we will begin with general features common to all devices. Each SCI module has a *baud rate* control register, which we use to select the transmission rate. There is a mode bit, **M**, which selects 8-bit (M=0) or 9-bit (M=1) data frames. Each device is capable of creating its own serial port clock with a period that is an integer multiple of the E clock period. The programmer will select the baud rate by specifying the integer divide-by used to convert the E clock into the serial port clock. A *frame* is the smallest complete unit of serial transmission. Figure 3.30 plots the signal versus time on a serial port, showing a single frame, which includes a start bit (0), 8 bits of data (least significant bit first), and a stop bit (1). This protocol is used for both transmitting and receiving. The information rate, or *bandwidth*, is defined as the amount of data or useful information transmitted per second. From Figure 3.30, we see that 10 bits are sent for every byte of usual data. Therefore, the bandwidth of the serial channel (in bytes/second) is the baud rate (in bits/sec) divided by 10.

Figure 3.30
A serial data frame with M=0.

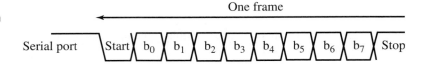

Common error: If you change the E clock frequency without changing the baud rate register, the SCI will operate at an incorrect baud rate.

Checkpoint 3.9: Assuming M=0 and a baud rate of 9600 bits/sec, what is the bandwidth in bytes/sec?

3.6.1 Transmitting in Asynchronous Mode

We will begin with transmission, because it is simpler than reception. The transmitter portion of the SCI includes a TxD data output pin, with TTL voltage levels (see Figure 3.31). The transmitter has a 10- or 11-bit shift register, which cannot be directly accessed by the programmer, and this shift register is separate from the receive shift register. To output data using the SCI, the software will write to the Serial Communications Data Register. On the 6811, the data register is called SCDR; on the 6812, it is called SCIDRL. The transmit data register is write only, which means the software can write to it (to start a new transmission), but cannot read from it. Even though the transmit data register is at the same address as the receive data register, the transmit and receive data registers are two separate registers. When using 9-bit data mode (M=1), we first set the T8 bit, then we write to the transmit data register to start transmission.

Figure 3.31
Data and shift registers implement the serial transmission.

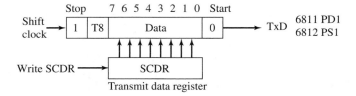

Four control bits affect transmission. We initialize the Transmit Enable control bit, TE, to 1 to enable the transmitter. Interrupt-driven software will be developed in Chapter 7. There are two status bits generated by transmitter activity. The Transmit Data Register Empty flag, TDRE, is set when the transmit SCDR is empty. The TDRE bit is cleared by reading the TDRE flag (with it set), then writing to the SCDR. The Transmit Complete flag, TC, is set when the transmit shift register is done shifting. The TC is cleared by reading the TC flag (with it set), then writing to the SCDR.

When new data (8 bits) is loaded in the SCDR, it is copied into the 10- or 11-bit transmit shift register. Next, the start bit, T8 (if M=1), and stop bits are added. Then the frame is shifted out one bit at a time at a rate specified by the baud rate register. If there is already data in the shift register when the SCDR is written, it will wait until the previous frame is transmitted before it too is transferred. The serial port hardware is actually controlled by a clock that is 16 times faster than the baud rate. The digital hardware in the SCI counts 16 times between changes to the TxD output line.

In essence, the SCDR and transmit shift register behave together like a two-element first-in-first-out queue (FIFO). In other words, the software can actually write two bytes to the SCDR, and the hardware will send them both, one at a time. In fact, the serial port interface chip used in most PC computers has a 16-byte hardware FIFO between the data register and the shift register. A PC that has a 16C550-compatible UART supports this hardware FIFO function. This FIFO reduces the software response time requirements of the operating system to service the serial port hardware.

3.6.2 Receiving in Asynchronous Mode

Receiving data frames is a little trickier than transmission, because we have to synchronize the receive shift register with the incoming data. The receiver portion of the SCI includes and RxD data input pin, with TTL voltage levels (see Figure 3.32). There is also a 10- or 11-bit shift register, which cannot be directly accessed by the programmer. Again, the receive shift register is separate from the transmit shift register. The receiver has a Serial Communications Data Register. Again, this register is called SCDR on the 6811, and SCIDRL on the 6812. The receive data register is read only, which means write operations to this address have no effect on this register. When operating in 9-bit mode (M=1), the ninth data bit is saved in the R8 bit.

Figure 3.32
Data register shift registers implement the receive serial interface.

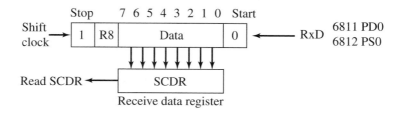

We will set the Receiver Enable control bit, RE, to 1 to enable the receiver. The Receive Data Register Full flag, RDRF, is set when new input data is available. The RDRF bit is cleared by reading the RDRF flag (with it set), then reading the SCDR. The Overrun flag, OR, is set when input data is lost because previous data frames had not been read. The OR bit is cleared by reading the OR flag (with it set), then reading the SCDR. The Noise flag, NF, is set when the input is noisy. The NF bit is cleared by reading the NF flag (with it set), then reading the SCDR. Each bit is sampled three times by the receiver. The NF bit is set when any of the groups of three samples do not all agree. NF errors can occur if there is indeed noise on the line, but more likely it is caused by a mismatch between the transmitter and receiver baud rates. The Framing Error, FE, is set when the stop bit is incorrect. The FE bit is cleared by reading the FE flag (with it set), then reading the SCDR. Framing errors are also probably caused by a mismatch in baud rate.

The receiver waits for the 1 to 0 edge signifying a start bit, then shifts in 10 or 11 bits of data one at a time from the RxD line. The start and stop bits are removed (checked for noise and framing errors), the 8 bits of data are loaded into the SCDR, the 9th data bit is put in R8 (if M=1), and the RDRF flag is set. If there is already data in the SCDR when the shift register is finished, it will wait until the previous frame is read by the software before it is transferred.

> *Observation:* If the receiving SCI device has a baud rate mismatch of more than 5%, then a framing error can occur when the stop bit is incorrectly captured.

An overrun occurs when there is one receive frame in the SCDR, one receive frame in the receive shift register, and a third frame comes into RxD. In order to avoid overrun, we can design a real-time system (i.e., one with a maximum latency). The latency of a SCI receiver is the delay between the time when new data arrives in the receiver SCDR and the time the software reads the SCDR. If the latency is always less than 10 (11 if M=1) bit times, then overrun will never occur.

> *Observation:* With a serial port that has a shift register and one data register (no additional FIFO buffering), the latency requirement of the input interface is the time it takes to transmit one data frame.

In the example illustrated in Figure 3.33, assume the SCI receive shift register and receive data register are initially empty. Three incoming serial frames occur one right after another, but the software does not respond. At the end of the first frame, the $31 goes into the receive SCDR and the RDRF flag is set. In this scenario, the software is busy doing other things and does not respond to the setting of RDRF. Next, the second frame is entered into the receive shift register. At the end of the second frame, there is the $31 in the SCDR and the $32 in the shift register. If the software were to respond at this point, then both characters would be properly received. If the third frame begins before the first is read by the software, then an overrun error occurs and a frame is lost. We can see from this worst-case scenario that the software must read the data from SCDR within 10 bit times of the setting of RDRF.

Figure 3.33
Three receive data frames result in an overrun (OR) error.

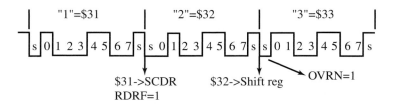

Next we will overview the specific SCI functions on particular Freescale microcomputers. This section is intended to supplement rather than replace the Freescale manuals. When designing systems with a SCI, please also refer to the reference manual of your specific Freescale microcomputer.

3.6.3
6811 SCI Details

Most versions of the 6811 have one asynchronous serial port using Port D bits 1,0. Table 3.12 shows some of the 6811 I/O ports that implement the SCI functions. The full details can be found in the Freescale data sheets. We will assume the 6811 SCI is running in the default 8-bit data mode (M=0).

Address	Bit 7	6	5	4	3	2	1	Bit 0	Name
$102B	TCLR	SCP2	SCP1	SCP0	RCKB	SCR2	SCR1	SCR0	BAUD
$102C	R8	T8	0	M	Wake	0	0	0	SCCR1
$102D	TIE	TCIE	RIE	ILIE	TE	RE	RWU	SBK	SCCR2
$102E	TDRE	TC	RDRF	IDLE	OR	NF	FE	0	SCSR
$102F	R7/T7	R6/T6	R5/T5	R4/T4	R3/T3	R2/T2	R1/T1	R0/T0	SCDR

Table 3.12
6811 SCI ports.

We use the BAUD register to select the baud rate. There are two integer factors that divide the E clock to get the baud rate, shown in Table 3.13. The three bits **SCP2 SCP1** and **SCP0** determine the prescale factor. **SCP2** is not present in some of the original 6811s. The three bits **SCR2, SCR1,** and **SCR0** specify the other factor. The baud rate can be calculated using the equation

$$\text{SCI Baud Rate} = \frac{\text{ECLK}}{(16 \bullet P \bullet BR)}$$

SCP2	SCP1	SCP0	Prescaler, P	SCR2	SCR1	SCR0	Divider, BR
0	0	0	1	0	0	0	1
0	0	1	3	0	0	1	2
0	1	0	4	0	1	0	4
0	1	1	13	0	1	1	8
1	0	0	39	1	0	0	16
				1	0	1	32
				1	1	0	64
				1	1	1	128

Table 3.13
Baud rate selection bits for the 6811.

Checkpoint 3.10: What is the baud rate on a 2 MHz 6811 if BAUD equals 0?

Checkpoint 3.11: What value should BAUD be for a baud rate of 9600 bits/sec on a 2 MHz 6811?

The **SCCR2** control register contains the bits that turn on the SCI and the interrupt arm bits. Interrupts will be discussed in Chapter 7. **TE** is the Transmitter Enable bit. You turn this bit off to disable the transmitter and set this bit to enable the transmitter. **RE** is the Receiver Enable bit. This bit should be set to one to activate the receiver. If RE is 0, then the receiver is disabled.

The flags in the **SCSR** register can be read by the software, but cannot be modified by writing to this register. **TDRE** is the Transmit Data Register Empty Flag. TDRE is set if transmit data can be written to SCDR. If TDRE is zero, the transmit data register contains previous data that has not yet been moved to the transmit shift register. TDRE is cleared by

SCSR read with TDRE set followed by SCDR write. **RDRF** is the Receive Data Register Full Flag. It is set if a received character is ready to be read from SCDR. RDRF is cleared by an SCSR read with RDRF set, followed by a SCDR read.

3.6.4
6812 SCI Details

The MC68HC812A4 has two asynchronous serial ports using Port S bits 3,2 and bits 1,0. On the other hand, most 6812 versions have only one serial port using Port S bits 1,0. Except for the specific addresses listed in Table 3.14, the MC9S12C32 SCI details in the section will apply to most 6812 chips. Each serial port has a 16-bit baud rate register. The full details can be found in the Freescale data sheet. We will assume the 6812 SCI is running in the default 8-bit data mode (SCICR1=0).

Address	msb														lsb	Name
$00C8	–	–	–	12	11	10	9	8	7	6	5	4	3	2	1	0	SCIBD

Address	Bit 7	6	5	4	3	2	1	Bit 0	Name
$00CA	LOOPS	SWAI	RSRC	M	WAKE	ILT	PE	PT	SCICR1
$00CB	TIE	TCIE	RIE	ILIE	TE	RE	RWU	SBK	SCICR2
$00CC	TDRE	TC	RDRF	IDLE	OR	NF	FE	PF	SCISR1
$00CF	R7T7	R6T6	R5T5	R4T4	R3T3	R2T2	R1T1	R0T0	SCIDRL

Table 3.14
MC9S12C32 SCI ports.

The **SCIBD** register is 16 bits, occupying two bytes. The least significant 13 bits of SCIBD determine the baud rate for the SCI port. If BR is the value written to bits 12:0 and MCLK is the module clock (typically this is the same as the E clock), then the baud rate is

$$\text{SCI Baud Rate} = \frac{\text{MCLK}}{(16 \cdot \text{BR})}$$

Checkpoint 3.12: Assume MCLK is 4 MHz. What is the baud rate if SCIBD equals 13?

Checkpoint 3.13: Assume MCLK is 25 MHz. What value should SCIBD be for a baud rate of 38400 bits/sec?

The **SCICR2** control register contains the bits that turn on the SCI and the interrupt arm bits. Interrupts will be covered in Chapter 7. **TE** is the Transmitter Enable bit, and **RE** is the Receiver Enable bit. We set both TE and RE equal to 1 in order to activate the SCI device.

The flags in the **SCISR1** register can be read by the software, but cannot be modified by writing to this register. **TDRE** is the Transmit Data Register Empty Flag. It is set by the SCI hardware if transmit data can be written to SCIDRL. If TDRE is zero, the transmit data register contains previous data that has not yet been moved to the transmit shift register. Writing into the SCIDRL when TDRE is zero will result in a loss of data. On the other hand, when this bit is set, the software can begin another output transmission by writing to SCIDRL. This flag is cleared by first reading SCISR1 with TDRE set, followed by a SCIDRL write. **RDRF** is the Receive Data Register Full bit. RDRF is set if a received character is ready to be read from SCDR. We clear the RDRF flag by reading SCISR1 with RDRF set followed by reading SCIDRL.

The **SCIDRL** register contains the data transmitted out and received in by the SCI device. Even though there are separate transmit and receive data registers, these two registers exist at the same I/O port address. Reads to SCIDRL access the eight bits of the read-only SCI receive data register. Writes to SCIDRL access the eight bits of the write-only SCI transmit data register.

3.6.5
SCI Device Driver

Software that sends and receives data must implement a mechanism to synchronize the software with the hardware. In particular, the software should read data from the input device only when data is indeed ready. Similarly, software should write data to an output device only when the device is read to accept new data. *Busy-waiting*, or *gadfly* are two equivalent names for the same synchronization method. In this scheme, the software continuously checks the hardware status waiting for it to be ready. In this section, we will use busy-waiting to write I/O programs that send and receive data using the SCI port. When a new 8-bit character is received into the serial port, RDRF is set and the 8-bit data is available in the serial data register, SCDR. To get new data from the serial port, the software first waits for the flag to be set, then reads the result. Recall that when the software reads SCDR (SCIDRL on the 6812), it accesses the receive data register. This operation is illustrated in Figure 3.34 and shown in Programs 3.31 and 3.32. In a similar fashion, when the software wishes to output via the serial port, it first waits for TDRE to be set, then performs the output. When the software writes SCDR (SCIDRL on the 6812), it sets the transmit data register. An interrupt synchronization method will be presented in Chapter 7.

Figure 3.34
Flowcharts of InChar and OutChar using busy-waiting.

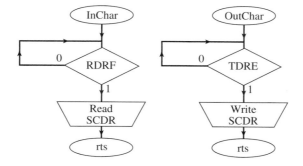

The initialization program, `SCI_Init`, enables the SCI device and selects the baud rate. The input routine waits in a loop until RDRF (receive data register full) is set, then reads the data register, SCDR. The function returns the new ASCII character by value in RegA. The output routine first waits in a loop until TDRE (transmit data register empty) is set, then writes data to the data register, SCDR. The function accepts the new ASCII character as a call by value in RegA. As we saw previously in Section 3.3, this is an efficient way to implement an output gadfly loop.

Checkpoint 3.14: Rewrite the SCI_OutChar, removing one instruction.

Checkpoint 3.15: How does the software clear RDRF?

Checkpoint 3.16: How does the software clear TDRE?

The **TExaS** simulator has example serial port functions for the 6811 and 6812 written in both assembly and C. Once **TExaS** is installed, the busy-waiting SCI functions can be found in the TUT2.* files. The TUT4.* files implement similar serial port operations using interrupt synchronization.

Program 3.31
Assembly functions that
implement serial I/O.

```
; Initalize 6811 SCI
; Inputs: none    Outputs: none
SCI_Init
      ldaa #$0C
      staa SCCR2  ;enable SCI
      ldaa #$32
      staa BAUD    ;9600 bps
      rts
; Read 8 bits from serial port
; Return 8-bit byte in RegA
RDRF equ  $20
SCI_InChar
      ldaa SCSR   ;serial status
      anda #RDRF ;available?
      beq  SCI_InChar ;wait RDRF
      ldaa SCDR   ;read ASCII
      rts
; Write 8 bits to serial port
; Input 8-bit byte in RegA
TDRE equ  $80
SCI_OutChar
      ldab SCSR    ;serial status
      andb #TDRE  ;output ready?
      beq  SCI_OutChar ;wait TDRE
      staa SCDR    ;write ASCII
      rts
```

```
; Initalize 6812 SCI
; Inputs: none    Outputs: none
SCI_Init
      movb #$0c,SCICR2 ;enable SCI
      movw #13,SCIBD   ;19200 bps
      rts


; Read 8 bits from serial port
; Return 8-bit byte in RegA
RDRF equ  $20
SCI_InChar
      ldaa SCISR1 ; serial status
      anda #RDRF  ; available?
      beq  SCI_InChar ; wait RDRF
      ldaa SCIDRL ; read ASCII
      rts
; Write 8 bits to serial port
; Input 8-bit byte in RegA
TDRE equ  $80
SCI_OutChar
      ldab SCISR1  ; serial status
      andb #TDRE    ; output ready?
      beq  SCI_OutChar ; wait TDRE
      staa SCIDRL  ; write ASCII
      rts
```

Program 3.32
C functions that
implement serial I/O.

Not right should be 13b) for Dragon12 board

```
// 6811 initialize SCI
void SCI_Init(void){
  BAUD = 0x32;  // 9600 bits/sec
  SCCR2= 0x0C; // enable
}

#define RDRF 0x20
// Wait for new input,
// then return ASCII code
char SCI_InChar(void){
  while((SCSR&RDRF) == 0){};
  return(SCDR);
}
#define TDRE 0x80
// Wait for buffer to be empty,
// then output
void SCI_OutChar(char data){
  while((SCSR&TDRE) == 0){};
  SCDR = data;
}
```

```
// 6812 initialize SCI
void SCI_Init(void){
  SCIBD  = 13;  // 19200 bits/sec
  SCICR2 = 0x0C; // enable
}

#define RDRF 0x20
// Wait for new input,
// then return ASCII code
char SCI_InChar(void){
  while((SCISR1&RDRF) == 0){};
  return(SCIDRL);
}
#define TDRE 0x80
// Wait for buffer to be empty,
// then output
void SCI_OutChar(char data){
  while((SCISR1&TDRE) == 0){};
  SCIDRL = data;
}
```

Checkpoint 3.17: Describe what happens if the receiving computer is operating on a baud rate that is twice as fast as the transmitting computer?

Checkpoint 3.18: Describe what happens if the transmitting computer is operating on a baud rate that is twice as fast as the receiving computer?

3.7 Exercises

3.1 a) In this problem you will write a function that outputs data to the printer in Figure 3.35 using a gadfly handshake protocol. You may write the software in assembly or C. The following sequence will print one ASCII character:

Figure 3.35
Simple printer interface.

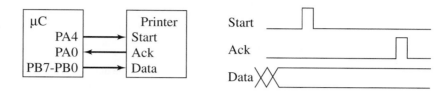

1. The microcomputer puts the 8-bit ASCII on the Data lines
2. The microcomputer issues a Start pulse (does not matter how wide)
3. The microcomputer waits for the Ack pulse (printer is done)

You do not have to save/restore registers. You may assume the Ack pulse is larger than 10 μs. The 8-bit ASCII data to print is *passed by value on the stack*. An example calling sequence is

```
ldab  #V
pshb          The ASCII  V  is pushed on the stack
jsr   Output
ins           The  V  is discarded
```

Full credit will be given to the solution that includes the appropriate use of the bset bclr brset and brclr instructions. If you write in C, pass by value with the following prototype:

```
void Output(unsigned char);
```

b) How long is your Start pulse? Explain your calculation.

3.2 a) What is a trap?
b) Why do we use traps? Give an example.

3.3 What are five different methods for I/O synchronization?

3.4 What does DMA stand for?

3.5 What is the biggest disadvantage of blind cycle counting?

3.6 What is "gadfly"?

3.7 What is the difference between regular interrupts and periodic polling?

3.8 What are the interrupt-related bits in CCR?

3.9 Mention three instructions that can be used to disable or enable interrupts.

3.10 Can you explain what is meant by "saving program context"? Show the stack contents after context is saved, if an interrupt is recognized while executing a LSLA instruction located at address $e010. The contents of the processor registers are: X=$ab12, Y=$cd34, A=$55, B=$66, CCR=$04.

3.11 In all your programs, you included some assembler directives depositing the program start address to locations $FFFE & $FFFF. For example:

```
org $FFFE
fdb Start
```

where Start is the starting address of your program. Explain the purpose of doing this.

Figure 3.36
An input device
interfaced to a 6811.

3.12 The objective of this problem is to interface an input device to a 6811 (Figure 3.36) and implement a binary tree command interpreter. To solve this problem with a different microcomputer, you could substitute any available I/O port for STRA STRB PC7-PC0. You may write the software in assembly or C. The sequence to input an ASCII character from the input device is as follows. When a new character is available, the input device puts its ASCII code on the 8-bit Data, then the input device makes Ready=1. Next your software should read the 8-bit ASCII code, then acknowledge receipt of the data by pulsing Acknowledge. After the pulse, the input device will make Ready=0 again (Figure 3.37).

Figure 3.37
Timing diagram of the
interface between an
input device and a 6811.

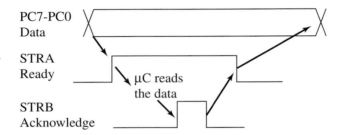

a) Show bit by bit your choice (what and why) for the parallel I/O control register.
b) Show the ritual that initializes the microcomputer.
c) Show the subroutine that inputs one data byte. The sequence of steps is described above. The input data are returned by value using RegB. Use gadfly synchronization. If you write in C, implement call by reference with the following prototype.

```
unsigned char Input(void);
```

To improve search time, your command interpreter will use a binary tree (Figure 3.38) instead of a table or single-linked list. The six subroutines NEXT, FORWARD, BACK, LEFT, RIGHT, STOP are given (you do not write these six subroutines.) Each node of the tree has one ASCII character (e.g., 'B', 'L', . . .), a 16-bit subroutine address (e.g., BACK, LEFT, . . .), and two pointers to its children. Notice that the tree is defined in alphabetical order. The command interpreter will input one character from the input device, search the tree, and if a match is found, the interpreter will execute the appropriate subroutine.

Figure 3.38
A linked binary tree data
structure.

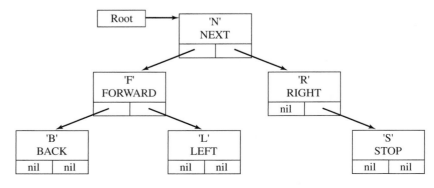

d) Show the pseudo-operations (e.g., FCC, FCB, FDB) that define the above binary tree. You may define a 'nil' pointer as any address that can not be a valid node pointer (e.g., 0, $1000, $FFFF)

e) Show the command interpreter that initializes Port C (calls the subroutine you wrote in part b, then repeats over and over the following sequence:

1. It inputs one ASCII character (calls the subroutine you wrote in part c)
2. It looks up that letter in the binary tree
3. If found then it executes the appropriate subroutine.

In this first example assume the input letter is 'L'. Your program initializes a pointer with the ROOT (it points to node N.) Since 'L'<'N', your program updates the pointer in the direction of nodes that are alphabetically before 'N' (now it points to node F). Since 'L'>'F', your program updates the pointer in the direction of nodes that are alphabetically after 'F' (now it points to node L). Since 'L'='L', your program executes the subroutine LEFT.

In this second example assume the input letter is 'P'. Your program initializes a pointer with the ROOT (it points to node N). Since 'P'>'N', your program updates the pointer in the direction of nodes that are alphabetically after 'N' (now it points to node R). Since 'P'<'R', your program updates the pointer in the direction of nodes that are alphabetically before 'R' (now it points to nil). Since the pointer is nil, there is no match.

3.13 The objective of this problem is to interface an input device to a 6811 single-chip computer using Port C (Figure 3.39). To solve this problem with a different microcomputer, you could substitute any available I/O port for STRA STRB PC7-PC0. You may write the software in assembly or C.

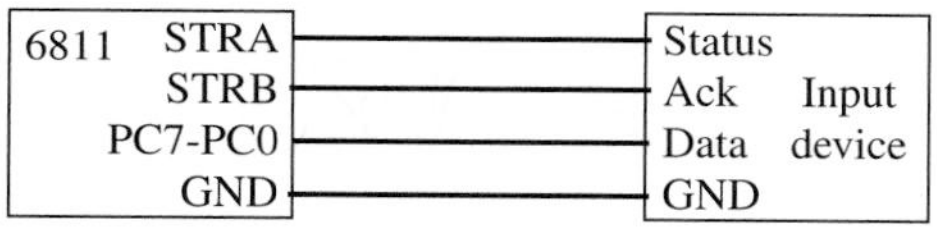

Figure 3.39
An input device interfaced to a 6811.

```
6811  STRA  ──────────  Status
      STRB  ──────────  Ack    Input
   PC7-PC0  ──────────  Data   device
      GND   ──────────  GND
```

The timing shown in Figure 3.40 occurs when 1 byte is transferred.
a) Show bit by bit your choice (what and why) for the parallel I/O control register.
b) Show the ritual that initializes the microcomputer.
c) Show the subroutine that inputs one data byte. Use gadfly synchronization. If you write in assembly return the new data by reference using RegY. RegY points to the place to store the next data byte. If you write in C, implement call by reference with the following prototype:

```
void Input(unsigned char *);
```

Figure 3.40
Timing diagram of the interface between an input device and a 6811.

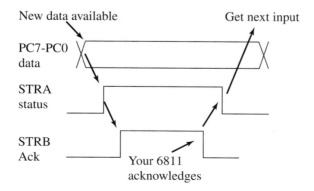

3.14 The objective of this problem is to interface an output device to a 6811 single-chip computer using Port C (Figure 3.41). To solve this problem with a different microcomputer, you could substitute any available I/O port for STRA STRB PC7-PC0. You may write the software in assembly or C.

Figure 3.41
An output device
interfaced to a 6811.

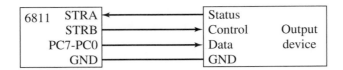

First, the microcomputer outputs data on PC7-PC0. Then, the microcomputer makes STRB=1; 1 ms later the output device will make STRA=1 (the microcomputer waits for STRA=1). Then, the microcomputer makes STRB=0; 1 μs later the output device will make STRA=0 (the microcomputer need not check for STRA=0) (Figure 3.42).

a) Show bit by bit your choice (what and why) for the parallel I/O control register.

b) Show the ritual that initializes the microcomputer. Use assembly or C.

c) Show the subroutine that outputs one data byte. The sequence of steps is described above. If you write in assembly, the output data are passed by value using Reg B. Use gadfly synchronization. If you write in C, implement call by value with the following prototype:

```
void OUTPUT(unsigned char);
```

Figure 3.42
Timing diagram of the
interface between an
output device and a
6811.

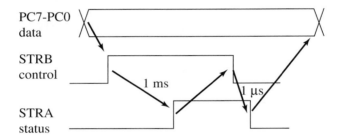

3.15 The objective of this problem is to interface a position transducer array to a microcomputer (Figure 3.43). You may use any available I/O port. Use gadfly synchronization. You may write the software in assembly or C. The sequence to read a 5-bit sensor position from the input device is as follows:

The microcomputer specifies which of eight sensors is to be read (sets N2,N1,N0)
The microcomputer tells the sensor array to read the position by a negative logic pulse on
    START
The sensor signals it is working by setting BUSY=1
After about 1 ms the 5-bit position is available on the lines P4, P3, P2, P1, P0
The sensor signals it is done by setting BUSY=0
The microcomputer reads the 5-bit position

Figure 3.43
Timing diagram for a
position sensor.

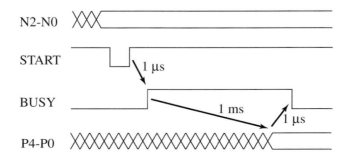

a) Show the connections between the position transducer array and the microcomputer. Label chip numbers but not pin numbers of any required digital logic (Figure 3.44).

b) Show bit by bit your choice (what and why) for the parallel I/O control register. Specify 0 for bits that must be 0. Specify 1 for bits that must be 1. Specify X for bits that can be set in the ritual to be either 0 or 1. The following program illustrates the calling sequence of the RITUAL and SENSOR functions:

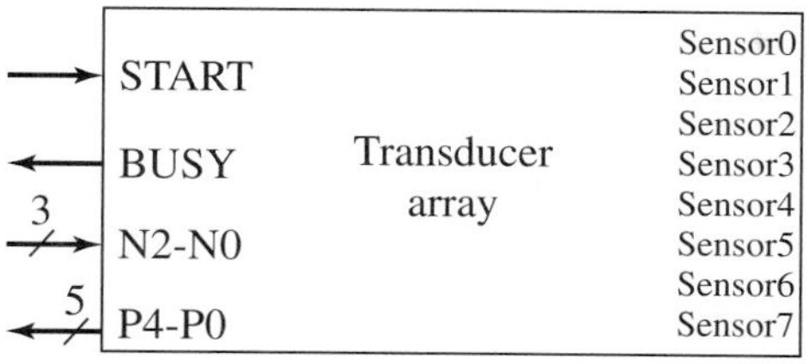

Figure 3.44
An eight-channel sensor array.

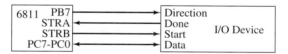

```
unsigned char data[8];              /* Room for all 8 results        */
void main(void) { unsigned char N;  /* Sensor position 0,1,2..., 7   */
      RITUAL();                     /* Initialize uC Part c)         */
      for (N=0;N<8;N++)
            data[N]=SENSOR(N);      /* Read sensor N Part d)         */
}
```

c) Show the RITUAL() procedure that initializes the microcomputer.

d) Show the SENSOR(N) function that reads one data byte. The sequence of steps is described above. Use gadfly synchronization.

3.16 The objective of this problem is to interface an I/O device to a 6811 (Figure 3.45). The same interface must be capable of both input and output. To solve this problem with a different microcomputer, you could substitute any available I/O port for STRA STRB PC7-PC0. You may write the software in assembly or C.

Figure 3.45
An I/O device interfaced to a 6811.

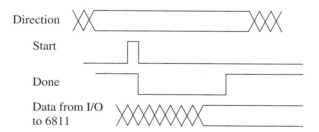

The sequence to receive an 8-bit data value from the input device to the microcomputer is as follows. First the microcomputer sets Direction=1 and issues a Start pulse. When the input device has the data ready, it will drive its data lines and signal with Done=1. The microcomputer should then read the data (Figure 3.46).

Figure 3.46
Input timing diagram of the interface between an I/O device and a 6811.

a) Show the C (or assembly language) procedure that initializes the microcomputer.

b) Show the C (or assembly language) procedure that inputs one data byte. Assume the ritual in part a has been called. The sequence of steps is described above. The input data is returned by value. Use gadfly synchronization. If you write in C, implement call by reference with the following prototype:

```
void Input(unsigned char *);
```

The sequence to transmit an 8-bit data value from the microcomputer to the output device is as follows. First the microcomputer sets `Direction=0`, places the desired 8-bit information on the data lines, and issues a `Start` pulse. When the output device is complete, it signals with `Done=1`. The microcomputer should then make its data output hiZ (tristate) (Figure 3.47).

Figure 3.47
Output timing diagram of the interface between an I/O device and a 6811.

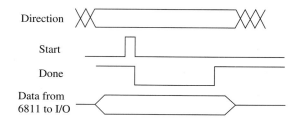

c) Show the C (or assembly language) procedure that outputs one data byte. Assume the ritual in part a has been called. The sequence of steps is described above. The output data is passed by value. Use gadfly synchronization. If you write in C pass by value with the following prototype:

```
void Output(unsigned char);
```

3.17 The objective of this problem is to interface an input device to a 6811 (Figure 3.48). To solve this problem with a different microcomputer, you could substitute any available I/O port for STRA STRB PC7-PC0. You may write the software in assembly or C.

Figure 3.48
An input device interfaced to a 6811.

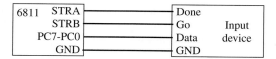

The sequence to input an ASCII character from the input device is as follows. When a new character is available, the input device puts its ASCII code on the 8-bit `Data`, then the input device makes `Done=1`. The hardware of the 6811 Port C should make `Go=0` almost immediately. Recognizing that `Done=1`, your software should read the 8-bit ASCII code, then acknowledge receipt of the data by making `Go=1` again. After the `Go=1`, the input device will make `Done=0` again (Figure 3.49).

Figure 3.49
Timing diagram of the interface between an input device and a 6811.

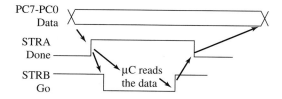

a) Show bit by bit your choice (what and why) for the parallel I/O control register. Specify that 1 for this bit must be 1, 0 for this bit must be 0, and X for this bit can be written with either 0 or 1.

b) Show the C program (ritual) that initializes the microcomputer.

c) Show the C procedure that inputs one data byte. The sequence of steps is described above. The input data is returned by value. Use gadfly synchronization. If you write in C, implement a function return value with the following prototype:

```
unsigned char Input(void);
```

3.18 The objective of this problem is to interface the following input device to a microcomputer single-chip computer using any available I/O port (Figure 3.50). The figure shows the connections to Port C. You may write the software in assembly or C.

Figure 3.50
An input device
interfaced to a
microcomputer.

First, your microcomputer sets `Control=1`, meaning input is requested. Then, the input device will create a new 6-bit input and place it on `Data`. The input device will then make `Status=1` signifying new data available. Next, your microcomputer program should read the 6-bit input value. Your program should then make `Control=0`, meaning it has read the data. The input device will then make `Status=0`, signifying the transfer is complete. Your software should wait until `Status=0` before returning (Figure 3.51).

Figure 3.51
Timing diagram of the
interface between an
input device and a
microcomputer.

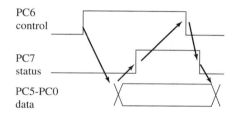

a) Show your choice in hexadecimal for data direction register, e.g., DDRC.
b) Show the assembly language subroutine that inputs one 6-bit data. The sequence of steps is described above. The input data is returned by value in RegA. Use gadfly synchronization. Full credit will be given to the subroutine that appropriately uses the `bset bclr brset` and/or `brclr` assembly language instructions. If you write in C, implement a function return value with the following prototype:

```
unsigned char Input(void);
```

3.19 These questions refer specifically to the 6811 PIOC.
a) What are the different bits in the 6811 PIOC? Name each bit and describe the purpose of each bit.
b) What is the address of PIOC?
c) Which bit in PIOC enables handshaking mode?
d) Which bit in PIOC is not writable? What operation sets this bit? What operation clears this bit?
e) Which two pins on the 6811 are used specifically for handshaking?
f) How can you generate a strobe pulse on the STRB pin on the 6811?

3.8 Lab Assignments

Lab 3.1. The overall objective is to create a serial port device driver that supports fixed-point input/output. The format will be 16-bit unsigned decimal fixed-point with a resolution of 0.001. You will be able to find on the CD implementations of a serial port driver that supports character, string, and integer I/O, which will be similar to those presented in Section 3.6. In particular, you will design, implement, and test two routines called `SCI_FixIn` and `SCI_FixOut`. `SCI_FixIn` will accept input from the SCI similar to the function `SCI_InUDec`. `SCI_FixOut` will transmit output to the SCI similar to the function `SCI_OutUDec`. During the design phase of this lab, you should define the range of features available. You should design, implement, and test a main program that illustrates the range of capabilities.

Lab 3.2. The same as Lab 3.1, except the format will be 16-bit signed binary fixed-point with a resolution of 2^{-8}.

Lab 3.3. The same as Lab 3.1, except the format will be 32-bit unsigned integer.

Lab 3.4. Design a four-function (add, subtract, multiply, divide) calculator using fixed-point math and the SCI for input/output. The format should be 16-bit signed decimal fixed-point with a resolution of 0.01. The syntax of the calculator should be Reverse Polish with a data stack.

Lab 3.5. The overall objective of this lab is to design, implement, and test a parallel port expander. Using less than 16 I/O pins of your microcontroller, you will design hardware and software that supports two **8-bit latched input** ports and two strobed output ports. Each input port has 8 data lines and one **latch** signal. On the rising each of the **latch**, your system should capture (latch), the data lines. Each output port has 8 data lines and one **strobe** signal. The hardware/software system should generate a pulse out on **strobe** whenever new output is sent. The output ports do not need to be readable. E.g., consider using four 74HC374 registers.

Lab 3.6. The overall objective of this lab is to design, implement, and test a parallel output port expander. Using just three I/O pins of your microcontroller, you will design hardware and software that supports four 8-bit output ports. The output ports do not need to be readable. E.g., consider using four 74HC595 registers.

Lab 3.7. The overall objective of this lab is to design, implement, and test a parallel input port expander. Using just three I/O pins of your microcontroller, you will design hardware and software that supports four 8-bit input ports. The input ports do not need to be latched. E.g., consider using four 74HC597 registers.

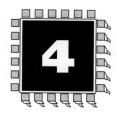

4 Interrupt Synchronization

Chapter 4 objectives are to:

❑ Introduce the concept of interrupt synchronization
❑ Discuss the issues involved in reentrant programming
❑ Implement and apply the FIFO circular queue
❑ Discuss the specific details of using interrupts on the 6811, and 6812
❑ Interface a keyboard and printer using external IRQ interrupts
❑ Show a high-priority/low-latency XIRQ power failure interface
❑ Define mechanisms to establish priority including round robin
❑ Design and implement background I/O for simple devices using periodic polling

There are many reasons to consider interrupt synchronization. The first consideration is that the software in a real-time system must respond to hardware events within a prescribed time. To illustrate the need for interrupts, consider a keyboard interface where the time between new keyboard inputs might be as small as 10 ms. In this situation, the software latency is the time from when the new keyboard input is ready until the time the software reads the new data. To prevent loss of data in this case, the software latency must be less than 10 ms. We can implement real-time software using gadfly synchronization only when the size and complexity of the system are very small. Interrupts are important for these real-time systems because they provide a mechanism to guarantee an upper bound on the software response time. Interrupts also give us a way to respond to infrequent but important events. Alarm conditions like low battery power and error conditions can be handled with interrupts. Periodic interrupts, generated by the timer at a regular rate, will be necessary to implement data acquisition and control systems. In the unbuffered interfaces of Chapter 3, the hardware and software took turns waiting for each other. Interrupts provide a way to buffer the data so that the hardware and software spend less time waiting. In particular, the buffer we will use is a FIFO queue placed between the interrupt routine and the main program to increase the overall bandwidth. We will begin our discussion with general issues, then present the specific details about the 6811, and 6812 microcomputers. After that, a number of simple interrupt examples will be presented, and at the end of the chapter we will discuss some advanced concepts like priority, round-robin polling, and periodic polling.

4.1   What Are Interrupts?

4.1.1
Interrupt
Definition

An *interrupt* is the automatic transfer of software execution in response to hardware that is asynchronous with the current software execution. The hardware can be either an external I/O device (like a keyboard or printer) or an internal event (like an opcode fault or a periodic timer). When the hardware needs service (busy-to-done state transition), it will request an interrupt. Recall from Chapter 2 that a *thread* is defined as the path of action of software as it executes. The execution of the interrupt service routine is called a *background* thread. This thread is created by the hardware interrupt request and is killed when the interrupt service routine executes the `rti` instruction. A new thread is created for each interrupt request. It is important to consider each individual request as a separate thread because local variables and registers used in the interrupt service routine are unique and separate from one interrupt event to the next. In a multithreaded system we consider the threads as cooperating to perform an overall task. Consequently we will develop ways for the threads to communicate (see Section 4.3 on FIFO queues) and synchronize (see the discussion of semaphores in Section 5.2) with each other. Most embedded systems have a single common overall goal. On the other hand, general-purpose computers can have multiple unrelated functions to perform. A *process* is also defined as the action of software as it executes. The difference is processes do not necessarily cooperate toward a common shared goal. Threads share access to global variables, while processes have separate globals.

The software has dynamic control over aspects of the interrupt request sequence. First, each potential interrupt source has a separate arm bit that the software can activate or deactivate. The software will set the arm bits for those devices it wishes to accept interrupts from, and it will deactivate the arm bits within those devices from which interrupts are not to be allowed. In other words, it uses the arm bits to individually select which devices will and which devices will not request interrupts. The second aspect that the software controls is the interrupt enable bit, I, which is in the condition code register. The software can enable all armed interrupts by setting I=0, or it can disable all interrupts by setting I=1. The disabled interrupt state (I=1) does not dismiss the interrupt requests; rather it postpones them until a later time, when the software deems it convenient to handle the requests. We will pay special attention to these enable/disable software actions. In particular we will need to disable interrupts when executing nonreentrant code, but disabling interrupts will have the effect of increasing the response time of software.

There are two general methods with which we configure external hardware so that it can request an interrupt. The first method is a shared negative-logic-level-active request like $\overline{\text{IRQ}}$. All the devices that need to request interrupts have an open-collector negative-logic interrupt request line. The hardware requests service by pulling the interrupt request $\overline{\text{IRQ}}$ line low. The line over the IRQ signifies negative logic. In other words, an interrupt is requested when $\overline{\text{IRQ}}$ is zero. Because the request lines are open-collector, a pull-up resistor is needed to make $\overline{\text{IRQ}}$ high when no devices need service (Figure 4.1).

Figure 4.1
Wire- or negative-logic
interrupt request line.

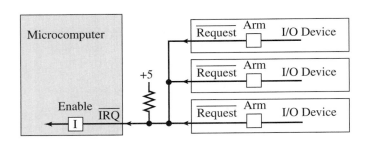

Figure 4.2
Dedicated edge-triggered
interrupt request lines.

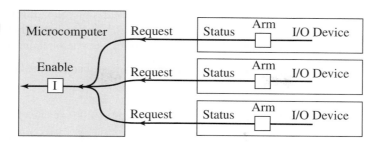

Normally these interrupt requests share the same interrupt vector. This means that whichever device requests an interrupt, the same interrupt service routine is executed. Therefore the interrupt service routine must first determine which device requested the interrupt. The 6811 and 6812 have two of these shared negative-logic-level-active interrupts, IRQ and XIRQ. XIRQ interrupts on the 6811 and 6812 are nonmaskable. In other words, once they are enabled (the X bit in the CCR is 0), they cannot be disabled.

The second method uses multiple dedicated *edge-triggered* requests (Figure 4.2). In this method, each device has its own interrupt request line that is usually connected to a status signal in the I/O device. In this way, an interrupt is requested on the *busy-to-done* state transition. Edge-triggered means the interrupt is requested on the rise, the fall, or both the rise and the fall of the *Status* signal. With *vectored* interrupts these individual requests have a unique interrupt vector. This means there can be a separate interrupt service routine for each device. In this way the microcomputer will automatically execute the appropriate software interrupt handler when an interrupt is requested. Therefore the interrupt service routine does not need to determine which device requested the interrupt.

The original IBM PC had only eight dedicated edge-triggered interrupt lines, and the current IBM PC I/O bus has only 15. This small number can be a serious limitation in a computer system with many I/O devices.

Observation: Microcomputer systems running in expanded mode often use shared negative-logic-level-active interrupts for their external I/O devices.

Observation: Microcomputer systems running in single-chip mode often use dedicated edge-triggered interrupts for their I/O devices.

Observation: The number of interrupting devices on a system using dedicated edge-triggered interrupts is limited when compared to a system using shared negative-logic-level-active interrupts.

Observation: Most Freescale microcomputers support both shared negative-logic and dedicated edge-triggered interrupts.

The advantages of a wire- or negative-logic interrupt request are (1) additional I/O devices can be added without redesigning the hardware, (2) there is no fundamental limit to the number of interrupting I/O devices you can have, and (3) the microcomputer hardware is simple. The advantages of the dedicated edge-triggered interrupt request are (1) the software is simpler, therefore it will be easier to debug and it will run faster, (2) there will be less coupling between software modules, making it easier to debug and to reuse code, and (3) it will be easier to implement priority such that higher-priority requests are handled quickly, while lower-priority interrupt requests can be postponed. Because the Freescale microcomputers support both types of interrupt, the designer can and must address these considerations.

**4.1.2
Interrupt Service
Routines**

The interrupt service routine (ISR) is the software module that is executed when the hardware requests an interrupt. From Section 4.1.1 we see that there may be one large ISR that handles all requests (polled interrupts) or many small ISRs specific for each potential

source of interrupt (vectored interrupts). The design of the ISR requires careful consideration of many factors that will be discussed in this chapter. When an interrupt is requested
(and the device is armed and the I bit is zero), the microcomputer will service an interrupt:

1. The execution of the main program is suspended (the current instruction is
    finished)[1]
2. The ISR, or background thread, is executed
3. The main program is resumed when the ISR executes `rti`

When the microcomputer accepts an interrupt request, it will automatically save the
execution state of the main thread by pushing its registers on the stack (i.e., PC, Y, X, A,
B, CCR). After the ISR provides the necessary service, it will execute a `rti` instruction. This
instruction pulls the registers from the stack, which returns control to the main program.
Execution of the main program will then continue with the exact stack and register values
that existed before the interrupt. Although interrupt handlers can allocate, access, then deallocate local variables, parameters passing between threads must be implemented using
global memory variables. Global variables are also required if an interrupt thread wishes to
pass information to itself, (e.g., from one interrupt instance to another). The execution of the
main program is called the *foreground thread,* and the executions of ISRs are called *background threads* (Figure 4.3).

Checkpoint 4.1: What value of the I bit enables interrupts to occur?

Figure 4.3
An interrupt causes the
main thread to be
suspended, and the
interrupt thread is run.

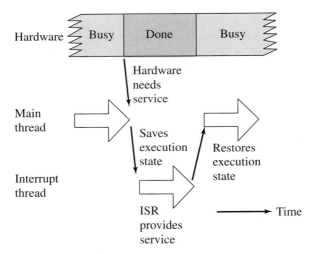

4.1.3 When to Use Interrupts

The factors listed in Table 4.1 should be considered when deciding the most appropriate
mechanism to synchronize hardware and software. One should not always use gadfly just
because one is too lazy to implement the complexities of interrupts. On the other hand, one
should not always use interrupts just because they are fun and exciting. Direct Memory
Access (DMA) will be covered in Chapter 10.

4.1.4 Interthread Communication

For regular function calls we use the registers and stack to pass parameters, but interrupt threads have logically separate resisters and stack. In particular, all registers are
automatically saved by the microcomputer as it switches from the main program (foreground thread) to the ISR (background thread). The `rti` instruction will restore the registers (including the interrupt enable bits and the PC) back to their previous values.

---

[1]The 6812 instructions `rev`, `revw`, and `wav` can be interrupted in the middle of their execution.

Table 4.1
Each synchronization method has motivations for its use.

Gadfly	Interrupts	DMA
Predicable	Variable arrival times	Low latency
Simple I/O	Complex I/O, different speeds	High bandwidth
Fixed load	Variable load	
Dedicated, single thread	Other functions to do	
Single process	Multithread or multiprocess	
Nothing else to do	Infrequent but important alarms	
	Program errors	
	Overflow, invalid opcode	
	Illegal stack or memory access	
	Machine errors	
	Power failure, memory fault	
	Breakpoints for debugging	
	Real-time clocks	
	Data acquisition and control	

Thus, all parameter passing must occur through global memory. One cannot pass data from the main program to the ISR using registers or the stack. The classic producer/consumer problem has two threads. One thread produces data, and the other consumes data. For an input device, the background thread is the producer because it generates new data, and the foreground thread is the consumer because it uses the data up. For an output device, the data flows in the other direction so that the producer/consumer roles are reversed. It is appropriate to pass data from the producer thread to the consumer thread using a FIFO queue (Figure 4.4).

Observation: For systems with interrupt-driven I/O on multiple devices, there will be a separate FIFO for each device.

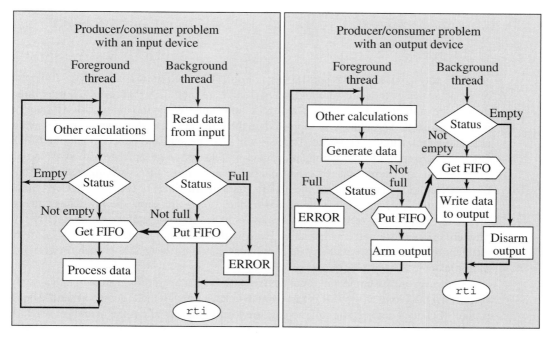

Figure 4.4
FIFO queues can be used to pass data between threads.

An input device needs service (busy-to-done state transition) when new data are available. The ISR (background) will accept the data and process it. If more data is desired, the ISR will restart the input hardware (done-to-busy state transition).

An output device needs service (busy-to-done state transition) when the device is idle, ready to output more data. The ISR (background) will get more data and output it. The output function will restart the hardware (done-to-busy state transition). For this output device example, it would be appropriate to GET more data from a FIFO queue and send it to the device. Two particular problems with output device interrupts are:

1. How does one generate the first interrupt? In other words, how does one start the output thread?
2. What does one do if an output interrupt occurs (device is idle) but there are no more data currently available (e.g., FIFO is empty)?

These problems can be solved by arming and disarming when needed.

The foreground thread (main program) executes a loop and accesses the FIFO when it needs to input or output data. The background threads (interrupts) are executed when the hardware needs service. For an input device, an interrupt is requested (causing the ISR to be executed) when new input data are available. The busy-to-done state transition causes an interrupt. These hardware state transitions are illustrated in Figures 4.5 and 4.8.

Figure 4.5
The input device interrupts the computer when it has new data.

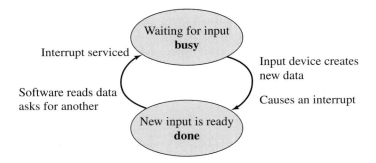

One way to visualize the interrupt synchronization is to draw a state versus time plot of the activities of the hardware and the two software modules (Figure 4.6). For an input device, the main thread begins by waiting for new input. When the input device is busy, it is in the process of creating new input. When the input device is done, new data are available and an interrupt is requested. The ISR will read the data and put them into the FIFO. Once data are in the FIFO, the main program is released to go on. The arrows from one graph to the other represent the synchronizing events. In this first example, the time for the software to read and process the data is less than the time for the input device to create new input. This situation is called *I/O-bound*. In this situation, the FIFO has either 0 or 1 entry, and the use of interrupts does not enhance the bandwidth over the gadfly implementations presented in Chapter 3. Even with an I/O-bound device it may be more efficient to utilize interrupts because it provides a straightforward approach to servicing multiple devices.

In this second example, the input device starts with a burst of high-bandwidth activity (Figure 4.7). As long as the ISR is fast enough to keep up with the input device and as long as the FIFO does not become full, no data are lost. In this situation, the overall bandwidth is higher than it would be with a gadfly implementation, because the input device does not have to wait for each data byte to be processed. This is the classic example of a

Figure 4.6
Hardware/software
timing of an I/O-bound
input interface.

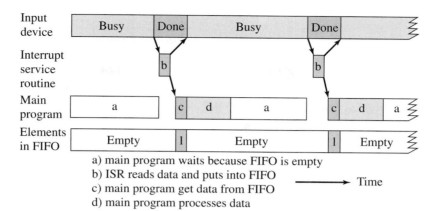

a) main program waits because FIFO is empty
b) ISR reads data and puts into FIFO
c) main program get data from FIFO
d) main program processes data

→ Time

Figure 4.7
Hardware/software
timing of an input
interface during a
high-bandwidth burst.

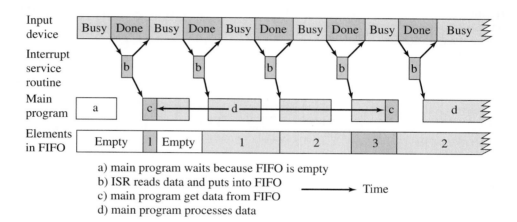

a) main program waits because FIFO is empty
b) ISR reads data and puts into FIFO
c) main program get data from FIFO
d) main program processes data

→ Time

buffered input, because data enter the system (via the interrupts), are temporarily stored in a buffer (put into the FIFO), and are processed later (by the main program, get from the FIFO). When the I/O device is faster than the software, the system is called *CPU-bound.* As we will see later, this system can work if the producer rate temporarily exceeds the consumer rate (a short burst of high-bandwidth input). If the external device sustained the high-bandwidth input rate, then the FIFO would become full and data would be lost.

For an output device, the interrupt is requested when the output is idle and ready to accept more data (Figure 4.8). The busy-to-done state transition causes an interrupt. The ISR gives the output device another piece of data to output.

Again, we can visualize the interrupt synchronization by drawing a state versus time plot of the activities of the hardware and the two software modules. For an output device interface, the output device is initially disarmed and the FIFO is empty. The main thread begins by generating new data. After the main program puts the data into the FIFO, it arms the output interrupts. This first interrupt occurs immediately, and the ISR gets the data from the FIFO and outputs it to the external device. The output device becomes busy because it is in the process of outputting data. It is important to realize that it only takes the software on the order of 1 μs to write data to one of its output ports, but usually it takes the output device much longer to fully process the data. When the output device is done, it is ready to accept more data and an interrupt is requested. If the FIFO is empty at this point, the ISR will disarm the output device. If the FIFO is not empty, the ISR will get data from the FIFO

Figure 4.8
The output device
interrupts the computer
when it is idle and needs
new data.

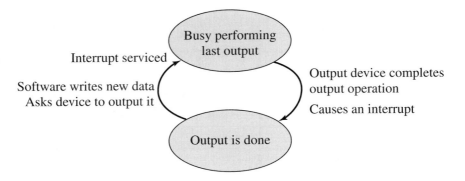

Figure 4.9
Hardware/software
timing of a CPU-bound
output interface.

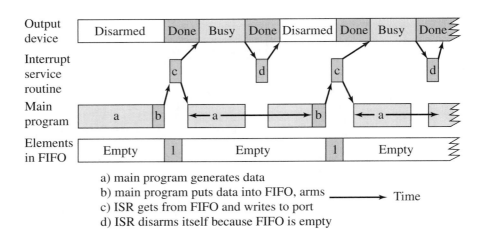

a) main program generates data
b) main program puts data into FIFO, arms
c) ISR gets from FIFO and writes to port
d) ISR disarms itself because FIFO is empty

and write them out to the output port. Once data are written to the output port, the output device is released to go on. In this first example, the time for the software to generate data is longer than the time for the external device to output it. This is an example of a CPU-bound system (Figure 4.9). In this situation, the FIFO has either 0 or 1 entry, and the use of interrupts does not enhance the bandwidth over the gadfly implementations presented in Chapter 3. Nevertheless, interrupts provide a well-defined mechanism for dealing with complex systems.

In this second example, the software starts with a burst of high-bandwidth activity (Figure 4.10). As long as the FIFO does not become full, no data are lost. In this situation, the overall bandwidth is higher than it would be with a gadfly implementation, because the software does not have to wait for each data byte to be processed by the hardware. This is the classic example of a *buffered* output, because data enter the system (via the main program), are temporarily stored in a buffer (put into the FIFO), and are processed later (by the ISR, get from the FIFO, write to external device). When the I/O device is slower than the software, the system is called I/O bound. Just like the input scenario, the FIFO might become full if the producer rate is too high for too long.

There are other types of interrupts that are not an input or an output. For example, we will configure the computer to request an interrupt on a periodic basis. This means an interrupt handler will be executed at fixed time intervals. This periodic interrupt will be essential for the implementation of real-time data acquisition and real-time control systems. For example, if we are implementing a digital controller that executes a control algorithm 100 times a second, then we will set up the timer hardware to request an interrupt every

Figure 4.10
Hardware/software
timing of an I/O-bound
output interface.

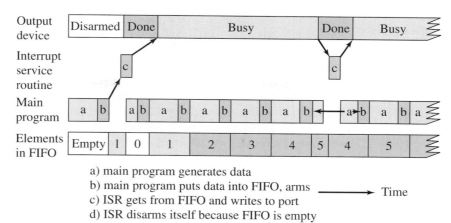

a) main program generates data
b) main program puts data into FIFO, arms
c) ISR gets from FIFO and writes to port
d) ISR disarms itself because FIFO is empty

Time

10 ms. The ISR will execute the digital control algorithm and return to the main thread. We will learn how to create the period interrupt in Section 4.15 and use it in Chapters 12 and 13.

An axiom with interrupt synchronization is that the interrupt program should execute as fast as possible. The interrupt should occur when it is time to perform a needed function, and the ISR should come in clean, perform that function, and return right away. Placing backward branches (gadfly loops, iterations) in the interrupt software should be avoided if possible. The percentage of time spent executing interrupt software should be minimized. For an input device, the *interface latency* is the time between when new input are available and the time when the software reads the input data. We can also define *device latency* as the response time of the external I/O device. For example, if we request that a certain sector be read from a disk, then the device latency is the time it takes to find the correct track and spin the disk (seek) so that the proper sector is positioned under the read head. For an output device, the interface latency is the time between when the output device is idle and the time when the software writes new data. A *real-time* system is one that can guarantee an upper bound interface latency.

4.2 Reentrant Programming

A program segment is reentrant if it can be concurrently executed by two (or more) threads. This issue is very important when using interrupt programming. To implement reentrant software, we place local variables on the stack and avoid storing into global memory variables. When writing in assembly, we use *registers* or the *stack* for parameter passing to create reentrant subroutines. Typically each thread will have its own set of registers and stack. A nonreentrant subroutine will have a section of code called a *vulnerable window* or *critical section*. An error occurs if:

1. One thread calls the nonreentrant subroutine
2. That thread is executing in the "vulnerable" window when interrupted by a second thread
3. The second thread calls the same subroutine

A number of scenarios can happen next. In the first scenario, the second thread is allowed to complete the execution of the subroutine, control is then returned to the first thread, and the first thread finishes the subroutine. This first scenario is the usual case with interrupt programming. In the second scenario, the second thread executes part of it, is interrupted and then reentered by a third thread, the third thread finishes, the control is returned to the second thread and it finishes, and last the control is returned to the first thread and it finishes. This second scenario can happen in interrupt programming if interrupts are reenabled during the execution of the ISR. In the third scenario, the second thread executes part of it, is interrupted and the first thread continues, the first thread finishes, the control is returned to the second thread and it finishes. This third scenario will occur only when you use a thread scheduler like the ones described in the next chapter. Since most embedded systems do not have a thread scheduler, we will focus on the first two scenarios. A vulnerable window may exist when two different subroutines access and modify the same memory-resident data structure.

In Figure 4.4, we saw two concurrent threads communicating via a FIFO queue. When processing input, it is possible for the main program to start to execute `Fifo_Get`, be interrupted, and the input interrupt routine calls `Fifo_Put`. To verify correctness of our system, we must consider what would happen if the `Fifo_Put` subroutine is executed in between any two assembly instructions of the `Fifo_Get` routine. A similar problem arises when processing outputs. In this situation the main program may start to execute `Fifo_Put`, be interrupted, and the output interrupt routine calls `Fifo_Get`. To verify correctness of this system, we must consider what would happen if the `Fifo_Get` subroutine is executed in between any two assembly instructions of the `Fifo_Put` routine. If we are processing both input and output, then two FIFOs would be used, so none of the individual functions can be reentered. Nevertheless, the `Fifo_Put` and `Fifo_Get` routines share access to global variables, so we must search for potential critical sections. The three functions `Fifo_Init` `Fifo_Put` and `Fifo_Get` all manipulate the same global variables; therefore, to operate properly we need a mechanism to implement mutual exclusion, which simply means only one thread at a time is allowed access to the FIFO module. For most of this book, we will implement mutual exclusion by disabling interrupts. In the next chapter, we will describe how to create a preemptive thread scheduler and use semaphores to implement mutual exclusion.

This first example (Program 4.1) is a nonreentrant assembly subroutine that uses a memory variable `Second`. This subroutine could have been made reentrant by implementing `Second` as a local variable, but the purpose of the example is to illustrate what can go wrong when a nonreentrant subroutine is reentered.

Program 4.1
This subroutine is nonreentrant because of the read-modify-write access to a global.

```
Second   rmb   2     Temporary global variable
* Input parameters: Reg X,Y contain 2 16 bit numbers
* Output parameter: Reg X is returned with the average
Ave      sty   Second  Save the second number in memory
         xgdx            Reg D contains first number
         addd  Second    Reg D=First+Second
         lsrd            (First+Second)/2
         adcb  #0        round up?
         adca  #0
         xgdx
         rts
```

A *vulnerable window* exists between the `sty` and the `addd` instructions. Assume there are two concurrent threads (`Main`, `InterruptService`) that both call this subroutine. Concurrent means that both threads are ready to run. Because there is only one computer, exactly one thread will be active (running) at a time. Typically, the operating system (OS) switches execution control back and forth using interrupts. For example, the `Main` thread might be executing when an interrupt causes the computer to switch over and execute the

InterruptService. When the InterruptService is done, it executes a rti, and the control returns back to the Main thread. An error occurs if (Figure 4.11):

1. The Main thread calls Ave
2. The Main thread executes the sty instruction, saving its second number in Second
3. The OS halts the Main thread (using an interrupt) and starts the Interrupt-Service thread
4. The InterruptService thread calls Ave

   The InterruptService thread executes the sty, saving its second number in Second

   The InterruptService thread finishes Ave

5. The OS returns control back to the Main thread
6. The Main thread executes the addd instruction but gets the wrong Second number

Figure 4.11
One sequence of operations that results in the reentering of the subroutine Ave.

A *vulnerable window* exists in the following program between the Result=y and the Result+x instructions.

```
int Result;    /* Temporary global variable */
int Ave(int x,y);{
     Result=y;    /* Save the second number in global memory */
     Result=(Result+x)>>1;   /* (First+Second)/2 */
     return(Result);}
```

An error occurs if (Figure 4.12):

1. The Main thread calls Ave
2. The Main thread executes the Result=y instruction, saving its second number in Result
3. The OS halts the Main thread (using an interrupt) and starts the InterruptService thread
4. The InterruptService thread calls Ave

   The InterruptService thread executes the Result=y, saving its number over other copy

   The InterruptService thread finishes Ave

5. The OS returns control back to the Main thread
6. The Main thread calculates (Result+x)>>1 but gets the wrong Result number

The solution to this simple problem is to implement Result as a local variable either in a register or on the stack. But we saw earlier in the chapter that global variables are needed for interthread communication, so not all nonreentrant activity can be solved by using local variables. In this chapter we will disable interrupts during the vulnerable window. In the

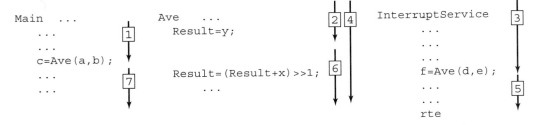

Figure 4.12
One sequence of operations that results in the reentering of the C function `Ave`.

next chapter semaphores are used to provide mutual exclusive access to certain memory-resident data structures.

An *atomic operation* is one that once started is guaranteed to finish. In most computers, once an instruction has begun, the instruction must be finished before the computer can process an interrupt. Therefore, the following read-modify-write sequence is atomic because it cannot be halted in the middle of its operation:

```
inc counter    where counter is a global variable
```

On the other hand, this read-modify-write sequence is *nonatomic* because it can start, then be interrupted:

```
ldaa counter    where counter is a global variable
inca
staa counter
```

In general, nonreentrant code can be grouped into three categories, all involving nonatomic writes to global variables. We will classify I/O ports as global variables for the consideration of reentrancy. We will group registers into the same category as local variables because each thread will have its own registers and stack.

The first group is the *read-modify-write* sequence:

1. The software reads the global variable, producing a copy of the data.
2. The software modifies the copy (at this point the original variable is still unmodified).
3. The software writes the modification back into the global variable.

An assembly example is shown in Program 4.2. In C, this read-modify-write could be implemented as shown in Program 4.3.

Program 4.2
This subroutine is also nonreentrant because of the read-modify-write access to a global.

```
Money rmb   2      bank balance implemented as a global
* add $100 to the account
more  ldd   Money  where Money is a global variable
      addd  #100
      std   Money  Money=Money+100
      rts
```

Program 4.3
This C function is nonreentrant because of the read-modify-write access to a global.

```
unsigned int Money;   /* bank balance implemented as a global */
/* add 100 dollars */
void more(void){
      Money += 100;}
```

In the second group is the *write followed by read,* where the global variable is used for temporary storage:

1. The software writes to the global variable (this becomes the only copy of important information).
2. The software reads from the global variable, expecting the original data to still be there.

An assembly example is shown in Program 4.4. In C, Program 4.5 shows an illustrative example of this write-read.

Program 4.4
This assembly subroutine is nonreentrant because of the write-read access to a global.

```
temp   rmb   2         temporary result implemented as a global
* calculate RegX=RegX+2*RegD
mac    stx  temp    Save X so that it can be added
       lsld           RegD=2*RegD
       addd temp      RegD=RegX+2*RegD
       xgdx           RegX=RegX+2*RegD
       rts
```

Program 4.5
This C function is nonreentrant because of the write-read access to a global.

```
int temp;  /* global temporary */
/* calculate x+2*d */
int mac(int x, int d){
       temp = x+2*d;    /* write to a global variable */
       return (temp);}  /* read from global */
```

In the third group, we have a *nonatomic multistep write* to a global variable:

1. The software writes part of the new value to a global variable.
2. The software writes the rest of the new value to a global variable.

Program 4.6 shows an assembly example, and in C, this multistep write sequence could be implemented as shown in Program 4.7.

Program 4.6
This assembly subroutine is nonreentrant because of the multistep write access to a global.

```
Info   rmb   4         32-bit data implemented as a global
* set the variable using RegX and RegY
set    stx  Info    Info is a 32 bit global variable
       sty  Info+2
       rts
```

Program 4.7
This C function is nonreentrant because of the multistep write access to a global.

```
int info[2];  /* 32-bit global */
void set(int x, int y){
    info[0]=x;
    info[1]=y;}
```

Observation: When considering reentrant software and vulnerable windows, we classify accesses to I/O ports the same as accesses to global variables.

We can make a subroutine reentrant by using a stack variable

If we can eliminate the global variables, then the subroutine becomes reentrant. The example of Program 4.1 is redesigned by placing the temporary on the stack, as shown in Program 4.8. There are no "vulnerable" windows because each thread has its own registers and stack.

We can make a subroutine reentrant by disabling interrupts

Sometimes one must access global memory to implement the desired function. Consider the example of a message mailbox (Program 4.9). This mailbox has a status flag (0 means empty) and an 8-bit value. The subroutine send will check the status, and if empty it will store the message (passed in RegB) into the mailbox. In this example there are multiple concurrent threads that may call Send. The mailbox will be covered in more detail in Chapter 5.

Program 4.8
This assembly subroutine is reentrant because it does not write to any globals.

```
* Input parameters: Reg X,Y contain 2 16 bit numbers
* Output parameter: Reg X is returned with the average
Ave       pshy          Save the second number on the stack
          tsy           Reg Y points the Second number
          xgdx          Reg D contains first number
          addd 0,Y      Reg D=First+Second
          lsrd          (First+Second)/2
          adcb #0       round up?
          adca #0
          xgdx
          puly
          rts
```

Program 4.9
This assembly subroutine is nonreentrant because of the read-modify-write access to a global.

```
Status    rmb  1     0 means empty, -1 means it contains something
Message   rmb  1     data to be communicated
* Input parameter: Reg B contains an 8 bit message
* Output parameter: Reg CC (C bit) is 1 for OK, 0 for busy error
Send      tst  Status   check if mailbox is empty
          bmi  Busy     full, can't store, so return with C=0
          stab Message store
          dec  Status   signify it now contains a message
          sec           stored OK, so return with C=1
Busy      rts
```

Clearly there is a *vulnerable window* or *critical section* between the tst and dec instructions that makes this subroutine nonreentrant. One can make this subroutine reentrant by disabling interrupts during the vulnerable window. It is important not to disable interrupts too long so as not to affect the interface latency of the other threads. Notice also that the interrupts are not simply disabled then enabled, but rather the interrupt status is saved, the interrupts disabled, then the interrupt status is restored. Consider what would happen if you simply added a sei at the beginning and a cli at the end of the above subroutine, then called it with the interrupts disabled.

Program 4.10
This assembly subroutine is reentrant because it disables interrupts during the critical section.

```
Status    rmb  1     0 means empty, -1 means it contains something
Message   rmb  1     data to be communicated
* Input parameter: Reg B contains an 8 bit message
* Output parameter: Reg CC (C bit) is 1 for OK, 0 for busy error
Send      clc           Initialize carry=0
          tpa           save current interrupt state
          psha
          sei           disable interrupts during vulnerable window
          tst  Status   check if mailbox is empty
          bmi  Busy     full, so return with C=0
          staa Message store
          dec  Status   signify it now contains a message
          pula
          oraa #1       OK, so return with C=1
          psha
Busy      pula          restore interrupt status
          tap
          rts
```

Program 4.11
This C function is reentrant because it disables interrupts during the critical section.

```
int Empty;            // -1 means empty, 0 means it contains something
short Message;        //  data to be communicated
int Send(short data){ int OK;
unsigned char SaveCCR;
   asm tpa              // previous interrupt enable
   asm staa SaveCCR     // save previous
   asm sei              // make atomic, start critical section
   OK = 0;               // Assume it is not OK
   if(Empty){
       Message = data;
       Empty = 0;       // signify it is now contains a message
       OK = -1;         // Successfull
   }
   asm ldaa SaveCCR     // recall previous
   asm tap              // end critical section
   return(OK);
}
```

Some machines provide a *test and set* function to solve this reentrant problem. This single (nondivisible) operation is equivalent to the subroutine in Program 4.12. The test and set operation, once started, must be allowed to complete. More details about semaphores can be found in Chapter 5.

Program 4.12
This assembly subroutine can be used as part of a binary semaphore.

```
* Global parameter: Semi4 is the memory location to test and set
* If the location is zero, it will set it (make it -1)
*      and return Reg CC (Z bit) is 1 for OK
* If the location is nonzero, it will return Reg CC (Z bit) = 0
Semi4 fcb  0             Semaphore is initially free
Tas   tst  Semi4         check if already set
      bne  Out           busy, operation failed, so return with Z=0
      dec  Semi4         signify it is now busy
      bita #0            operation successful, so return with Z=1
Out   rts
```

Reentrant programming is very important when writing high-level language software, too. Obviously, we minimize the use of global variables. But when global variables are necessary, we must be able to recognize potential sources of bugs due to nonreentrant code. We must study the assembly language output produced by the compiler. For example, we cannot determine whether the following read-modify-write operation is reentrant without knowing if it is atomic:

```
time++;
```

If the compiler generates the following object code, then `time++;` is atomic (therefore not critical):

```
inc time
```

If the compiler generates the following object code, then `time++;` is not atomic (therefore critical):

```
ldd  time
addd #1
std  time
```

Similarly, it is possible that your compiler might generate reentrant code for Program 4.5, so we must always examine assembly listings when considering reentrancy.

4.3 First-In–First-Out Queue

4.3.1 Introduction to FIFOs

As we saw earlier, the FIFO circular queue is quite useful for implementing a buffered I/O interface. It can be used for both buffered input and buffered output. The order-preserving data structure temporarily saves data created by the source (producer) before they are processed by the sink (consumer). The class of FIFOs studied in this section will be statically allocated global structures. Because they are global variables, it means they will exist permanently and can be carefully shared by more than one thread. The advantage of using a FIFO structure for a data flow problem is that we can decouple the producer and consumer threads. Without the FIFO we would have to produce one piece of data, then process it, produce another piece of data, then process it. With the FIFO, the producer thread can continue to produce data without having to wait for the consumer to finish processing the previous data. This decoupling can significantly improve system performance (Figure 4.13).

Figure 4.13
The FIFO is used to buffer data between the producer and consumer.

You have probably already experienced the convenience of FIFOs. For example, you can continue to type another command into the Windows command interpreter while it is still processing a previous command. The ASCII codes are put (calls *Fifo-Put*) in a FIFO whenever you hit the key. When the Windows command interpreter is free, it calls *Fifo-Get* for more keyboard data to process. A FIFO is also used when you ask the computer to print a file. Rather than waiting for the actual printing to occur character by character, the print command will *PUT* the data in a FIFO. Whenever the printer is free, it will *GET* data from the FIFO. The advantage of the FIFO is it allows you to continue to use your computer while the printing occurs in the background. To implement this magic of background printing we will need interrupts.

There are many producer/consumer applications. In Table 4.2 the tasks on the left are producers that create or input data, while the tasks on the right are consumers that process or output data.

Table 4.2
Producer-consumer examples.

Source/producer		Sink/consumer
Keyboard input	→	Program that interprets
Program with data	→	Printer output
Program sends message	→	Program receives message
Microphone and ADC	→	Program that saves sound data
Program that has sound data	→	DAC and speaker

The producer puts data into the FIFO. The `Fifo_Put` operation does not discard information already in the FIFO. If the FIFO is full and the user calls `Fifo_Put`, the `Fifo_Put` routine will return a full error, signifying the last (newest) data were not properly saved. The sink process removes data from the FIFO. The `Fifo_Get` routine will modify the FIFO. After a get, the particular information returned from the `Fifo_Get` routine is no longer saved on the FIFO. If the FIFO is empty and the user tries to get, the `Fifo_Get` routine will return an empty error, signifying no data could be retrieved. The FIFO is order-preserving, such that the information is returned by repeated calls of `Fifo_Get` in the same order as the data were saved by repeated calls of `Fifo_Put`.

There are many ways to implement a statically allocated FIFO. We can use either a pointer or an index to access the data in the FIFO. We can use either two pointers (or two indices) or two pointers (or two indices) and a counter. The counter specifies how many entries are currently stored in the FIFO. There are even hardware implementations of FIFO queues. We begin with the two-pointer implementation. It is a little harder to implement but does have some advantages over the other implementations.

4.3.2 Two-Pointer FIFO Implementation

We will begin with the two-pointer implementation. If we were to have infinite memory, a FIFO implementation is easy (Figure 4.14). `GetPt` points to the data that will be removed by the next call to `Fifo_Get`, and `PutPt` points to the empty space where the data will be stored by the next call to `Fifo_Put` (Program 4.13).

Figure 4.14
The FIFO implementation with infinite memory.

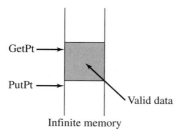

GetPt →

PutPt →

Valid data

Infinite memory

```
GetPt rmb 2   ;Pointer to oldest data
PutPt rmb 2   ;Pointer to free memory
* Call by value with RegA
Fifo_Put ldx   PutPt ;place to put next
         staa 1,x+  ;Store data into FIFO
         stx  PutPt ;Update pointer
         ldaa #-1   ;success
         rts
* Call by reference with RegX
Fifo_Get ldy   GetPt ;place to remove next
         ldaa 1,y+  ;Read data from FIFO
         staa 0,x   ;return by reference
         sty  GetPt ;Update pointer
         ldaa #-1   ;success
         rts
```

```
char static volatile *PutPt;  // put next
char static volatile *GetPt;  // get next
// call by value
int Fifo_Put(char data){
   *PutPt = data;   // Put
   PutPt++;         // next
   return(1);       // true if success
}
// call by reference
int Fifo_Get(char *datapt){
   *datapt = *GetPt; // return by reference
   GetPt++;          // next
   return(1);        // true if success
}
```

Program 4.13
Code fragments showing the basic idea of a FIFO.

Three modifications are required to the above subroutines. If the FIFO is full when `Fifo_Put` is called, then the subroutine should return a full error. Similarly, if the FIFO is empty when `Fifo_Get` is called, then the subroutine should return an empty error.

The `PutPt` and `GetPt` must be wrapped back up to the top when they reach the bottom (Figures 4.15, 4.16).

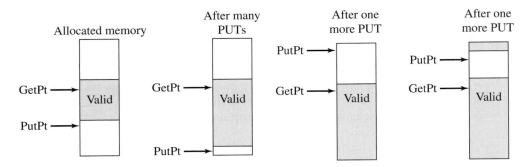

Figure 4.15
The `Fifo_Put` operation showing the pointer wrap.

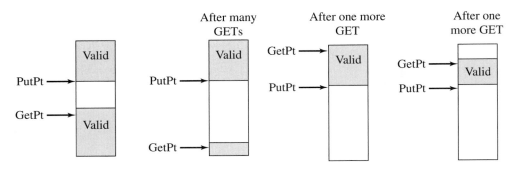

Figure 4.16
The `Fifo_Get` operation showing the pointer wrap.

Two mechanisms can be used to determine whether the FIFO is empty or full. A simple method is to implement a counter containing the number of bytes currently stored in the FIFO. `Fifo_Get` would decrement the counter, and `Fifo_Put` would increment the counter. The second method is to prevent the FIFO from being completely full. For example, if the FIFO had 10 bytes allocated, then the `Fifo_Put` subroutine would allow a maximum of 9 bytes to be stored. If there were 9 bytes already in the FIFO and another `Fifo_Put` were called, then the FIFO would not be modified and a full error would be returned. In this way if `PutPt` equals `GetPt` at the beginning of `Fifo_Get`, then the FIFO is empty. Similarly, if `PutPt+1` equals `GetPt` at the beginning of `Fifo_Put`, then the FIFO is full. Be careful to wrap the `PutPt+1` before comparing it to `Fifo_Get`. This second method does not require the length to be stored or calculated. The implementation of this FIFO module is shown in Program 4.14.

To check for FIFO full, the `Fifo_Put` routine in Program 4.14 attempts to put using a temporary `PutPt`. If putting makes the FIFO look empty, then the temporary `PutPt` is discarded and the routine is exited without saving the data. This is why a FIFO with 10 allocated bytes can hold only 9 data points. If putting doesn't make the FIFO look empty, then the temporary `PutPt` is stored into the actual `PutPt`, saving the data as desired.

```
;Two-pointer implementation of the FIFO          // Two-pointer implementation of the FIFO
FIFOSIZE   equ  10    ;can hold 9 elements         #define FIFOSIZE 10      // can hold 9
PutPt      rmb  2     ;where to put next           char static volatile *PutPt;  // put next
GetPt      rmb  2     ;where to get next           char static volatile *GetPt;  // get next
Fifo       rmb  FIFOSIZE                           char static Fifo[FIFOSIZE];
Fifo_Init ldx #Fifo                                void Fifo_Init(void){
           tpa        ;make atomic                 unsigned char SaveCCR;
           sei        ;critical section            asm   tpa
           stx PutPt                               asm   staa SaveCCR
           stx GetPt ;Empty                        asm   sei               // make atomic
           tap        ;end critical section          PutPt=GetPt=&Fifo[0]; // Empty
           rts                                     asm   ldaa SaveCCR
;******Put a byte into the FIFO******              asm   tap           // end critical section
;Input  RegA is 8-bit data to put                  }
;Output RegA is -1 if ok, 0 if full                int Fifo_Put(char data){
Fifo_Put   ldx  PutPt  ;Temporary                  char volatile *tempPt;
           staa 1,x+    ;Try to put data             tempPt = PutPt;
           cpx  #Fifo+FIFOSIZE                        *(tempPt++) = data;      // try to Put
           bne  PutNoWrap                            if(tempPt==&Fifo[FIFOSIZE]){
           ldx  #Fifo    ;Wrap                         tempPt = &Fifo[0];          // wrap
PutNoWrap clra                                        }
           cpx  GetPt    ;Full if same               if(tempPt == GetPt ){
           beq  PutDone                                return(0);     // Failed, fifo full
           coma          ;-1 means OK                  }
           stx  PutPt                                 else{
PutDone    rts                                          PutPt = tempPt;  // Success, update
;*******Get a byte from the FIFO*******                 return(1);
; Input  RegX place for 8-bit data                     }
; Output RegA is -1 if ok, 0 if empty               }
Fifo_Get   clra         ;assume it will fail        int Fifo_Get(char *datapt){
           ldy  GetPt                                 if(PutPt == GetPt ){
           cpy  PutPt ;Empty if the same               return(0);    // Empty if PutPt=GetPt
           beq  GetDone                               }
           coma       ;RegA=-1 means OK               else{
           ldab 1,y+ ;Data from FIFO                    *datapt = *(GetPt++);
           stab 0,x  ;Return by reference              if(GetPt==&Fifo[FIFOSIZE]){
           cpy  #Fifo+FIFOSIZE                            GetPt = &Fifo[0]; // wrap
           bne  GetNoWrap                               }
           ldy  #Fifo     ;Wrap                         return(1);
GetNoWrap sty  GetPt                                    }
GetDone    rts                                      }
```

Program 4.14

Two-pointer implementation of a FIFO.

To check for FIFO empty, the `Fifo_Get` routine in Program 4.14 simply checks to see if `GetPt` equals `PutPt`. If they match at the start of the routine, then `Fifo_Get` returns with the "empty" condition signified.

Since `Fifo_Put` and `Fifo_Get` have read-modify-write accesses to global variables, they themselves are not reentrant. Similarly, `Fifo_Init` has a multiple-step write-access to global variables. Therefore `Fifo_Init` is not reentrant. Consequently, interrupts could be temporarily disabled to prevent one thread from reentering these FIFO functions. Notice that at the end of the critical section, interrupts are not enabled, but rather the interrupt status is restored to its previous state (enabled or disabled). This method of disabling interrupts allows you to nest a function call from one nonreentrant function to another.

One advantage of this pointer implementation is that if you have a single thread that calls the Fifo_Get (e.g., the main program) and a single thread that calls the Fifo_Put (e.g., the serial port receive interrupt handler) as shown in Figure 4.4, then this Fifo_Put function can interrupt this Fifo_Get function without loss of data. So in this particular situation, interrupts would not have to be disabled. It would also operate properly if there were a single interrupt thread calling Fifo_Get (e.g., the serial port transmit interrupt handler) and a single thread calling Fifo_Put (e.g., the main program). On the other hand, if the situation is more general, and multiple threads could call Fifo_Put or multiple threads could call Fifo_Get, then the interrupts would have to be temporarily disabled.

4.3.3
Two-Pointer/
Counter FIFO
Implementation

The other method to determine if a FIFO is empty or full is to implement a counter. In the code of Program 4.15, Size contains the number of bytes currently stored in the FIFO. The advantage of implementing the counter is that FIFO quarter-full and three-quarter-full conditions are easier to implement. If you were studying the behavior of a system, it might be informative to measure the values of Size as a function of time.

```
; Pointer, counter implementation
FIFOSIZE  equ  10   ;can hold 10
PutPt     rmb  2    ;where to put next
GetPt     rmb  2    ;where to get next
Size      rmb  1    ;empty if Size=0
Fifo      rmb  FIFOSIZE
Fifo_Init ldx  #Fifo
          tpa        ;make atomic
          sei        ;critical section
          stx  PutPt
          stx  GetPt ;Empty
          clr  Size
          tap        ;end critical section
          rts
********Put a byte into the FIFO*******
* Input  RegA contains 8-bit data to put
* Output RegA is -1 if ok, 0 if full
Fifo_Put  pshc       ;save old CCR
          sei        ;make atomic
          ldab Size
          cmpb #FIFOSIZE ;Full ?
          bne  PNotFull
          clra
          bra  PutDone
PNotFull  inc  Size   ;one more element
          ldx  PutPt
          staa 1,x+   ;Put data into fifo
          cpx  #Fifo+FIFOSIZE
          bne  PutNoWrap
          ldx  #Fifo  ;Wrap
PutNoWrap ldaa #-1    ;success means OK
          stx  PutPt
PutDone   pulc        ;end critical section
          rts
```

```
// Pointer, counter implementation
#define FIFOSIZE 10    // can hold 10
char static volatile *PutPt;  // put next
char static volatile *GetPt;  // get next
char static Fifo[FIFOSIZE];
unsigned char Size;  // Number of elements
void Fifo_Init(void){
unsigned char SaveCCR;
asm  tpa
asm  staa SaveCCR
asm  sei              // make atomic
   PutPt=GetPt=&Fifo[0]; // Empty
   Size = 0;
asm  ldaa SaveCCR
asm  tap           // end critical section
}
int Fifo_Put(char data){
unsigned char SaveCCR;
   if(Size == FIFOSIZE){
      return(0);    // fifo was full
   }
   else{
asm  tpa
asm  staa SaveCCR
asm  sei              // make atomic
      Size++;  // one more element in FIFO
      *(PutPt++) = data;  // put data
      if(PutPt == &Fifo[FIFOSIZE]){
         PutPt = &Fifo[0]; // Wrap
      }
asm  ldaa SaveCCR
asm  tap           // end critical section
      return(-1);   // Successful
   }
}
```

```
*****Get a byte from the FIFO********        int Fifo_Get(char *datapt) { char SaveSP;
* Input  RegX points to place for data       unsigned char SaveCCR;
* Output RegA is -1 if ok, 0 if empty          if(Size == 0){
Fifo_Get  pshc    ;save old CCR                   return(0);  // Empty if Size=0
          sei     ;critical section            }
          clra    ;assume it will fail         else{
          tst  Size                        asm   tpa
          beq  GetDone                      asm   staa SaveCCR
          dec  Size   ;one less element     asm   sei            // make atomic
          ldy  GetPt                            *datapt=*(GetPt++);
          coma      ;RegA=-1 means OK           Size--;   // one less element in FIFO
          ldab 1,y+  ;Data from FIFO            if(GetPt == &Fifo[FifoSize]){
          stab 0,x   ;Return by reference         GetPt = &Fifo[0];  // wrap
          cpy  #Fifo+FIFOSIZE                    }
          bne  GetNoWrap                    asm   ldaa SaveCCR
          ldy  #Fifo ;Wrap                  asm   tap            // end critical section
GetNoWrap sty  GetPt                             return(-1);  // Successful
GetDone   pulc      ;end critical section      }
          rts                                }
```

Program 4.15
Implementation of a two-pointer with counter FIFO.

To check for FIFO full, the `Fifo_Put` routine in Program 4.15 simply compares `Size` to the maximum allowed value. If the FIFO is already full, then the routine is exited without saving the data. With this implementation a FIFO with 10 allocated bytes actually can hold 10 data points.

To check for FIFO empty, the `Fifo_Get` routine of Program 4.15 simply checks to see if `Size` equals 0. If `Size` is zero at the start of the routine, then `Fifo_Get` returns with the "empty" condition signified.

4.3.4 FIFO Dynamics

As you recall, the FIFO passes the data from the producer to the consumer. In general, the rates at which data are produced and consumed can vary dynamically. Human beings do not enter data into a keyboard at a constant rate. Even printers require more time to print color graphics versus black and white text. Let t_p be the time (in seconds) between calls to `Fifo_Put` and r_p be the arrival rate (producer rate in bytes per second) into the system. Similarly, let t_g be the time (in seconds) between calls to `Fifo_Get` and r_g be the service rate (consumer rate in bytes per second) out of the system.

$$r_g = \frac{1}{t_g} \qquad r_p = \frac{1}{t_p}$$

If the minimum time between calls to `Fifo_Put` is greater than the maximum time between calls to `Fifo_Get`,

$$\min t_p \geq \max t_g$$

then a FIFO is not necessary and the data flow program could be solved with a simple global variable. On the other hand, if the time between calls to `Fifo_Put` becomes less than the time between calls to `Fifo_Get` because either:

- The arrival rate temporarily increases
- The service rate temporarily decreases

then information will be collected in the FIFO. For example, a person might type very fast for a while, followed by a long pause. The FIFO could be used to capture without loss all the data as they come in very fast. Clearly on average the system must be able to process the data (the consumer thread) at least as fast as the average rate at which the data arrives (producer thread). If the average producer rate is larger than the average rate the consumer is capable of handling data,

$$\bar{r}_p > \bar{r}_g$$

then the FIFO will eventually overflow no matter how large the FIFO. If the producer rate is temporarily high and that causes the FIFO to become full, then this problem can be solved by increasing the FIFO size.

There is a fundamental difference between an empty error and a full error. Consider the application of using a FIFO between your computer and its printer. This is a good idea because the computer can temporarily generate data to be printed at a very high rate, followed by long pauses. The printer is like a turtle. It can print at a slow but steady rate. The computer will put a byte into the FIFO that it wants printed. The printer will get a byte out of the FIFO when it is ready to print another character. A full error occurs when the computer calls `Fifo_Put` at too fast a rate. A full error is serious, because if ignored data will be lost. On the other hand, an empty error occurs when the printer is ready to print but the computer has nothing in mind. An empty error is not serious, because in this case the printer just sits there doing nothing.

Checkpoint 4.2: If the FIFO becomes full, can the situation be solved by increasing the size?

Figure 4.17 shows a data flow graph with buffered input and buffered output. FIFOs implemented in this section are statically allocated global structures. The system shown in Figure 4.17 has two channels, one for input and one for output, and each channel employs a separate FIFO queue. An assembly language implementation of this system can be found as TUT4 within TExaS, and a C implementation as project SCIA on the CD. The details of this implementation will be presented in Chapter 7.

Figure 4.17
A data flow graph showing two FIFOs that buffer data between producers and consumers.

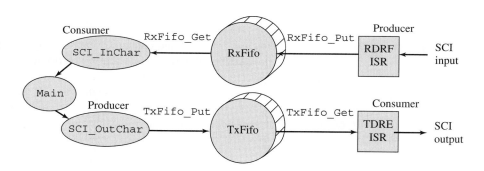

Checkpoint 4.3: What does it mean if the RxFifo in Figure 4.17 is empty?

Checkpoint 4.4: What does it mean if the TxFifo in Figure 4.17 is empty?

4.4 General Features of Interrupts on the 6811/6812

In this section we will present the specific details for each of the 6811/6812 microcomputers. As you develop experience using interrupts, you will come to notice a few common aspects that most computers share. The following paragraphs outline three essential mechanisms that are needed to utilize interrupts. Although every computer that uses interrupts includes all three mechanisms, there is a wide spectrum of implementation methods.

All interrupting systems have the *ability for the hardware to request action from the computer.* The interrupt requests can be generated by using a separate connection to the microprocessor for each device or by using a negative-logic wire-or requests employing open-collector logic. The Freescale microcomputers support both types.

All interrupting systems must have the *ability for the computer to determine the source.* A vectored interrupt system employs separate connections for each device so that the computer can give automatic resolution. You can recognize a vectored system because each device has a separate interrupt vector address. With a polled interrupt system, the software must poll each device, looking for the device that requested the interrupt.

The third necessary component of the interface is the *ability for the computer to acknowledge the interrupt.* Normally there is a flag in the interface that is set on the busy-to-done state transition. Acknowledging the interrupt involves clearing the flag that caused the interrupt. It is important to shut off the request so that the computer will not mistakenly request a second (and inappropriate) interrupt for the same condition. Some Intel systems use a hardware acknowledgment that automatically clears the request. Most Freescale microcomputers use a software acknowledge. So when we identify the flag, it will be important to know exactly what conditions will set the flag (and request an interrupt) and how the software will clear it (acknowledge) in the ISR.

Even though there are no standard definitions for the terms *mask, enable,* and *arm* in the computer science or computer engineering communities, in this book we will assign the following specific meanings. To *arm (disarm)* a device means to enable (shut off) the source of interrupts. One arms (disarms) a device if one is (is not) interested in interrupts at all. For example, the TOI bit is the arm bit for the TOF flag. Similarly the 6812 has eight arm bits (C7I–C0I) for the output compare and input capture interrupts. The Freescale literature calls the arm bit an "interrupt enable mask." To *enable (disable)* means to allow interrupts at this time (postpone interrupts until a later time). We disable interrupts if it is currently not convenient to process. In particular, to disable interrupts we set the I bit in 6811 and 6812 condition code register using the `sei` instruction. There are some interrupts that cannot be disabled, such as XIRQ, SWI, illegal opcode, reset. For most of this book (except in Chapter 5) we will disable interrupts while executing in a vulnerable window.

> **Common error:** The system will crash if the ISR does not either acknowledge or disarm the device requesting the interrupt.

> **Common error:** The ISR software does not have to explicitly disable interrupts at the beginning (`sei`) or explicitly reenable interrupts at the end (`cli`).

The sequence of events that occurs during an interrupt service is quite similar on the 6811 and 6812. The sequence begins with the *Hardware needs service (busy-to-done) transition.* This signal is connected to an input of the microcomputer that can generate an interrupt. For example, the STRA (6811 only), key wakeup, input capture, SCI, and SPI systems support interrupt requests. Some interrupts are internally generated like output compare, real-time interrupt (RTI), and timer overflow.

The second event is the *setting of a flag* in one of the I/O status registers of the microcomputer. This is the same flag that a gadfly interface would be polling on. Examples include the STRA (*STAF*), key wakeup (*KWIFJ2*), input capture (*IC3F*), SCI (*RDRF*), SPI (*SPIF*), output compare (*OC5F*), RTI (*RTIF*), and timer overflow (*TOF*). For an interrupt to be requested the appropriate *flag* bit must be *armed.* Examples include the STRA (*STAI*), key wakeup (*KWIEJ2*), input capture (*IC3I*), SCI (*RIE*), SPI (*SPIE*), output compare (*OC5I*), RTI (*RTII*), and timer overflow (*TOI*). The sequence order of these three conditions does not matter, as long as all three become true:

The interrupting event occurs that sets the flag   TOF$=$1
The device is armed   TOI$=$1
The microcomputer interrupts are enabled   I$=$0

The third event in the interrupt processing sequence is the *thread switch*. The thread switch is performed by the microcomputer hardware automatically. The specific steps include:

1. The microcomputer will finish the current instruction.[2]
2. All the registers are pushed on the stack (i.e., PC, Y, X, A, B, CCR) the **CCR** is on top, with the **I** bit still equal to 0.
3. The microcomputer will get vector address from memory and put it into the **PC**.
4. The microcomputer will set $I = 1$.

The fourth event is the software *execution of the ISR*. For a polled interrupt configuration the ISR must poll each possible device and branch to a specific handler for that device. The polling order establishes device priority. For a vectored interrupt configuration, you could poll anyway to check for run-time hardware/software errors. The ISR must either acknowledge or disarm the interrupt. We acknowledge an interrupt by clearing the flag that was set in the second event shown above. We will see later in the chapter that there is an optional step at this point for low-priority interrupt handlers. After we acknowledge a low-priority interrupt, we may reenable interrupts (`cli`) to allow higher-priority devices to go first. All ISRs must perform the necessary operations (read data, write data, etc.) and pass parameters through global memory (e.g., FIFO queue).

The last event is another thread switch to *return control back to the thread that was running* when the interrupt was processed. In particular, the software executes a `rti` at the end of the ISR, which will pull all the registers off the stack (i.e., CCR, B, A, X, Y, PC). Since the CCR was pushed on the stack in step 2 above with I=0, the execution of `rti` automatically reenables interrupts. After the ISR executes `rti`, the stack is restored to the state it was before the interrupt (there may be one more or less entry in the FIFO however).

4.4.1
6811 Interrupts

The 6811 interrupt hardware will automatically save all registers on the stack during the thread switch. The thread switch is the process of stopping the foreground (main) thread and starting the background (interrupt handler). The "oldPC" value on the stack points to the place in the foreground thread to resume once the interrupt is complete. At the end of the interrupt handler, another thread switch occurs as the `rti` instruction restores registers from the stack (including the PC) (Figure 4.18).

The 6811 has two external requests $\overline{IRQ}$ and $\overline{XIRQ}$ that are level-zero active. Many of the internal I/O devices can generate interrupt requests based on external events (e.g.,

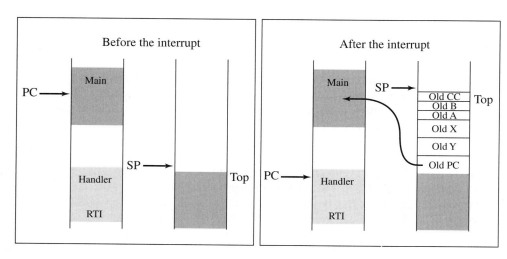

Figure 4.18
6811 stack before and after an interrupt.

---
[2]Again, the 6812 allows the instructions `rev`, `revw`, and `wav` to be interrupted.

STRA, input capture, SCI, SPI). Most of the interrupt requests will temporarily set the I bit in the CCR during the interrupt program to prevent other interrupts (including itself). On the other hand, the XIRQ request temporarily sets both the I and X bits in the CCR during the interrupt program to postpone all other interrupts sources. The 6811 can support:

A STRA interrupt
Three or four input capture interrupts
Five output compare interrupts
Three timer interrupts (timer overflow, RTI, pulse accumulator)
Two serial port interrupts (SCI and SPI)

The interrupts have a fixed priority, but you can elevate one request to highest priority using the hardware priority interrupt (HPRIO) register ($103C).

We typically use XIRQ to interface a single highest-priority device. XIRQ has a separate interrupt vector ($FFF4) and a separate enable bit (X). Once the X bit is cleared (enabled), the software cannot disable it. A XIRQ interrupt is requested when the external XIRQ pin is low and the X bit in the CCR is 0. XIRQ processing will automatically set **X = I = 1** (an IRQ cannot interrupt an XIRQ service) at the start of the XIRQ handler. Just like regular interrupts, the X and I bits will be restored to their original values by the `rti` instruction.

4.4.2
6812 Interrupts

CPU12 exceptions include resets and interrupts. Each exception has an associated 16-bit vector that points to the memory location where the routine that handles the exception is located. Vectors are stored in the upper 128 bytes of the standard 64-kbyte address map.

A hardware priority hierarchy determines which reset or interrupt is serviced first when simultaneous requests are made. Six sources are not maskable. The remaining sources have a mask bit that can be enabled (armed) or turned off (disarmed). The priorities of the non-maskable sources are:

1. Power-on-reset (POR) or regular hardware RESET pin
2. Clock monitor reset
3. COP watchdog reset
4. Unimplemented instruction trap
5. Software interrupt instruction (`swi`)
6. XIRQ signal (if X bit in CCR = 0)

Maskable interrupt sources include on-chip peripheral systems and external interrupt service requests. Interrupts from these sources are recognized when the global interrupt enable bit (I) in the CCR is cleared. The default state of the I bit out of reset is 1, but it can be written at any time.

The 6812 interrupt hardware also automatically saves all registers on the stack during the thread-switch (i.e., PC, Y, X, A, B, CCR). The "oldPC" value on the stack points to the place in the foreground thread to resume once the interrupt is complete. At the end of the interrupt handler, another thread switch occurs as the `rti` instruction restores registers from the stack (including the PC) (Figure 4.19).

Checkpoint 4.5: During the startup of an ISR, which happens first: the CCR is pushed on the stack, or the I bit is set?

Like the 6811, the 6812 has two external requests $\overline{\text{IRQ}}$ and $\overline{\text{XIRQ}}$ that are level-zero active. Many of the internal I/O devices can generate interrupt requests based on external events (e.g., key wakeup, input capture, SCI, SPI). These interrupt requests will temporarily set the I bit in the CCR during the interrupt program to prevent other interrupts (including itself). On the other hand, the XIRQ request temporarily sets both the I and X bits in the CCR during the interrupt program to postpone all other interrupts sources. The MC68HC812A4 can support:

24 key wakeup interrupts (Ports D, H, and J)
Eight input capture/output compare interrupts

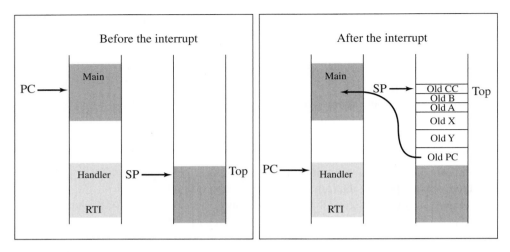

Figure 4.19
6812 stack before and after an interrupt.

An ADC interrupt
Three timer interrupts (timer overflow, RTI, pulse accumulator),
Three serial port interrupts (two SCIs and SPI)

The MC9S12C32 can support

10 key wakeup interrupts (Ports J and P)
8 input capture/output compare interrupts
An ADC interrupt
4 timer interrupts (timer overflow, real-time interrupt, 2 pulse accumulator)
2 serial port interrupts (SCI and SPI)
4 CAN interrupts

The interrupts have a fixed priority, but you can elevate one request to highest priority using the HPRIO register ($001F). The relative priorities of the other interrupt sources remain the same.

The 6812 XIRQ interrupt system is similar to the one on the 6811. We typically use XIRQ to interface a single highest-priority device. XIRQ has a separate interrupt vector ($FFF4) and a separate enable bit (X). Once the X bit is cleared (enabled), the software cannot disable it. A XIRQ interrupt is requested when the external XIRQ pin is low and the X bit in the CCR is 0. XIRQ processing will automatically set **X=I=1** (an IRQ cannot interrupt an XIRQ service) at the start of the XIRQ handler. Just like regular interrupts, the X and I bits will be restored to their original values by the `rti` instruction.

Checkpoint 4.6: What makes XIRQ interrupts high priority?

4.5 Interrupt Vectors and Priority

4.5.1
MC68H11E
Interrupt Vectors
and Priority

The priority is fixed in the order shown in Table 4.3, with *SCI* having the lowest priority and Reset having the highest. For some interrupt sources, such as the SCI interrupts, flags are automatically cleared during the response to the interrupt requests. For example, the RDRF flag in the SCI system is cleared by the automatic clearing mechanism, consisting of a read of the SCI status register while RDRF is set, followed by a read of the SCI data

Table 4.3

Interrupt vectors for the 6811.

Vector	Interrupt source	Enable	Arm
Start:			
$FFFE	Power on reset	Always	Always highest priority
$FFFE	Hardware reset	Always	Always
COP:			
$FFFC	COP clock monitor fail	Always	OPTION.CME=1
$FFFA	COP failure	Always	CONFIG.NOCOP=0
External interrupts:			
$FFF4	Nonmaskable XIRQ	X=0	External hardware
$FFF2	External IRQ	I=0	External hardware (e.g., 6821, 6840, 6850)
Parallel I/O:			
$FFF2	Parallel I/O handshake, STAF	I=0	PIOC.STAI=1
Programmable timer:			
$FFF0	Real time interrupt, RTIF	I=0	TMSK2.RTII=1
$FFEE	Timer input capture 1, IC1F	I=0	TMSK1.IC1I=1
$FFEC	Timer input capture 2, IC2F	I=0	TMSK1.IC2I=1
$FFEA	Timer input capture 3, IC3F	I=0	TMSK1.IC3I=1
$FFE8	Timer output compare 1, OC1F	I=0	TMSK1.OC1I=1
$FFE6	Timer output compare 2, OC2F	I=0	TMSK1.OC2I=1
$FFE4	Timer output compare 3, OC3F	I=0	TMSK1.OC3I=1
$FFE2	Timer output compare 4, OC4F	I=0	TMSK1.OC4I=1
$FFE0	Timer output compare 5, OC5F	I=0	TMSK1.OC5I=1
$FFDE	Timer overflow, TOF	I=0	TMSK2.TOI=1
Pulse accumulator:			
$FFDC	Pulse accumulator overflow	I=0	TMSK2.PAOVI=1
$FFDA	Pulse accumulator input edge	I=0	TMSK2.PAII=1
SPI:			
$FFD8	SPI serial transfer complete, SPIF	I=0	SPCR.SPIE=1
SCI:			
$FFD6	Receive data register full, RDRF	I=0	SCCR2.RIE=1
$FFD6	Receive overrun, OVRN	I=0	SCCR2.RIE=1
$FFD6	Transmit data register empty, TDRE	I=0	SCCR2.TIE=1
$FFD6	Transmit complete, TC	I=0	SCCR2.TCIE=1
$FFD6	Idle line detect, IDLE	I=0	SCCR2.ILIE=1
Software interrupts:			
$FFF8	Illegal opcode trap	Always	Always
$FFF6	Software interrupt SWI	Always	Always lowest priority

register. The normal response to an RDRF interrupt request is to read the SCI status register to check for receive errors, then to read the received data from the SCI data register. These two steps satisfy the automatic clearing mechanism without requiring any special instructions. Many of the potential interrupt requests share the same interrupt vector (e.g., there are multiple possible interrupt sources [STAF and all external connections to IRQ] that all use the vector at $FFF2). Therefore, when this request is processed, the software must determine which of the possible signals caused the interrupt. The entire list of potential 6811 interrupts is included in Table 4.3.

How we establish the vector depends on the memory configuration. For a stand-alone single-chip 6811 system or multichip systems with your own PROM at $C000 to $FFFF, we can use assembly language to set the vectors.

```
org    $FFF0
fdb    RTIHAN      Pointer to real time interrupt handler
org    $FFF2
fdb    IRQHAN      Pointer to external IRQ and STRA handler
```

```
org   $FFF4
fdb   XIRQHAN    Pointer to external XIRQ handler
org   $FFFE
fdb   RESETHAN   Pointer to hardware reset and power on reset handler
```

With the 6811 EVB, we usually use a $C000 to $FFFF 8K PROM containing the debugger called BUFFALO. Since the interrupt vector addresses are in this PROM (something you do not wish to change while developing 6811 EVB software), BUFFALO creates jumps into RAM. You must place a 3-byte `jmp` instruction at these reserved RAM memory locations. Table 4.4 lists all BUFFALO RAM interrupt jump locations.

Table 4.4
Interrupt vectors for the 6811 EVB using BUFFALO.

Interrupt source	6811 vector	BUFFALO jump location
SCI	$FFD6	$00C4–$00C6
SPI	$FFD8	$00C7–$00C9
Pulse accumulator input edge	$FFDA	$00CA–$00CC
Pulse accumulator overflow	$FFDC	$00CD–$00CF
Timer overflow	$FFDE	$00D0–$00D2
Timer output compare 5	$FFE0	$00D3–$00D5
Timer output compare 4	$FFE2	$00D6–$00D8
Timer output compare 3	$FFE4	$00D9–$00DB
Timer output compare 2	$FFE6	$00DC–$00DE
Timer output compare 1	$FFE8	$00DF–$00E1
Timer input capture 3	$FFEA	$00E2–$00E4
Timer input capture 2	$FFEC	$00E5–$00E7
Timer input capture 1	$FFEE	$00E8–$00EA
RTI	$FFF0	$00EB–$00ED
IRQ (6821,6840,6850, and STAF)	$FFF2	$00EE–$00F0
XIRQ (external hardware)	$FFF4	$00F1–$00F3
Software interrupt (SWI)	$FFF6	$00F4–$00F6
Illegal opcode	$FFF8	$00F7–$00F9
COP	$FFFA	$00FA–$00FC
Clock monitor	$FFFC	$00FD–$00FF
Reset	$FFFE	

In each case, a 3-byte value is stored in the jump location similar to the following examples. The jump opcode ($7E) is stored in the first byte, followed by the 16-bit address of the device handler. For systems with the BUFFALO monitor add this code to the ritual

```
ldaa  #$7E      Op code for JMP
staa  $00EB
ldx   #RTIHAN
stx   $00EC     JMP RTIHAN
ldaa  #$7E      Op code for JMP
staa  $00EE
ldx   #IRQHAN
stx   $00EF     JMP IRQHAN
ldaa  #$7E      Op code for JMP
staa  $00F1
ldx   #XIRQHAN
stx   $00F2     JMP XIRQHAN
```

**4.5.2
MC68HC812A4
Interrupt Vectors
and Priority**

The priority is fixed in the order shown in Table 4.5, with *Key Wakeup H* having the lowest priority and *Reset* having the highest. Any one particular system usually uses just a few interrupts, but the entire list of potential MC68HC812A4 interrupts is included in Table 4.5.

Table 4.5
Interrupt vectors for the MC68HC812A4.

Vector address	CodeWarrior number	Interrupt source	CCR mask	Local arm
$FFFE	0	Reset	None	None
$FFFC	1	COP clock monitor fail reset	None	COPCTL.CME
			None	COPCTL.FCME
$FFFA	2	COP failure reset	None	COP rate selected
$FFF8	3	Unimplemented instruction trap	None	None
$FFF6	4	SWI	None	None
$FFF4	5	XIRQ	X bit	None
$FFF2	6	IRQ or	I bit	INTCR.IRQEN
		Key wakeup D	I bit	KWIED,[7:0]
$FFF0	7	RTI, RTIF	I bit	RTICTL.RTIE
$FFEE	8	Timer channel 0, C0F	I bit	TMSK1.C0I
$FFEC	9	Timer channel 1, C1F	I bit	TMSK1.C1I
$FFEA	10	Timer channel 2, C2F	I bit	TMSK1.C2I
$FFE8	11	Timer channel 3, C3F	I bit	TMSK1.C3I
$FFE6	12	Timer channel 4, C4F	I bit	TMSK1.C4I
$FFE4	13	Timer channel 5, C5F	I bit	TMSK1.C5I
$FFE2	14	Timer channel 6, C6F	I bit	TMSK1.C6I
$FFE0	15	Timer channel 7, C7F	I bit	TMSK1.C7I
$FFDE	16	Timer overflow, TOF	I bit	TMSK2.TOI
$FFDC	17	Pulse accumulator overflow	I bit	PACTL.PAOVI
$FFDA	18	Pulse accumulator input edge	I bit	PACTL.PAI
$FFD8	19	SPI serial transfer complete, SPIF	I bit	SP0CR1.SPI0E
$FFD6	20	SCI 0 transmit buffer empty, TDE	I bit	SC0CR2.TIE
		SCI 0 transmit complete, TC	I bit	SC0CR2.TCIE
		SCI 0 receiver buffer full, RDF	I bit	SC0CR2.RIE
		SCI 0 receiver idle	I bit	SC0CR2.ILIE
$FFD4	21	SCI 1 transmit buffer empty, TDE	I bit	SC1CR2.TIE
		SCI 1 transmit complete, TC	I bit	SC1CR2.TCIE
		SCI 1 receiver buffer full, RDF	I bit	SC1CR2.RIE
		SCI 1 receiver idle	I bit	SC1CR2.ILIE
$FFD2	22	ATD sequence complete	I bit	ATDCTL2.ASCIE
$FFD0	23	Key wakeup J (stop wakeup)	I bit	KWIEJ.[7:0]
$FFCE	24	Key wakeup H (stop wakeup)	I bit	KWIEH.[7:0]
$FF80–$FFCD		Reserved	I bit	

For some interrupt sources, such as the SCI interrupts, flags are automatically cleared during the response to the interrupt requests. For example, the RDRF flag in the SCI system is cleared by the automatic clearing mechanism, consisting of a read of the SCI status register while RDRF is set, followed by a read of the SCI data register. The normal response to an RDRF interrupt request is to read the SCI status register to check for receive errors, then to read the received data from the SCI data register. These two steps satisfy the automatic clearing mechanism without requiring any special instructions. Many of the potential interrupt requests share the same interrupt vector. (e.g., there are eight possible interrupt sources [PJ7-PJ0] that all use the vector at $FFD0). Therefore, when this request is processed, the software must determine which of the eight possible signals caused the interrupt.

4.5.3 MC68HC912B32 Interrupt Vectors and Priority

The priority is fixed in the order shown in Table 4.6, with *BDLC* having the lowest priority and Reset having the highest. Many of the potential interrupt requests share the same interrupt vector. For example, there are four possible SCI interrupt sources (TDRE, TC, RDRF, IDLE) that all use the vector at $FFD6. Therefore, when this request is processed, the software must determine which of the four possible signals caused the interrupt. Any one partic-ular system usually uses just a few interrupts, but the entire list of potential MC68HC912B32 interrupts is included in Table 4.6.

Table 4.6
Interrupt vectors for the MC68HC912B32.

Vector address	CodeWarrior number	Interrupt source	CCR mask	Local enable
$FFFE	0	Reset	None	None
$FFFC	1	COP clock monitor fail reset	None	COPCTL.CME
			None	COPCTL.FCME
$FFFA	2	COP failure reset	None	COP rate selected
$FFF8	3	Unimplemented instruction trap	None	None
$FFF6	4	SWI	None	None
$FFF4	5	XIRQ	X bit	None
$FFF2	6	IRQ	I bit	INTCR.IRQEN
$FFF0	7	RTI, RTIF	I bit	RTICTL.RTIE
$FFEE	8	Timer channel 0, C0F	I bit	TMSK1.C0I
$FFEC	9	Timer channel 1, C1F	I bit	TMSK1.C1I
$FFEA	10	Timer channel 2, C2F	I bit	TMSK1.C2I
$FFE8	11	Timer channel 3, C3F	I bit	TMSK1.C3I
$FFE6	12	Timer channel 4, C4F	I bit	TMSK1.C4I
$FFE4	13	Timer channel 5, C5F	I bit	TMSK1.C5I
$FFE2	14	Timer channel 6, C6F	I bit	TMSK1.C6I
$FFE0	15	Timer channel 7, C7F	I bit	TMSK1.C7I
$FFDE	16	Timer overflow, TOF	I bit	TMSK2.TOI
$FFDC	17	Pulse accumulator overflow	I bit	PACTL.PAOVI
$FFDA	18	Pulse accumulator input edge	I bit	PACTL.PAI
$FFD8	19	SPI serial transfer complete	I bit	SP0CR1.SPIE
$FFD6	20	SCI 0 transmit buffer empty, TDE	I bit	SC0CR2.TIE
		SCI 0 transmit complete, TC	I bit	SC0CR2.TCIE
		SCI 0 receiver buffer full, RDF	I bit	SC0CR2.RIE
		SCI 0 receiver idle	I bit	SC0CR2.ILIE
$FFD4	21	Reserved	I bit	—
$FFD2	22	ATD	I bit	ATDCTL2.ASCIE
$FFD0	23	BDLC	I bit	BCR1.IE
$FF80–$FFCF		Reserved	I bit	—

4.5.4 MC9S12C32 Interrupt Vectors and Priority

The priority is fixed in the order shown in Table 4.7, with *Key Wakeup P* having the lowest priority and *Reset* having the highest. Any one particular system usually uses just a few interrupts, but the entire list of potential MC9S12C32 interrupts is included.

For many of the interrupts, the flag is cleared by writing a one to it. On the other hand, for some interrupt sources, such as the SCI interrupts, flags are automatically cleared during the response to the interrupt requests. For example, the RDRF flag in the SCI system is cleared by the automatic clearing mechanism, consisting of a read of the SCI status register while RDRF is set, followed by a read of the SCI data register. The normal response to an RDRF interrupt request is to read the SCI status register to check for receive errors, then to read the received data from the SCI data register. These two steps satisfy the automatic clearing mechanism without requiring any special instructions. Many of the potential interrupt requests share the same interrupt vector. For example, there are eight possible interrupt sources (PP7-PP0) that all use the vector at $FF8E. Therefore, when this request is processed the software must determine which of the eight possible signals caused the interrupt.

Vector address	CodeWarrior number	Interrupt source	CCR mask	Local arm	HPRIO value to elevate
$FFFE	0	Reset	None	None	—
$FFFC	1	COP Clock Monitor Fail Reset	None	COPCTL.CME	—
			None	COPCTL.FCME	—
$FFFA	2	COP Failure Reset	None	COP rate selected	—
$FFF8	3	Unimplemented Instruction Trap	None	None	—
$FFF6	4	SWI	None	None	—
$FFF4	5	XIRQ	X bit	None	—
$FFF2	6	IRQ	I bit	INTCR.IRQEN	$F2
$FFF0	7	Real Time Interrupt, RTIF	I bit	CRGINT.RTIE	$F0
$FFEE	8	Timer Channel 0, C0F	I bit	TIE.C0I	$EE
$FFEC	9	Timer Channel 1, C1F	I bit	TIE.C1I	$EC
$FFEA	10	Timer Channel 2, C2F	I bit	TIE.C2I	$EA
$FFE8	11	Timer Channel 3, C3F	I bit	TIE.C3I	$E8
$FFE6	12	Timer Channel 4, C4F	I bit	TIE.C4I	$E6
$FFE4	13	Timer Channel 5, C5F	I bit	TIE.C5I	$E4
$FFE2	14	Timer Channel 6, C6F	I bit	TIE.C6I	$E2
$FFE0	15	Timer Channel 7, C7F	I bit	TIE.C7I	$E0
$FFDE	16	Timer Overflow, TOF	I bit	TIE.TOI	$DE
$FFDC	17	Pulse Acc. Overflow, PAOVF	I bit	PACTL.PAOVI	$DC
$FFDA	18	Pulse Acc. Input Edge, PAIF	I bit	PACTL.PAI	$DA
$FFD8	19	SPI Serial Transfer Complete, SPIF	I bit	SPICR1.SPIE	$D8
		SPI Transmit Empty, SPTEF		SPICR1. SPTIE	
$FFD6	20	SCI Transmit Buffer Empty, TDRE	I bit	SCICR2.TIE	$D6
		SCI Transmit Complete, TC		SCICR2.TCIE	
		SCI Receiver Buffer Full, RDRF		SCICR2.RIE	
		SCI Receiver Idle, IDLE		SCICR2.ILIE	
$FFD2	22	ATD Sequence Complete, ASCIF	I bit	ATDCTL2.ASCIE	$D2
$FFCE	24	Key Wakeup J, PIFJ.[7:6]	I bit	PIEJ.[7:6]	$CE
$FFB6	36	CAN wakeup	I bit	CANRIER.WUPIE	$B6
$FFB4	37	CAN errors	I bit	CANRIER.CSCIE	$B4
				CANRIER.OVRIE	
$FFB2	38	CAN receive	I bit	CANRIER.RXFIE	$B2
$FFB0	39	CAN transmit	I bit	CANTIER.TXEIE[2:0]	$B0
$FF8E	56	Key Wakeup P, PIFP[7:0]	I bit	PIEP.[7:0]	$8E

Table 4.7
Interrupt vectors for the MC9S12C32.

4.6 External Interrupt Design Approach

This short section deals with the design steps that occur when interfacing an I/O device with interrupt synchronization. Because the details of specific I/O devices can vary considerably, this section only serves as a framework, listing the issues we should consider when using interrupts. Computer engineering design, like all disciplines, is an iterative process. We will begin with the external hardware design.

First, we identify the busy-to-done state transition to cause interrupt. We ask the question, what status signal from the I/O device signifies when the input device has new data and needs the software to read them. For an output device, we look for a signal that specifies when it is finished and needs the software to give it more data. As shown in Figure 4.20 this signal will be an output of the I/O device and an input to the microcomputer.

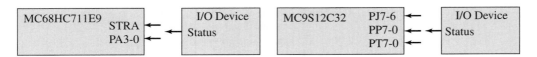

Figure 4.20
We connect external devices to microcomputer lines that can generate interrupts.

Next, we connect I/O status signal to a microcomputer input that can generate interrupts. On the MC68HC11, we could use STRA or input capture on Port A. On the MC68HC812A4, we could use the Key wakeup on Ports D, J, H, or input capture on Port T. On the MC68HC912B32, we could use input capture on Port T. On the MC9S12C32, we could use Key wakeup on Ports J, P or input capture on Port T.

There are four major components to the interrupting software system. In no particular order, they are the ritual, the main program, the ISR(s), and the interrupt vectors.

The *ritual* is executed once on start-up. It is probably a good idea to disable interrupts at the beginning of the ritual so that interrupts are not requested before the entire initialization sequence is allowed to finish. The ritual usually initializes the global data structures (e.g., Fifo_Init), sets the I/O port direction register(s) as needed, and sets the I/O port interrupt hardware control register, specifying the proper conditions to request the interrupt. An optional step is to clear the flag (the one that when set will request an interrupt). We like to add this optional step so that the first interrupt to occur will be the result of activity that occurred after the ritual, and not the result of activities prior to executing the ritual. For example, it is possible for the power-on sequence to different hardware modules to be different, causing edges to occur on digital lines that falsely mimic actual I/O activity. The last steps of the ritual are to arm the device and enable interrupts.

The *main program* executes in the foreground. At the start the SP is initialized, and the ritual is executed. The main program generally performs tasks that are not time-critical. The main program interacts with the ISRs via global memory data structures (e.g., FIFO queue).

The *interrupt handler* executes in the background. For a polled interrupt, it must determine the source by polling potential devices. The polling order will establish a priority, although later in the chapter we discuss more sophisticated methods for implementing priority. The ISR must acknowledge (clear the flag that requested the interrupt) or disarm. We acknowledge if we are interested in more interrupts and disarm if we are no longer interested in interrupts. Information is exchanged with the main program and other interrupt handlers (including itself) via global memory. The ISR executes rti to return control back to the program previously executing. Because of the real-time nature of interrupting devices, debugging tools must be minimally intrusive. Examples of good debugging tools to use for interrupts are (1) instrumentation that dumps into an array, (2) instrumentation that dumps into an array with filtering, and (3) instrumentation using an output port. The following are examples of bad debugging tools because they significantly affect the dynamic response of the interface (i.e., they require too much time to execute): (1) single-stepping, (2) breakpoints, and (3) print statements. These techniques were discussed in Section 2.11.

The last component is the *interrupt vectors*. On general-purpose computers, interrupt vectors are in RAM, and the software dynamically attaches and unattaches interrupt handlers to these vectors. On most embedded systems, the interrupt vectors reside in PROM or ROM and their values are determined at compile time and initialized by the ROM programmer. Nevertheless, we must establish interrupt vectors to point to the appropriate ISR. The syntax for setting vectors is compiler-dependent. Program 4.16 establishes an interrupt vector using the ImageCraft ICC12 and Metrowerks CodeWarrior compilers.

```
// ImageCraft ICC12 C for MC9S12C32        // Metrowerks CodeWarrior C for MC9S12C32
unsigned short Count;                       unsigned short Count;
#pragma interrupt_handler RTIHan()
void RTIHan(void){                          void interrupt 7 RTIHan(void){
  CRGFLG = 0x80;  // ack, clear RTIF          CRGFLG = 0x80;  // ack, clear RTIF
  Count++;         // number of interrupts    Count++;          // number of interrupts
}                                           }
#pragma abs_address:0xfff0
void (*RTI_vector[])() = { RTIHan };
#pragma end_abs_address
void RTI_Init(void){                        void RTI_Init(void){
  asm(" sei"); // Make ritual atomic          asm sei        // Make ritual atomic
  CRGINT = 0x80; // RTIE=1 enable rti         CRGINT = 0x80; // RTIE=1 enable rti
  RTICTL = 0x33; // 4096us or 244.14Hz        RTICTL = 0x33; // 4096us or 244.14Hz
  Count = 0;     // interrupt counter         Count = 0;     // interrupt counter
  asm(" cli");                                asm cli
}                                           }
```

Program 4.16
ICC12 and CodeWarrior C syntax to set an interrupt vector.

4.7 Polled versus Vectored Interrupts

As we defined earlier, when more than one source of interrupt exists, the computer must have a reliable method to determine which interrupt request has been made. There are two common approaches, and the computers studied in this book (6811/6812) apply a combination of both methods. The first approach is called vectored interrupts. With a vectored interrupt system each potential interrupt source has a unique interrupt vector address (Program 4.17). You simply place the correct handler address in each vector, and the hardware automatically calls the correct software when an interrupt is requested. One external interrupt request is shown in Figure 4.21 and another interrupt is generated by a timer overflow.

```
;MC68HC711E9                                ;MC9S12C32
TimeHan ldaa #$80  ;TOF is bit 7            TimeHan movb #$80,TFLG2 ;clear TOF
        staa TFLG2 ;clear TOF               ;*Timer interrupt calculations*
;*Timer interrupt calculations*                     rti
        rti
                                            ExtHan  movb #$80,PIFJ ;clear flag7
ExtHan  ldaa PIOC                           ;*External interrupt calculations*
        ldaa PORTCL ;clear STAF                     rti
;*External interrupt calculations*
        rti                                         org  $FFDE  ;timer overflow
        org  $FFDE  ;timer overflow                 fdb  TimeHan
        fdb  TimeHan                                org  $FFD0  ;Key wakeup J
        org  $FFF2  ;IRQ external                   fdb  ExtHan
        fdb  ExtHan
```

Program 4.17
Example of a vectored interrupt.

Figure 4.21
An external I/O device is connected to the microcomputer.

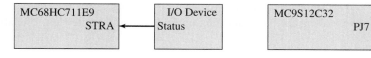

Since the two sources have separate vectors, the timer interrupt hardware (timer overflow TOF) will automatically activate `TimeHan` and the external interrupt hardware will automatically activate `ExtHan`.

The second approach is called polled interrupts. With a polled interrupt system, multiple interrupt sources share the same interrupt vector address (Program 4.18). Once the interrupt has occurred, the ISR software must poll the potential devices to determine which device needs service. Two external interrupt requests are shown in Figure 4.22.

```
;MC68HC711E9                          ;MC9S12C32
ExtHan  ldaa PIOC    ;which one       ExtHan brset PIFJ,$80,KJ7Han
        bita #$80   ;STAF?                   brset PIFJ,$40,KJ6Han
        bne  STAFHan                         swi       ;error
        ldaa OtherStatus
        bita #$80   ;External?
        bne  OtherHan
        swi         ;error
STAFHan ldaa PORTCL ;clear STAF        KJ7Han movb #$80,PIFJ ;clear flag7
;*STAF interrupt calculations*         ;*KJ7 interrupt calculations*
        rti                                   rti
OtherHan ldaa OtherData                KJ6Han movb #$40,PIFJ ;clear flag6
;*Other interrupt calculations*        ;*KJ6 interrupt calculations*
        rti                                   rti
        org  $FFF2  ;IRQ external              org  $FFCE   ;Key wakeup J
        fdb  ExtHan                           fdb  ExtHan
```

Program 4.18
Example of a polled interrupt.

Figure 4.22
When two or more devices share an interrupt line, then the ISR must poll.

Since the two sources have the same vector, the ISR software must first determine which one caused the interrupt. The 6811/6812 systems have a separate acknowledgment so that if both interrupts are pending, acknowledging one will not satisfy the other, so the second device[3] will request a second interrupt and get serviced.

The sequence of requests to the 6811/6812 does not matter because each request has a separate acknowledgement.

Common error: If two interrupts were requested, it would be a mistake to service just one and acknowledge them both.

Observation: External events are often asynchronous to program execution, so careful thought is required to consider the effect if an external interrupt request were to come in between each pair of instructions.

Observation: The computer automatically sets the I bit during processing so that an interrupt handler will not interrupt itself.

Two polling techniques are presented in this book. The first is *minimal* polling. This method performs a minimally sufficient check to determine the source. Usually this entails simply checking the particular flag bit that caused the interrupt. The above interrupt polling are

---

[3]This other device could be a 6821 parallel port connected to the 6811 in expanded mode.

examples of minimal polling. A more robust polling method is called *polling for 0s and 1s.* This verifies as much information in the control/status register as possible and usually entails checking for the presence of both 1s and 0s in the control/status register. For example, assume the interrupt flag bit is in bit 7, bit 6 is unknown, and bits 5–0 should be 000111. We write this expected condition as 1x000111. Simple polling only checks the flag bit, while polling for 0s and 1s compares the actual 7 bits with their expected values. One implementation is

```
ldaa CSR        read control/status register
anda #%1011111  clear bits that are indeterminate
cmpa #%1000111  expected value if this device active
beq  Handler    execute if this device is requesting
```

It is good software engineering practice to install consistency checks during the prototype and debugging stages. Once debugged, the system can be optimizing by removing them.

4.8 Keyboard Interface Using Interrupts

The interface in Figure 4.23 uses interrupts to input characters from the keyboard. When the user types a key on this keyboard, the 7-bit ASCII code becomes available on the **DATA**, followed by a rise in the signal **STROBE** that causes an interrupt. The data remain available until the next key is typed (Figure 4.24). The 6811 Port C hardware will latch the data in on the rising edge of **STROBE**. The software will poll to verify the interrupt has occurred, then call the handler that inputs the data, acknowledges the interrupt, and puts the character into a FIFO (Figure 4.25).

The ritual and interrupt programs in assembly language are shown in Program 4.19.

Performance tip: In the early stages of a development project, inserting consistency checks (like polling in the above example when it was not necessary) can identify

Figure 4.23
A keyboard is interfaced to the microcomputer.

Figure 4.24
The keyboard timing diagram.

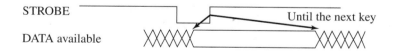

Figure 4.25
A flowchart of the interrupting keyboard interface.

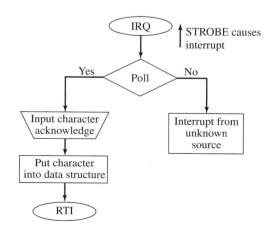

```
; MC68HC711E9                          ; MC9S12C32
; PC6-PC0 inputs = keyboard DATA       ; PT6-PT0 inputs = keyboard DATA
; STRA=STROBE interrupt on rise        ; PJ7=STROBE interrupt on rise
; 6 STAI 1  Interrupts armed           Key_Init sei          ; make atomic
; 5 CWOM 0  Normal outputs                    movb #$80,DDRT  ;PT7 unused output
; 4 HNDS 0  No handshake                      bclr DDRJ,#$80  ;PJ7 input
; 3 OIN  0                                     bset PPSJ,#$80  ;rise on PJ7
; 2 PLS  0  STRB not used                      bset PIEJ,#$80  ;arm PJ7
; 1 EGA  1  STAF set on rise of READY          movb #$80,PIFJ  ;clear flag7
; 0 INVB 0  STRB not used                      clr  PTT        ;PT7=0
Key_Init sei        ;Mmake atomic             jsr  Fifo_Init
        ldaa #$80   ;PC7 is an output         cli             ;Enable IRQ
        staa DDRC   ;PC6-0 inputs             rts
        ldaa #$42
        staa PIOC
        ldaa PIOC   ;clears STAF
        ldaa PORTCL
        clr  PORTC  ;Make PC7=0
        jsr  Fifo_Init
        cli         ;Enable IRQ
        rts
ExtHan ldaa PIOC    ;poll STAF        ExtHan brset PIFJ,$80,KeyHan
        bmi  KeyHan                          swi          ;error
        swi         ;error            KeyHan movb #$80,PIFJ  ;ack, clear flag7
KeyHan ldaa PORTCL  ;clear STAF              ldaa PTT        ;read key
        jsr  Fifo_Put                        jsr  Fifo_Put
        rti                                  rti
        org  $FFF2  ;IRQ external            org  $FFCE  ;Key wakeup J
        fdb  ExtHan                          fdb  ExtHan
```

Program 4.19
Interrupting keyboard software.

hardware and software bugs. Once the system is thoroughly tested, you could remove the checks for enhanced speed.

Observation: When power is applied to a system, each device turns on separately; therefore, it is possible to get initial rising and/or falling edges that do not represent actual I/O events. Thus, it is good software practice to clear interrupt flags in the initialization so that the first interrupt represents the first I/O event.

The latency for this system is the time between the rise of **STROBE**, and the time when the software reads the input data. The logical analyzer output during the 6811 interrupt request would be as follows. *The 14 italicized cycles represent the thread switch performed automatically in hardware by the 6811.* Latency is 28 cycles (+1 instruction) for the 6811. When running on a 6811 EVB with BUFFALO, you have to add the jmp instruction.

STRA	R/W	Address	Data	Comment
0	1	F100	B7	STAA $1234 instruction in MAIN
1	1	F101	12	Key typed rise of STROBE
1	1	F102	34	Finish instruction (IRQ requested)
1	0	1234	55	Write to memory
1	*1*	*F103*		*Start to process IRQ, ??dummy PC??*
1	*1*	*F104*		*??dummy PC read??*
1	*0*	*SP*	*03*	*PCL push all registers*
1	*0*	*SP-1*	*F1*	*PCH*

1	0	SP-2		YL
1	0	SP-3		YH
1	0	SP-4		XL
1	0	SP-5		XH
1	0	SP-6		A
1	0	SP-7		B
1	0	SP-8		CC (with I=0)
1	1	SP-8		??dummy stack read??, then I=1
1	1	FFF2	00	High byte of IRQ vector fetch
1	1	FFF3	EE	Low byte of IRQ vector fetch
1	1	00EE	7E	op code jmp INTHAN (if using BUFFALO)
1	1	00EF	81	operand high byte of INTHAN (if using BUFFALO)
1	1	00F0	23	operand low byte of INTHAN (if using BUFFALO)
1	1	8123	B6	op code for LDAA
1	1	8124	10	operand high byte of 1002 extended addr
1	1	8125	02	operand low byte of 1002
1	1	1002	C2	Read PIOC with STAF set
1	1	8126	28	Op code for BMI
1	1	8127	03	PC relative offset
1	1	FFFF		Null
1	1	812B	B6	Op code for LDAA
1	1	812C	10	operand high byte of 1005 extended addr
1	1	812D	05	operand low byte of 1005
1	1	1005	55	Read CL data, STAF=0, Acknowledge

Observation: The CycleView mode of the TExaS simulator allows you to observe the bus activity during an interrupt service, both the thread switch from main to ISR and from ISR to main.

The interrupt programs, for the 6811/6812 are shown in Program 4.20.

Observation: Data are lost when the FIFO gets full.

```
// MC68HC711E9 ICC11 C
// PC6-PC0 inputs = keyboard DATA
// STRA=STROBE interrupt on rise
void Key_Init(void){
unsigned char dummy;
asm(" sei");
  PIOC = 0x42;    // EGA=1, STAI
  DDRC = 0x80;    // STRA=STROBE
  PORTC = 0x00;   // PC7=0
  dummy = PIOC;  dummy=PORTCL;
  Fifo_Init();
asm(" cli");
}
#pragma interrupt_handler ExtHan()
void ExtHan(void){
  if((PIOC&STAF)==0)asm(" swi");
  Fifo_Put(PORTCL); // ack
}
#pragma abs_address:0xfff2
void (*IRQ_vector[])() = { ExtHan };
#pragma end_abs_address
```

```
// MC9S12C32 CodeWarrior C
// PT6-PT0 inputs = keyboard DATA
// PJ7=STROBE interrupt on rise
void Key_Init(void){
asm sei
  DDRT = 0x80;  // PT6-0 DATA
  DDRJ &= ~0x80;
  PPSJ|= 0x80;  // rise on PJ7
  PIEJ|= 0x80;  // arm PJ7
  PIFJ = 0x80;  // clear flag7
  Fifo_Init();
asm cli
}

void interrupt 24 ExtHan(void){
  if((PIFJ&0x80)==0){
    asm swi
  }
  PIFJ =0x80;  // clear flag
  Fifo_Put(PTT);
}
```

Program 4.20
Interrupting keyboard software.

4.9 Printer Interface Using IRQ Interrupts

To output a character on this printer (Figure 4.26), the user first outputs the 7-bit ASCII code to the **DATA**, followed by a pulse on the signal **START**. The completion of the output operation is signified by the rise of **READY**, which causes an interrupt. This printer interrupt example will output one line of characters, then stop. A FIFO could have been used to create a continuous data flow.

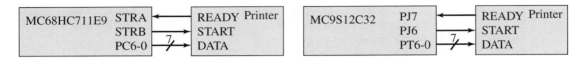

Figure 4.26
The hardware interface of a printer to the microcomputer.

For the first character, which is started in the ritual, the timing (Figure 4.27) looks similar to the gadfly solution presented in Chapter 3. But for subsequent characters, the timing (Figure 4.28) begins with the busy-to-done state transition, which in this case is the rise of **READY**. The rise of **READY** causes an interrupt. The interrupt software outputs new data to the printer and issues another **START** pulse. For the 6811, the acknowledge function is the same as the service function (STAA PORTCL). The other implementations will have an explicit acknowledge operation.

Figure 4.27
Timing diagram of a printer showing the steps to print the first character.

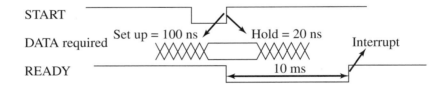

Figure 4.28
Timing diagram of a printer showing the steps during an interrupt service.

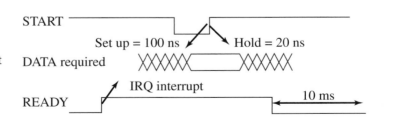

We begin this interface with a single-buffer data structure, which contains ASCII characters terminated by the null (0) character. Fill is used to copy new data into the Line buffer during initialization, and Get will be called within the interrupt handler to get more data. This data structure could be enhanced by checking for pointer overflow within both the Fill and Get routines (Program 4.21).

Next, we have the ritual Init that initializes the data structure, arming the device so that the rise of READY causes an interrupt. The system is started by writing the first data from the ritual (Program 4.22).

```
; MC68HC711E9                          ; MC9S12C32
;*****goes in RAM*************          ;*****goes in RAM*************
OK    rmb  1   ;0=busy, 1=done          OK    rmb  1    ;0=busy, 1=done
Line  rmb  20  ;ASCII, end with 0       Line  rmb  20   ;ASCII, end with 0
Pt    rmb  2   ;pointer to Line         Pt    rmb  2    ;pointer to Line
;*****goes in ROM*************          ;*****goes in ROM*************
;Input RegX=>string                     ;Input RegX=>string
Fill  ldy  #Line ;RegX=>string          Fill  ldy  #Line ;RegX=>string
      sty  Pt    ;initialize pointer          sty  Pt    ;initialize pointer
Floop ldaa 0,X   ;copy data             Floop ldaa 1,x+ ;copy data
      staa 0,Y                                staa 1,y+
      inx                                     tsta       ;end?
      iny                                     bne  Floop
      tsta       ;end?                        clr  OK
      bne  Floop                              rts
      clr  OK
      rts


;Return RegA=data                       ;Return RegA=data
Get   ldx  Pt                           Get   ldx  Pt
      ldaa 0,X   ;read data                   ldaa 1,x+ ;read data
      inx                                     stx  Pt
      stx  Pt                                 rts
      rts
```

Program 4.21
Helper routines for the printer interface.

```
; MC68HC711E9                          ; MC9S12C32
; PC6-PC0 outputs = printer DATA       ; PT6-PT0 outputs = printer DATA
; STRA=READY interrupt on rise         ; PJ7=READY interrupt on rise
; 6 STAI 1  Interrupts armed           ;Input RegX=>string
; 5 CWOM 0  Normal outputs             Print_Init
; 4 HNDS 1  Output handshake               sei         ; make atomic
; 3 OIN  1                                 bsr  Fill   ;Init global
; 2 PLS  1  START=STRB pulse               movb #$FF,DDRT ;PT6-0 outputs
; 1 EGA  1  STAF set on rise of READY       movb #$40,DDRJ ;PJ6=START, PJ7=READY
; 0 INVB 0  STRB not used                  bset PPSJ,#$80 ;rise on PJ7
;Input RegX=>string                        bset PIEJ,#$80 ;arm PJ7
Print_Init sei     ; atomic                movb #$80,PIFJ ;clear flag7
      bsr  Fill    ;Init global             bsr  Get
      ldaa #$FF    ;PC7 is an output        bsr  Out    ;start first
      staa DDRC    ;PC6-0 outputs           cli         ;Enable IRQ
      ldaa #$5E                             rts
      staa PIOC                        ;RegA is data to print
      bsr  Get     ;start first        Out   bclr PTJ,#$40 ;START=0
      staa PORTCL                            staa PTT        ;write DATA
      cli          ;Enable IRQ              bset PTJ,#$40 ;START=1
      rts                                   rts
```

Program 4.22
Initialization routines for the printer interface.

Last, we have the interrupt handler that gets called on the rise of READY (Figure 4.29). The handler will first poll, executing an error routine if the interrupt is from some other unknown source. It will then acknowledge the interrupt, get a character from the data structure, and output it to the printer. If the data are the null (0) character, then the global flag OK is set and the system is disarmed (Program 4.23).

Figure 4.29
Flowchart of the
printer ISR.

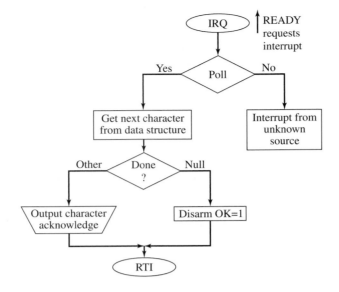

Program 4.23
ISR routines for the printer interface.

```
; MC68HC711E9
ExtHan ldaa PIOC      ;poll STAF?
       bmi  PrtHan
       swi             ;error
PrtHan bsr  Get
       tsta
       beq  Disarm
       staa PORTCL    ;Ack,start next
       bra  Done
Disarm ldaa #$1E      ;STAI=0
       staa PIOC
       inc  OK        ;line complete
Done   rti
       org  $FFF2     ;IRQ
       fdb  ExtHan
```

```
; MC9S12C32
ExtHan brset PIFJ,$02,PrtHan
       swi             ;error

PrtHan movb #$80,PIFJ ;Ack, clear flag7
       bsr  Get
       tsta
       beq  Disarm
       bsr  Out       ;start next
       bra  Done
Disarm bclr PIEJ,#$80 ;disarm PJ7
       inc  OK        ;line complete
Done   rti
       org  $FFCE     ;Key wakeup J
       fdb  ExtHan
```

In C, the 6811/6812 solutions are shown in Program 4.24.

```
// MC68HC711E9, ICC11 C
// PC6-PC0 outputs = printer DATA
// STRA=READY interrupt on rise
// STRB=START pulse out
unsigned char OK;  // 0=busy, 1=done
unsigned char Line[20]; //ASCII data
unsigned char *Pt; // pointer to line
void Fill(unsigned char *p){
  Pt = &Line[0];
  while((*Pt++)=(*p++)); // copy
  Pt = &Line[0];     // initialize
  OK = 0;            // busy
}
```

```
// MC9S12C32 CodeWarrior C
// PT6-PT0 outputs = printer DATA
// PJ7=READY interrupt on rise
// PJ6=START pulse out
unsigned char OK;  // 0=busy, 1=done
unsigned char Line[20]; //ASCII data
unsigned char *Pt; // pointer to line
void Fill(unsigned char *p){
  Pt = &Line[0];
  while((*Pt++)=(*p++)); // copy
  Pt = &Line[0];    // initialize pointer
  OK = 0;           // busy
}
```

```
unsigned char Get(void){
  return(*Pt++);
}
void Print_Init(unsigned char *thePt){
asm(" sei");        // make atomic
  Fill(thePt);      // copy data
  DDRC = 0xFF;      // Port C outputs
  PIOC = 0x5E;      // arm out handshake
  PORTCL = Get();   // start first
asm(" cli");
}
#pragma interrupt_handler ExtHan()
void ExtHan(void){ unsigned char data;
  if((PIOC & STAF)==0)asm(" swi");
  if(data=Get())
    PORTCL=data;    // start next
  else{
    PIOC=0x1E;      // disarm
    OK=1;           // line complete
  }
}
#pragma abs_address:0xfff2
void (*IRQ_vector[])() = { ExtHan };
#pragma end_abs_address
```

```
unsigned char Get(void){
  return(*Pt++);
}
void Out(unsigned char data){
  PTJ &=~0x40;    // START=0
  PTT  = data;    // write DATA
  PTJ |= 0x40;    // START=1
}
void Print_Init(unsigned char *thePt){
asm sei    // make atomic
  Fill(thePt); // copy data into global
  DDRT = 0xFF;   // PT6-0 output DATA
  DDRJ = 0x40;   // PJ7=START
  PPSJ|= 0x80;   // rise on PJ7
  PIEJ|= 0x80;   // arm PJ7
  PIFJ = 0x80;   // clear flag7
  Out(Get()); // start first
asm cli
}
void interrupt 24 ExtHan(void){
  if((PIFJ&0x02)==0)asm(" swi");
  PIFJ = 0x80;   // clear flag7
  if(data=Get())
    Out(data);   // start next
  else{
    PIEJ &=~0x80; // disarm
    OK=1;}}       // line complete
```

Program 4.24
C language software interrupt for the printer.

Two problems with output interrupts were introduced earlier: (1) How does one generate the first interrupt? (2) What does one do if an output interrupt occurs (device is idle) but there are no more data currently available (e.g., the data structure is empty)? In this example, the ritual could start the output because the first character to print was available at that time. If the first data were not available at the time of the ritual, arming the output initiation would have to be postponed until the first data are available. Also, in this case, the system disarms when there are no more data to print. The system is rearmed (by calling the ritual again) when more output is desired.

Another solution involves the use of nonprinting dummy characters. In this technique, a special nonprinting character like NULL ($00) or SYN ($16) is transmitted when the device is ready but there is nothing to print. This method is a little simpler because one does not have to arm, disarm, and rearm. The disadvantage is that software overhead is required to process interrupts when no real function is being performed. In other words, the main thread will execute slower. In addition, the printer must discard these dummy characters. The arm/disarm technique was presented at the beginning of the chapter during the discussion of the producer/consumer problem. It will be presented again in Chapter 7 on the serial communications chapter.

4.10 Power System Interface Using $\overline{\text{XIRQ}}$ Synchronization

The objective of this section is to use XIRQ interrupts to create a very low latency interface (Figure 4.30). This power system will monitor the voltage level from the regular power supply. If the level drops below a safe threshold, it will trigger an XIRQ interrupt by signaling the problem with a rising edge on **TooLow**. This rising edge will clear the 74HC74 flip-flop, making its output, XIRQ, low. The XIRQ handler will service the crisis by enabling the backup power system by making PB1=1. The handler can acknowledge the XIRQ interrupt by toggling PB0=0, then PB0=1 again. Assuming the program is running with

XIRQ interrupts enabled, the latency of this interface is quite low (Program 4.25). The C language implementation is shown in Program 4.26.

Figure 4.30
Hardware interface of an
XIRQ interrupting device.

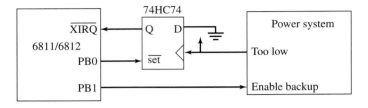

Program 4.25
Assembly software for
the XIRQ interrupt.

```
* Called to initialize the power system
RITUAL ldaa #$FF
       staa DDRB      Port B outputs (6812 only)
       ldaa #0        Backup power initially off
       staa PORTB     Set the flip flop, make XIRQ=1
       ldaa #1
       staa PORTB     Flip flop ready to receive rising edge of TooLow
       ldaa #$10      Enable XIRQ, Disable IRQ
       tap
       rts            Back to main, foreground thread

*Note that the software can only enable XIRQ and cannot disable XIRQ.
* In this way, XIRQ is nonmaskable.
XIRQHAN ldaa #2
       staa PORTB     Enable BackUp power, acknowledge XIRQ
       ldaa #3
       staa PORTB     Will allow another rising edge of TooLow
       rti


       org  $FFF4
       fdb  XIRQHAN  XIRQ interrupt vector
```

Program 4.26
C language software for
the XIRQ interrupt.

```
/* Power System interface
   XIRQ requested on a rise of TooLow
   PB0, negative logic pulse, will acknowledge XIRQ
   PB1=1 will activate backup power */
#pragma interrupt_handler PowerLow()
void PowerLow(void){ PORTB=2; PORTB=3; }  /* Ack, turn on backup power */
void Ritual(void){
     DDRB=0xFF;         // Port B outputs (6812 only)
     PORTB=0; PORTB=1; // Make XIRQ=1
asm(" ldaa #0x10\n"
    "  tap");
}
```

Observation: Some older computers have a nonmaskable interrupt (NMI) that is always enabled. These computers have problems when a NMI is requested before the system has been initialized, because the system is not yet ready to handle the NMI request.

Common error: It is a mistake to enable XIRQ interrupts before the system has been initialized to handle the XIRQ request.

The latency of this interface is defined as the time from the rising edge of **TooLow** to when the software enables the backup power. On the 6811 it includes the 2–41 cycles required to finish the current instruction, the 14 cycles to process the XIRQ, and the six cycles needed to execute the first two instructions of the XIRQHAN. The logic analyzer output during the interrupt request would be as follows. *The italicized portion is the thread switch performed automatically in hardware by the 6811.*

TooLow	$\overline{\text{XIRQ}}$	R/W	Address	Data	Comment
0	1	1	F200	B7	STAA $5678 instruction in MAIN
1	1	1	F201	56	Power Failure occurs
1	0	1	F202	78	Finish instruction (XIRQ requested)
1	0	0	5678	12	Write to memory
1	*0*	*1*	*F203*		*Start to process XIRQ, dummy PC*
1	*0*	*1*	*F204*		*??dummy PC read??*
1	*0*	*0*	*SP*	*03*	*PCL push all registers*
1	*0*	*0*	*SP-1*	*F2*	*PCH*
1	*0*	*0*	*SP-2*		*YL*
1	*0*	*0*	*SP-3*		*YH*
1	*0*	*0*	*SP-4*		*XL*
1	*0*	*0*	*SP-5*		*XH*
1	*0*	*0*	*SP-6*		*A*
1	*0*	*0*	*SP-7*		*B*
1	*0*	*0*	*SP-8*		*CC (with X=0 and the old I)*
1	*0*	*1*	*SP-8*		*??dummy stack read??, then , X=1, I=1*
1	*0*	*1*	*FFF4*	*E6*	*High byte of XIRQHAN (vector fetch)*
1	*0*	*1*	*FFF5*	*00*	*Low byte of XIRQHAN*
1	0	1	E600	86	Op code for LDAA
1	0	1	E601	02	Immediate operand
1	0	1	E602	B7	Opcode for STAA
1	0	1	E603	10	Extended mode operand
1	0	1	E604	04	PORTB address is $1004
1	0	0	1004	02	**set**=0 acknowledge, enable backup
1	1	1	E605	86	Opcode for LDAA
1	1	1	E606	03	Immediate operand
1	1	1	E607	B7	Opcode for STAA
1	1	1	E608	10	Extended mode operand
1	1	1	E604	04	PORTB address is $1004
1	1	0	1004	03	**set**=1
1	1	1	E609	3B	Opcode for RTI
1	1	1	E60A		Dummy PC read
1	1	1	SP-9		Dummy stack read
1	1	1	SP-8		CC (with X=0 and the old I)
1	1	1	SP-7		B
1	1	1	SP-6		A
1	1	1	SP-5		XH
1	1	1	SP-4		XL
1	1	1	SP-3		YH
1	1	1	SP-2		YL
1	1	1	SP-1	F2	PCH
1	1	1	SP	03	PCL pull all registers
1	1	1	F203		Continue execution of previous function

4.11 Interrupt Polling Using Linked Lists

The process of polling involves software at the beginning of the ISR that checks one by one the list of possible devices that might have caused the interrupt. When the interrupt polling software detects a device that needs service, it executes the appropriate device driver to service the interrupt. There are two considerations when we decide whether or not to poll at the beginning of the interrupt handler. First, if there are two or more potential sources of interrupt that operate through the same interrupt vector, then we must poll to determine which one is interrupting. Examples of this situation are (1) external IRQ and STAF and (2) the asynchronous serial port, SCI. Second, even if we do not have to poll because our interrupting device has its own dedicated vector, sometimes we will poll anyway just to add a layer of robustness to our software. In this way, if the software arrives at the first location of the interrupt handler due to a software or hardware error, our software can determine an error has occurred because the device that should have been ready is not.

Figure 4.31
Linked list data structure used to implement polling.

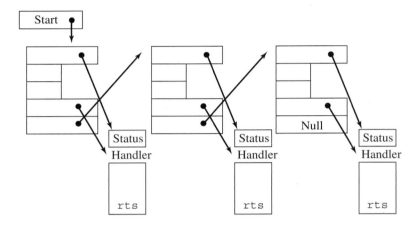

In this example, we implement polling using linked lists (Figure 4.31). The purpose of using linked lists is to make it easier to debug, change the polling order, add devices, or subtract devices. Although this example uses statically allocated linked list, if the linked list was created at run time, then changing polling order or adding/subtracting devices could be performed dynamically.

4.11.1
6811 Interrupt
Polling Using
Linked Lists

In this 6811 implementation, we have three potential interrupt sources using the same interrupt vector. The first is STAF, the second is Port A of an external 6821, and the third is Port B of the 6821. The details of the 6821 are not important for this example. Rather the purpose of this example is to illustrate the use of linked lists when implementing interrupt polling. There will be one node for each potential device. Each node of the statically allocated linked list has enough information to poll the device. If successful, the node also has a handler address that will be called. Program 4.27 shows the 6811 assembly language.

Program 4.27
6811 assembly structure for interrupt polling using linked lists.

```
start   fdb   11STAF   place to start polling
Sreg    equ   0        Index to Status Register
Amask   equ   2              and mask
Cmask   equ   3              compare mask
DevHan  equ   4              device handler
NextPt  equ   6              next pointer
num     fcb   3        number of devices
11STAF  fdb   $1002    address of PIOC
        fcb   $ff      look at all the bits in PIOC
        fcb   $C0      expect exact match with $C0
        fdb   STAFhan  device handler
        fdb   11CA1    pointer to next device to poll
```

```
llCA1    fdb    $2011        address of 6821 Port A Control/Status
         fcb    $87          look at bits 7,2,1,0
         fcb    $85          expect bits 7,2,1,0 to be 1,1,0,1
         fdb    CA1han       device handler for CA1
         fdb    llCB2        pointer to next device to poll
llCB2    fdb    $2013        address of 6821 Port B Control/Status
         fcb    $7C          look at bits 6,5,4,3,2
         fcb    $4C          expect bits 6,5,4,3,2 to be 1,0,0,1,1
         fdb    CB2han       device handler for CB2
         fdb    0            no more
```

The interrupt handler performs the following four steps: (1) read status register (`ldaa,y`), (2) eliminate irrelevant bits (`anda`), (3) determine if this device is requesting (`cmpa`), and (4) execute the handler (`jsr`) if it is requesting. This particular implementation does not report an error if the interrupt is from an unknown source. We will see in Section 4.14 how to modify this solution to implement round-robin polling. We assume each device handler terminates with a RTS instruction and saves registers B and X (Program 4.28).

Program 4.28
6811 assembly implementation of interrupt polling using linked lists.

```
IrqHan   ldx    start        Reg X points to linked list place to start
         ldab   num          number of possible devices
next     ldy    Sreg,x       Reg Y points to status reg
         ldaa   ,y           read status
         anda   Amask,x      clear bits that are indeterminate
         cmpa   Cmask,x      expected value if this device active
         bne    Notyet       skip if this device not requesting
         ldy    DevHan,x     Reg Y points to device handler
         jsr    ,y           call device handler, will return here
Notyet   ldx    NextPt,x     Reg X points to next entry
         decb                device counter
         bne    next         check next device
         rti
```

In C, this linked list polling system is shown in Program 4.29.

```c
const struct  Node{
      unsigned char *StatusPt;       /* Pointer to status register */
      unsigned char Amask;           /* And Mask */
      unsigned char Cmask;           /* Compare Mask */
      void (*Handler)(void);         /* Handler for this task */
      const struct Node *NextPt;     /* Link to Next Node */
};
unsigned char CLdata,PIAAdata,PIABdata;
void STRAHan(void){    // regular functions that return (rts) when done
    CLdata=PORTCL;}
void PIAHanA(void){
    PIAAdata=ADATA;}
void PIAHanB(void){
    PIABdata=BDATA;}
typedef const struct Node NodeType;
typedef NodeType * NodePtr;
```

continued on p. 234

Program 4.29
C language implementation of interrupt polling on the 6811 using linked lists.

continued from p. 233

```
NodeType sys[3]={
    {&PIOC, 0xFF, 0xC0, STRAHan, &sys[1]},
    {&ACNT, 0x87, 0x85, PIAHanA, &sys[2]},
    {&BCNT, 0x7C, 0x4C, PIAHanB, 0} };
#pragma interrupt_handler IRQHan()
void IRQHan(void){ NodePtr Pt; unsigned char Status;
  Pt=&sys[0];
  while(Pt){              // executes device handlers for all requests
    Status=*(Pt->StatusPt);
    if((Status&(Pt->Amask))==(Pt->Cmask)){
        (*Pt->Handler)();}      /* Execute handler */
    Pt=Pt->NextPt; } }  // returns after all devices have been polled
```

Program 4.29
C language implementation of interrupt polling on the 6811 using linked lists.

4.11.2
6812 Interrupt
Polling Using
Linked Lists

In this 6812 implementation, we have three potential interrupt sources using the same interrupt vector. In this system we will have three key wakeup interrupts on PJ2, PJ1, and PJ0. The purpose of this example is to illustrate the use of linked lists when implementing interrupt polling. There will be one node for each potential device. Each node of the statically allocated linked list has enough information to poll the device. If successful, the node also has a handler address that will be called. This solution is simpler than in the 6811 because all key wakeups on Port J have the same status register, and we only need to test for 1s and no 0s. Program 4.30 shows the 6812 assembly language.

Program 4.30
6812 assembly structure for interrupt polling using linked lists.

```
start   fdb   llPJ2     place to start polling
Mask    equ   0               and mask
DevHan  equ   1               device handler
NextPt  equ   3               next pointer
num     fcb   3         number of devices
llPJ2   fcb   $04       look at bit 2
        fdb   PJ2han    device handler
        fdb   llPJ1     pointer to next device to poll
llPJ1   fcb   $02       look at bit 1
        fdb   PJ1han    device handler
        fdb   llPJ0     pointer to next device to poll
llPJ0   fcb   $01       look at bit 0
        fdb   PJ0han    device handler
        fdb   0         end of list
```

The interrupt handler performs the following four steps: (1) read status register (ldaa ,y), (2) determine if this device is requesting (anda), and (3) execute the handler (jsr) if it is requesting (Program 4.31). This particular implementation does not report an error if the interrupt is from an unknown source. We will see in Section 4.14 how to modify this solution to implement round-robin polling. We assume each device handler terminates with a RTS instruction and saves registers B and X. In C, this linked list polling system is shown in Program 4.32.

Program 4.31
6812 assembly implementation of interrupt polling using linked lists.

```
IrqHan   ldx    start        Reg X points to linked list place to start
         ldab   num          number of possible devices
next     ldaa   KWIFJ        read status
         anda   Mask,x       check if proper bit is set
         beq    Notyet       skip if this device not requesting
         jsr    [DevHan,x]   call device handler, will return here
Notyet   ldx    NextPt,x     Reg X points to next entry
         decb                device counter
         bne    next         check next device
         rti
```

```c
const struct Node{
     unsigned char Mask;          /* And Mask */
     void (*Handler)(void);       /* Handler for this task */
     const struct Node *NextPt;      /* Link to Next Node */
};
unsigned char Counter2,Counter1,Counter0;
void PJ2Han(void){    // regular functions that return (rts) when done
    KWIFJ=0x04;       // acknowledge
    Counter2++;}
void PJ1Han(void){    // regular functions that return (rts) when done
    KWIFJ=0x02;       // acknowledge
    Counter1++;}
void PJ0Han(void){    // regular functions that return (rts) when done
    KWIFJ=0x01;       // acknowledge
    Counter0++;}
typedef const struct Node NodeType;
typedef NodeType * NodePtr;
NodeType sys[3]={
    {0x04, PJ2Han, &sys[1]},
    {0x02, PJ1Han, &sys[2]},
    {0x01, PJ0Han,   0    } };
void interrupt 23 IRQHan(void){ NodePtr Pt; unsigned char Status;
 Pt=&sys[0];
 while(Pt){              // executes device handlers for all requests
    if(KWIFJ&(Pt->Mask)){
        (*Pt->Handler)();}     /* Execute handler */
    Pt=Pt->NextPt; } }  // returns after all devices have been polled
```

Program 4.32
C language implementation of interrupt polling on the MC68HC814A4 using linked lists.

4.12 Fixed Priority Implemented Using One Interrupt Line

In this example (Figure 4.32), Device 1 has higher priority than Device 2. Since they share a common interrupt vector, this is an example of a polled interrupt. The hardware is shown with external parallel ports, but the discussion would be identical if two key wakeups were used.

There are two steps that we need to perform to let Device 1 go ahead of Device 2. First, if both simultaneously request interrupts, then obviously Device 1 goes first. The more difficult situation is when Device 2 requests an interrupt, is allowed to start its function, then Device 1 requests service. For this scenario, we need a mechanism to postpone the service of Device 2, serve Device 1 right now, then finish the service for Device 2 after we are done with Device 1. We can accomplish this effect by reenabling interrupts in the

Figure 4.32
A polled interrupt where Device 1 has higher priority than Device 2.

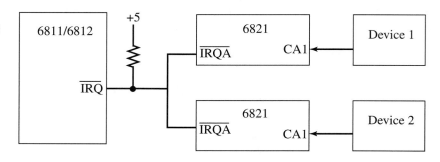

Device 2 handler after the Device 2 interrupt has been acknowledged. The software will crash if we reenable interrupts in the Device 2 handler before we acknowledge it (the three conditions flag set, flag armed, I=0 would cause Device 2 to interrupt itself over and over). Obviously we would not want to reenable interrupts while servicing the higher-priority device. This *selective reenabling* approach works for two devices but does not work quite the same way for three or more devices. If we were to have three devices, clearly we would reenable interrupts for the lowest-priority device and not reenable interrupts for the highest-priority device. The difficult question is whether to reenable interrupts while servicing the middle-priority device. If we do reenable interrupts while servicing the middle device, then we implement a two-level priority system, where 1 is higher than 2 and 3 but 2 and 3 essentially have the same priority. If we do not reenable interrupts while servicing the middle device, then we implement a different two-level priority system, where 1 and 2 are higher than 3 but 1 and 2 have the same priority. Luckily, when we are interested in priority on an embedded system, it usually fits the two-level situation, where one or two devices need to have priority over all the rest. So, the flowchart shown in Figure 4.33 can be applied in general to solve a two-level priority system. We enable interrupts for the lower-priority devices and do not reenable interrupts for the higher-priority devices.

Figure 4.33
A flowchart of the ISR where Device 1 has higher priority than Device 2.

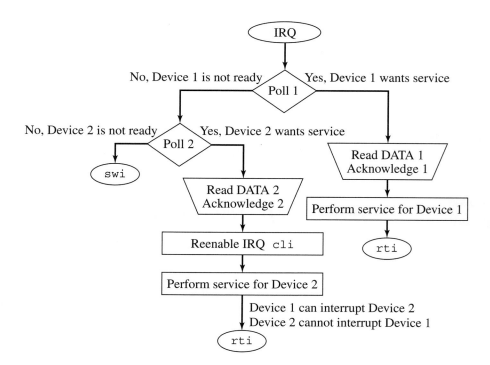

4.13 Fixed Priority Implemented Using XIRQ

One of the disadvantages of the priority implementation of Section 4.12 is that it takes some time to poll and acknowledge a low-priority device. If we wish to reduce the latency and increase the priority of one device over all the other interrupt sources, then XIRQ can be used (Figure 4.34). Once the X bit is 0, the software including ISRs cannot postpone an XIRQ request. It doesn't matter which request comes first, an XIRQ interrupt will be processed at the completion of the next instruction.

Figure 4.34
A vectored interrupt where Device 1 has higher priority than Device 2.

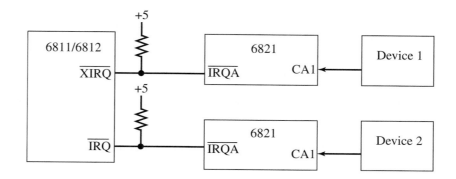

The flowchart in Figure 4.35 shows the operations of the XIRQ and IRQ interrupt service routines.

Figure 4.35
Two flowcharts of the ISRs where Device 1 has higher priority than Device 2.

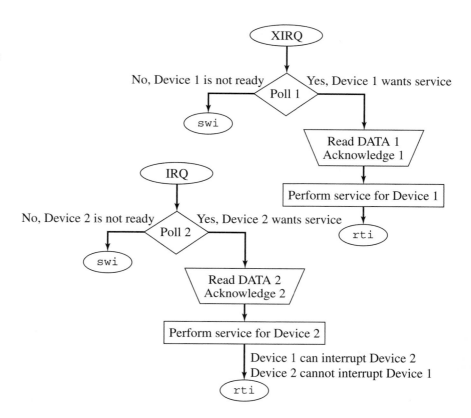

4.14 Round-Robin Polling

Rather than implement priority, sometimes we wish to implement *no priority*. This means we can guarantee service under heavy load for equally important devices. For interrupts that occur on an infrequent basis, priority doesn't really matter. But as the number of interrupts increase and the interrupt request rates increase, we may wish to use round-robin polling. This implementation works only for polled interrupts and does not apply to vectored interrupts. Two round-robin schemes are shown for a situation with three devices called A, B, C. The particular example sequence of events is shown for the situation that B and C always want service and A never does.

Scheme 1 Start polling after device that last got service. Example sequence of events:
  Interrupt, poll A, B, C (B, C need service)
  Service B
  Interrupt, poll C, A, B (B, C need service)
  Service C
  Interrupt, poll A, B, C (B, C need service)
  Service B
  Interrupt, poll C, A, B
Scheme 2 Cycle through list of devices independent of which device last got service.
  Example sequence of events:
  Interrupt, poll A, B, C
  Interrupt, poll B, C, A
  Interrupt, poll C, A, B
  Interrupt, poll A, B, C, etc.

Our previous polling examples in Section 4.11 can be modified to implement either round-robin scheme. Programs 4.33 and 4.34 implement the second simple method that rotates the polling order independent of which devices request service.

```
NodeType sys[3]={
    {&PIOC, 0xFF, 0xC0, STRAHan, &sys[1]},
    {&ACNT, 0x87, 0x85, PIAHanA, &sys[2]},
    {&BCNT, 0x7C, 0x4C, PIAHanB, &sys[0]} };
NodePtr Pt=&sys[0];  // points to the one that got polled first at last interrupt
#pragma interrupt_handler IRQHan()
void IRQHan(void){  unsigned char Counter,Status;
 Counter=3;         // quit after three devices checked
 Pt=Pt->NextPt;     // rotates ABC BCA CAB polling orders
 while(Counter--){
    Status=*(Pt->StatusPt);
    if((Status&(Pt->Amask))==(Pt->Cmask)){
        (*Pt->Handler)();}     /* Execute handler */
    Pt=Pt->NextPt; } }
```

Program 4.33
C language implementation of round-robin polling on the 6811.

4.15 Periodic Polling

The purpose of this section is to present alternative methods to create a real-time periodic interrupt (RTI). A RTI is one that is requested on a fixed time basis. This interfacing technique is required for data acquisition and control systems, because software servicing must be performed at accurate time intervals. For a data acquisition system, it is important to establish

```
NodeType sys[3]={
    {0x04, PJ2Han, &sys[1]},
    {0x02, PJ1Han, &sys[2]},
    {0x01, PJ0Han, &sys[0]} };
NodePtr Pt=&sys[0];  // points to the one that got polled first at last interrupt
void interrupt 23 IRQHan(void){ unsigned char Counter,Status;
 Counter=3;          // quit after three devices checked
 Pt=Pt->NextPt;      // rotates ABC BCA CAB polling orders
 while(Counter--){
    if(KWIFJ&(Pt->Mask)){
        (*Pt->Handler)();}      /* Execute handler */
    Pt=Pt->NextPt; } }  // returns after all devices have been polled
```

Program 4.34
C language implementation of round-robin polling on the MC68HC812A4.

an accurate sampling rate. The time in between ADC samples must be equal (and known) for the digital signal processing to function properly. Similarly for microcomputer-based control systems, it is important to maintain both the ADC and DAC timing.

Another application of RTIs is called *intermittent* or *periodic polling*. In regular gadfly, the main program polls the I/O devices continuously (refer back to Section 3.3). With intermittent polling, the I/O devices are polled on a regular basis (established by the RTI). If no device needs service, then the interrupt simply returns. This method frees the main program from the I/O tasks. Older IBM PC computers use an 18-Hz RTI to interface its keyboard. We use periodic polling (Figure 4.36) if the following two conditions apply:

1. The I/O hardware cannot generate interrupts directly.
2. We wish to perform the I/O functions in the background.

Figure 4.36
An ISR flowchart that implements periodic polling.

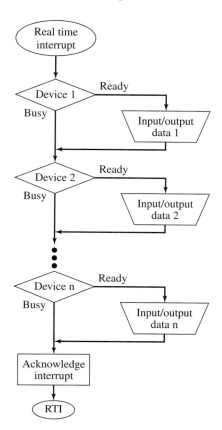

In each of the examples of this section, we will design a real-time periodic interrupt software system that increments a global variable. It is real-time because we can put an upper bound on the latency, which is the time between when the software task is supposed to run (in this case at fixed time intervals), and when the ISR increments the global variable Time. In general, the latency of interrupt-driven tasks includes

1. the maximum time the system runs with interrupts disabled
2. the time to finish the current instruction in the main program
3. the time to perform the thread switch (push registers, fetch interrupt vector)
4. the time to execute the interrupt service routine

The initialization program will arm the device and establish the periodic interrupt rate. A flag will be set by the timer hardware at a periodic rate. Because the flag is armed, an interrupt will be requested each time the flag is set. The interrupt service routine will acknowledge the interrupt by clearing the flag, then it will increment the global variable.

Checkpoint 4.7: In a typical system, which of the four components of latency is largest?

Observation: The TCNT timer is very accurate because of the stability of the crystal clock.

4.15.1
MC68HC711E9
Periodic Interrupts

There are three mechanisms on the 6811 that generate periodic interrupts: real-time interrupt, timer overflow, and output compare. Table 4.8 shows the 6811 registers used in periodic interrupts. The entries shown in bold will be used in this section.

Address	msb															lsb	Name
$100E	**15**	**14**	**13**	**12**	**11**	**10**	**9**	**8**	**7**	**6**	**5**	**4**	**3**	**2**	**1**	**0**	TCNT
$1016	15	14	13	12	11	10	9	8	7	6	5	4	3	2	1	0	TOC1
$1018	**15**	**14**	**13**	**12**	**11**	**10**	**9**	**8**	**7**	**6**	**5**	**4**	**3**	**2**	**1**	**0**	TOC2
$101A	15	14	13	12	11	10	9	8	7	6	5	4	3	2	1	0	TOC3
$101C	15	14	13	12	11	10	9	8	7	6	5	4	3	2	1	0	TOC4
$101E	15	14	13	12	11	10	9	8	7	6	5	4	3	2	1	0	TI4/O5

Address	Bit 7	6	5	4	3	2	1	Bit 0	Name
$1022	OC1I	**OC2I**	OC3I	OC4I	IC4/OC5I	IC1I	IC2I	IC3I	TMSK1
$1023	OC1F	**OC2F**	OC3F	OC4F	IC4/OC5F	IC1F	IC2F	IC3F	TFLG1
$1024	**TOI**	RTII	PAOVI	PAII	0	0	**PR1**	**PR0**	TMSK2
$1025	**TOF**	RTIF	PAOVF	PAIF	0	0	0	0	TFLG2
$1026	DDRA7	PAEN	PAMOD	PEDGE	DDRA3	I4/O5	**RTR1**	**RTR0**	PACTL

Table 4.8
6811 registers used to configure periodic interrupts.

First, the real-time interrupt feature can be used to generate interrupts at a fixed rate. Two bits (RTR1 and RTR0) in the PACTR register determine the interrupt rate. Program 4.35 will increment a global variable, Time, every 32.767 ms. Four possible interrupt periods are possible with the 6811, as shown in Table 4.9.

The background ISR thread is invoked at 30.517 Hz. To clear the RTIF flag (acknowledge the interrupt), the software writes a one to it. After acknowledging, the ISR increments the Time variable.

Checkpoint 4.8: How would you modify Program 4.35 to count at 61.035 Hz?

```
; MC68HC711E9 assembly                      // MC68HC711E9 ICC11 C
        org   $0000  ;RAM                   unsigned short Time;
Time    rmb   2
        org   $D000                         void RTI_Init(void){
RTI_Init sei         ;make atomic             asm(" sei");    // Make ritual atomic
        ldaa  #3     ;32.768ms                PACTL  = 3;     // 30.517Hz
        staa  PACTL                           TMSK2  = 0x40;  // Arm
        ldaa  #$40                            Time = 0;       // Initialize
        staa  TMSK2  ;arm RTI                 asm(" cli");
        ldd   #0                            }
        std   Time
        cli          ;enable IRQ
        rts                                 #pragma interrupt_handler RTIHan()
RTIHan  ldaa  #$40   ;acknowledge           void RTIHan(void){
        staa  TFLG2  ;about 30.517Hz          TFLG2 = 0x40;     // Acknowledge
        ldd   Time                            Time++;
        addd  #1                            }
        std   Time
        rti
        org   $FFF0                         #pragma abs_address:0xfff0
        fdb   RTIHan ;vector                void (*RTI_vector[])() ={
                                              RTIHan};  // fff0 RTI
                                            #pragma end_abs_address
```

Program 4.35
Implementation of a periodic interrupt using the real-time interrupt feature.

Table 4.9
Real-time interrupt rates on a 2 MHz 6811.

RTR1	RTR0	Divide E by	Period (µs)	Frequency (Hz)
0	0	2^{13}	4096	244.14
0	1	2^{14}	8192	122.07
1	0	2^{15}	16384	61.035
1	1	2^{16}	32768	30.517

The timer overflow interrupt feature can also be used to generate interrupts at a fixed rate (see Table 4.10). The 16-bit TCNT register is incremented at a fixed rate. The TOF flag is set when the counter overflows and wraps back around (automatically) to zero. If armed, the TOF flag will generate an interrupt. Two bits (PR1 and PR0) in the TMSK2 register determine the rate at which the counter will increment, hence will determine the TOF interrupt rate. To create a TOF periodic interrupt, we arm the timer overflow (TOI), and set the rate (PR1-0) (see Program 4.36). To clear the TOF flag (acknowledge the interrupt), the software writes a one to it.

The third mechanism to generate periodic interrupts is output compare. Program 4.37 shows a 1 kHz periodic interrupt using output compare 2. When the TCNT register matches TOC2, the output compare flag, OC2F, is set. If armed (OC2I=1), then it will request an interrupt. To clear the OC2F flag (acknowledge the interrupt), the software writes a one to it. The ISR will acknowledge the interrupt and set TOC2 to the time for the next interrupt. The interrupting period is determined by the TCNT period (set by TMSK2) multiplied by the constant PERIOD. Program 4.37 will increment a global variable, Time, every 1 ms.

Checkpoint 4.9: How would you modify Program 4.37 to count at 100 Hz?

Table 4.10
Timer overflow rates assuming a 2 MHz 6811.

PR1	PR0	Divide by	TCNT Clock	TOF Interrupt	TOF Interrupt
0	0	1	500 ns	32.768 ms	30.517 Hz
0	1	4	2 µs	131.072 ms	7.63 Hz
1	0	8	4 µs	262.144 ms	3.81 Hz
1	1	16	8 µs	524.288 ms	1.91 Hz

```
; MC68HC711E9 assembly                    // MC68HC711E9 ICC11 C
        org   $0000 ;RAM                  unsigned short Time;
Time    rmb   2
        org   $D000
TOF_Init sei        ;make atomic          void TOF_Init(void){
        ldaa  #$80  ;arm, 32.768ms          asm(" sei");    // Make ritual atomic
        staa  TMSK2 ;Set TOI=1 arm TOF       TMSK2 = 0x80 ;  // Arm, 30.517Hz
        ldd   #0                             Time = 0;       // Initialize
        std   Time                           asm(" cli");
        cli         ;enable IRQ           }
        rts                               #pragma interrupt_handler TOFHan()
TOFHan  ldaa  #$80  ;acknowledge          void TOFHan(void){
        staa  TFLG2 ;about 30.517Hz          TFLG2 = 0x80;    // Acknowledge
        ldd   Time                           Time++;
        addd  #1                          }
        std   Time                        #pragma abs_address:0xffde
        rti                               void (*TOF_vector[])() ={
        org   $FFDE                          TOFHan};  // ffde TOF
        fdb   TOFHan  ;vector             #pragma end_abs_address
```

Program 4.36
Implementation of a 6811 periodic interrupt using timer overflow.

```
; MC68HC711E9 assembly                    // MC68HC711E9 ICC11 C
PERIOD  equ   1000                        #define PERIOD 1000
        org   $0000 ;RAM                  unsigned short Time;
Time    rmb   2
        org   $D000
OC2_Init sei        ;make atomic          void OC2_Init(void){
        ldaa  #$01                          asm(" sei");     // Make ritual atomic
        staa  TMSK2 ;1 MHz TCNT             TMSK1 |= 0x40;   // arm OC2
        ldaa  #$40                          TMSK2 = 0x01;    // 1 MHz TCNT
        staa  TMSK1 ;arm OC2                TOC2= TCNT+PERIOD; // first in 1ms
        ldd   #0                            Time=0;          // Initialize
        std   Time                          asm(" cli");     // enable IRQ
        ldd   TCNT  ;time now             }
        addd  #PERIOD ;first in 1ms
        std   TOC2
        cli         ;enable IRQ
        rts
OC2Han  ldaa  #$40  ;acknowledge          #pragma interrupt_handler TOC2handler
        staa  TFLG1                        void TOC2handler(void){
        ldd   TOC2                           TOC2 = TOC2+PERIOD; // next in 1 ms
        addd  #PERIOD                        TFLG1 = 0x40;       // acknowledge C6F
        std   TOC2  ;next in 1 ms            Time++;
        ldd   Time                         }
        addd  #1
        std   Time                        #pragma abs_address:0xffe2
        rti                               void (*TOC2_vector[])() ={
        org   $FFE2                          TOC2handler};  // ffe2 TOC2
        fdb   OC2Han  ;vector             #pragma end_abs_address
```

Program 4.37
Implementation of a 6811 periodic interrupt using output compare.

4.15.2
MC68HC812A4
Periodic Interrupts

There are also three mechanisms on the 6812 that generate periodic interrupts: real-time interrupt, timer overflow, and output compare. The solutions in this section will also apply to the MC68HC912B32 and MC68HC912D60. Table 4.11 shows the 6812 registers used in periodic interrupts. The entries shown in bold will be used in this section.

Address	msb															lsb	Name
$0084	**15**	**14**	**13**	**12**	**11**	**10**	9	8	7	6	5	4	3	2	1	**0**	TCNT
$0090	15	14	13	12	11	10	9	8	7	6	5	4	3	2	1	0	TC0
$0092	15	14	13	12	11	10	9	8	7	6	5	4	3	2	1	0	TC1
$0094	15	14	13	12	11	10	9	8	7	6	5	4	3	2	1	0	TC2
$0096	15	14	13	12	11	10	9	8	7	6	5	4	3	2	1	0	TC3
$0098	15	14	13	12	11	10	9	8	7	6	5	4	3	2	1	0	TC4
$009A	15	14	13	12	11	10	9	8	7	6	5	4	3	2	1	0	TC5
$009C	**15**	**14**	**13**	**12**	**11**	**10**	**9**	**8**	**7**	**6**	**5**	**4**	**3**	**2**	**1**	**0**	TC6
$009E	15	14	13	12	11	10	9	8	7	6	5	4	3	2	1	0	TC7

Address	Bit 7	6	5	4	3	2	1	Bit 0	Name
$0014	**RTIE**	RSWAI	RSBCK	0	RTBYP	**RTR2**	**RTR1**	**RTR0**	RTICTL
$0015	**RTIF**	0	0	0	0	0	0	0	RTIFLG
$0047	LCK	PLLON	PLLS	BCSC	BCSB	BCSA	MCSB	MCSA	CLKCTL
$0080	IOS7	**IOS6**	IOS5	IOS4	IOS3	IOS2	IOS1	IOS0	TIOS
$0086	**TEN**	TSWAI	TSBCK	TFFCA	PAOQE	T7QE	T1QE	T0QE	TSCR
$008C	C7I	**C6I**	C5I	C4I	C3I	C2I	C1I	C0I	TMSK1
$008D	**TOI**	0	TPU	TDRB	TCRE	**PR2**	**PR1**	**PR0**	TMSK2
$008E	C7F	**C6F**	C5F	C4F	C3F	C2F	C1F	C0F	TFLG1
$008F	**TOF**	0	0	0	0	0	0	0	TFLG2

Table 4.11
MC68HC812A4 registers used to configure periodic interrupts.

First, the real-time interrupt feature can be used to generate interrupts at a fixed rate. Three bits (RTR2, RTR1, and RTR0) in the RTICTL register ($0014) determine the interrupt rate. To clear the RTIF flag (acknowledge the interrupt), the software writes a one to it. Program 4.38 will increment a global variable, Time, every 32.767 ms. Seven possible interrupt periods are possible with the 6812, see Table 4.12.

```
; MC68HC812A4 assembly
        org   $0800   ;RAM
Time    rmb   2
        org   $F000
RTI_Init sei                   ;make atomic
        movb  #$86,RTICTL ;32.768ms arm
        movw  #0,Time
        cli                    ;enable IRQ
        rts
RTIHan  movb  #$80,RTIFLG ;ack
        ldd   Time
        addd  #1
        std   Time
        rti
        org   $FFF0
        fdb   RTIHan      ;vector
```

```
// MC68HC812A4 CodeWarrior C
unsigned short Time;
void RTI_Init(void){
  asm sei         // Make atomic
  RTICTL = 0x86;  // Arm, 30.517Hz
  Time = 0;       // Initialize
  asm cli
}



void interrupt 7 RTIHan(void){
  RTIFLG = 0x80;     // Acknowledge
  Time++;
}
```

Program 4.38
Implementation of a MC68HC812A4 periodic interrupt using the real-time clock feature.

Table 4.12
8MHz MC68HC812A4
real-time interrupt rates.

RTR2	RTR1	RTR0	Divide E by	Period (µs)	Frequency (Hz)
0	0	0	off	off	off
0	0	1	2^{13}	1024	976.56
0	1	0	2^{14}	2048	488.28
0	1	1	2^{15}	4096	244.14
1	0	0	2^{16}	8192	122.07
1	0	1	2^{17}	16384	61.035
1	1	0	2^{18}	32768	30.517
1	1	1	2^{19}	65536	15.259

Checkpoint 4.10: How would you modify Program 4.38 to count at 61.035 Hz?

The timer overflow interrupt feature can also be used to generate interrupts at a fixed rate (see Table 4.13). The 16-bit TCNT register is incremented at a fixed rate. The TOF flag is set when the counter overflows and wraps back around (automatically) to zero. If armed, the TOF flag will generate an interrupt. Three bits (PR2, PR1, and PR0) in the TMSK2 register determine the rate at which the counter will increment, hence will determine the TOF interrupt rate. To clear the TOF flag (acknowledge the interrupt), the software writes a one to it. To create a TOF periodic interrupt, we enable the timer (TEN=1), arm the timer overflow (TOI), and set the rate (PR2-0) (see Program 4.39).

Table 4.13
Timer overflow rates,
assuming an 8 MHz
6812. The divide by
64 and the divide by
128 modes are available
only on the
MC68HC912D60.

PR2	PR1	PR0	Divide by	TCNT Clock	TOF Interrupt	TOF Interrupt
0	0	0	1	125 ns	8.192 ms	122.07 Hz
0	0	1	2	250 ns	16.384 ms	61.035 Hz
0	1	0	4	500 ns	32.768 ms	30.517 Hz
0	1	1	8	1 µs	65.536 ms	15.259 Hz
1	0	0	16	2 µs	131.072 ms	7.63 Hz
1	0	1	32	4 µs	262.144 ms	3.81 Hz
1	1	0	64	8 µs	524.288 ms	1.91 Hz
1	1	1	128	16 µs	1048.576 ms	0.95 Hz

```
; MC68HC812A4 assembly
        org   $0800   ;RAM
Time    rmb   2
        org   $F000
TOF_Init sei               ;make atomic
        movb  #$82,TMSK2  ;arm, 32.768ms
        movb  #$80,TSCR   ;enable TCNT
        movw  #0,Time
        cli                ;enable IRQ
        rts
TOFHan  movb  #$80,TFLG2   ;acknowledge
        ldd   Time
        addd  #1
        std   Time
        rti
        org   $FFDE
        fdb   TOFHan   ;vector
```

```
// MC68HC812A4 CodeWarrior C
unsigned short Time;
void TOF_Init(void){
  asm sei        // Make atomic
  TMSK2 = 0x82;  // Arm, 30.517Hz
  TSCR = 0x80;   // enable counter
  Time = 0;      // Initialize
  asm cli        // enable interrupts
}


interrupt 16 void TOFHan(void){
  TFLG2 = 0x80;  // Acknowledge
  Time++;
}
```

Program 4.39
Implementation of a periodic interrupt using timer overflow.

The third mechanism to generate periodic interrupts is output compare. Program 4.40 shows a 1 kHz periodic interrupt using output compare 6. To enable output compare the corresponding bit in the TIOS register must be set. When the TCNT register matches TC6, the output compare flag, C6F, is set. If armed (C6I=1), then it will request an interrupt. To clear the C6F flag (acknowledge the interrupt), the software writes a one to it. The ISR will acknowledge the interrupt and set TC6 to the time for the next interrupt. The interrupting period is determined by the TCNT period (set by TMSK2) multiplied by the constant PERIOD. Program 4.40 will increment a global variable, Time, every 1 ms.

```
; MC68HC812A4 assembly
PERIOD  equ  1000   ;in usec
        org  $0800  ;RAM
Time    rmb  2
        org  $F000
OC6_Init sei             ;make atomic
        movb #$03,TMSK2 ;1us
        movb #$80,TSCR  ;enable TCNT
        bset TIOS,#$40  ;activate OC6
        bset TMSK1,#$40 ;arm OC6
        movw #0,Time
        ldd  TCNT   ;time now
        addd #50    ;first in 50us
        std  TC6
        cli         ;enable IRQ
        rts
OC6Han  movb #$40,TFLG1 ;acknowledge
        ldd  TC6
        addd #PERIOD
        std  TC6     ;next in 1 ms
        ldd  Time
        addd #1
        std  Time
        rti
        org  $FFE2
        fdb  OC6Han ;vector
```

```c
// MC68HC812A4 CodeWarrior C
#define PERIOD 1000
unsigned short Time;


void OC6_Init(void){
  asm sei          // Make atomic
  TSCR = 0x80;
  TIOS |= 0x40;    // activate OC6
  TMSK1 |= 0x40;   // arm OC6
  TMSK2 = 0x03;    // 1 MHz TCNT
  TC6 = TCNT+50;   // first in 50us
  Time = 0;        // Initialize
  asm cli          // enable IRQ
}


interrupt 14 void OC6handler(void){
  TC6 = TC6+PERIOD; // next in 1 ms
  TFLG1 = 0x40;     // ack C6F
  Time++;
}
```

Program 4.40
Implementation of a periodic interrupt using output compare.

Checkpoint 4.11: How would you modify Program 4.40 to count at 100 Hz?

4.15.3 MC9S12C32 Periodic Interrupts

There are also three mechanisms on the MC9S12C32 that generate periodic interrupts: real-time interrupt, timer overflow, and output compare. Table 4.14 shows the 6812 registers used in periodic interrupts. The entries shown in bold will be used in this section.

First, the real-time interrupt mechanism can generate interrupts at a fixed rate. Seven bits (RTR6-0) in the RTICTL register ($003B$) specify the interrupt rate. The 7-bit value is composed of two parts:

Let **RTR6**, **RTR5**, **RTR4** be **n**, which is a 3-bit number ranging from 0 to 7.
Let **RTR3**, **RTR2**, **RTR1**, **RTR0** be **m**, which is a 4-bit number ranging from 0 to 15.

Address	msb															lsb	Name
$0044	15	14	13	12	11	10	9	8	7	6	5	4	3	2	1	0		TCNT
$0050	15	14	13	12	11	10	9	8	7	6	5	4	3	2	1	0		TC0
$0052	15	14	13	12	11	10	9	8	7	6	5	4	3	2	1	0		TC1
$0054	15	14	13	12	11	10	9	8	7	6	5	4	3	2	1	0		TC2
$0056	15	14	13	12	11	10	9	8	7	6	5	4	3	2	1	0		TC3
$0058	15	14	13	12	11	10	9	8	7	6	5	4	3	2	1	0		TC4
$005A	15	14	13	12	11	10	9	8	7	6	5	4	3	2	1	0		TC5
$005C	15	14	13	12	11	10	9	8	7	6	5	4	3	2	1	0		TC6
$005E	15	14	13	12	11	10	9	8	7	6	5	4	3	2	1	0		TC7

Address	Bit 7	6	5	4	3	2	1	Bit 0	Name
$0037	RTIF	PROF	0	LOCKIF	LOCK	TRACK	SCMIF	SCM	CRGFLG
$0038	RTIE	0	0	LOCKIE	0	0	SCMIE	0	CRGINT
$003B	0	RTR6	RTR5	RTR4	RTR3	RTR2	RTR1	RTR0	RTICTL
$0046	TEN	TSWAI	TSBCK	TFFCA	0	0	0	0	TSCR1
$004D	TOI	0	0	0	TCRE	PR2	PR1	PR0	TSCR2
$0040	IOS7	IOS6	IOS5	IOS4	IOS3	IOS2	IOS1	IOS0	TIOS
$004C	C7I	C6I	C5I	C4I	C3I	C2I	C1I	C0I	TIE
$004E	C7F	C6F	C5F	C4F	C3F	C2F	C1F	C0F	TFLG1
$004F	TOF	0	0	0	0	0	0	0	TFLG2

Table 4.14
MC9S12C32 registers used to configure periodic interrupts.

If **n** is zero, then the RTI system is off. A MC9S12C32 with an 8 MHz crystal will have an OSCCLK frequency of 8 MHz and a default E clock frequency of 4 MHz. Table 4.15 shows the available interrupt periods, assuming an 8 MHz crystal. To clear the RTIF flag (acknowledge the interrupt), the software writes a one to it. Program 4.41 will increment a

		n [6:4] of the RTICTL							
		000	001	010	011	100	101	110	111
	0000	off	0.128	0.256	0.512	1.024	2.048	4.096	8.192
	0001	off	0.256	0.512	1.024	2.048	4.096	8.192	16.384
	0010	off	0.384	0.768	1.536	3.072	6.144	12.288	24.576
	0011	off	0.512	1.024	2.048	4.096	8.192	16.384	**32.768**
	0100	off	0.640	1.280	2.560	5.120	10.240	20.480	40.960
	0101	off	0.768	1.536	3.072	6.144	12.288	24.576	49.152
	0110	off	0.896	1.792	3.584	7.168	14.336	28.672	57.344
m [3:0]	0111	off	1.024	2.048	4.096	8.192	16.384	32.768	65.536
	1000	off	1.152	2.304	4.608	9.216	18.432	36.864	73.728
	1001	off	1.280	2.560	5.120	10.240	20.480	40.960	81.920
	1010	off	1.408	2.816	5.632	11.264	22.528	45.056	90.112
	1011	off	1.536	3.072	6.144	12.288	24.576	49.152	98.304
	1100	off	1.664	3.328	6.656	13.312	26.624	53.248	106.496
	1101	off	1.792	3.584	7.168	14.336	28.672	57.344	114.688
	1110	off	1.920	3.840	7.680	15.360	30.720	61.440	122.880
	1111	off	2.048	4.096	8.192	16.384	32.768	65.536	131.072

Table 4.15
MC9S12C32 real-time interrupt period in ms.

global variable, Time, every 32.767 ms. The interrupt rate is determined by the crystal clock and the RTICTL value

$$\text{RTI interrupt frequency (Hz)} = 15625*2^{-n}/(m+1)$$
$$\text{RTI interrupt period (ms)} = 0.064*(m+1)*2^n$$

Observation: The phase-lock-loop (PLL) on the 9S12 will not affect the RTI rates.

Checkpoint 4.12: How would you modify Program 4.41 to count at 61.035 Hz? (there are four answers).

```
; MC9S12C32 assembly
         org   $3800   ;RAM
Time     rmb   2
         org   $4000
RTI_Init sei           ;make atomic
         movb  #$73,RTICTL ;32.768ms
         movb  #$80,CRGINT ;arm RTI
         movw  #0,Time
         cli           ;enable IRQ
         rts
RTIHan   movb  #$80,CRGFLG   ;ack
         ldd   Time
         addd  #1
         std   Time
         rti
         org   $FFF0
         fdb   RTIHan  ;vector
```

```
// MC9S12C32 CodeWarrior C
unsigned short Time;
void RTI_Init(void){
  asm sei          // Make atomic
  RTICTL = 0x73;   // 30.517Hz
  CRGINT = 0x80;   // Arm
  Time = 0;        // Initialize
  asm cli
}



void interrupt 7 RTIHan(void){
  CRGFLG = 0x80;  // Acknowledge
  Time++;
}
```

Program 4.41
Implementation of a periodic interrupt using the real-time clock feature.

The timer overflow interrupt feature can also be used to generate interrupts at a fixed rate (see Table 4.16). The 16-bit TCNT register is incremented at a fixed rate. The TOF flag is set when the counter overflows and wraps back around (automatically) to zero. If armed, the TOF flag will generate an interrupt. Three bits (PR2, PR1, and PR0) in the TSCR2 register determine the rate at which the counter will increment, hence will determine the TOF interrupt rate. To clear the TOF flag (acknowledge the interrupt), the software writes a one to it. To create a TOF periodic interrupt, we enable the timer (TEN=1), arm the timer overflow (TOI), and set the rate (PR2-0) (see Program 4.42).

Table 4.16
Timer overflow rates, assuming a 4 MHz MC9S12C32.

PR2	PR1	PR0	Divide by	TCNT Clock	TOF Period	TOF Rate
0	0	0	1	250 ns	16.384 ms	61.035 Hz
0	0	1	2	500 ns	32.768 ms	30.517 Hz
0	1	0	4	1 μs	65.536 ms	15.259 Hz
0	1	1	8	2 μs	131.072 ms	7.629 Hz
1	0	0	16	4 μs	262.144 ms	3.815 Hz
1	0	1	32	8 μs	524.288 ms	1.907 Hz
1	1	0	64	16 μs	1048.576 ms	0.954 Hz
1	1	1	128	32 μs	2097.152 ms	0.477 Hz

Observation: The phase-lock-loop (PLL) on the 9S12 will affect the TOF and output compare rates.

The third mechanism to generate periodic interrupts is output compare. Program 4.43 shows a 1 kHz periodic interrupt using output compare 6. To enable output compare, the corresponding bit in the TIOS register must be set. When the TCNT register matches TC6,

```
; MC9S12C32 assembly                        // MC9S12C32 CodeWarrior C
      org  $3800  ;RAM                      unsigned short Time;
Time  rmb  2                                void TOF_Init(void){
      org  $4000                              asm sei      // Make atomic
TOF_Init sei       ;make atomic               TSCR1 = 0x80;  // enable counter
      movb #$80,TSCR1 ;enable TCNT            TSCR2 = 0x81;  // Arm, 30.517Hz
      movb #$81,TSCR2 ;arm, 32.768ms          Time = 0;      // Initialize
      movw #0,Time                            asm cli        // enable interrupts
      cli        ;enable IRQ                 }
      rts
TOFHan movb #$80,TFLG2  ;acknowledge
      ldd  Time                             interrupt 16 void TOFHan(void){
      addd #1                                 TFLG2 = 0x80;  // Acknowledge
      std  Time                               Time++;
      rti                                   }
      org  $FFDE
      fdb  TOFHan  ;vector
```

Program 4.42
Implementation of a periodic interrupt using timer overflow.

the output compare flag, C6F, is set. If armed (C6I=1), then it will request an interrupt. To clear the C6F flag (acknowledge the interrupt), the software writes a one to it. The ISR will acknowledge the interrupt and set TC6 to the time for the next interrupt. The interrupting period is determined by the TCNT period (set by TSCR2) multiplied by the constant PERIOD. Program 4.43 will increment a global variable, Time, every 1 ms.

Checkpoint 4.13: How would you modify Program 4.43 to count at 1 Hz?

```
; MC9S12C32 assembly                        // MC9S12C32 CodeWarrior C
PERIOD    equ  1000   ;in usec              #define PERIOD  1000
          org  $3800  ;RAM                  unsigned short Time;
Time      rmb  2
          org  $4000
OC6_Init sei          ;make atomic          void OC6_Init(void){
          movb #$80,TSCR1 ;enable TCNT        asm sei        // Make atomic
          movb #$02,TSCR2 ;1us                TSCR1 = 0x80;
          bset TIOS,#$40  ;activate OC6       TSCR2 = 0x02;  // 1 MHz TCNT
          bset TIE,#$40   ;arm OC6            TIOS |= 0x40;  // activate OC6
          movw #0,Time                        TIE  |= 0x40;  // arm OC6
          ldd  TCNT  ;time now                TC6 = TCNT+50; // first in 50us
          addd #50   ;first in 50us           Time = 0;      // Initialize
          std  TC6                            asm cli        // enable IRQ
          cli        ;enable IRQ            }
          rts
OC6Han    movb #$40,TFLG1 ;acknowledge      interrupt 14 void OC6handler(void){
          ldd  TC6                            TC6 = TC6+PERIOD; // next in 1 ms
          addd #PERIOD                        TFLG1 = 0x40;    // acknowledge C6F
          std  TC6   ;next in 1 ms            Time++;
          ldd  Time                         }
          addd #1
          std  Time
          rti
          org  $FFE2
          fdb  OC6Han ;vector
```

Program 4.43
Implementation of a periodic interrupt using output compare.

4.16 Exercises

4.1 The objective of this problem is to use interrupts to interface the input device in Figure 4.37 to a 6811. To solve this problem with a different microcomputer, you could substitute any available I/O port for STRA STRB PC7-PC0. You may write the software in assembly or C.

Figure 4.37
Input device interface.

The sequence to input an ASCII character from the input device is as follows. When a new character is available, the input device puts its ASCII code on the 8-bit **Data**, then the input device makes **Ready**=1. Next your interrupt software should read the 8-bit ASCII code, then acknowledge receipt of the data by pulsing **Acknowledge**. After the pulse, the input device will make **Ready**-0 again (Figure 4.38).

Figure 4.38
Input device timing.

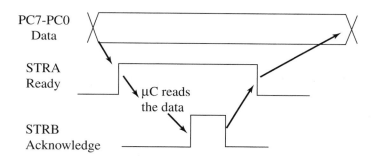

a) Show bit by bit your choice (what and why) for the PIOC register.
b) Show the ritual that initializes the 6811. You should call the function `InitFifo()`, which initializes the FIFO. You may assume the FIFO routines already exist. Do not write the FIFO routines. You may write your answers in assembly or C.
c) Show the interrupt handler that inputs one data byte. The sequence of steps is described above. Store the input data in the FIFO, call `Put(unsigned char)`. Ignore FIFO full errors. No polling is required. You do not have to write the main program that gets and processes the data.

4.2 The objective of this problem is to interface an output device (Figure 4.39) to a 6811 single-chip computer using Port C using interrupt synchronization. To solve this problem with a different microcomputer, you could substitute any available I/O port for STRA STRB PC7-PC0. You may write the software in assembly or C.

Figure 4.39
Output device interface.

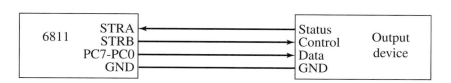

Figure 4.40
Output device timing.

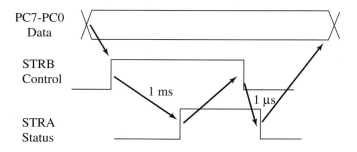

First, your 6811 outputs data on PC7-PC0. Then, your 6811 makes STRB=1; 1 ms later the output device will make STRA=1 (causes an interrupt). Then, your 6811 makes STRB=0; 1 µs later the output device will make STRA=0 (your 6811 need not check for STRA=0) (Figure 4.40).

a) Show bit by bit your choice (what and why) for the PIOC register.
b) Write a C program with a prototype `void Print(unsigned char *pt);` that begins the operations required to print the null (0) terminated ASCII string pointed to by `pt` on the output device. Clear a global variable `DONE` (which will be set in the interrupt routine when the system is done printing). This routine includes the ritual that initializes the 6811.
c) Show the ISR that outputs one data byte. The sequence of steps is described above. Disarm and set the `DONE` flag when the string is finished. Do not output the null. Execute `swi` if you get an illegal interrupt. You do not have to write the main program that generates the data and calls `Print()`.

4.3 *Read the entire question before starting.* The objective is to interface a standard IBM PC keyboard to a microcomputer using STRA, Port C, and output compare (Figure 4.41). The TTL level **Clock** I/O is connected to both **STRA** and **PC0**. The **Clock** is sometimes an input and sometimes an open-collector output of the 6811. The TTL level **Data** is connected to **PC7**. The **Data** also is sometimes an input and sometimes an open-collector output of the 6811. To solve this problem with a different microcomputer, you could substitute any available I/O port for STRA PC7-PC0. You may write the software in assembly or C.

Figure 4.41
Keyboard interface.

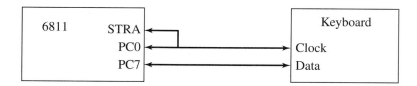

You will *not* use FIFOs in this problem, although your system could be significantly enhanced with FIFOs that decouple the main and background threads. Rather, there will be two global variables (which in actuality behave like a 1-byte FIFO):

```
unsigned char info;    // 8 bit data from the keyboard
unsigned char flag;    // true means info contains valid information
```

You will implement a simple half-duplex solution. In other words, your microcomputer software will be operating in either receive or transmit mode, but never both at the same time. You will use `info` and `flag` (no other globals are allowed). You can assume the main program leaves the system mostly in receive mode, and the operator does not type a key during the short intervals when the main program activates transmit mode. A real IBM PC keyboard interface must handle the situation where communication is attempted in both directions simultaneously (but you don't have to worry about it here).

Receive mode is where data is being received from the keyboard into the microcomputer. When the operator types a key on the keyboard, one or more (up to six) *scan codes* (8 bits each) are sent from the keyboard to the microcomputer. You will write a background interrupt activated by STRA that performs communication with the keyboard. You will also write a C function with the following prototype:

```
void ReceiveMode(void);    // place the keyboard system in receive mode
```

The main program, which you do not write, is free to access the info and flag to collect and process the scan codes. In the Figure 4.42 timing diagram, the thin lines represent keyboard outputs that are open-collector with resistor pull-up, and the thick lines represent computer outputs that are also open-collector.

Figure 4.42
Keyboard receive timing.

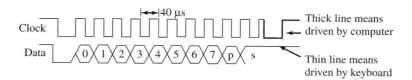

To receive a byte, your microcomputer first places both the **Clock** and **Data** in input mode. In the idle state, the keyboard makes both **Clock** and **Data** high (actually the keyboard signals are also open-collector, and the high levels are generated by resistor pull-ups on the keyboard side). When the keyboard wishes to send a scan code (remember one to six scan codes are sent after the operator types/releases a key), the keyboard will first set the **Data** low, then create a high-to-low transition on the **Clock**. This is like the start bit we saw in the asynchronous serial interface. You will accept and ignore the start bit. Then, every 40 μs the keyboard will put the next information bit on the **Data** line and create another high-to-low transition on **Clock**. Notice the **Data** changes on the low-to-high transition, and you should input the binary bits on the high-to-low edges. It will send 8 bits of information (bit 0 first), which you will read and then create an 8-bit byte from the 8 individual bits. It also sends an odd parity bit (p) and stop bit (s) that you will read and ignore. It would be a good idea to do some error checking (returning with an error on a timeout, making sure the start bit is low, the odd parity is correct, and the stop bit is high), but these features are *not* required in this problem. Since the main program is performing other unrelated tasks, you will implement these receive functions in a background thread. If the flag is 0 when the stop bit is received (meaning the info is currently empty), put the new byte into info and set the flag (like a FIFO put). If the flag is still set when the stop bit is received (because the main program hasn't read the previous transmission), simply discard the new data (like an overrun on the SCI serial port or like a FIFO full error). As an acknowledgment back to the keyboard, your computer should make the **Clock** an open-collector output and drive **Clock** low for 40 μs. Use OC2 (gadfly) to create this 40-μs timing. At the end of the STRA interrupt handler, your microcomputer system should leave **Clock** and **Data** in input mode, ready to accept the next scan code. There will be exactly one interrupt request for each scan code received.

Figure 4.43
Keyboard transmit timing.

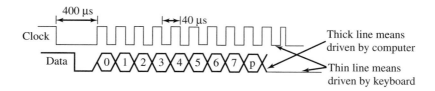

Transmit mode is where commands are sent from the microcomputer to the keyboard. When the main program wishes to modify the keyboard functions (turn on LEDs, enable/disable certain functions) it sends a *command code* (8 bits each) to the keyboard (Figure 4.43). You will write a gadfly C function (with *no* background interrupt threads and *no* globals) with the following prototype:

```
void SendCommand(unsigned char);    // transmit one command to the keyboard
```

The main program, which you do not write, can call this function to transmit command codes to the keyboard. *Notice that this main program does not disable interrupts as it accesses the shared globals.* For example, the main program could first initialize the keyboard using transmit mode, then switch to receive mode to input and process scan codes:

```
void main(void){ unsigned char ScanCode;
   OtherInitialization();    // unrelated to your keyboard (don't write this)
   SendCommand(0x25);        // specify keyboard operation mode
   SendCommand(0x42);        // specify keyboard operation mode
   ReceiveMode();            // your function that initializes the keyboard
   while(1){
     if(flag) {
        ScanCode=info;        // read next byte (like a FIFO get)
        flag=0;               // means can accept another receive transmission
        Process(ScanCode);} // don't write this Process
      OtherProcess();}}        // unrelated to your keyboard (don't write this)
```

To transmit a byte, the microcomputer first disables the receive mode functions, places **Clock** and **Data** in open-collector output mode with the output high (floating), and waits for **Clock** to be high. Next the computer sets **Clock** low for 400 μs. Again use OC2 (gadfly) to create this timing delay. Then you pull the **Data** low (while the **Clock** is still low). Next, you release the **Clock** by making it an input (it should float high because of the pull-up in the keyboard). The keyboard now should capture control of the **Clock** and drive it low. On the next nine low-to-high transitions you will set a new binary bit on the **Data** line. On these nine high-to-low transitions, the binary bit is latched into the keyboard. The first 8 bits are the command code (bit 0 first), and the ninth bit is odd parity. After the odd parity bit is received by the keyboard, both the **Clock** and **Data** lines are held low by the keyboard until it is ready to accept another command. At the end, your microcomputer system should leave both **Clock** and **Data** in input mode.

4.4 You will design and implement a FSM using the 6812 *key wakeup interrupts* (Figure 4.44). One of the limitations of the previous FSM implementations is that they require 100% of the processor time and run in the foreground. In this system, there are two inputs and two outputs. We will implement a FSM where state transitions only occur on the rising edges of one of the two inputs. These rising edges should cause a key wakeup interrupt, and the FSM controller will be run in the interrupt handler. Your system will use a statically allocated linked data structure, with a one-to-one correspondence to the FSM. machine. You should be able to change the data structure to accommodate other two-input, two-output FSMs of similar structure without changing the program controller. Even though you could shut off the controller once it gets to the "End" state, continue to accept key wakeup interrupts. You can assume that the inputs occur independently, and many occur at the same time.

Figure 4.44
Finite-state machine.

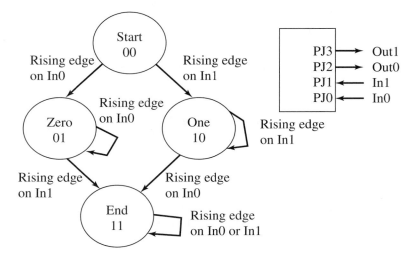

The FSM controller sequence is:

1. Wait for a rising edge on **In1** or **In0** (causing key wakeup interrupt)
2. Go to the next state depending on the current state and which rising edge occurred
3. Output the pattern of the new state on **Out1** and **Out0**
   a) Show the linked data structure. You may write this in C or assembly.
   b) Show the ritual that is executed once at the beginning. There will be a main program (which you will not write) and possibly other interrupts, but this is the only device using Port J. The initial state is **Start** with its initial output of 00.
   c) Show the key wakeup interrupt handler. Don't worry about how the interrupt vector is set.

4.5 Design the hardware/software interface that receives data from the following fully interlocked sensor device. The sensor has **status** and **data** outputs and an **ack** input.

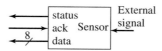

The following timing diagram illustrates the sequence of events required to input a sensor reading into the computer. The procedure begins with the software setting its **ack** output high. Second, the sensor will perform a conversion, place the 8-bit result on its **data** lines, and set its **status** low. Third, the software will read the **data**, then set its **ack** low. Fourth, the sensor will bring its **status** high again. The fifth step (really the same as the first step for the next data transfer) is for the software to set its **ack** high, meaning it wishes to collect another sensor reading.

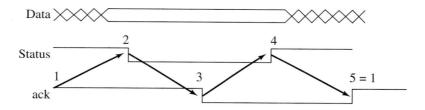

The interface timing has the following additional constraints/observations:

- Sensor **data** outputs are valid when **status** is low (software reads **data** after 2 and before 3).
- Timing events 1, 2, 3, 4 occur exactly in this order.
- Delays from 1 to 2 and from 3 to 4 are determined by the sensor, varying from 10 μs to 1 s.
- Delays from 2 to 3 and from 4 to 1 are system response times that should be minimized.

Your software must follow these conditions:

- Both the fall and rise of **status** will cause interrupts.
- These two interrupts will have separate vectors (vectored interrupt).
- If your computer doesn't support vectored interrupt, then you may use polled interrupts.
- A FIFO queue will link your producer (background interrupt) with a consumer (main program).
- You will not write the main program that gets from the FIFO and processes the data.
- You may use any one of the FIFOs in Chapter 4 without showing its implementation.
- You may ignore COP and TOF interrupts.
- Don't worry about how the interrupt vector is set.
- You may ignore FIFO full errors (although it would have been possible to prevent them).

a) Show the hardware interface between the sensor and your computer. You may use any available port, but make sure two interrupts can be requested that have separate vectors.
b) Show the ritual that will be called at the beginning of the main program.
c) Show the interrupt handler that is executed on the fall of status (timing event 2).
d) Show the interrupt handler that is executed on the rise of status (timing event 4).

4.17 Lab Assignments

Lab 4.1. The overall objective is to create an alarm clock. A periodic interrupt establishes the time of day. Input/output of the system uses the gadfly SCI serial port developed in Chapter 3. Input is used to set the time and the alarm. Design a command interpreter that performs the necessary operations. Output is used to display the current time. An LED or sound buzzer can be used to signify the alarm.

Lab 4.2. The overall objective is to create a software-driven pulse-width–modulated output. A periodic interrupt will set an output pin high/low. Input/output of the system uses the gadfly SCI serial port developed in Chapter 3. Input is used to set the period and duty-cycle for the wave. Design a command interpreter that performs the necessary operations. Connect the pulse-width modulated output to an oscilloscope.

Lab 4.3. The overall objective is to create an interrupt-driven LED light display. Interface 8 to 16 colored LEDs to individual output pins. A periodic interrupt will change the LED pattern. Connect one or two switches to the system, and use them to control which LED light pattern is being displayed. Design a linked data structure that contains the light patterns. Create device drivers for the LED outputs, switch inputs, and periodic interrupt. The main program will initialize the LED outputs, switch inputs, and periodic interrupt. All of the input/output will be performed in the ISR. You should create a general purpose timer system that accepts a function to execute and a period. For example,

```
void main(void){
  LED_Init();     // Initialize LED output system
  Switch_Init();  // Initialize switch input system
  Timer_Init(&FSMcontroller,250);   // Run FSMcontroller() every 250ms
  while(1){}
}
```

Lab 4.4. The overall objective is to create an **interrupt-driven traffic light controller**. The system has three digital inputs and seven digital outputs. You can simulate the system with three switches and seven LEDs. The inputs are North, East, and Walk. The outputs are six for the traffic light and one for a walk signal. Implement the controller using a finite state machine. Choose a Moore or Mealy data structure as appropriate. A periodic interrupt will run the FSM. The main program will initialize the LED outputs, switch inputs, and periodic interrupt. All of the input/output will be performed in the ISR. You should create a general-purpose timer system that accepts a function to execute and a period. For example,

```
void main(void){
  Traffic_Init();     // Initialize switches and LED
  Timer_Init(&Traffic_Controller,100);   // Run FSM every 100ms
  while(1){}
}
```

5 Threads

Chapter 5 objectives are to:

- ❑ Define a thread control block
- ❑ Design and implement a preemptive thread scheduler
- ❑ Design and implement spin-lock semaphores
- ❑ Design and implement blocking semaphores
- ❑ Present applications that employ semaphores

In Chapter 4 we used interrupts to create a multithreaded environment. In that configuration we had a single foreground thread (the main program) and multiple background threads (the ISRs). These threads are run using a simple algorithm. The ISR of an input device is invoked when new input is available. The ISR of an output device is invoked when the output device is idle and needs more data. Last, the ISR of a periodic task is run at a regular rate. The main program runs in the remaining intervals. Because most embedded applications are small in size, and static in nature, this configuration is usually adequate. The limitation of a single foreground thread comes as the size and complexity of the system grows. As we saw in Chapter 2, it is appropriate to partition a large project into modules, then piece together those modules in either a layered or hierarchical manner. It is during this "piece together" phase of a large project that the simple single foreground thread environment breaks down. It is easy to combine each module's initialization routine into one common initialization sequence. The only difficulty during initialization is to prevent the initialization of one module from undoing the initialization of a previously initialized module. For example, you must be careful when two modules initialize the same I/O control register. Similarly, it is straightforward to combine the ISRs of the modules. If the modules use separate vectors, then no additional effort is required. If two modules use the same interrupt vector, then one of the polling schemes described in Chapter 4 can be used. The difficulty arises when combining the foreground functions of the modules. With a single foreground thread, we are forced into a sequential program structure similar to Figure 3.10. For the projects where the modules are tightly coupled (interdependent), this sequential model is both natural and appropriate. On the other hand, the projects where the modules are more loosely coupled (independent) may more naturally fit a multiple foreground thread configuration. The goal of this chapter is to develop the software techniques to implement multiple foreground threads (the scheduler) and provide synchronization tools (semaphores) that allow the threads to interact with each other. Systems that implement a thread scheduler still may employ regular I/O driven interrupts. In this way, the system supports multiple foreground threads and multiple background threads.

5.1 Multithreaded Preemptive Scheduler

We define a thread as the execution of a software task that has its own stack and registers (Figure 5.1). Another name for thread is lightweight process. Since each thread has a separate stack, its local variables are private, which means it alone has access. Multiple threads cooperate to perform an overall function. Since threads interact for a common goal, they do share resources, such as global memory, and I/O devices (Figure 5.2).

Figure 5.1
Each thread has its own registers and stack.

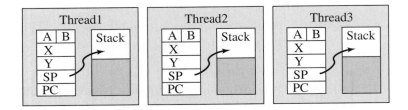

Figure 5.2
Threads share global memory and I/O ports.

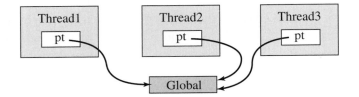

Some simple examples of multiple threads are the interrupt-driven I/O examples of Chapter 4. In each of these examples, the background thread (ISR) executes when the I/O device is done performing the required I/O operation. The foreground thread (main program) executes during the times when no interrupts are needed. A global data structure is used to communicate between threads. Notice that the information stored on the stack or in the microcomputer registers by one thread is not accessible by another thread.

Checkpoint 5.1: What is the difference between a program and a thread?

Checkpoint 5.2: Why can't threads pass parameters to each other on the stack?

A thread can be in one of three states (Figure 5.3). A thread is in the *blocked state* when it is waiting for some external event like I/O (keyboard input available, printer ready, I/O device available.) If a thread communicates with other threads, then it can be blocked waiting for an input message or waiting for another thread to be ready to accept its output message. If a thread wishes to output to the display, but another thread is currently outputting, it will block. We will use a semaphore mechanism to implement the sharing of the display output among multiple threads. If a thread needs information from a FIFO (calls `Fifo_Get`), then it will be blocked if the FIFO is empty (because it cannot retrieve any information). On the other hand, if a thread outputs information to a FIFO (calls `Fifo_Put`), then it will be blocked if the FIFO is full (because it cannot save its information).

Figure 5.3
A thread can be in one
of three states.

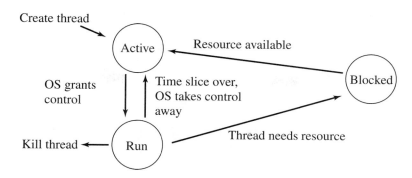

A thread is in the *active state* if it is ready to run but waiting for its turn. A thread is in the *run state* if it is currently executing. With a single instruction stream computer like the 6811 and 6812, at most one thread can be in the run state at a time. A good implementation is to use linked list data structures to hold the ready and blocked threads. We can create a separate blocked linked list for each reason why the thread cannot execute. For example, one blocked list for waiting for the output display to be free, one for full during a call to Fifo_Put, and one for empty during a call to Fifo_Get. In general, we will have one blocked list with each blocking semaphore.

In Figure 5.4, thread 5 is running, threads 1 and 2 are ready to run, thread 6 is blocked waiting for the printer, and threads 3 and 4 are blocked because the FIFO is empty.

Figure 5.4
Thread 5 is running,
threads 1 and 2 are read
to run, and the rest are
blocked.

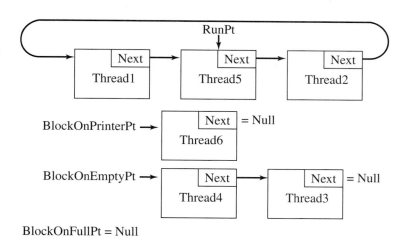

Checkpoint 5.3: Why is it efficient to block a thread that needs a resource that is not available?

Checkpoint 5.4: If a thread is blocked because the output display is not available, when should you wake it up?

**5.1.1
Round-Robin
Scheduler**

In this section we will develop a simple multithreaded round-robin scheduler. In particular, there will be three statically allocated threads that are each allowed to execute 10 ms in a round-robin fashion. Even though there are two programs, ProgA and ProgB, there will be three threads (Figure 5.5). Recall that a thread is not simply the software but the execution of the software. In this way, we will have two threads executing the same program, ProgA.

Figure 5.5
The circular linked list allows the scheduler to run all three threads equally.

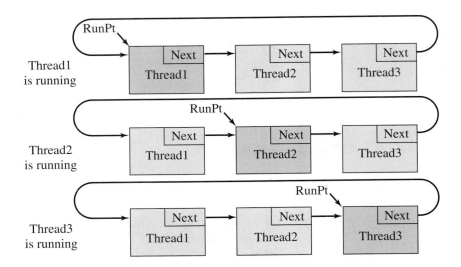

The thread control block (TCB) will store the information private to each thread. There will be a 64-byte TCB structure for each thread. The TCB must contain:

1. A pointer so that it can be chained into a linked list
2. The value of its stack pointer
3. A stack area that includes local variables and other registers

While a thread is running, it uses the actual 6811 hardware registers CCR, B, A, X, Y, PC, SP (Figure 5.6). In addition to these necessary components, the TCB might also contain:

4. Thread number, type, or name
5. Age, or how long this thread has been active
6. Priority
7. Resources that this thread has been granted

Figure 5.6
The running thread uses the actual registers, while the other threads have their register values saved on the stack.

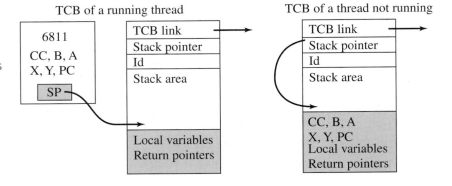

To illustrate the concept of a preemptive scheduler, we will implement a simple system first in 6811 assembly then in 6811 C. The three statically allocated threads are arranged in a circular linked list (Program 5.1). This example illustrates the difference between a

Program 5.1
Assembly code that
statically allocates three
threads.

TCB1	fdb	TCB2	
	fdb	IS1	
	fcb	1	
	rmb	49	
IS1	rmb	1	
	fcb	$40	
	fdb	0,0,0	
	fdb	ProgA	

TCB2	fdb	TCB3	
	fdb	IS2	
	fcb	2	
	rmb	49	
IS2	rmb	1	
	fcb	$40	
	fdb	0,0,0	
	fdb	ProgA	

TCB3	fdb	TCB1	link
	fdb	IS3	SP
	fcb	4	Id
	rmb	49	
IS3	rmb	1	
	fcb	$40	CCR
	fdb	0,0,0	DXY
	fdb	ProgB	PC

program (e.g., ProgA and ProgB) and a thread (e.g., Thread1, Thread2, and Thread3). Notice that Threads 1 and 2 both execute ProgA. There are many applications where the same program is being executed multiple times.

Even though the thread has not yet been allowed to run, it is created with an initial stack area that "looks like" it had been previously suspended by an OC5 interrupt. Notice that the initial value loaded into the CCR ($40) when the thread runs for the first time has XIRQ disabled (X=1) and IRQ enabled (I=0). When the thread is launched for the first time, it will execute the program specified by the value in the "initial PC" location. The Id is used to visualize the active process. On the 6812, change rmb 49, rmb 1 to rmb 50, rmb 0.

The round-robin preemptive scheduler simply switches to a new thread every 10 ms (Programs 5.2, 5.3).

```
Next   equ  0        pointer to next TCB
SP     equ  2        Stack pointer for this thread
Id     equ  4        Used to visualize which thread is current running
RunPt  rmb  2        pointer to thread that is currently running
Main   ldaa #$FF
       staa DDRC     PortC displays which program is executing
       ldx  #TCB1    First thread to run
       jmp  Start
* Suspend thread which is currently running
OC5Han ldx  RunPt
       sts  SP,x     save Stack Pointer in TCB
* launch next thread
       ldx  Next,x
Start  stx  RunPt
       ldaa Id,x
       staa PORTB    visualizes running thread
       lds  SP,x     set SP for this new thread
       ldd  TOC5
       addd #20000   interrupts every 10 ms
       std  TOC5
       ldaa #$08     ($20 on the 6812)
       staa TFLG1    acknowledge OC5
       rti
```

Program 5.2
Assembly code that implements the preemptive thread switcher.

Checkpoint 5.5: What does RunPt point to?

Checkpoint 5.6: What is the purpose of the sts lds instructions?

Program 5.3
Assembly code for the two main programs and shared subroutine.

```
ProgA pshx              ProgB pshx              Sub pshx
      tsx                     tsx                   tsx
      ldd   #5                ldd   #5              std   0,x
      std   0,x              std   0,x             ldaa #1
LoopA ldaa #2         LoopB ldaa #4              staa PORTC
      staa PORTC            staa PORTC            ldd   0,x
      ldd   0,x             ldd   0,x             addd #1
      jsr   sub             jsr   sub             pulx
      std   0,x             std   0,x             rts
      bra   LoopA           bra   LoopB
```

Next we will implement a simple multithreaded system in 6811 C, also with three statically allocated threads running two programs (Figure 5.7). The syntax of these programs, like all 6811 C programs in this book, conforms to the ImageCraft ICC11 version 4.5 or later. The thread.c example that is installed with the TExaS simulator (look for it in the ICC11 subfolder) conforms with the freeware version ICC11 and has slightly different syntax (Program 5.4).

Figure 5.7
Profile of the three threads running two main programs.

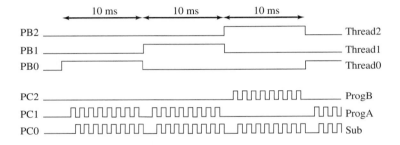

Program 5.4
C code for the two main programs and shared subroutine.

```
short Sub(short j){ short i;
    PORTC=1;  /* Port C=program is being executed */
    i=j+1;
    return(i);}
void ProgA(){ short i;
    i=5;
    while(1) { PORTC=2; i=Sub(i);}}
void ProgB(){ short i;
    i=6;
    while(1) { PORTC=4; i=Sub(i);}}
```

One of the tricky parts about defining the TCB is to choose the size of MoreStack so that RegA and RegX are contiguous. Most compilers will try to allocate 16-bit variables (e.g., RegX) on an even address. Notice that TCB starts at an even address CCR and RegA will be at odd addresses and the rest will be even. If you change the 49 to 50, then the compiler will leave a 1-byte gap between RegA and RegX, and the program will not work (Programs 5.5, 5.6).

Checkpoint 5.7: What happens if a thread executes **TOC5=TCNT+10;**

The 6812 implementation is virtually identical, except the 6812 SP points to the top stack entry, whereas the 6811 SP+1 points to the top stack entry. Assembly and C implementations of this example for both the 6811 and 6812 can be found as part of the TExaS simulator. Look for the Thread.* files after the simulator has been installed. The TExaS examples include a logic analyzer connection so that you can create profiles like the one presented in Figure 5.7.

Program 5.5
6811 C code for the thread control block.

```c
struct TCB
{   struct TCB *Next;        /* Link to Next TCB */
    unsigned char *SP;       /* Stack Pointer when not running  */
    unsigned short  Id;      /* output to PortB visualizing active thread */
    unsigned char MoreStack[49];  /* more stack */
    unsigned char CCR;       /* Initial CCR */
    unsigned char RegB;      /* Initial RegB */
    unsigned char RegA;      /* Initial RegA */
    unsigned short RegX;     /* Initial RegX */
    unsigned short RegY;     /* Initial RegY */
    void (*PC)(void);        /* Initial PC */
};
typedef struct TCB TCBType;
typedef TCBType * TCBPtr;
TCBType sys[3]={
  {  &sys[1],               /* Pointer to Next */
     &sys[0].MoreStack[49],    /* Initial SP */
     1,                    /* Id */
     { 0},
     0x40,0,0,0,0,         /* CCR,B,A,X,Y */
     ProgA, },             /* Initial PC */
  {  &sys[2],               /* Pointer to Next */
     &sys[1].MoreStack[49],    /* Initial SP */
     2,                    /* Id */
     { 0},
     0x40,0,0,0,0,         /* CCR,B,A,X,Y */
     ProgA, },             /* Initial PC */
  {  &sys[0],               /* Pointer to Next */
     &sys[2].MoreStack[49],    /* Initial SP */
     4,                    /* Id */
     { 0},
     0x40,0,0,0,0,         /* CCR,B,A,X,Y */
     ProgB, } };           /* Initial PC */
```

```c
TCBPtr RunPt;   /* Pointer to current thread  */
#pragma interrupt_handler ThreadSwitch()
void ThreadSwitch(){
asm(" ldx _RunPt\n"
    " sts 2,x");
    RunPt=RunPt->Next;
    PORTB=RunPt->Id;  /* PortB=active thread */
asm(" ldx _RunPt\n"
    " lds 2,x");
    TOC3=TCNT+20000;  /* Thread runs for 10 ms */
    TFLG1=0x20; }     /* ack by clearing TOC3F */
void main(void){ DDRC=0xFF;      /* PortC outputs specify that program is running */
    RunPt=&sys[0];   /* Specify first thread */
asm(" sei");
    TOC3vector=&ThreadSwitch;
    TFLG1 = 0x20;   /* Clear OC3F */
    TMSK1 = 0x20;   /* Arm TOC3 */
    TOC3=TCNT+20000;
    PORTB=RunPt->Id;
asm(" ldx _RunPt\n"
    " lds 2,x\n"
    " cli\n"
    " rti");}    /* Launch First Thread */
```

Program 5.6
6811 C code for the thread switcher.

5.1.2
Other Scheduling
Algorithms

A non-preemptive (cooperative) scheduler trusts each thread to voluntarily release control on a periodic basis. Although easy to implement, because it doesn't require interrupts, it is not appropriate for real-time systems.

A priority scheduler assigns each thread a priority number (e.g., 1 is the highest). Normally, we add priority to a system that implements blocking semaphores and not to one that uses spin-lock semaphores. Two or more threads can have the same priority. A priority 2 thread is run only if no priority 1 threads are ready to run. Similarly, we run a priority 3 thread only if no priority 1 or priority 2 threads are ready. If all threads have the same priority, then the scheduler reverts to a round-robin system. The advantage of priority is that we can reduce the latency (response time) for important tasks by giving those tasks a high priority. The disadvantage is that on a busy system, low-priority threads may never be run. This situation is called *starvation.*

5.1.3
Dynamic
Allocation of
Threads

In the above examples, the number of threads was specified at assembly/compile time. We can use the compiler's memory manager to dynamically allocate the deallocate threads. The Program 5.7 code fragment dynamically allocates space for a new TCB and initializes its contents. For the 6812, change CCR-1 to CCR.

Program 5.7
6811 C function to
create a new thread.

```
void create(void (*program)(void), short TheId){
  TCBPtr NewPt;      // pointer to new thread control block
  NewPt=(TCBPtr)malloc(sizeof(TCBType)); // space for new TCB
  if(NewPt==0)return;
  NewPt->SP=&(NewPt->CCR-1);   /* 6811 Stack Pointer when not running  */
  NewPt->Id=TheId;             /* used to visualize active thread */
  NewPt->CCR=0x40;             /* Initial CCR, I=0 */
  NewPt->RegB=0;               /* Initial RegB */
  NewPt->RegA=0;               /* Initial RegA */
  NewPt->RegX=0;               /* Initial RegX */
  NewPt->RegY=0;               /* Initial RegY */
  NewPt->PC=program;           /* Initial PC */
  if(RunPt){
    NewPt->Next=RunPt->Next;
    RunPt->Next=NewPt;}        /* will run Next */
  else
    RunPt=NewPt;               /* the first and only thread */
}
```

5.2 Semaphores

We will use semaphores to implement synchronization, sharing, and communication between threads. A semaphore is a counter and has two atomic functions (methods) apart from initialization that operate on the counter. The operations are called P or wait (derived from the Dutch *proberen,* "to test"), and V or signal (derived from the Dutch *verhogen,* "to increment"). When we use a semaphore, we usually can assign a meaning or significance to the counter value. A binary semaphore has a global variable that can be 1 for free and 0 for busy.

**5.2.1
Spin-Lock
Semaphore
Implementation**

There are many implementations of semaphores, but the simplest is called *spin-lock* (Figure 5.8). If the thread calls `wait` with the counter less than or equal to 0 it will "spin" (do nothing) until the counter goes above zero (Program 5.8). In the context of the previous round-robin scheduler, a thread that is "spinning" will perform no useful work but eventually will be suspended by the OC5 handler, and other threads will execute. It is important to allow interrupts to occur while the thread is spinning so that the computer does not hang up. The read-modify-write operation on the global semaphore must be made atomic, because the scheduler might switch threads in between any two instructions that execute with the interrupts enabled.

Figure 5.8
Flowcharts of a spin-lock counting semaphore.

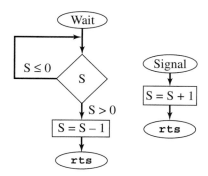

```
;assembly implementation
S      fcb   1    ;semaphore, initialized to 1

; spin if zero, otherwise decrement
wait sei        ;read-modify-write atomic
     ldaa S     ;value of semaphore
     bhi  OK    ;available if >0
     cli
     bra  wait ;**interrupts occur here**
OK   deca
     staa S       ;S=S-1
     cli
     rts

;increment semaphore
signal
     inc  S      ;S=S+1, this is atomic
     rts
```

```c
// ******** OS_Wait ***********
// decrement and spin if less than 0
// input:  pointer to a semaphore
// output: none
void OS_Wait(short *semaPt){
  asm sei        // Test and set is atomic
  while(*semaPt <= 0){   // disabled
    asm cli                // disabled
    asm nop                // enabled
    asm sei                // enabled
  }
  (*semaPt)--;             // disabled
  asm cli                  // disabled
}                          // enabled
// ******** OS_Signal ***********
// increment semaphore
// input:  pointer to a semaphore
// output: none
void OS_Signal(short *semaPt){
unsigned char SaveCCR;
  asm tpa
  asm staa SaveCCR    // save previous
  asm sei             // make atomic
  (*semaPt)++;
  asm ldaa SaveCCR    // recall previous
  asm tap             // end critical
}
```

Program 5.8
A spin-lock counting semaphore.

Checkpoint 5.8: What happens if a spin-lock **wait** loops with interrupts disabled?

Checkpoint 5.9: What happens if the **sei cli** instructions are removed entirely from the spin-lock wait?

The 6812 minm instruction can be used to implement a binary semaphore without disabling interrupts (Program 5.9). In C, these functions are shown in Program 5.10.

```
S       fcb   1        semaphore flag initialized to 1
bWait   clra           new value
loop    minm  S        in either case S is now 0, carry set if S used to 1
        bcc   loop      loop until S=1
        rts
bSignal ldaa  #1
        staa  S         S=1
        rts
```

Program 5.9
6812 assembly code for a spin-lock binary semaphore.

Program 5.10
6812 C code for a spin-lock binary semaphore.

```
void bWait(char *semaphore){
asm(" clra\n"            // new value for semaphore
"loop: minm [2,x]\n"    // test and set (ICC12 version 5)
"      bcc loop\n"); }
void bSignal(char *semaphore){
     (*semaphore)=1;} // compiler makes this atomic
```

Observation: If the semaphores can be implemented without disabling interrupts, then the latency in response to external events will be improved.

We can use three binary semaphores and a counter to implement a counting semaphore again without disabling interrupts (Program 5.11).

Program 5.11
C code for a counting semaphore.

```
struct sema4       // counting semaphore based on 3 binary semaphores
{   short value;   // semaphore value
    char s1;       // binary semaphore
    char s2;       // binary semaphore
    char s3;       // binary semaphore
};
typedef struct sema4 sema4Type;
typedef sema4Type * sema4Ptr;
void Wait(sema4Ptr semaphore){
    bWait(&semaphore->s3);  // wait if other caller to Wait gets here first
    bWait(&semaphore->s1);   // mutual exclusive access to value
    (semaphore->value)--;   // basic function of Wait
    if((semaphore->value)<0){
       bSignal(&semaphore->s1); // end of mutual exclusive access to value
       bWait(&semaphore->s2);   // wait for value to go above 0
       }
    else
       bSignal(&semaphore->s1); // end of mutual exclusive access to value
    bSignal(&semaphore->s3);       // let other callers to Wait get in
}
void Signal(sema4Ptr semaphore){
    bWait(&semaphore->s1);   // mutual exclusive access to value
    (semaphore->value)++;   // basic function of Signal
```

```
        if((semaphore->value)<=0)
            bSignal(&semaphore->s2);    // allow S2 spinner to continue
        bSignal(&semaphore->s1); // end of mutual exclusive access to value
}
void Initialize(sema4Ptr semaphore, short initial){
    semaphore->s1=1;   // first one to bWait(s1) continues
    semaphore->s2=0;   // first one to bWait(s2) spins
    semaphore->s3=1;   // first one to bWait(s3) continues
    semaphore->value=initial;}
```

5.2.2
Blocking
Semaphore
Implementation

For this implementation, each semaphore has an integer global variable (an up/down counter) and a blocked **tcb** linked list. This linked list contains the threads that are blocked (not ready to run) because they called the Wait function when the semaphore counter was less than or equal to 0. Our semaphore "object" also has three "methods": Initialize, Wait, and Signal (Figure 5.9). The proper way to make a multistep sequence atomic is to first save the interrupt status (I bit) on the stack, then disable interrupts. At the end of the atomic sequence, we restore the I bit back to its original value rather than simply assuming it was enabled to begin with. This procedure to disable/enable interrupts handles the situation of nested critical sections. We must be careful to block and wake up the correct number of threads.

Figure 5.9
Flowcharts of a blocking counting semaphore.

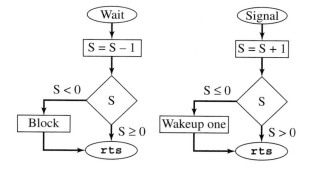

Initialize (Program 5.12):

1. Set the counter to its initial value
2. Clear the associated blocked **tcb** linked list

Wait:

1. Disable interrupts to make atomic:

   ```
   tpa
   psha    save old interrupt status
   sei
   ```

2. Decrement the semaphore counter, $S=S-1$
3. If the semaphore counter is less than 0, then
       Block this thread onto the associated **tcb** blocked linked list by executing SWI
           The SWI operation performs stack operations similar to an OC5 interrupt
           The SWI handler suspends the current thread
               Moves the **tcb** of this thread from the active list to the blocked list
               Launches another thread from the active list

4. Restore interrupt status:

```
pula
tap
```

Signal:

1. Disable interrupts to make atomic:

```
tpa
psha    save old interrupt status
sei
```

2. Increment the semaphore counter, S=S+1
3. If the semaphore counter is less than or equal to 0, then
    Wake up one thread from the **tcb** linked list
    Do not suspend execution of this thread
    Simply move the **tcb** of the "wakeup" thread from the blocked list to the active list

4. Restore interrupt status:

```
pula
tap
```

Program 5.12
Assembly code to initialize a blocking semaphore.

```
S       rmb  1    semaphore counter
BlockPt rmb  2    Pointer to threads blocked on S
Init    tpa
        psha             Save old value of I
        sei              Make atomic
        ldaa #1
        staa S           Init semaphore value
        ldx  #Null
        stx  BlockPt     empty list
        pula
        tap              Restore old value of I
        rts
```

The implementation in Program 5.13 assumes the system only has one semaphore. In a more general implementation, there would be multiple semaphores, each with its own counter and blocked list. In this case, a pointer (call by reference) to the semaphore structure (counter and blocked list) would be passed to the SWI handler.

Program 5.13
Assembly code helper function to block a thread, used to implement a blocking semaphore.

```
;To block a thread, execute SWI
;RunPt points to the active thread, which will be blocked
;BlockPt points to the blocked list for this semaphore
SWIhan ldx  RunPt   ;running process "to be blocked"
       sts  SP,x    ;save Stack Pointer in its TCB
; Unlink "to be blocked" thread from RunPt list
       ldy  Next,x  ;find previous thread (see Figure 5.10)
       sty  RunPt   ;next one to run
look   cpx  Next,y  ;search to find previous
       beq  found
       ldy  Next,y
       bra  look
found  ldd  RunPt   ;one after blocked (see Figure 5.11)
       std  Next,y  ;link previous to next to run
* Put "to be blocked" thread on block list
       ldy  BlockPt ;(see Figure 5.12)
       sty  Next,x  ;link "to be blocked"
       stx  BlockPt
```

```
* Launch next thread
        ldx   RunPt
        lds   SP,x     ;set SP for this new thread
        ldd   TCNT     ;Next thread gets a full 10ms time slice
        addd  #20000   ;interrupt after 10 ms
        std   TOC5
        ldaa  #$08     ;($20 on the 6812)
        staa  TFLG1    ;clear OC5F
        rti
```

Figure 5.10
Linked list of threads ready to run before the thread is blocked.

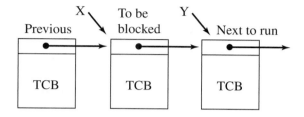

Figure 5.11
Linked list at this point in the program.

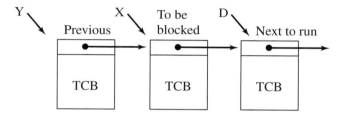

Figure 5.12
TCBs after the thread is blocked (ready-to-run TCBs are not shown).

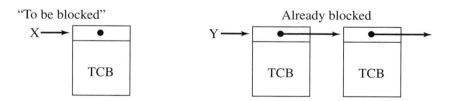

5.3 Applications of Semaphores

This section can be used in two ways. First, it provides a short introduction to the kinds of problems that can be solved using semaphores. In other words, if you have a problem similar to one of these examples, then you should consider a thread scheduler with blocking semaphores as one possible implementation. Second, this section provides the basic approach to solving these particular problems. As stated in the introduction to this chapter, we will generally find that large software systems with more-or-less independent modules can use semaphores when synchronization is needed. On the other hand, for smaller systems with more tightly coupled modules, then the overhead of a thread scheduler will probably not be justified.

5.3.1
Thread Synchronization or Rendezvous

The objective of this example is to synchronize threads 1 and 2. In other words, whichever thread gets to this part of the code first will wait for the other. Initially semaphores S1 and S2 are both 0. The two threads are said to *rendezvous* at the code following the signal and wait calls. The significance of the two semaphores is illustrated in the following table:

S1	S2	Meaning
0	0	Neither thread has arrived at the rendezvous location or both have passed
−1	+1	Thread 2 arrived first and is waiting for thread 1
+1	−1	Thread 1 arrived first and is waiting for thread 2

Thread 1	Thread 2
`signal(&S1);`	`signal(&S2);`
`wait(&S2);`	`wait(&S1);`

5.3.2
Resource Sharing, Nonreentrant Code or Mutual Exclusion

The objective of this example is to share a common resource on a one-at-a-time basis. The critical section (or vulnerable window) of nonreentrant software is that region that should be executed only by one thread at a time. Mutual exclusion means that once a thread has begun executing in its critical section (the `printf();` in this example), then no other thread is allowed to execute its critical section. In other words, whichever thread starts to print first will be allowed to finish printing. Either a binary or a counting semaphore can be used. Initially, the semaphore is 1. If S=1, it means the printer is free. If S=0, it means the printer is busy and no thread is waiting. If S<0, it means the printer is busy and one or more threads are blocked.

Thread 1	Thread 2	Thread 3
`bwait(&S);`	`bwait(&S);`	`bwait(&S);`
`printf("bye");`	`printf("tchau");`	`printf("ciao");`
`bsignal(&S);`	`bsignal(&S);`	`bsignal(&S);`

5.3.3
Thread Communication Between Two Threads Using a Mailbox

The objective of this example is to send mail from thread 1 to thread 2. The `send` semaphore allows the producer to tell the consumer that new mail is available. The `ack` semaphore is a mechanism for the consumer to tell the producer that the mail was received. Initially, semaphores `send` and `ack` are both 0. The significance of the two semaphores is illustrated in the following table. In this example, the two threads will rendezvous at this point.

send	ack	Meaning
0	0	No mail available, consumer is not waiting
−1	0	No mail available, consumer is waiting for mail
+1	−1	Mail available and producer is waiting for acknowledgment

Producer thread	Consumer thread
`Mail=4;`	`wait(&send);`
`signal(&send)`	`read(Mail);`
`wait(&ack)`	`signal(&ack);`

5.3.4
Thread Communication Between Many Threads Using a FIFO Queue

In the *bounded buffer* problem, we can have multiple threads putting data into a finite-size FIFO and multiple threads getting data out of this FIFO. Bounded buffer is simply another name for the standard FIFO queue we used in Chapter 4 to solve the producer/consumer application. We need two counting semaphores called `CurrentSize` and `RoomLeft` that contain the number of entries currently stored in the FIFO and the number of empty spaces left in the FIFO, respectively. `CurrentSize` is initialized to zero, and `RoomLeft` is initialized to the maximum allowable number of elements in the FIFO. The `Fifo_Put` routine executes the following steps:

```
Wait(&RoomLeft);          /* This will block the thread when FULL */
asm(" sei");              /* mutually exclusive access to FIFO */
Enter information into the FIFO structure
asm(" cli");
Signal(&CurrentSize);     /* Wake up a thread if blocked on empty */
```

The `Fifo_Get` routine executes the following steps:

```
Wait(&CurrentSize);              /* This will block the thread when EMPTY */
asm(" sei");                     /* mutually exclusive access to FIFO */
Remove information from the FIFO structure
asm(" cli");
Signal(&RoomLeft);              /* Wake up a thread if blocked on full */
```

Another implementation of this bounded buffer problem adds the semaphore called `Mutex` to implement the mutually exclusive access to the FIFO. `Mutex` is initialized to 1. The significance of this semaphore is illustrated in the following table:

Mutex	Meaning
1	No thread is currently entering or removing data
0	One thread is entering or removing data, none are blocked
−1	One thread is entering or removing data, one is blocked

The `Fifo_Put` routine executes the following steps:

```
Wait(&RoomLeft);         /* This will block the thread when FULL */
Wait(&Mutex);            /* Access to FIFO is critical code */
Enter information into the FIFO structure
Signal(&Mutex);
Signal(&CurrentSize);    /* Wake up a thread blocked on empty? */
```

The `Fifo_Get` routine executes the following steps:

```
Wait(&CurrentSize);      /* This will block the thread when EMPTY */
Wait(&Mutex);            /* Access to FIFO is critical code */
Remove information from the FIFO structure
Signal(&Mutex);
Signal(&RoomLeft);       /* Wake up a thread blocked on full? */
```

> **Checkpoint 5.10:** On average over the long term, what is the relationship between the number of times `Wait` is called compared to the number of times `Signal` is called?

5.4 Fixed Scheduling

In the round-robin scheduler of Section 5.1, the threads were run one at a time and each was given the same time slice. When using semaphores, the thread scheduler dynamically runs or blocks threads depending on conditions at that time. There is another application of thread scheduling sometimes found in real-time embedded systems, which involves a fixed scheduler. In this scheduler, the thread sequence and the allocated time-slices are determined a priori, during the design phase of the project. This class of problems is like creating the city bus schedule, or routing packages through a warehouse. What we do first is create a list of tasks to perform, as follows:

1. Assigning a priority to each task,
2. Defining the resources required for each task,
3. Determining how often each task is to run, and
4. Estimating how long each task will require to complete.

Next, we consider the available resources. Since this chapter deals with thread scheduling, the only resource we will consider here is processor cycles. In more complex systems, we could consider other resources such as memory and I/O channels. For real-time tasks we want to guarantee performance, so we must consider the worst-case estimate of how long each task will take, so the schedule can be achieved 100% of the time. On the other hand, if it is acceptable to meet the scheduling requirement most of the time, we could consider the average time it takes to perform each task. Lastly, we schedule the run times for each task by assigning times for the highest priority tasks first and then shuffle the assignments like placing pieces in a puzzle until all real-time tasks are scheduled as required. The tasks that are not real-time can be scheduled in the remaining slots. If all real-time tasks cannot be scheduled, then a faster microcontroller will be required. The design of this type of fixed scheduler is illustrated with a design example.

The goal of this design example is to schedule three real-time tasks: a finite state machine (FSM), a proportional-integral-derivative controller (PID), and a data acquisition system (DAS). There will also be one non-real-time task, PAN, which will input/output with the front panel. Figure 5.13 shows that each real-time task in the example has a required period of execution, a maximum execution time, and a minimum execution time.

Figure 5.13
Real-time specifications for these three tasks.

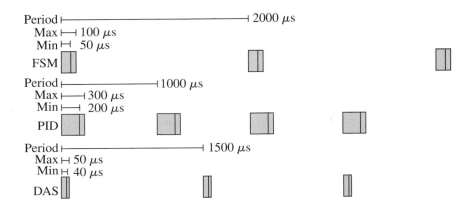

Because we wish to guarantee that tasks will always be started on time, we will consider the maximum times. If a solution were to exist, then we will be able find one with a repeating 6000μs pattern, because 6000 is the least common multiple of 2000, 1000, and 1500. The basic approach to scheduling periodic tasks is to time-shift the second and third tasks so that when the three tasks are combined, there are no overlaps, as shown in Figure 5.14. We start with the most frequent task, which in this example is the PID controller; then we schedule the FSM task immediately after it. In this example, about 41% of the time is allocated to real-time tasks. A solution is possible for this case, because the number of tasks is small and there is a simple 1/1.5/2 relationship between the required periods. Then, we schedule non–real-time tasks in the remaining intervals.

Figure 5.14
Repeating pattern to schedule these three real-time tasks.

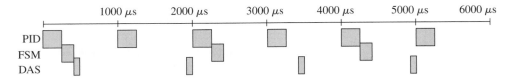

Program 5.14 shows the four threads for this system. The real-time threads execute OS_Sleep when it completes its task, which will suspend the thread and run the non–real-time thread. In this way, each thread will run one time through the for loop at the period requirement specified in Figure 5.13. When the treads explicitly release control (in this case

by calling OS_Sleep), the system is called **cooperative multitasking**. The non–real-time thread (PAN) will be suspended by the timer interrupt, in a manner similar to the preemptive schedule described previously in Section 5.1.

Program 5.14
Four user threads.

```
//******************FSM************************
void FSM(void){ StatePtr Pt;
  Pt = SA;                    // Initial State
  DDRT = 0x03;                // PT1,PT0 outputs, PT3,PT2 inputs
  PTT = Pt->Out;              // Output depends on the current state
  for(;;) {
    OS_Sleep();              // Runs every 2ms
    Pt = Pt->Next[PTT>>2];   // Next state depends on the input
    PTT = Pt->Out;           // Output depends on the current state
  }
}
//******************PID************************
void PID(void){ unsigned char speed,power;
  PID_Init();                 // Initialize
  for(;;) {
    OS_Sleep();              // Runs every 1ms
    speed = PID_In();        // read tachometer
    power = PID_Calc(speed);
    PID_Out(power);          // adjust power to motor
  }
}
//******************DAS************************
void DAS(void){ unsigned char raw;
  DAS_Init();                 // Initialize
  for(;;) {
    OS_Sleep();              // Runs every 1.5ms
    raw = DAS_In();          // read ADC
    Result = DAS_Calc(raw);
  }
}
//******************PAN************************
void PAN(void){ unsigned char input;
  PAN_Init();                 // Initialize
  for(;;) {
    input = PAN_In();        // front panel input
    if(input){
      PAN_Out(input);        // process
    }
  }
}
```

Program 5.15 creates the four thread control blocks. The initial CCR has I=0, enabling interrupts. In this system the TCBs are not linked together, but rather exist as a table of four entries, one for each thread. Each thread will have a total of 100 bytes of stack, and the stack itself exists inside the TCB. The RunPt will point to the TCB of the currently running thread. On the 6811, you will have to replace InitialReg[0] with MoreStack[90] in the initialization of sys[4], because the SP points the free memory just above the stack.

Program 5.16 defines the data structure containing the details of the fixed scheduler. This structure is a circular linked list, because the schedule repeats. In particular, the 22 entries explicitly define the schedule drawn in Figure 5.14. The front panel thread (PAN) is assigned to run in the gaps when no real-time thread requires execution.

Program 5.15
The thread control blocks.

```
struct TCB{
  unsigned char *StackPt;        // Stack Pointer
  unsigned char MoreStack[91];   // 100 bytes of stack
  unsigned char InitialReg[7];   // initial CCR,B,A,X,Y
  void (*InitialPC)(void);       // starting location
};
typedef struct TCB TCBType;
TCBType *RunPt;                // thread currently running
#define TheFSM &sys[0]         // finite state machine
#define ThePID &sys[1]         // proportional-integral-derivative
#define TheDAS &sys[2]         // data acquisition system
#define ThePAN &sys[3]         // front panel
TCBType sys[4]={
  { TheFSM.InitialReg[0],{ 0},{0x40,0,0,0,0,0,0},FSM},
  { ThePID.InitialReg[0],{ 0},{0x40,0,0,0,0,0,0},PID},
  { TheDAS.InitialReg[0],{ 0},{0x40,0,0,0,0,0,0},DAS},
  { ThePAN.InitialReg[0],{ 0},{0x40,0,0,0,0,0,0},PAN}
};
```

Program 5.16
The scheduler defines both the thread and the duration.

```
struct Node{
  struct Node *Next;        // circular linked list
  TCBType *ThreadPt;        // which thread to run
  unsigned short TimeSlice; // how long to run it
};
typedef struct Node NodeType;
NodeType *NodePt;
NodeType Schedule[22]={
{ &Schedule[1], ThePID, 300}, // interval      0,  300
{ &Schedule[2], TheFSM, 100}, // interval    300,  400
{ &Schedule[3], TheDAS,  50}, // interval    400,  450
{ &Schedule[4], ThePAN, 550}, // interval    450, 1000
{ &Schedule[5], ThePID, 300}, // interval   1000, 1300
{ &Schedule[6], ThePAN, 600}, // interval   1300, 1900
{ &Schedule[7], TheDAS,  50}, // interval   1900, 1950
{ &Schedule[8], ThePAN,  50}, // interval   1950, 2000
{ &Schedule[9], ThePID, 300}, // interval   2000, 2300
{ &Schedule[10],TheFSM, 100}, // interval   2300, 2400
{ &Schedule[11],ThePAN, 600}, // interval   2400, 3000
{ &Schedule[12],ThePID, 300}, // interval   3000, 3300
{ &Schedule[13],ThePAN, 100}, // interval   3300, 3400
{ &Schedule[14],TheDAS,  50}, // interval   3400, 3450
{ &Schedule[15],ThePAN, 550}, // interval   3450, 4000
{ &Schedule[16],ThePID, 300}, // interval   4000, 4300
{ &Schedule[17],TheFSM, 100}, // interval   4300, 4400
{ &Schedule[18],ThePAN, 500}, // interval   4400, 4900
{ &Schedule[19],TheDAS,  50}, // interval   4900, 4950
{ &Schedule[20],ThePAN,  50}, // interval   4950, 5000
{ &Schedule[21],ThePID, 300}, // interval   5000, 5300
{ &Schedule[0], ThePAN, 700}  // interval   5300, 6000
};
```

The thread scheduler is shown in Program 5.17. The software interrupt creates the cooperative multitasking and is used by the real-time threads when their task is complete. In this example, there is only one non–real-time thread, but it would be straightforward to implement a round-robin scheduler for these threads in the software interrupt handler.

Program 5.17
The scheduler defines both the thread and the duration.

```
void OS_Sleep(void){  // cooperative multitasking
  asm swi            // suspend this tread and run another
}
interrupt 4 void swiISR(void){
asm ldx RunPt        // cooperative multitasking
asm sts 0,x          // thread goes to sleep when it is done
  RunPt = ThePAN;    // non-real time thread
asm ldx RunPt
asm lds 0,x
}
interrupt 11 void threadSwitchISR(void){
asm ldx RunPt
asm sts 0,x
  NodePt = NodePt->Next;
  RunPt = NodePt->ThreadPt;    // which thread to run
  TC3 = TC3+NodePt->TimeSlice; // Thread runs for a unit of time
  TFLG1 = 0x08;                // acknowledge by clearing TC3F
asm ldx RunPt
asm lds 0,x
}
void main(void) {
  NodePt = &Schedule[0];    // first thread to run
  RunPt = NodePt->ThreadPt;
  TIOS |= 0x08;     // activate OC3
  TSCR1 = 0x80;     // enable TCNT
  TSCR2 = 0x02;     // usec TCNT
  TIE |= 0x08;      // Arm TC3
  TC3 = TCNT+NodePt->TimeSlice; // Thread runs for a unit of time
  TFLG1 = 0x08;     // Clear C3F
asm ldx RunPt
asm lds 0,x
asm rti     // Launch First Thread
}
```

Checkpoint 5.11: Why does the thread switch Program 5.17 execute sts 0,x, while the thread switches in Programs 5.2 and 5.6 execute sts 2,s?

We could have attempted to implement this system with regular periodic interrupts. In particular, we could have created three independent periodic interrupts and performed each task in a separate ISR. Unfortunately, there would be situations when one or more tasks would overlap. In other words, one interrupt might be requested while we are executing one of the other two ISRs. Although all tasks would run, some would be delayed. This delay is called **time-jitter**, which is defined as the difference between the time a thread is supposed to run (see comments of Program 5.16) and the time it does run. This solution presented in Programs 5.14 to 5.17 will create a situation in which the interrupts are always enabled when the C3F flag gets set; therefore, no real-time tasks are delayed. The measured time-jitter for this system is always less than 1 µsec. The origin of the error is the variability in which instruction of PAN() is being executed at the time of the output compare interrupt request.

Checkpoint 5.12: What would be the effect on time-jitter by activating the PLL in the MC9S12C32 so that the computer runs six times faster (24MHz) and multiplying the TimeSlice constants in Program 5.16 by six?

5.5   Exercises

5.1 In this exercise you will extend the preemptive scheduler to support priority. This system should support three levels of priority: 1 will be the highest. You can solve this problem on the 6811 or 6812 using either assembly or C.

a) Redesign the TCB to include a 16-bit integer for the priority (although the values will be restricted to 1, 2, 3). Show the static allocation for the three threads from the example in this chapter, assuming the first two are priority 2 and the last is priority 3. There are no priority 1 threads in this example, but there might be in the future.

b) Redesign the scheduler to support this priority scheme.

c) In the chapter it said "*Normally, we add priority to a system that implements blocking semaphores and not to one that uses spin-lock semaphores.*" What specifically will happen here if the system is run with spin-lock semaphores?

d) Even when the system supports blocking semaphores, starvation might happen to the low-priority threads. Describe the sequence of events that causes starvation.

e) Suggest a solution to the starvation problem.

5.2 We can use semaphores to limit access to resources. In the following example, both threads need access to a printer and a SPI port. The binary semaphore `sPrint` provides mutual exclusive access to the printer, and the binary semaphore `sSPI` provides mutual exclusive access to the SPI port. The following scenario has a serious flaw:

Thread 1	**Thread 2**
`bwait(&sPrint);`	`bwait(&sSPI);`
`bwait(&sSPI);`	`bwait(&sPrint);`
`printf("bye");`	`OutSPI(5);`
`OutSPI(6);`	`printf("tchau");`
`bsignal(&sPrint);`	`bsignal(&sSPI);`
`bsignal(&sSPI);`	`bsignal(&sPrint);`

a) Describe the sequence of events that would lead to the condition where both threads are stuck forever.

b) How could you change the above program to prevent the deadlock?

5.3 You are given four identical I/O ports to manage on the MC68HC12A4: Port A, Port B, Port C, and Port D. You may assume there is a preemptive thread scheduler and blocking semaphores.

a) Look up the address of each port and its direction register.

b) Create a data structure to hold an address of the port and the address of the data direction register. Assume the type of this structure is called `PortType`.

c) Design and implement a manager that supports two functions. The first function is called `NewPort`. Its prototype is

```
PortType *NewPort(void);
```

If a port is available when a thread calls `NewPort`, then a pointer to the structure, defined in part b, is returned. If no port is available, then the thread will block. When a port becomes available, this thread will be awakened and the pointer to the structure will be returned. You may define and use blocking semaphores without showing the implementation of the semaphore or scheduler. The second function is called `FreePort`, and its prototype is

```
void FreePort(PortType *pt);
```

This function returns a port so that it can be used by the other threads. Include a function that initializes the system, where all five ports are free. (*Hint:* The solution is very similar to the FIFO queue example shown in Section 5.3.4.)

5.4 Some computers have a `swap` instruction. Modify the subroutines shown in Program 5.9, written in 6812 assembly language, to implement a binary spin-lock semaphore using this fictional instruction. The system must provide for mutual exclusion. You may not disable interrupts. Do not use the `minm` instruction. The syntax of this fictional atomic operation is

```
swap        Operand,r
```

where `Operand` is any standard 6812 addressing mode and `r` is any of the 6812 registers. This instruction exchanges the 8- or 16-bit contents of the `Operand` and data register. It does not set any condition code bits.

5.6   Lab Assignments

Lab 5.1. The overall objective is to create a preemptive thread scheduler with blocking semaphores. There is a preemptive scheduler with spinlock semaphores on the CD in the `Lab17` folder. You will first adjust the FIFO sizes so no data is lost. Then, you will redesign the system to implement blocking semaphores.

Lab 5.2. The overall objective is to create a fixed thread scheduler for three real-time tasks and two non–real-time tasks. An implementation of the fixed scheduler developed in Section 5.4 can be found as the FixedScheduler example at http://www.ece.utexas.edu/~valvano/metrowerks/. First, you will create a second non–real-time thread similar to `PAN`, but using other I/O pins. Next, you will calculate the maximum time to execute each loop of the real-time threads. Then, you will develop a fixed scheduler with the following specification:

```
FSM period is 100 ms
PID period is 75 ms
DAS period is 10 ms
```

The `swi` handler should alternate starting the two non–real-time threads. You can use switches for inputs and LEDs for outputs so that you can interact with the system.

6 Timing Generation and Measurements

Chapter 6 objectives are to use:

❏ Input capture to generate interrupts and measure period or pulse width
❏ Output compare to create periodic interrupts, generate square waves, and measure frequency
❏ Both input capture and output compare to make flexible and robust measurement systems
❏ Pulse accumulator to measure frequency and period
❏ Pulse-width modulator to generate wave forms

The timer systems on the Freescale microcomputers are very versatile. Over the last 25 years, the evolution of these timer functions has paralleled the growth of new applications possible with these embedded computers. In other words, inexpensive yet powerful embedded systems have been made possible by the capabilities of the timer system. In this chapter we will introduce these functions, then use them throughout the remainder of the book. If we review the applications introduced in the first chapter (see Section 1.1), we will find that virtually all of them make extensive use of the timer functions developed in this chapter.

6.1 Input Capture

6.1.1 Basic Principles of Input Capture

We can use input capture to measure the period or pulse width of TTL-level signals. The input capture system can also be used to trigger interrupts on rising or falling transitions of external signals. TCNT is a 16-bit counter incremented at a fixed rate. Each input capture module has

An external input pin, ICn,
A flag bit,
Two edge control bits, EDGnB EDGnA,
An interrupt mask bit (arm), and
A 16-bit input capture register.

The various members of the Freescale HC11 and HC12 families have from zero to eight input capture modules. For example, the MC68HC11A8 has three, the MC68HC711E9 has four, and the MC9S12C32 has eight. In this book we use the term **arm** to describe the bit that allows or denies a specific flag from requesting an interrupt. The Freescale manuals refer to this bit as a **mask**. That is, the device is armed when the mask bit is 1. Typically, there is a separate arm bit for every flag that can request an interrupt.

Figure 6.1
Basic components of
input capture.

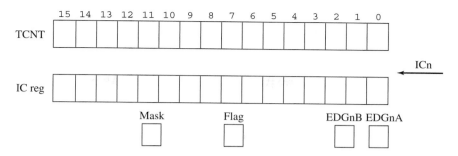

An external input signal is connected to the input capture pin. The EDGnB, EDGnA bits specify whether the rising, falling, or both rising and falling edges of the external signal will trigger an input capture event. Two or three actions result from an input capture event: 1) the current TCNT value is copied into the input capture register, 2) the input capture flag is set, and 3) an interrupt is requested if the mask bit is 1. This means an interrupt can be requested on a capture event. The input capture mechanism has many uses. Three of common applications are:

1. An interrupt service routine is executed on the active edge of the external signal,
2. Perform two rising edge input captures and subtract the two measurements to get period,
3. Perform a rising edge then a falling edge capture and subtract the two measurements to get pulse width.

The flag bit does not behave like a regular memory location. In particular, the flag cannot be set by software. Rather, an input capture or output compare hardware event will set the flag. The other peculiar behavior of the flag is that the software must write a one to the flag in order to clear it. If the software writes a zero to the flag, no change will occur.

Checkpoint 6.1: When does an input capture event occur?

Checkpoint 6.2: What happens during an input capture event?

Observation: The TCNT timer is very accurate because of the stability of the crystal clock.

Observation: When measuring period or pulse width, the measurement resolution will equal the TCNT period.

6.1.2
Input Capture
Details

Next we will overview the specific input capture functions on two Freescale microcomputers. This section is intended to supplement rather than replace the Freescale manuals. When designing systems with input capture, please refer to the reference manual of your specific Freescale microcomputer.

6.1.2.1
MC68HC711E9 Input
Capture Details

Table 6.1 lists the registers associated with input capture on the 6811. The entries shown in bold will be used in this section. **TCNT** is a 16-bit unsigned counter that is incremented at a rate determined by two bits (PR1 and PR0) in the **TMSK2** register, as defined in Table 6.2. This counter cannot be stopped or reset. The fundamental approach to input capture involves connecting an external TTL-level signal(s) to PA3, PA2, PA1, and/or PA0. Figure 6.2 shows the mapping from port pin to input capture channel. For example, input capture channel 3 is controlled by pin PA0. Since there are no direction register bits associated with PA2-0, these bits are always inputs. PA3 can be used as output compare channel 5 or input capture channel 4. To use PA3 as an input capture, we have to clear bit DDRA3 and set I4/O5 bit in the **PACTL** register. The input capture event occurs on the rising, falling, or both rising and falling edge of this external signal. The 6811 will set the input capture flag (IC1F, IC2F, IC3F, or IC4F) and latch the current 16-bit TCNT value into the input capture latch (TIC1,

TIC2, TIC3, or TIC4). If the input capture flag is armed (IC1I, IC2I, IC3I, IC4I) then an interrupt will be requested when the flag is set. We can arm or disarm the input capture interrupts by initializing the **TMSK1** register. Our software can determine whether an input capture event has occurred by reading the **TFLG1** register. Every time the TCNT register overflows from $FFFF to 0, the TOF flag in the **TFLG2** register is set. The TOF flag will cause an interrupt if the mask TOI equals 1, which is described in Section 4.15.1.

Figure 6.2
Input capture interface on the MC68HC711E9.

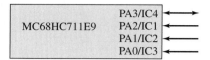

Address	msb														lsb		Name
$100E	15	14	13	12	11	10	9	8	7	6	5	4	3	2	1	0	TCNT
$1010	15	14	13	12	11	10	9	8	7	6	5	4	3	2	1	0	TIC1
$1012	15	14	13	12	11	10	9	8	7	6	5	4	3	2	1	0	TIC2
$1014	15	14	13	12	11	10	9	8	7	6	5	4	3	2	1	0	TIC3
$101E	15	14	13	12	11	10	9	8	7	6	5	4	3	2	1	0	TIC4/TOC5

Address	Bit 7	6	5	4	3	2	1	Bit 0	Name
$1000	PA7	PA6	PA5	PA4	PA3	PA2	PA1	PA0	PORTA
$1021	EDG4B	EDG4A	EDG3B	EDG3A	EDG2B	EDG2A	EDG1B	EDG1A	TCTL2
$1022	OC1I	OC2I	OC3I	OC4I	IC4/OC5I	IC1I	IC2I	IC3I	TMSK1
$1023	OC1F	OC2F	OC3F	OC4F	IC4/OC5F	IC1F	IC2F	IC3F	TFLG1
$1024	TOI	RTII	PAOVI	PAII	0	0	PR1	PR0	TMSK2
$1026	DDRA7	PAEN	PAMOD	PEDGE	DDRA3	I4/O5	RTR1	RTR0	PACTL

Table 6.1
MC68HC711E9 registers used for input capture.

Table 6.2
Two control bits define the TCNT clock rate assuming a 2 MHz E clock.

PR1	PR0	Divide by	TCNT Clock
0	0	1	500 ns
0	1	4	2 μs
1	0	8	4 μs
1	1	16	8 μs

Checkpoint 6.3: Which Port A pins are input only, which are output only, and which are bidirectional?

We specify the active edge (i.e., the edge that latches TCNT and sets the flag) by initializing the TCTL2 register, as shown in Table 6.3.

Table 6.3
Two control bits define the active edge used for input capture.

EDGnB	EDGnA	Active edge
0	0	None
0	1	Capture on rising
1	0	Capture on falling
1	1	Capture on both rising and falling

Checkpoint 6.4: List the steps to initialize PA0 as an armed rising edge input capture.

Checkpoint 6.5: List the steps to initialize PA3 as an armed falling edge input capture.

The flags in the TFLG1 register are cleared by writing a 1 into the specific flag bit we wish to clear. For example, writing a $FF into TFLG1 will clear all eight flags. The following is a valid method for clearing IC3F. In other words, this acknowledge sequence clears the IC3F flag without affecting the other seven flags in the TFLG1 register.

```
TFLG1=0x01;
```

It is inappropriate to use the `bset` instruction (e.g., `bset $23,X,#$01`) to clear flags, because this read/modify/write instruction will also clear all the other bits in the flag register. For example, assume the TFLG1 initially equals $19. The execution of the "`bset $23,X,#$01`" instruction produces the following two accesses:

```
read  $1023 $19    ;the $19 is ored with $01 to get $19
write $1023 $19    ;that clears OC4F, OC5F and IC3F
```

Common error: Executing `TFLG1 |= 0x01;` will mistakenly clear all the bits in the TFLG1 register.

Checkpoint 6.6: Write assembly or C code to clear IC2F.

6.1.2.2
MC9S12C32 Input Capture Details

Table 6.4 shows the 6812 registers used for input capture. The entries shown in bold will be used in this section. **TCNT** is a 16-bit unsigned counter that is incremented at a rate determined by three bits (PR2, PR1, and PR0) in the **TSCR2** register. To use any of the features involving TCNT, such as input capture, we have to set the TEN bit in the **TSCR1** register. Table 6.5 shows the available TCNT periods for two different E clock periods. The fundamental approach to input capture involves connecting an external TTL-level signal to one of the eight input capture pins on Port T, as shown in Figure 6.3. The pin is selected as input capture by placing a 0 in the corresponding bit of the **TIOS** register. There is a direction register, **DDRT**, for port T, which means we should clear the corresponding bits for the

Address	msb															lsb	Name
$0044	15	14	13	12	11	10	9	8	7	6	5	4	3	2	1	0	TCNT
$0050	15	14	13	12	11	10	9	8	7	6	5	4	3	2	1	0	TC0
$0052	15	14	13	12	11	10	9	8	7	6	5	4	3	2	1	0	TC1
$0054	15	14	13	12	11	10	9	8	7	6	5	4	3	2	1	0	TC2
$0056	15	14	13	12	11	10	9	8	7	6	5	4	3	2	1	0	TC3
$0058	15	14	13	12	11	10	9	8	7	6	5	4	3	2	1	0	TC4
$005A	15	14	13	12	11	10	9	8	7	6	5	4	3	2	1	0	TC5
$005C	15	14	13	12	11	10	9	8	7	6	5	4	3	2	1	0	TC6
$005E	15	14	13	12	11	10	9	8	7	6	5	4	3	2	1	0	TC7

Address	Bit 7	6	5	4	3	2	1	Bit 0	Name
$0240	PT7	PT6	PT5	PT4	PT3	PT2	PT1	PT0	PTT
$0242	DDRT7	DDRT6	DDRT5	DDRT4	DDRT3	DDRT2	DDRT1	DDRT0	DDRT
$0046	**TEN**	TSWAI	TSBCK	TFFCA	0	0	0	0	TSCR1
$004D	TOI	0	0	0	TCRE	**PR2**	**PR1**	**PR0**	TSCR2
$0040	**IOS7**	**IOS6**	**IOS5**	**IOS4**	**IOS3**	**IOS2**	**IOS1**	**IOS0**	TIOS
$004C	**C7I**	**C6I**	**C5I**	**C4I**	**C3I**	**C2I**	**C1I**	**C0I**	TIE
$004E	**C7F**	**C6F**	**C5F**	**C4F**	**C3F**	**C2F**	**C1F**	**C0F**	TFLG1
$004F	TOF	0	0	0	0	0	0	0	TFLG2
$004A	**EDG7B**	**EDG7A**	**EDG6B**	**EDG6A**	**EDG5B**	**EDG5A**	**EDG4B**	**EDG4A**	TCTL3
$004B	**EDG3B**	**EDG3A**	**EDG2B**	**EDG2A**	**EDG1B**	**EDG1A**	**EDG0B**	**EDG0A**	TCTL4

Table 6.4
MC9S12C32 registers used for input capture.

Table 6.5
Three control bits define
the TCNT clock period.

PR2	PR2	PR0	Divide by	TCNT Period (4 MHz E Clock)	TCNT Period (24 MHz E Clock)
0	0	0	1	250 ns	41.7 ns
0	0	1	2	500 ns	83.3 ns
0	1	0	4	1 μs	166.7 ns
0	1	1	8	2 μs	333.3 ns
1	0	0	16	4 μs	666.7 ns
1	0	1	32	8 μs	1.333 μs
1	1	0	64	16 μs	2.667 μs
1	1	1	128	32 μs	5.333 μs

Figure 6.3
Input capture interface
on the 6812.

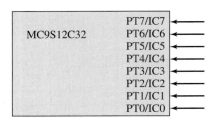

input capture inputs. The input capture event occurs on the rising, falling, or both rising and falling edge of this external signal. The 6812 will set the input capture flag (CnF in the **TFLG1** register) and latch the current 16-bit TCNT value into the input capture latch (TCn). If the input capture flag is armed (CnI in the **TIE** register), then an interrupt will be requested when the flag is set. There is a separate 16-bit input capture register for each of the eight input capture modules. We specify the active edge (i.e., the edge that latches TCNT and sets the flag) by initializing the **TCTL3** and **TCTL4** registers, as described in Table 6.6. We can arm or disarm the input capture interrupts by initializing the TIE register. Our software can determine whether an input capture event has occurred by reading the TFLG1 register. Every time the TCNT register overflows from $FFFF to 0, the TOF flag in the **TFLG2** register is set. The TOF flag will cause an interrupt if the mask TOI equals 1 which is described in Section 4.15.3.

Table 6.6
Two control bits define
the active edge used for
input capture.

EDGnB	EDGnA	Active edge
0	0	None
0	1	Capture on rising
1	0	Capture on falling
1	1	Capture on both rising and falling

Observation: The phase-lock-loop (PLL) on the 9S12 will affect the TCNT period.

Observation: The MC68HC812A4 does not support the two slowest TCNT prescale options.

Checkpoint 6.7: List the steps to initialize PT0 as an armed rising edge input capture.

Checkpoint 6.8: List the steps to initialize PT3 as an armed falling edge input capture.

Fast clear (TFFCA) is a mode that attempts to automatically clear flag bits. We will not use fast clear mode, in order to make our software clearer and to make it friendly when using multiple timer features. TCNT can be reset with a special mode of output compare 7 (TCRE=1). But in this book, we will consider TCNT as simple 16-bit counter clocked at a continuous rate (we will always set TCRE=0).

The flags in the TFLG1 and TFLG2 registers are cleared by writing a 1 into the specific flag bit we wish to clear. For example, writing a $FF into TFLG1 will clear all eight flags. The following is a valid method for clearing C3F. That is, this acknowledge sequence clears the C3F flag without affecting the other seven flags in the TFLG1 register.

```
TFLG1 = 0x08;
```

Checkpoint 6.9: Write assembly or C code to clear C6F.

Common error: Executing `TFLG1 |= 0x08;` will mistakenly clear all the bits in the TFLG1 register.

6.1.3
Real-Time
Interrupt Using an
Input Capture

In this example, we create a periodic interrupt using an external clock and input capture (Figure 6.4). Although most microcomputers have internal clocks that can be used to create periodic interrupts, there are applications where it is desirable to use an external clock. For example, if you had multiple microcomputers and wished to create simultaneous interrupt requests, you could use this approach. The clock frequency of a TLC555 timer is determined by the resistor and capacitor values. The data sheet for various 555 timers can be found on the accompanying CD. This astable multivibrator creates a 1-kHz square wave. The period of a TLC555 timer is $0.693 \times C_T \times (R_A + 2R_B)$. In our circuit, R_A is 4.4 kΩ, R_B is 5 kΩ, and C_T is 0.1 μF. The stability of this circuit is determined by the stability of R_A R_B and C_T.

Figure 6.4
An external signal is connected to the input capture.

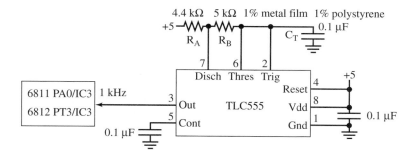

An input capture interrupt occurs on each rise of the square wave generated by the TLC555. The latency of the system is defined as the time delay between the rise of the input capture signal to the increment of `Time`. Assuming there are no other interrupts, and assuming the Main program does not disable interrupts, the delay includes the time to (1) finish the current instruction, (2) process the interrupt, and (3) execute the interrupt handler up to and including changing `Time` (Table 6.7).

Table 6.7
Components of latency and their values for the assembly language implementation.

Component	6811	6812
Longest instruction (cycles, μs)	41=20.5 μs	13=3.25 μs
Process the interrupt (cycles, μs)	14=7 μs	9=2.25 μs
Execute handler (cycles, μs)	19=9.5 μs	11=2.75 μs
Max latency (μs)	37	8.25

The latency may be larger if there are other sections of code that execute with the interrupts disabled, like other interrupt handlers. The ritual `Init` sets input capture to interrupt on the rise and initializes the global `Time` (Program 6.1).

```
; MC68HC711E9                          ; MC9S12C32
;external signal to PA0/IC3            ;external signal to PT3/IC3
Time rmb  2       ;every 1 ms          Time rmb  2          ;every 1 ms
Init sei          ;make atomic         Init sei             ;make atomic
     ldaa TCTL2   ;Old value                bclr TIOS,#$08  ;PT3=input capture
     anda #$FC    ;Clear EDG3B=0             bclr DDRT,#$08  ;PT3 is input
     oraa #$01    ;EDG3BA =01               movb #$80,TSCR1 ;enable TCNT
     staa TCTL2   ;on rise of PA0           movb #$01,TSCR2 ;500 ns clock
     ldaa TMSK1   ;Old value                bclr TCTL4,#$80 ;EDG3BA =01
     oraa #$01    ;IC3I=1                   bset TCTL4,#$40 ;on rise of IC3
     staa TMSK1   ;Arm IC3F                 bset TIE,#$08   ;Arm C3F
     ldd  #0                                movw #0,Time    ;init global
     std  Time    ;init global              movb #$08,TFLG1 ;clear C3F
     ldaa #$01    ;clear IC3F               cli             ;enable
     staa TFLG1                             rts
     cli          ;enable
     rts
IC3Han ldaa #$01  ;clear IC3F   [2]     IC3Han movb #$08,TFLG1 ;ack C3F   [4]
       staa TFLG1 ;Acknowledge  [4]            ldx  Time                  [3]
       ldx  Time               [5]            inx                        [1]
       inx                     [3]            stx  Time                  [3]
       stx  Time               [5]            rti
       rti
       org  $FFEA                             org  $FFE8  ;timer channel 3
       fdb  IC3Han                            fdb  IC3Han
```

Program 6.1
Periodic interrupt using input capture and an external clock.

The interrupt software acknowledges the interrupt, and increments the global variable. The interrupt handlers cannot "poll for 0s and 1s" because none of the bits in the status register is guaranteed to be zero. The same algorithm written in C is presented in Program 6.2.

```
// MC68HC711E9, ICC11 C                 // MC9S12C32, CodeWarrior C
// PA0 input = external signal          // PT3 input = external signal
unsigned short Time;  // incremented     unsigned short Time;   // incremented
void Init(void){                         void Init(void){
asm(" sei");       // make atomic          asm sei        // make atomic
  TCTL2 = (TCTL2&0xFC)|0x01;                TIOS &=~0x08; // PT3 input capture
  TMSK1 |= 0x01;   // Arm IC3, rising       DDRT &=~0x08; // PT3 is input
  TFLG1 = 0x01;    // initially clear       TSCR1 = 0x80; // enable TCNT
  Time = 0;                                 TSCR2 = 0x01; // 500ns clock
asm(" cli");                                TCTL4 = (TCTL4&0x3F)|0x40;
}                                           TIE |= 0x08;  // Arm IC3, rising
#pragma interrupt_handler TIC3handler       TFLG1 = 0x08; // initially clear
void TIC3handler(void){                     Time = 0;
  TFLG1 = 0x01;    // acknowledge           asm cli
  Time++;                                 }
}                                        void interrupt 11 IC3Han(void){
#pragma abs_address:0xffea                 TFLG1 = 0x08;  // acknowledge
void (*IC3_vector[])() = {TIC3handler};    Time++;
#pragma end_abs_address                  }
```

Program 6.2
Periodic interrupt using input capture and an external clock.

Common Error: When two software modules both need to set the same configuration register, a poorly written initialization by one software module undoes the initialization performed by the other.

6.1.4
Period
Measurement

Before one implements a system that measures period, it is appropriate to consider the issues of resolution, precision, and range. The *resolution* of a period measurement is defined as the smallest change in period that can reliably be detected. In the first example, if the period increases by 500 ns, then there will be one more TCNT clock between the first rising edge and the second rising edge. In this situation, the period calculated by `TIC1-First` will increase by 1; therefore, the period measurement resolution is 500 ns. The resolution is also the significance or the units of the measurement. In this first example, if the calculation of `Period` results in 1000, then it represents a period of 1000 · 500 ns, or 500 μs. In the example presented later in Section 6.5.1, the period must increase at least 1 ms for the measurement to be able to reliably detect the change. In the example of Section 6.5.1, the period measurement resolution is 1 ms. The *precision* of the period measurement is defined as the number of separate and distinguishable measurements. In the first example, a 16-bit counter is used, so there are about 65,536 different periods that can be measured. We can specify the precision in alternatives (e.g., 65536) or in bits (e.g., 16 bits). The precision of the example in Section 6.1.4.2 is 32 bits. The last issue to consider is the *range* of the period measurement, which is defined as the minimum and maximum values that can reliably be measured. We are concerned what happens if the period is too small or too large. A good measurement system should be able to detect overflows and underflows. In addition, we would not like the system to crash, or hang up if the input period is out of range. Similarly, it is desirable if the system can detect when there is no period.

The default TCNT rate for the 6811 is 2 MHz. For the 6812 implementations, we will set the three bits (PR2, PR1, and PR0) in the TSCR2 register to 001, making TCNT clock every 500 ns. In this way, the two implementations will be equivalent.

6.1.4.1
16-bit Period
Measurement with a
500-ns Resolution

In this example, the TTL-level input signal is connected to an input capture pin. Each rising edge will generate an input capture interrupt (Figure 6.5). The period is calculated as the difference in TIC1 latch values from one rising edge to the other (Figure 6.6). For example, if the period is 8192 μs, the IC1 interrupts will be requested every 16,384 cycles, and the difference between TIC1 latch values will be 16384=$4000. This subtraction remains valid even if the TCNT overflows and wraps around in between IC1 interrupts. On the other hand, this method will not operate properly if the period is larger than 65,535 cycles, or 32,767 μs.

Figure 6.5
To measure period we connect the external signal to an input capture.

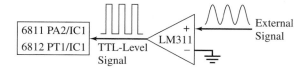

The resolution is 500 ns because the period must increase by at least this amount before the difference between TIC1 measurements will reliably change. Even though a 16-bit counter is used, the precision is a little less than 16 bits, because the shortest period that can be handled with this interrupt-driven approach is shown in Table 6.8. This factor is determined by counting all the cycles of the assembly version of `IC1han` and adding the time needed for the microcomputer to process the interrupt. In other words, if the

Figure 6.6
Timing example showing counter rollover during 16-bit period measurement.

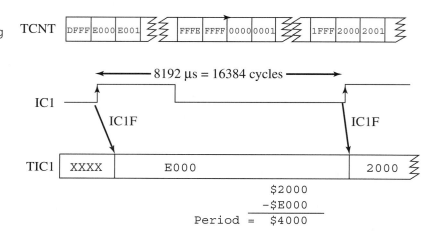

$$\$2000$$
$$-\$E000$$
$$\text{Period} = \quad \$4000$$

interrupts are requested at a rate faster than the minimum period specified in Table 6.8, then some interrupts will be lost. Also notice that as the period approaches this minimum, a higher and higher percentage of the computer execution is utilized just in the `IC1han` itself. For example, on the 6811 Table 6.9 shows that if the period is 64 μs, then 50% of the time is used executing the ISR.

Table 6.8
Calculation of the smallest period that can be handled by the assembly language input capture interrupt.

Component	6811	6812
Process the interrupt (cycles/μs)	14=7 μs	9=2.25 μs
Execute entire handler (cycles/μs)	50=25 μs	31=7.75 μs
Minimum period (cycles/μs)	64=32 μs	40=10 μs

Table 6.9
Calculation of the percentage time in the 6811 interrupt handler as a function of input period.

Period (μs)	Cycles/interrupt	Percentage time in handler (%)
32	64	100
64	64	50
320	64	10
P	64	3200/P

Depending on the complier, C code may execute faster or slower than the assembly, but the general trend (percentage overhead approaches 100% as the period approaches zero) still holds. As mentioned earlier, this implementation cannot detect if the period is larger than 32 ms. For example, a signal with period of 40,960 μs gives the same result as a signal with a period of 8192 μs. This limitation will be corrected in the other two period measurement examples presented later in this chapter.

The period measurement system written in assembly is presented in Program 6.3. The 16-bit subtraction calculates the number of TCNT clocks between rising edges. Since the ritual does not wait for the first edge, the first period measurement will be incorrect and should be neglected. Because the input capture interrupt has a separate vector, the software does not poll. An interrupt is requested on each rising edge of the input signal.

The period measurement system written in C is presented in Program 6.4. The code **Period=TIC1-First;** calculates the number of E clocks between rising edges. The first period measurement will be incorrect and should be neglected.

```
; MC68HC711E9                              ; MC9S12C32
;external signal to PA2/IC1                ;external signal to PT1/IC1
Period rmb  2  ;units 500 ns              Period rmb  2    ;units 500 ns
First  rmb  2  ;TCNT at first edge        First  rmb  2    ;TCNT at first edge
Done   rmb  1  ;set each rising           Done   rmb  1    ;set each rising
Init sei        ;make atomic              Init sei          ;make atomic
     ldaa TCTL2  ;Old value                   bclr TIOS,#$02  ;PT1=input capture
     anda #$CF   ;Clear EDG1B=0                bclr DDRT,#$02  ;PT1 is input
     oraa #$10   ;EDG1BA =01                   movb #$80,TSCR1 ;enable TCNT
     staa TCTL2  ;on rise of PA2               movb #$01,TSCR2 ;500ns clk
     ldd  TCNT                                 bclr TCTL4,#$08 ;EDG1BA =01
     std  First  ;init global                  bset TCTL4,#$04 ;on rise of PT1
     clr  Done                                 movw TCNT,First ;init global
     ldaa #$04   ;clear IC1F                    clr  Done
     staa TFLG1                                movb #$02,TFLG1 ;clear C1F
     ldaa TMSK1  ;Old value                    bset TIE,#$02   ;Arm C1F
     oraa #$04   ;IC1I=1                        cli             ;enable
     staa TMSK1  ;Arm IC1F                      rts
     cli         ;enable
     rts
IC1Han ldaa #$01  ;clear IC3F    [2]       IC1Han movb #$02,TFLG1 ;clear C1F [4]
       staa TFLG1 ;Acknowledge   [4]              ldd  TC1                    [3]
       ldd  TIC1                 [5]              subd First                  [3]
       subd First                [6]              std  Period                 [3]
       std  Period               [5]              movw TC1,First              [6]
       ldd  TIC1                 [5]              movb #$FF,Done              [4]
       std  First                [5]              rti                         [8]
       ldaa #$FF  ;set flag      [2]
       staa Done                 [4]
       rti                       [12]
       org  $FFEE                                 org  $FFEC  ;timer channel 1
       fdb  IC1Han                                fdb  IC1Han
```

Program 6.3
Assembly language 16-bit period measurement.

```
// MC68HC711E9, ICC11 C                    // MC9S12C32, CodeWarrior C
// PA2/IC1 input = external signal         // PT1/IC1 input = external signal
// rising edge to rising edge              // rising edge to rising edge
// resolution = 500ns                      // resolution = 500ns
// Range = 36 us to 32 ms,                 // Range = 36 us to 32 ms,
// no overflow checking                    // no overflow checking
unsigned short Period;  // 500 ns units    unsigned short Period;  // 500 ns units
unsigned short First;   // TCNT first edge unsigned short First;   // TCNT first edge
unsigned char Done;     // Set each rising unsigned char Done;     // Set each rising
void Init(void){                           void Init(void){
  asm(" sei");   // make atomic              asm sei          // make atomic
  TCTL2 = (TCTL2&0xCF)|0x10; // rising       TIOS &=~0x02;    // PT1 input capture
  First = TCNT;  // first will be wrong      DDRT &=~0x02;    // PT1 is input
  Done = 0;      // set on subsequent        TSCR1 = 0x80;    // enable TCNT
  TFLG1 = 0x04;  // Clear IC1F               TSCR2 = 0x01;    // 500ns clock
  TMSK1 |= 0x04; // Arm IC1                   TCTL4 = (TCTL4&0xF3)|0x04; // rising
  asm(" cli");                               First = TCNT;    // first will be wrong
}                                            Done = 0;        // set on subsequent
```

continued on p. 286

Program 6.4
C language period measurement.

continued from p. 285

```
#pragma interrupt_handler TIC1handler()        TFLG1 = 0x02;    // Clear C1F
void TIC1handler(void){                         TIE |= 0x02;    // Arm IC1
  Period = TIC1-First; // 500ns                  asm cli
  First = TIC1;   // Setup for next            }
  TFLG1 = 0x04;   // ack by clearing IC1F       void interrupt 9 TC1handler(void){
  Done = 0xFF;                                    Period = TC1-First; // 500ns resolution
}                                                 First = TC1;    // Setup for next
#pragma abs_address:0xffee                        TFLG1 = 0x02;   // ack by clearing C1F
void (*IC1_vector[])() = {TIC1handler};           Done = 0xFF;
#pragma end_abs_address                         }
```

Program 6.4
C language period measurement.

> **Checkpoint 6.10:** How would you modify Program 6.3 or 6.4 to implement a 2 μs measurement resolution?

For the 6811 implementation, we can adjust the two bits (PR1 and PR0) in the TMSK2 register to specify one of four possible period measurement resolutions varying from 500 ns to 4 μs. For the 6812 implementation, we can adjust the three bits (PR2, PR1, and PR0) in the TSCR2 register to specify one of eight possible period measurement resolutions varying from 250 ns to 32 μs.

6.1.4.2
32-bit Period
Measurement with a
500-ns Resolution

In this example, we measure the period of an external signal with a precision of 32 bits and a resolution of 500-ns. The TTL-level signal connected to an input capture pin (Figure 6.5). Every time the TCNT register overflows from $FFFF to 0, the TOF flag is set. We can increase the precision of the period measurement as well as implementing period-too-long error checking by counting the number of TOF flag setting events during one period. To implement the period measurement task in the background, we will arm both the input capture and the timer overflow interrupts.

If we use a 16-bit counter, `Count`, for the number of times the TOF is set during one period, the precision of the period measurement is 32 bits. Let T_1 be the 32-bit time of the first rising edge of the input signal. The high 16 bits of T_1 are 0, and the low 16 bits are the input capture latch value at the time of the first rising edge. Let T_2 be the 32-bit time of the second rising edge of the input signal. The high 16 bits of T_2 are derived from the value of `Count`, and the low 16 bits are the input capture latch value at the time of the second rising edge. The period is calculated from the 32-bit subtraction:

$$\text{Period} = T_2 - T_1$$

Figure 6.7 illustrates a typical 32-bit measurement of period that is 73728=$00012000 cycles long. On the first rising edge of IC1, the TIC1 value of $4000 is copied into the global variable `First`, and `Count` is cleared. In this situation, T_1 is $00004000 cycles. Next, the TCNT overflows causing a TOF interrupt. The `TOhandler()` increments `Count` to 1. On the second rising edge of IC1, the time T_2 is measured as `Count` concatenated with TIC1. In this case, $T_2 = $00016000. The period is the difference between T_2 and T_1, that is, 73728 = $12000. Notice, how the C program below handles the borrow from the least significant word into the most significant word during the 32-bit subtraction, using the statement `if (TIC1<First) MsPeriod--;`.

The tricky part about implementing this 32-bit precision measurement is when the TOF and IC1F flags are both set approximately at the same time. At the time of the first rising edge of IC1, if TOF is not set, then the time T_1 is simply TIC1. If TOF is set, then it could have occurred just before IC1, in which case the next TOF interrupt should not increment `Count` or it could have occurred just after IC1, in which case the next TOF interrupt should increment `Count`.

Figure 6.7
Simple illustration of
the 32-bit period
measurement.

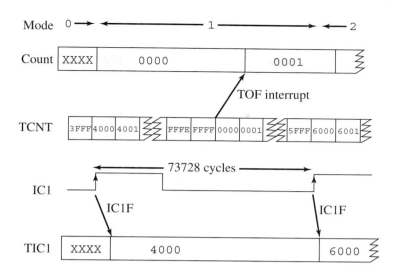

Figure 6.8 illustrates the situation when the TOF is set just before the IC1F is set. Even though the TOF flag is set before the IC1F flag, they both could occur during the same instruction. Since IC1F is a higher-priority interrupt than TOF, the `TIC1handler()` will be executed before the `TOhandler()`. If the TOF flag setting occurred just before the IC1, then the most significant bit of TIC1 will be 0. In this situation, the statement `if (((TIC1&0x8000)==0)&&(TFLG2&0x80))Count--;}` will initialize `Count` to $FFFF so that the first TOF interrupt will effectively not be counted. In this case, T_1 is $00000001 cycles and T_2 is $00011006 cycles, and the period is $11005, or 69,637 cycles.

Figure 6.8
Situation when TOF is
set just before the first
rising edge of IC1.

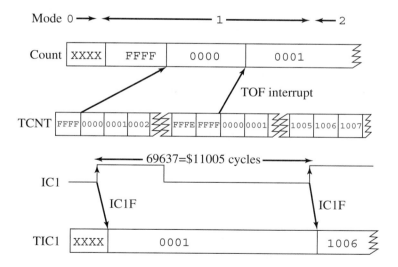

Figure 6.9 illustrates the situation when a TOF occurs just after the IC1F. We assume the TOF flag is set before the line `if(((TIC1&0x8000)==0)&&(TFLG2&0x80))Count--;}` is executed. If the TOF flag is set after this line, then we have a case like Figure 6.7, and there is no problem. Since the TOF flag setting occurred just after the IC1, then the most significant bit of TIC1 will be 1. In this situation, the statement `if (((TIC1&0x8000)==0)&&(TFLG2&0x80))Count--;}` will leave `Count` at 0 so that the first TOF interrupt will be properly counted. In this case, T_1 is $0000FFFE cycles and T_2 is $00021006 cycles, and the period is $11008, or 69,640 cycles.

Figure 6.9
Situation when TOF is
set just after the first
rising edge of IC1.

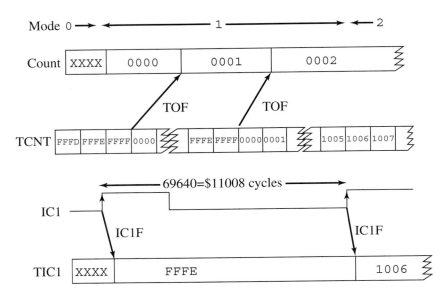

There is a similar problem that exists if the TOF is set just before or just after the second setting of the IC1F. In this case, we will count the TOF, by incrementing `Count` if the TOF occurs just before IC1F. The statement `if(((TIC1&0x8000)==0)&& (TFLG2&0x80))` `Count++;` will correct for this situation.

Notice also that if the `Count` reaches 65535 either in `Mode=0` (searching for first rise) or in `Mode=1` (search for second rise), the measurement is terminated. This mechanism allows the system to detect a period-too-long. This measurement system does not measure every period, but rather it could measure every other one. This fairly complicated solution was derived from a 6811 assembly language program found in the Freescale 68HC11 Reference Manual, Example 10-3.

Program 6.5 measures period from rising edge to rising edge with a resolution of 500 ns (E clock). The precision is 32 bits. The range varies from 0 to 35 min (E clock*4billion). An input capture interrupt is generated on each rising edge. The signal is connected to IC1 (Figure 6.5). The timer overflow interrupt counts the global variable `Count`. The globals `MsPeriod` and `LsPeriod` compose the 32-bit period measurement. `First` is the TCNT at the first rising edge. The global `Mode` means:

```
0 means looking for first edge of IC1
1 means looking for second edge
2 means measurement is done
```

6.1.5
Pulse-Width
Measurement

The basic idea of pulse-width measurement is to cause an input capture event on both the rising and falling edges of an input signal. The difference between these two times will be the pulse width. Just like period measurement, the resolution is determined by the rate at which TCNT is incremented.

6.1.5.1
Gadfly Pulse-Width
Measurement

The objective is to use *input capture pulse-width measurement* to measure resistance (Figure 6.10). This basic approach is employed by most joystick interfaces. The resistance measurement range is $0 \leq R \leq 1$ MΩ. The desired resolution is 1 kΩ. We will use gadfly synchronization.

Figure 6.10 shows the hardware interface between the unknown resistance R and the input capture pin of the microcomputer. A rising edge on PB7 causes a monostable positive

```c
// MC68HC711E9, ICC11 C
unsigned short MsPeriod,LsPeriod;
unsigned short First;
unsigned short Count;
unsigned char Mode;
#pragma interrupt_handler TOhandler()
void TOhandler(void){
  TFLG2 = 0x80;    // ack
  Count++;
  if(Count==65535){ // 35 minutes
    MsPeriod=LsPeriod=65535;
    TMSK1=0x00; TMSK2=0x00;  // Disarm
    Mode=2;
  }
}
#pragma abs_address:0xffde
void (*TO_vector[])() = {TOhandler};
#pragma end_abs_address
#pragma interrupt_handler TIC1handler()
void TIC1handler(void){
  if(Mode==0){  // first edge
    First = TIC1; Count=0; Mode=1;
    if(((TIC1&0x8000)==0)
        &&(TFLG2&0x80)) Count--;
  }
  else{        // second edge
    if(((TIC1&0x8000)==0)
        &&(TFLG2&0x80))Count++;
    MsPeriod = Count; Mode=2; // done
    LsPeriod = TIC1-First;
    if (TIC1<First) MsPeriod--;
    TMSK1=0x00; TMSK2=0x00;  // Disarm
  }
  TFLG1 = 0x04;  // ack, clear IC1F
}
#pragma abs_address:0xffee
void (*IC1_vector[])() ={TIC1handler};
#pragma end_abs_address
void Init(void){
  asm(" sei");      // make atomic
  TFLG1 = 0x04;     // Clear IC1F
  TMSK1 |= 0x04;    // Arm  IC1
  TCTL2 = (TCTL2&0xCF)|0x10; // rising
  TFLG2 = 0x80;     // Clear TOF
  TMSK2 |= 0x80;    // Arm TOF
  Mode = 0;         // searching for first
  asm(" cli"); }
```

```c
// MC9S12C32, CodeWarrior C
unsigned short MsPeriod,LsPeriod;
unsigned short First;
unsigned short Count;
unsigned char Mode;
void interrupt 16 TOhandler(void){
  TFLG2 = 0x80;  // ack
  Count++;
  if(Count==65535){ // 35 minutes
    MsPeriod=LsPeriod=65535;
    TIE=0x00; TSCR2=0x00;  // Disarm
    Mode = 2;       // done
  }
}
void interrupt 9 TIC1handler(void){
  if(Mode==0){ // first edge
    First = TC1; Count=0;
    Mode=1;
    if(((TC1&0x8000)==0)
        &&(TFLG2&0x80)) Count--;
  }
  else{          // second edge
    if(((TC1&0x8000)==0)
        &&(TFLG2&0x80)) Count++;
    Mode = 2;  // measurement done
    MsPeriod = Count;
    LsPeriod = TC1-First;
    if(TC1<First){
      MsPeriod--; // borrow
    }
    TIE=0x00; TSCR2=0x00; // Disarm
  }
  TFLG1 = 0x02;   // ack, clear C1F
}
void Init(void){
  asm sei       // make atomic
  TIOS &=~0x02;  // PT1 input capture
  DDRT &=~0x02;  // PT1 is input
  TSCR2 = 0x81 ; // Arm, TOF 30.517Hz
  TSCR1 = 0x80;  // enable counter
  TFLG1 = 0x02;  // Clear C1F
  TIE |= 0x02;   // Arm IC1, C1I=1
  TCTL4 = (TCTL4&0xF3)|0x04; // rising
  TFLG2 = 0x80;  // Clear TOF
  Mode = 0;      // searching for first
  asm cli
}
```

Program 6.5
C language 32-bit period measurement.

Figure 6.10
To measure resistance using pulse width we connect the external signal to an input capture.

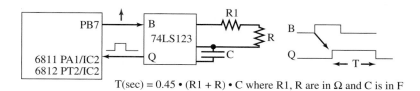

T(sec) = 0.45 • (R1 + R) • C where R1, R are in Ω and C is in F

logic pulse on the "Q" pin of the 74LS123. We choose R_1 and C so that the resistance resolution maps into a pulse-width measurement resolution of 500 ns, and the resistance range $0 \le R \le 1$ MΩ maps into $500 \le T \le 1000$ μs. The following equation describes the pulse width generated by the 74LS123 monostable as a function of the resistances and capacitance.

$$T = 0.45 \times (R + R_1) \times C$$

For a linear system, with x as input and y as output, we can use calculus to relate the measurement resolution of the input and output.

$$\Delta y = \frac{\partial y}{\partial x} \Delta x$$

Therefore, the relationship between the pulse-width measurement resolution ΔT and the resulting resistance measurement resolution is determined by the value of the capacitor.

$$\Delta T = 0.45 \Delta R \cdot C$$

To make a ΔT of 500 ns correspond to a ΔR of 1 kΩ, we choose

$$C = \Delta T / (0.45 \cdot \Delta R) = 500 \text{ ns} / 0.45 \text{ kΩ} = 1111 \text{ pF}$$

To study the range of pulse widths, we look at R=0,

$$T = 0.45 \cdot R_1 \cdot C$$

so

$$R_1 = T / (0.45 \cdot C) = 500 \text{ μs} / (0.45 \cdot 1111 \text{ pF}) = 1 \text{ MΩ}$$

As a check, notice at R=1 MΩ, T= 0.45 · (2 MΩ) · 1111 pF =1 ms.

The measurement subroutine including ritual returns in RegD the current resistance R in kΩ. For example, if the resistance R is 123 kΩ, then RegD will be 123. We will not worry about resistances R greater than 1 MΩ or if R is disconnected.

The Init function is the ritual that will configure the system in Program 6.6. Next, the Meas subroutine used to perform the input capture measurement. This algorithm written in C is presented in Program 6.7. Again the function returns resistance in units of kΩ.

6.1.5.2
Interrupt-Driven Pulse-Width Measurement

In this example, the TTL-level input signal is connected to an input capture (Figure 6.5). Both the rising and falling edges will generate an input capture interrupt. The pulse width is calculated as the difference in TIC1 latch values from a rising edge to the next falling edge (Figure 6.11). In this example the background thread sets the global variable Rising on the rising edge and calculates the pulse width PW on the falling edge. The interrupt handler simply reads the current value on the input capture pin to determine if this interrupt is a rising or falling edge. If the first interrupt is a falling edge, then the first pulse width measured will be inaccurate.

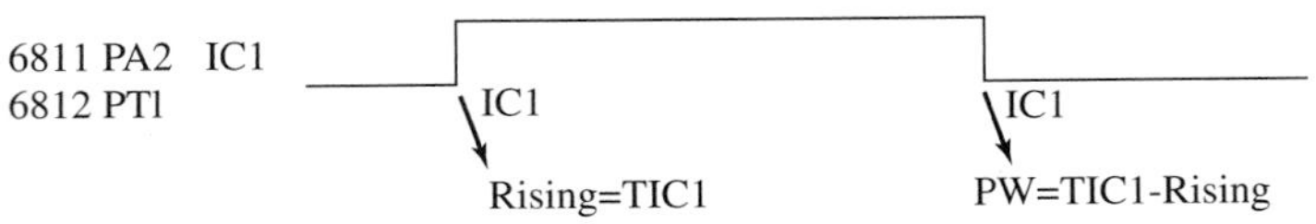

Figure 6.11
Pulse width is the time between the rising and falling edges.

6811 PA2   IC1
6812 PTI

IC1   IC1
Rising=TIC1   PW=TIC1-Rising

The pulse-width measurement is performed from rising edge to falling edge. The resolution is 500 ns (determined by E clock). The range is about 50 μs to 32 ms, with no overflow checking. The low end of the range is determined by the software overhead to process the interrupt (see the discussion about minimum period in Section 6.1.4.1). IC1 interrupts occur on both the rising and falling edges. The global PW contains the most recent measurement. Rising is a private[1] global variable containing the TCNT value at the rising edge.

---

[1]A global variable only accessed by one module.

```
; MC68HC711E9
; B=PB7,  Q=PA1/IC2
Init    ldaa  #$00   ;gadfly
        staa  TMSK1  ;IC2I=0
        rts
; return Reg D as R in Kohm
Rising equ   0      ;First TCNT
Meas    ldx   #$1000 ;I/O registers
        ldaa  #$04   ;Rising edge
        staa  $21,X  ;Set TCTL2
        bclr  $23,X,#$FD  ;IC2F=0
        bclr  $04,X,#$80  ;PB7=0
        bset  $04,X,#$80  ;PB7=1
First   brclr $23,X,#$02,First
;Wait for first rising edge
        ldy   $12,X  ;TCNT at rising
        ldaa  #$08   ;Falling edge
        staa  $21,X  ;Set TCTL2
        bclr  $23,X,#$FD  ;IC2F=0
        pshy         ;Save on stack
Second  brclr $23,X,#$02,Second
;Wait for next falling edge
        ldd   $12,X  ;TCNT at falling
        tsy
        subd  Rising,Y
;RegD=pulse width 1000 to 2000 cyc
        subd  #1000  ;0<=R<=1000Kohm
        puly
        rts
```

```
; MC9S12C32
; B=PB7,   Q=PT2/IC2
Init    bclr  TIOS,#$04  ;PT2=input capture
        movb  #$80,TSCR1 ;enable
        movb  #$01,TSCR2 ;500 ns clock
        clr   TIE        ;gadfly, C2I=0
        bset  DDRB,#$80
        rts

; return Reg D as R in Kohm
Meas    movb  #$10,TCTL4 ;Rising edge
        movb  #$04,TFLG1 ;C2F=0
        bclr  PORTB,#$80 ;PB7=0
        bset  PORTB,#$80 ;PB7=1
First   brclr TFLG1,#$04,First
;Wait for first rising edge
        ldy   TC2  ;TCNT at rising
        movb  #$04,TFLG1 ;C2F=0
        movb  #$20,TCTL4 ;Falling edge
        pshy         ;Save on stack
Second  brclr TFLG1,#$04,Second
;Wait for next falling edge
        ldd   TC2 ;TCNT at falling
        subd  2,SP+
;RegD=pulse width 1000 to 2000 cyc
        subd  #1000   ;0<=R<=1000Kohm
        rts
```

Program 6.6
Assembly language pulse-width measurement.

```
// MC68HC711E9
unsigned short Measure(void) {
unsigned short Rising;
  TCTL2 = (TCTL2&0xF3)|0x04; // Rising
  TFLG1 = 0x02; // clear IC2F
  PORTB&=~0x80;
  PORTB|= 0x80; // rising edge on PB7
  while(TFLG1&0x02==0){}; // wait
  Rising = TIC2;   // TCNT at rising
  TCTL2 = (TCTL2&0xF3)|0x08; // Falling
  TFLG1 = 0x02;            // clear IC2F
  while(TFLG1&0x02==0){}; // wait
  return(TIC2-Rising-1000);
}
void Init(void){
  TMSK1 = 0x00;      // no interrupts
}
```

```
// MC9S12C32
unsigned short Measure(void) {
unsigned short Rising;
  TCTL4 = (TCTL4&0xCF)|0x10; // Rising
  TFLG1 = 0x04;    // clear C2F
  PORTB&=~0x80;
  PORTB|= 0x80;    // rising edge on PB7
  while(TFLG1&0x04==0){}; // wait for rise
  Rising = TC2;    // TCNT at rising edge
  TFLG1 = 0x04;    // clear C2F
  TCTL4 = (TCTL4&0xCF)|0x20; // Falling
  while(TFLG1&0x04==0){}; // wait for fall
  return(TC2-Rising-1000); }
void Init(void){
  DDRB |= 0x80;  // PB7 is output
  TIOS &=~0x04;  // clear bit 2
  DDRT &=~0x04;  // PT2 is input capture
  TSCR1 =0x80;   // enable
  TSCR2 =0x01;   // 500 ns clock
  TIE = 0x00;}   // no interrupts
```

Program 6.7
C language pulse-width measurement.

Done is set at the falling edge of IC1, signifying a new measurement is available. The main program reads Done and PW to process the measurements. The main program can clear Done after it has processed the measurement. The interrupt handler could (but doesn't in this simple example) check Done and trigger an error if a new measurement is complete, but the previous measurement has not been processed yet. Because the Ritual() sets Rising equal to the current time (and not the time of a rising edge), if the first interrupt is falling edge, then the first measurement will be incorrect. If the first interrupt after the Ritual() is executed is a rising edge, then the first measurement will be correct (Programs 6.8).

```
// MC68HC711E9, ICC11 C
unsigned short PW;      // units of 500 ns
unsigned short Rising; // TCNT at rising
unsigned char Done;   // Set each falling
#pragma interrupt_handler TIC1handler()
void TIC1handler(void){
  if(PORTA&0x04){     // PA2=1 if rising
    Rising = TIC1;   // Setup for next
  }
  else{
    PW = TIC1-Rising; // measurement
    Done = 0xFF;
  }
  TFLG1 = 0x04;           // ack, IC1F=0
}
#pragma abs_address:0xffee
void (*IC1_vector[])() ={TIC1handler};
#pragma end_abs_address
void Init(void){
  asm(" sei");      // make atomic
  TCTL2 |= 0x30;    // both edges
  TMSK1|= 0x04;    // Arm IC1
  TFLG1 = 0x04;    // Clear IC1F
  Done = 0;           // set on falling
  asm(" cli");
}
```

```
// MC9S12C32, CodeWarrior C
unsigned short PW;      // units of 500 ns
unsigned short Rising; // TCNT at rising
unsigned char Done;   // Set each falling
void interrupt 9 TC1handler(void){
  if(PTT&0x02){     // PT1=1 if rising
    Rising = TC1;   // Setup for next
  }
  else{
    PW = TC1-Rising; // measurement
    Done = 0xFF;
  }
  TFLG1 = 0x02;     // ack, clear C1F
}
void Init(void) {
  asm sei      // make atomic
  TIOS &=~0x02;    // clear bit 1
  DDRT &=~0x02;    // PT1 is input capture
  TSCR1 =0x80;     // enable
  TSCR2 =0x01;     // 500 ns clock
  TCTL4|=0x0C;     // Both edges IC1
  TIE |= 0x02;     // arm IC1
  TFLG1 = 0x02;    // clear C1F
  Done = 0;
  asm cli
}
```

Program 6.8
C language pulse-width measurement.

Checkpoint 6.11: What is the relationship between the resolution and the maximum pulse width that can be measured?

6.1.5.3
Pulse-Width
Measurement Using Two
Input Capture Channels

One of the limitations of the previous measurement technique is that the lower bound on the range is determined by the software overhead to process the input capture interrupt. If the pulse-width is below that minimum, the error goes undetected. To solve this problem, the TTL-level input signal is connected to both IC1 and IC2 (Figure 6.12). The rising-edge

Figure 6.12
To measure pulse we could connect the external signal to two input captures.

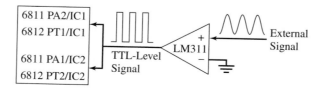

time will be measured by IC2 without the need of an interrupt, and the falling-edge interrupts will be handled by IC1.

The pulse width is calculated as the difference in TIC1–TIC2 latch values. In this example the IC1 interrupt handler simply sets the global variable PW at the time of the falling edge. Because no software is required to process the IC2 measurement, there is no minimum pulse width. On the other hand, software processing is required to handle the IC1 signal, so there is a minimum period (e.g., there must be more than 50 μs from one falling edge to the next falling edge). Again, the first measurement may or may not be accurate (Figure 6.13).

Figure 6.13
The rising edge is measured with IC2, and the falling edge is measured with IC1.

6811 PA2
6812 PT1

6811 PA1
6812 PT2

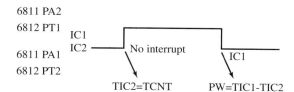

The pulse-width measurement is performed from rising edge to falling edge. The resolution is 500 ns (determined by TCNT). The range is about 500 ns to 32 ms, with no overflow checking. IC1 interrupts only occur on the falling edges. The global PW contains the most recent measurement. Done is set at the falling edge of IC1, signifying a new measurement is available. If the first edge after the Ritual() is executed is a falling edge, then the first measurement will be incorrect (because TIC2 is incorrect). If the first edge after the Ritual() is executed is a rising edge, then the first measurement will be correct. Compared to the last example, notice how little software overhead is required to perform these measurements (Program 6.9).

Notice how the C code in Program 6.9 sets two bits of TCTL2 (6811) and TCTL4 (6812) without modifying the other six bits. We call this a "friendly" ritual because it does not undo other initializations. On the other hand, the rituals all have to get together and agree on a common value for the 6812 TSCR2.

```
// MC68HC711E9, ICC11 C
unsigned short PW;    // units of 500 ns
unsigned char Done;  // Set each falling
#pragma interrupt_handler TIC1handler()
void TIC1handler(void){
  PW = TIC1-TIC2; // from rise to fall
  Done = 0xFF;
  TFLG1 = 0x04;   // ack by clearing IC1F
}
#pragma abs_address:0xffee
void (*IC1_vector[])() ={TIC1handler};
#pragma end_abs_address
void Init(void){
  asm(" sei");       // make atomic
  TCTL2=(TCTL2&0xCF)|0x20;
// falling edges of IC1, TCNT->TIC1
  TCTL2=(TCTL2&0xF3)|0x04;
// rising edges of IC2, TCNT->TIC2
  Done=0;   // set on the falling edge
  TFLG1 = 0x04;    // Clear IC1F
  TMSK1|= 0x04;    // Arm IC1, not IC2
  asm(" cli");}
```

```
// MC9S12C32, CodeWarrior C
unsigned short PW;   // units of 500 ns
unsigned char Done;  // Set each falling

void interrupt 9 TIC1handler(void){
  TFLG1 = 0x02;  // ack C1F
  PW = TC1-TC2;  // from rise to fall
  Done = 0xFF;
}
void Init(void) {
  asm sei       // make atomic
  TIOS &=~0x06;   // clear bits 2,1
  DDRT &=~0x06;   // PT2,PT1 input captures
  TSCR1 = 0x80;   // enable
  TSCR2 = 0x01;   // 500 ns clock
  TCTL4 =(TCTL4&0xCF)|0x10; // IC2 Rise
  TCTL4 =(TCTL4&0xF3)|0x08; // IC1 Fall
  Done = 0;       // set on the falling edge
  TIE |= 0x02;  // arm IC1, not IC2
  TFLG1 = 0x02; // clear C1F
  asm cli
}
```

Program 6.9
C language pulse-width measurement using two input captures.

6.2 Output Compare

**6.2.1
General Concepts**

As with our introduction to input capture, we begin our discussion of output compare with some general comments. Output compare will be used to create squarewaves, generate pulses, implement time delays, and execute periodic interrupts. A common output technique used in computer-based control systems is a variable duty-cycle actuator. Another name for this is pulse-width modulation (PWM). In Figure 6.14, the shaded areas represent times when power is applied. The computer generates a squarewave with variable duty-cycle using the output compare interface. To apply a large amount of power to the control system it issues a squarewave that is mostly high. To apply a small amount of power, it outputs a squarewave that is mostly low.

Figure 6.14
The output power is controlled by varying the duty cycle of a squarewave.

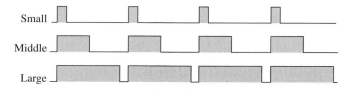

Figure 6.15
The basic components of output compare.

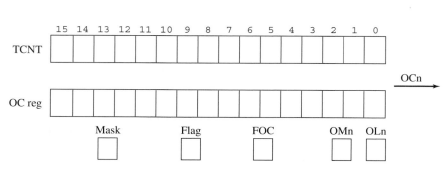

We will also use output compare together with input capture to measure frequency. Output compare and input capture can also be combined to measure period and frequency over a wide range of ranges and resolutions. The same 16-bit TCNT, incremented at a fixed rate, is used for input capture and output compare. As shown in Figure 6.15, each output compare module has

> An external output pin, OCn
> A flag bit
> A force compare control bit, FOCn
> Two control bits, OMn OLn
> An interrupt mask bit
> A 16-bit output compare register

The various microcomputers from Freescale have from zero to eight output compare modules. For example, the MC68HC711E9 has five and the MC9S12C32 has eight. The maximum number of input capture modules plus output compare modules on the 6812 is eight.

The output compare pin is an output of the computer; hence it will be used to control an external device. An output compare event occurs, setting the flag bit, when either

> **1.** The 16-bit TCNT matches the 16-bit OC register, or
> **2.** The software writes a one to the FOC bit.

The OMn, OLn bits specify what effect the output compare event will have on the output pin. Two or three actions result from an output compare event: 1) the OCn output bit changes, 2) the output compare flag is set and 3) an interrupt is requested if the mask bit is 1. Just like the input capture, the output compare flag is cleared by writing a one to it. The module is armed when the mask bit is 1. One simple application of output compare is to create a fixed time delay. Let `delay` be the number of cycles you wish to wait. The steps to create the delay are as follows:

1. Read the current 16-bit TCNT
2. Calculate `TCNT+delay`
3. Set the 16-bit output compare register to `TCNT+delay`
4. Clear the output compare flag,
5. Wait for the output compare flag to be set

If we perform the addition `TCNT+delay` with 16-bit integer math (e.g., addd instruction), then this approach will function properly even if the counter rolls over from $FFFF to 0. For example, assume the current TCNT is $F000, and we wish to delay 8192 ($2000) cycles. The 16-bit addition $F000+$2000 will result in a sum of $1000. If we put the $1000 into the output compare register, then it will take the correct number (i.e., 8192 cycles), for the TCNT to go from $F000 to $1000. The time for the software to execute steps 1–4, will determine the minimum delay that can be created with this approach. For example, if the value of `delay` is so small that the TCNT hits `TCNT+delay` before the flag is cleared in step 4, then the delay will be a `65536+delay` delay. For obvious reasons, the maximum delay with this simple method is 65536 cycles.

One of the output compare modules, PA7/OC1 on the 6811 and PT7/OC7 on the 6812, can be configured such that an output compare event on it will cause changes on some or all of the other output compare pins. This coupled behavior can be used to create synchronized signals. For example, we can create pulses that start together or end together.

Checkpoint 6.12: When does an output compare event occur?

Checkpoint 6.13: What happens during an output compare event?

6.2.2 Output Compare Details

In the following subsections we overview the specific output compare functions on particular Freescale microcomputers. This section is intended to supplement rather than replace the Freescale manuals. When designing systems with output compare, please refer to the reference manual of your specific Freescale microcomputer.

6.2.2.1 MC68HC711E9 Output Compare Details

Table 6.10 lists the registers associated with output compare on the 6811. Figure 6.16 shows the 6811 pins available for output compare. The entries shown in bold will be used in this section. **TCNT** is a 16-bit unsigned counter that is incremented at a rate determined by two bits (PR1 and PR0) in the **TMSK2** register. Since there are no direction register bits associated with PA6-4, these bits are always outputs. There are direction register bits for PA3 and PA7 (**DDR3 DDR7** bits in the PACTL register) that should be set to one if that pin is to be used as an output. In addition, to use PA3 as output compare channel 5, we should clear the **I4/O5** bit. An output compare event occurs when one of the 16-bit output compare registers (TOC1, TOC2, TOC3, TOC4, or TOC5) matches the 16-bit TCNT register. This event will set the corresponding output compare flag (OC1F, OC2F, OC3F, OC4F, or OC5F) in the **TFLG1** register. If the output pin is activated, as specified in the **TCTL1** register, then this output compare event can also modify the output value (set to one, clear to

Figure 6.16
Available signals for output compare on the MC68HC711E9.

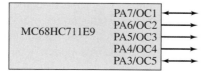

zero, or toggle.) We can arm or disarm the individual output compare interrupts by initializing the **TMSK1** register. Our software can determine if an output compare event has occurred by reading the TFLG1 register.

Address	msb															lsb	Name
$100E	15	14	13	12	11	10	9	8	7	6	5	4	3	2	1	0	TCNT
$1016	15	14	13	12	11	10	9	8	7	6	5	4	3	2	1	0	TOC1
$1018	15	14	13	12	11	10	9	8	7	6	5	4	3	2	1	0	TOC2
$101A	15	14	13	12	11	10	9	8	7	6	5	4	3	2	1	0	TOC3
$101C	15	14	13	12	11	10	9	8	7	6	5	4	3	2	1	0	TOC4
$101E	15	14	13	12	11	10	9	8	7	6	5	4	3	2	1	0	TIC4/TOC5

Address	Bit 7	6	5	4	3	2	1	Bit 0	Name
$1000	**PA7**	**PA6**	**PA5**	**PA4**	**PA3**	PA2	PA1	PA0	PORTA
$100B	FOC1	FOC2	FOC3	FOC4	FOC5	0	0	0	CFORC
$100C	OC1M7	OC1M6	OC1M5	OC1M4	OC1M3	0	0	0	OC1M
$100D	OC1D7	OC1D6	OC1D5	OC1D4	OC1D3	0	0	0	OC1D
$1020	**OM2**	**OL2**	**OM3**	**OL3**	**OM4**	**OL4**	**OM5**	**OL5**	TCTL1
$1022	**OC1I**	**OC2I**	**OC3I**	**OC4I**	**IC4/OC5I**	IC1I	IC2I	IC3I	TMSK1
$1023	**OC1F**	**OC2F**	**OC3F**	**OC4F**	**IC4/OC5F**	IC1F	IC2F	IC3F	TFLG1
$1024	TOI	RTII	PAOVI	PAII	0	0	**PR1**	**PR0**	TMSK2
$1026	**DDRA7**	PAEN	PAMOD	PEDGE	**DDRA3**	**I4/O5**	RTR1	RTR0	PACTL

Table 6.10
MC68HC711E9 registers used for output compare.

Recall that the flags in the TFLG1 and TFLG2 registers are cleared by writing a one into the specific flag bit we wish to clear. This acknowledge sequence clears the OC3F flag without affecting the other seven flags in the TFLG1 register.

```
TFLG1=0x20;      // Clears OC3F in TFLG1
```

Checkpoint 6.14: Write assembly or C code to clear the OC4F flag.

The output compare event (when the output compare latch equals TCNT) can be configured to affect an output pin. The TCTL1 register determines what effect the OC2, OC3, OC4, and OC5 events will have (none, toggle, clear, or set) on the output pin, as listed in Table 6.11.

Table 6.11
Two bits determine the action caused by a 6811 output compare event.

OMn	OLn	Effect of When TOCn=TCNT
0	0	Does not affect OCn
0	1	Toggle OCn
1	0	Clear OCn=0
1	1	Set OCn=1

The output compare 1, OC1, operates differently. PA7 can also be used the pulse accumulator mechanism. On a successful OC1 event (TOC1=TCNT), the 6811 can be programmed to set or clear any of the output compare pins. The **OC1M** register selects which pin(s) will be affected by the OC1 event. Set the corresponding bit(s) of this register to zero to disconnect OC1 from the output pin(s). Conversely, set the corresponding bit(s) of this register to one to attach the OC1 event to the other output pin(s). For every bit in the OC1M register that is one, the corresponding bit in the **OC1D** register will specify the resulting value of the output pin(s) after an OC1 event. This mode can be used to create synchronized outputs, or short pulses.

6.2.2.2
MC9S12C32 Output
Compare Details

Because the 6812 input capture and output compare modules share I/O pins (Figure 6.17) and many registers, we have already introduced much of the output compare components. Table 6.12 summarizes the 6812 registers used for output compare. The entries shown in bold will be used in this section. The TEN bit in the TSCR1 register must be enabled. TCNT is a 16-bit unsigned counter that is incremented at a rate determined by three bits (PR2, PR1, and PR0) in the TSCR2 register. The fundamental approach to output compare involves connecting one of the eight output compare pins on Port T to an external TTL-level device. The pin is selected as output compare by placing a 1 in the corresponding bit of the TIOS register. With input capture, we must set the direction bit in DDRT to zero, but selecting output compare in the TIOS register automatically makes the bit an output. An output compare event occurs when one of the Output Compare Registers (TCn) matches the TCNT register. This event will set the corresponding output compare flag (CnF) in the TFLG1 register. If the output pin is activated, as specified in the TCTL1 and TCTL2 registers, then this output compare event can also modify the output value (set to one, clear to zero, or toggle.)

Figure 6.17
Available signals for output compare on the 6812.

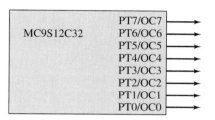

Address	msb															lsb	Name
$0044	15	14	13	12	11	10	9	8	7	6	5	4	3	2	1	0	TCNT
$0050	15	14	13	12	11	10	9	8	7	6	5	4	3	2	1	0	TC0
$0052	15	14	13	12	11	10	9	8	7	6	5	4	3	2	1	0	TC1
$0054	15	14	13	12	11	10	9	8	7	6	5	4	3	2	1	0	TC2
$0056	15	14	13	12	11	10	9	8	7	6	5	4	3	2	1	0	TC3
$0058	15	14	13	12	11	10	9	8	7	6	5	4	3	2	1	0	TC4
$005A	15	14	13	12	11	10	9	8	7	6	5	4	3	2	1	0	TC5
$005C	15	14	13	12	11	10	9	8	7	6	5	4	3	2	1	0	TC6
$005E	15	14	13	12	11	10	9	8	7	6	5	4	3	2	1	0	TC7

Address	Bit 7	6	5	4	3	2	1	Bit 0	Name
$0240	PT7	PT6	PT5	PT4	PT3	PT2	PT1	PT0	PTT
$0242	DDRT7	DDRT6	DDRT5	DDRT4	DDRT3	DDRT2	DDRT1	DDRT0	DDRT
$0046	**TEN**	TSWAI	TSBCK	TFFCA	0	0	0	0	TSCR1
$004D	TOI	0	0	0	TCRE	**PR2**	**PR1**	**PR0**	TSCR2
$0040	**IOS7**	**IOS6**	**IOS5**	**IOS4**	**IOS3**	**IOS2**	**IOS1**	**IOS0**	TIOS
$004C	**C7I**	**C6I**	**C5I**	**C4I**	**C3I**	**C2I**	**C1I**	**C0I**	TIE
$004E	**C7F**	**C6F**	**C5F**	**C4F**	**C3F**	**C2F**	**C1F**	**C0F**	TFLG1
$004F	TOF	0	0	0	0	0	0	0	TFLG2
$0048	**OM7**	**OL7**	**OM6**	**OL6**	**OM5**	**OL5**	**OM4**	**OL4**	TCTL1
$0049	**OM3**	**OL3**	**OM2**	**OL2**	**OM1**	**OL1**	**OM0**	**OL0**	TCTL2
$0041	FOC7	FOC6	FOC5	FOC4	FOC3	FOC2	FOC1	FOC0	CFORC
$0042	OC7M7	OC7M6	OC7M5	OC7M4	OC7M3	OC7M2	OC7M1	OC7M0	OC7M
$0043	OC7D7	OC7D6	OC7D5	OC7D4	OC7D3	OC7D2	OC7D1	OC7D0	OC7D
$0047	TOV7	TOV6	TOV5	TOV4	TOV3	TOV2	TOV1	TOV0	TTOV

Table 6.12
MC9S12C32 registers used for output compare.

We can arm or disarm the individual output compare interrupts by initializing the TIE register. Our software can determine whether an output compare event has occurred by reading the TFLG1 register. Recall, the flags in the TFLG1 and TFLG2 registers are cleared by writing a one into the specific flag bit we wish to clear. For example, writing a $FF into TFLG1 will clear all eight flags. The following is valid method for clearing C3F. That is, these acknowledge sequences clear the C3F flag without affecting the other seven flags in the TFLG1 register.

```
TFLG1=0x08;
```

>   *Checkpoint 6.15:* Write assembly or C code to clear the C4F flag.

>   *Checkpoint 6.16:* Explain why `TFLG1|=0x20;` is unfriendly.

The output compare event (when the 16-bit output compare register equals the 16-bit TCNT) can be configured to affect an output pin. The TCTL1 and TCTL2 registers determine what effect each of the four possible output compare events will have (none, toggle, clear or set) on the output pin, as shown in Table 6.13.

Table 6.13
Two bits determine the action caused by 6812 output compare event.

OMn	OLn	Effect of When TOCn=TCNT
0	0	does not affect OCn
0	1	Toggle OCn
1	0	Clear OCn=0
1	1	Set OCn=1

Like the 6811, there is one output compare, OC7, which operates differently. On a successful OC7 event (TC7=TCNT), the 6812 can be programmed to set or clear any of the other output compare pins. The OC7M register selects which pin(s) will be affected by the OC7 event. Clear the corresponding bit(s) of the OC7M register to zero to disconnect OC7 from the output pin(s). Conversely, set the corresponding bit(s) of the OC7M register to one to attach the OC7 event to the output pin(s). For every bit in the OC7M register that is one, the corresponding bit in the OC7D register will specify the resulting value of the output pin(s) after an OC7 event.

>   *Observation:* The phase-lock-loop (PLL) on the 9S12 will affect the TCNT period.

>   *Checkpoint 6.17:* List the steps to initialize PT0 as an armed output compare with toggled output.

**6.2.3
Square-Wave
Generation**

This example generates a 50% duty cycle square wave using output compare. The output is high for `Period` cycles, then low for `Period` cycles. Output compare interrupts will be requested at a rate twice as fast as the resulting square-wave frequency. One interrupt is required for the rising edge on the output compare pin and another for the falling edge. Toggle mode is used to create the 50% duty cycle square wave (Figure 6.18). In this mode, the output compare pin is toggled whenever the output compare latch matches TCNT.

Figure 6.18
Square-wave generation using output compare.

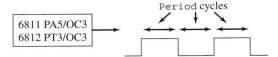

The output compare interrupt handler simply acknowledges the interrupt and calculates the time for the next signal transition. This implementation will create an exact square wave independent of software execution delays as long as the interrupt is serviced within `Period` cycles. If the interrupt latency (i.e., the time between the setting of the output compare flag and the running of the ISR) is more than `Period` cycles, then the next edge will not occur for another 65,536 cycles.

In assembly, Program 6.10 contains the ritual and the interrupt handlers.

```
; MC68HC711E9                              ; MC9S12C32
Period rmb  2      ;units usec             Period rmb  2      ;units usec
Init   sei         ;make atomic            Init   sei         ;make atomic
       ldaa TMSK1  ;Old value                     bset TIOS,#$08  ;OC3
       oraa #$20   ;TMSK1 OC3I=1                  bset DDRT,#$08  ;PT3 output
       staa TMSK1  ;Arm OC3F                      movb #$80,TSCR1  ;enable
       ldaa TCTL1                                 movb #$01,TSCR2  ;500 ns clock
       anda #$CF   ;OM3=0, toggle                 bset TIE,#$08   ;Arm OC3
       oraa #$10   ;OL3=1                         bset TCTL2,#$40  ;OL3=1, toggle
       staa TCTL1                                 bclr TCTL2,#$80  ;OM3=0
       ldaa #$20   ;clear OC3F                    movb #$08,TFLG1  ;clear C3F
       staa TFLG1                                 ldd  TCNT  ;current time
       ldd  TCNT   ;current time                  addd #50   ;first in 25us
       addd #50    ;first in 25us                 std  TC3
       std  TOC3                                  cli        ;enable
       cli         ;enable                        rts
       rts
OC3Han ldaa #$20   ;clear OC3F [2]         OC3Han movb #$08,TFLG1 ;Ack    [4]
       staa TFLG1  ;Ack        [4]                ldd  TC3                 [3]
       ldd  TOC3               [5]                addd Period ;next        [3]
       addd Period ;next       [6]                std  TC3                 [2]
       std  TOC3               [5]                rti                      [8]
       rti                     [12]
       org  $FFE4                                 org  $FFE8
       fdb  OC3Han                                fdb  OC3Han
```

Program 6.10
Assembly language square wave using output compare.

To determine the fastest square wave that can be created we count the time it takes to process an interrupt. The fastest square wave will occupy 100% of the computer execution and request interrupts at twice the frequency (Table 6.14). In C, this square-wave software is shown in Program 6.11.

Table 6.14
Total time in the handler calculated for different assembly language implementations.

Component	6811	6812
Process the interrupt (cycles, μs)	$14 = 7 \, \mu$s	$9 = 2.25 \, \mu$s
Execute entire handler (cycles, μs)	$34 = 17 \, \mu$s	$20 = 5 \, \mu$s
Total time (μs)	24	7.25

```
// MC68HC711E9, ICC11 C                      // MC9S12C32, CodeWarrior C
unsigned short Period;  // in usec           unsigned short Period;    // in usec
// Number of Cycles Low and High            // Number of Cycles Low and High
#pragma interrupt_handler TOC3handler()      void interrupt 11 TC3handler(void){
void TOC3handler(void){                        TFLG1 = 0x08;      // ack C3F
  TOC3 = TOC3+Period; // calculate Next        TC3 = TC3+Period; // calculate Next
  TFLG1 = 0x20;        // ack,  OC3F=0       }
}                                            void ritual(void) {
#pragma abs_address:0xffe4                     asm sei         // make atomic
void (*OC3_vector[])()={TOC3handler};          TIOS |= 0x08;   // enable OC3
#pragma end_abs_address                        DDRT |= 0x08;   // PT3 is output
void ritual(void){                             TSCR1 = 0x80;   // enable
  asm(" sei");       // make atomic            TSCR2 = 0x01;   // 500 ns clock
  TFLG1 = 0x20;      // clear OC3F             TCTL2 = (TCTL2&0x3F)|0x40; // toggle
  TMSK1 |= 0x20;     // arm OC3                TIE |= 0x08;    // Arm output compare 3
  TCTL1 = (TCTL1&0xCF)|0x10; // toggle         TFLG1 = 0x08;   // Initially clear C3F
  TOC3  = TCNT+50; // first in 25 us           TC3 = TCNT+50; // First one in 25 us
  asm(" cli");                                 asm cli
}                                            }
```

Program 6.11
C language square wave using output compare.

The following is the assembly listing output of the TOC3handler() generated by the ICC11 compiler (Version 4.5). We later added the cycles shown on the right.

```
                      13 ; void TOC3handler(void){
                      14 ;       TOC3=TOC3+Period;
0000 FC 10 1A         15        ldd 4122          [5]
0003 F3 00 00         16        addd _Period      [6]
0006 FD 10 1A         17        std 4122          [5]
0009                  18 SW.12:: ;    TFLG1=0x20;}
0009 C6 20            19     ldab #32             [2]
000B F7 10 23         20     stab  4131           [4]
000E 3B               21     rti                  [12]
```

The total time to process this C language OC3 interrupt includes (1) 14 cycles for the 6811 to process the interrupt and (2) 34 cycles to complete the TOC3handler(). Therefore, each OC3 interrupt requires 48 cycles to process. In this case, we did not include the time to finish the current instruction because we will calculate the overhead required to generate this square wave (Table 6.15). Since an OC3 interrupt is requested every Period cycle and each one takes exactly 48 cycles to complete, the overhead required to produce the square wave can be calculated as a percentage of the total available 6811 execution cycles.

Table 6.15
Percentage overhead calculated for a 6811 C language implementation versus frequency.

Frequency	Period	Interrupt every (cycles)	Time to process (cycles)	Overhead (%)
10 Hz	100 ms	50,000	48	0.1
100 Hz	10 ms	5,000	48	1
1 kHz	1 ms	500	48	10
5 kHz	200 μs	200	48	24
1/P	P (μs)	P	48	4800/P

Just like the previous example of the real-time clock, as long as no other software executes with the interrupts disabled for longer than Period cycles, this solution will create the perfect square wave. Remember the toggle event (i.e., when PA5/OC3 changes) occurs automatically in hardware at the time when TOC3=TCNT, and not during the software execution of the `TOC3handler()`.

The slowest square wave that can be generated by this implementation has a period of 65,535 μs—that is, about 15.3 Hz. On the 6811 you can set up the TCNT to clock every 4 μs. Thus, the slowest square wave you can generate is about 2 Hz. On the MC9S12C32 you can use a 32 μs TCNT to create a squarewave with a 4-second period. Less accurate square waves can be made using an interrupt real-time clock and a software counter. Using the real-time clock, we could interrupt at a fixed interval **T** and decrement a software counter. When that counter hits 0, we could toggle an output bit and reset the counter to **N**. The period of the resulting square wave would be 2·**N**·**T**. These functions could be added to the implementations presented earlier in Programs 6.10 and 6.11.

6.2.4
Pulse-Width
Modulation

This example generates a variable duty cycle square wave using output compare (Figure 6.19). Pulse-width modulation is an effective and thus popular mechanism for the embedded microcomputers to control external devices. The output is 1 for High cycles, then 0 for Low cycles. Output compare interrupts will again be requested at a rate twice as fast as the resulting square-wave frequency. One interrupt is required for the rising edge and another for the falling edge. Toggle mode is used to create the variable duty cycle square wave. In the examples below, High plus Low will always equal 10,000, so in each case the square-wave period will be 10,000 cycles, or 5 ms. By adjusting the ratio of

$$\text{Duty cycle} = \frac{\text{High}}{\text{High} + \text{Low}}$$

the software can control the duty cycle. This implementation cannot generate waves close to 0 or 100% duty cycle. The upper and lower limits can be calculated by counting the cycles required to process the output compare interrupt like in the previous example. If **T** is the maximum number of cycles to process the output compare interrupt, then both High and Low must be greater than **T**.

Figure 6.19
Pulse-width modulation using output compare.

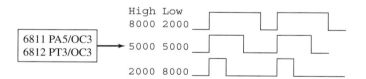

In assembly language, the ritual and the interrupt handler are in Program 6.12 (there will be an interrupt for both the rise and fall).

```
; MC68HC711E9
High rmb  2 ;number of cycles high
Low  rmb  2 ;number of cycles low
Init  sei         ;make atomic
      ldaa TMSK1 ;Old value
      oraa #$20   ;TMSK1 OC3I=1
      staa TMSK1 ;Arm OC3F
      ldaa TCTL1
      anda #$CF   ;OM3=0, toggle
```

```
; MC9S12C32
High rmb  2  ;number of cycles high
Low  rmb  2  ;number of cycles low
Init  sei         ;make atomic
      bset TIOS,#$08   ;OC3
      bset DDRT,#$08   ;PT3 output
      movb #$80,TSCR1  ;enable
      movb #$01,TSCR2  ;500 ns clock
      bset TIE,#$08    ;Arm OC3
```

continued on p. 302

Program 6.12
Assembly language pulse-width-modulated square wave using output compare.

continued from p. 301

```
          oraa #$10   ;OL3=1                              bset TCTL2,#$40   ;OL3=1, toggle
          staa TCTL1                                      bclr TCTL2,#$80   ;OM3=0
          ldaa #$20   ;clear OC3F                         movb #$08,TFLG1   ;clear C3F
          staa TFLG1                                      ldd  TCNT   ;current time
          ldd  TCNT   ;current time                       addd #50    ;first in 25us
          addd #50    ;first in 25us                      std  TC3
          std  TOC3                                       cli         ;enable
          cli         ;enable                             rts
          rts
OC3Han ldaa #$20   ;clear OC3F  [2][2]            OC3Han movb #$08,TFLG1 ;Ack      [4][4]
       staa TFLG1 ;Ack          [4][4]                   ldaa PTT     ;rise/fall? [3][3]
       ldaa PORTA ;rise/fall?   [4][4]                   bita #$08    ;PT3        [1][1]
       bita #$20  ;PA5          [2][2]                   beq  zero               [1][3]
       beq  zero                [3][3]            one     ldd  TC3                [3]
one    ldd  TOC3                [5]                       addd High ;now PT3 is 1 [3]
       addd High ;PA5 is 1      [6]                       std  TC3                [2]
       std  TOC3                [5]                       bra  done               [3]
       bra  done                [3]               zero    ldd  TC3                   [3]
zero   ldd  TOC3                   [5]                    addd Low  ;now PT3 is 0    [3]
       addd Low   ;PA5 is 0        [6]                    std  TC3                   [2]
       std  TOC3                   [5]            done    rti                       [8][8]
done   rti                     [12][12]
       org  $FFE4                                        org  $FFE8
       fdb  OC3Han                                       fdb  OC3Han
```

Program 6.12
Assembly language pulse-width-modulated square wave using output compare.

To determine the minimum and maximum pulse widths that can be created we count the time it takes to process an interrupt. The cycle times shown in brackets, e.g., [2], in Program 6.12 are executed for each interrupt. The two columns in Table 6.16 represent the two possible branch patterns of the interrupt execution. This table presents the **T** parameter discussed previously.

Table 6.16
Total time in the handler calculated for different assembly language implementations.

Component	6811	6812
Process the interrupt (cycles)	14	9
Execute entire handler (cycles)	43–46	27–28
Total time **T** (μs)	30 μs	9.25 μs

In C, this PWM software is shown in Program 6.13. To determine the **T** parameter for the C implementation, you could observe the assembly listing from the compiler, count cycles, and perform a calculation like Table 6.16.

Checkpoint 6.18: How do you change program 6.13 so that the period is 20 ms instead of 5 ms?

Observation: The 6812 ISR produced by Metrowerks executes in 24 to 27 cycles, which is actually faster than the assembly program written in Program 6.12.

```
// MC68HC711E9, ICC11 C                          // MC9S12C32, CodeWarrior C
unsigned short High; // Cycles High              unsigned short High; // Cycles High
unsigned short Low;  // Cycles Low               unsigned short Low;  // Cycles Low
// Period is High+Low Cycles                     // Period is High+Low Cycles
#pragma interrupt_handler TOC3handler()          void interrupt 11 TC3handler (void){
void TOC3handler(void){                            TFLG1 = 0x08;      // ack C3F
  if(PORTA&0x20){      // PA5 is now high           if(PTT&0x08){      // PT3 is now high
    TOC3 = TOC3+High; // 1 for High                   TC3 = TC3+High; // 1 for High cyc
  }                                                  }
  else{                // PA5 is now low             else{              // PT3 is now low
    TOC3 = TOC3+Low;  // 0 for Low                    TC3 = TC3+Low;  // 0 for Low cycles
  }                                                  }
  TFLG1=0x20;          // ack, clear OC3F          }
}                                                void Init(void){
#pragma abs_address:0xffe4                         asm sei         // make atomic
void (*OC3_vector[])()={TOC3handler};              TIOS  |= 0x08;  // enable OC3
#pragma end_abs_address                            DDRT  |= 0x08;  // PT3 is output
void Init(void){                                   TSCR1 = 0x80;   // enable
  asm(" sei");         // make atomic              TSCR2 = 0x01;   // 500 ns clock
  TFLG1 = 0x20;        // initially OC3F=0          TIE  |= 0x08;   // Arm output compare 3
  TMSK1 |= 0x20;       // arm OC3                   TFLG1 = 0x08;   // Initially clear C3F
  TCTL1 = (TCTL1&0xCF)|0x10; // toggle             TCTL2 = (TCTL2&0x3F)|0x40; // toggle
  TOC3 = TCNT+50; // first right away              TC3 = TCNT+50; // first right away
  asm(" cli");                                     asm cli
}                                                }
```

Program 6.13
C language pulse-width modulated square wave using output compare.

6.2.5 Delayed Pulse Generation

One application of the coupled output compare mechanism is a delayed output pulse (Figure 6.20). One pulse will be generated each time the function Pulse() is called. We will supply two parameters to this pulse generation function, Delay and Width. The first parameter is the delay (in cycles), and the second is the width of the pulse. Delayed pulse generation could have been created with a single output compare module, but the use of two output compare modules allows the Width to be as short as 1 cycle. We can set 6811 TOC1 (or 6812 TC7) to the time we want the pulse to go high and set 6811 TOC3 (or 6812 TC3) to the time we want the pulse to go low. The OC3 interrupt is disarmed in the interrupt handler so that the output pulse occurs only once and is not repeated every 65,536 counts of TCNT. Although the Width can be as short as 1 cycle, the Delay must be big enough so that TCNT does not reach the TOC3 value before the OC3F flag is cleared (i.e., the execution time of this function must be less than the Delay time).

Figure 6.20
Delayed pulse generation using output compare.

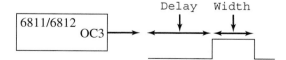

On the 6811, we set both OC1M5 and OC1D5 to 1. When TCNT equals TOC1, PA5/OC3 will go high. We set OM3 to 1 and OL3 to 0. When TCNT=TOC3, PA5/OC3 will go low. Similarly on the 6812, we set both OC7M3 and OC7D3 to 1. When TCNT equals TC7, PT3/OC3 will go high. We set OM3 to 1 and OL3 to 0. When TCNT=TC3, PT3/OC3 will go low (Program 6.14).

```
// MC68HC711E9, ICC11 C              // MC9S12C32
void Pulse(unsigned short Delay,     void Pulse(unsigned short Delay,
      unsigned short Width){               unsigned short Width){
  asm(" sei");  // make atomic         asm sei        // make atomic
  TOC1 = TCNT+Delay;                   TIOS |= 0x08;  // enable OC3
  TOC3 = TOC1+Width;                   DDRT |= 0x08;  // PT3 is output
  OC1M = 0x20;  // connect OC1 to PA5  TSCR1 = 0x80;  // enable
  OC1D = 0x20;  // PA5=1 when TOC1=TCNT TSCR2 = 0x01;  // 500 ns clock
  TCTL1 = (TCTL1&0xCF)|0x20;          TC7 = TCNT+Delay;
// PA5=0 when TOC3=TCNT                TC3 = TC7+Width;
  TFLG1 = 0x20;       // Clear OC3F    OC7M = 0x08;     // connect OC7 to PT3
  TMSK1 |= 0x20;      // Arm OC3F      OC7D = 0x08;     // PT3=1 when TC7=TCNT
  asm(" cli");                        TCTL2=(TCTL2&0x3F)|0x80;
}                                    // PT3=0 when TC3=TCNT
#pragma interrupt_handler TOC3handler() TFLG1 = 0x08;    // Clear C3F
void TOC3handler(void){               TIE |= 0x08;     // Arm C3F
  OC1M = 0;  // disconnect OC1 from PA5 asm cli
  OC1D = 0;                          }
  TCTL1 &=~0x30;  // disable OC3      void interrupt 11 TC3handler(void){
  TMSK1 &=~0x20;  // disarm OC3F       OC7M = 0;  // disconnect OC7 from PT3
}                                      OC7D = 0;
#pragma abs_address:0xffe4             TCTL2 &=~0xC0;   // disable OC3
void (*OC3_vector[])()={TOC3handler};  TIE &= ~0x08;    // disarm C3F
#pragma end_abs_address              }
```

Program 6.14
C language delayed pulse output using output compare.

Observation: We acknowledge an interrupt (clear its flag) within the interrupt handler when we are interested in subsequent interrupts, and we disarm an interrupt (clear its mask) when we are not interested in any more interrupts from this source.

6.3   Frequency Measurement

6.3.1 Frequency Measurement Concepts

The direct measurement of frequency involves counting input pulses for a fixed amount of time. The basic idea is to use Input Capture to count pulses and use Output Compare to create the fixed time interval. For example, we could initialize Input Capture to interrupt on every rising edge of our input signal. During the Input Capture handler, we could increment a `Counter`. At the beginning of our fixed time interval, the `Counter` is initialized to zero, and at the end of the interval, we can calculate frequency:

$$f = \frac{counter}{fixed\ time}$$

The frequency resolution, Δf, is defined to be the smallest change in frequency that can be reliably measured by the system. For the system to detect a change, the frequency must increase (or decrease) enough so that there is one more (or one less) pulse during the fixed time interval. Therefore, the frequency resolution is

$$\Delta f = \frac{1}{fixed\ time}$$

This frequency resolution also specifies the units of the measurement.

6.3.2
Frequency
Measurement
with Δf=100 Hz

If we count pulses in a 10-ms time interval, then the number of pulses represents the signal frequency with units 1/10 ms, or 100 Hz. For example, if there are 5 pulses during the 10-ms interval, then the frequency is 500 Hz. For this system, the measurement resolution is 100 Hz, so the frequency would have to increase to 600 Hz (or decrease to 400 Hz) for the change to be reliably detected (Figure 6.21).

Figure 6.21
Basic timing involved in frequency measurement using both input capture and output compare.

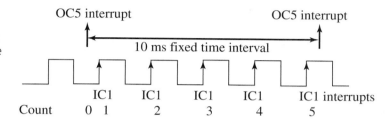

The highest frequency that can be measured will be determined by how fast the Input Capture interrupt handler can count pulses. We should select a counter precision (in this case only 8 bits is needed) to hold this maximum number. In this example, the TTL-level input signal is connected to IC1 (PA2 on the 6811 or PT1 on the 6812). The rising edge will generate an input capture interrupt (Figure 6.22).

Figure 6.22
Frequency measurement using both input capture and output compare.

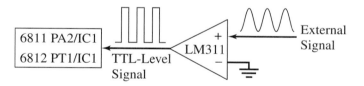

In C, the frequency measurement software is as follows. The frequency measurement counts the number of rising edges in a 10-ms interval. The measurement resolution is 100 Hz (determined by the 10-ms interval). An IC1 interrupt occurs on each rising edge, and an OC5 interrupt occurs every 10 ms. The TTL-level input signal is connected to IC1 (6811 PA2 or 6812 PT1). The foreground/background threads communicate via two shared globals. The background thread will update Freq with a new measurement and set Done. When Done is set, the foreground thread will read the global Freq and clear Done.

```
unsigned short Freq;    /* Frequency with units of 100 Hz */
unsigned char Done;     /* Set each measurement, every 10 ms */
```

There is a private global (only accessed by the background threads and not the foreground).

```
unsigned short Count;   /* Number of rising edges */
```

If we assume the fastest period is 50 μs, then the largest frequency will be 20 kHz. Therefore, the maximum values of Count and Freq will be 200, so 8-bit variables could have been used. A frequency of 0 will result in no input capture interrupts, and the system will properly report the frequency of 0. The 6811 and 6812 implementation are shown in C in Program 6.15.

```
// MC68HC711E9, ICC11 C                      // MC9S12C32, CodeWarrior C
#pragma interrupt_handler TIC1handler()      void interrupt 9 TC1handler(void){
void TIC1handler(void){                        Count++;      // number of rising edges
  Count++;       // number of rising edges     TFLG1 = 0x02; // ack, clear C1F
  TFLG1 = 0x04; // ack, clear IC1F           }
}                                            #define Rate 20000  //  10 ms
#pragma abs_address:0xffee                   void interrupt 13 TC5handler(void){
void (*IC1_vector[])() ={TIC1handler};         TFLG1= 0x20;       // Acknowledge
#pragma end_abs_address                        TC5 =  TC5+Rate;   // every 10 ms
#define Rate 20000  //  10 ms                   Freq = Count;     // 100 Hz units
#define OC5F 0x08                               Done = 0xff;
#pragma interrupt_handler TOC5handler()        Count = 0;         // Setup for next
void TOC5handler(void){                      }
  TFLG1= OC5F;        // Acknowledge         void Init(void) {
  TOC5 = TOC5+Rate; // every 10 ms            asm sei       // make atomic
  Freq = Count;       // 100 Hz units         TIOS  |= 0x20; // enable OC5
  Done = 0xff;                                TSCR1 = 0x80; // enable
  Count = 0; }        // Setup for next       TSCR2 = 0x01; // 500 ns clock
#pragma abs_address:0xffe0                     TIE  |= 0x22;  // Arm OC5 and IC1
void (*OC5_vector[])() ={TOC5handler};         TC5 = TCNT+Rate;   // First in 10 ms
#pragma end_abs_address                        TCTL4 = (TCTL4&0xF3)|0x04;
void Init(void) {                            /* C1F set on rising edges */
  asm(" sei");        // make atomic           Count = 0;     // Set up for first
  TMSK1 |= 0x0C;     // Arm OC5 and IC1        Done = 0;      // Set on measurements
  TOC5 = TCNT+Rate; // First in 10 ms          TFLG1 = 0x22; // clear C5F, C1F
  TCTL2 = (TCTL2&0xCF)|0x10; // rising         asm cli
  Count = 0;          // Set up for first   }
  Done = 0;       // Set on measurements
  TFLG1 = 0x0C;   // clear OC5F, IC1F
  asm(" cli");
}
```

Program 6.15
C language frequency measurement.

6.4 Conversion Between Frequency and Period

6.4.1
Using Period Measurement to Calculate Frequency

Period and frequency are obviously related, so when faced with a problem that requires frequency information we could measure period and calculate frequency from the period measurement.

$$f = \frac{1}{p}$$

Assume we use the 16-bit period measurement interface described earlier in Section 6.1.4.1. In this system a global variable Period contains the measured period, with a range from 36 μs^2 to 32 ms and a resolution of 500 ns. This corresponds to a frequency range of 31 to 27,778 Hz. If we add another global variable, Freq, that will hold the calculated frequency in hertz, then a 32-bit by 16-bit divide (a 6811 subroutine or 6812 opcode) can be used to calculate

$$\text{Freq} = \frac{2000000}{\text{Period}}$$

---

[2]Each microcomputer has a separate lower bound on period; 36 μs was the fastest period that could be measured by the 2-MHz 6811.

It is easy to see how the 36-μs to 32-ms period range maps into the 31- to 27,778-Hz frequency range, but mapping the 500-ns period resolution into an equivalent frequency resolution is a little more tricky. If the frequency is f, then the frequency must change to f+Δf such that the period changes by at least Δp=500 ns. 1/f is the initial period, and 1/(f+Δf) is the new period. These two periods differ by 500 ns. In other words,

$$\Delta p = \frac{1}{f} - \frac{1}{f + \Delta f}$$

We can rearrange this equation to relate Δf as a function of Δp and f.

$$\Delta f = \frac{1}{(1/f) - \Delta p} - f$$

This very nonlinear relationship, shown in Table 6.17, illustrates that although the period resolution is fixed at 500 ns, the equivalent frequency resolution varies from 500 Hz to 0.0005 Hz. If the signal frequency is restricted to values below 1413 Hz, then we can say the frequency resolution will be better than 1 Hz.

Table 6.17
Relationship between frequency resolution and frequency when calculated using period measurement.

Frequency (Hz)	Period (μs)	Δf (Hz)
31,250	32	500
20,000	50	200
10,000	100	50
5,000	200	13
2,000	500	2
1,000	1,000	0.5
500	2,000	0.13
200	5,000	0.02
100	10,000	0.005
50	20,000	0.001
31.25	32,000	0.0005

6.4.2
Using Frequency Measurement to Calculate Period

Similarly, when faced with a problem that requires a period measurement, we could measure frequency and calculate period from the frequency measurement.

$$p = \frac{1}{f}$$

A similar nonlinear relationship exists between the frequency resolution and period resolution. In general, the period measurement approach will be faster, but the frequency measurement approach will be more robust in the face of missed edges or extra pulses. See Exercise 6.3 for a comparison between frequency and period measurement.

6.5 Measurements Using Both Input Capture and Output Compare

6.5.1
Period Measurement with Δp = 1 ms

The objective is to measure period with a resolution of 1 ms. The TTL-level input signal is connected to an input capture. Each rising edge will generate an input capture interrupt. In addition, output compare is used to increment a software counter, Time, every 1 ms. The period is calculated as the number of 1-ms output compare interrupts between one rising edge of the input capture pin to the other rising edge of the input capture pin. For example, if the period is 8192 μs, there will be eight output compare interrupts between successive input capture interrupts (Figure 6.27).

C language implementations for the 6811/6812 are presented. The period measurement counts the number of 1-ms intervals between successive rising edges. The period measurement resolution is 1 ms, because the period must increase by at least 1 ms for there to be a different number of counts. The range is 0 to 65 s, and the precision of 16 bits is determined by the size of the Time counter. If the gadfly loop in the ritual is removed, the first measurement will be inaccurate, but the subsequent measurements will be okay. There is an IC1 interrupt each period, and the external signal is connected to IC1. A real-time clock with a 1-ms rate is created with OC3. The foreground/background threads communicate via three shared globals. The background thread will update Period with a new measurement and set Done. When Done is set, the foreground thread will read the global Period and clear Done. OverFlow is set by the background and read by the foreground if the period is larger than 65 s. The ritual will wait for the first rising edge. In this way the first measurement will be correct. The only problem is that if no signal exists (the input capture pin is a constant level), this ritual software will hang at this wait.

```
unsigned short Period;    /* Period in msec */
unsigned char OverFlow;   /* Set if Period is too big*/
unsigned char Done;       /* Set each rising edge of IC1 */
```

There is a private global (only accessed by the background threads and not the foreground).

```
unsigned short Cnt;       /* number of msec in one period */
```

The output compare handler simply counts 1-ms time intervals. The input capture interrupt occurs on the external input. This ISR occurs when a new measurement is ready (Program 6.16).

```
// MC68HC711E9, ICC11 C
#define RESOLUTION 2000
#pragma interrupt_handler TOC3handler()
void TOC3handler(void){
  TOC3 = TOC3+RESOLUTION;  // every 1 ms
  TFLG1 = 0x20;        // ack, clear OC3F
  Cnt++;
  if(Cnt==0) OverFlow = 0xFF;
}
#pragma abs_address:0xffe4
void (*OC3_vector[])()={TOC3handler};
#pragma end_abs_address
#pragma interrupt_handler TIC1handler()
void TIC1handler(void){
  TFLG1 = 0x04;    // ack, clear IC1F
  if(OverFlow){
    Period = 65535; // greater than 65535
    OverFlow = 0;}
  else
    Period = Cnt;
  Cnt = 0;    // start next measurement
  Done = 0xFF;
}
#pragma abs_address:0xffee
void (*IC1_vector[])() ={TIC1handler};
```

```
// MC9S12C32, CodeWarrior C
#define RESOLUTION 2000
void interrupt 11 TC3handler(void){
  TFLG1 = 0x08;         // Acknowledge
  TC3 = TC3+RESOLUTION; // every 1 ms
  Cnt++;
  if(Cnt==0) OverFlow=0xFF;
}
void interrupt 9 TC1handler(void){
  TFLG1 = 0x02;   // ack, clear C1F
  if(OverFlow){
    Period = 65535; // greater than 65535
    OverFlow = 0;
  }
  else
    Period = Cnt;
  Cnt = 0;   // start next measurement
  Done = 0xFF;
}
void Ritual(void){
  asm sei        // make atomic
  TIOS |= 0x08;  // enable OC3
  TSCR1 = 0x80;  // enable
  TSCR2 = 0x01;  // 500 ns clock
  TFLG1 = 0x0A;  // Clear C3F,C1F
```

```
#pragma end_abs_address              TFLG1 = 0x0A;   // Clear C3F,C1F
void Ritual(void){                   TIE = 0x0A;     // Arm OC3 and IC1
  asm(" sei");  // make atomic        TCTL4 = (TCTL4&0xF3)|0x04;
  TFLG1 = 0x24;      // Clear OC3F,IC1F  /* C1F set on rising edges */
  TMSK1 = 0x24;      // Arm OC3 and IC1    while((TFLG1&0x02)==0); // wait rising
  TCTL2 = 0x10;      // rising edges       TFLG1 = 0x02;     // Clear C1F
  while((TFLG1&0x04)==0);               TC3 = TCNT+RESOLUTION;
// wait for first rising               Cnt=0; OverFlow=0; Done=0;
  TFLG1 = 0x04;      // Clear IC1F        asm cli
  TOC3 = TCNT+RESOLUTION;             }
  Cnt=0; OverFlow=0; Done=0;
  asm(" cli"); }
```

Program 6.16
C language implementation of period measurement.

6.5.2 Frequency Measurement with Δf = 0.1 Hz

If we count pulses in a 10-s time interval, then the number of pulses represents the signal frequency with units 1/10 s, or 0.1 Hz (e.g., if there are 12,345 pulses during the 10-s interval, then the frequency is 1234.5 Hz). For this system, the measurement resolution is 0.1 Hz. For example, the frequency would have to increase to 1234.6 Hz (or decrease to 1234.4 Hz) for the change (a different number of IC1 interrupts) to be reliably detected. OC5 interrupts every 25 ms, and it takes 400 OC5 interrupts to create the 10-s time delay. The number of IC1 interrupts in the 10-s interval represents the input frequency with units of 0.1 Hz (Figure 6.23).

Figure 6.23
Basic timing involved in frequency measurement.

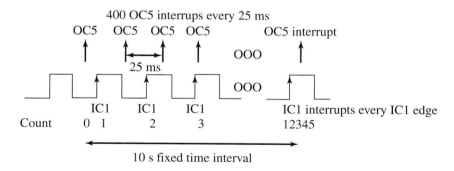

The highest frequency that can be measured will be determined by how fast the Input Capture interrupt handler can count pulses. We should select a counter precision (e.g., 8, 16, or 24 bits) so that the highest frequency does not overflow the counter. The TTL-level input signal is connected to IC1 (PA2 on the 6811 or PT1 on the 6812). Each rising edge will generate an input capture interrupt.

C language implementations for the 6811/6812 are presented. To simplify the implementation, we will assume the maximum frequency is 6.5 kHz so that a 16-bit counter can be used. The frequency measurement is the number of rising edges in a 10-s interval. A periodic interrupt is generated with OC5. The foreground/background threads communicate via two shared globals. The background thread will update Freq with a new measurement and set Done. When Done is set, the foreground thread will read the global Freq and clear Done. Just like the other frequency measurement system, a frequency of 0 will result in no input capture interrupts, and the system will properly report the frequency of 0.

```c
unsigned short Freq;   /* Frequency with units of 0.1 Hz */
unsigned char Done;    /* Set each measurement, every 10 s */
```

There are two private globals (only accessed by the background threads and not the foreground.)

```c
unsigned short FourHundred; // Used to create 10sec interval
unsigned short Count;       // Number of rising edges
```

The input capture interrupt counts the number of cycles, and the output compare interrupt is used to mark the end of the fixed time measurement (Program 6.17).

```c
// MC68HC711E9, ICC11 C
#define PERIOD 50000  // 25 ms
#pragma interrupt_handler TIC1handler()
void TIC1handler(void){
  Count++;         // number of rising edges
  TFLG1 = 0x04;} // ack, clear IC1F
#pragma abs_address:0xffee
void (*IC1_vector[])() ={TIC1handler};
#pragma end_abs_address
#pragma interrupt_handler TOC5handler()
void TOC5handler(void){
  TFLG1 = 0x08;        // Acknowledge
  TOC5 = TOC5+PERIOD; // every 25 ms
  if(++FourHundred==400){
    Freq = Count;      //  0.1 Hz units
    FourHundred=0;
    Done = 0xff;
    Count = 0; }}  // Setup for next
#pragma abs_address:0xffe0
void (*OC5_vector[])() ={TOC5handler};
#pragma end_abs_address
void Init(void) {
asm(" sei");       // make atomic
  TMSK1 |= 0x0C; // Arm OC5,IC1
  TCTL2 = (TCTL2&0xCF)|0x10; // rising
  Count = 0;       // Set up for first
  Done=0;          // Set on measurement
  FourHundred = 0;
  TOC5 = TCNT+PERIOD; // First in 25 ms
  TFLG1 = 0x0C;       // Clear OC5F,IC1F
asm(" cli"); }
```

```c
// MC9S12C32, CodeWarrior C
#define PERIOD 50000  //  25 ms
void interrupt 9 TC1handler(void){
  Count++;         // number of rising edges
  TFLG1 = 0x02;  // ack, clear C1F
}
void interrupt 13 TC5handler(void){
  TFLG1 = 0x20;        // Acknowledge
  TC5 = TC5+PERIOD;  // every 25 ms
  if (++FourHundred==400){
    Freq = Count;    //  0.1 Hz units
    FourHundred = 0;
    Done = 0xff;
    Count = 0;     // Setup for next
  }
}
void Init(void) {
asm sei          // make atomic
  TIOS| = 0x20;  // enable OC5
  TSCR1 = 0x80; // enable
  TSCR2 = 0x01;  // 500 ns clock
  TIE = 0x22;    // Arm OC5 and IC1
  TCTL4 = (TCTL4&0xF3)|0x04; // rising
  Count = 0;       // Set up for first
  Done = 0;        // Set on measurement
  FourHundred = 0;
  TC5 = TCNT+PERIOD; // First in 25 ms
  TFLG1 = 0x22;      // Clear C5F,C1F
asm cli
}
```

Program 6.17
C language implementation of frequency measurement.

6.6 Pulse Accumulator

6.6.1 MC68HC711E9 Pulse Accumulator Details

The 6811 pulse accumulator is an 8-bit read/write counter that can operate in either of two modes. External event counting mode can be used for counting events or frequency measurement. We will use gated time accumulation mode for pulse width measurement. In the event-counting mode, the 8-bit counter is incremented on either the rising edge or the falling edge of PA7. The maximum clocking rate for the external event-counting mode is the E clock frequency divided by two. In the gated time accumulation mode, a free-running E clock/64 signal increments the 8-bit counter, but only while the PA7 input is enabled. For a 2 MHz 6811, the period of this clock will be 32 μs. The I/O ports involved in the 6811 pulse accumulator are shown in Table 6.18.

Address	Bit 7	6	5	4	3	2	1	Bit 0	Name
$1000	**PA7**	PA6	PA5	PA4	PA3	PA2	PA1	PA0	PORTA
$1024	TOI	RTII	**PAOVI**	**PAII**	0	0	PR1	PR0	TMSK2
$1025	TOF	RTIF	**PAOVF**	**PAIF**	0	0	0	0	TFLG2
$1026	**DDRA7**	**PAEN**	**PAMOD**	**PEDGE**	DDRA3	I4/O5	RTR1	RTR0	PACTL
$1027	**Bit 7**	**6**	**5**	**4**	**3**	**2**	**1**	**Bit 0**	PACNT

Table 6.18
MC68HC711E9 I/O ports used by the pulse accumulator.

DDRA7 is the Data Direction bit for PA7. Normally, the DDRA7 bit is cleared so PA7 is an input, but even if it is configured for output, PA7 still drives the pulse accumulator. **PAEN** is the Pulse Accumulator System Enable bit. We set **PAEN** to one in order to activate the pulse accumulator. The PAMOD and PEDGE bits select the operation mode, as shown in Table 6.19.

PAMOD	PEDGE	Mode	Action on Clock	Sets PAIF
0	0	Event counting	PA7 falling edge increments PACNT	Falling edge PA7
0	1	Event counting	PA7 rising edge increments PACNT	Rising edge PA7
1	0	Gated time accumulation	Counts when PA7=1	Falling edge PA7
1	1	Gated time accumulation	Counts when PA7=0	Rising edge PA7

Table 6.19
6811 pulse accumulator operation modes.

The **PAOVF** status bit is set each time the pulse accumulator count rolls over from $FF to $00. To clear this status bit, we write a one to the TFLG2 register bit 5. The **PAOVI** will arm the device so that a pulse accumulator interrupt is requested when PAOVF is set. When **PAOVI** is zero, pulse accumulator overflow interrupts are disarmed. The **PAIF** status bit is automatically set each time a selected edge is detected at the PA7 pin (PAMOD=0 means falling edge, and PAMOD=1 means rising edge). To clear this status bit, write to the TFLG2 register bit 4. The **PAII** will arm the device so that a pulse accumulator interrupt is requested when PAIF is set. When PAII is zero, pulse accumulator input interrupts are disarmed.

6.6.2 MC9S12C32 Pulse Accumulator Details

The 6812 pulse accumulator is a 16-bit read/write counter that can operate in either of two modes. External event counting mode can be used for counting events or frequency measurement. We will use gated time accumulation mode for pulse width measurement. The I/O ports involved in the 6812 pulse accumulator are shown in Table 6.20.

Address	msb															lsb	Name
$0062	15	14	13	12	11	10	9	8	7	6	5	4	3	2	1	0	PACNT

Address	Bit 7	6	5	4	3	2	1	Bit 0	Name
$0046	**TEN**	TSWAI	TSBCK	TFFCA	0	0	0	0	TSCR1
$0060	0	**PAEN**	**PAMOD**	**PEDGE**	CLK1	CLK0	**PAOVI**	**PAI**	PACTL
$0061	0	0	0	0	0	0	**PAOVF**	**PAIF**	PAFLG
$0240	**PT7**	PT6	PT5	PT4	PT3	PT2	PT1	PT0	PTT
$0242	**DDRT7**	6	5	4	3	2	1	Bit 0	DDRT

Table 6.20
MC9S12C32 I/O ports used by the pulse accumulator.

DDRT7 is the Data Direction bit for PT7. Normally, the DDRT7 bit is cleared so that PT7 is an input, but even if it is configured for output, PT7 still drives the pulse accumulator. **PAEN** is the Pulse Accumulator System Enable bit. We set PAEN to one in order to activate the pulse accumulator. The PAMOD and PEDGE bits select the operation mode, as shown in Table 6.21.

PAMOD	PEDGE	Mode	Action on Clock	Sets PAIF
0	0	Event counting	PT7 falling edge increments PACNT	Falling edge PT7
0	1	Event counting	PT7 rising edge increments PACNT	Rising edge PT7
1	0	Gated time accumulation	Counts when PT7=1	Falling edge PT7
1	1	Gated time accumulation	Counts when PT7=0	Rising edge PT7

Table 6.21
6812 pulse accumulator operation modes.

In the event counting mode, the 16-bit counter (**PACNT**) is incremented on either the rising edge or the falling edge of PT7. The maximum clocking rate for the external event counting mode is the E clock frequency divided by two. Event counting mode does not require the timer to be enabled.

In the gated time accumulation mode, a free-running clock (E clock divided by 64) increments the 16-bit counter. In this mode, the E clock divided by 64 increments PACNT while the PT7 input is active. Gated accumulation mode does require the TEN in the TSCR1 register to be set.

The **PAOVF** status bit is set each time the pulse accumulator count rolls over from $FFFF to $0000. To clear this status bit, we write a one to the PAFLG register bit 1. The **PAOVI** will arm the device so that a pulse accumulator interrupt is requested when PAOVF is set. When **PAOVI** is zero, pulse accumulator overflow interrupts are disarmed. The **PAIF** status bit is automatically set each time a selected edge is detected at the PT7 pin (PAMOD=0 means falling edge, and PAMOD=1 means rising edge). To clear this status bit, write to the PAFLG register bit 1. The **PAII** will arm the device so that a pulse accumulator interrupt is requested when PAIF is set. When PAII is zero, pulse accumulator input interrupts are disarmed.

> **Observation:** The PACNT input and timer channel 7 use the same pin PT7. To use the pulse accumulator, disconnect PT7 from the output compare logic by clearing bits, OM7 and OL7. Also clear the channel 7 output compare 7 mask bit, OC7M7.

6.6.3 Frequency Measurement

The goal is to measure frequency in Hz, which in this case is defined as the number of falling edges that occur in one second. The signal to be measured will be connected to the pulse accumulator input, which is PA7 on the 6811, and PT7 on the 6812. The frequency measurement function, shown in Program 6.18, enables the pulse accumulator and selects event counting mode. When measuring frequency it usually doesn't matter whether we count rising or falling edges. However, in this case, falling edges will be counted. The approach will be to initialize the pulse accumulator to event counting, clear the count, wait one second, then read the counter. Since frequency is defined as the number of edges in one second, the value in the PACNT after the one-second time delay will be frequency in Hz. The 6811 frequency measurement range is 0 to 255 Hz, and the 6812 can measure 0 to 65535 Hz. In both cases, the frequency resolution (which is the smallest change in frequency that can be distinguished) will be 1 Hz. In general, the frequency resolution will be one divided by the fixed time during which counts are measured. The PAOVF bit will be set if the input frequency exceeds the measurement range.

```
;6811 measures 0 to 255 Hz              ;6812 measures 0 to 65535 Hz
;returns Reg A = freq in Hz             ;returns Reg D = freq in Hz
Freq ldaa #$40  ;PA7 is input, PEN=1    Freq bclr DDRT,#$80  ;PT7 is input
     staa PACTL ;count falling edges         movb #$40,PACTL ;count falling
     clr  PACNT                              movw #0,PACNT
     ldaa #$20                               movb #$02,PAFLG ;clear PAOVF
     staa TFLG2 ;clear PAOVF                 ldy  #100
     ldy  #100                               bsr  Timer_Wait10ms ;Program 2.6
     bsr  Timer_Wait10ms ;Program 2.6        brclr PAFLG,#$02,ok ;check PAOVF
     ldaa TFLG2                         bad  ldd  #65535          ;too big
     bita #$20  ;check PAOVF                 bra  out
     beq  ok                           ok   ldd  PACNT  ;units in Hz
bad  ldaa #255  ;too big               out  rts
     bra  out
ok   ldaa PACNT ;units in Hz
out  rts
```

Program 6.18
Frequency measurement using the pulse accumulator.

Checkpoint 6.19: What is the frequency resolution of the system implemented in Program 6.18? What does it mean?

Checkpoint 6.20: How do you modify Program 6.18 so that it measures frequency with a resolution of 1 kHz?

6.6.4
Pulse-Width
Measurement

The goal is to measure pulse width, which in this case will be defined as the time the input signal is high. Again, the input signal will be connected to the pulse accumulator input, which is PA7 on the 6811, and PT7 on the 6812. The pulse-width measurement function, shown in Program 6.19, enables the pulse accumulator and selects gated accumulation mode. In this case, PEDGE is set to zero, so the PACNT will accumulate when the input is high. With PEDGE equal to zero, the PAIF will be set on the falling edge of the input, signaling that the pulse-width measurement is complete. The approach will be to initialize the pulse accumulator to gated accumulation mode, clear the count, wait for PAIF to be set, and then read the counter. Since PACNT counts while the input is high, the value in this counter will represent the width of the pulse. The pulse-width resolution is the smallest change in pulse width that can be distinguished. In general, the pulse-width resolution will be the period of the free-running clock used to increment the counter. For the 6811, the pulse-width resolution will be 32 μs. Assuming the 6812 E clock is the default value of 4 MHz, its pulse-width resolution will be 16 μs. The 6811 pulse-width measurement range is 32 μs to 8.16 ms, whereas the 6812 can measure 16 μs to 1.05 s. The PAOVF bit will be set if the input pulse width exceeds the measurement range. If the input signal has a pulse width of 1 ms, then the 6811 function will return a value of 1000/32 or 31, whereas the 6812 function will give a result of 1000/16 or 62.

```
;6811 measures 32us to 8.16ms          ;6812 measures 16us to 1.05s
;returns Reg A = pulse width in 32us    ;returns Reg D = pulse width in 16us
Puls ldaa #$60   ;PA7 is input, PEN=1   Puls bclr DDRT,#$80  ;PT7 is input
     staa PACTL ;measure when high           movb #$60,PACTL ;measure high
     clr  PACNT                               movw #0,PACNT
     ldaa #$20                                movb #$02,PAFLG ;clear PAOVF
     staa TFLG2 ;clear PAOVF            loop brclr PAFLG,#$01,loop
loop ldaa TFLG2                              brclr PAFLG,#$02,ok ;check PAOVF
     bita #$10  ;check PAIF            bad  ldd  #65535          ;too big
     beq  loop  ;wait for falling edge      bra  out
     bita #$20  ;check PAOVF           ok   ldd  PACNT  ;units in 16us
     beq  ok                          out  rts
     ldaa #255  ;bad measurement
     bra  out
ok   ldaa PACNT ;units in 32us
out  rts
```

Program 6.19
Pulse-width measurement using the pulse accumulator.

Checkpoint 6.21: What is the pulse-width resolution of the system implemented in Program 6.19? What does it mean?

Checkpoint 6.22: What will be the output of Program 6.19 if the pulse width is 1234.5 μsec?

6.7 Pulse-Width Modulation on the MC9S12C32

Earlier in the chapter we created a pulse-width modulated output using output compare. The disadvantages of that approach include the software overhead to service the periodic interrupts and its inability to create 0% and 100% signals. Appreciating the importance of pulse-width modulation, Freescale added dedicated hardware to handle PWM, not previously available in the 6811. Table 6.22 shows the MC9S12C32 registers used to create pulse-width modulated outputs. Each of three PWM devices can be configured either as two 8-bit channels or one 16-bit channel. In particular, each of the 16-bit registers in Table 6.22 could be considered as two separate 8-bit registers. For example, the 16-bit register **PWMPER01** could be considered as the two 8-bit registers **PWMPER0** (at address $00F2) and **PWMPER1** (at address $00F3). Some versions of the MC9S12C32 do not have all the Port P pins available, as described in Table 1.17 back in Chapter 1. The PWM channel 5 always uses output PP5 (which is available on all MC9S12C32 packages), but the other channels can be connected to either Port P or Port T. If a bit in the **MODRR** register is 1, the corresponding Port T pin is connected to the PWM system. If the bit is 0, the Port P pin is connected to the PWM and the corresponding Port T pin is connected to the timer system.

Address	msb															lsb	Name
$00F2	15	14	13	12	11	10	9	8	7	6	5	4	3	2	1	0	PWMPER01
$00F4	15	14	13	12	11	10	9	8	7	6	5	4	3	2	1	0	PWMPER23
$00F6	15	14	13	12	11	10	9	8	7	6	5	4	3	2	1	0	PWMPER45
$00F8	15	14	13	12	11	10	9	8	7	6	5	4	3	2	1	0	PWMDTY01
$00FA	15	14	13	12	11	10	9	8	7	6	5	4	3	2	1	0	PWMDTY23
$00FC	15	14	13	12	11	10	9	8	7	6	5	4	3	2	1	0	PWMDTY45

Address	Bit 7	6	5	4	3	2	1	Bit 0	Name
$0247	0	0	0	MODRR4	MODRR3	MODRR2	MODRR1	MODRR0	MODRR
$00E0	0	0	PWME5	PWME4	PWME3	PWME2	PWME1	PWME0	PWME
$00E1	0	0	PPOL5	PPOL4	PPOL3	PPOL2	PPOL1	PPOL0	PWMPOL
$00E2	0	0	PCLK5	PCLK4	PCLK3	PCLK2	PCLK1	PCLK0	PWMCLK
$00E3	0	PCKB2	PCKB1	PCKB0	0	PCKA2	PCKA1	PCKA0	PWMPRCLK
$00E4	0	0	CAE5	CAE4	CAE3	CAE2	CAE1	CAE0	PWMCAE
$00E5	0	CON45	CON23	CON01	PSWAI	PFRZ	0	0	PWMCTL
$00E8	Bit 7	6	5	4	3	2	1	Bit 0	PWMSCLA
$00E9	Bit 7	6	5	4	3	2	1	Bit 0	PWMSCLB

Table 6.22
MC9S12C32 registers used to configure pulse-width modulated outputs.

The **PWME** register allows you to enable/disable individual PWM channels. The **PWMCTL** register is used to concatenate two 8-bit channels into one 16-bit PWM. For example, if the **CON23** is 1, then channels 2 and 3 become one 16-bit channel with the output generated on PP3 (if MODRR3=0) or PT3 (if MODRR3=1). Concatenated channels are controlled using the higher of the two channels. For example, concatenated channel 23 is enabled with bit PWME3. The **PWMPOL** register specifies the polarity of the output. Figure 6.24 shows a PWM output for the case when the $PPOL_x$ bit is 1. The output will be high for the number of counts in the **PWMDTY** register. The **PWMPER** register contains the number of counts in one complete cycle. The duty cycle is defined as the fraction of time the signal is high, calculated as a percent, depending on PWMPER and PWMDTY.

$$DutyCycle = 100\% * PWMDTY_x / PWMPER_x$$

Figure 6.24
PWM output generated when PPOL=1.

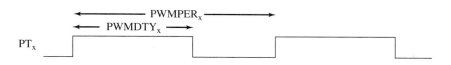

If the **PPOL**$_x$ bit is 0, the output will be low for the number of counts in the **PWMDTY** register, as illustrated in Figure 6.25. The duty cycle, defined as a fraction of time the signal is high, is

$$DutyCycle = 100\% * (PWMPER_x - PWMDTY_x)/ PWMPER_x$$

Figure 6.25
PWM output generated when PPOL=0.

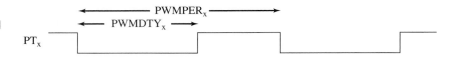

There are many possible choices for the clock. The base clock is derived from the E clock. Activating the PLL affects the E clock and hence will affect the PWM generation. Channels 0, 1, 4, and 5 use either clock A or clock SA. Channels 2 and 3 use either clock B or clock SB. The six bits in the **PWMPRCLK** register, as shown in Table 6.23, determine the relationship between clocks A,B and the E clock.

Table 6.23
Clock A and Clock B prescale in PWMCLK.

PCKB2	PCKB1	PCKB0	Clock B	PCKA2	PCKA1	PCKA0	Clock A
0	0	0	E	0	0	0	E
0	0	1	E/2	0	0	1	E/2
0	1	0	E/4	0	1	0	E/4
0	1	1	E/8	0	1	1	E/8
1	0	0	E/16	1	0	0	E/16
1	0	1	E/32	1	0	1	E/32
1	1	0	E/64	1	1	0	E/64
1	1	1	E/128	1	1	1	E/128

It is possible to divide the A and B clocks further using the **PWMSCLA** and **PWMSCLB** registers. The period of the SA clock is the period of the A clock divided by two times the value in the PWMSCLA register. Similarly, the period of the SB clock is the period of the B clock divided by two times the value in the PWMSCLB register. If the value in PWMSCLA(B) is 0, then a divide by 512 is selected. The clock used for each channel is determined by the **PWMCLK** register. The period of the PWM output is the period of the selected clock times the value in the **PWMPER** register.

PCLK5	=1	Clock SA is the clock source for PWM channel 5
	=0	Clock A is the clock source for PWM channel 5
PCLK4	=1	Clock SA is the clock source for PWM channel 4
	=0	Clock A is the clock source for PWM channel 4
PCLK3	=1	Clock SB is the clock source for PWM channel 3
	=0	Clock B is the clock source for PWM channel 3
PCLK2	=1	Clock SB is the clock source for PWM channel 2
	=0	Clock B is the clock source for PWM channel 2
PCLK1	=1	Clock SA is the clock source for PWM channel 1
	=0	Clock A is the clock source for PWM channel 1
PCLK0	=1	Clock SA is the clock source for PWM channel 0
	=0	Clock A is the clock source for PWM channel 0

Let **n** be the 3-bit value for PCKA2-0 in the PWMCLK register. Let the E clock period be **Period**$_E$. Then if the A clock is selected for channel x, the periods of the A clock and PWM output will be

$$Period_A = 2^n * Period_E$$
$$Period_{PTx} = 2^n * PWMPER_x * Period_E$$

If the SA clock is selected for channel x, the periods of the SA clock and PWM output will be

$$Period_{SA} = 2^n * 2 * PWMSCLA * Period_E$$
or $\quad Period_{SA} = 2^n * 512 * Period_E$ (if PWMSCLA equals 0)

$$Period_{PTx} = 2^n * 2* PWMSCLA * PWMPER_x * Period_E$$
or $\quad Period_{PTx} = 2^n * 512 * PWMPER_x * Period_E$ (if PWMSCLA equals 0)

The design of a PWM system considers three factors. The first factor is period of the PWM output. Most applications choose a period, initialize the waveform at that period, and adjust the duty cycle dynamically. The second factor is precision, which is the total number of duty cycles that can be created. An 8-bit PWM channel may have up to 256 different outputs, while a 16-bit channel can potentially create up to 65536 different duty cycles. More specifically, since the duty cycle register must be less than or equal to the period register (e.g., $PWMDTY_x \leq PWMPER_x$), the precision of the system will equal $PWMPER_x + 1$ in alternatives. The final consideration is the number of channels. The MC9S12C32 supports up to six 8-bit channels or three 16-bit channels. It is possible to mix and match, creating, for example, four 8-bit channels and one 16-bit channel. Different versions of the 6812 will have from 0 to 8 channels of PWM.

In this first design example, we will create a 10ms period 8-bit PWM output using channel 0 generated on the PT0 output. In order to maximize precision, it is best to create the 10 ms period using as large a value in PWMPER0 as possible. We have the limitation that the prescale and PWMPER0 factors will be integers. Since 10 ms/256 equals 39.0625 μs, we need a clock just larger than 39 μs. The fastest clock that can be used is 40 μs, resulting in PWMPER0 equal to 250. Assuming the E clock period is 250 ns, the prescale needs to be 40/0.25 or 160. There are a number of ways to make this happen, but one way is to select Clock A to be E/16, create SA=A/10, then select the SA clock for channel 0, as shown in Program 6.20.

```
;MC9S12C32 assembly                    // MC9S12C32 C
PWM_Init0          ;10ms PWM on PT0    // 10ms PWM on PT0
   bset MODRR,#$01  ;PT0 with PWM      void PWM_Init(void){
   bset PWME,#$01   ;enable chan 0       MODRR  |= 0x01; // PT0 with PWM
   bset PWMPOL,#$01 ;high then low       PWME   |= 0x01; // enable channel 0
   bset PWMCLK,#$01 ;Clock SA            PWMPOL |= 0x01; // PT0 high then low
   ldaa PWMPRCLK                         PWMCLK |= 0x01; // Clock SA
   anda #$F8                             PWMPRCLK = (PWMPRCLK&0xF8)|0x04; // A=E/16
   oraa #$04                             PWMSCLA = 5;     // SA=A/10, 0.25*160=40us
   staa PWMPRCLK     ;A=E/16             PWMPER0 = 250;   // 10ms period
   movb #5,PWMSCLA   ;SA=A/10            PWMDTY0 = 0;     // initially off
   movb #250,PWMPER0 ;10ms period      }
   clr  PWMDTY0      ;initially off    // Set the duty cycle on PT0 output
   rts                                 void PWM_Duty0(unsigned char duty){
PWM_Duty0          ;RegA is duty cycle   PWMDTY0 = duty;  // 0 to 250
   staa PWMDTY0     ;0 to 250          }
   rts
```

Program 6.20
Implementation of an 8-bit PWM output.

Checkpoint 6.23: State another way to create a prescale of 160 on channel 0.

Checkpoint 6.24: How would you modify Program 6.20 to have a 100ms period?

In this second design example, we will create a 1s period 16-bit PWM output using concatenated channel 23 generated on the PT3 output. In order to maximize precision, it is best to create the 1s period using as large a value in PWMPER23 as possible. Since 1s/65536 equals 15.2587890625 μs, we need a clock just larger than 15 μs. The fastest clock that can be used is 16 μs, resulting in PWMPER23 equal to 62500. Assuming the E clock period is 250 ns, the prescale needs to be 16/0.25 or 64. There are a number of ways to make this happen, but one way is to make Clock B to be E/64, then select the B clock for channel 23, as shown in Program 6.21.

```
;MC9S12C32 assembly                          // MC9S12C32 C
PWM_Init3           ;1s PWM on PT3            // 1s PWM on PT3
    bset MODRR,#$08  ;PT0 with PWM            void PWM_Init(void){
    bset PWME,#$08   ;enable chan 3             MODRR  |= 0x08; // PT3 with PWM
    bset PWMPOL,#$08 ;high then low            PWME   |= 0x08; // enable channel 3
    bclr PWMCLK,#$08 ;Clock B                  PWMPOL |= 0x08; // PT3 high then low
    bset PWMCTL,#$20 ;concat 2+3               PWMCLK &=~0x08; // Clock B
    ldaa PWMPRCLK                              PWMCTL |= 0x20; // Concatenate 2+3
    anda #$8F                                  PWMPRCLK = (PWMPRCLK&0x8F)|0x60; // B=E/64
    oraa #$60                                  PWMPER23 = 62500; // 1s period
    staa PWMPRCLK          ;B=E/64             PWMDTY23 = 0;     // initially off
    movw #62500,PWMPER23 ;1s period          }
    movw #0,PWMDTY23      ;off                // Set the duty cycle on PT3 output
    rts                                       void PWM_Duty(unsigned short duty){
PWM_Duty3           ;RegD is duty cycle        PWMDTY23 = duty;  // 0 to 62500
    std  PWMDTY0  ;0 to 62500                }
    rts
```

Program 6.21
Implementation of an 8-bit PWM output.

Checkpoint 6.25: What would be the effect of creating the 1s output using a 1ms SB clock and a PWMPER23 value of 1000?

Checkpoint 6.26: Are programs 6.20 and 6.21 friendly enough to be used together?

6.8   Exercises

6.1 The objective of this problem is to measure the frequency of a square wave connected to IC1. You may only use IC1 and OC5 (i.e., no other 6811/6812 I/O feature or external device is allowed). The frequency range is 0 to 200 Hz, and the *resolution is 0.01 Hz*. For example, if the frequency is 56.783 Hz, then your software will set the global `Freq` to 5678. The C program in Section 6.5.2 measures frequency with units of 0.1 Hz. Make modifications to this program so that the resolution is improved to 0.01 Hz. Don't worry about frequencies above 200 Hz.

6.2 When a debugger wishes to single-step a program that exists in RAM, it can replace opcodes one at a time with SWI (it will have to replace two opcodes when single-stepping a conditional branch). This approach is not feasible when testing software stored in ROM or PROM. In this exercise you will use output compare interrupts to implement single-step debugging. Your approach will work for programs in RAM or ROM.

a) Write a debugging function that initializes the OC5 interrupt then calls the `UserRoutine`. The first OC5 interrupt should occur after exactly one instruction of the `UserRoutine` has been executed. You may assume the `UserRoutine` has no I/O parameters. You may also assume that `UserRoutine` has no interrupts of its own and it does not disable interrupts. When the `UserRoutine` returns back to your function, you should shut off OC5. You can start with the following syntax and add the OC5 code. You may write your answer in assembly or C.

```
// Single step in C              * Single Step in assembly
void debug(void (*UserRoutine)(void)){  debug   * X points to UserRoutine
// add stuff here to initialize OC5     * stuff here to initialize OC5
   (*UserRoutine)();                          jsr 0,X   call UserRoutine
// add stuff here to stop OC5          * add stuff here stop OC5
}                                      rts
```

You are given two functions that you will call but do not need to write.

```
char GetChar(void); // waits for keyboard input and returns the ASCII code
void Display(void); // displays debugging information like registers
```

> **b)** Write the OC5 interrupt handler that calls `Display` then `GetChar`. In a real debugger we would process the keyboard input and interact with the user, but in this simple solution the keyboard input is used only to pause. Before returning from interrupt you should set up OC5 so that the 6811/6812 will execute exactly one more instruction of `UserRoutine` before another OC5 interrupt is generated.

6.3 A microcomputer-based PID controller requires a frequency measurement to be performed every 20 ms. This problem will investigate three techniques:

> **A.** Direct frequency measurement
> **B.** Period measurement
> **C.** A combined frequency/period measurement

The frequency range is 100 to 1000 Hz, and we wish to obtain the best frequency resolution possible under the constraint that a new frequency measurement must be available every 20 ms for the PID controller. Each method has two interrupt handlers:

> **1.** A 20-ms real time clock using OC5
> **2.** Falling edge of the square wave using 6811 STRA, or 6812 PJ0 key wake up.

There are five 16-bit unsigned global variables used by the three techniques:

> FREQ is the frequency in hertz calculated during the OC5 handler   used by A, B, C
> CNT is the number of falling edges in the current 20-ms interval   used by A, C
> FIRST is the TCNT value at the time of the first falling edge   used by C
> PREVIOUS is the TCNT value at the time of the previous falling edge   used by B
> LAST is the TCNT value at the time of the last falling edge   used by B, C

A. Direct frequency measurement   The ritual performs:

```
CNT=0;
```

The 6811 STRA or 6812 key wake-up interrupt handler performs:

```
// 6811 code                        // 6812 code
   Read PIOC,Read Port CL /* Ack */    KWIFJ=0×01; // ack
   CNT=CNT+1;                          CNT=CNT+1;
```

The OC5 interrupt handler performs:

```
Write OC5F=1;                       /* Acknowledge OC5 */
TOC5=TOC5+40000;
Calculates FREQ=50·CNT;
CNT=0;
```

The frequency resolution is 50 Hz regardless of the square-wave frequency.

B. Period measurement   The ritual performs:

```
PREVIOUS =LAST =0;
```

The 6811 STRA or 6812 key wake-up interrupt handler performs:

// 6811 code	// 6812 code
Read PIOC,Read Port CL /* Ack */	KWIFJ=0X01; // ack
PREVIOUS=LAST	PREVIOUS=LAST
LAST=TCNT;	LAST=TCNT;

The OC5 interrupt handler performs:

```
Write OC5F=1;                            /* Acknowledge OC5 */
TOC5=TOC5+40000;
Calculates FREQ=????(answered in part a)
PREVIOUS =LAST =0;
```

C. Combined frequency/period measurement   The ritual performs:

```
FIRST=LAST=0;
CNT=-1;
```

The 6811 STRA or 6812 key wake-up interrupt handler performs:

// 6811 code	// 6812 code
Read PIOC,Read Port CL /* Ack */	KWIFJ=0X01; // ack
If (CNT==-1) FIRST=TCNT;	If (CNT==-1) FIRST=TCNT;
else LAST=TCNT;	else LAST=TCNT;
CNT=CNT+1;	CNT=CNT+1;

The OC5 interrupt handler performs:

```
Write OC5F=1;   /* Acknowledge OC5 */
TOC5=TOC5+40000
Calculates FREQ=????(answered in part d)
FIRST=LAST=0;
CNT=-1;
```

The hardware for each method is shown in Figure 6.26.

Figure 6.26
Interface for
Exercise 6.3.

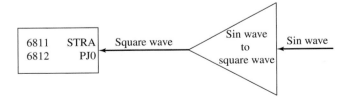

a) Specify the equation used to calculate FREQ from PREVIOUS and LAST in the **Period measurement** OC5 handler (no software is required).
b) What is the frequency resolution with units when the square-wave frequency is 100 Hz?
c) What is the frequency resolution with units when the square-wave frequency is 1000 Hz?
d) Specify the equation used to calculate FREQ from FIRST LAST and CNT in the **Combined frequency/period measurement** OC5 handler (no software is required).
e) What is the frequency resolution with units when the square-wave frequency is 100 Hz?
f) What is the frequency resolution with units when the square-wave frequency is 1000 Hz?

6.4 The objective of this problem is to measure body temperature using input capture (pulse-width measurement) (Figure 6.27). A shunt resistor is placed in parallel with a thermistor. The thermistor-shunt combination R has the following linear relationship for temperatures from 90 to 110°F.

$$R = 100k\Omega - (T - 90°F) \cdot 1k\Omega/°F \quad \text{where R is the resistance of the thermistor-shunt}$$

Figure 6.27
Interface for
Exercise 6.4.

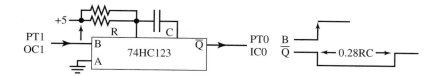

In other words, the resistance varies from 100kΩ to 80kΩ as the temperature varies from 90 to 110°F. The range of your system is 90 to 110°F and the resolution should be better than 0.01°F. You will use a 74HC123. On the rise of the B trigger input, the 123 creates a negative logic pulse on its output. The width of the pulse is about 0.28·R·C. Temperature will be measured 100 times a second. For the 6811 connect to PA3/OC5 and PA2/IC1.

a) The pulse-width measurement resolution will be 500 ns for the 6811 and 125 ns for the 6812. Choose the capacitor value so that this pulse width measurement resolution matches the desired temperature resolution of 0.01°F.

b) Given this value of C, what is the pulse width at 90°F? Give the answer in both microseconds and E clock cycles.

c) Given this value of C, what is the pulse width at 110°F? Give the answer in both microseconds and E clock cycles.

d) Write the ritual that configures an output compare on PT1 as a periodic interrupt. A rising edge should occur every 10 ms. Configure an input capture on PT0 to measure the pulse width.

e) Show the interrupt handler(s) that perform the temperature measurement tasks in the background and sets a global, `Temperature`, 100 times a second. The units of this unsigned decimal fixed point number are 0.01°F. For example, `Temperature` will vary from 9000 to 11000 as temperature varies from 90 to 110°F. The output compare interrupt will start the pulse (You may assume the interrupt vectors are properly established, but please use simple interrupt handler names like `TOC1handler()` and `TIC0handler()`.

6.5 Design a wind direction measurement instrument using the input capture technique. Again, you are given a transducer that has a resistance that is linearly related to the wind direction. As the wind direction varies from 0 to 360 degrees, the transducer resistance varies from 0 to 1000Ω. The frequencies of interest are 0 to 0.5 Hz, and the sampling rate will be 1 Hz. One way to interface the transducer to the computer is to use an astable multivibrator like the 555. The period of a 555 timer is $0.693 \cdot C_T \cdot (R_A + 2R_B)$. The 555 output could be connected to the Input Capture Port channel 7. (See Exercise 12.10.)

a) Show the hardware interface.

b) Write the ritual and gadfly function/subroutine that measures the wind direction and returns a 16-bit unsigned result with units of degrees (i.e., the value varies from 0 to 359). (You do not have to write software that samples at 1 Hz, simply a function that measures wind direction once.)

6.6 The objective of this problem is to design an underwater ultrasonic ranging system. The distance to the object, **d**, can vary from 1 to 100 m. The ultrasonic transducer will send a short 5-μs sound pulse into the water in the direction of interest. The sound wave will travel at 1500 m/s and reflect off the first object it runs into. The reflected wave will also travel at 1500 m/s back to the transducer. The reflected pulse is sensed by the same transducer. Your system will trigger the electronics (give a 5-μs digital pulse), measure the time of flight, then calculate the distance to the object. Using periodic interrupts, the software will issue a 5-μs pulse out PT1 once a second. Using interrupting input capture, the software will measure the time of flight, Δt. The input capture interrupt handler will calculate distance **d** as a decimal fixed-point value with units of 0.01 m and enter it into a FIFO queue. The main program will call the ritual, then get data out of the FIFO queue. The main program will call `Alarm()` if the distance is less than 15 m. You do not have to give the implementation of `Alarm()`. You may use any of the FIFOs in Chapter 4 without showing its implementation. To solve this problem on other microcomputers, you may modify the hardware in Figure 6.28 to conform to the available ports. As shown in the figure the time of flight, Δt, is measured from the rise of PT1 (6812 output) to the rise of PT0 (6812 input).

Figure 6.28
Interface for Exercise 6.6.

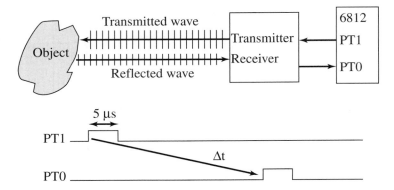

a) Derive an equation that relates the distance **d** to the time of flight **Δt**.

b) Use this equation to calculate the minimum and maximum possible time of flight **Δt**.

c) Choose the TCNT rate that will satisfy the range and resolution requirements of this problem.

d) Give the ritual that initializes the interface, including PORTT timer channels 0 and 1. Don't worry about the other six timer channels.

e) Give the main program that first calls the ritual. The main program will empty the FIFO and call `Alarm()` if the distance drops below 15 m. Decide whether it is better to convert time (TCNT counts) to distance (decimal fixed point, 0.01m) here in the main program or in the interrupt handler.

f) Give the `TC1handler()` periodic interrupt handler that issues pulses of about 5-μs duration on PT1 every 1 sec. Good interrupt software has no backward jumps.

g) Give the `TC0handler()` interrupt handler that measures the time of flight and puts the result (either the count or the calculated distance) into the FIFO. Good interrupt software has no backward jumps. There may be additional sonic echoes, so only calculate the range of the first one and ignore the others.

6.7 The objective of this exercise is to interface a silicon-controlled rectifier (SCR) using input capture and/or output compare. A 120-Hz digital logic waveform (**Sync**) is available that is synchronized to the zero crossings of the 60-Hz alternating current wave. A 100-μs pulse on the digital logic **Control** signal will turn on the SCR. The SCR will automatically shut off on the next zero crossing of the 60-Hz wave. **Sync** will be an input to the microcomputer, and **Control** will an output (Figure 6.29).

The software controls the amount of delivered power by adjusting the time **T**, which is the delay from the rising edge of the **Sync** input to the rising edge of **Control** output. To solve this problem on other microcomputers, you may modify the hardware in Figure 6.30 to conform to the available ports. On the 6811 Sync is connected to IC2/PA1 and Control is connected to OC2/PA6.

The delay **T** will vary from 300 to 8000 μs. When **T** is 300 μs, full power is being delivered. When **T** is 8000 μs, almost no power is delivered to the load. A 16-bit unsigned global variable

```
unsigned short T; // delay 2400 to 64000 in 125 ns clock cycles
```

will be set by the main program (which you will not write) and read by the interrupt software (which you will write).

Figure 6.29
Timing for Exercise 6.7.

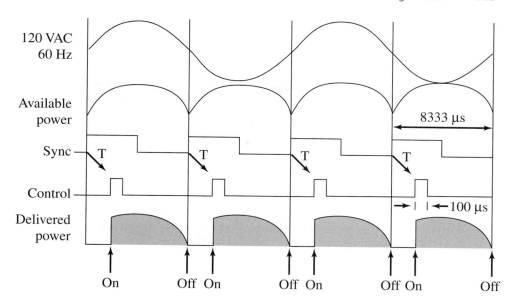

Figure 6.30
Interface for
Exercise 6.7.

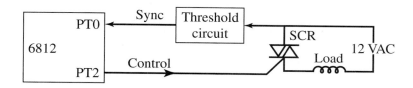

a) Give the ritual which initializes the interface. You may use any available feature of your microcomputer.

b) Give the `TC2handler()` and `TC0handler()` interrupt handlers that implement this interface. Good interrupt software has no backward jumps.

c) This is indeed an example of a real-time system. Give the upper bound on the latency of the `TC0handler()`. Don't calculate what it actually is but rather determine theoretically how fast it must be to work properly.

6.9 **Lab Assignments**

Lab 6.1. The overall objective is to interface a joystick to the microcontroller. The joystick is made with two potentiometers. You can use two astable multivibrators (LM555) to convert the two resistances into two periods. Use two period measurement channels to estimate the X–Y position of the joystick. Organize the software interface into a device driver, and write a main program to test the interface.

Lab 6.2. The overall objective is to measure temperature. A thermistor is a transducer with a resistance that is a function of its temperature. You can use an astable multivibrator (LM555) to convert the resistance into a period. Use a period measurement channel to estimate the resistance of the thermistor. Use a table lookup with linear interpolation to convert resistance to temperature. Organize the software interface into a device driver, and write a main program that outputs temperature to the SCI channel.

Lab 6.3. The overall objective is to measure linear position. A slide-pot is a transducer with a resistance that is a function of the linear position of the slide. You can use an astable multivibrator (LM555) to convert the resistance into a period. Use a period measurement channel to estimate the resistance of the potentiometer. Use a table lookup with linear interpolation to convert resistance to position. Organize the software interface into a device driver, and write a main program that outputs position to the SCI channel.

Lab 6.4. The objective of this lab is to control a servo motor (see Figure 6.31). Interface a servo motor to the PWM output pin of the microcontroller. The desired angle is input from the SCI channel (connected to a PC running HyperTerminal.) Organize the software interface into a device driver, and write a main program that inputs from the SCI channel and maintains the PWM output to the servo. Servos are a popular mechanism to implement steering in robotics. Ranging from micro servos with 15 oz-in torque to powerful heavy-duty sailboat servos, they all share several common characteristics. Servos allow you to control the position of the shaft, or **horn**. The servo senses where it is (the actual position). You specify where it should be (the desired position). When the servo receives a position, it attempts to move the servo horn to the desired position. The task of the servo, then, is to make the actual position the desired position. The first step to understanding how servos work is to understand how to control them. Power is usually between 4.8V and 6V and should be separate from system power (as servos are electrically noisy). Even small servos can draw over an amp under heavy load, so the power supply should be appropriately rated. Servos are commanded through "pulse-width modulation" (PWM), signals sent through the command wire. Essentially, the width of a pulse defines the position. For example, sending a 1.5 ms pulse to the servo, tells the servo that the desired position is at the midpoint, as shown in Figure 6.32. In order for the servo to hold this position, the command must be sent at about 50 Hz, or every 20 ms. If you were to send a pulse longer than 2.1 ms or shorter than 0.9 ms, the servo would attempt to overdrive (and possibly damage) itself. Once the servo has received the desired position (via the PWM signal) the servo must attempt to match the desired and actual positions. It does this by turning a small, geared motor left or right. If, for example, the desired position is less than the actual position, the servo will turn to the left. On the other hand, if the desired position is greater than the actual position, the servo will turn to the right. In this manner, the servo "zeros-in" on the correct position. Should a load force the servo horn to the right or left, the servo will attempt to compensate. Note that there is no control mechanism for the speed of movement and, for most servos, the speed is specified in degrees per second. For more information refer to the data sheet of your servo.

Figure 6.31
HS-311 servo
http://www.hitecrcd.com.

Figure 6.32
Typical servo command signals.

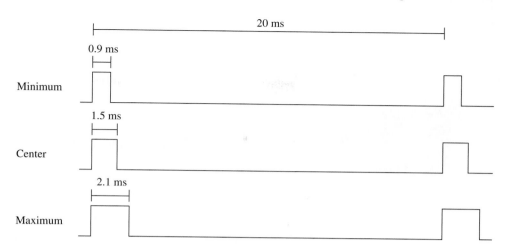

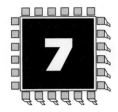

7 Serial I/O Devices

Chapter 7 objectives are to:

❏ Discuss fundamental concepts associated with serial communication systems: asynchronous, synchronous, bandwidth, full-duplex, half-duplex and simplex
❏ Present the RS232 and RS422 communication protocols, signals, and interface chips
❏ Write low-level device drivers that perform basic I/O with serial ports
❏ Discuss interfacing issues associated with performing the I/O as a background interrupt thread

In many applications, a single dedicated microcomputer is insufficient to perform all the tasks of the embedded system. One solution would be to use a larger and more powerful microcomputer, and another approach would be to distribute the tasks among multiple microcomputers. This second approach has the advantages of modularity and expandability. To implement a distributed system, we need a mechanism to send and receive information between the microcomputers. A second scenario that requires communication is a central general-purpose computer linked to multiple remote embedded systems for the purpose of distributed data collection or distributed control. For situations where the required bandwidth is less than about 1000 bytes/s, the built-in serial ports of the microcomputer can be used.

Chapter 7 deals with external devices that we connect to the serial I/O ports of our computer. In particular, we will interface terminals, keyboards, displays, printers, and other computers. In this chapter we will focus on serial channels that employ a direct physical connection between the microcomputers; later (in Chapter 14) we expand the communication system to include networks and modems.

7.1 Introduction and Definitions

Serial communication involves the transmission of one bit of information at a time. One bit is sent, a time delay occurs, then the next bit is sent. This section will introduce the use of serial communication as an interfacing technique for various microcomputer peripherals. Since many peripheral devices such as printers, keyboards, scanners, and mice have their own computers, the communication problem can be generalized to one of transmitting information between two computers. The *universal asynchronous receiver/transmitter (UART)* is the hardware port that implements the serial data transmission. Freescale calls it an *asynchronous communications interface adapter (ACIA)* or *SCI*. The *serial channel* is the collection of signals (or wires) that implement the communication. To improve bandwidth, remove noise, and increase range, we place interface logic between the digital logic UART port and the serial channel. We define the *data terminal equipment (DTE)* as the computer or a terminal and the *data communication equipment (DCE)* as the modem or printer (Figure 7.1).

Figure 7.1
A serial channel connects a DTE to a DCE.

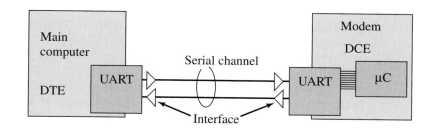

The interface logic (e.g., DS275, MAX232, or MC145407) converts between TTL/MOS/CMOS logic levels and RS232 logic levels. In this protocol, a typical bidirectional channel requires three wires (RxD, TxD, ground). We use RS422 voltage levels when we want long cable lengths and high bandwidths. The binary signal is encoded on the RS422 line as a voltage difference (MC3486/3487, MAX485, 3691, 3695, 8921/8922/8923, 75176, or 78120). In this protocol, a typical bidirectional channel requires five wires (RxD^+, RxD^-, TxD^+, TxD^-, ground). Typical voltage levels are shown in Table 7.1.

Table 7.1
Voltage levels for the CMOS RS232 and RS422 protocols.

		Typical CMOS Level	**Typical RS232 Level**	**Typical RS422 Level**
True	Mark	+5 V	TxD = −12 V	$(TxD^+ - TxD^-) = -3$ V
False	Space	+0.1 V	TxD = +12 V	$(TxD^+ - TxD^-) = +3$ V

Figure 7.2
A RS232 frame showing one start, seven data, one parity, and two stop bits.

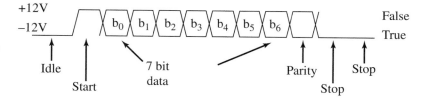

A *frame* is a complete and nondivisible packet of bits. A frame includes both information (e.g., data, characters) and overhead (start bit, error checking, and stop bits). A frame is the smallest packet that can be transmitted. The RS232 and RS422 protocols have one *start bit,* seven/eight data bits, no/even/odd *parity,* and one/1.5/two *stop bits* (Figure 7.2). The RS232 idle level is true (−12 V). The start bit is false (+12 V). A true data bit is −12 V, and a false data bit is +12 V. Stop bits are true (−12V).

Observation: The RS232 protocol always has one start bit and at least one stop bit.

Observation: RS232 and RS422 data channels are in negative logic because the true voltage is less than the false voltage.

Checkpoint 7.1: If the RS232 protocol has eight data bits, no parity, and one stop bit, how many total bits are in a frame?

Parity is generated by the transmitter and checked by the receiver. Parity is used to detect errors. For *even parity,* the number of 1s in the data plus parity is an even number. For *odd parity,* the number of 1s in the data plus parity is an odd number. If errors are unlikely, then operating without parity is faster and simpler. The *bit time* is the basic unit of time used in serial communication. It is the time between each bit. The transmitter outputs a bit, waits one bit time, then outputs the next bit. The start bit is used to synchronize the receiver with the transmitter. The receiver waits on the idle line until a start bit is first detected. After the true-to-false transition, the receiver waits a half a bit time. The half a bit time wait places the input sampling time in the middle of each data bit, giving the best tolerance to variations between the transmitter and receiver clock rates. We will discuss the detailed timing later in the chapter. Next, the receiver reads one bit every bit time. The *baud rate* is the total number of bits (information, overhead, and idle) per time that is transmitted in the serial communication. Later in Chapter 14, we will define the baud rate of a modem as the number of sounds per second. But for now, we define,

$$\text{Baud rate} = \frac{1}{\text{bit time}}$$

We will define *information* as the data that the "user" intends to be transmitted by the communication system. Examples of information include:

- Characters to be printed on your printer
- A picture file to be transmitted to another computer
- A digitally encoded voice message communicated to your friend
- The object code file to be downloaded from the PC to the 6811/6812

We will define *overhead* as signals added by the "operating system" to the communication to affect reliable transmission. Examples of overhead include:

- Start bit(s), start byte(s), or start code(s)
- Stop bit(s), stop byte(s), or stop code(s)
- Error checking bits such as parity, cyclic redundancy check (CRC), and checksum
- Synchronization messages like ACK, NAK, XON, XOFF

Although, in a general sense overhead signals contain "information," overhead signals are not included when calculating bandwidth or considering full-duplex, half-duplex, or simplex.

An important parameter in all communication systems is *bandwidth.* We will use the three terms bandwidth, *bit rate,* and *throughput* interchangeably to specify the number of information bits per time that is transmitted. These terms apply to all forms of communication:

- Parallel
- Serial
- Mixed parallel/serial

For serial communication systems, we can calculate:

$$\text{Bandwidth} = \frac{\text{number of information bits/frame}}{\text{total number of bits/frame}} \times \text{baud rate}$$

A *full-duplex communication system* allows information (data, characters) to transfer simultaneously in both directions. A *full-duplex channel* allows bits (information, error

Figure 7.3
A full-duplex serial channel connects two DTEs (computers).

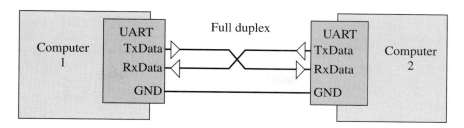

checking, synchronization, or overhead) to transfer simultaneously in both directions (Figure 7.3).

A *half-duplex communication* system allows information to transfer in both directions, but in only one direction at a time. Half-duplex is a term usually defined for modem communications, but in this book we will expand its meaning to include any serial protocol that allows communication in both directions, but in only one direction at a time. A fundamental problem with half-duplex is the detection and recovery from a collision. A collision occurs when both computers simultaneously transmit data. Fortunately, every transmission frame is echoed back into its own receiver. The transmitter program can output a frame, wait for the frame to be transmitted (which will be echoed into its own receiver), then check the incoming parity and compare the data to detect a collision. If a collision occurs, then it probably will be detected by both computers. After a collision, the transmitter can wait awhile and retransmit the frame. The two computers need to decide which one will transmit first after a collision so that a second collision can be avoided. The first hardware mechanism to implement half-duplex utilizes tristate logic. In this system, the transmitter will enable the driver ($\overline{\text{RTS}}$ =0) before transmission, then disable the driver after complete transmission ($\overline{\text{RTS}}$ =1).

Observation: People communicate in half-duplex.

When using the SCI (built into most of the 6811, and 6812 computers), complete transmission occurs when the SCI **Receive Data Register Full** flag is set (RDRF=1) and not when the SCI **Transmit Data Register Empty** flag is set (TDRE=1). The SCI Transmit Complete Flag (TC) should be set approximately at the same time as the RDRF, because TC=1 means a frame has been shifted out, and RDRF=1 means a frame has been shifted in (Figure 7.4).

Figure 7.4
A half-duplex serial channel can be implemented with tristate logic.

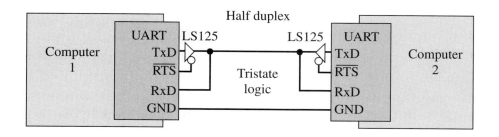

Another hardware mechanism for half-duplex utilizes open-collector logic (Figure 7.5). The 7407 driver has two output states: zero and hiZ. The logic high is created with the passive pull-up. With open collector, the half-duplex channel is the logical AND of the two TxD outputs. In this system, the transmitter simply transmits its frame without needing to enable or disable the driver.

Figure 7.5
A half-duplex serial channel can be implemented with open-collector logic.

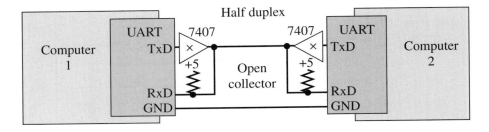

Figure 7.6
A desktop network is created using a half-duplex serial channel implemented with open-collector logic.

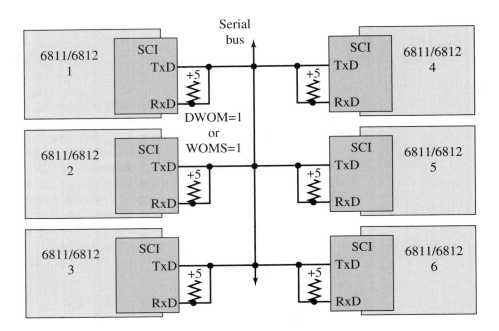

One application of the half-duplex protocol is the desktop serial bus, or multidrop network (Figure 7.6). In the following example, each microcomputer can send a message to another microcomputer. The 6811 Port D outputs can be made open-collector by setting DWOM=1. The 6812 serial port output can be made open-collector by setting WOMS=1. All the grounds are tied together. Assume there are less than 128 microcomputers in this network so that each microcomputer can have a unique 7-bit address. A frame with bit 7 equal to 1 is defined as an address; otherwise, the frame is data. The transmitting microcomputer first sends the address of the destination microcomputer, followed by multiple data frames. All the receiver microcomputers listen for their particular address. Once a microcomputer receiver recognizes its address, it will accept the data frames that follow. There are various options for determining the end of the message:

1. Define each message to be a fixed length
2. Send the message length as the second character
3. Define a special character that specifies end of message

Checkpoint 7.2: What is the difference between full duplex and half duplex?

A *simplex communication* system allows information to transfer in only one direction. The XON/XOFF protocol that we will cover later is an example of a communication system that has a full-duplex channel but implements simplex communication. This is because with XON/XOFF information (characters) are transmitted from the computer to the printer, but only XON/XOFF (error checking) flags are sent from the printer back to the computer. In this case, no information (data, characters) is sent from the printer to the computer. Figure 7.7 shows a simplex serial channel.

Figure 7.7
A simplex serial channel between two computers.

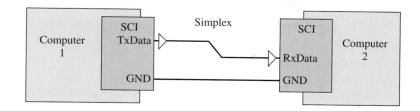

One application of simplex channels is the ring network. Rather than using a common bus like Figure 7.6, the six microcomputers could be connected in a ring topology using six simplex channels like Figure 7.7. The system could use the same address/data format described for the multidrop half-duplex connection. If a microcomputer receives a message addressed to a different node, it simply retransmits it along the ring. This system is slower than the multidrop system, but it does not have to contend with collision errors.

Figure 7.8
Nodes on an asynchronous channel run at the same frequency but have separate clocks.

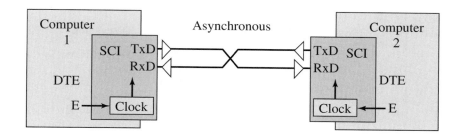

To transfer information correctly, both sides of the channel must operate at the same baud rate. In an *asynchronous* communication system, the two devices have separate and distinct clocks (Figure 7.8). Because these two clocks are generated separately (one on each side), they will not have exactly the same frequency or be in phase. If the two baud rate clocks have different frequencies, the phase between the clocks will also drift over time. Transmission will occur as long as the periods of the two baud rate clocks are close enough. The $-12\,V$ to $+12\,V$ edge at the beginning of the start bit is used to synchronize the receiver with the transmitter. If the two periods in a RS232 system differ by less than 5%, then after 10 bits the receiver will be off by less than half a bit time (and no error will occur). Any larger difference between the two periods will cause an error.

In a *synchronous* communication system, the two devices share the same clock (Figure 7.9). Typically a separate wire in the serial cable carries the clock. In this way, very high baud rates can be obtained. Another advantage of synchronous communication is that very long frames can be transmitted. Larger frames reduce the OS overhead for long transmissions because

Figure 7.9
Nodes on a synchronous channel operate off a common clock.

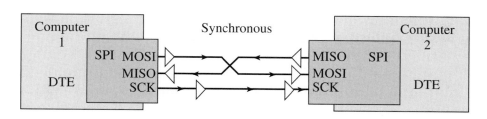

fewer frames need be processed per message. Even though in this chapter we will design various low-bandwidth synchronous systems using the SPI, synchronous communication is best applied to systems that require bandwidths above 1 Mb/s. The cost of this increased performance is the additional wire in the cable. The clock must be interfaced with channel drivers (e.g., RS232, RS422, optocouplers) similar to the transmit and receiver data signals.

Checkpoint 7.3: What is the difference between synchronous and asynchronous serial?

7.2 RS232 Specifications

The baud rate in a RS232 system can be as high as 20,000 bits/s. Because a typical cable has 50 pF/ft, the maximum distance for RS232 transmission is limited to 50 ft. There are 21 signals defined for full modem (*MO*dulate/*DEM*odulate) communication (Figure 7.10).

Figure 7.10
DB25 (RS232), DB9 (EIA-574), and RJ45 (EIA-561) connectors used in many serial applications.

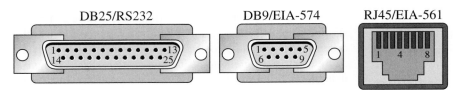

DB25 Pin	RS232 Name	DB9 Pin	EIA-574 Name	RJ45 Pin	EIA-561 Name	Signal	Description	True (V)	DTE	DCE
1						FG	Frame Ground/Shield			
2	BA	3	103	6	103	TxD	Transmit Data	−12	Out	In
3	BB	2	104	5	104	RxD	Receive Data	−12	In	Out
4	CA	7	105/133	8	105/133	RTS	Request to Send	+12	Out	In
5	CB	8	106	7	106	CTS	Clear to Send	+12	In	Out
6	CC	6	107			DSR	Data Set Ready	+12	In	Out
7	AB	5	102	4	102	SG	Signal Ground			
8	CF	1	109	2	109	DCD	Data Carrier Detect	+12	In	Out
9							Positive Test Voltage			
10							Negative Test Voltage			
11							Not Assigned			
12						sDCD	Secondary DCD	+12	In	Out
13						sCTS	Secondary CTS	+12	In	Out
14						sTxD	Secondary TxD	−12	Out	In
15	DB					TxC	Transmit Clk (DCE)		In	Out
16						sRxD	Secondary RxD	−12	In	Out
17	DD					RxC	Receive Clock		In	Out
18	LL						Local Loopback			
19						sRTS	Secondary RTS	+12	Out	In
20	CD	4	108	3	108	DTR	Data Terminal Rdy	+12	Out	In
21	RL					SQ	Signal Quality	+12	In	Out
22	CE	9	125	1	125	RI	Ring Indicator	+12	In	Out
23						SEL	Speed Selector DTE		In	Out
24	DA					TCK	Speed Selector DCE		Out	In
25	TM					TM	Test mode	+12	In	Out

Table 7.2
Pin assignments for the RS232 EIA-574 and EIA-561 protocols.

Table 7.2 shows the entire set of RS232 signals. The RS232 standard uses a DB25 connector that has 25 pins. The EIA-574 standard uses RS232 voltage levels and a DB9 connector that has only nine pins. The EIA-561 standard also uses RS232 voltage levels but with a RJ45 connector that has only eight pins. The most commonly used signals of the full RS232 standard are available with the EIA-561/EIA-574 protocols.

The frame ground is connected on one side to the *ground shield* of the cable. The shield will provide protection from electric field interference. The *twisted cable* has a small area between the wires. The smaller the area, the less the magnetic field pickup. There is one disadvantage to reducing the area between the connectors. The capacitance to ground is inversely related to the separation distance between the wires. Thus as the area decreases, the capacitance will increase. This increased capacitive load will limit both the distance and the baud rate (Figure 7.11).

Figure 7.11
The simplest RS232 cable uses just TxD, RxD, and ground (with an optional ground shield).

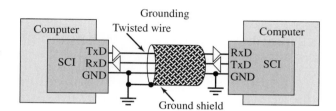

The signal ground is connected on both sides to the supply return. A separate wire should be used for signal ground (do not use the ground shield to connect the two signal grounds). The noise immunity will be degraded if the ground shield is connected on both sides. There are many available RS232 driver chips, but the Maxim MAX232 (Figure 7.33) and DS275 (Figure 7.12) are popular devices because of their low cost and simple implementation. The MAX232 employs a charge pump (using 100-nF capacitors) to create the standard +12 and −12 output voltages with only a +5-V supply. The DS275 operates on +5-V and requires no external capacitors.

Figure 7.12
RS232 interface to Freescale microcomputers.

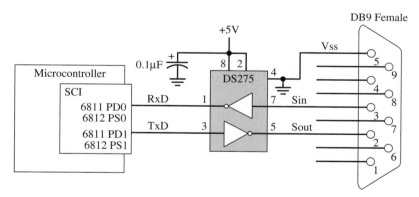

Figure 7.13
RS232 output specifications.

Must withstand
    Short to ground
    Short to any other wire
Operating range
    True $-15 \leq V_{out} \leq -5V$
    False $+5 \leq V_{out} \leq +15V$
Maximum output voltage
    $-25 \leq V_{out} \leq +25V$
Short circuit current
    $I_{out} \leq 0.5$ A
Transition (-3 to 3) range time $\leq 4\%$

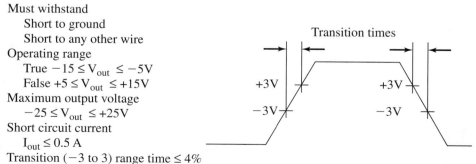

Figure 7.13 overviews the RS232 specifications for an output signal. Similarly, Figure 7.14 overviews the RS232 specifications for an input signal.

Figure 7.14
RS232 input
specifications.

Maximum Slew Rate
  $dV_{in}/dt \leq 30V/\mu s$
Operating Range
  True $-15 \leq V_{in} \leq -3V$
  Translation $-3 < V_{in} < +3V$
  False $+3 \leq V_{in} \leq +15V$
Input Resistance
  $3000\Omega \leq R_{in} \leq 7000\Omega$
Input Capacitance including cable
  $C_{in} \leq 2500pF$
Input open circuit voltage
  $E_{in} \leq 2V$

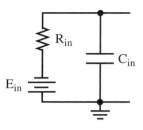

On the data lines, a true or mark voltage is negative. Conversely, for the control lines, a true signal has a positive voltage. *Request to Send (RTS)* is a signal from the computer to the modem requesting transmission be allowed. *Clear to Send (CTS)* is the acknowledge signal back from the modem signifying transmission can proceed. *Data Set Ready (DSR)* is a modem signal specifying that I/O can occur. *Data Carrier Detect (DCD)* is a modem signal specifying that the carrier frequencies have been established on its telephone line. *Ring Indicator (RI)* is a modem signal that is true when the phone rings. The *Receive Clock* and *Transmit Clock* are used to establish synchronous serial communication. *Data Terminal Ready (DTR)* is a printer signal specifying the status of the printer. This signal is sometimes called $\overline{Busy}$. When DTR is $+12$ V, the printer is ready and can accept more characters. Conversely when DTR is -12 V, the printer is busy and cannot accept more characters. None of the built-in serial ports of the Freescale microcomputer discussed in this book supports these control lines explicitly. On the other hand, it is straightforward to implement these hardware handshaking signals using simple I/O lines. If we wish to generate interrupts on edges of these lines, then we would use input capture or key wake-up features.

A typical sequence for initiating communication between a computer (DTE) and a modem (DCE) begins with the turning on of the power in both devices. The computer activates the DTR line, and the modem responds by activating the DSR signal. Next, the modem establishes a link across the telephone line with the other modem. A functional link requires a separate carrier frequency from both modems. See Section 14.9 for additional details about the modem/modem link. The modem signals to the computer that a proper link has been established by activating DCD. When the computer wishes to transmit it, it activates RTS. If okay, the modem responds by activating CTS. After receiving the CTS, the computer can transmit data. The computer can continuously activate RTS if multiple frames are to be sent, but it must postpone transmission if CTS becomes false.

Faced with the design of a RS232 interface, we should use one of the many RS232 chips available. The MC1488/MC1489 require ±12-V supplies. Some devices (e.g., MAX232, MC145407) use a charge-pump mechanism so that they can operate on a single $+5$-V power supply; other devices (e.g., MC145406) have multiple bidirectional interfaces (Figure 7.15).

7.3 RS422/USB/RS423/RS485 Balanced Differential Lines

To increase the baud rate and maximum distance, the balanced differential line protocols were introduced. The RS422 signal is encoded in a differential signal, A−B. There are many RS422 interface chips, such as SP301, SP304, MAX486, MAX488, MC3486/3487,

Figure 7.15
More RS232 interface chips.

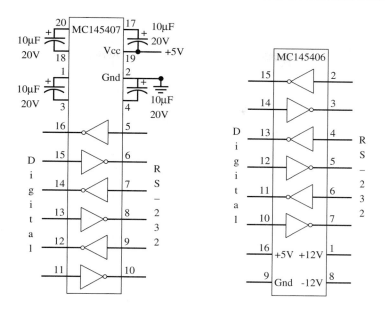

Figure 7.16
RS422 serial channel.

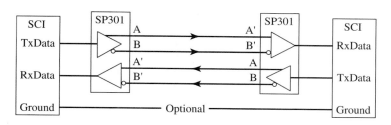

MC3691, MC3695, 8921/8922/8923, 75176, or 78120. A full-duplex RS422 channel is implemented in Figure 7.16 with Sipex SP301 drivers.

Because each signal requires two wires, five wires (ground included) are needed to implement a full-duplex channel. With RS232 one typically connects one receiver to one transmitter. But with RS422, up to ten receivers can be connected to one transmitter. Table 7.3 summarizes four common EIA standards.

Table 7.3
Specifications for the RS232, RS423A, RS422, and RS485 protocols.

Specification	RS232D	RS423A	RS422	RS485
Mode of operation	Single-ended	Single-ended	Differential	Differential
Drivers on one line	1	1	1	32
Receivers on one line	1	10	10	32
Maximum distance (ft)	50	4,000	4,000	4,000
Maximum data rate	20 kbits/sec	100 kbits/sec	10 Mbits/sec	10 Mbits/sec
Maximum driver output	±25 V	±6 V	-0.25 to $+6$ V	-7 to $+12$ V
Driver output (loaded)	±5 V	±3.6 V	±2 V	±1.5 V
Driver output (unloaded)	$\pm$ 15V	±6 V	±5 V	±5 V
Driver load impedance	3kΩ to 7kΩ	450Ω min	100Ω	54Ω
Receiver input voltage	±15 V	±12 V	±7 V	-7 to $+12$ V
Receiver input sensitivity	±3 V	±200 mV	±200 mV	±200 mV
Receiver input resistance	3kΩ to 7kΩ	4kΩ min	4kΩ min	12kΩ min

The maximum baud rate at 40 ft is 10 Mbits/sec. At 4000 ft, the baud rate can only be as high as 100 kbits/sec. Table 7.4 shows two implementations of the RS422 protocol.

Table 7.4
Pin assignments for the
EIA-530 and RS449
protocols.

DB25 Pin	EIA-530 Name	DB37 Pin	RS449 Name	Signal	Description	DTE	DCE
1		1		FG	Frame Ground/Shield		
2	BA (A)	4	SD (A)	TxD	Transmit Data	Out	In
14	BA (B)	22	SD (B)	TxD	Transmit Data	Out	In
3	BB (A)	6	RD (A)	RxD	Receive Data	In	Out
16	BB (B)	24	RD (B)	RxD	Receive Data	In	Out
4	CA (A)	7	RS (A)	RTS	Request to Send	Out	In
19	CA (B)	25	RS (B)	RTS	Request to Send	Out	In
5	CB (A)	9	CS (A)	CTS	Clear to Send	In	Out
13	CB (B)	27	CS (B)	CTS	Clear to Send	In	Out
6	CC (A)	11	DM (A)	DSR	Data Set Ready	In	Out
22	CC (B)	29	DM (B)	DSR	Data Set Ready	In	Out
20	CD (A)	12	TR (A)	DTR	Data Terminal Rdy	Out	In
23	CD (B)	30	TR (B)	DTR	Data Terminal Rdy	Out	In
7	AB	19	SG	SG	Signal Ground		
8	CF (A)	13	RR (A)	DCD	Data Carrier Detect	In	Out
10	CF (B)	31	RR (B)	DCD	Data Carrier Detect	In	Out
15	DB (A)	5	ST (A)	TxC	Transmit Clk (DCE)	In	Out
12	DB (B)	23	ST (B)	TxC	Transmit Clk (DCE)	In	Out
17	DD (A)	8	RT (A)	RxC	Receive Clock	In	Out
9	DD (B)	26	RT (B)	RxC	Receive Clock	In	Out
18	LL	10	LL		Local Loopback	Out	In
21	RL	14	RL	RL	Remote Loopback	Out	In
24	DA (A)	17	TT (A)	TCK	Speed Selector DCE	Out	In
11	DA (B)	35	TT (B)	TCK	Speed Selector DCE	Out	In
25	TM	18	TM	TM	Test mode	In	Out

The EIA-530 standard uses a DB25 connector that has 25 pins. The RS449 standard uses a DB37 connector that has 37 pins (Figure 7.17).

Figure 7.17
DB25 (EIA-530), and
DB37 (RS-449)
connectors used in
RS422 applications.

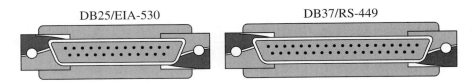

DB25/EIA-530   DB37/RS-449

In this section we will introduce the electrical specifications of the **Universal Serial Bus** (USB). Figure 7.18 shows the two types of USB connectors. A single host computer controls the USB, and there can only be one host per bus. The host controls the scheduling of all transactions using a token-based protocol. The USB architecture is a tiered star topology, similar to 10BaseT Ethernet. The host is at the center of the star and devices are attached to the host. The number of nodes on the bus can be extended using USB hubs. Up to 127 devices can be connected to any one USB bus at any one given time. USB plug 'n'plug is implemented with dynamically loadable and unloadable drivers. When the user plugs the device into the USB bus, the host will detect the connection, interact with the

Figure 7.18
USB connectors.

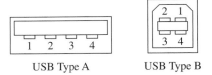

USB Type A   USB Type B

newly inserted device, and load the appropriate driver. The USB device can be used without explicitly installing drivers or rebooting. When the device is unplugged, the host will automatically unload its driver.

Table 7.5
USB signals.

Pin Number	Color	Function
1	Red	VBUS (5 V)
2	White	D−
3	Green	D+
4	Black	Ground

USB uses four shielded wires, +5V power, GND, and twisted-pair differential data signals, as listed in Table 7.5. It uses a NRZI (non-return-to zero-invert) encoding scheme to send data with a sync field to synchronize the host and receiver clocks. The D+ signal has a 15 kΩ pull-down resistor to ground, and the D− signal has a 1.5 kΩ pull-up resistor to +3.6V. Like the other protocols in this section, the data is encoded as a differential signal between D+ and D−. In general, a differential '1' exists when D+ is greater than D−. More specifically, a differential '1' is transmitted by pulling D+ over 2.8V and D− under 0.3V. The transmitter creates a differential '0' by making D− greater than 2.8V and D+ less than 0.3V. The receiver recognizes the differential '1' when D+ is 0.2V greater than D−. The receiver will consider the input as a differential '0' when D+ 0.2V less than D−. The polarity of the signal is inverted depending on the speed of the bus. Therefore the terms 'J' and 'K' states are used in signifying the logic levels. At low speed, a 'J' state is a differential '0'. At high speed, a 'J' state is a differential '1'. USB interfaces employ both differential and single-ended outputs. Certain bus states are indicated by single-ended signals on D+, D− or both. For example, a single-ended zero (SE0) signifies device reset when held for more than 10mS. More specifically, SE0 is generated by holding both D− and D+ low (< 0.3V). USB can operate at three speeds. The low speed/full speed bus has a characteristic impedance of 90Ω. High-speed mode uses a a constant current protocol to reduce noise.

High-speed data is clocked at 480Mb/s
Full-speed data is clocked at 12Mb/s
Low-speed data is clocked at 1.5Mb/s

7.3.1 RS422 Output Specifications

The output voltage levels are shown in Table 7.6. A key RS422 specification is that the output impedances should be balanced. If the I/O impedances are balanced, then added noise in the cable creates a common-mode voltage, and the common-mode rejection of

Table 7.6
Output voltage levels for the RS422 differential line protocol.

	Output Voltage
True or Mark	$-6 \leq A-B \leq -2\,V$
Transition	$-2 \leq A-B \leq +2\,V$
False or Space	$+2 \leq A-B \leq +6\,V$

the input will eliminate it. More details about common mode are presented later in Chapters 11 and 12.

$$R_{Aout} = R_{Bout} \approx 100\Omega$$

The time in the transition region must be less than

10% for baud rates above 5 Mb/s
20 ns for baud rates below 5 Mb/s

The input voltage levels are as shown in Table 7.7.

Table 7.7
Input voltage thresholds
for the RS422
differential line protocol.

	Input Voltage
True or Mark	$A-B \leq -0.2\,V$
Transition	$-0.2\,V \leq A-B \leq +0.2\,V$
False or Space	$+0.2\,V \leq A-B$

As mentioned earlier, to provide noise immunity the common-mode input impedances must also be balanced

$$4k\Omega \leq R_{A'in} = R_{B'in}$$

The balanced nature of the interface produces good noise immunity. The differential input impedance is specified by the plot in Figure 7.19. Any point within the shaded region is allowed. Even though the ground connection in the RS422 cable is optional, it is assumed the grounds are connected somewhere. In particular, the interface will operate with a common-mode voltage up to 7 V.

$$\frac{|A' + B'|}{2} \leq + 7\,V$$

Figure 7.19
RS422 input current
versus input voltage
relationship.

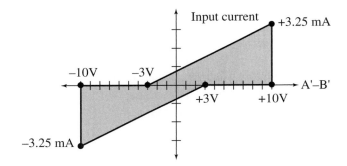

7.3.3
RS485 Half-Duplex
Channel

RS485 can be either half-duplex or full-duplex. The RS485 protocol, illustrated in Figure 7.20, implements a half-duplex channel using differential voltage signals. The Sipex SP483 or Maxim MAX483 implements the half-duplex RS485 channel. One of the advantages of RS485 is that up to 32 devices can be connected onto a single serial bus. When more than one transmitter can drive the serial bus, the protocol is also called *multidrop*. To transmit

Figure 7.20
A half-duplex serial
channel is implemented
with RS485 logic.

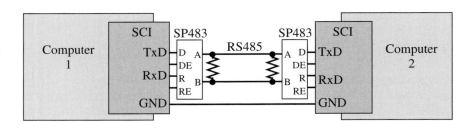

the computer enables the driver by making DE active, then sends the serial frame from the TxD output of the SCI port. If RE is also active during transmission, the transmitted frame is echoed into the serial receiver of the SCI RxD line. To receive a frame the computer simply enables its receiver (by making RE active) and accepts a serial frame on the RxD line in the usual manner. Be careful when selecting the resistances on a half-duplex network so that the total driver impedance is about 54 Ω.

7.4 Other Communication Protocols

There are a variety of communication protocols for implementing communications channels. They vary in cost, distance, bandwidth, and noise immunity. Some protocols require the two computers to have a common ground, while others are isolated from each other. Isolation is important to prevent noise on one system from creating errors on another.

7.4.1 Current Loop Channel

Current loop is an old standard where true is encoded as a 20-mA current and false is signified by no current. The advantage of current loop is its inherent electrical isolation between the two computers. We could use current loop in applications where we wished to prevent noise in one computer from coupling into the electronics of the other (Figure 7.21).

Figure 7.21
Current loop serial interface.

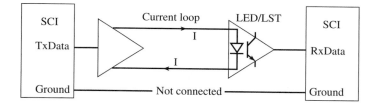

7.4.2 Introduction to Modems

Detailed information about modems is presented later in Chapter 14. *Frequency-shift keying* (FSK) modems encode the binary data as frequencies that are transmitted as sounds on standard phone lines (Figure 7.22). The modulator converts the TxD data from the transmission computer into sounds. The phone line transmits the sounds to the receiver modem. The demodulator on the receiver converts the sound frequencies back into a regular true/false digital line. The SCI on the receiver accepts the serial frame in the usual way (start bit, data bits, stop bit).

Figure 7.22
Modem serial interface.

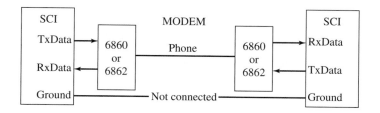

Two pairs of frequencies implement full-duplex at 300 bits/s (Table 7.8).

Table 7.8
Frequency parameters for a 300 bits FSK modem protocol.

Logic	Originate	Answer
True	1270 Hz	2225 Hz
False	1070 Hz	2025 Hz

The *phase-encoded modem* also uses standard phone lines. Phase-encoded modems provide for increased bandwidth by encoding multiple bits into periodic phase shifts of the sound signal. More details can be found in Section 14.9. Even higher bandwidth is implemented by combining phase and amplitude information. A significant amount of digital signal processing is required to produce reliable transmission at high speeds.

7.4.3 Optical Channel

A *fiber-optic light* channel uses a LED transmitter, a fiber-optic cable, and a light sensor. Binary information is encoded as the presence or absence of light in the cable (Figure 7.23). Fiber-optic cable is used in applications requiring long distances and/or high bandwidth. Similar to the current loop channel, fiber-optic cables provide electrical isolation between the two computers. Techniques for interfacing LEDs can be found in Section 8.2. Another advantage of the optical channel is its noise immunity. All the other protocols are to some degree susceptible to electric field noise. Communication using fiber optics is not affected by electric fields along the cable.

Figure 7.23
Fiber-optic serial interface.

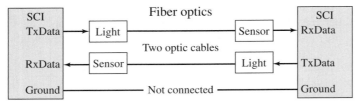

7.4.4 Digital Logic Channel

The simplest channel to implement uses standard *digital logic*. When the two computers are located in the same box, it is appropriate to send information directly from one to the other without special hardware circuits (Figure 7.24). Although standard digital logic will not be appropriate for long distances and/or in the presence of strong electric fields, it certainly is cheap and simple and should at least be considered for communication across short distances.

Figure 7.24
Simple digital logic serial interface.

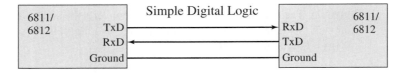

If isolation is required, then optoisolators like the 6N139 can be used (Figure 7.25).

Observation: When using an optocoupler such as the 6N139 in Figure 7.25, the power and ground signals are NOT connected between the two computers. This greatly reduces noise coupling.

Figure 7.25
Isolated digital logic serial interface.

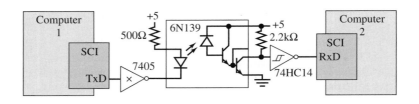

7.5 Serial Communications Interface

Most of the Freescale embedded microcomputers support at least one SCI. Before discussing the detailed operation of particular devices, we will begin with general features common to all devices. Common features that affect the entire SCI module include:

A baud rate control register used to select the transmission rate
A mode bit M used to select 8-bit (M=0) or 9-bit (M=1) data frames

Each device is capable of creating its own serial port clock with a period that is an integer multiple of the E clock period. The programmer will select the baud rate by specifying the integer divide-by used to convert the E clock into the serial port clock.

Common error: If you change the E clock frequency without changing the baud rate register, the SCI will operate at an incorrect baud rate.

Table 7.9 lists the I/O port locations of the serial ports for the various microcomputers discussed in this book. The MC68HC812A4 has two serial ports.

Table 7.9
Serial port pins available on various Motorola microcomputers.

Microcomputer	Pin for TxD	Pin for RxD
MC68HC711E9	PD1	PD0
MC68HC812A4	PS3/PS1	PS2/PS0
MC68HC912B32	PS1	PS0
MC9S12C32	PS1	PS0

7.5.1
Transmitting in Asynchronous Mode

We will begin with transmission, because it is straightforward. Common features that affect the transmitter portion of the SCI include (Figure 7.26):

TxD data output pin, with TTL voltage levels
10- or 11-bit shift register, which cannot be directly accessed by the programmer; this shift register is separate from the receive shift register
Serial Communications Data Register (SCDR), which is write only, even though at the same address this data register is separate from the receive data register
T8 data bit that you set before writing to SCDR when 9-bit data mode (M=1) is used

The control bits that affect the transmitter are:

Transmit Enable control bit (TE), which you initialize to 1 to enable the transmitter
Send Break control bit (SBK), which you set to 1 to send blocks of 10 or 11 zeros
Transmit Interrupt Enable control bit (TIE), which you set to 1 to arm the TDRE flag
Transmit Complete Enable control bit (TCIE), which you set to 1 to arm the TC flag

Figure 7.26
Data register and shift register used to implement the transmit serial interface.

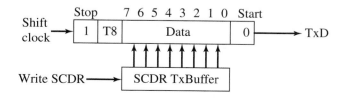

The status bits generated by transmitter activity are:

Transmit Data Register Empty flag (TDRE), which is set when the transmit SCDR is empty; TDRE is cleared by reading the TDRE flag (with it set), then writing to the SCDR

Transmit Complete flag (TC), which is set when the transmit shift register is done shifting; TC is cleared by reading the TC flag (with it set), then writing to the SCDR

When new data (8 bits) are loaded in the SCDR, they are copied into the 10- or 11-bit transmit shift register. Next, the start bit, T8 (if M=1), and stop bits are added. Then, the frame is shifted out one bit at a time at a rate specified by the baud rate register. If there are already data in the shift register when the SCDR is written, it will wait until the previous frame is transmitted before it, too, is transferred. In the timing diagrams of Figures 7.27 and 7.28, the "T" arrows refer to times when the transmit shift register is shifted, causing a change in the TxD output pin.

Figure 7.27
Transmit data frames for M=0 and M=1.

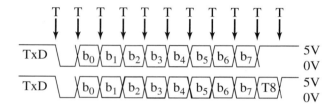

Figure 7.28
Start bit timing during a transmit data frame.

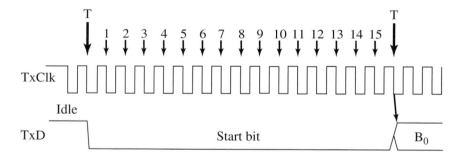

The serial port hardware is actually controlled by a clock that is 16 times faster than the baud rate. The digital hardware in the SCI counts 16 times in between changes to the TxD output line. Pseudocode for the transmission process is shown below (these operations occur automatically in hardware).

```
TRANSMIT  Set TxD=0              Output start bit
          Wait 16 clock times    Wait 1 bit time
          Set n=0                Bit counter
TLOOP     Set TxD=b_n            Output data bit
          Wait 16 clock times    Wait 1 bit time
          Set n=n+1
          Goto TLOOP if n≤7
          Set TxD=T8             Output T8 bit (optional if M=1)
          Wait 16 clock times    Wait 1 bit time (optional if M=1)
          Set TxD=1              Output a stop bit
          Wait 16 clock times    Wait 1 bit time
```

In essence, the SCDR and transmit shift register behave together like a two-element FIFO, with writing into the SCDR analogous to the PutFifo operation and the shifting data out analogous to the GetFifo operation. In fact, the serial port interface chip used in most PCs has a 16-byte hardware FIFO between the data register and the shift register. A PC that has a 16C550-compatible UART supports this hardware FIFO function. This FIFO reduces the latency requirements of the OS to service the serial port hardware.

7.5.2 Receiving in Asynchronous Mode

Receiving data frames is a little trickier than transmission because we have to synchronize the receive shift register with the incoming data. Common features that affect the receiver portion of the SCI include:

RxD data input pin, with TTL voltage levels

10- or 11-bit shift register, which cannot be directly accessed by the programmer; this shift register is separate from the transmit shift register

Serial Communications Data Register (SCDR), which is read only, even though at the same address this data register is separate from the transmit data register

R8 bit that you can read after receiving a frame in 9-bit data mode (M=1)

The control bits that affect the receiver are:

Receiver Enable control bit (RE), which you initialized to 1 to enable the receiver

Receiver Wakeup control bit (RWU), which you set to 1 to allow a receiver input to wake up the computer

Receiver Interrupt Enable control bit (RIE), which you set to 1 to arm the RDRF flag

Idle Line Interrupt Enable control bit (ILIE), which you set to 1 to arm the IDLE flag

The status bits generated by receiver activity are (Figure 7.29):

Receive Data Register Full flag (RDRF), which is set when new input data is available; RDRF is cleared by reading the RDRF flag (with it set), then reading the SCDR

Receiver Idle flag (IDLE), which is set when the receiver line becomes idle; IDLE is cleared by reading the IDLE flag (with it set), then reading the SCDR

Overrun flag (OR), which is set when input data is lost because previous data frames had not been read; OR is cleared by reading the OR flag (with it set), then reading the SCDR

Noise flag (NF), which is set when the input is noisy; NF is cleared by reading the NF flag (with it set), then reading the SCDR

Framing Error (FE), which is set when the stop bit is incorrect; FE is cleared by reading the FE flag (with it set), then reading the SCDR.

Figure 7.29
Data register and shift register used to implement the receive serial interface.

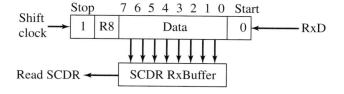

The receiver waits for the 1 to 0 edge signifying a start bit, then shifts in 10 or 11 bits of data one at a time from the RxD line. The start and stop bits are removed (checked for noise and framing errors), the 8 bits of data are loaded into the SCDR, the ninth data bit is put in R8 (if M=1), and the RDRF flag is set. If there are already data in the SCDR when the shift register is finished, it will wait until the previous frame is read by the software before it is transferred. An overrun occurs when there is one receive frame in the SCDR, one receive frame in the receive shift register, and a third frame comes into RxD. In the timing diagrams of Figures 7.30 and 7.31, the "S" arrow refers to the time when the receiver detects the 1 to 0 edge of the start bit, and the "R" arrows refer to the times the receiver shift register is shifted, causing the current value of the RxD input pin to be recorded.

Figure 7.30
Receive data frames for
M=0 and M=1.

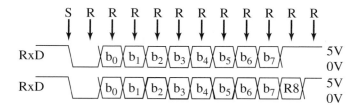

Figure 7.31
Start bit timing during a
receive data frame.

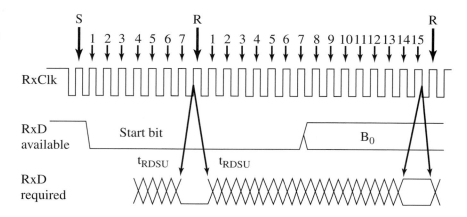

If "R" refers to the time the RxD pin is recorded, the setup time is the time before "R" the input must be valid. The hold time is the time after the "R" data must continue to be valid. Pseudocode for the receive process is shown below (these operations occur automatically in hardware). Because the receiver must wait a half a bit time after the start 1 to 0 edge, it requires a clock faster than the baud rate. In particular, with a serial clock 16 times faster than the baud rate, it waits a half a bit time by waiting 8 serial clock periods. All of the operations (e.g., waiting for RxD=0) are synchronized to the serial clock.

```
RECEIVE   Goto RECEIVE if RxD=1    Wait for start bit
          Wait 8 clock times       Wait half a bit time
          Goto RECEIVE if RxD=1    False start? (NF)
          Set n=0                  Bit counter
RLOOP     Wait 16 clock times      Wait 1 bit time
          Set b_n=RXD              Input data bit
          Set n=n+1
          Goto RLOOP if n≤7
          Wait 16 clock times      Wait 1 bit time (optional if M=1)
          Set R8= RxD              Read R8 bit (optional if M=1)
          Wait 16 clock times      Wait 1 bit time
          Set FE=1 if RxD =0       Framing error if no stop bit
```

An overrun occurs when there is one receive frame in the SCDR, one receive frame in the receive shift register, and a third frame comes into RxD. To avoid overrun, we can design a real-time system (i.e., one with a maximum latency). The latency of a SCI receiver is the delay between the time when new data arrives in the receiver SCDR and the time the software reads the SCDR. If the latency is always less than 10 (11 if M=1) bit times, then overrun will never occur.

> **Observation:** With a serial port that has a shift register and one data register (no additional FIFO buffering), the latency requirement of the input interface is the time it takes to transmit one data frame.

In the following example, assume the SCI receive shift register and receive data register are initially empty. Three incoming serial frames occur one right after another, but the software does not respond. At the end of the first frame, the $31 goes into the receive SCDR and the RDRF flag is set. In this scenario, the software is busy doing other things and does not respond to the setting of RDRF. Next, the second frame is entered into the receive shift register. At the end of the second frame, there is the $31 in the SCDR and the $32 in the shift register. If the software were to respond at this point, then both characters would be properly received. If the third frame begins before the first is read by the software, then an overrun error occurs and a frame is lost (Figure 7.32). We can see from this worst-case scenario that the software must read the data from SCDR within 10 bit times of the setting of RDRF.

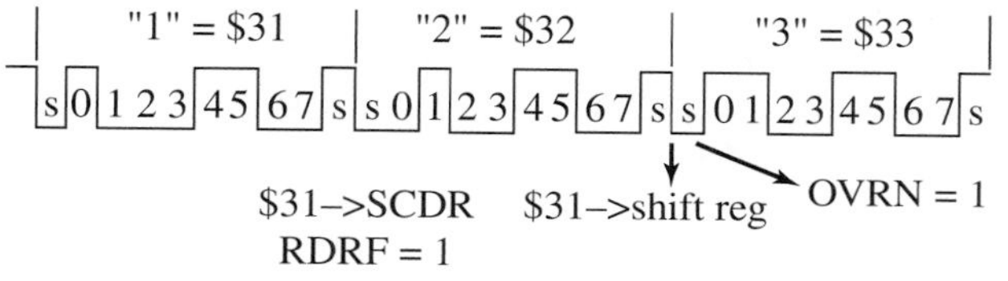

Figure 7.32
Three receive data frames result in an overrun (OR) error.

7.5.3 MC68HC711E9 SCI Details

The MC68HC711E9 has one asynchronous serial port using Port D bits 1,0. Table 7.10 shows the I/O ports that implement the SCI functions. It has a RS232 **non-return-to-zero** (NRZ) format with one start bit, eight or nine data bits, and one stop bit, as shown in Figure 7.27. The SCI transmitter and receiver are independent but use the same data format and bit rate.

Table 7.10
6811 SCI ports.

Address	Bit 7	6	5	4	3	2	1	Bit 0	Name
$102B	TCLR	SCP2	SCP1	SCP0	RCKB	SCR2	SCR1	SCR0	BAUD
$102C	R8	T8	0	M	Wake	0	0	0	SCCR1
$102D	TIE	TCIE	RIE	ILIE	TE	RE	RWU	SBK	SCCR2
$102E	TDRE	TC	RDRF	IDLE	OR	NF	FE	0	SCSR
$102F	R7/T7	R6/T6	R5/T5	R4/T4	R3/T3	R2/T2	R1/T1	R0/T0	SCDR

We use the BAUD register to select the baud rate, previously described in Table 3.13. The three bits **SCP2:0** determine the prescale factor 0,1,2,3,4 specify a prescale of $P=1,3,4,13,39$ respectively, where P=39 mode is available only on the MC68HC(7)11E20. Let **n** be the number specified by the three bits **SCR2:0**. The baud rate can be calculated using the equation

$$\text{SCI Baud Rate} = \frac{\text{Eclk}}{16*P*2^n}$$

The **SCCR2** control register contains most of the important bits that configure the SCI. **TE** is the Transmitter Enable bit. We turn this bit off to disable the transmitter, and set this bit to enable the transmitter. **RE** is the Receiver Enable bit. This bit should be set to one to activate the receiver. If RE is 0, then the receiver is disabled. **TIE** is the Transmit Interrupt Enable bit. We set TIE=1 to arm the transmitter so that an interrupt is requested when TDRE is set. We clear TIE=0 to disarm the TDRE-triggered interrupts. **TCIE** is the Transmit

Complete Interrupt Enable bit. We set TCIE=1 to arm the transmitter so that an interrupt is requested when TC is set. We clear TCIE=0 to disarm the TC-triggered interrupts. **RIE** is the Receiver Interrupt Enable bit. We set RIE=1 to arm the receiver so that an interrupt is requested when RDRF is set. We clear RIE=0 to disarm the RDRF-triggered interrupts. **ILIE** is the Idle Line Interrupt Enable bit. We set ILIE=1 to arm the receiver so that an interrupt is requested when IDLE is set. We clear ILIE=0 to disarm the IDLE-triggered interrupts. **RWU** is the Receiver Wake Up Control bit. We set RWU=1 to enable wake up and inhibit receiver interrupts. We can place the 6811 in a low-power sleep mode, and wake it up from the SCI input (see Wake bit). **SBK** is the Send Break bit. We set SBK=1 to send a break, which is a continuous TxD low. TxD will remain low for as long as the SBK is 1.

The **SCCR1** control register contains the rest of the bits we use to configure the SCI. **M** is the Mode bit. We set M=0 to create a 10-bit frame with 1 start bit, 8 data bits, 1 stop bit. We set M=1 to create an 11-bit frame with 1 start bit, 9 data bits, 1 stop bit (the ninth data bit is in T8/R8). If **Wake**=0, then the SCI will wake up by an IDLE line recognition. If Wake=1, then the SCI will wake up by address mark (most significant data bit set).

The flags in the **SCSR** register can be read by the software, but cannot be modified by writing to this register. **TDRE** is the Transmit Data Register Empty Flag. TDRE is set if the transmit data can be written to SCDR. If TDRE is zero, the transmit data register contains previous data that has not yet been moved to the transmit shift register. TDRE is cleared by SCSR read with TDRE set followed by SCDR write. **TC** is the Transmit Complete Flag. It is set if the transmitter is idle (no data, preamble, or break transmission in progress). We can clear TC by reading SCSR with TC set followed by writing to SCDR. **RDRF** is the Receive Data Register Full Flag. It is set if a received character is ready to be read from SCDR. RDRF is cleared by a SCSR read with RDRF set followed by a SCDR read. **IDLE** is the Idle Line Detected Flag. It is set if the RxD line is idle (10 or 11 consecutive logic ones). The IDLE flag is inhibited when RWU is set to one. It is cleared by reading SCSR with IDLE set followed by reading SCDR. Once cleared, IDLE is not set again until the RxD line has been active and becomes idle again. **R8** is Receive Data Bit 8. It is a read-only bit that contains the ninth bit of the receive data when M=1. **T8** is Transmit Data Bit 8. If M bit is set, T8 stores the ninth bit in transmit data character. T8 can be set to one to implement a second stop bit. It can also be used to implement even or odd parity. To calculate an even parity, we can perform a bit-wise exclusive-or of the data bits.

Three error conditions (overrun, noise, and framing) can occur during SCI receiver activity. Three bits (OR, NF, and FE) in the serial communications status register (SCSR) indicate whether one of these error conditions exists. The overrun error (**OR**) bit is set when the next byte is ready to be transferred from the receive shift register to the SCDR and the SCDR is already full (RDRF bit is set) (see Figure 7.32). When an overrun error occurs, the data that caused the overrun is lost and the data that was already in SCDR is not disturbed. The OR is cleared when the SCSR is read (with OR set), followed by a read of the SCDR. The noise flag (**NF**) bit is set if there is noise on any of the received bits, including the start and stop bits. In particular, each data bit is sample three times and the NF bit is set if the three samples are not all the same. The NF bit is not set until the RDRF flag is set. The NF bit is cleared when the SCSR is read (with FE equal to 1) followed by a read of the SCDR. When no stop bit is detected in the received data character, the framing error (**FE**) bit is set. FE is set at the same time as the RDRF. If the byte received causes both framing and overrun errors, the processor recognizes only the overrun error. The framing error flag inhibits further transfer of data into the SCDR until it is cleared. The FE bit is cleared when the SCSR is read (with FE equal to 1) followed by a read of the SCDR.

7.5.4 MC9S12C32 SCI Details

The MC9S12C32 has one serial port using Port S bits 1,0. Table 7.11 shows the I/O ports that implement the SCI functions. It has a RS232 **Non-Return-to-Zero** (NRZ) format with one start bit, eight or nine data bits, and one stop bit, as shown in Figure 7.27. The SCI transmitter and receiver are independent, but use the same data format and bit rate.

Table 7.11
MC9S12C32 SCI ports.

Address	msb														lsb	Name
$00C8	-	-	-	12	11	10	9	8	7	6	5	4	3	2	1	0	SCIBD

Address	Bit 7	6	5	4	3	2	1	Bit 0	Name
$00CA	LOOPS	SWAI	RSRC	M	WAKE	ILT	PE	PT	SCICR1
$00CB	TIE	TCIE	RIE	ILIE	TE	RE	RWU	SBK	SCICR2
$00CC	TDRE	TC	RDRF	IDLE	OR	NF	FE	PF	SCISR1
$00CD	0	0	0	0	0	BRK13	TXDIR	RAF	SCISR2
$00CE	R8	T8	0	0	0	0	0	0	SCIDRH
$00CF	R7T7	R6T6	R5T5	R4T4	R3T3	R2T2	R1T1	R0T0	SCIDRL

The least significant 13 bits of **SCIBD** determine the baud rate for the SCI port. If **BR** is the value written to bits 12:0, and **Mclk** is the module clock (typically this is the same as the E clock), then the baud rate is

$$\text{SCI Baud Rate} = \frac{\text{Mclk}}{16 * \text{BR}}$$

The **SCICR2** control register contains the bits that turn on the SCI, and contains the interrupt arm bits. **TE** is the Transmitter Enable bit, and **RE** is the Receiver Enable bit. We set both TE and RE equal to 1 in order to activate the SCI device. **TIE** is the Transmit Interrupt Enable bit. We set TIE=1 to arm the transmitter so that an interrupt is requested when TDRE is set. We clear TIE=0 to disarm the TDRE-triggered interrupts. **TCIE** is the Transmit Complete Interrupt Enable bit. We set TCIE=1 to arm the transmitter so that an interrupt is requested when TC is set. We clear TCIE=0 to disarm the TC-triggered interrupts. **RIE** is the Receiver Interrupt Enable bit. We set RIE=1 to arm the receiver so that an interrupt is requested when RDRF is set. We clear RIE=0 to disarm the RDRF-triggered interrupts. **ILIE** is the Idle Line Interrupt Enable bit. We set ILIE=1 to arm the receiver so that an interrupt is requested when IDLE is set. We clear ILIE=0 to disarm the IDLE-triggered interrupts. **RWU** is the Receiver Wake Up Control bit. We set RWU=1 to enable wake up and inhibit receiver interrupts. **SBK** is the Send Break bit. We set SBK=1 to send a break (continuous TxD low) as long as the SBK is 1.

The **SCICR1** control register contains the bits that handle special modes of the SCI. **LOOPS** is the SCI LOOP Mode/Single Wire Mode Enable bit. We set it to 1 to enable loop mode. When loop mode is active, the SCI receive section is disconnected from the RxD pin and the RxD pin is available as general purpose I/O. The receiver input is determined by the RSRC bit. The transmitter output is controlled by the associated DDRS bit. Both the transmitter and the receiver must be enabled to use the LOOP or the single wire mode. **RSRC** is the Receiver Source when LOOPS=1, the RSRC bit determines the internal feedback path for the receiver. If RSRC equals 0, the receiver input is connected to the transmitter internally (not TxD pin). If RSRC is 1, then the receiver input is connected to the TxD pin. **M** is the Mode bit. We set M=0 to create a 10 bit frame with 1 start bit, 8 data bits, 1 stop bit. We set M=1 to create an 11-bit frame with 1 start bit, 9 data bits, 1 stop bit (the ninth data bit is in T8/R8). If **Wake**=0, then the SCI will wake up by an IDLE line recognition. If Wake=1, then the SCI will wake up by address mark (most significant data bit set). **ILT** is the Idle Line Type specifying which of two types of idle line detection will be used by the SCI receiver. ILT determines when the receiver starts counting logic 1s as idle character bits. The counting begins either after the start bit or after the stop bit. If the count begins after the start bit, then a string of logic 1s preceding the stop bit may cause false recognition of an idle character. Beginning the count after the stop bit avoids false idle character recognition, but requires properly synchronized transmissions. If ILT is 1, then the idle character bit count begins after the stop bit. If ILT is 0, then the idle character bit count begins after start bit. To enable parity we set **PE** to 1. If parity is enabled, the SCI will insert a parity bit into the most significant position, and we specify the parity type with PT. If PT is 0, then an even number of ones in the data character causes the parity bit to be zero, and an odd number of ones causes the parity bit to be one. If PT is 1, then odd parity is selected. An odd number of ones in the data character causes the parity bit to be zero and an even number of ones causes the parity bit to be one. If parity is enabled, the receiver will test the parity of each incoming frame. Typically, we set $M=1$ along with $PE=1$ to create an 11-bit frame (1 start, 8 data, 1 parity, and 1 stop). Alternatively, we count set $M=0$ and with $PE=1$ to create a 10-bit frame (1 start, 7 data, 1 parity, and 1 stop).

The flags in the **SCISR1** register can be read by the software, but cannot be modified by writing to this register. **TDRE** is the Transmit Data Register Empty Flag. It is set by the SCI hardware if transmit data can be written to SCIDRL. If TDRE is zero, transmit data register contains previous data that has not yet been moved to the transmit shift register. Writing into the SCIDRL when TDRE is set will result in a loss of data. On the other hand, when this bit is set, the software can begin another output transmission by writing to SCIDRL. This flag is cleared by first reading SCISR1 with TDRE set followed by a SCIDRL write. **TC** is the Transmit Complete Flag. It is set if transmitter is idle (no data, preamble, or break transmission in progress). We can clear TC by reading SCISR1 with TC set followed by writing to SCIDRL. **RDRF** is the Receive Data Register Full bit. RDRF is set if a received character is ready to be read from SCIDR. We clear the RDRF flag by reading SCISR1 with RDRF set followed by reading SCIDRL. **IDLE** is the Idle Line Detected Flag. It is set if the RxD line is idle (10 or 11 consecutive logic ones). The IDLE flag is inhibited when RWU is set to one. It is cleared by reading SCISR1 with IDLE set followed by reading SCIDRL. Once cleared, IDLE is not set again until the RxD line has been active and becomes idle again.

Four error conditions can occur during generation of SCI input/output. Four bits (OR, NF, FE and PE) in the serial communications status register (SCISR1) indicate whether one of these error conditions exists. The overrun error (**OR**) bit is set when the next byte is ready to be transferred from the receive shift register to the SCDR and the SCDR is already full (RDRF bit is set), see Figure 7.32. When an overrun error occurs, the data that caused the overrun is lost and the data that was already in SCIDRL is not disturbed. The OR is cleared when the SCISR1 is read (with OR set), followed by a read of the SCIDRL. The noise flag (**NF**) bit is set if there is noise on any of the received bits, including the start and stop bits. In particular, each data bit is sample three times and the NF bit is set if the three samples are not all the same. The NF bit is not set until the RDRF flag is set. The NF bit is cleared when the SCISR1 is read (with FE equal to 1) followed by a read of the SCIDRL. When no stop bit is detected in the received data character, the framing error (**FE**) bit is set. FE is set at the same time as the RDRF. If the byte received causes both framing and overrun errors, the processor only recognizes the overrun error. The framing error flag inhibits further transfer of data into the SCIDRL until it is cleared. The FE bit is cleared when the SCISR1 is read (with FE equal to 1) followed by a read of the SCIDRL. The parity flag (**PF**) bit is set when the parity enable bit (PE) is set and the parity of the received data does not match the parity type bit (PT). The PF bit is set during the same cycle as the RDRF flag but does not get set in the case of an overrun. We can clear PF by reading SCISR1 and then reading SCIDRL.

The **SCIDRL** register contains the data transmitted out and received in by the SCI device. Even though there are separate transmit and receive data registers, these two registers exist at the same I/O port address. Reads to SCIDRL access the eight bits of the read-only SCI receive data register. Writes to SCIDRL access the eight bits of the write-only SCI transmit data register. **R8** is Receive Data Bit 8. It is a read-only bit that contains the ninth bit of the receive data when $M=1$. **T8** is Transmit Data Bit 8. If M bit is set, T8 stores ninth bit in transmit data character. T8 can be set to 1 to implement a second stop bit. When using 9-bit data, it is necessary to read and write the SCI data register as one 16-bit register.

Common error: If we read or write the SCIDRL and SCIDRH registers in two separate 8-bit accesses, it is possible to confuse the MSbyte and the LSbyte between sequential frames.

The **SCISR1** register contains two mode control bits and one status bit. **BRK13** is the Break Transmit character length bit, which determines the transmit break length. If BRK13 is 1, then the break character is 13 or 14 bit long, and if it is 0, then the break Character is 10 or 11 bit long. **TXDIR** specifies the transmitter pin data direction in single-wire mode. It determines whether the TxD pin is going to be used as an input or output, in the single-wire mode of operation. If TXDIR is 1, then the TxD pin is an output in Single-Wire mode. If it is 0, then the TXD pin is an input in Single-Wire mode. RAF is the Receiver Active Flag. This flag is read only and is controlled by the receiver front end. It is set during the detection of a start bit. It is cleared when an idle state is detected or when the receiver circuitry detects a false start bit (generally due to noise or baud rate mismatch). If RAF is one, then a frame is being received. It is useful in half-duplex systems to avoid a collision.

> **Observation:** The advances in RS232 interface hardware have reduced the probability of transmission errors to such an extent that most serial channels no longer implement parity checking.

> **Observation:** There is no simple way for the transmitter to know whether the baud rates are not matched.

> **Checkpoint 7.4:** In what simple way can the receiver software tell whether the baud rates are not matched?

7.6 SCI Software Interfaces

7.6.1
Full Duplex
Serial Channel

The objective of this section is to develop software to support bidirectional data transfer using interrupt synchronization. We could connect the microcontroller to another computer and use this channel to transfer data. For example, if we connect the DB9 cable to a serial port on a PC, we could run HyperTerminal on the PC and communicate with the microcontroller. A Maxim converter chip is used to generate the RS232 voltage levels, as shown in Figure 7.33. The RS232 timing is generated automatically by the SCI. A simple device driver for this interface was developed in Chapter 3, and in this section we will redesign the device driver to use interrupts.

The data flow graph for this interface was shown previously as Figure 4.17. This example can be found as tut4 within **TExaS**. A portion of that system is presented as Program 7.1.

Figure 7.33
Hardware interface implementing an asynchronous RS232 channel.

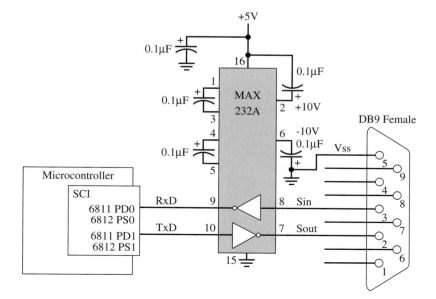

In order for data to be properly received, the baud rate must match with the other module, which in this interface will be 9600 bits/sec. Initially, the two FIFOs are cleared, and just the receiver is armed. The transmitter will be armed when data is available within the SCI_OutChar routine. A flowchart of this system is show in Figure 7.34. An interrupt occurs when new incoming data arrives in the receiver data register (RDRF=1). An interrupt also occurs when the transmit data register is empty (TDRE=1). TDRE is one, when the output channel is idle, needing the software to supply additional data. Notice that the transmit channel is disarmed when the TxFifo is empty and rearmed when new data is put into the TxFifo. When the RxFifo becomes full, then data is lost, but when the TxFifo becomes full, the main program simply waits for space to become available. The assembly language implementation of the FIFO was shown previously as Program 4.14. A C implementation of this system is presented as Program 7.2.

Figure 7.34
FIFO queues can be used to pass data between threads.

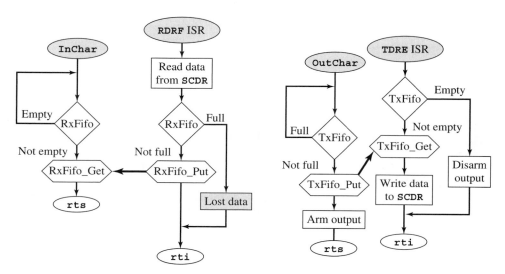

Checkpoint 7.5: How does the software clear the RDRF flag?

Checkpoint 7.6: How does the software clear the TDRE flag?

```
*MC68HC711E9, 2MHz
SCI_Init jsr RxFifo_Init ;empty
      jsr TxFifo_Init ;empty
      ldaa #$2c          ;arm just RDRF
      staa SCCR2         ;enable SCI
      ldaa #$30
      staa BAUD          ;baud rate=9600
      cli
      rts
* Inputs: none Outputs: RegA is ASCII
SCI_InChar pshb
iloop  jsr  RxFifo_Get ;B=0 if empty
      tstb
      beq  iloop
      pulb
      rts                ;A=character
```

```
*MC9S12C32, 4MHz
SCI_Init jsr RxFifo_Init ;FIFO is empty
      jsr  TxFifo_Init   ;FIFO is empty
      movb #$2c,SCICR2   ;arm just RDRF
      movw #26,SCIBD     ;baud rate=9600
      cli
      rts


* Inputs: none Outputs: RegA is ASCII
SCI_InChar pshb
iloop  jsr  RxFifo_Get   ;B=0 if empty
      tbeq B,iloop
      pulb
      rts                ;A=character
```

continued on p. 351

continued from p. 350

```
* Inputs: RegA is ASCII Outputs: none
SCI_OutChar pshb        ;A=character
oloop  jsr  TxFifo_Put ;save in FIFO
       tstb
       beq  oloop    ;FIFO full, wait
       ldab #$AC
       stab SCCR2     ;arm TDRE,RDRF
       pulb
       rts
SCIhandler ldaa SCSR1
       bita #$20
       beq  CkTDRE    ;Not RDRF set
       ldaa SCDR      ;ASCII character
       bsr  RxFifo_Put
CkTDRE ldaa SCSR1
       bpl  sdone     ;Not TDRE set
       ldaa SCCR2     ;bit 7 is TIE
       bpl  sdone     ;disarmed?
       bsr  TxFifo_Get
       tstb
       beq  nomore
       staa SCDR      ;start next output
       bra  sdone
nomore ldaa #$2C
       staa SCCR2     ;disarm TDRE
sdone  rti
       org  $FFD6
       fdb  SCIhandler
```

```
* Inputs: RegA is ASCII Outputs: none
SCI_OutCh pshb              ;A=character
oloop  jsr  TxFifo_Put  ;save in FIFO
       tbeq B,oloop       ;B=0 if full
       movb #$AC,SCICR2 ;arm TDRE
       pulb
       rts


SCIhandler ldaa SCISR1
       bita #$20
       beq  CkTDRE    ;Not RDRF set
       ldaa SCIDRL    ;ASCII character
       bsr  RxFifo_Put
CkTDRE ldaa SCISR1
       bpl  sdone     ;Not TDRE set
       ldaa SCICR2    ;bit 7 is TIE
       bpl  sdone     ;disarmed?
       bsr  TxFifo_Get
       tbeq B,nomore
       staa SCIDRL    ;start output
       bra  sdone
nomore movb #$2C,SCICR2 ;disarm TDRE
sdone  rti


       org  $FFD6
       fdb  SCIhandler
```

Program 7.1
Assembly language implementation of an interrupting SCI interface.

```
// MC68HC711E9, ICC11 C
void SCI_Init(void){
  RxFifo_Init(); // empty FIFOs
  TxFifo_Init();
  BAUD = 0x30;   // 9200 bits/sec
  SCCR1 = 0;     // M=0
  SCCR2 = 0x2c;  // enable, arm RDRF
asm(" cli");     // enable interrupts
}
// Input ASCII character from SCI
// spin if RxFifo is empty
char SCI_InChar(void){ char letter;
  while(RxFifo_Get(&letter) == 0);
  return(letter);
}
// Output ASCII character to SCI
// spin if TxFifo is full
void SCI_OutChar(char data){
  while(TxFifo_Put(data) == 0);
  SCCR2 = 0xAC;     // arm TDRE
}
```

```
// MC9S12C32, CodeWarrior C
void SCI_Init(void){
  RxFifo_Init(); // empty FIFOs
  TxFifo_Init();
  SCIBD = 26;    // 9600 bits/sec
  SCICR1 = 0;    // M=0, no parity
  SCICR2 = 0x2C; // enable, arm RDRF
asm cli          // enable interrupts
}
// Input ASCII character from SCI
// spin if RxFifo is empty
char SCI_InChar(void){ char letter;
  while (RxFifo_Get(&letter) == 0){};
  return(letter);
}
// Output ASCII character to SCI
// spin if TxFifo is full
void SCI_OutChar(char data){
  while (TxFifo_Put(data) == 0){};
  SCICR2 = 0xAC; // arm TDRE
}
```

continued on p. 352

Program 7.2
C language implementation of an interrupting SCI interface.

```
continued from p. 351
// RDRF set on new receive data
// TDRE set on empty transmit register
#pragma interrupt_handler SciHandler()
void SciHandler(void){ char data;
  if(SCSR&RDRF){
    RxFifo_Put(SCDR); // clear RDRF
  }
  if((SCCR2&0x80)&&(SCSR&TDRE)){
    if(TxFifo_Get(&data)){
      SCDR = data;    // clear TDRE
    }
    else{
      SCCR2 = 0x2c;   // disarm TDRE
    }
  }
}
#pragma abs_address:0xffd6
void (*sci_vector[])() ={&SciHandler};
#pragma end_abs_address
```

```
// RDRF set on new receive data
// TDRE set on empty transmit register
interrupt 20 void SciHandler(void){
char data;
  if(SCISR1 & RDRF){
    RxFifo_Put(SCIDRL); // clears RDRF
  }
  if((SCICR2&0x80)&&(SCISR1&TDRE)){
    if(TxFifo_Get(&data)){
      SCIDRL = data;    // clears TDRE
    }
    else{
      SCICR2 = 0x2c;    // disarm TDRE
    }
  }
}
```

Program 7.2
C language implementation of an interrupting SCI interface.

Observation: Data is lost when the RxFifo gets full.

Common error: Notice that the foregoing transmit device driver either acknowledges the interrupt by sending another character or disarms itself because the TxFifo is empty. The software will crash (infinite loop) if it returns from interrupt without acknowledging or disarming.

Checkpoint 7.7: Why didn't the initialization software arm TDRE?

Checkpoint 7.8: What bad thing would happen if the RDRF ISR waited for there to be room in the RxFifo, just as SCI_OutChar waits for there to be room in the TxFifo?

Checkpoint 7.9: Modify either Program 7.1 or 7.2 so the baud rate is 1200 bits/sec.

**7.6.2
Use of Data
Terminal Ready
(DTR) to Interface
a Printer**

One problem with printers is that the printer bandwidth (the actual number of characters per second that can be printed) may be less than the maximum bandwidth supported by the serial channel. There are five conditions that might lead to a situation where the computer outputs serial data to the printer, but the printer isn't ready to accept the data. First, special characters may require more time to print (e.g., carriage return, line feed, tab, formfeed, and graphics). Second, most printers have internal FIFOs that could get full. If the FIFO is not full, then it can accept data as fast as the channel will allow, but when the FIFO becomes full, the computer should stop sending data. Third, the printer cable may be disconnected. Fourth, the printer may be deselected. Fifth, the printer power may be off. The output interfaces shown previously provide no feedback from the printer that could be used to detect and correct these five problems. There are two mechanisms, called flow control, to synchronize the computer with a variable rate output device. These two flow control protocols are called DTR and XON/XOFF.

The first method uses a hardware signal, DTR (pin 4 on the DB9 connector or pin 20 on the DB25 connector), as feedback from the printer to the microcomputer (see Figure 7.35). DTR is −12V if the printer is busy and is not currently able to accept transmission. DTR is +12V if the printer is ready and able to accept transmission. This mechanism can handle all five of the above situations. The computer input mechanism will handle the DTR protocol using additional software checking. With a standard RS232 interface when DTR is −12V,

Figure 7.35
Hardware interface implementing a RS232 simplex channel with DTR handshaking.

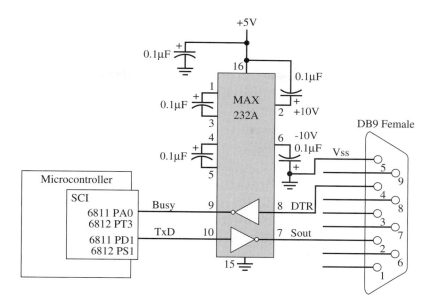

the input line will be +5V (which will stop the transmission if the TDR register is empty). Thus, when DTR is –12V, transmission is temporarily suspended. When DTR is +12V, the input line will be 0V and transmission can proceed normally. In this design, input capture will be used to detect changes in the DTR signal.

The DTR signal from the printer provides feedback information about the printer status. When DTR is −12V (6811 PA0 or 6812 PT3 is high), the printer is not ready to accept more data. In this case, our computer will postpone transmitting more frames. When DTR is +12V (6811 PA0 or 6812 PT3 is low), the printer is ready to accept more data. At this point, the computer will resume transmission. The `Printer_OutChar(data)`, shown below, is called by the main program when it wishes to print.

```
// MC68HC711E9, ICC11 C
void Printer_Init(void){
  TxFifo_Init(); // empty FIFOs
  BAUD = 0x30;   // 9200 bits/sec
  SCCR1 = 0;     // M=0
  SCCR2 = 0x2c;  // enable, arm RDRF
  TCTL2|= 0x03;  // both rise and fall
  TMSK1 = 0x01;  // Arm IC3
  TFLG1 = 0x01;  // clear IC3F
asm(" cli");     // enable interrupts
}
void checkIC3(void){
  if(PORTA&0x01)    // PA0=1 if DTR=-12
    SCCR2 = 0x0C;   // busy, so disarm
  else
    SCCR2 = 0x8C;   // not busy, so arm
}
```

```
// MC9S12C32, CodeWarrior C
void Printer_Init(void){
  TxFifo_Init(); // empty FIFOs
  SCIBD = 26;    // 9600 bits/sec
  SCICR1 = 0;    // M=0, no parity
  SCICR2 = 0x0C; // enable disarm TDRE
  TIOS &=~0x08;  // PT3 input capture
  DDRT &=~0x08;  // PT3 is input
  TSCR1 = 0x80;  // enable TCNT
  TCTL4 |= 0xC0; // both rise and fall
  TIE |= 0x08;   // Arm IC3
  TFLG1 = 0x08;  // initially clear
asm cli          // enable interrupts
}
```

continued on p. 354

Program 7.3
C language implementation of a printer interface with DTR synchronization.

continued from p. 353

```
// Output ASCII character to Printer
// spin if TxFifo is full
void Printer_OutChar(char data){
  while(TxFifo_Put(data) == 0);
  SCCR2 = 0xAC;      // arm TDRE
}
// TDRE set on empty transmit register
#pragma interrupt_handler SciHandler()
void SciHandler(void){ char data;
  if((SCCR2&0x80)&&(SCSR&TDRE)){
    if(TxFifo_Get(&data)){
      SCDR = data;      // clear TDRE
    }
    else{
      SCCR2 = 0x2c;     // disarm TDRE
    }
  }
}
#pragma abs_address:0xffd6
void (*sci_vector[])() ={&SciHandler};
#pragma end_abs_address
#pragma interrupt_handler IC3Han()
void IC3Han(void) {
  TFLG1 = 0x01; // Ack, clear IC3F
  checkIC3();}  // Arm SCI if DTR=+12
#pragma abs_address:0xffea
void (*IC3_vector[])() ={&IC3Han};
#pragma end_abs_address
```

```
void checkIC3(void){
  if(PTT&0x08)      // PT3=1 if DTR=-12
    SCICR2 = 0x0C; // busy, so disarm
  else
    SCICR2 = 0x8C; // not busy, so arm
}
// Output ASCII character to Printer
// spin if TxFifo is full
void Printer_OutChar(char data){
  while (TxFifo_Put(data) == 0){};
  checkIC3();
}
// TDRE set on empty transmit register
interrupt 20 void SciHandler(void){
char data;
  if((SCICR2&0x80)&&(SCISR1&TDRE)){
    if(TxFifo_Get(&data)){
      SCIDRL = data;   // clears TDRE
    }
    else{
      SCICR2 = 0x2c;    // disarm TDRE
    }
  }
}
// IC3 interrupt on any change of DTR
void interrupt 11 IC3Han(void) {
  TFLG1 = 0x08; // Ack, clear C3F
  checkIC3();   // Arm SCI if DTR=+12
}
```

Program 7.3
C language implementation of a printer interface with DTR synchronization.

7.6.3
Use of XON/XOFF
to Interface a
Printer

Another flow control technique is the XON/XOFF protocol. The hardware, which is a standard full duplex serial channel, is described in Figure 7.36. This technique allows the printer to signal back to the computer that it cannot currently accept more data. The advantage of XON/XOFF is that it uses a standard full duplex serial channel. This is an example of a simplex communication *system* constructed on top of a full duplex *channel*.

Figure 7.36
Hardware interface
implementing a serial
channel with XON/XOFF
handshaking.

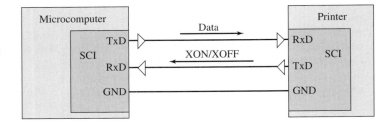

In this system we must assume the printer power is on, and the cable is properly connected. Most printers maintain a FIFO between the incoming serial data from the computer, and the actual printing hardware. When the printer FIFO is almost full (e.g., 80% full) or paper out, the printer will send a single XOFF ($13) transmission back to the computer. Once the FIFO has more room (e.g., 20% full) or more paper, the printer will send a single XON ($11) to the computer. Typically the computer assumes the printer FIFO is empty, so it can initially begin transmission without first receiving an XON. But, once the computer receives an XOFF, it must wait for an XON before it continues transmission. One disadvantage of XON/XOFF is that it cannot handle cable disconnection or printer power-off problems. Figure 7.37 shows a flowchart to implement the XON/XOFF protocol using busy-wait synchronization.

Figure 7.37
Software flowchart implementing a serial channel with XON/XOFF handshaking.

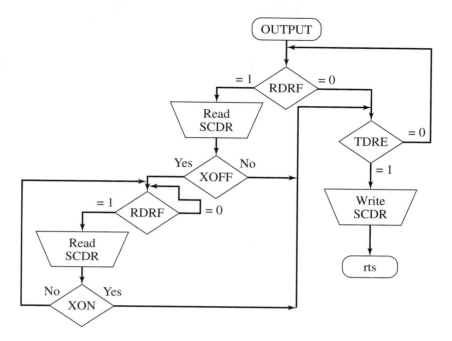

7.7 Synchronous Transmission and Receiving Using the SPI

7.7.1
SPI Fundamentals

Many of the Freescale embedded microcomputers include a SPI. The fundamental difference between a SCI, which implements an asynchronous protocol, and a SPI, which implements a synchronous protocol, is the manner in which the clock is implemented. Two devices communicating with asynchronous serial interfaces (SCI) operate at the same frequency (baud rate) but have two separate (not synchronized) clocks. Two devices communicating with synchronous serial interfaces (SPI) operate from the same clock (synchronized). Typically, the master device creates the clock, and the slave device(s) uses the clock to latch the data in or out. Before discussing the detailed operation of particular devices, we will begin with general features common to all devices. The Freescale SPI includes four I/O lines. The slave select ($\overline{SS}$) is an optional negative logic control signal from master to slave signifying the channel is active. The second line, SCK, is a 50% duty cycle clock generated by the master. The master-out slave-in (MOSI) is a data line driven by the master and received by the slave. The master-in slave-out (MISO) is a data line driven by the slave and received by the master. To work properly, the transmitting device uses one edge of the clock to change its output, and the receiving device uses the other edge to accept the data. Table 7.12

Table 7.12
Synchronous serial port pins on various Freescale microcomputers.

Microcomputer	Pin for $\overline{SS}$	Pin for SCK	Pin for MOSI	Pin for MISO
MC68HC11	PD5	PD4	PD3	PD2
MC68HC812A4	PS7	PS6	PS5	PS4
MC68HC912B32	PS7	PS6	PS5	PS4
MC9S12C32	PM3	PM5	PM4	PM2

lists the I/O port locations of the synchronous serial ports for the various microcomputers discussed in this book.

The SPI allows the computer to communicate synchronously with peripheral devices and other microprocessors. The SPI system in the microcomputer can operate as a master or as a slave. In the SPI system the 8-bit data register, SPDR, in the master and the 8-bit data register in the slave, also SPDR, are linked to form a distributed 16-bit register. Figure 7.38 illustrates communication between master and slave. The interface logic shown in the figure can implement any of the physical channels discussed earlier in the chapter (e.g., simple CMOS digital logic, RS232, RS422, optically isolated).

Figure 7.38
A synchronous serial interface between two Freescale microcomputers.

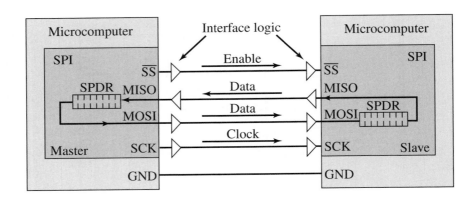

When a data transfer operation is performed, this 16-bit register is serially shifted eight bit positions by the SCK clock from the master so that the data are effectively exchanged between the master and the slave. Data written to the SPDR register of the master are transmitted to the slave. Data written to the SPDR register of the slave are transmitted to the master. The SPI is also capable of interprocessor communications in a multiple master system.

Common control features for the SPI module include:

A baud rate control register used to select the transmission rate
A mode bit in the control register to select master versus slave, clock polarity, clock phase
Interrupt arm bit
Ability to make the outputs open-drain (open-collector)

Common status bits for the SPI module include:

SPIF, transmission complete
WCOL, write collision
MODF, mode fault

Observation: Because the clocks are shared, if you change the E clock frequency, the transfer rate will change, but the SPI still should operate properly.

The key to proper transmission is to select one edge of the clock (shown as T in Figure 7.39) to be used by the transmitter to change the output, and use the other edge (shown as R) to latch the data in the receiver. In this way data are latched during the time when they are stable. Data available is the time when the output data is actually valid, and data required is the time when the input data must be valid. For the communication to occur without error, the data available from the device that is driving the data line must overlap (start before and end after) the data required by the other device that is receiving the data. It is this overlap

Figure 7.39
Synchronous serial timing showing that the data available interval overlaps the data required interval.

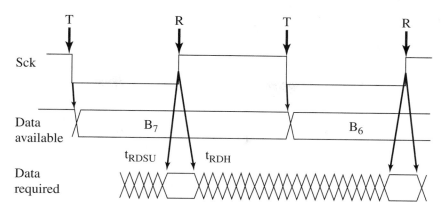

that will determine the maximum frequency at which synchronous serial communication can occur. More discussion of the concepts of data available and data required can be found in Section 9.3.

The general idea of SPI communication can be illustrated by the following pseudocodes. Although it is possible to implement synchronous serial transmission on any computer with simple I/O pins in software using the bit-banging approach (see the DS1620 example in Chapter 3), these operations are implemented in hardware by the SPI interface.

```
TRANSMIT   Set n=7                              Bit counter
TLOOP          On the fall of Sck, set Data=b_n  Output data bit
               Set n=n-1
               Goto TLOOP if n≥0
               set Data=1                        Idle output

RECEIVE    Set n=7                              Bit counter
RLOOP          On the rise of Sck, read Data
               Set b_n=Data                      Input data bit
               Set n=n-1
               Goto RLOOP if n≥0
```

Observation: Because the clocks are shared, if you change the E clock frequency, the transfer rate will change.

The SPI timing is shown in Figure 7.40. The SPI transmits 8-bit data at the same time as it receives input. In all modes, the SPI changes its output on the opposite edge of the clock as it uses to shift data in. There are three mode control bits (MSTR, CPOL, CPHA) that affect the transmission protocol. If the device is a master (**MSTR**=1), it generates the **SCLK**; data is output on the **MOSI** pin and input on the **MISO** pin. If the device is a slave (MSTR=0), the SCLK is an input, and data is received on the MOSI pin and transmitted on the MISO pin. The **CPOL** control bit specifies the polarity of the SCLK. In particular, the CPOL bit specifies the logic level of the clock when data is not being transferred. The **CPHA** bit affects the timing of the first bit transferred and received. If CHPA is 0, then the device will shift data in on the first (and 3rd, 5th, 7th, . . . etc.) clock edge. If CHPA is 1, then the device will shift data in on the second (and 4th, 6th, 8th, . . . etc.) clock edge. In Figure 7.40, the data is shown with MSB transferred first, but the 6812 has an option where the bits are transferred in the other order (LSB first.)

Next we will overview the specific SPI functions on particular Freescale microcomputers. This section is intended to supplement rather than replace the Freescale manuals. When designing systems with a SPI, it is important to refer to the reference manual of your specific Freescale microcomputer.

Figure 7.40
Synchronous serial
modes of the Freescale
SPI interface.

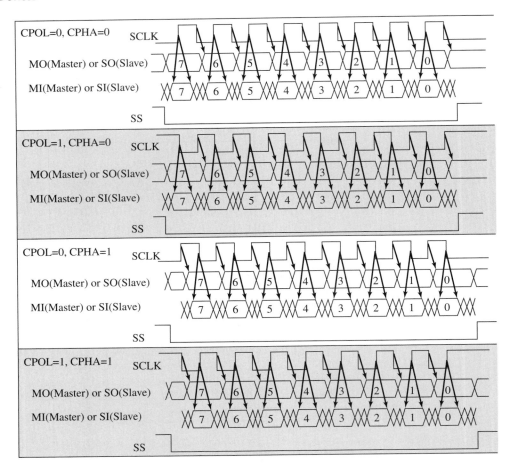

7.7.2
**MC68HC711E9 SPI
Details**

The SPI port on the MC68HC711E9 uses the four pins PD5 = $\overline{SS}$, PD4 = SCLK, PD3 = MOSI, and PD2 = MISO. If the 6811 is the master, then we should set the DDRD register to make PD5, PD4, and PD3 outputs. PD2 should be an input. Table 7.13 shows the 6811 SPI ports.

Table 7.13
6811 SPI ports.

Address	Bit 7	6	5	4	3	2	1	Bit 0	Name
$1009	0	0	DDRD5	DDRD4	DDRD3	DDRD2	DDRD1	DDRD0	DDRD
$1028	SPIE	SPE	DWOM	MSTR	CPOL	CPHA	SPR1	SPR0	SPCR
$1029	SPIF	WCOL	0	MODF	0	0	0	0	SPSR
$102A	Bit 7	6	5	4	3	2	1	Bit 0	SPDR

The **SPCR** register contains the configuration bits that determine the mode of operation. **SPIE** is the Serial Peripheral Interrupt Enable bit. We will turn this bit off to use busy-waiting synchronization. **SPE** is the Serial Peripheral System Enable bit. We will turn this bit on to enable the SPI device. The **DWOM** bit in the SPCR register is used to activate the Port D wired-OR mode. This one bit affects all six port D pins. If DWOM is 0, then PD5–PD0 are normal CMOS signals. A normal CMOS output has two states, either driven low (0V or logic zero) or driven high (+5V or logic one). If DWOM is 1, then any port D output is open-drain. Open-drain means the output is either driven low or floats. Alternative names for same behavior are *open-collector* and *wire-or-mode*. **MSTR** is the Master Mode Select bit. We will set this bit to one to make the 6811 the master. **CPOL** and **CPHA** determine the SPI Clock Polarity and Clock Phase. These two bits are used to specify the protocol, as shown in Figure 7.40. The SPR1 and SPR0 control bits determine the transfer rate. Assuming an E clock frequency of 2 MHz, Table 7.14 shows the possible transmission rates.

Table 7.14
Bit rate selection for the synchronous serial port on the MC68HC711E9.

SPR1	SPR0	Divisor	Transfer Frequency	Bit Time
0	0	2	1 MHz	1 μs
0	1	4	500 kHz	2 μs
1	0	16	125 kHz	8 μs
1	1	32	62.5 kHz	16 μs

The **SPSR** register contains the SPI status bits. SPIF is the SPI Transfer Complete Flag. It is set when an SPI transfer is complete. SPIF is cleared by reading SPSR with SPIF set followed by a SPDR access. **WCOL** is the Write Collision bit. It is set when SPDR is written while transfer is in progress. It is cleared by SPSR with WCOL set followed by a SPDR access. **MODF** is the mode fault bit. A mode fault will terminates SPI operation. It is set when $\overline{SS}$ is pulled low while MSTR=1. It is cleared by SPSR read with MODF set followed by SPCR write.

7.7.3 MC9S12C32 SPI Details

The SPI port on the MC9S12C32 uses the four pins PM3=$\overline{SS}$, PM5=SCLK, PM4=MOSI, and PM2=MISO. When the SPI is enabled (**SPE**=1), all pins that are defined by the configuration as inputs will be inputs regardless of the state of the DDRM bits for those pins. All pins that are defined as SPI outputs will be outputs only if the DDRM bits for those pins are set. If the 6812 is the master, then we should set the DDRM register to make PM5, PM4, and PM3 outputs. PM2 will automatically be an input. A bidirectional serial pin is possible using the **BIDIROE** as the direction control. Table 7.15 shows the 6812 SPI ports.

Address	Bit 7	6	5	4	3	2	1	Bit 0	Name
$00D8	SPIE	SPE	SPTIE	MSTR	CPOL	CPHA	SSOE	LSBF	SPICR1
$00D9	0	0	0	MODFEN	BIDIROE	0	SPISWAI	SPC0	SPICR2
$00DA	0	0	0	0	0	SPR2	SPR1	SPR0	SPIBR
$00DB	SPIF	0	SPTEF	MODF	0	0	0	0	SPISR
$00DD	Bit 7	6	5	4	3	2	1	Bit 0	SPIDR
$0250	0	0	PM5	PM4	PM3	PM2	PM1	PM0	PTM
$0252	0	0	DDRM5	DDRM4	DDRM3	DDRM2	DDRM1	DDRM0	DDRM

Table 7.15
MC9S12C32 SPI ports.

The SPI functions in three modes: run, wait, and stop. **Run mode** is the basic mode of operation. The SPI operation in **wait mode** is a configurable low-power mode, controlled by the **SPISWAI** bit. In wait mode, if the SPISWAI bit is clear, the SPI operates as in Run Mode. If the SPISWAI bit is set, the SPI goes into a power-conservative state, with the SPI clock generation turned off. If the SPI is configured as a master, any transmission in progress stops but is resumed after CPU goes into Run Mode. If the SPI is configured as a slave, reception and transmission of a byte continues, so that the slave stays synchronized to the master. The SPI is inactive in **stop mode** for reduced power consumption. If the SPI is configured as a master, any transmission in progress stops, but is resumed after CPU goes into Run Mode. If the SPI is configured as a slave, reception and transmission of a byte continues, so that the slave stays synchronized to the master.

The module clock (same frequency as the E clock) is input to a divider series and the resulting SPI clock rate may be selected to be divided by 2, 4, 8, 16, 32, 64, 128, or 256. Three bits in the SPIBR register control the SPI clock rate. The SPIBR register determines the transfer rate. Table 7.16 shows the possible transmission rates for two different module clocks.

Table 7.16
Bit rate selection for the
synchronous serial port
on the 6812.

SPR2	SPR1	SPR0	Divisor	Module Clock=4 MHz		Module Clock=24 MHz	
				Frequency	**Bit Time**	**Frequency**	**Bit Time**
0	0	0	2	2 MHz	500 ns	12 MHz	83.3 ns
0	0	1	4	1 MHz	1 μs	6 MHz	166.7 ns
0	1	0	8	500 kHz	2 μs	3 MHz	333.3 ns
0	1	1	16	250 kHz	4 μs	1.5 MHz	666.7 ns
1	0	0	32	125 kHz	8 μs	750 kHz	1.33 μs
1	0	1	64	62.5 kHz	16 μs	375 kHz	2.67 μs
1	1	0	128	31.25 kHz	32 μs	187.5 kHz	5.33 μs
1	1	1	256	15.625 kHz	64 μs	93.75 kHz	10.67 μs

We use the **SPICR1** register to specify the SPI mode of operations. **SPE** is the SPI System Enable bit. We will turn this bit on whenever we wish to use the SPI. We set the **SSOE** bit to enable the $\overline{SS}$ signal as shown in Figure 7.40. We clear the **SSOE** bit when we want to use PM3 as a regular I/O pin. The **SPIE** bit is the arm bit for SPIF, and **SPTIE** bit is the arm bit for SPTEF. These arm bits will be cleared because interrupts are not needed. We will set the **MSTR** to one so that the 6812 becomes the master. **CPOL** and **CPHA** determine the SPI Clock Polarity and Clock Phase. These two bits are used to specify the protocol, as shown in Figure 7.40. All other bits not specifically mentioned will be cleared. **SPIF** is the SPI Interrupt Request bit. Even though we won't be using interrupts, this bit gets set after each byte is transferred, and will be used to implement the busy-waiting synchronization. In particular, SPIF is set after the eighth SCK cycle in a data transfer, and it is cleared by reading the **SPISR** register (with SPIF set) followed by an access (read or write) to the SPI data register. **SPTEF** is the Transmit Empty Interrupt Flag. If set, this bit indicates that the transmit data register is empty. To clear this bit and place data into the transmit data register, SPISR has to be read with SPTEF=1, followed by a write to SPIDR. Any write to the SPI Data Register without reading SPTEF=1, is effectively ignored.

The **SPIDR** 8-bit register is both the input and output register. A write to SPIDR allows a data byte to be queued and transmitted. For a SPI configured as a master, a queued data byte is transmitted immediately after the previous transmission has completed. The SPTEF in the SPISR register indicates when the SPI Data Register is ready to accept new data. Reading the data can occur anytime from after the SPIF is set to before the end of the next transfer. If the SPIF is not serviced by the end of the successive transfers, those data bytes are lost, and the data within the SPIDR retains the first byte until SPIF is serviced. **LSBFE** is the LSB-First Enable bit. This bit does not affect the position of the MSB and LSB in the data register. Reads and writes of the data register always have the MSB in bit 7. In master mode, a change of this bit will abort a transmission in progress and force the SPI system into idle state. If LSBFE is 1, data is transferred least significant bit first. If LSBFE is 0, data is transferred most significant bit first.

The control bits **MODFEN** and **SSOE** affect the operation of the PM3 pin as defined in Table 7.17. **MODF** is the Mode Error Interrupt Status Flag. This bit is set if the $\overline{SS}$ input becomes low when the SPI is configured as a master and mode fault detection is enabled, MODFEN bit of SPICR2 register is set. The flag is cleared automatically by a read of the SPI Status Register (with MODF set) followed by a write to the SPICR1.

Table 7.17
Mode selection for the
synchronous serial port
on the MC9S12C32.

MODFEN	SSOE	Master Mode (MSTR=1)	Slave Mode (MSTR=0)
0	0	PM3 not used with SPI	PM3 is $\overline{SS}$ input
0	1	PM3 not used with SPI	PM3 is $\overline{SS}$ input
1	0	PM3 is $\overline{SS}$ input with MODF feature	PM3 is $\overline{SS}$ input
1	1	PM3 is $\overline{SS}$ output	PM3 is $\overline{SS}$ input

Bidirectional modes are controlled by the bits **SPC0**, **BIDIROE**, and **MSTR**. These control bits determine the input/output configuration of the PM2 and PM4 as illustrated in Table 7.18 and Figure 7.41.

Table 7.18
Bidirectional modes for the SPI on the MC9S12C32.

Pin Mode	MSTR	SPC0	BIDIROE	MISO	MOSI
Normal	1	0	X	Master In	Master Out
Bidirectional	1	1	0	MISO not used	Master In
			1		Master I/O
Normal	0	0	X	Slave Out	Slave In
Bidirectional	0	1	0	Slave In	MOSI not used
			1	Slave I/O	

Figure 7.41
Synchronous serial modes of the Freescale 6812 SPI interface.

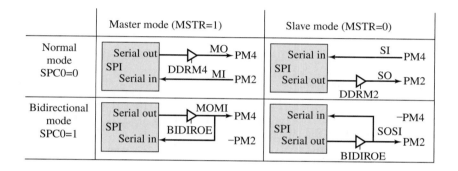

Checkpoint 7.10: How does the software clear the SPIF flag?

Checkpoint 7.11: What hardware interface and which modes do we use to create a bidirectional communication channel between two microcontrollers using their SPI ports?

7.7.6
SPI Applications

7.7.6.1
Digital-to-Analog
Converter

This first example shows the synchronous serial interface between the computer and an Analog Devices DAC8043 DAC. A DAC accepts a digital input (in our case a number between 0 and 4095) and creates an analog output (in our case a voltage between 0 and −5). Detailed discussion of DACs will be presented later in Chapter 11. Here in this section we will focus on the digital hardware and software aspects of the serial interface (Figure 7.42).

Figure 7.42
A 12-bit DAC interfaced to the SPI port.

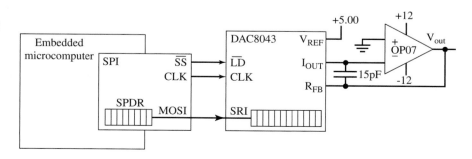

As with any SPI interface, there are basic interfacing issues to consider.

1. *Word size.* In this case we need to transmit 12 bits to the DAC. The DAC8043 data sheet suggests that we can embed the 12-bit data right-justified (it will ignore the first 4 bits that are sent) into a 16-bit transmission.
2. *Bit order.* The DAC8043 requires the most significant bits first.
3. *Clock phase, clock polarity.* There are two issues to resolve. Since the DAC8043 samples its serial input data on the rising edge of the clock, the SPI must change the data on the falling edge. CPOL=CPHA=0 and CPOL=CPHA=1 both satisfy this requirement. The second issue is which edge comes first, the rise or the fall. In this interface it probably doesn't matter.
4. *Bandwidth.* We look at the timing specifications of the DAC8043. The minimum clock low width of 120 ns means the shortest SPI period we can use is 250 ns.

The first 4 bits sent will be 0, followed by the 12 data bits that specify the analog output (Figure 7.43).

Figure 7.43
DAC8043 DAC serial timing.

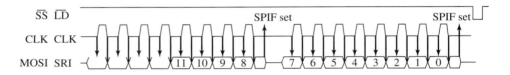

Because we want the $\overline{LD}$ signal to remain high for the entire 16-bit transfer then pulse low, we will implement it using the regular I/O pin functions. The ritual initializes the direction register, SPI mode, and bandwidth (Program 7.4). To change the digital-to-analog (D/A) output, two 8-bit transmissions are sent (Program 7.5).

```
// MC68HC711E9                              // MC9S12C32
void DAC_Init(void){ // PD5=LD=1           void DAC_Init(void){ // PM3=LD=1
  DDRD |= 0x38; // PD4=CLK=SPI clock out      DDRM |= 0x38; // PM5=CLK=SPI clock out
  PORTD|= 0x20; // PD3=SRI=SPI master out     PTM  |= 0x08;  // PM4=SRI=SPI master out
/* bit SPCR                                 /* bit SPICR1
 7 SPIE = 0    no interrupts                 7 SPIE = 0    no interrupts
 6 SPE  = 1    enable SPI                    6 SPE  = 1    enable SPI
 5 DWOM = 0    regular outputs              5 SPTIE= 0    no interrupts
 4 MSTR = 1    master                       4 MSTR = 1    master
 3 CPOL = 0    output changes on fall,       3 CPOL = 0    output changes on fall,
 2 CPHA = 0    clock normally low           2 CPHA = 0    clock normally low
 1 SPR1 = 0    1 MHz operation              1 SSOE = 0    PM3 regular output, LD
 0 SPR0 = 0     */                          0 LSBF = 0    most sign bit first */
  SPCR = 0x50;                                SPICR1 = 0x50;
}                                             SPICR2 = 0x00; // normal mode
                                              SPIBR = 0x00;} // 2MHz
```

Program 7.4
C language initialization for a DAC interface using the SPI.

```
// MC68HC711E9
#define SPIF 0x80
void DAC_out(unsigned short code){
unsigned char dummy;
  SPDR = (code>>8);        // msbyte
  while((SPSR&SPIF)==0); // gadfly wait
  dummy = SPDR;            // clear SPIF
  SPDR = code;             // lsbyte
  while((SPSR&SPIF)==0); // gadfly wait
  dummy = SPDR;            // clear SPIF
  PORTD &= ~0x20;          // PD5=LD=0
  PORTD |= 0x20; }         // PD5=LD=1
```

```
// MC9S12C32
#define SPIF 0x80
void DAC_out(unsigned short code){
unsigned char dummy;
  SPIDR = (code>>8);        // msbyte
  while((SPISR&SPIF)==0); // gadfly wait
  dummy = SPIDR;            // clear SPIF
  SPIDR = code;             // lsbyte
  while((SPISR&SPIF)==0); // gadfly wait
  dummy = SPIDR;            // clear SPIF
  PTM &= ~0x08;             // PM3=LD=0
  PTM |= 0x08; }            // PM3=LD=1
```

Program 7.5
C language function for a DAC interface using the SPI.

7.7.6.2
Analog-to-Digital
Converter

This second SPI example shows the synchronous serial interface between the computer and a Maxim MAX1247 ADC. An ADC accepts an analog input (in our case a voltage between 0 and +2.5 V on one of the four analog inputs CH3–CH0) and creates a digital output (in our case a number between 0 and 4095). Detailed discussion of ADCs will also be presented later in Chapter 11. Here in this section we will focus on the hardware and software aspects of the serial interface (Figure 7.44).

Figure 7.44
A four-channel 12-bit ADC interfaced to the SPI port.

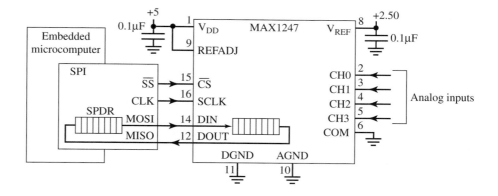

Again, there are basic interfacing issues to consider for this interface:

1. *Word size.* In this case we need to first transmit 8 bits to the ADC, then receive 12 bits back from the ADC. The MAX1247 data sheet suggests that it will embed the 12-bit data into two 8-bit transmissions.
2. *Bit order.* The MAX1247 requires the most significant bits first.
3. *Clock phase, clock polarity.* Since the MAX1247 samples its serial input data on the rising edge of the clock, the SPI must change the data on the falling edge. CPOL=CPHA=0 and CPOL=CPHA=1 both satisfy this requirement. We will use the CPOL=CPHA=0 mode as suggested in the Maxim data sheet.
4. *Bandwidth.* We look at the timing specifications of the MAX1247. The maximum SCLK frequency is 2 MHz, and the minimum clock low/high widths is 200 ns, so the shortest SPI period we can use is 500 ns.

The first 8 bits sent will specify the channel and ADC mode. After conversion, the 12 data bits result is returned. Notice that bit 7 of the mode select is always high, and the 12-bit

Figure 7.45
DAC8043 D/A serial timing.

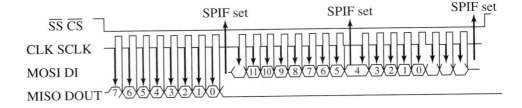

ADC result is embedded into the middle of the two 8-bit transmissions. The software will shift the 16-bit data 3 bits to the right to produce the 0 to 4095 result (Figure 7.45).

Because we want the $\overline{CS}$ signal to remain low for the entire 24-bit transfer, we will implement it using the regular I/O pin functions. The ritual initializes the direction register, SPI mode, and bandwidth (Program 7.6).

```
// MC68HC711E9
void ADC_Init(void){ // PD5=CS=1
  DDRD |=0x38; // PD4=SCLK=SPI clock out
  PORTD|=0x20; // PD3=DIN=SPI master out
/* bit SPCR       PD2=DOUT=SPI master in
7 SPIE = 0   no interrupts
6 SPE  = 1   enable SPI
5 DWOM = 0   regular outputs
4 MSTR = 1   master
3 CPOL = 0   output changes on fall,
2 CPHA = 0   clock normally low
1 SPR1 = 0   1 MHz operation
0 SPR0 = 0     */
  SPCR = 0x50;
}
```

```
// MC9S12C32
void ADC_Init(void){ // PM3=CS=1
  DDRM |=0x38;  // PM5=SCLK=SPI clock out
  DDRM &=~0x04; // PM2=DOUT=SPI master in
  PTM |= 0x08;  // PM4=DIN=SPI master out
/* bit SPICR1
7 SPIE = 0   no interrupts
6 SPE  = 1   enable SPI
5 SPTIE= 0   no interrupts
4 MSTR = 1   master
3 CPOL = 0   output changes on fall,
2 CPHA = 0   clock normally low
1 SSOE = 0   PM3 regular output, LD
0 LSBF = 0   most sign bit first */
  SPICR1 = 0x50;
  SPICR2 = 0x00; // normal mode
  SPIBR = 0x00;} // 2MHz
```

Program 7.6
C language initialization for an ADC interface using the SPI.

Recall that when the software outputs to the SPI data register, the 8-bit register in the SPI is exchanged with the 8-bit register in the ADC. To read the ADC, three 8-bit transmissions are exchanged (Program 7.7). On the first exchange, the software specifies the channel and ADC mode, then the 12-bit ADC data are returned during the second and third transmission. All the ADC modes in the following #define statements implement unipolar voltage range, single-ended, and external clock:

```
#define CH0 0x9F
#define CH1 0xDF
#define CH2 0xAF
#define CH3 0xEF
```

7.7.6.3
Temperature
Sensor/Controller
One of the basic building components of a microprocessor-based control system is the sensor. In this bang-bang control example, we will interface a DS1620 to the computer and use it as part of the temperature controller. The DS1620 chip automatically implements a bang-bang digital control system that applies heat to the room to maintain the temperature

```
// MC68HC711E9
unsigned short ADC_in(unsigned char code){
unsigned short data;
  PORTD &= ~0x20; // PD5=CS=0
  SPDR = code;      // set channel,mode
  while((SPSR&SPIF)==0); // gadfly wait
  data = SPDR;      // clear SPIF
  SPIDR = 0;        // start SPI
  while((SPSR&SPIF)==0); // gadfly wait
  data = SPDR<<8;   // msbyte of ADC
  SPDR = 0;         // start SPI
  while((SPSR&SPIF)==0); // gadfly wait
  data += SPDR;     // lsbyte of ADC
  PORTD |= 0x20;    // PD5=CS=1
  return data>>3;}  // right justify
```

```
// MC9S12C32
unsigned short ADC_in(unsigned char code){
unsigned short data;
  PTM &= ~0x08;     // PM3=CS=0
  SPIDR = code;     // set channel,mode
  while((SPISR&0x80)==0); // gadfly wait
  data = SPIDR;     // clear SPIF
  SPIDR = 0;        // start SPI
  while((SPISR&0x80)==0); // gadfly wait
  data = SPIDR<<8;  // msbyte of ADC
  SPIDR = 0;        // start SPI
  while((SPISR&0x80)==0); // gadfly wait
  data += SPIDR;    // lsbyte of ADC
  PTM |= 0x08;      // PM3=CS=1
  return data>>3;}  // right justify
```

Program 7.7
C language function for an ADC interface using the SPI.

as close to a desired temperature setpoint T* as possible. The microcomputer will communicate with the DS1620 to set the two threshold temperatures, T_{HIGH} and T_{LOW}.

Once programmed with the two setpoint temperatures T_{HIGH} and T_{LOW}, the DS1620 will perform the above bang-bang algorithm automatically. Figure 7.46 shows the DS1620-based controller. The DS1620 can be programmed at the factory before installing the chip into a system. Here it is shown with a microcomputer, which allows the operator to adjust the setpoints.

Figure 7.46
DS1620-based
temperature controller.

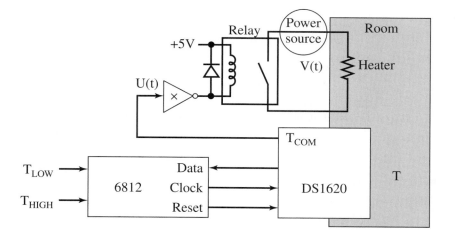

To run the controller, the microcomputer simply sends the two setpoints to the DS1620 and puts it in continuous mode. The DS1620 was previously interfaced to the computer in Chapter 3 using simple output commands; in this section we will interface it using the 6812 SPI (Program 7.8). The 6812 SPI module has a few unique features that allow it to be used with the DS1620. First, it allows data to be transferred least significant bit first. Second, and more important, it has a bidirectional half-duplex mode that

Program 7.8

C language initialization of a temperature sensor interface using the SPI.

```
// Interface between MC9S12C32 and DS1620 using the SPI
// bitstatus  Configuration/Status Register meaning
// 7   DONE    1=Conversion done, 0=conversion in progress
// 6   THF     1=temperature above TH, 0=temperature below TH
// 5   TLF     1=temperature below TL, 0=temperature above TL
// 1   CPU     1=CPU control, 0=stand alone operation
// 0   1SHOT   1=one conversion and stop, 0=continuous conversions
// temperature         digital value (binary) digital value (hex)
// +125.0 C    011111010               $0FA
//   +64.0 C   010000000               $080
//    +1.0 C   000000010               $002
//    +0.5 C   000000001               $001
//       0 C   000000000               $000
//    -0.5 C   111111111               $1FF
//   -16.0 C   111100000               $1E0
//   -55.0 C   110010010               $192
void DS1620_Init(void){ // PM3=RST=0
  DDRM |= 0x38; // PM5=CLK=SPI clock out
  PTM &=~0x08;  // PM4=DQ=SPI bidirectional data
/* bit SPICR1
 7 SPIE = 0    no interrupts
 6 SPE  = 1    enable SPI
 5 SPTIE= 0    no interrupts
 4 MSTR = 1    master
 3 CPOL = 1    output changes on fall
 2 CPHA = 1    and input clocked in on rise
 1 SSOE = 0    PM3 regular output DS1620 RST
 0 LSBF = 1    least significant bit first */
  SPICR1 = 0x5D;
  SPICR2 = 0x01; // bidirectional mode
  SPIBR = 0x01;} // 1MHz could be 2MHz
```

allows data to be transmitted and received on the same pin. The basic idea is that to transmit 9 bits of temperature data to the DS1620 (Program 7.9), the 6812:

1. Sets the RST signal=1
2. Outputs the 8-bit command code for write TH (or write TL) to the DS1620 using one SPI transfer
3. Outputs the 9-bit data value to the DS1620 using two SPI transfers, least significant bit first and right-justified (i.e., most significant 7 bits are 0)
4. Sets the RST signal=0

Program 7.9

C language helper functions for a temperature sensor interface using the SPI.

```
#define SPIF 0x80
void out8(char code){ unsigned char dummy;
// assumes DDRM bit 4 is 1, output
  SPIDR = code;
  while((SPISR&SPIF)==0);    // gadfly wait for SPIF
  dummy = SPIDR;}            // clear SPIF
void out9(short code){ unsigned char dummy;
  SPIDR = code;              // lsbyte
  while((SPISR&SPIF)==0);    // gadfly wait for SPIF
  dummy = SPIDR;             // clear SPIF
  SPIDR = (code>>8);         // msbyte
  while((SPISR&SPIF)==0);    // gadfly wait for SPIF
  dummy = SPIDR;}            // clear SPIF
```

```
unsigned char in8(void){ short n; unsigned char result;
  DDRM &=~0x10; // PM4=DQ input
  SPIDR = 0;      // start shift register
  while((SPISR&SPIF)==0); // gadfly wait for SPIF
  result = SPIDR;          // get data, clear SPIF
  DDRM |= 0x10;            // PM4=DQ output
  return result;}
 short in9(void){ short result;
  DDRM &=~0x10; // PM4=DQ input
  SPIDR = 0;      // start shift register
  while((SPISR&SPIF)==0); // gadfly wait for SPIF
  result = SPIDR;          // get LS data, clear SPIF
  SPIDR = 0;               // start shift register
  while((SPISR&SPIF)==0); // gadfly wait for SPIF
  if(SPIDR&0x01)           // get MS data, clear SPIF
    result |= 0xFF00;      // negative
  else
    result &=~0xFF00;      // positive
  DDRM |= 0x10; // PM4=DQ output
  return result;}
```

The basic idea is that to receive 9 bits of temperature data from the DS1620, the 6812:

1. Sets the RST signal=1
2. Outputs the 8-bit command code for read T to the DS1620 using one SPI transfer
3. Switches the direction bit for the data line to input
4. Inputs the 9-bit data value to the DS1620 using two SPI transfers, least significant bit first and right-justified (i.e., most significant 7 bits are sign extended)
5. Switches the direction bit for the data line to output
6. Sets the RST signal=0

On top of these low-level DS1620 driver functions, we implement the higher-level routines of Program 7.10.

Program 7.10
C language functions for a temperature sensor interface using the SPI.

```
void DS1620_Start(void){
  PTM |= 0x08;      // RST=1
  out8(0xEE);
  PTM &=~0x08;}     // RST=0
void DS1620_WriteConfig(char data){
  PTM |= 0x08;      // RST=1
  out8(0x0C);
  out8(data);
  PTM &=~0x08;}     // RST=0
void DS1620_WriteTH(short data){
  PTM |= 0x08;      // RST=1
  out8(0x01);
  out9(data);
  PTM &=~0x08;}     // RST=0
void DS1620_WriteTL(short data){
  PTM |= 0x08;      // RST=1
  out8(0x02);
  out9(data);
  PTM &=~0x08;}     // RST=0
```

continued on p. 368

continued from p. 367

```
unsigned char DS1620_ReadConfig(void){ unsigned char value;
  PTM |= 0x08;      // RST=1
  out8(0xAC);
  value = in8();
  PTM &=~0x08;      // RST=0
  return value;}
unsigned short DS1620_ReadT(void){ unsigned short value;
  PTM |= 0x08;      // RST=1
  out8(0xAA);
  value = in9();
  PTM &=~0x08;      // RST=0
  return value;}
```

7.8 Exercises

7.1 The objective is to build the equivalent of the SCI receive function using input capture and output compare. Assume the baud rate is 1000 bits/s, resulting in a bit time of 1 ms. The TTL-level serial input is connected to an input capture pin like the 6811 PA1/IC2. You can also use an output compare as well. The hardware in Figure 7.47 is one possibility. To implement this problem on another computer, simply connect to an available input capture.

Figure 7.47
Interface used to create a serial input channel using input capture.

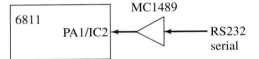

You are given (use without writing these three functions) a FIFO with prototypes:

```
int InitFifo(unsigned short);  // Create and initialize a FIFO,
// pass in the size, returns true if successful
int PutFifo(char);  // Enter 8 bit data into FIFO,
// returns true if successful, false if full and data not stored
int GetFifo(&char);   // Remove 8 bit data from FIFO,
// returns true if successful, false if empty, no data removed
```

The main program, which you do not write, will call `GetFifo` to receive data from your serial interface. Your input capture/output compare interrupt handler(s) will accept the one start, eight data, one stop bit transmission. There is a global error flag, FLAG, which should be set on any of the following three conditions. Set the flag if the FIFO becomes full and data are lost (overrun). Also set the flag if the start bit is incorrect (false start). Also set the flag if the stop bit is incorrect (framing error). If there are no false start or framing errors, put the 8-bit data into the FIFO using `PutFifo`. The single-chip 6811 has 256 bytes of RAM, 2 kbytes of EEPROM, and an E clock of 2 MHz.

   a) Carefully draw the input signal on PA1/IC2 when the ASCII character '5' is received. Draw the signal to scale. Assume the TCNT value is 1000 (decimal) when the first 1 to 0 transition occurs. Give the TCNT value for each transition of the input signal. Place arrows (↑) on the drawing exactly when each IC and OC interrupt will occur, giving the TCNT value at each interrupt.

   b) Define all global variables you need. Assume the compiler will place them in the 6811 single-chip RAM.

```
char FLAG;  // set when a serial input error occurs
```

c) Show the ritual that initializes global variables, input capture IC2, and output compare OC2. Assume that the 6811 is running in single-chip mode and that the two interrupt vectors are already initialized in ROM. Carefully choose an appropriate size when calling `InitFifo()` to create/initialize the FIFO.

d) Show the IC2 and OC2 interrupt handlers. *No backward jumps* are allowed.

7.2 The objective of this exercise is to design a printer interface that handles DTR hardware synchronization. To implement this problem on another computer, simply connect to an available input capture. For the 6811 connect: (1) SCI transmit to the printer receive, (2) the DTR pin 20 to input capture IC3/PA0, and (3) ground to ground. Your software must be *interrupt-driven and written in C*. The baud rate is 1200 bits/s. The RS232 protocol is one start bit, seven bit, no parity, and two stop bits. There is a DTR signal from the printer. When DTR is −12 V (PA0=high), your 6811 should not begin transmitting any more frames because the printer is not ready to accept it. When DTR is +12 V (PA0=low), the printer is ready to accept more data. There are three FIFO functions given (you do *not* write them). Your `OutChar(data)` is called by the main program (again, you do not write the main program). To implement this problem on another computer, simply connect DTR to an available input capture.

`InitFifo()`	initializes the FIFO
`flag=PutFifo(data)`	enters a byte `data` into the FIFO
	returns a `flag`=true if the FIFO is full, and the data was not saved.
`flag=GetFifo(&data)`	returns with a byte and `flag` is false if successful
	returns `flag` true if the FIFO was empty at the time of the call

a) Show the ritual software that is executed once. Assume the interrupt vectors will be initialized automatically by the compiler.

b) Show the C function `OutChar(data)` that initiates an output. The actual output occurs in the background. `OutChar` should return a true flag if the operation cannot be performed.

c) Give the interrupt routine(s) that performs the serial output operations in the background. Give careful thought as to when and how to interrupt. Poll anyway you want. *Good software has* no *backward jumps*. Use labels `IC3handler()` and `SCIhandler()` for the two interrupt routines. Comments will be graded.

7.3 Design an automatic baud rate selector. The input capture system will be used to measure the first three positive logic pulse widths and the first three negative logic widths. To implement this problem on another computer, simply connect to an available input capture. Your software will then choose the smallest pulse width as the communication bit time. For example, if the smallest bit time is 400 E cycles (which falls between 625 and 313), then the baud rate should be 4800. Finally, your software will initialize the SCI to communicate at that correct baud rate (eight bit data, one start, one stop and no parity). Specify the E clock rate. You may assume the baud rate is 600, 1200, 2400, 4800, or 9600.

a) Show the hardware connections between a three-wire ±12 V RS232 full-duplex line (TxD, RxD, and Gnd) and the microcomputer. Any RS232 interface chip may be used. Give chip numbers but not pin numbers.

b) Show the appropriate data structure that holds the information of the above table. The organization of this information should be easy to understand and easy to change (e.g., add more baud rates, change of E clock frequency). This data structure will eliminate the need for a complex `if then` or `switch code` in part c.

c) Show the C software function that automatically selects the correct baud rate and initializes the SCI. Gadfly synchronization will be used. Input/output functions for the SCI are not required.

7.4 You will build a low-cost computer-based logic analyzer. It will be able to collect 2048 8-bit digital samples at 80 Mhz triggered by your software. You are given the following hardware components (you may use other 74HC digital devices, too).

A 20-MHz clock generator (black box with 50% duty cycle digital output)
A 2048 9-bit hardware FIFO (CY7C429, IDT7203, AM7203, or LH540203)

The FIFO contains a 2048 by 9 bit RAM and implements the classic FIFO functions. You will only use eight of the nine data pins. To reset the FIFO, you pull its ***RS** line low. This will clear the

FIFO. When the reset ***RS** is high, you can perform a put operation by toggling the ***W** line low, then high. The 9 bits on the input data lines **D8–D0** are stored (put) in the FIFO on the rising edge of the ***W** line. When the reset ***RS** is high, you can perform a get operation by toggling the ***R** line low, then high. The 9 bits are removed (get) from the FIFO on the rising edge of the *R* line. These 9 bits are available on the output data lines **O8–O0** when the ***R** line is low. There are three negative logic FIFO status outputs that are available:

full flag	***FF**	0 means full
half flag	***HF**	0 means more than half full
empty flag	***EF**	0 means empty

a) Show the hardware connections to your single-chip microcomputer. A RS232 channel will connect your logic analyzer system to a personal computer. The only external connections to your logic analyzer system are the eight FIFO data inputs. For some cool features that are possible with this approach, see the Circuit Cellar, *INK,* vol. 89, December 1997, pp. 46–49. Other features discussed in this article, but not to be implemented here, are computer-controlled sampling rate, optional external clock, and external reset to the device under test so that the FIFO and external circuit are started together.

b) Show the main program that initializes the SCI to 9600 baud, one stop, no parity, then loops:

- Resets the FIFO
- Waits for any character to be received on the SCI input (start command from the PC)
- Allows the FIFO to fill with 8-bit data at 20 MHz
- Waits for the FIFO to be full
- Reads all 2048 bytes from the FIFO and transmits them out the SCI output one at a time

This computer is dedicated to this task, so you must show all the software for the computer. The program never quits, repeats the loop over and over. Use gadfly or interrupt synchronization, whichever is most appropriate.

7.9   Lab Assignments

Lab 7.1. The overall objective is to redesign the Lab 6.2 or 6.3 data acquisition system to employ two microcontrollers. The slave microcontroller will perform the data acquisition (position or temperature), and the master microcontroller will interface to the PC. The two microcontrollers will be linked using their SPI ports, as shown in Figure 7.48. The master microcontroller will fetch data from the slave and transmit it to the PC, using SCI interrupt synchronization. The user will interact with a PC running HyperTerminal.

Figure 7.48
Data flow graph for
Lab 7.1

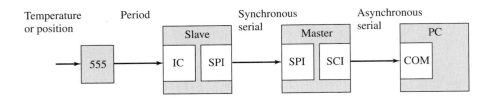

Lab 7.2. The overall goal is to develop an interrupting SCI device driver that implements fixed-point input/output. The fixed-point constant is 0.001. The full-scale range is from 0 to 65.534. The Fix_InDec function should provide for flexible operation. The input/output specifications of Fix_InDec are illustrated in Table 7.19. The operator creates the input by typing on the keyboard, and the output is a number passed back as the return parameter of the function.

Table 7.19
Examples of 16-bit unsigned decimal fixed point input with $\Delta = 10^{-3}$.

User Types in to `Fix_InDec`	Actual Value	Return Parameter of `Fix_InDec()`
0	0.000	0
.2	0.200	200
50.5	50.500	50500
12.48	12.480	12480
0.0023	0.002	2
1.4595	1.460	1460
1.4604	1.460	1460
6	6.000	6000
65.534	65.534	65534

Notice that `Fix_InDec` rounds the input to the closest fixed-point result (e.g., 1.4595 rounds to 1.460, and 1.4604 also rounds to 1.460). Some numbers such as 1.2345678 might be considered illegal because they cause overflow of intermediate results. In the comments of your software, please discuss why you chose your particular implementation method over the other available choices. Please handle the backspace character, allowing the operator to erase characters. In particular, you are free to use iterative or recursive algorithms. You are free to modify the prototypes as well as handle illegal inputs in any way you feel is appropriate. You must detect illegal input, but you have a choice as to how your system responds to the illegal input. One possibility for handling an illegal number would be to return 65535, which you could define as an illegal number. A second possibility for when an illegal number is typed is to output an error message, and require the operator to enter the number again. The input/output specifications for `Fix_OutDec` are illustrated in Table 7.20.

Table 7.20
Examples of 16-bit unsigned decimal fixed-point output with $\Delta = 10^{-3}$.

Input to `Fix_OutDec`	Output of `Fix_OutDec`
$0000 0	0.000
$0032 50	0.050
$00FA 250	0.250
$0781 1921	1.921
$3039 12345	12.345
$7FFF 32767	32.767
$FFFE 65534	65.534

Lab 7.3. The overall goal is to develop a solid-state disk. Your system will be able to create files, append data to the end of a file, print out the entire contents of a file, and delete files. In addition, your system will be able to list the names and sizes of the available files. Basically, you will interface a large serial EEPROM, such as the 25LC640, using the SPI port, then write a series of software functions that make it appear as a disk. Solid-state disks can be also made from battery-backed RAM or flash EEPROM. A personal computer uses disks made with magnetic storage media and moving parts. While this hard disk technology is acceptable for the personal computer because of its large size (>gigabyte) and low cost (<$100 OEM), it is not appropriate for an embedded system, because it its physical dimensions, electrical power requirements, noise, sensitivity to motion (maximum acceleration), and weight. Embedded applications that might require disk storage include data acquisition, data base systems, and signal generation. The personal desk accessory (PDA) devices currently employ solid-state disks because of their small physical size and low power requirements. Unfortunately, solid-state disks have smaller sizes and higher cost per bit than the traditional magnetic storage disk. In particular, you will implement a 64-Kbyte disk that costs about $1. The cost per bit is therefore about $15/Mbyte. Compare this cost to a 100-Gbyte hard drive that costs about $100. This cost per bit is only $0.001/Mbyte.

Every byte of the serial EEPROM must be accessible. There must be a low-level device driver, and this software alone can directly access the serial EEPROM. Using a driver means there are separate header and code files. Space is allocated to a file in fixed-size blocks (e.g., two files do not store data into the same block.) The high-level structure should include a directory that supports multiple logical files. Your system must support at least 10 files. The files are dynamically created and can grow in size (shrinking is easy to do but not necessary in this Lab.) There must be three software layers, including a high-level command interpreter (e.g., a `main.c` file with the editor program), a middle-level logical file system (e.g., `file.h` and `file.c`), and a low-level memory-access system (e.g., `disk.h` and `disk.c`). Each layer should have its own code file, and careful thought should go into deciding which components are private and which are public. All information (directory, linking, and data) must be stored on the serial EEPROM. The EEPROM is nonvolatile. This means if the microcontroller were to lose power, all information would be accessible when power is restored back to the microcontroller.

The specific function prototypes are included for illustration purposes. You are free to change and function names and parameters, as long as the basic operations are supported. You may choose any size for the fixed-size disk blocks. You are free to adjust the specific syntax of the interpreter, as long as similar functions are available. You may implement any disk allocation method, as long as files can grow in size.

The first step is to interface the 64-Kbyte serial EEPROM to the microcontroller using the SPI interface. The second step is to develop a low-level device driver for the disk. Disks are partitioned into fixed-size blocks. The function of the low-level is to allow read/write access to the solid-state disk (i.e., the serial EEPROM.) There should be separate `Disk.c` and `Disk.h` files containing software that implements the low-level disk operations. The following prototypes illustrate these operations. The `Disk_Open` command should enable the SPI interface to your serial EEPROM. All routines return a 1 if successful and a 0 on failure.

```
int Disk_Open(void);  // initialize disk interface
int Disk_ReadByte(unsigned char *bytePt,  // result returned by reference
  unsigned short blockNum,  // which block to read from
  unsigned short byteNum);  // which byte within this block
int Disk_ReadWord(unsigned short *wordPt,  // result returned by reference
  unsigned short blockNum,  // which block to read from
  unsigned short wordNum);  // which word within this block
int Disk_WriteByte(unsigned char byte,  // value to be saved
  unsigned short blockNum,  // which block to write into
  unsigned short byteNum);  // which byte within this block
int Disk_WriteWord(unsigned short word,  // value to be saved
  unsigned short blockNum,  // which block to write into
  unsigned short wordNum);  // which word within this block
```

The third step is to develop a file system. There should be separate `File.c` and `File.h` files containing software that implements the file system. There are three components of the file system: directory, allocation, and free-space management. The first component is the **directory**, as shown in Figure 7.49. The directory contains a mapping between the symbolic filename and the physical address of the data. Specific information contained in the directory might include the file name, the number of the first block containing data, and the total number of bytes stored in the file. Other information that one often finds in a directory entry includes a pointer to the last block of the file, access rights, date of creation, date of last modification, and file type. One possible implementation places the directory in block 0 (again, you are free to develop your own method). In this simple system, all files are listed in this one directory (there are no subdirectories). There is one fixed-size directory

Figure 7.49
Linked file allocation with 256-byte blocks.

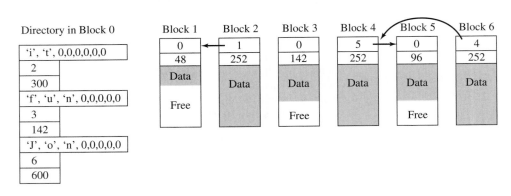

entry for each file. Each entry contains the file name, the block number of the first block containing data (0 means there is no first block), the block number of the last block containing data (0 means there is no last block), and the total number of bytes stored in the file. A filename is stored as a null-terminated ASCII string in a fixed-size array. A null-string (first byte 0) means no file. Because the directory itself is located in block 0, 0 can be used as a null-block pointer. Given that the entire directory must fit into the block 0, the maximum number of files can be calculated by dividing the block size by the number of bytes used for each directory entry.

The second component of the file system is the **logical to physical address translation**. Logically, the data in the file are addressed in a simple linear fashion. The logical address ranges from the first to the last. There are many algorithms one could use to keep track of where all the data for a file belongs. One simple mechanism is called **linked allocation**, which is also illustrated in Figure 7.49. Recall that the directory contains the block number of the first block containing data for the file. The start of every block contains a link (the block number) of the next block, and a byte count (the number of data bytes in this block). If the link is zero, this is last block of the file. If the byte count is zero, this block is empty (contains no data). Once the block is full, the file must request a free block (empty and not used by another file) to store more data. Linked allocation is effective for systems like this that employ sequential access. Sequential read access involves two functions similar to a magnetic tape: rewind (start at beginning), and read the next data. Sequential write access simply involves appending data to the end of the file. The following figure assumes the block size is 256 bytes and the filename has a maximum of 7 characters. All counters and pointers have 16-bit precision. Since each data block has a 2-byte link and a 2-byte counter, each block can store up to 252 bytes of data. The file "it" has 300 bytes, 252 of them in block 2 and the rest in block 1. All of the 142 bytes for file "fun" are stored in block 3.

The third component of the file system is **free-space management**. Initially all blocks except the one used for the directory are free, available for files to store data. To store data into a file, blocks must be allocated to the file. When a file is deleted, its blocks must be made available again. One simple free-space management technique uses **linked allocation**, similar to the way data is stored (see Figure 7.50). You could assign the last directory entry for free space management. This entry is hidden from the user. That is, this free space file can't be opened, printed, or deleted. It doesn't use any of the byte count fields, but does use the links to access all the free blocks. Initially, all the blocks (except the directory itself) are linked together, with the special directory entry pointing to the first one and the last one having a null pointer. When a file requests a block, it is unlinked from the free space and linked to the file. When a file is deleted, all its blocks are linked to the free space again.

Figure 7.50

Free-space management.

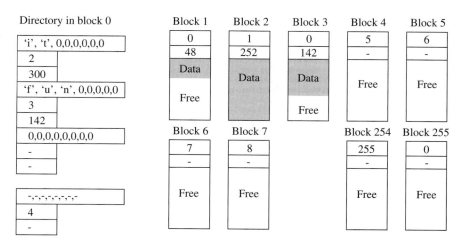

The following prototypes illustrate the operations to be performed by the file system. It is OK to make copies of pointers and counters into regular memory while a command is being executed. But after each operation, all counters and pointers should be written back onto the disk. Similarly it is OK for the Open command to make copies of pointers and counters to be used by the Write command, but the Close command should leave the disk in a consistent state (ready for a power loss). All routines return a 1 if successful and a 0 on failure.

```
int File_Format(void);                    // initialize file system
int File_Create(unsigned char name[8]);   // create new file, makeit empty
int File_Open(unsigned char name[8]);     // open a file for appending
int File_Write(unsigned char byte);       // save at end of the open file
int File_Close(void);                     // close the file
int File_Print(unsigned char name[8]);    // print entire contents
int File_Directory(void);                 // print directory contents
int File_Delete(unsigned char name[8]);   // delete this file
```

File_Format should return an error if the call to the low-level Disk_Open returns an error. File_Create should return an error if the directory is full or if the file already exists. File_Open should return an error if another file is already open (only one file can be open at a time in this simple system), or if the file doesn't exist. File_Write should return an error if the disk is full or if no file is open. File_Close should return an error if no file is open. File_Print and File_Delete should return an error if the file doesn't exist.

The fourth step is to develop a simple interpreter/editor that illustrates the features of your file system. This software includes the main program, which first initializes, then implements the interpreter/editor. The following commands illustrate the types of features you need to develop:

f	format the disk, erasing all data and all files
d	display the directory, including names and sizes of the files
c mine	create a new empty file called "mine"
p yours	display the contents of file "yours"
a hers	open the file "hers", and add characters
	subsequently typed characters are added to the file
	close the file and return to the interpreter when an <esc> is typed ($1B)
e his	erase file "his"

8 Parallel Port Interfaces

Chapter 8 objectives are to:

❑ Develop fundamental concepts associated with the physical behavior of the device
❑ Design the hardware interface between the device and the parallel port
❑ Write low-level device drivers that perform basic I/O with our device
❑ Discuss interfacing issues associated with performing the I/O as a background interrupt process

Chapter 8 deals with external devices that we connect to the parallel I/O port of our computer. In particular, we will interface switches, keyboards, single LEDs, LED displays, single LCDs, LCD displays, relays, and motors. Starting with this chapter and running through Chapter 11, we make the shift away from the details of the microcomputer itself and discuss devices external to the computer. Later, in Chapters 12 through 14, the pieces (microcomputer and external devices) are combined to design embedded systems.

8.1 Input Switches and Keyboards

8.1.1 Interfacing a Switch to the Computer

We begin with the simple switch interface (Figure 8.1). To convert the mechanical signal into an electric signal, a resistor pull-up is used. When the switch is open, the output is pulled to +5 V. The amount of current this output can source (its equivalent I_{OH}) is determined by the resistor value. The smaller the resistor, the larger the I_{OH}. When the switch is closed, the output is forced to ground. The amount of current this output can sink (its equivalent I_{OL}) is huge, limited only by the capacity of the switch to conduct current. The resistor does not affect the equivalent I_{OL}. A smaller resistor does waste current when the switch contact is closed.

In TTL digital logic we usually pull-up to +5 V rather than pulling down to zero because in most situations I_{OL} needs to be larger than I_{OH}. CMOS digital logic is an exception to this rule. Either pull-up or pull-down could be used for CMOS. When the switch in the pull-down circuit of Figure 8.2 is open, the output is pulled to ground. The amount of current this output can sink (its equivalent I_{OL}) is determined by the resistor value. The smaller the resistor, the larger the I_{OL}. When the switch is closed, the output is forced to +5 V. The amount of current this output can source (its equivalent I_{OH}) is huge, limited only by the capacity of the switch to conduct current. The resistor does not affect the equivalent I_{OH}. Notice that the polarity of the interface is reversed in the pull-down interface as compared to the pull-up case.

Ports AD, J, M, P, S and T on the MC9S12C32 support both internal pull-ups and pull-downs. The functions of Ports J and P were presented previously in Table 3.7. To use

Figure 8.1
A simple switch
interface.

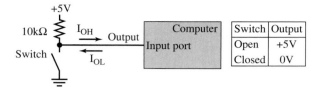

Figure 8.2
Another simple switch
interface.

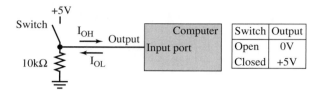

internal pull registers, the pin must be selected as an input or a wire-or output. To use Port AD as a digital port, the corresponding bits in the ATDDIEN must be set. Positive or negative logic switch interfaces can be implemented on the 6812 without a resistor, as shown in Figure 8.3, where PAD1 is configured to have a pull-down resistor, and PAD0 is configured to have a pull-up resistor.

Figure 8.3
The MCOS12C32
supports internal pull-up
or pull-down.

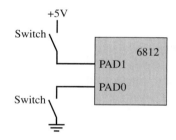

The software initialization sets bits in the Port Pull Select Register (PPSAD, PPSJ, PPSP, PPSM, PPSS, PPST) to select pull-up or pull-down. For each port pin that is enabled for pull-up or pull-down, the corresponding bit in the Port Pull Select Register determines whether it is pull-up(0) or pull-down(1). For each of these port pins there is a bit in the Pull Enable Register (PERAD, PERJ, PERP, PERM, PERS, PERT), which we set to enable the pull-up or pull-down function. It is good programming practice to first set the PPSx register, then set the PERx register so that temporary glitches are avoided. Program 8.1 will initialize Port AD with pull-up on PAD0 and pull-down on PAD1, as needed for the interface shown in Figure 8.3.

Program 8.1
The MC9S12C32
Port AD initialization.

```
// MC9S12C32
// PortAD bit 1 is connected to a switch to +5, using internal pull-down
// PortAD bit 0 is connected to a switch to 0, using internal pull-up
void PortAD_Init(void){
  ATDDIEN |= 0x03;    // PAD1-0 digital I/O
  DDRAD &= ~0x03;     // PAD1-0 inputs
  PPSAD |=  0x02;     // pull-down on PAD1
  PPSAD &= ~0x01;     // pull-up on PAD0
  PERAD |=  0x03;     // enable pull-up and pull-down
}
```

Checkpoint 8.1: Write the software that configures all of Port M as inputs with internal pull-ups.

8.1.2
Hardware
Debouncing Using
a Capacitor

The mechanical properties of a switch strongly affect the mechanical response to an applied force according to the following second-order differential equation. The three terms arise from Newton's Second Law internia, friction, and the spring constant, respectively.

$$F = m\frac{d^2x}{dt^2} + K_d\frac{dx}{dt} + Kx$$

where m is the switch mass, K_d is the friction coefficient, and K is the spring constant. The damping ratio can be calculated as

$$\xi = \frac{K_d}{2\sqrt{K \cdot m}}$$

Depending on the relative values for m, K_d, and K, the step response (i.e., what happens when you put your finger on the button) will be underdamped ($\xi > 1$ causes ringing) or overdamped ($\xi < 1$ causes a delayed but smooth rise.) Most inexpensive switches are underdamped. Hence, they will bounce both when touched and when released. Typical bounce times range from 1 to 25 ms. Ideally, the switch resistance is zero (actually about 0.1Ω) when closed and infinite when open (Figure 8.4).

Figure 8.4
Switch timing showing
bounce on touch and
release.

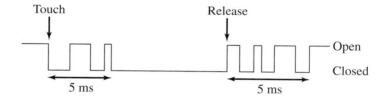

Checkpoint 8.2: What is the natural solution to the above differential equation with no friction ($K_d = 0$)?

More expensive switches can be purchased with proper damping. These oil-filled switches do not bounce. Because it is easy to add software to solve the bounce problem, most keyboard devices use inexpensive switches that bounce. With the circuit having just a pull-up resistor, the electric output will bounce because the mechanical input bounces (Figure 8.5).

Figure 8.5
Switch bounce can be
seen on the voltage
signal.

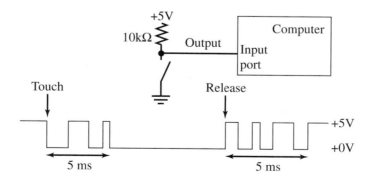

It may or may not be important to debounce the switch. For example, if we are counting the number of times the operator pushes the button, then we must debounce so that one push results in incrementing just once. We will debounce our standard computer keyboard so that when the operator types the letter A into her word processor, only one A is entered into the file. On the other hand, if the switch position specifies some static condition, and the operator sets the switch before she turns on the computer, then debouncing is not necessary.

We will study both hardware and software mechanisms to debounce the switch. The simplest hardware method is to use a capacitor across the switch (Figure 8.6). The capacitor value is chosen large enough so that the input voltage does not exceed the 0.7-V threshold of the NOT gate while it is bouncing. A detailed discussion about selecting the capacitor value will be presented later in this section.

Figure 8.6
Switch bounce removed with a capacitor (this is a bad circuit).

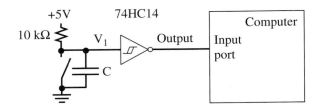

The touch timing with and without the capacitor is as shown in Figure 8.7. Notice that there is no delay between the touching of the switch and the transition of the signal output. The release timing with and without the capacitor is shown in Figure 8.8. There is a significant delay from the release of the switch until the fall of the output. To repeat, the capacitor is chosen such that the input voltage does not exceed threshold during the bouncing (Figure 8.9).

Figure 8.7
Switch touch bounce is removed by the capacitor.

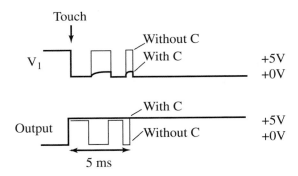

Figure 8.8
Switch release bounce is also removed by the capacitor.

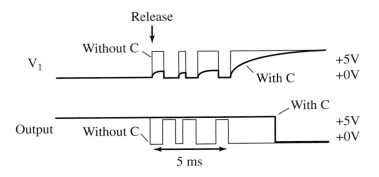

Figure 8.9
Timing used to calculate
capacitor value.

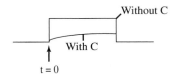

The voltage rise during a bounce interval when the switch is open is given by

$$V \geq 5 - 5e^{-t/RC}$$

In this example, R=10 kΩ, and the bounce time is 5 ms. Thus, we will choose C such that

$$0.7 \geq 5 - 5e^{-5 \text{ ms}/(10 \text{ k}\Omega \cdot C)}$$
$$0.86 \leq e^{-5 \text{ ms}/(10 \text{ k}\Omega \cdot C)}$$
$$\ln(1.16) \geq 5 \text{ ms}/(10 \text{ k}\Omega \cdot C)$$
$$C \geq 5 \text{ ms}/(10 \text{ k}\Omega \cdot \ln(1.16)) = 3.3 \text{ }\mu\text{F}$$

One problem with the above interface is the instantaneous current that occurs when the switch bounces closed. At $t = 0^-$, there is a charge on the capacitor. At $t = 0^+$, the energy has been discharged and the voltage is zero. Theoretically, this can only occur if the current at $t = 0$ is infinite (Figure 8.10).

Figure 8.10
A spark will occur
because a large current
occurs when the switch
is touched.

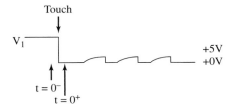

In reality, the current is not infinite, but it is large enough to cause a spark. These sparks will produce carbon deposits on the switch that will build up until the switch no longer works. To limit the current (thereby eliminating the sparks) a 22Ω resistor is placed in series with the switch. The value 22Ω was chosen to be much smaller than the 10 kΩ, but much larger than the 0.1Ω contact resistance of the switch.

If an input switch is closed, its resistance will be about 0.1Ω, and the output of the 74HC14 will be high (a logic 1). If an input switch is open, its resistance will be infinite, and the output of the 74HC14 will be low (a logic 0). If you connect the output of this switch interface to a computer parallel port input, then the computer can read the state (on/off) of the switch (Figure 8.11).

Figure 8.11
A hardware interface
that removes the bounce
(good circuit compared
to Figure 8.6).

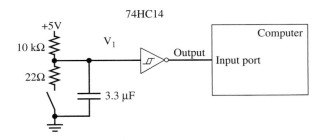

The Schmitt trigger 74HC14 is used to prevent multiple transitions when the switch is released. The difference between a regular NOT gate (74HC04) and the Schmitt trigger

NOT gate (74HC14) is hysteresis (Figure 8.12). Hysteresis is required because the input of the NOT gate in the transition range is

$$0.7 \le V_1 \le 2.0 \text{ V}$$

Figure 8.12
Input/output relationship showing the difference between a 74HC04 and a 74HC14.

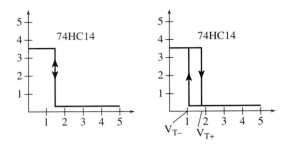

for a long period. Normally the time during which a digital input is in transition is on the order of a few nanoseconds. Recall that the rise and fall time for a 2-MHz 6811 E clock is 20 to 25 ns. For faster clocks, the transition times become even shorter. But in this application because of the 3.3 μF capacitor, the time in the transition range, Δt, is a huge 12 ms! In particular,

$$\Delta t = t_2 - t_1$$

where

$$0.7 = 5 - 5\, e^{-t_1/(10\text{k}\Omega \cdot 3.3\ \mu\text{F})}$$

and

$$2.0 = 5 - 5 e^{-t_2/(10\text{k}\Omega \cdot 3.3\ \mu\text{F})}$$

Thus

$$t_1 = 10\text{k}\Omega \cdot 3.3\ \mu\text{F} \cdot \ln(1.16) = 5 \text{ ms}$$

and

$$t_2 = 10\text{k}\Omega \cdot 3.3\ \mu\text{F} \cdot \ln(1.67) = 17 \text{ ms}$$

Thus

$$\Delta t = 12 \text{ ms}$$

If the input stays in the transition range for a long time with just a little noise, then the output of a regular digital gate will toggle with the noise. The hysteresis removes the extra transitions that might occur with a regular gate (Figure 8.13).

Figure 8.13
Timing showing why a 74HC14 is used instead of a regular digital gate.

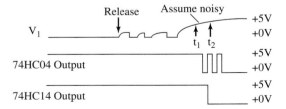

Observation: With a capacitor-based debounced switch, there is no delay between the closing of the switch and the rising edge at the computer input.

Observation: With a capacitor-based debounced switch, there is a large delay (over 10 ms) between the opening of the switch and the falling edge at the computer input.

Checkpoint 8.3: Is the time duration of the bounce a function of the value of the pull-up resistor?

8.1.3
Software
Debouncing

8.1.3.1
Software Debouncing
with Gadfly
Synchronization

It will be less expensive to remove the bounce using software methods. It is appropriate to use a software approach because the software is fast compared to the bounce time. Typically we use the pull-up resistor to convert the switch position into a TTL-level digital signal. The 6812 also supports internal pull-up resistors, which can reduce the component count and simplify manufacturing. The 6812 key wake up could have been used in place of the input capture (Figure 8.14).

Figure 8.14
Switch interface.

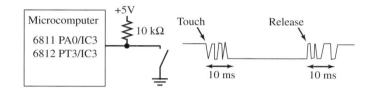

In these examples, we assume the switch bounce is less than 10 ms. The first two routines implement software debouncing using gadfly synchronization. The computer is dedicated to the interface and does not perform any other functions while the routines are running. The first routine (Program 8.2) waits for the switch to be pressed (PA0/PT3 low) and returns 10 ms after the switch is pressed (Figure 8.15). The advantages of using output compare for implementing the wait loop are (1) it is easier to understand and change if needed and (2) it is more accurate if background interrupts are occurring. No 6811 initialization is required because PA0 is always an input and OC5 output compare is active by default. In C, these routines could be written as shown in Program 8.3.

```
; MC68HC711E9                              ; MC9S12C32
;loop until switch is pressed             ;loop until switch is pressed
Key_WaitPress                             Key_WaitPress
    ldaa PORTA  ;PA0=0 if pressed             brset PTT,#$08,*  ;PT3=0 if pressed
    anda #$01                                 ldy  #1
    bne  WaitPress                            bsr  Timer_Wait10ms ;Program 2.6
    ldy  #1                                   rts
    bsr  Timer_Wait10ms ;Program 2.6      ;loop until switch is released
    rts                                   Key_WaitRelease
;loop until switch is released                brclr PTT,#$08,*  ;PT3=1 if released
Key_WaitRelease                               ldy  #1
    ldaa PORTA ;PA0=1 if released             bsr  Timer_Wait10ms ;Program 2.6
    anda #$01                                 rts
    beq  WaitRelease                      Key_Init
    ldy  #1                                   bsr  Timer_Init
    bsr  Timer_Wait10ms ;Program 2.6          bclr DDRT,#$08    ;PT3 is input
    rts                                       rts
```

Program 8.2
Switch debouncing using assembly software.

Figure 8.15
Software flowcharts for
debouncing the switch.

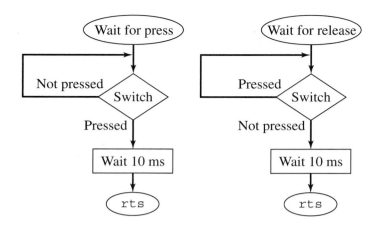

```
// MC68HC711E9                               // MC9S12C32
void Key_WaitPress(void){                    void Key_WaitPress(void){
  while(PORTA&0x01); // PA0=0->pressed          while(PTT&0x08);    // PT3=0 when pressed
  Timer_Wait10ms(1); // debouncing              Timer_Wait10ms(1); // debouncing
}                                            }
void Key_WaitRelease(void){                  void Key_WaitRelease(void){
  while((PORTA&0x01)==0); //PA0=1->released     while((PTT&0x08)==0); // PT3=1 -> released
  Timer_Wait10ms(1);        // debouncing       Timer_Wait10ms(1);    // debouncing
}                                            }
                                             void Key_Init(void){
// no ritual required                          Timer_Init();
                                               DDRT &=~0x08; // PT3 is input
                                             }
```

Program 8.3
Switch debouncing using C software.

This second routine reads the current value of the switch (Program 8.4). If the switch is currently bouncing, it will wait for stability (Figure 8.16). A return value of 0 means pressed (PA0/PT3=0), and 1 means not pressed (PA0/PT3=1). Notice that the software always waits in a "do nothing" loop for 10 ms. This inefficiency can be eliminated by placing the switch I/O in a background interrupt-driven thread. In C, these routines could be written as shown in Program 8.5.

> ***Observation:*** With a software-based debounced switch, the signal arrives at the computer input without delay, but software delays may occur at either touch or release.

8.1.3.2
Software Debouncing
with Interrupt
Synchronization

Input capture, which is available on most Freescale microcomputers, is a convenient mechanism to detect changes on the digital signal. The key wake-up mechanism on the 6812 provides an alternative to input capture. First we will present the interface using input capture. The input capture can be configured to interrupt either on the rise, the fall, or both the rise and the fall. Because of the bounce, any of these modes will generate an interrupt request when the key is touched or released (Figure 8.17).

```
; MC68HC711E9, PA0 is input          ; MC9S12C32, PT3 is input
; RegB is the return value           ; RegB is the return value
Key_Read ldd  TCNT                   Key_Read ldd  TCNT
         addd #20000 ;10ms delay              addd #20000 ;10ms delay
         std  TOC5   ;start timer              std  TC5    ;start timer
         ldaa #$08                            movb #$20,TFLG1 ; clear flag
         staa TFLG1 ;clear OC5F                ldab PTT   ;0 if pressed
         ldab PORTA ;0 if pressed             andb #$08  ;B=old value
         andb #$01  ;B=old value      same    ldaa TFLG1 ;10ms bouncing
same     ldaa TFLG1 ;10ms bouncing            anda #$20  ;C5F set?
         anda #$08  ;OC5F set?                bne done
         bne  done                            ldaa PTT   ;0 if pressed
         ldaa PORTA ;0 if pressed             anda #$08  ;A=new value
         anda #$01  ;A=new value               cba        ;same as before
         cba        ;same as before           beq  same
         beq  same                            bra  Key_Read ; different
         bra  Key_Read               done     rts
;start over if different              Key_Init bset TIOS,#$20  ;enable OC5
done     rts  ; same value for 10ms            movb #$80,TSCR1 ;enable
                                               movb #$01,TSCR2 ;500 ns clk
; no initialization needed                     bclr DDRT,#$08  ;PT3 is input
                                               rts
```

Program 8.4
Another example of switch debouncing using assembly software.

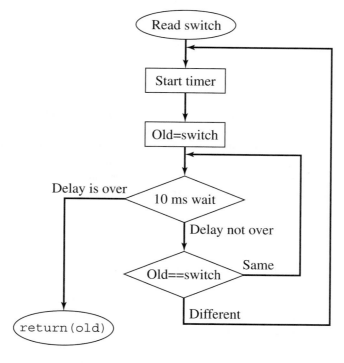

Figure 8.16
Another software
flowchart for
debouncing the switch.

```
// MC68HC711E9, PA0 is input          // MC9S12C32, PT3 is input
// waits until unchanged for 10ms     // waits until unchanged for 10ms
unsigned char Key_Read(void){         unsigned char Key_Read(void){
unsigned char old;                    unsigned char old;
  old = PORTA&0x01;   // Current value   old = PTT&0x08;       // Current value
  TOC5 = TCNT+20000;  // 10ms delay      TC5 = TCNT+20000;     // 10ms delay
  TFLG1 = 0x08;       //  Clear OC5F     TFLG1 = 0x20;         // Clear C5F
  while((TFLG1&0x08)==0){ // 10ms?       while((TFLG1&0x20)==0){ // 10ms?
    if(old != (PORTA&0x01)){ // changed?   if(old != (PTT&0x08)){ // changed?
      old = PORTA&0x01;  // New value        old = PTT&0x08;     // New value
      TOC5 = TCNT+20000; // delay           TC5 = TCNT+20000;}  // restart delay
      TFLG1 = 0x08; }   // Clear OC5F
  }                                     }
  return(old);                          return(old);
}                                     }
                                      void Key_Init(void) {
                                        TIOS |= 0x20;      // enable OC5
                                        TSCR1 = 0x80;      // enable
                                        TSCR2 = 0x01;      // 500 ns clock
                                        DDRT &=~0x08;}     // PT3 is input

// no ritual required
```

Program 8.5
Another example of switch debouncing using C software.

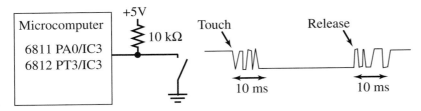

Figure 8.17
Switch interface.

A combination of input capture and output compare interrupts allows the switch interface to be performed in the background. This example simply counts the number of times the switch is pressed. The IC3 interrupt occurs immediately after the switch is pressed and released. Because the IC3 handler disarms itself, the bounce will not cause additional interrupts. The OC5 interrupt occurs 10 ms after the switch is pressed and 10 ms after the switch is released. At this time the switch position is stable (no bounce) (Figure 8.18).

The first IC3 interrupt occurs when the switch is first touched. The first OC5 interrupt occurs 10 ms later. At this time the global variable, Count, is incremented. The second IC3 interrupt occurs when the switch is released. The second OC5 interrupt does not increment the Count but simply rearms the input capture system. The ritual initializes the system with IC3 armed and OC5 disarmed (Program 8.6).

The latency of this interface is defined as the time when the switch is touched until the time when the count is incremented. Because of the delay introduced by the OC5 interrupt, the latency is 10 ms. If we assume the switch is not bouncing (currently being touched or released) at the time of the ritual, we can reduce this latency to less than 50 µs by introducing

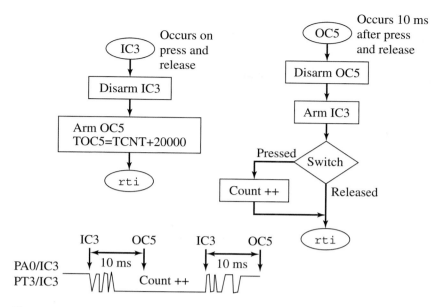

Figure 8.18
Another software flowchart for debouncing the switch with interrupts.

```
// MC68HC711E9, ICC11 C
// counts the number of button pushes
// button connected to IC3=PA0
unsigned short Count; // times pushed
#pragma interrupt_handler TOC5handler()
void TOC5handler(void){
  TMSK1 |= 0x01; // arm IC3
  TMSK1 &=~0x08; // disarm OC5
  if(PORTA&0x01==0) Count++;
  TFLG1 = 0x09;  // clear OC5F,IC3F
}
#pragma interrupt_handler TIC3handler()
void TIC3handler(void){
  TMSK1 &=~0x01;     // disarm IC3
  TMSK1 |= 0x08;     // arm OC5
  TOC5 = TCNT+20000; // 10 ms
  TFLG1 = 0x09;}     // clear OC5F,IC3F
void Key_Init(void){
  asm(" sei");     // make atomic
  TMSK1 |= 0x01;   // arm IC3
  TMSK1 &=~0x08;   // disarm OC5
  TFLG1 = 0x09;    // clear OC5F,IC3F
  TCTL2 |= 0x03;   // both edges
  Count = 0;
  asm(" cli");
}
```

```
// MC9S12C32, CodeWarrior C
// counts the number of button pushes
// button connected to IC3=PT3
unsigned short Count; // times pushed
void interrupt 13 TC5handler(void){
  TIE |= 0x08;    // arm IC3,
  TIE &=~0x20;    // disarm OC5
  if(PTT&0x08==0) Count++;
  TFLG1 = 0x28;   // clear C5F,C3F
}
void interrupt 11 TC3handler(void){
  TIE &=~0x08;     // disarm IC3,
  TIE |= 0x20;     // arm OC5
  TC5 = TCNT+20000; // 10 ms
  TFLG1 = 0x28;    // clear C5F,C3F
}
void Key_Init(void){
  asm sei         // make atomic
  TIOS=(TIOS&0xF7)|0x20; // enable OC5,IC3
  TSCR1 = 0x80;    // enable
  TSCR2 = 0x01;    // 500 ns clock
  DDRT &=~0x08;    // PT3 is input
  TIE |= 0x08;     // arm IC3,
  TIE &=~0x20;     // disarm OC5
  TFLG1 = 0x28;    // clear C5F,C3F
  TCTL4 |= 0x20;   // both edges
  Count = 0;
  asm cli }
```

Program 8.6
Switch debouncing using interrupts in C software.

a global variable. If `LastState` is true, then the switch is currently not pressed and the software is searching for a touch. If `LastState` is false, then the switch is currently pressed and the software is searching for a release (Program 8.7).

```
// MC68HC711E9, ICC11 C
// counts the number of button pushes
// signal connected to IC3=PA0
unsigned short Count; // times pushed
char LastState; // looking for touch?
// true means open, looking for a touch
// false means closed, looking for release
#pragma interrupt_handler TOC5handler()
void TOC5handler(void){
  TMSK1 |= 0x01; // arm IC3
  TMSK1 &=~0x08; // disarm OC5
  TFLG1 = 0x09;  // clear OC5F,IC3F
}
#pragma interrupt_handler TIC3handler()
void TIC3handler(void){
  if(LastState){ // every other
    Count++;      // a touch has occurred
    LastState=0;}
  else
    LastState=1; // release occurred
  TMSK1 &=~0x01; // disarm IC3
  TMSK1 |= 0x08; // arm OC5
  TOC5 = TCNT+20000; // 10ms
  TFLG1 = 0x09;} // clear OC5F,IC3F
void Key_Init(void){
  asm(" sei");    // make atomic
  TMSK1 |= 0x01; // arm IC3
  TMSK1 &=~0x08; // disarm OC5
  TFLG1 = 0x09;   // clear OC5F,IC3F
  TCTL2 |= 0x03; // set on both edges
  Count = 0;
  LastState = PORTA&0x01;
  asm(" cli");
}
```

```
// MC9S12C32, CodeWarrior C
// counts the number of button pushes
// signal connected to IC3=PT3
unsigned short Count; // times pushed
char LastState; // looking for touch?
// true means open, looking for a touch
// false means closed, looking for release
void interrupt 13 TC5handler(void){
  TIE |= 0x08;   // arm IC3,
  TIE &=~0x20;   // disarm OC5
  TFLG1 = 0x28;  // clear C5F,C3F
}
void interrupt 11 TC3handler(void){
  if(LastState){ // every other
    Count++;      // a touch has occurred
    LastState=0;}
  else
    LastState=1; // release occurred
  TIE &=~0x08;   // disarm IC3,
  TIE |= 0x20;   // arm OC5
  TC5 = TCNT+20000; // 10 ms
  TFLG1 = 0x28;   // clear C5F,C3F
}
void Key_Init(void){
  asm sei         // make atomic
  TIOS=(TIOS&0xF7)|0x20; // enable OC5,IC3
  TSCR1 = 0x80;   // enable
  TSCR2 = 0x01;   // 500 ns clock
  DDRT &=~0x08;   // PT3 is input
  TIE |= 0x08;    // arm IC3,
  TIE &=~0x20;    // disarm OC5
  TFLG1 = 0x28;   // clear C5F,C3F
  TCTL4 |= 0xC0; // both edges
  Count = 0;
  LastState = PTT&0x08;
  asm cli }
```

Program 8.7
Low latency switch debouncing using interrupts in C software.

Now the latency is simply the time required for the microcomputer to recognize and process the input capture interrupt. Assuming there are no other interrupts, this time is less than 20 cycles.

**8.1.4
Basic Approaches to Interfacing Multiple Keys**

In this section we attempt to interface as many switches as possible to a single 8-bit parallel port, and we will consider three interfacing schemes (Figure 8.19). In a *direct interface* we connect each switch to a separate microcomputer input pin. For example, using just one 8-bit parallel port, we can connect eight switches using the direct scheme. An advantage of this interfacing approach is that the software can recognize all 256 (2^8) possible switch patterns. If the switches were remote from the microcomputer, we would need a nine-wire

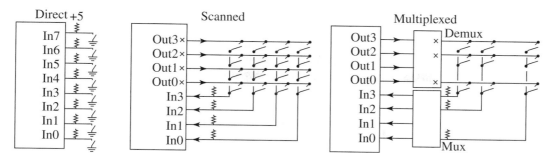

Figure 8.19
Three approaches to interfacing multiple keys.

cable to connect it to the microcomputer. In general, if there are n switches, we would need $n/8$ parallel ports and $n+1$ wires in our remote cable. This method will be used when there are a small number of switches or when we must recognize multiple and simultaneous key presses (Figure 8.19). Examples include music keyboards and the shift and control option keys. As illustrated in the Music Keyboard example in Section 8.15, implementing an interrupt-driven interface requires a lot of hardware. Therefore, if interrupt synchronization is required, it may be more appropriate to utilize periodic polling interrupt synchronization.

Figure 8.20
Multiple keys are implemented by placing the switches in a matrix. (Notice there are fewer wires in the cable than there are keys.)

In a *scanned interface* the switches are placed in a row/column matrix (Figure 8.20). The × at the four outputs signifies an open collector (an output with two states, HiZ and low). The computer drives one row at a time to zero while leaving the other rows at HiZ. By reading the column, the software can detect if a key is pressed in that row. The software "scans" the device by checking all rows one by one. For the 6811, the CWOM = 1 feature of Port C can be used to implement this HiZ/0 binary output. For the other microcomputers, the open-collector functionality will be implemented by toggling the direction register. Table 8.1 illustrates the sequence to scan the four rows.

Table 8.1
Scanning patterns for a 4 by 4 matrix keyboard.

Row	Out3	Out2	Out1	Out0
3	0	HiZ	HiZ	HiZ
2	HiZ	0	HiZ	HiZ
1	HiZ	HiZ	0	HiZ
0	HiZ	HiZ	HiZ	0

For computers without an open-collector output mode, the direction register can be toggled to simulate the two output states, HiZ/0, of open-collector logic (see Exercise 1.10). This method can interface many switches with a small number of parallel I/O pins. In our example situation, the single 8-bit I/O port can handle 16 switches with only an eight-wire cable. The disadvantage of the scanned approach over the direct approach is that it can only handle situations where zero, one, or two switches are simultaneously pressed. This method is used for most of the switches in our standard computer keyboard. Recall that the shift, alt, and control keys are interfaced with the direct method. We can "arm" this interface for interrupts by driving all the rows to 0. The key wake-up or input capture mechanism can be used to generate interrupts on touch and release. Because of the switch bounce, an interrupt will occur when any of the keys changes. For microcomputers like the 6811 without a key wake-up, we can add a simple NAND gate connected to all columns. For the 6811 interface, the output of the NAND gate will change when any of the keys are pressed.

In a *multiplexed interface,* the computer outputs the binary value defining the row number, and a hardware demultiplexer will output the 0 on the selected row and HiZ's on the other rows. The demultiplexer must have open-collector outputs (illustrated again by the × in the Figure 8.19 circuit). The computer outputs the sequence $00, $10, $20, $30 . . . , $F0 to scan the 16 rows, as shown in Table 8.2.

Table 8.2
Scanning patterns for a multiplexed 16 by 16 matrix keyboard.

	Computer output				Demultiplexer output			
Row	Out3	Out2	Out1	Out0	15	14	. . .	0
15	1	1	1	1	0	HiZ	. . .	HiZ
14	1	1	1	0	HiZ	0	. . .	HiZ
. . .	. . .	. . .	. . .	. . .	. . .	. . .	. . .	. . .
1	0	0	0	1	HiZ	HiZ	. . .	HiZ
0	0	0	0	0	HiZ	HiZ	. . .	0

In a similar way, the column information is passed to a hardware multiplexer that calculates the column position of any 0 found in the selected row. One additional signal is necessary to signify the condition that no keys are pressed in that row. Since this interface has 16 rows and 16 columns, we can interface up to 256 keys! We could sacrifice one of the columns to detect the no key pressed in this row situation. In this way, we can interface 240 keys (15 · 16) on the single 8-bit parallel port. If more than one key is pressed in the same row, this method will detect only one of them. Therefore, we classify this scheme as being able to handle only 0 or one key pressed. Applications that can utilize this approach include touch screens and touch pads because they have a lot of switches but are interested only in the 0 or 1 touch situation. Implementing an interrupt-driven interface would require too much additional hardware. In this case, periodic polling interrupt synchronization would be appropriate.

8.1.5
Sixteen-Key
Electronic Piano

In this direct interface, we connect each key up to a separate computer input pin so that we can distinguish all 2^{16} possible combinations (Figure 8.21). Again, we assume a 10-ms switch bounce time (Figure 8.17).

The computer inputs require an external pull-up resistor. Each bit in the global, KEY, is affected by a key (1 for not pressed, 0 for pressed) (Program 8.8).

The 6811 interface connects the 16 individual switches to Ports E and C (Figure 8.22). 6811 interrupts will be requested with STRA (Figure 8.22). To generate interrupts on the touch and release, the 6811 hardware requires a 16-bit latch and a 16-bit "equals" gate (Figure 8.23). In this way, a 6811 STRA interrupt is requested whenever the switch pattern changes.

Figure 8.21
Multiple keys are interfaced directly to input ports.

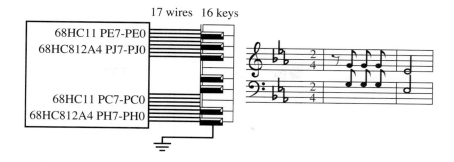

17 wires 16 keys

68HC11 PE7-PE0
68HC812A4 PJ7-PJ0

68HC11 PC7-PC0
68HC812A4 PH7-PH0

```
; MC68HC11A8
KEY:    ds  2      ;current value
RITUAL: clr DDRC ; all inputs
        ldaa #$54
; Bit signal value comment
;  6  STAI     1    arm STAF
;  5  CWOM     0    not applicable
;  4  HNDS     1    Input handshake
;  3  OIN      0
;  2  PLS      1    pulse out
;  1  EGA      0    falling edge
;  0  INVB     0    negative pulse
        staa PIOC
        bsr  KeyBoard ;set 74LS374's
        cli           ;Enable IRQ
        rts
KeyBoard: ldaa PIOC ;part of clear
        ldaa PORTE   ;Read MSB
        ldab PORTCL  ;Read LSB, ack
        std  KEY
        rts
IRQHan: ldaa #$14   ;Disarm STAF
        staa PIOC
        ldd  TCNT
        addd #20000  ;OC5 10ms later
        std  TOC5
        ldaa #$08    ;Arm OC5
        staa TMSK1
        staa TFLG1   ;Clear OC5F
        rti
OC5Han: clr  TMSK1   ;Disarm OC5
        ldaa #$54    ;Rearm STAF
        staa $PIOC
        bsr  KeyBoard ;read keyboard
        rti
```

```
; MC68HC812A4
KEY:    ds  2      ;current value
RITUAL: bset #$20,TIOS ;enable OC5
        movb #$90,TSCR  ;enable fast clr
        movb #$32,TMSK2 ;500 ns clk
        bclr #$20,TMSK1 ;disarm OC5
        clr  DDRH       ;key inputs
        clr  DDRJ       ;key inputs
        movb #$FF,KWIEH ;arm key wakeup
        movb #$FF,KWIEJ ;arm key wakeup
        clr  KPOLJ      ;falling edge
        clr  PUPEJ      ;regular inputs
        bsr  KeyBoard ;
        cli             ;Enable IRQ
        rts
KeyBoard: movb #$FF,KWIFH ;clr flags
        movb #$FF,KWIFJ ;clear flags
        ldaa PORTJ   ;Read MSB
        ldab PORTH   ;Read LSB
        std  KEY
        rts
KeyHHan: clr KWIEH ;disarm key wakeup
        clr  KWIEJ    ;disarm key wakeup
        bset #$20,TMSK1 ;arm OC5
        ldd  TCNT    ;TCNT
        addd #20000  ;OC5 10ms later
        std  TC5     ;fast clear
        rti
KeyJHan: clr KWIEH ;disarm key wakeup
        clr  KWIEJ    ;disarm key wakeup
        bset #$20,TMSK1 ;arm OC5
        ldd  TCNT    ;TCNT
        addd #20000  ;OC5 10ms later
        std  TC5     ;fast clear
        rti
OC5Han: bclr #$20,TMSK1 ;disarm OC5
        movb #$FF, KWIEH ;arm key wakeup
        movb #$FF, KWIEJ ;arm key wakeup
        bsr  KeyBoard    ;read keyboard
        rti
```

Program 8.8
Assembly software interface of a direct connection keyboard.

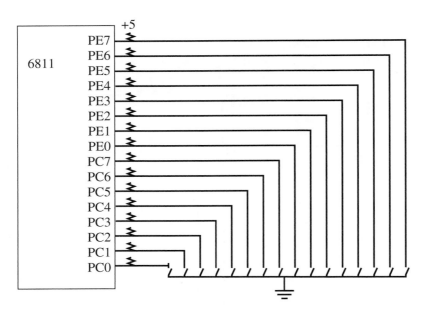

Figure 8.22
Multiple keys are interfaced directly to input ports.

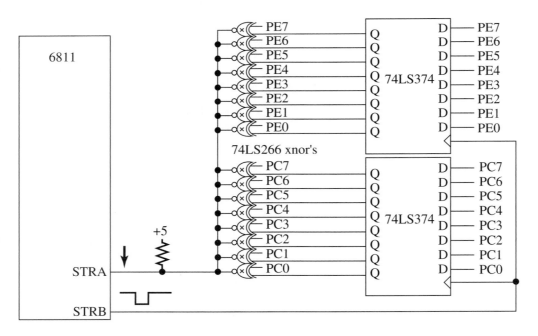

Figure 8.23
Hardware for generating interrupts on the 6811.

The read PIOC followed by the read CL data will cause a STRB pulse that will save the current keyboard values into the 16 flip-flips built with two 74LS374s. The 74LS266 has open-collector outputs that are "wire-ANDed" together. Four 74LS266 chips have 16 exclusive-NOR gates that create a 16 by 16 "equals" function. If the current keyboard values equal the values set by the interrupt handler, then STRA will equal 1. If any key changes since the last interrupt, STRA falls, causing an IRQ interrupt.

The 6812 interface does not require the 16-bit latch and 16-bit equals gate, because key wake-up interrupts can be generated on the touch or release of any of the 16 switches (Figure 8.24).

In C, the interface software would be as shown in Program 8.9. It would be possible to set both the 6812 Key WakeupH and Key WakeupJ interrupt vectors to point to the same ISR because they perform identical functions.

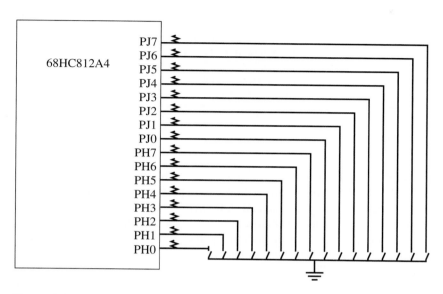

Figure 8.24
Hardware interface for the 6812.

```
// MC68HC711E9, ICC11 C
unsigned short Key; // pattern
void KeyBoard(void){
  dummyRead = PIOC;  // read status
  Key = (PORTE<<8)+PORTCL;}
// Set global, clear STAF


void Key_Init(void){
  asm(" sei");    // make atomic
  TMSK1 = 0x00;  // Disarm OC5
  DDRC = 0;
// PortC are inputs from the keyboard
  PIOC=0x54;
// Input hndshk, fall, arm, neg pulse
  KeyBoard();     // Initially read
  asm(" cli");
}


#pragma interrupt_handler IRQhandler()
void IRQhandler(void){
```

```
// MC68HC812A4, ICC12 C
unsigned short Key;  // current pattern
void KeyBoard(void){
  KWIFH = KWIFJ = 0xFF;     // clear flags
  Key = (PORTJ<<8)+PORTH;}  // Set global
void Key_Init(void){
  asm(" sei");       // make atomic
  TIOS |= 0x20;      // enable OC5
  TSCR = 0x80;       // enable
  TMSK2 = 0x32;      // 500 ns clock
  TMSK1 &= 0xDF;     // Disarm OC5
  DDRH = DDRJ = 0;   // H,J are key inputs
  KWIEH=KWIEJ=0xFF;  // all 16 are armed
  KPOLJ = 0;         // falling edge
  PULEJ = 0;         // regular input
  KeyBoard();        // Initially read
  asm(" cli"); }
#pragma interrupt_handler KeyHhandler()
void KeyHhandler(void){
  KWIEH=KWIEJ=0x00; // all 16 are disarmed
```

continued on p. 392

Program 8.9
C software interface of a direct connection keyboard.

continued from p. 391

```
  PIOC = 0x14;        // Disarm STAF
  TMSK1 = 0x08;       // Arm OC5
  TOC5 = TCNT+20000;  // 10 ms
  TFLG1 = 0x08;       // clear OC5F
}
#pragma interrupt_handler TOC5handler()
void TOC5handler(void){
  TMSK1 = 0x00;  // Disarm OC5
  PIOC = 0x54;   // Rearm STAF
  KeyBoard();    // Read, set LS374's
}
```

```
  TMSK1 |= 0x20;      // Arm OC5
  TC5 = TCNT+20000;  // 10 ms
  TFLG1 = 0x20;}      // clear C5F
#pragma interrupt_handler KeyJhandler()
void KeyJhandler(void){
  KWIEH=KWIEJ=0x00; // all 16 are disarmed
  TMSK1 |= 0x20;      // Arm OC5
  TC5 = TCNT+20000;  // 10 ms
  TFLG1 = 0x20;}      // clear C5F
#pragma interrupt_handler TOC5handler()
void TOC5handler(void){
  TMSK1 &= 0xDF;     // Disarm OC5
  KWIEH=KWIEJ=0xFF;  // all 16 are rearmed
  KeyBoard();}       // Read keys
```

Program 8.9

C software interface of a direct connection keyboard.

The four 74LS266 and two 74LS374 chips can be eliminated if periodic polling interrupt synchronization is used instead. With periodic polling, we establish a periodic interrupt (one that occurs at a regular fixed rate). In this example we will interrupt every 10 ms whether or not there has been a change in the keyboard. To handle the bounce, we set the periodic interrupt interval longer than the bounce time. If the switch happens to be bouncing at the time of the interrupt, global variable Key may change during this interrupt or maybe the next. The advantage of the previous hardware-based interrupt interface is that the global variable is changed exactly 10 ms after the key is pressed (or released). In this example there is a ±10-ms time jitter between the key press and when the global variable is changed. This interface, like all periodic polling systems, continuously generates interrupts that are wasted (inefficient) when no key change has occurred (Program 8.10).

```
// MC68HC711E9, ICC11 C
unsigned short Key;      // pattern
unsigned short KeyBoard(void){
  return((PORTE<<8)+PORTC); // pattern
}
void Key_Init(void){
 asm(" sei");        // make atomic
 DDRC = 0;           // inputs from keyboard
 Key = KeyBoard();   // read 16 keys
 TMSK1 = 0x08;       // Arm OC5
 TOC5 = TCNT+20000;
 TFLG1 = 0x08;       // clear OC5F
 asm(" cli");
}
#pragma interrupt_handler TOC5handler()
void TOC5handler(void){
  Key = KeyBoard();   // Current pattern
  TOC5 = TOC5+20000; // every 10ms
  TFLG1 = 0x08;       // ack OC5F
}
```

```
// MC68HC812A4, ICC12 C
unsigned short Key;      // current pattern
unsigned short KeyBoard(void){
  return((PORTH<<8)+PORTJ);} // pattern
void Key_Init(void){
 asm(" sei");        // make atomic
 TIOS |= 0x20;       // enable OC5
 TSCR = 0x80;        // enable
 TMSK2 = 0x32;       // 500 ns clock
 TMSK1 |= 0x20;      // Arm OC5
 DDRH = DDRJ = 0;    // H,J are key inputs
 PULEJ=0;            // regular input
 Key = KeyBoard();   // read 16 keys
 TC5 = TCNT+20000;
 asm(" cli"); }
#pragma interrupt_handler TOC5handler()
void TOC5handler(void){
  Key = KeyBoard(); // Current pattern
  TC5 = TC5+20000; // every 10ms
  TFLG1 = 0x20;}      // ack OC5F
```

Program 8.10

C software interface of a direct connection keyboard using periodic polling.

8.1.6
4 by 4 Scanned Keyboard

To save I/O lines, the scanned approach is used. In this approach the keys are divided into rows and columns. The computer can drive the rows with open-collector outputs and read the columns. The 6811 interface will use open-collector logic on PC7-PC4 by setting CWOM=1 and DDRC=$F0. Interrupts will be requested on both the touch and release using the STRA mechanism (Figure 8.25).

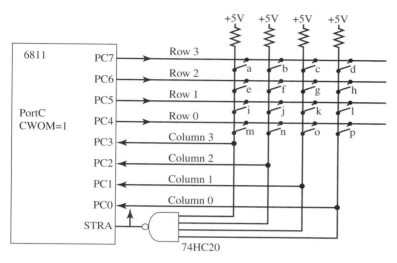

Figure 8.25
A matrix keyboard interfaced to a 6811.

The 6812 interface will produce the open-collector function (outputs have two states, HiZ, 0) by toggling the direction register. The resistor pull-ups on the PT3-PT0 inputs will be configured internally. Interrupts will be requested on the touch and release using the input capture mechanism (Figure 8.26).

Figure 8.26
A matrix keyboard interfaced to the microcomputer.

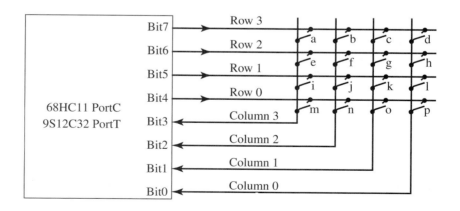

There are two steps to scan a particular row:

1. Select that row by driving it low, shown in Table 8.3 (make output 0), while the other rows are left not driven (make output HiZ)
On the 6811 we have CWOM=1, so the Port C outputs will be HiZ when output is 1
On the 6812 we set the direction register, so only one row is output

2. Read the columns to see if any keys are pressed in that row (Table 8.3)
0 means the key is pressed
1 means the key is not pressed

Table 8.3
Patterns for a 4 by 4
matrix keyboard.

Row 3	Row 2	Row 1	Row 0	Column 3	Column 2	Column 1	Column 0
0	1	1	1	a	b	c	d
1	0	1	1	e	f	g	h
1	1	0	1	i	j	k	l
1	1	1	0	m	n	o	p

When all rows are 0, the 6811 STRA input will rise when any of the 16 keys is pressed. When all rows are 0, one of the 6812 PTT inputs will fall when any of the 16 keys is pressed. The scanned keyboard operates properly if:

1. No key is pressed
2. Exactly one key is pressed
3. Exactly two keys are pressed

This method can not be used to detect three or more keys simultaneously pressed. If three keys are pressed in an L shape, then the fourth key that completes the rectangle will appear to be pressed. Therefore special keys like shift, control, option, and alt are not placed in the scanned matrix but rather are interfaced directly, each to a separate input port, like the piano keyboard example. The 6812 implementation sets the direction register to inputs to create the open-collector output. The assembly language software to scan this keyboard is shown in Program 8.11. The C language software to scan this keyboard is shown in Program 8.12.

A debounced scanned interface is created by combining the scanning software from this section with the debouncing software from the last section. In particular remove the simple KeyBoard() function in Program 8.9 and replace it with the Scan() function of Program 8.12.

```
; MC68HC711E9
Key_Init ldaa #$40
         staa PIOC   ;CWOM=1
         ldaa #$F0   ;PC7-PC4 outputs
         staa DDRC   ;PC3-PC0 inputs
         rts
ScanTab  fcb  $70,"abcd" ;top row
         fcb  $B0,"efgh"
         fcb  $D0,"ijkl"
         fcb  $E0,"mnop" ;bottom row
         fcb  0
; Returns RegA ASCII key pressed,
;      RegY number of keys pressed
;      Y=0 if no key pressed
Key_Scan ldy  #0    ;Number pressed
         ldx  #ScanTab
loop     ldab 0,x
         beq  done
         stab PORTC ;select row
         ldab PORTC ;read columns
         lsrb        ;PC0 into carry
         bcs  notPC0
         ldaa 4,x
         iny
notPC0   lsrb        ;PC1 into carry
         bcs  notPC1
         ldaa 3,x
         iny
```

```
; MC9S12C32
Key_Init clr  DDRT  ;PT3-PT0 inputs
         clr  PTT   ;PT7-PT4 oc output
         clr  PPST  ;pull-up on PT3-PT0
         movb #$0F,PERT
         rts
ScanTab  fcb  $80,"abcd" ;top row
         fcb  $40,"efgh"
         fcb  $20,"ijkl"
         fcb  $10,"mnop" ;bottom row
         fcb  0
; Returns RegA ASCII key pressed,
;      RegY number of keys pressed
;      Y=0 if no key pressed
Key_Scan ldy  #0     ;Number pressed
         ldx  #ScanTab
loop     ldab 0,x    ;row select
         beq  done
         stab DDRT   ;select row
         brset PTT,#$01,notPT0
         ldaa 4,x    ;code for column 0
         iny
notPT0   brset PTT,#$02,notPT1
         ldaa 3,x    ;code for column 1
         iny
notPT1   brset PTT,#$04,notPT2
         ldaa 2,x    ;code for column 2
         iny
```

```
notPC1    lsrb        ;PC2 into carry        notPT2    brset PTT,#$08,notPT3
          bcs  notPC2                                  ldaa 1,x   ;code for column 3
          ldaa 2,x                                     iny
          iny                                notPT3    leax 5,x   ;Size of entry
notPC2:   lsrb        ;PC3 into carry                  bra  loop
          bcs  notPC3                        done      rts
          ldaa 1,x
          iny
notPC3    ldab #5     ;Size of entry
          abx
          bra  loop
done      rts
```

Program 8.11
Assembly software interface of a matrix scanned keyboard.

```
// MC68HC711E9                             // MC9S12C32
const struct Row                          const struct Row
{   unsigned char out;                    { unsigned char direction;
    unsigned char keycode[4];}              unsigned char keycode[4];}
typedef const struct Row RowType;         typedef const struct Row RowType;
RowType ScanTab[5]={                       RowType ScanTab[5]={
{   0x70, "abcd" },                        {   0x80, "abcd" }, // row 3
{   0xB0, "efgh" },                        {   0x40, "efgh" }, // row 2
{   0xD0, "ijkl" },                        {   0x20, "ijkl" }, // row 1
{   0xE0, "mnop" },                        {   0x10, "mnop" }, // row 0
{   0x00, "    " }};                       {   0x00, "    " }};
void Key_Init(void){ // PC3-PC0 are inputs  void Key_Init(void){
  PIOC = 0x40;   // CWOM=1                    DDRT = 0x00; // PT3-PT0 inputs
  DDRC = 0xF0;}  // PC7-PC4 are outputs      PTT = 0;     // PT7-PT4 oc output
/* Returns ASCII code for key pressed,       PPST = 0     // pull-up on PT3-PT0
   Num is the number of keys pressed         PERT = 0x0F;}
   both equal zero if no key pressed */    /* Returns ASCII code for key pressed,
unsigned char Key_Scan(short *Num){           Num is the number of keys pressed
RowType *pt; unsigned char column,key;        both equal zero if no key pressed */
short j;                                    unsigned char Key_Scan(short *Num){
  (*Num)=0; key=0;    // default values    RowType *pt; unsigned char column,key;
  pt=&ScanTab[0];                          short j;
  while(pt->out){                            (*Num)=0; key=0;    // default values
    PORTC = pt->out;  // select row          pt=&ScanTab[0];
    column = PORTC;   // read columns        while(pt->direction){
    for(j=3; j>=0; j--){                        DDRT = pt->direction;  // one output
      if((column&0x01)==0){                     column = PTT;   // read columns
        key = pt->keycode[j];                   for(j=3; j>=0; j--){
        (*Num)++;}                                if((column&0x01)==0){
      column>>=1;} // shift into position           key = pt->keycode[j];
    pt++;                                            (*Num)++;}
  }                                               column>>=1;} // shift into position
  return key;                                 pt++; }
}                                            return key;}
```

Program 8.12
C software interface of a matrix scanned keyboard.

Observation: Because of capacitance, a short delay (5 ms) may need to be inserted after setting the row before reading the column. This delay can be shortened by reducing the value of the pullup resistor.

8.1.7
Multiplexed/
Demultiplexed
Scanned
Keyboard

To interface many switches, the multiplexed interface approach is used. In this approach the keys are again divided into rows and columns. In this interface there are 16 rows and 9 columns, giving 144 keys. The computer specifies the row number (0 to 15) by outputting to the four most significant bits. The 4 to 16 demultiplexer, 74159, has open-collector outputs and will drive exactly one row to 0 and the other rows will be HiZ. The 74147 is a 10 to 4 line priority encoder. If all nine inputs of the 74147 are high (i.e., no key pressed in this row), then the least significant 4 bits signals will be 1111. If exactly one key is pressed in the row, then exactly one 0 will exist on the 74147's inputs, and the 74147 output will be the negative logic location of the column number. For example, if the key in column 4 is pressed, then the 74147 "4" input will be 0 and bit 3–0 will be 1011. If two or more keys are pressed in the same row, then bits 3–0 will be the column number (again in negative logic), the column with the highest number. For example, if the keys in columns 3 and 7 are both pressed, then the 74147 "3" and "7" inputs will both be 0 and bits 3–0 will only signify the "7" 1000. As mentioned earlier, a multiplexed interface cannot handle more than one key pressed simultaneously (Figure 8.27). The assembly language software to scan this keyboard is shown in Program 8.13.

Figure 8.27
A multiplexed matrix keyboard interfaced to the microcomputer.

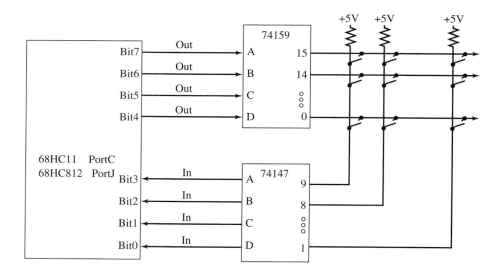

```
; MC68HC11A8
Key:       ds 1   ;current key
PrevKey: ds 1   ;previous key
Ritual: sei
        ldaa #$F0   ;PC7-PC4 outputs
        staa DDRC   ;PC3-PC0 inputs
        bsr KeyScan
        staa PrevKey
        staa Key
        ldaa TMSK1
        oraa #$08   ; arm OC5
        staa TMSK1
        ldd  TCNT
        addd #20000
        std  TOC5   ; first in 10ms
        ldaa #$08
        staa TFLG1 ; clear OC5F
```

```
; MC68HC812A4
Key:       ds 1   ;current key
PrevKey: ds 1   ;previous key
Ritual: sei
        ldaa #$F0   ;PJ7-PJ4 outputs
        staa DDRJ   ;PJ3-PJ0 inputs
        bsr KeyScan
        staa PrevKey
        staa Key
        bset #$20,TIOS   ;activate OC5
        bset #$20,TMSK1 ;arm OC5
        ldaa #$32
        staa TMSK2   ;500ns clock
        ldd  TCNT
        addd #20000
        std  TC5    ; first in 10ms
        ldaa #$20
```

```
              cli
              rts
; returns RegA=code (0 for none)
KeyScan: clra      ;=0 means no
              clr  PORTC ;row=0
loop:   ldab PORTC ;read columns
              andb #$0F
              cmpb #$0F  ;$0F means no
              beq  none
              eorb #$0F  ;code 1-9
              tba         ;found one
none:   ldab PORTC
              addb #$10  ;next row
              stab PORTC
              bcc  loop
              rts
TOC5handler:
              bsr  KeyScan
              cmpa PrevKey ;same as last?
              bne  skip
              staa Key   ;new value
skip:   staa PrevKey
              ldd  TOC5
              addd #20000
              std  TOC5  ;every 10ms
              ldaa #$08
              staa TFLG1 ;ack OC5
              rti
```

```
              staa TFLG1 ; clear C5F
              cli
              rts
; returns RegA=code (0 for none)
KeyScan: clra      ;=0 means no
              clr  PORTJ ;row=0
loop:   ldab PORTJ ;read columns
              andb #$0F
              cmpb #$0F  ;$0F means no
              beq  none
              eorb #$0F  ;code 1-9
              tba           ;found one
none:   ldab PORTJ
              addb #$10   ;next row
              stab PORTJ
              bcc  loop
              rts
TOC5handler:
              bsr  KeyScan
              cmpa PrevKey  ;same as last?
              bne  skip
              staa Key        ;new value
skip:   staa PrevKey
              ldd  TC5
              addd #20000
              std  TC5        ;every 10ms
              ldaa #$08
              staa TFLG1     ;ack OC5
              rti
```

Program 8.13
Assembly software interface of a multiplexed keyboard.

Periodic polling interrupt synchronization is appropriate for this interface. The background thread will maintain an 8-bit global variable, Key, storing in it the current key code from the matrix. To handle the bounce, we set the periodic interrupt interval longer than the bounce time. If the switch happens to be bouncing at the time of the interrupt, the 74147 outputs may contain incorrect patterns (e.g., representing patterns that are neither the old nor the new status). Therefore we will update Key only after the KeyScan() result is the same for two consecutive interrupts. In this example, there is a 10- to 20-ms time delay (latency) between the key press and when the global variable is changed. The C language software to scan this keyboard is shown in Program 8.14.

```
// MC68HC11A8
// PC7-PC4 row output
// PC3-PC0 column inputs
unsigned char Key;     // current pattern
unsigned char PreviousKey;  // 10ms ago
#define period 20000        // 10 ms
unsigned char KeyScan(void){
    unsigned char key,row;
    key=0;     // means no key pressed
    for(row=0;row<16;row++){
```

```
// MC68HC812A4
// PJ7-PJ4 row output
// PJ3-PJ0 column inputs
unsigned char Key;     // current pattern
unsigned char PreviousKey;  // 10ms ago
#define period 20000        // 10 ms
unsigned char KeyScan(void){
    unsigned char key,row;
    key=0;     // means no key pressed
    for(row=0;row<16;row++){
```

continued on p. 398

Program 8.14
C software interface of a multiplexed keyboard.

continued from p. 397

```
        PORTC=row<<4;   // Select row
        if((PORTC&0x0F)!=0x0F){
             key=PORTC^0x0F; }}
   return(key);}
void Ritual(void){
   asm(" sei");    // make ritual atomic
   DDRC=0xF0;
   PreviousKey=Key=KeyScan(); // read
   TMSK1|=0x08;         // Arm OC5
   TOC5=TCNT+period;
   TFLG1=0x08;          // clear OC5F
   asm(" cli"); }
#pragma interrupt_handler TOC5handler()
void TOC5handler(void){
unsigned char NewKey;
   NewKey=KeyScan(); // Current pattern
   if(NewKey==PreviousKey) Key=NewKey;
   PreviousKey=NewKey;
   TOC5=TOC5+period;
   TFLG1=0x08;}         // ack OC5F
```

```
        PORTJ=row<<4;   // Select row
        if((PORTJ&0x0F)!=0x0F){
             key=PORTJ^0x0F; }}
   return(key);}
void Ritual(void){
   asm(" sei");    // make ritual atomic
   DDRJ=0xF0;
   PreviousKey=Key=KeyScan(); // read
   TMSK1|=0x20;         // Arm OC5
   TIOS|=OC5;      // enable OC5
   TSCR|=0x80;     // enable
   TMSK2=0x32;     // 500 ns clock
   TC5=TCNT+period;
   TFLG1=0x20;         // clear OC5F
   asm(" cli"); }
#pragma interrupt_handler TOC5handler()
void TOC5handler(void){
unsigned char NewKey;
   NewKey=KeyScan();  // Current pattern
   if(NewKey==PreviousKey) Key=NewKey;
   PreviousKey=NewKey;
   TOC5=TOC5+period;
   TFLG1=0x20;}            // ack OC5F
```

Program 8.14
C software interface of a multiplexed keyboard.

Checkpoint 8.4: How should the Ctrl, Alt, and Shift keys be interfaced within a standard PC keyboard? How should the remaining 100 keys be interfaced?

Checkpoint 8.5: How would you interface a touchpad that has the equivalent of over one million keys?

8.2   Output LEDs

Similar to the keyboard interfaces in the previous section, we will develop LED interfaces in the direct, scanned, and multiplexed categories (Figure 8.28). A direct interface has a unique computer output pin for each LED. In this way, a single output port can control eight LED segments. Once the output port is set, no software action is required to maintain the direct interface. All possible "on/off" patterns can be generated by the direct interface. The scanned interface organizes the LEDs in a matrix with rows and columns, and each row and each column has a unique output pin. In Figure 8.29 current sources are used to drive the rows and current sinks are connected to the columns. If the LEDs are configured in a 4 by 4 matrix, a single output port can control 16 LED segments. Software executed on a continuous and regular basis will be required to maintain the display. All possible on/off patterns can be generated by the scanned interface. The multiplexed display is also organized in a matrix with rows and columns but has multiplexing hardware so that there are more rows and columns than there are computer output pins. Theoretically, as shown in Figure 8.29, it is possible for a single output port to control $16 \cdot 16$, or 256, LED segments. Again, software executed on a continuous and regular basis will be required to maintain the display. Depending on the configuration of the multiplexer, not all possible on/off patterns can be generated by the multiplexed interface. Because of the maximum allowable LED current limitations (the details to be presented later in the section), it will not be practical to have a 16 by 16 LED matrix. Nevertheless, the

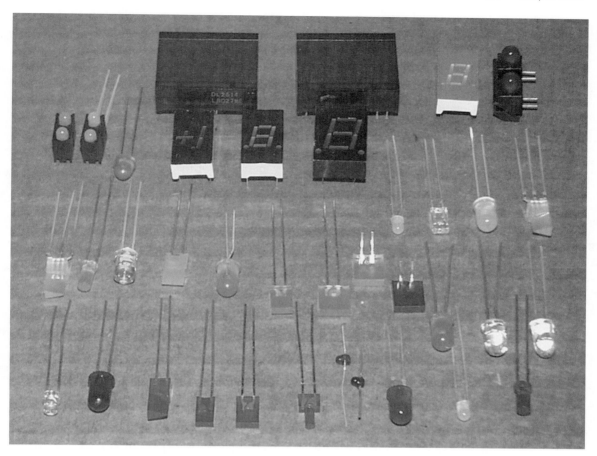

Figure 8.28
LEDs come in a wide variety of shapes, sizes, colors and configurations.

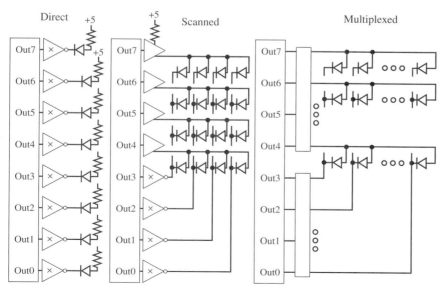

Figure 8.29
Three approaches to interfacing multiple LEDs.

multiplexed approach is common, especially for 7-segment and 15-segment LED displays. We will also develop an LED interface that uses a multiplexed approach but performs the scanning operations in hardware so that periodic software maintenance is not required.

8.2.1
Single LED
Interface

If you connect a microcomputer output pin to the LED interface, then the microcomputer can control the state (on/off) of the LED. If you make the microcomputer output high (a logic 1), the output of the 7405 will be low, current will flow through the LED, and it will be lit. The LED voltage will be about 2 V when it is lit. If you make the computer output low (a logic 0), the output of the 7405 will be off (HiZ, not driven, high imped-ance, disconnected), no current will flow through the LED, and it will be dark. The use of the 7405 provides the necessary current (about 10 mA) to activate the light. The 250Ω resis-tor is selected to control the brightness of the light (Figure 8.30).

Figure 8.30
A single LED interface.

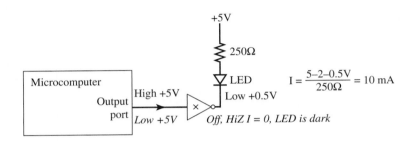

We can use either open-collector logic or open-emitter logic as current switches. The current on some LEDs is so low (e.g., 1 mA) that they can be connected directly to an output port, as shown in Figure 8.31. When the open-collector output is 0, it will sink current to ground (turning on the LED, relay, stepper coil, solenoid, etc.). When the open-collector output is floating, it will not sink any current (turning off the LED, relay, stepper coil, sole-noid, etc.). Similarly, when the 75491 input is high, the open-emitter output transistor is active, sourcing current into the LED. When the 75491 input is low, the open-emitter output transistor is off, making the LED current zero. Table 8.4 provides the output low currents for some typical open-collector devices. Table 8.5 provides the output source currents for some typical open-emitter devices.

Figure 8.31
Three approaches to controlling the current to an LED.

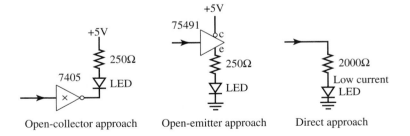

Darlington switches like the ULN-2061 through ULN-2077 and MOSFETs like the IRF-540 can be used either in open-collector mode to sink current or in open-emitter mode to source current. For all the devices the actual output voltage depends on the output current. What is shown in Tables 8.4 and 8.5 is the output voltage at maximum output current. For the transistor devices, the output voltages/currents also depend on the input currents.

Table 8.4
Output parameters for various open-collector gates.

Family	Example	V_{OL}*	I_{OL}
Standard TTL	7405	0.4 V	16 mA
Schottky TTL	74S05	0.5 V	20 mA
Low-power Schottky TTL	74LS05	0.5 V	8 mA
High-speed CMOS	74HC05	0.33 V	4 mA
High-voltage output TTL	7406	0.7 V	40 mA
High-voltage output TTL	7407	0.7 V	40 mA
Silicon monolithic IC	75492	0.9 V	250 mA
Silicon monolithic IC	75451-75454	0.5 V	300 mA
Darlington switch	ULN-2074	1.4 V	1.25 A
MOS field-effect transistor (MOSFET)	IRF-540	Varies	28 A

* Voltage at maximum I_{OL}.

Table 8.5
Output parameters for various open-emitter gates.

Family	Example	V_{CE}	I_{CE}
Silicon monolithic IC	75491	0.9 V	50 mA
Darlington switch	ULN-2074	1.4 V	1.25 A
MOSFET	IRF-540	Varies	28 A

For scanned and multiplexed LED interfaces, we will use both current sources and sinks. In the following interface circuits we assume the desired LED setpoint is 2 V and 10 mA. Since the LED is a diode, the voltage and current relationship is quite nonlinear. The voltage/current relationship for the LTP-1057A and LTP-1157A dot matrix LED displays is plotted in Figure 8.32.

Figure 8.32
Typical voltage/current response of a LED.

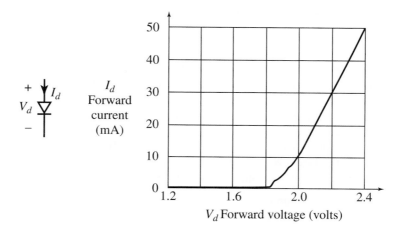

To prevent the LED from overheating, we must limit the electric power by using a current-limiting series resistor as shown in Figure 8.33. Table 8.6 illustrates typical LED specifications.

Table 8.6
Absolute maximum rating for LTP-1057A and LTP-1157A 5 by 7 dot matrix displays.

Parameter	Red	Green	Yellow	Orange	Units
Maximum power	55	75	60	75	mW
Peak forward current	160	100	80	100	mA
Max continuous current	25	25	20	25	mA

The LED power can be calculated from its voltage and current.

$$P_d = V_d \cdot I_d$$

The resistor value is chosen to establish the desired voltage/current (V_d/I_d) operating point for the LED. For the 75492 open-collector circuit, the resistor is calculated as (Figure 8.33)

$$R = (5 - V_d - V_{OL})/I_d = (5 - 2 - 0.9)/10 \text{ mA} = 210 \ \Omega$$

Similarly for the 75491 open-emitter circuit, the resistor is calculated as (Figure 8.33)

$$R = (5 - V_d - V_{ce})/I_d = (5 - 2 - 0.9)/10 \text{ mA} = 210 \ \Omega$$

Figure 8.33
Calculating the resistor used in the LED interface.

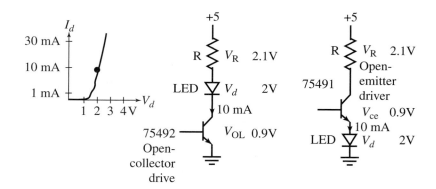

Checkpoint 8.6: What resistor value do you need in the direct approach LED interface of Figure 8.31 if the LED voltage is 2V and current is 1 mA, assuming VOH is 4.5V?

8.2.2 Seven-Segment LED Interfaces

When creating LEDs that display numbers, it is appropriate to use common-cathode or common-anode seven-segment LED modules. Some modules have an eighth or ninth segment to display a right-side and/or left-side decimal point. The circuit in Figure 8.34 shows a direct interface to a single seven-segment common-cathode LED display. It is called common cathode because the seven cathodes are connected. We label it direct because there is a separate computer output pin for each LED segment. The software simply writes to the output port to control the seven segments. A current sink is required to interface a common-anode display. It is called common anode because the seven anodes are connected (Figure 8.35).

Common error: If you try to replace the seven individual resistors in either of the two circuits in Figures 8.36 and 8.37 with a single resistor on the "common" side, then the brightness of each LED segment will be a function of how many LEDs are on. For example, if only one segment is on, it will be very bright; if all seven are on, then they will be dim.

Observation: The current-limiting resistors should be placed on the individual LED connections and not on the "common" connection.

8.2.3 Scanned Seven-Segment LED Interface

We will design a three-digit LED display capable of displaying the numbers from 000 to 999. The 21 LED segments will be interfaced in a 3 column by 7 row rectangular matrix. There will be a column for each digit and a row for each of the segments a, b, c, d, e, f, g. The open-collector outputs of the 7406 will sink current from the seven rows. The ULN2074,

Figure 8.34
Seven-segment
common-cathode LED
interface.

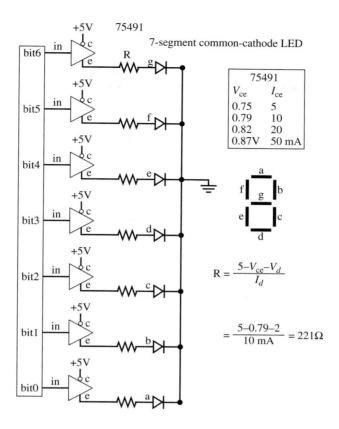

$$R = \frac{5 - V_{ce} - V_d}{I_d}$$

$$= \frac{5 - 0.79 - 2}{10 \text{ mA}} = 221\Omega$$

Figure 8.35
Seven-segment
common-anode LED
interface.

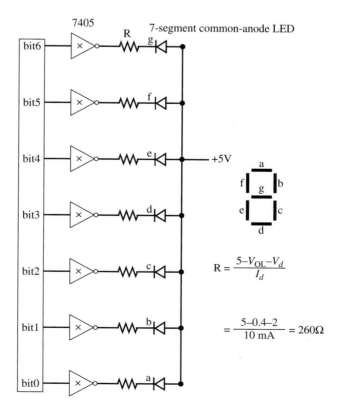

$$R = \frac{5 - V_{OL} - V_d}{I_d}$$

$$= \frac{5 - 0.4 - 2}{10 \text{ mA}} = 260\Omega$$

in open-emitter mode, will source current for the three columns. The software will utilize a simple 3-byte variable to contain the current value to be displayed. For example, if the value "456" is to be displayed, the main program will set the 24-bit global to $666D7C, as explained in Table 8.7. Our interrupt software will read this global and output the appropriate signals to the output ports (Figure 8.36).

Figure 8.36
Seven-Segment
common-anode LED's.

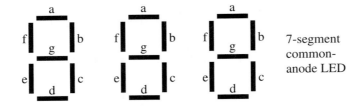

7-segment
common-
anode LED

Table 8.7 gives the conversion between decimal digit and seven-segment binary code assuming segments g, f, e, d, c, b, a are mapped into bits 6, 5, 4, 3, 2, 1, 0. This interface can also be used to create hexadecimal displays.

Table 8.7
Patterns for a
seven-segment display.

Digit	Segments	Binary code
0	f, e, d, c, b, a	%00111111=$3F
1	b, c	%00000110=$06
2	g, e, d, b, a	%01011011=$5B
3	g, d, c, b, a	%01001111=$4F
4	g, f, c, b	%01100110=$66
5	g, f, d, c, a	%01101101=$6D
6	g, f, e, d, c	%01111100=$7C
7	c, b, a	%00000111=$07
8	g, f, e, d, c, b, a	%01111111=$7F
9	g, f, c, b, a	%01100111=$67
A	g, f, e, c, b, a	%01101111=$6F
b	g, f, e, d, c	%01111100=$7C
C	f, e, d, a	%00111001=$39
d	g, e, d, c, b	%01011110=$5E
E	g, f, e, d, a	%01111001=$79
F	g, f, e, a	%01110001=$71

Assume the desired LED operating point is 10 mA, 2 V. Since the display will be scanned, we will operate each active LED at 30 mA with a 33% duty cycle. We will change the display every 5 ms. In this way, we will scan through the entire display every 15 ms, and each active LED will run at about 66 Hz, faster than the human eye can detect. The LED will "look like" it is continuously on at 10 mA (Figure 8.37).

The software will have to convert the decimal digits into the seven-segment codes using Table 8.7. The 7406 open-collector driver can sink the required 30 mA. If all seven segments are on, the source current will be 7 · 30 mA, or 210 mA. The ULN2074 can supply the necessary current (Figure 8.38). The 6812 system uses Port M and Port T.

Figure 8.37
Timing used to scan a
LED interface.

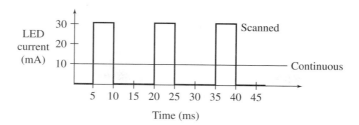

Figure 8.38
Circuit used to scan a LED interface. (For the MC9S1232 replace PB with PT and replace PC with PM.)

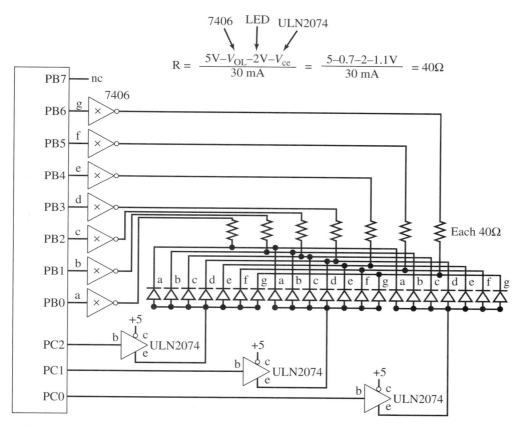

It is important to place the resistors on the segment select side rather than on the digit select side. In this way the current through each diode is not a function of the number of diodes on. The software that initializes/maintains the display is shown in Program 8.15.

**8.2.4
Scanned LED
Interface Using
the 7447 Seven-
Segment Decoder**

The purpose of the previous example was to illustrate the details involved in scanning a LED display. For practical designs we will use special LED display decoder logic. The simplest of the LED drivers is the 7447 seven-segment decoder. Again we will design a three-digit LED display interfaced only to a single output port. The 21 LED segments will again be interfaced in a 3 column by 7 row rectangular matrix. There will be a column for each digit and a row for each of the segments a, b, c, d, e, f, g. The open-collector outputs of the 7447 will sink current from the seven rows. The ULN2074, in open-emitter mode, will source current for the three columns. The software will utilize a 12-bit packed BCD global

```
// MC68HC711E9, ICC11 C                      // MC9S12C32, CodeWarrior C
// PB7-PB0 output, 7 bit pattern             // PT7-PT0 output, 7 bit pattern
// PC2-PC0 output, selects LED digit         // PM2-PM0 output, selects LED digit
unsigned char code[3]; // binary codes       unsigned char code[3]; // binary codes
static unsigned char select[3]={4,2,1};      static unsigned char select[3]={4,2,1};
unsigned short index;   // 0,1,2             unsigned short index;   // 0,1,2
#define OC5F 0x08                             void interrupt 13 TC5handler(void){
#pragma interrupt_handler TOC5handler()        TFLG1 = 0x20;        // Acknowledge
void TOC5handler(void){                         TC5 = TC5+10000;     // every 5 ms
  TFLG1 = 0x08;        // Acknowledge           PTM = select[index]; // which LED?
  TOC5 = TOC5+10000; // every 5 ms              PTT = code[index];   // enable
  PORTC = select[index]; // which LED?          if(++index==3) index=0;}
  PORTB = code[index];   // enable           void LED_Init(void) {
  if(++index==3) index=0;}                   asm sei              // make atomic
void LED_Init(void) {                          index = 0;
asm(" sei");           // make atomic          DDRT = 0xFF;     // outputs 7 segment code
  index = 0;                                    DDRM |= 0x07;    // outputs select LED
  DDRC = 0xFF;         // outputs               TIE  |=0x20;     // Arm OC5
  TMSK1 |= 0x08;   // Arm OC5                    TIOS |=0x20;     // enable OC5
  TFLG1 = 0x08;    // clear OC5F                 TSCR1 =0x80;     // enable
  TOC5 = TCNT+10000;                            TSCR2 =0x01;     // 500 ns clock
asm(" cli");                                    TC5 = TCNT+10000;
}                                            asm cli }
```

Program 8.15
C software interface of a scanned LED display.

variable to contain the current value to be displayed. For example, if the value "456" is to be displayed, the main program will set the 16-bit global to $0456. Our interrupt software will read this global and output the appropriate signals to the output port.

Again, we assume the desired LED operating point is 10 mA, 2 V. Since the display will be scanned, we will operate each active LED at 30 mA with a 33% duty cycle. The 7447A will simplify the generation of the seven-segment codes. This seven-segment LED driver can sink the required 30 mA. Just like the previous example, if all seven segments are on, the source current will be 7 • 30 mA, or 210 mA (Figure 8.39). The software that initializes/maintains the display is shown in Program 8.16. The 6812 system uses Port T.

If more digits are needed, it would be easy to extend this approach by adding additional ULN2074 current drivers. There are two issues to consider as the number of digits is added. The first is the scan frequency. For the display to "look" continuous, each digit must be updated faster than 60 Hz. (If you look closely into the specifications of your computer monitor and television sets, you will see this same constraint.) So as you increase the number of digits the interrupt rate must increase so that each individual digit is scanned faster than 60 Hz. If there are five digits, then the periodic interrupt rate must be increased to at least 300 Hz for each digit to be updated at a rate of 60 Hz. The second issue to consider is the duty cycle for each digit. As the number of digits is increased, the duty cycle for each digit decreases. This makes it necessary to increase the instantaneous current. For example, if there were five digits, the duty cycle would be 20% and the current would have to increase to 50 mA. This design would require an open-collector driver capable of sinking 50 mA and an open-emitter capable of sourcing 350 mA. Another limitation is the maximum instantaneous current allowed by the LED. There is an upper bound to the instantaneous LED current even if the duty cycle decreases, leaving the average power the same. Each LED is different, but this parameter is about 100 mA, as shown in Table 8.6.

Observation: The ratio of the maximum instantaneous current divided by the desired LED current determines the maximum number of columns in the LED matrix.

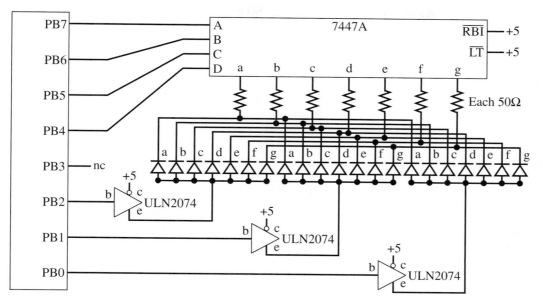

$$R = \frac{5V - V_{OL} - 2V - V_{ce}}{30 \text{ mA}} = \frac{5 - 0.4 - 2 - 1.1V}{30 \text{ mA}} = 50\Omega$$

Figure 8.39

An encoder used to interface three seven-segment common-anode LED digits. (For the MC9S12C32 replace PB with PT.)

```
// MC68HC711E9
unsigned short Global;// 12-bit packed BCD
const struct LED
{ unsigned char enable; // select
  unsigned char shift;  // bits to shift
  const struct LED *Next; };  // Link
typedef const struct LED LEDType;
typedef LEDType * LEDPtr;
LEDType LEDTab[3]={
{ 0x04, 8, &LEDTab[1] },  // Most sig
{ 0x02, 4, &LEDTab[2] },
{ 0x01, 0, &LEDTab[0] }}; // least sig
LEDPtr Pt;    // Points to current digit
#pragma interrupt_handler TOC5handler()
void TOC5handler(void){
  TFLG1 = 0x08;        // Acknowledge
  TOC5 = TOC5+10000;   // every 5 ms
  PORTB=(Pt->enable)
      +(Global>>(Pt->shift))<<4);
  Pt = Pt->Next; }
```

```
// MC9S12C32, CodeWarrior C
unsigned short Global; // 12-bit packed BCD

const struct LED
{ unsigned char enable; // select
  unsigned char shift;  // bits to shift
  const struct LED *Next; };  // Link
typedef const struct LED LEDType;
typedef LEDType * LEDPtr;
LEDType LEDTab[3]={
{ 0x04, 8, &LEDTab[1] },  // Most sig
{ 0x02, 4, &LEDTab[2] },
{ 0x01, 0, &LEDTab[0] }}; // least sig
LEDPtr Pt;    // Points to current digit
void interrupt 13 TC5handler(void){
  TFLG1 = 0x20;        // Acknowledge
  TC5 = TC5+10000;   // every 5 ms
  PTT = (Pt->enable)
      +(Global>>(Pt->shift))<<4);
  Pt = Pt->Next; }
```

continued on p. 408

Program 8.16

C software interface of a multiplexed LED display.

```
continued from p. 407

void LED_Init(void) {
asm(" sei");        // make atomic
  Global = 0;
  TMSK1 = 0x08;     // Arm OC5
  Pt = &LEDTab[0];
  TFLG1 = 0x08;     // clear OC5F */
  TOC5 = TCNT+10000;
asm(" cli"); }
```

```
void LED_Init(void) {
asm sei             // make atomic
  DDRT = 0xFF;    // outputs to LED's
  Global = 0;
  Pt=&LEDTab[0];
  TIE  |= 0x20;       // Arm OC5
  TIOS |= 0x20;       // enable OC5
  TSCR1 = 0x80;       // enable
  TSCR2 = 0x01;       // 500 ns clock
  TC5 = TCNT+10000;
asm(" cli"); }
```

Program 8.16
C software interface of a multiplexed LED display.

**8.2.5
Integrated LED
Interface Using
the MC14489
Display Driver**

For LED displays with many segments, one attractive solution is to use an integrated LED display driver. There are three advantages of a chip like the MC14489 over the other LED designs in this section. The first advantage is chip count. A single MC14489 20-pin chip and one resistor are all the components required to interface a five-digit LED display. The second advantage is that the scanning functions occur in hardware, eliminating the need for the software to execute periodic interrupts. Normally, when we think of the hardware/software trade-off, we often select the software solution because of its low cost and flexibility. This is a situation where a hardware solution could be cheaper. The third advantage is that it is easy to cascade multiple MC14489 drivers so that larger displays can be interfaced without requiring additional microcomputer output ports. Each of the five eight-segment LED devices has seven segments ("a–g") for creating the decimal digit plus one more segment ("h") for the decimal point. The packed BCD format is shifted serially into the MC14489. The SPI port is a convenient solution for the hardware/software interface between the computer and MC14489(s). Once the BCD pattern is loaded into the MC14489, the hardware will continuously generate the 8 by 5 matrix scanning signals that display the five digits. The **Rx** resistor controls the LED voltage/current operating point for the display. The software also has the ability to select the LED brightness: "regular" or "dim." The SPI SS pin (6811 PD5, 6812 PM3) will be configured as a simple output because the ENABLE pin will be low for 8 clock cycles when transmitting a command and for 24 clock cycles when transmitting data (Figure 8.40).

The MC14489 can handle multiple formats. This interface utilizes a packed BCD format, as shown in Figure 8.41. The clock signal (output of computer, input to MC14489)

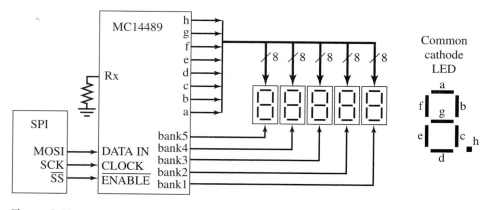

Figure 8.40
An integrated IC used to interface five seven-segment common-cathode LED digits.

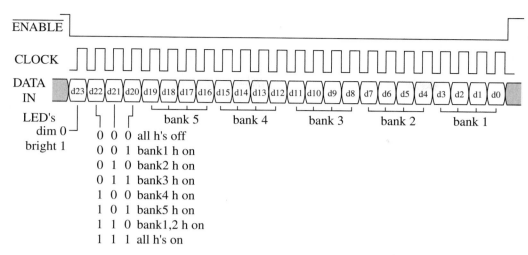

Figure 8.41
Data timing of an integrated LED controller.

is used to shift 24 bits of data into the LED display interface. For data to be properly transferred, the computer will change the data on the falling edge of the CLOCK and the MC14489 will shift the data on the rising edge. This timing can be achieved with the SPI in master mode and CPHA=0, CPOL=0. For ENABLE to be low for 24 bits, we will configure it as a simple output. The 20 bits that control the five banks are encoded as packed BCD. The MC14469 will also display BCD digits `Abcdef,` so the system can be used to show five decimal digits or five hexadecimal digits.

There is also an 8-bit command transmission that is used to configure the display. To implement the standard decimal (or hexadecimal) display, we will send a command of $01. This enables the device and puts all five banks in hexadecimal format (Figure 8.42). The software to control the interface is simplified when using the SPI. The transmission rate is configured for 1 MHz, but it actually could operate as fast as 2 MHz. Interrupt synchronization is not needed because each 8-bit data frame requires only 8 μs to complete (Program 8.17).

> ***Maintenance tip:*** It will be more reliable to design the system using a transmission rate, somewhat slower than the absolute maximum.

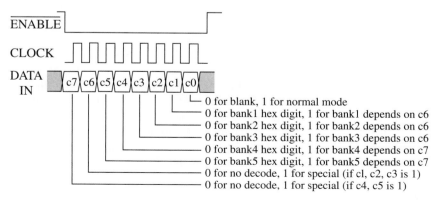

Figure 8.42
Configuration timing of an integrated LED controller.

```
// MC68HC711E9                            // MC68HC812A4
// PD3/MOSI = MC14489 DATA IN             // PM4/MOSI = MC14489 DATA IN
// PD4/SCLK = MC14489 CLOCK IN            // PM5/SCLK = MC14489 CLOCK IN
// PD5 (simple) = MC14489 ENABLE          // PM3 (simple output) = MC14489 ENABLE
void LED_Init(void) {                     void LED_Init(void) {
  DDRD |= 0x38;  // outputs to MC14489       DDRM |= 0x38;  // outputs to MC14489
  SPCR = 0x50;                              SPICR1 = 0x50;
// bit  meaning                           // bit  meaning
// 7 SPIE=0 no interrupts                 // 7 SPIE=0 no interrupts
// 6 SPE=1  SPI enable                    // 6 SPE=1  SPI enable
// 5 DWOM=0 regular outputs               // 5 SPTIE=0 no interrupts
// 4 MSTR=1 master                        // 4 MSTR=1 master
// 3 CPOL=0 match timing with MC14489     // 3 CPOL=0 match timing with MC14489
// 2 CPHA=0                               // 2 CPHA=0
// 1 SPR1=0 E/2 is 1Mhz SCLK              // 1 SSOE=0 PM3 is simple output
// 0 SPR0=0                               // 0 LSBF=0 MSB first
  PORTD |= 0x20;    // ENABLE=1             SPICR2 = 0x00; // regular drive
  PORTD &=~0x20;    // ENABLE=0             SPIBR = 0x01;  // 1MHz SCLK
  SPDR = 0x01;      // hex format           PTM |= 0x08;   // ENABLE=1
  while(SPSR&0x80)==0){};                   PTM &=~0x08;   // ENABLE=0
  PORTD |= 0x20;}   // ENABLE=1             SPIDR= 0x01;   // hex format
// 24 bit packed BCD                        while(SPISR&0x80)==0){};
void LED_out(unsigned char data[3]){        PTM |=0x08;}   // ENABLE=1
  PORTD &=~0x20;    // ENABLE=0           void LED_out(unsigned char data[3]){
  SPDR = data[2];  // send MSbyte           PTM &=~0x08;      // ENABLE=0
  while(SPSR&0x80)==0){};                   SPIDR = data[2];  // send MSbyte
  SPDR = data[1];  // send middle byte      while(SPISR&0x80)==0){};
  while(SPSR&0x80)==0){};                   SPIDR = data[1];  // send middle byte
  SPDR = data[0];  // send LSbyte           while(SPISR&0x80)==0){};
  while(SPSR&0x80)==0){};                   SPIDR = data[0];  // send LSbyte
  PORTD |= 0x20;   // ENABLE=1              while(SPISR&0x80)==0){};
}                                           PTM |=0x08;}      // ENABLE=1
```

Program 8.17
C software interface of an integrated LED display.

8.3   Liquid Crystal Displays

8.3.1
LCD Fundamentals

Liquid crystal displays are widely used in microcomputer systems (Figure 8.43). One advantage of LCDs over LEDs is their low power consumption. This allows the display, and perhaps the entire computer system, to be battery-operated. In addition, LCDs are more flexible in their sizes and shapes, permitting the combination of numbers, letters, words, and graphics to be driven with relatively simple interfaces. A LCD consists of a liquid-crystal material that behaves electrically as a capacitor. Whereas a LED converts electric power into emitted optical power, a LCD uses an alternating current (AC) voltage to change the light reflectivity (or sometimes transmittivity). The light energy is supplied by the room or a separate back light, and not by the electric power within the LCD (as with LEDs). The computer controls the display by altering the reflectivity of each segment. The disadvantage of LCDs is their slow response time. Fortunately, the bandwidth of most displays (both LEDs and LCDs) is limited by the human visual processing system (about 30 Hz).

It is important to provide the LCD with only AC and no DC signals. A DC signal above 50 mV will cause permanent damage to the LCD. Let V_{DD} be the supply voltage of the CMOS logic, and let V_{SS} be the ground voltage of the CMOS logic. The use of CMOS logic has two advantages. First, CMOS requires very little supply current. Second, CMOS logic has fairly

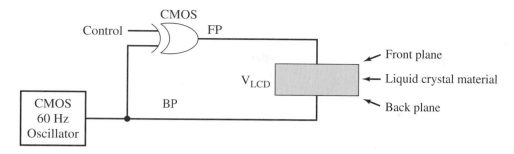

Figure 8.43
The basic idea of a liquid crystal interface.

stable high and low logic voltage levels. In other words, the V_{OH} of the CMOS EOR gate and CMOS 60-Hz oscillator will both be close to V_{DD} (hence close to each other). Similarly, the V_{OL} of the CMOS EOR gate and CMOS 60-Hz oscillator will also both be close to V_{SS} (hence close to each other). Therefore, for both **Control** high and low, $V_{LCD} = FP - BP$ will contain no DC component. Obviously, it will be important for the oscillator to have a 50% duty cycle.

The oscillator output **BP** is a square wave (V_{SS} to V_{DD}) with a frequency of 60 Hz. When the **Control** signal is low, the front-plane voltage, **FP**, is in phase with the back-plane voltage, **BP**. Hence, V_{LCD} will be zero. In this state, the display does not reflect light, and the display is blank. When the **Control** signal is high, the front-plane voltage, **FP**, is out of phase with the back-plane voltage, **BP**. Hence, V_{LCD} will be an AC square wave ($-V_{DD}$ to $+V_{DD}$). In this state, the display reflects light, and the display is visible (Figure 8.44).

Observation: LCDs are controlled with waveforms in the 40- to 60-Hz range.

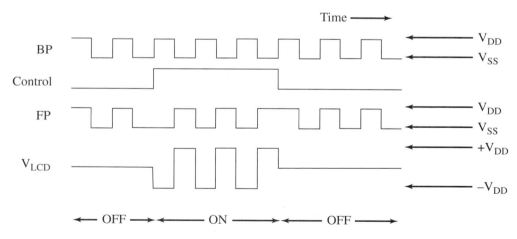

Figure 8.44
The basic timing of a liquid crystal interface.

One of the most important advantages of the LCD technology over lamps and LEDs is the flexibility in configuring the shapes and sizes of the segments. With LEDs the shapes of the segments are limited to simple regular shapes like circles and rectangles. Liquid crystal displays are created by sandwiching the liquid crystal material between a front and a back plane. The LCD segment is created in the overlap area of the front and back planes. Since the front and back planes are manufactured using techniques similar to PC board layout, there is a great flexibility in the sizes and shapes. In Figure 8.45 the entire phrase "**LCDs are**" represents a single segment, the overlap of FP1 and BP.

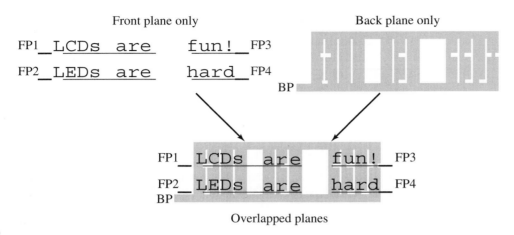

Figure 8.45
Example artwork for a LCD.

8.3.2
Simple LCD
Interface with the
MC14543

The LCD interface (hardware and software) is similar to that for LEDs. The direct-driven interface is simple but requires a large number of output bits and cable wires. The MC14543 is a CMOS BCD to seven-segment LCD driver. When **LD** = 1 **BI** = 0 and **Ph**=60 Hz square wave, one seven-segment LCD can be driven (Figure 8.46).

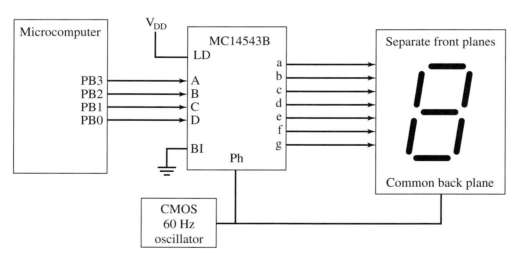

Figure 8.46
Direct interface of a LCD.

The ability to latch data into the driver allows multiple LCDs to be interfaced with a single microcomputer output port. To latch one digit, the software outputs the 4-bit BCD on PB3-0, then toggles one of PB7-4 high, then low, as, for example, shown in Program 8.18.

Program 8.18
Helper function for
a simple LCD display.

```
void LCDOutDigit(unsigned char position, unsigned char data) {
// position is 0x80, 0x40, 0x20, or 0x10  and data is the BCD digit
  PORTB = 0x0F&data;  // set BCD digit on the A-D inputs of the MC14543B
  PORTB |= position;  // toggle one of the LD inputs high
  PORTB = 0x0F&data;  // LD=0, latch digit into MC14543B
}
```

To set all four digits, the software calculates the BCD from an unsigned input, as, for example, in Program 8.19.

Program 8.19
C software interface
of a simple LCD display.

```
void LCD_OutNum(unsigned short data){ unsigned short digit,num,i;
unsigned char pos;
  num = min(data,9999); // data should be unsigned from 0 to 9999
  pos = 0x10;    // position of first digit (ones)
  for(i=0;i<4;i++){
    digit = num%10; num = num/10; // next BCD digit 0 to 9
    LCDOutDigit(pos,digit); pos = pos<<1;
  }
}
```

The 6812 requires a simple ritual that sets the direction register to outputs (e.g., DDRB=0xFF;). Port B on the 6811 is a fixed output port, so no ritual is required (Figure 8.47).

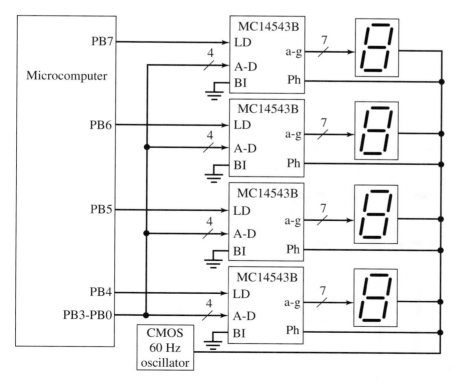

Figure 8.47
Latched interface of a LCD.

8.3.3
Scanned LCD
Interface with the
MC145000,
MC145001

Figure 8.48 shows a typical front-plane/back-plane configuration for an eight-segment LCD display. Different from LED displays, these LCD components do not have a common connection but rather utilize a 2 by 4 matrix. This configuration is compatible with LCD drivers like the MC145000 and MC145001.

Complex LCD artwork combines numbers, letters, words, and graphics on a single LCD. To simplify both the artwork and the interface, the front-plane and back-plane signals can be multiplexed. The concept is similar to the scanned LED displays described earlier. A four-digit display containing 32 segments can be multiplexed into four back-plane rows and eight front-plane columns (Figure 8.49). This reduces the number of cable wires from 32 (Figure 8.47) to 13 (Figure 8.49).

Figure 8.48
Artwork for an eight-segment liquid crystal digit.

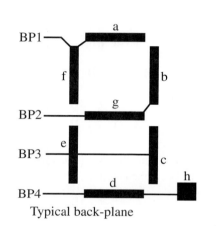

Typical back-plane

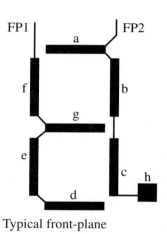

Typical front-plane

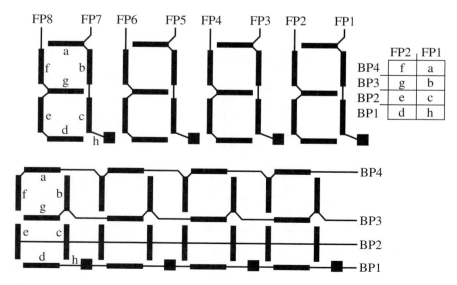

	FP2	FP1
BP4	f	a
BP3	g	b
BP2	e	c
BP1	d	h

Figure 8.49
Artwork for four eight-segment liquid crystal digits.

Consider the highlighted segment in Figure 8.50. To display this segment an AC voltage should be applied across BP3, FP5. To hide this segment, no AC voltage should be applied across BP3, FP5. It is impossible to control all 32 segments in this simple manner. Fortunately, Freescale has developed chips to simplify the interfacing of multiple LCDs. One MC145000 master and three MC145001 slaves accept serial input from the microcomputer and will directly drive 20 LCD digits or 180 segments. These same chips (MC145000 and MC145001) could be used to drive a complex LCD display with numbers, letters, words, and graphics. The MC14500 master can control up to 48 segments organized in 4 back planes and 12 front planes. Each additional MC145001 slave can control up to 44 segments organized in 4 back planes and 11 front planes.

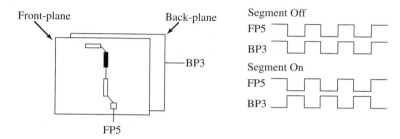

Figure 8.50
Each segment has a unique front-plane/back-plane combination.

The MC145000 master creates the frame-sync pulses in Figure 8.51 each time the display is updated. The time-multiplexed back-plane voltages used to create the scanned display are divided into four phases. During the first phase, the BP1 signal goes +5 V, then 0. If you wish any of the front-plane/BP1 segments to be active, the corresponding FP signal will go 0, then +5 V, during the first phase. This will create an AC signal large enough to activate the segment. Each of the BP signals has 25% of the time (when BP goes +5 V, then 0) to produce either an activation with the BP/FP pair (by making FP go 0, then +5 V) or a deactivation (by making FP go 3.33 V, then 1.67 V). In both the on and off cases then DC component is zero. In the on situation, the BP to FP differential voltage is 5 to 0 V, then 0 to 5 V—that is, 5 V then −5 V. In the off situation, the BP to FP differential voltage is 5 to 3.33 V, then 0 to 1.67 V—that is, 1.67 V then −1.67 V. The ±5-V AC signal is large enough to active the LCD, but the ±1.67-V AC signal is not (Figures 8.52, 8.53).

Figure 8.51
Synchronization pulses for the LCD display.

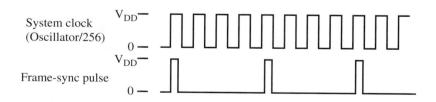

The interface between the microcomputer and the LCD driver(s) utilizes a synchronized serial format (both serial data and serial clock). Most 6811s and 6812 have a SPI that automatically creates these signals. The computer interface is responsible for sending 48 bits in serial (one bit at a time). The SPI software simply outputs 6 bytes (most

Figure 8.52
Fixed waveforms for the
four back-plane signals
for the LCD display.

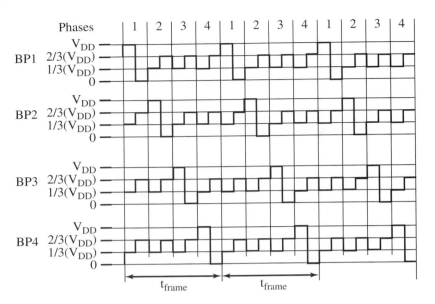

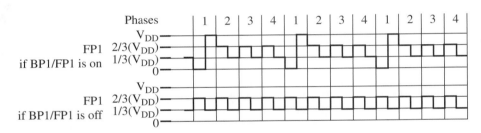

Figure 8.53
Variable waveforms for the front-plane signal for the LCD display.

significant first), and the SPI hardware creates the MOSI data output and SCK clock. Once the information is loaded, the latch is activated, and the MC145000 LCD driver will maintain the display without software overhead (until the software wishes to change the display) (Figure 8.54).

The details of the SPI port were discussed in Chapter 7. Even without a SPI port, the LCD interface would be possible with two ordinary output bits. Assume the MOSI, SCK outputs are implemented with a simple output port on PB0, PB1, respectively. For each bit, the software could output the data bit on the MOSI output, then toggle the SCK output. To fill the shift register and set all 48 segments, the software should output each bit one at a time, most significant bit first, as, for example, shown in Program 8.20. The SPI software to control this LCD interface is very similar to the LED interface using the MC14489 (Program 8.21).

8.3.4
Parallel Port LCD
Interface with the
HD44780
Controller

Microprocessor-controlled LCD displays are widely used, having replaced most of their LED counterparts, because of their low power and flexible display graphics. This example will illustrate how a handshaked parallel port of the microcomputer will be used to output to the LCD display. The hardware for the display uses an industry standard HD44780 controller (Figure 8.55). The low-level software initializes and outputs to the HD44780 controller.

Figure 8.54
Interface of a 48-segment LCD display.

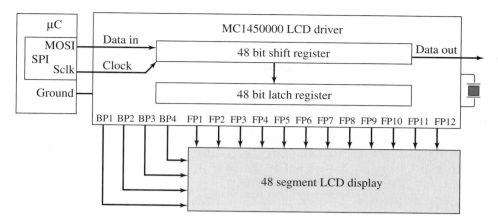

Program 8.20
Bit-banged interface to a scanned LCD display.

```
void LCD_Out (unsigned char *pt) {unsigned short i;  unsigned char mask;
  for(i=0;i<6;i++){
    for(mask=0x80;mask;mask=mask>>1){  // look at bits 7,6,5,4,3,2,1,0
      if((*pt)&mask) PORTB=1; else PORTB=0; // Serial data of the
      PORTB |= 0x02;   // toggle the serial clock first high
      PORTB &=~0x02;   // then low
    }
    pt++;
  }
}
```

```
// MC68HC711E9
// PD3/MOSI = MC145000 DATA IN
// PD4/SCLK = MC145000 CLOCK IN
void LCD_Init(void) {
  DDRD |= 0x18; // output->MC145000
  SPCR = 0x50; }
// bit  meaning
// 7 SPIE=0 no interrupts
// 6 SPE=1  SPI enable
// 5 DWOM=0 regular outputs
// 4 MSTR=1 master
// 3 CPOL=0 match timing with MC14489
// 2 CPHA=0
// 1 SPR1=0 E/2 is 1Mhz SCLK
// 0 SPR0=0
void LCD_out(unsigned char data[6]){
unsigned short j;
  for(j=5; j>=0 ; j--){
    SPDR = data[j];   // Msbyte first
    while(SPSR&0x80)==0){};
  }
}
```

```
// MC9S12C32
// PM4/MOSI = MC145000 DATA IN
// PM5/SCLK = MC145000 CLOCK IN
void LCD_Init(void) {
  DDRM |= 0x30;  // outputs to MC145000
  SPICR1 = 0x50;
// bit  meaning
// 7 SPIE=0 no interrupts
// 6 SPE=1  SPI enable
// 5 SPTIE=0 no interrupts
// 4 MSTR=1 master
// 3 CPOL=0 match timing with MC14489
// 2 CPHA=0
// 1 SSOE=0 PM3 is simple output
// 0 LSBF=0 MSB first
  SPICR2 = 0x00;  // regular drive
  SPIBR = 0x01;}  // 1MHz SCLK
void LCD_out(unsigned char data[6]){
unsigned short j;
  for(j=5; j>=0 ; j--){
    SPIDR = data[j];   // Msbyte first
    while(SPISR&0x80)==0){};}}
```

Program 8.21
SPI interface to a scanned LCD display using a MC145000.

Figure 8.55
Interface of a HD44780
LCD controller.

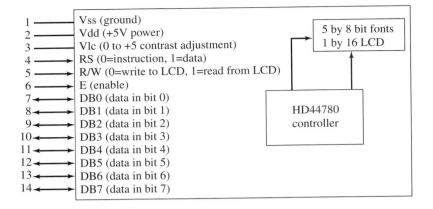

Table 8.8
Two control signals
specify the type of
access to the HD44780.

RS	R/W	Cycle
0	0	Write to Instruction Register
0	1	Read Busy Flag (bit 7)
1	0	Write data from microprocessor to the HD44780
1	1	Read data from HD44780 to the microprocessor

There are four types of access cycles to the HD44780, depending on RS and R/W (Table 8.8). Two types of synchronization can be used, blind cycle and gadfly. Most operations require 40 μs to complete, while some require 1.64 ms. This implementation uses Program 2.6 to create the blind cycle wait. A gadfly interface would have provided feedback to detect a faulty interface but has the problem of creating a software crash if the LCD never finishes. The best interface utilizes both gadfly and blind cycle so that the software can return with an error code if a display operation does not finish on time (because of a broken wire or damaged display.) First we present a low-level private helper function (Program 8.22). This function would not have a prototype in the LCD.H file because it is private.

```
// MC68HC711E9
// 1 by 16 char LCD Display (HD44780)
//   ground = pin 1   Vss
//   power  = pin 2   Vdd +5v
//   10kpot = pin 3   Vlc contrast adjust
//   PB2    = pin 6   E  enable
//   PB1    = pin 5   R/W 1=read, 0=write
//   PB0    = pin 4   RS 1=data, 0=control
//   PC0-7  = pins7-14 DB0-7 8-bit data
static void outCsr(unsigned char command){
  PORTC = command;
  PORTB &=~0x03;    // RS=0, R/W=0
  PORTB |= 0x04;    // E=1
  PORTB &=~0x04;    // E=0
  Timer_Wait(80);}  // Program 2.6
```

```
// MC9S12C32
// 1 by 16 char LCD Display (HD44780)
//   ground = pin 1   Vss
//   power  = pin 2   Vdd +5v
//   10kpot = pin 3   Vlc contrast adjust
//   PM2    = pin 6   E    enable
//   PM1    = pin 5   R/W 1=read, 0=write
//   PM0    = pin 4   RS 1=data, 0=control
//   PT0-7  = pins7-14 DB0-7 8-bit data
static void outCsr(unsigned char command){
  PTT = command;
  PTM &=~0x03;    // RS=0, R/W=0
  PTM |= 0x04;    // E=1
  PTM &=~0x04;    // E=0
  Timer_Wait(80);}  // Program 2.6
```

Program 8.22
Private functions for an HD44780 controlled LCD display.

Next we show the high-level public members (Program 8.23). These functions would have prototypes in the LCD.H file.

```
// MC68HC711E9
void LCD_OutChar(unsigned char letter){
  PORTC = letter;  // ASCII code
  PORTB |= 0x01;    // RS=1, R/W=0
  PORTB |= 0x04;    // E=1
  PORTB &=~0x04;    // E=0
  Timer_Wait(80);  // 40 us wait
}
void LCD_Clear(void){
  outCsr(0x01);        // Clear Display
  Timer_Wait(3280);  // 1.64ms wait
  outCsr(0x02);        // Cursor to home
  Timer_Wait(3280);} // 1.64ms wait
void LCD_Init(void){
  DDRC = 0xFF;
  PORTB &=~0x07; // R/W=0
  Timer_Wait(20000);
  outCsr(0x06);  // Incr,nodisplayshift
  outCsr(0x0C);  // on,cursoroff,blinkoff
  outCsr(0x14);  // cursormove,shiftright
  outCsr(0x30);  // 8bit,1line,5by7dots
  LCD_Clear();   // clear display
}
```

```
// MC9S12C32
void LCD_OutChar(unsigned char letter){
  PTT = letter; // ASCII code
  PTM |= 0x01;   // RS=1, R/W=0
  PTM |= 0x04;   // E=1
  PTM &=~0x04;   // E=0
  Timer_Wait(80);} // 40 us wait
void LCD_Clear(void){
  outCsr(0x01);    // Clear Display
  Timer_Wait(3280);   // 1.64ms wait
  outCsr(0x02);     // Cursor to home
  Timer_Wait(3280);} // 1.64ms wait
void LCD_Init(void){
  DDRT = 0xFF; DDRM |=0x07;
  PTM &=~0x07;  // R/W=0
  Timer_Init(); // Program 2.6
  Timer_Wait(20000);
  outCsr(0x06);  // Incr,nodisplayshift
  outCsr(0x0C);  // on,cursoroff,blinkoff
  outCsr(0x14);  // cursormove,shiftright
  outCsr(0x30);  // 8bit,1line,5by7dots
  LCD_Clear();   // clear display
}
```

Program 8.23
Public functions for an HD44780 controlled LCD display.

Checkpoint 8.7: What steps would you have to take to read the busy flag in the HD44780 controller?

8.4 Transistors Used for Computer-Controlled Current Switches

We can use individual transistors to source or sink current. In this chapter the transistors are used in saturated mode. This means that when the NPN transistors is on, current flows from the collector to the emitter. When the NPN transistor is off, no current flows from the collector to the emitter. Each transistor has an input and output impedance, h_{ie} and h_{oe}, respectively. The current gain is h_{fe}, or β. the model for the bipolar NPN transistor is shown in Figure 8.56.

Figure 8.56
NPN transistor model.

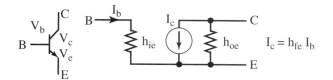

There are five basic design rules when using individual bipolar NPN transistors in saturated mode:

1. Normally $V_c > V_e$
2. Current can flow only in the following directions: from base to emitter (input current), from collector to emitter (output current), and from base to collector (doesn't usually happen but could if $V_b > V_c$)
3. Each transistor has maximum values for the following terms that should not be exceeded: I_b, I_c, V_{ce}, and $I_c \cdot V_{ce}$
4. The transistor acts like a current amplifier: $I_c = h_{fe} \cdot I_b$
5. The transistor will activate if $V_b > V_e + V_{be(SAT)}$, where $V_{be(SAT)}$ is typically above 0.6 V

The model for the bipolar PNP transistor is shown in Figure 8.57.

Figure 8.57
PNP transistor model.

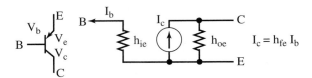

There are five basic design rules when using individual bipolar PNP transistors in saturated mode:

1. Normally $V_e > V_c$
2. Current can flow only in the following directions: from emitter to base (input current), from emitter to collector (output current), and from collector to base (doesn't usually happen but could if $V_c > V_b$)
3. Each transistor has maximum values for the following terms that should not be exceeded: I_b, I_c, V_{ce}, and $I_c \cdot V_{ce}$
4. The transistor acts like a current amplifier: $I_c = h_{fe} \cdot I_b$
5. The transistor will activate if $V_b < V_e - V_{be(SAT)}$, where $V_{be(SAT)}$ is typically above 0.6 V

Performance tip: A good transistor is one in which the I/O response is independent of h_{fe}. We can design the interface so that I_b can be twice as large as needed to supply the necessary I_c.

Table 8.9 illustrates the wide range of bipolar transistors that we can use.

Type	NPN	PNP	Package	$V_{be(SAT)}$	$V_{ce(SAT)}$	h_{fe} min/max	I_c
General purpose	2N3904	2N3906	TO-92	0.85 V	0.2 V	100	10 mA
General purpose	PN2222	PN2907	TO-92	1.2 V	0.3 V	100	150 mA
General purpose	2N2222	2N2907	TO-18	1.2 V	0.3 V	100/300	150 mA
Power transistor	TIP29A	TIP30A	TO-220	1.3 V	0.7 V	15/75	1 A
Power transistor	TIP31A	TIP32A	TO-220	1.8 V	1.2 V	25/50	3 A
Power transistor	TIP41A	TIP42A	TO-220	2.0 V	1.5 V	15/75	3 A
Power Darlington	TIP120	TIP125	TO-220	2.5 V	2.0 V	1000 min	3 A

Table 8.9
Parameters of typical transistors used by microcomputer to source or sink current.

Computer-Controlled Relays, Solenoids, and DC Motors

Relays, solenoids, and pulse-width modulated DC motors are grouped together because their electric interfaces are similar. In each case, there is a coil, and the computer must drive (or not drive) current through the coil.

8.5.1
Introduction
to Relays

A relay is a device that responds to a small current or voltage change by activating switches or other devices in an electric circuit. It is used to remotely switch signals or power. The input control is usually electrically isolated from the output switch. The input signal determines whether the output switch is open or closed. Figure 8.58 shows typical circuit drawings of classic general-purpose electromagnetic relays. In each, the input current affects the position of the output switch.

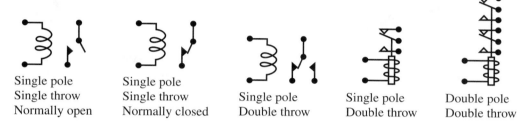

Single pole
Single throw
Normally open

Single pole
Single throw
Normally closed

Single pole
Double throw

Single pole
Double throw

Double pole
Double throw

Figure 8.58
Various types of relays.

The *American Heritage Dictionary* defines *relay* as "an act of passing something along from one person, group, or station to another." Electronic relays were originally used during the 19th century for extending the range of remote telegraph systems, *relaying* the signal from one station to another (Figure 8.59).

Figure 8.59
Original application
of relays.

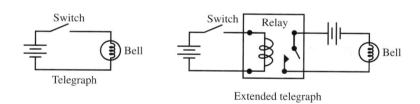

Switch

Bell

Telegraph

Switch   Relay

Bell

Extended telegraph

Relays are classified into four categories depending upon whether the output switches power (i.e., high currents through the switch) or electronic signals (i.e., low currents through the switch). Another difference is how the relay implements the switch. An electromagnetic (EM) relay uses a coil to apply EM force to a contact switch that physically opens and closes. The solid-state relay uses transistor switches made from solid-state components to electronically allow or prevent current flow across the switch. The four types (three shown in Figure 8.60) are:

- The classic general-purpose relay has an EM coil and can switch power
- The reed relay has an EM coil and can switch-low level DC electronic signals
- The solid-state relay (SSR) has an input-triggered semiconductor power switch
- The bilateral switch uses CMOS, FET, or biFET transistors

Figure 8.60
Photo of an EM, solid-state, and reed relay.

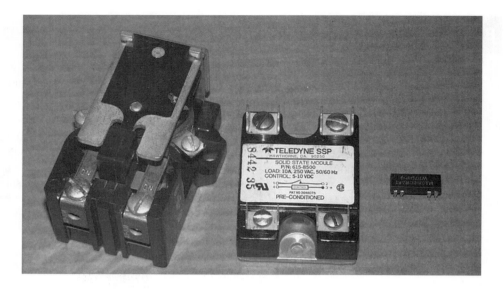

Actually, the bilateral switch is not a relay, but it is included in this discussion because it is a solid-state device that can be used to switch low-level signals.

8.5.2 Electromagnetic Relay Basics

Figure 8.61 illustrates a classical general-purpose EM relay. The input circuit is an EM coil with an iron core. The output switch includes two sets of silver or silver-alloy contacts (called *poles*). One set is fixed to the relay *frame,* and the other set is located at the end of leaf spring poles connected to the *armature.* The contacts are held in the "normally closed" position by the armature return spring. When the input circuit energizes the EM coil, a "pull-in" force is applied to the armature and the "normally closed" contacts are released (called *break*) and the "normally open" contacts are connected (called *make*). The armature pull-in can either energize or deenergize the output circuit, depending on how it is wired. The illustrated relay has its transparent protective polycarbonate cover removed. Relays are mounted in special sockets, or directly soldered onto a printed circuit board.

Figure 8.61
Drawing of an EM relay.

Double Pole Double Throw   (DPDT)

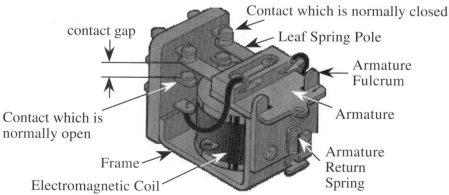

The number of poles (e.g., single-pole, double-pole, 3P, 4P) refers to the number of switches that are controlled by the input. *Single throw* means each switch has two contacts that can be open or closed. *Double throw* means each switch has three contacts. The common contact will be connected to one of the other two contacts (but not both at the same time). Figure 8.58 shows relays of different types.

The parameters of a relay are specified in terms of the input coil and output switch (Table 8.10). The input parameters include DC or AC excitation, coil resistance, pickup voltage, dropout voltage, and maximum coil power. The *pickup voltage* is the coil potential above in which activation is guaranteed. The *dropout voltage* is the coil potential below which deactivation is guaranteed. Typically the coil can handle sustained voltages a few volts above its nominal value before damage occurs. In microcomputer-based interfaces, DC coils are more convenient than AC coils. Unless there is an internal snubber diode, the polarity of the coil current does not matter.

Table 8.10
Parameters of four different types of computer-controlled switches.

Manufacturer Model number	Teledyne 712-5	Magnecraft W107DIP-2	Teledyne 611	PMI SW-01
Type	TO-5 relay	Reed	SSR	JFET
Coil resistance	50Ω	500Ω	500Ω	3-μA input
Pickup voltage	3.6 V	3.8 V	3.8 V	2 V
Dropout voltage		0.5 V	0.8 V	0.8 V
Contact load	AC/DC	DC	AC/DC	DC
Max contact power		10 W DC	2500 W AC	
Max contact voltage	250 V AC	100 V DC	250 V AC	+11 to −10 V
Max contact current	600 mA AC	0.5 A DC	10 A AC	5 mA DC
On resistance	0.2Ω	0.1Ω	0.15Ω	100Ω
Off resistance	Infinite	Infinite	28 kΩ	58 dB
Turn-on time	4 ms	600 μs	8.3 ms	400 ns
Turn-off time	3 ms	75 μs	16.6 ms	300 ns
Life expectancy	10^7	$25 \cdot 10^6$	Infinite	Infinite
Approximate cost	$2 to $4	$1 to $2	$10 to $20	$0.50–$2

The parameters of the output switch include maximum AC (or DC) power, maximum current, maximum voltage, on resistance, and off resistance. A DC signal welds the contacts together at a lower current value than an AC signal; therefore the maximum ratings for DC are considerably smaller than for AC. Other relay parameters include turn-on time, turn-off time, life expectancy, and I/O isolation. *Life expectancy* is measured in number of operations. Storage life for relays is usually quite long. Table 8.10 lists specifications for four devices.

Figure 8.62 illustrates the various configurations available. The sequence of operation is listed in Table 8.11.

Figure 8.62
Various relay configurations.

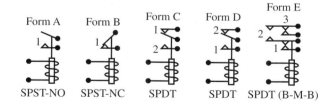

Form A — SPST-NO   Form B — SPST-NC   Form C — SPDT   Form D — SPDT   Form E — SPDT (B-M-B)

Table 8.11
Five relay configurations.

Form	Activation sequence	Deactivation sequence
A	Make 1	Break 1
B	Break 1	Make 1
C	Break 1, Make 2	Break 2, Make 1
D	Make 1, Break 2	Make 2, Break 1
E	Break 1, Make 2, Break 3	Make 3, Break 2, Make 1

Latching relays do not require continuous input current to remain in their present state. A short pulse on the two-input coil (1 to 50 ms) causes the switch to change state, after which it will remain in the new state (open or closed) indefinitely. These devices are suitable for low-power operation when the contact is infrequently switched. They are also convenient for low-noise applications, because they do not require continuous coil currents. Mercury-wetted relays provide for faster switching and eliminate contact bounce. Some devices have an internal snubber diode and/or electrostatic shielding. The shielding reduces the noise crosstalk from the input coil to the external circuits. A shielded relay should be used if the relay is switching simultaneously with critical low-noise functions.

8.5.3 Reed Relays

Reed relays (Figure 8.63) are often used in medical electronics, telecommunications, and automated test equipment (ATE) for signal-level switching (e.g., mode/gain/offset selection). The single-pole–single-throw (SPST) reed capsule has two contacts that are normally open. When the coil is activated, an EM force causes the contacts to close.

Figure 8.63
Drawing of a reed relay.

Single Pole Single Throw (SPST) Reed Relay

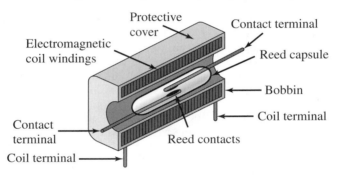

The coil is wound on a bobbin that surrounds the glass reed capsule. The reed capsule is hermetically sealed with inert gas. The contacts are typically made from precious metal like rhodium. Gold-cobalt contacts provide (at additional cost) reduced contact resistance (less than 0.05Ω) for applications that require higher accuracy. Reed relays are available as DIPs or single inline packages (SIPs). The relay can be purchased as SPST-NO (Form A), SPST-NC (Form B), DPST-NO (Form 2A), and SPDT (Form C).

8.5.4 Solenoids

Solenoids are used in door locks, automatic disk/tape ejectors, and liquid/gas flow control valves (on/off type). Much like an EM relay, there is a frame that remains motionless and an armature that moves in a discrete fashion (on/off). A solenoid has an electromagnet (Figure 8.64 and Figure 8.65). When current flows through the coil, a magnetic force is created, causing a discrete motion of the armature. When the current is removed, the magnetic force stops, and the armature is free to move. The motion in the opposite direction can be produced by a spring, gravity, or by a second solenoid.

8.5.5 Pulse-Width Modulated DC Motors

Similar to the solenoid and EM relay, the DC motor has a frame that remains motionless and an armature that moves. In this case, the armature moves in a circular manner (shaft rotation). A DC motor has an electromagnet as well. When current flows through the coil, a magnetic force is created, causing a rotation of the shaft. Brushes positioned between the frame and armature are used to alternate the current direction through the coil so that a DC

Figure 8.64
Mechanical drawing of
a solenoid showing that
the EM coil causes the
armature to move.

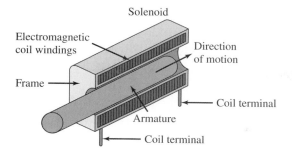

Figure 8.65
Photograph of two
solenoids.

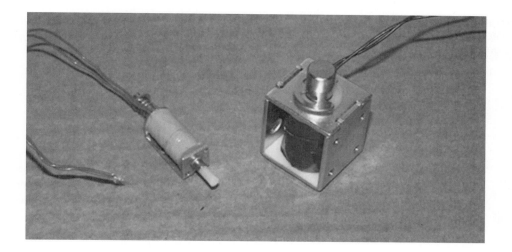

current generates a continuous rotation of the shaft. When the current is removed, the magnetic force stops and the shaft is free to rotate. In a pulse-width modulated DC motor, the computer activates the coil with a current of fixed magnitude but varies the duty cycle to control the motor speed. Software examples that generate variable duty-cycle square waves were presented in Chapter 6.

**8.5.6
Interfacing EM
Relays, Solenoids,
and DC Motors**

The interface circuit should provide sufficient current and voltage to activate the device. In the off state, the input current should be zero. Because of the inductive nature of the coil, huge back electromotive force (EMF) signals develop when the coil current is cut off. Because of the high-speed transistor switch, there is a large dI/dt when the computer deactivates the coil (some current to no current). There is also a dI/dt (but smaller because of the capacitances in the circuit) when the computer activates the coil (no current to some current). These voltages can range from 50 to 200 V, depending on the coil and driving circuit. To protect the driver electronics, a snubber diode is added to suppress the back EMF (Table 8.12). The choice of diode should consider the magnitude of the back EMF. For signals less than 75 V, the 1N914 is adequate. For larger and larger back EMFs the 1N4001 through 1N4007 can be used. All these diodes are fast enough to protect the driver electronics. The open-collector driver is used because the circuit requires two states: the need to sink current (during the activate state) and no current (during the off state). Because solenoids and DC motors (controlled with PWM) have coils similar to the EM relay, the interfaces shown in Figure 8.66 can be also used to control solenoids and DC motors.

The output low current (I_{OL}) of the driver should be sufficient to activate the relay. Other open-collector drivers (e.g., 7406, 75492, 75451, ULN2074, IRF540) have larger

Table 8.12
Maximum voltage
parameter for typical
snubber diodes.

Diode	Maximum voltage (V)
1N4001	50
1N914	75
1N4002	100
1N4003	200
1N4004	400
1N4005	600
1N4006	800
1N4007	1000

Figure 8.66 A relay
interface that sink
current through the coil.

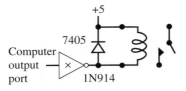

output low currents. Some designers suggest using two diodes to suppress both positive and
negative back EMF signals. Good engineering practice would be to measure the coil voltage and include both diodes if negative EMF signals are observed.

Transistors provide an alternative interfacing technique for relays. Choose a device that
can source (PNP) or sink (NPN) the necessary coil current. Speed is not an issue in transistor selection (Figure 8.67).

Figure 8.67
Two relay interfaces that
source current through
the coil.

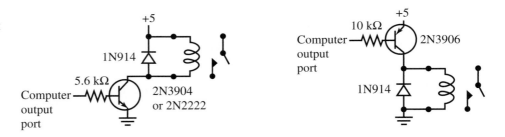

For coils that require currents above 500 mA, a MOSFET can be used. The microcomputer output sources the current to the gate (pin 1) to turn on the MOSFET. Although
the MOSFET gate does not need a lot of current to maintain the on state, large currents are
required to switch it from off to on and from on to off (Figure 8.68). The voltage V_m is
selected to match the specification of the motor.

Figure 8.68
Motor interface using
a high-current MOSFET.

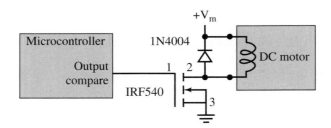

The relay by its nature isolates the computer and its driver electronics from the currents in the switch contacts. On the other hand, currents in the solenoid and DC motor must pass through the same ground as the computer. When these currents are large and noisy, we can decouple the coil currents from the computer ground using an optoisolator like the 6N139. Using electrical isolation can protect the computer electronics from surges in and around the motor. In the circuit of Figure 8.69, the computer (Vdd) and motor (Vpp) power are separate and their grounds are usually not connected. High-speed optoisolators, similar to these, can be used in computer networks to protect one computer from another. For more current, we can use the IRF540 MOSFET (Figures 8.70 and 8.71).

Figure 8.69
An isolated motor interface using a 6N139.

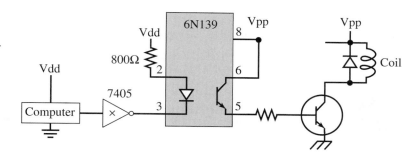

Figure 8.70
A high-current isolated motor interface using a 6N139 and a MOSFET.

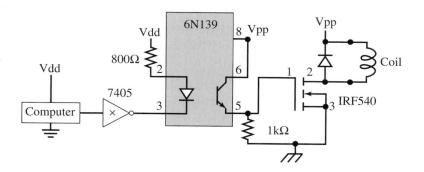

Figure 8.71
Another high-current isolated motor interface using a 6N139 and a MOSFET.

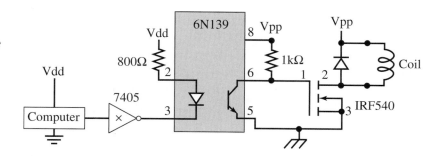

In some applications we wish to drive the motor forward and backward. To achieve this result we must be able to drive current both forward and backward through the motor coil. One approach to this interface is called the H-bridge. It is important not to simultaneously drive both Q1, Q2 or both Q3, Q4. The basic approach to the H-bridge is illustrated in Figure 8.72. If Q2 and Q3 are on, then current flows right to left across the coil. If Q1 and Q4 are on, then current flows in the opposite direction. PNP transistors

Figure 8.72
An H-bridge is used to drive current in both directions.

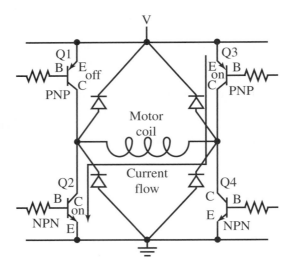

are used to source current into the coil, and NPN transistors are used to sink current out of the coil. Four diodes are required to prevent back EMF that occurs when the current is applied and removed.

The TIP125 PNP Darlington transistors can source up to 3 A. To activate the PNP transistor, the base voltage must be 0.6 V less than the emitter voltage (Vm). When light flows through its 6N139 isolation barrier, the 6N139 output goes to about 1 V, which drops the TIP125 base voltage low enough and sinks enough base current to activate the PNP source transistor. A low on the **Forward** signal will activate Q1 and Q4.

The TIP120 NPN Darlington transistors can sink up to 3 A. To activate the NPN transistor, the base voltage must be 0.6 V more than the emitter voltage (ground). When light flows through its 6N139 isolation barrier, the 6N139 output goes to about $V_m - 1$ V, which raises the TIP120 base voltage high enough and sources enough base current to activate the NPN sink transistor. A low on the **Reverse** signal will activate Q2 and Q3 (Figure 8.73).

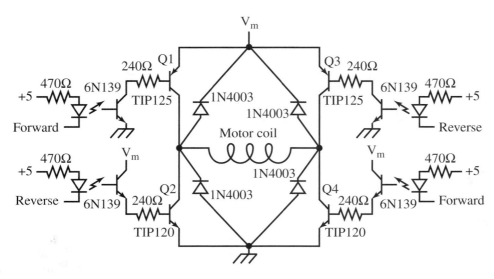

Figure 8.73
An isolated H-bridge can drive current in both directions.

As mentioned earlier it is important not to activate both **Forward** and **Reverse** at the same time. To prevent this situation even when the software crashes, the digital interface shown in Figure 8.74 could be used. To prevent temporary current paths through Q1+Q2 or Q3 + Q4, the software should change the **Direction** only while the motor has been stopped (**Go** = 0) for enough time to allow all transistors to turn off.

Figure 8.74
A digital circuit used with an H-bridge.

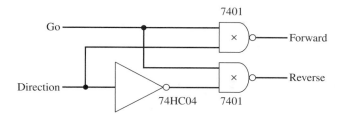

The classic approach to control a bidirectional DC motor is to use an H-bridge driver. We have the choice of creating an H-bridge with individual transistors, as shown in Figure 8.72, or using an integrated circuit. The advantage of using an integrated circuit is reduced cost, smaller size, and faster design. The TPIC0107B is an integrated H-bridge that is optimized for PWM input and reversible DC motor control. It can source/sink up to 3A. The TPIC0107B is made with a CMOS digital logic and Double-Diffused Metal-Oxide Semiconductor (DMOS) logic for switching power to the motor. It has an extremely low on resistance, 280 mΩ typical, to minimize system power dissipation. Power is applied to the motor when the PWM input is high. Figure 8.75 shows the interface such that the PWM pin is connected to a pulse-width-modulated output, DIR pin is connected to a regular output, and STATUS1, STATUS2 are connected to regular inputs. STATUS1 requires a pull-up resistor because it is open-collector. The DIR input controls the CW/CCW direction. Program 6.13 implements a PWM output that can be used to control the speed of this DC motor. The TPIC0107B provides protection against over-voltage, over-current, over-temperature, and cross-conduction faults. Fault diagnostics can be obtained by monitoring the STATUS1 and STATUS2 signals. The data sheet for this device can be found on the CD as file **tpic0107b.pdf**. Allegro also produces a complete line of DC motor controllers (e.g., A3936, A3949).

Checkpoint 8.8: What changes do you make to use the circuit in Figure 8.75 to control a 0.5A 12V DC motor?

Figure 8.75
Bidirectional DC motor interface.

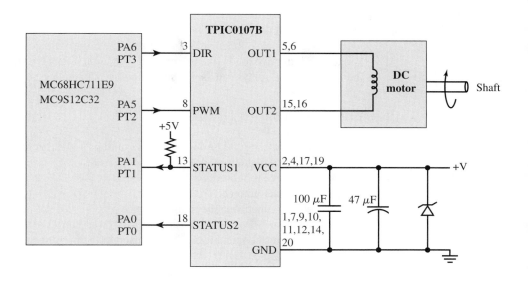

**8.5.7
Solid-State Relays**

To solve the limited life expectancy and contact bounce problems, SSRs were developed. Figure 8.76 illustrates the major components of a SSR. Figure 8.77 is a photograph of two SSRs. The SSR has no moving parts. The optocoupler provides isolation between the input circuit (pseudocoil) and the triac (pseudocontact). The switch function is constructed from either two inverse-parallel SCRs or an electrically equivalent triac. In the next section we will use silicon devices for switching DC signals that use power bipolar transistors or MOSFETs.

Figure 8.76
Internal components of a SSR.

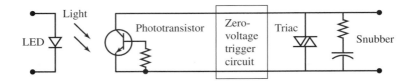

Figure 8.77
Photograph of two SSRs.

The signal from the phototransistor triggers the output triac so that it switches the load current. The zero-voltage detector triggers the triac only when the AC voltage is zero, reducing the surge currents when the triac is switched. Surge currents can occur when controlling capacitive loads like lamps. Once triggered the triac conducts until the next zero crossing. If the input signal continues, then the triac will continuously conduct. The RC circuit is called a snubber and is used to reduce the voltage transients that occur when switching inductive loads like motors and solenoids. SSRs cost five to ten times more than general-purpose relays but have the following advantages because there are no moving parts:

- Longer life
- Higher reliability
- Faster switching (although it occurs on the AC load zero crossing)
- Better mechanical stability
- Insensitive to vibrations and shock
- No contact bounce
- Reduces electromagnetic interference
- Quieter
- Eliminates contact arcing (can be used in the presence of explosive gases)

Interfacing the SSR to the microcomputer is identical to the LED interface (Figure 8.78).

Figure 8.78
Interface of a SSR.

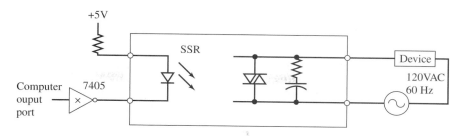

The value of the resistor, R, is chosen to select the V_d, I_d operating point of the LED.

$$R = \frac{+5 - V_d - V_{OL}}{I_d}$$

where V_{OL} is the output low voltage of the open-collector driver (7405). Again, the output low current (I_{OL}) of the driver should be sufficient to activate the LED (e.g., 7406, 75492, 75451, ULN2074).

8.6 ■ Stepper Motors

Stepper motors are very popular for microcomputer-controlled machines because of their inherent digital interface. It is easy for a microcomputer to control both the position and velocity of a stepper motor in an open-loop fashion. Although the cost of a stepper motor is typically higher than an equivalent DC permanent magnetic field motor, the overall system cost is reduced because stepper motors may not require feedback sensors. For these reasons, they are used in many computer peripherals such as hard disk drives, floppy disk drives, and printers. Stepper motors can also be used as shaft encoders, measuring both position and speed (Figure 8.79).

Figure 8.79
Three stepper motors.

8.6.1
Stepper Motor
Example

In this section, we will introduce stepper motors by showing a simple example. The computer can make the motor spin by outputting the sequence . . . 10, 9, 5, 6, 10, 9, 5, 6 . . . over and over. For a motor with 200 steps per revolution each new output will cause the motor to rotate 1.8°. If the time between outputs is fixed at Δt s, then the shaft rotation speed will be $0.005/\Delta t$ in revolutions per second (rps). In each system, we will connect the stepper motor to the least significant bits of output PORTB. In this first hardware interface, we will connect a unipolar stepper with four 5-V coils, A, A', B, B'. Because the coil resistance is about 70Ω, it will require about 5 V/70Ω, or 70 mA, to activate. The output low current of the 75492 is sufficient to sink the 70 mA. The 1N914 diodes will protect the 75492 from the back EMF that will develop across the coil when the current is shut off. Figure 8.80 illustrates the state when 10,(binary 1010) is output to PORTB. Bits PB3 and PB1 are high, making their 75492 outputs low and driving current through coils A and B. Bits PB2 and PB0 are low, making their 75492 outputs off (HiZ), and driving no current through coils A' and B'.

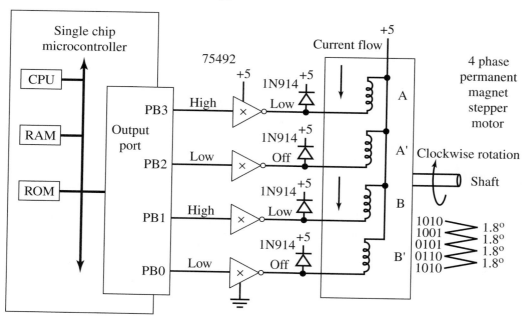

Figure 8.80
Simple stepper interface.

We define the active state of the coil when current is flowing. The basic operation is summarized in Table 8.13.

Table 8.13
Stepper motor sequence.

PORTB output	A	A'	B	B'
10	Activate	Deactivate	Activate	Deactivate
9	Activate	Deactivate	Deactivate	Activate
5	Deactivate	Activate	Deactivate	Activate
6	Deactivate	Activate	Activate	Deactivate

We will implement a linked list approach to the stepper motor control software. This approach yields a solution that is easy to understand and change. If the computer outputs the sequence backward, then the motor will spin in the other direction. To ensure proper operation, this . . . 10, 9, 5, 6, 10, 9, 5, 6 . . . sequence must be followed. For example, assume the computer outputs . . . 9, 5, 6, 10, and 9. Now it wishes to reverse direction, since the output is already at 9, then it should begin at 10, and continue with 6, 5, 9. . . . In other words, if the current output is "9," then the only two valid next outputs would be "5" if it wanted to spin clockwise or "10" if it wanted to spin counterclockwise. Maintaining this proper sequence will be simplified by implementing a double circular linked list. For each node in the linked list there are two valid next states, depending upon whether the computer wishes to spin clockwise or counterclockwise (Figure 8.81).

The sequence 10, 9, 5, 6 is called *full-stepping* and each output causes the motor to move $\Delta\theta$. The sequence 10, 8, 9, 1, 5, 4, 6, 2, is called *half-stepping* and each output causes the motor to move $\Delta\theta/2$.

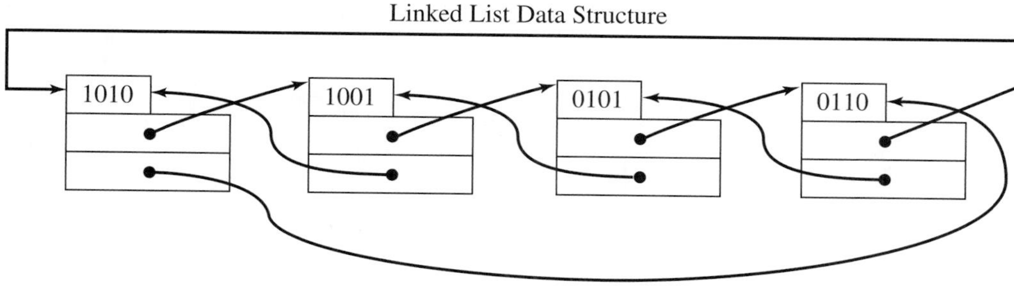

Linked List Data Structure

Figure 8.81
A double circular linked list used to control the stepper motor.

A slip is when the computer issues a sequence change, but the motor does not move. A slip can occur if the mechanical load on the shaft exceeds the available torque of the motor. A slip can also occur if the computer tries to change the outputs too fast. If the system knows the initial shaft angle, and the motor never slips, then the computer can control both the shaft speed and angle without a position sensor. The routines CW and CCW will step the motor once in the clockwise and counterclockwise directions, respectively. If every time the computer calls CW or CCW it were to wait for 5 ms, then the motor would spin at 1 rps. The 6811 and 6812 have similar solutions. In each case, the linked list data structure will be stored in EEPROM, and the variables are allocated into RAM and initialized at run time in the ritual (Program 8.24).

```
; 6811 or 6812
;Linked list stored in EEPROM
S10    fcb 10    ;Output pattern
       fdb S9    ;Next if CW
       fdb S6    ;Next if CCW
S9     fcb 9
       fdb S10
       fdb S5
S5     fcb 5
       fdb S9
       fdb S6
S6     fcb 6
       fdb S5
       fdb S10
;Global variables stored in RAM
Pos    ds  1   ;0<= Pos <=199
Pt     ds  2   ;to current state
```

```
// 6811 or 6812
const struct State{
  unsigned char Out;        // Output
  const struct State *Next[2]; // CW/CCW
};
typedef struct State StateType;
typedef StateType *StatePtr;
#define clockwise 0        // Next index
#define counterclockwise 1 // Next index
StateType fsm[4]={
  {10,{&fsm[1],&fsm[3]}},
  { 9,{&fsm[2],&fsm[0]}},
  { 5,{&fsm[3],&fsm[1]}},
  { 6,{&fsm[0],&fsm[2]}}
};
unsigned char Pos;  // between 0 and 199
StatePtr Pt;        // Current State
```

Program 8.24
A double circular linked list used to control the stepper motor.

The programs that step the motor also maintain the position in the global, Pos. If the motor slips, then the software variable will be in error. Also it is assumed the motor is initially in position 0 at the time of the ritual (Program 8.25).

```
; 6811 or 6812
;Move 1.8 degrees clockwise
CW:  ldx  Pt     ;current state
     ldx  1,x    ;Next clockwise
     stx  Pt     ;Update pointer
     ldaa ,x     ;Output pattern
     staa PORTB  ;Set phase control
     ldaa Pos    ;Update position
     inca         ;clockwise
     cmpa #200
     blo  OK1    ;0<= Pos <=199
     clra
OK1: staa Pos
     rts
;Move 1.8 degrees counterclockwise
CCW: ldx  Pt     ;current state
     ldx  3,x    ;Next CCW
     stx  Pt     ;Update pointer
     ldaa ,x     ;Output pattern
     staa PORTB  ;Set phase control
     ldaa POS    ;Update position
     deca         ;CCW direction
     cmpa #255
     bne  OK2    ;0<= Pos <=199
     ldaa #199
OK2: staa Pos
     rts
Init: clr Pos
     ldx  #S10
     stx  Pt
     movb #$FF,DDRB  ;6812 only
     rts
```

```c
// 6811 or 6812
// Move 1.8 degrees clockwise
void CW(void){
  Pt = Pt->Next[clockwise]; // circular
  PORTB = Pt->Out;          // step motor
  if(Pos==199){             // shaft angle
    Pos = 0;                // reset
  }
  else{
    Pos++;                  // CW
  }
}

// Move 1.8 degrees counterclockwise;
void CCW(void){
  Pt = Pt->Next[counterclockwise];
  PORTB = Pt->Out;          // step motor
  if(Pos==0){               // shaft angle
    Pos = 199;              // reset
  }
  else{
    Pos--;                  // CCW
  }
}

// Initialize Stepper interface
void Init(void){
  Pos = 0;
  Pt = &fsm[0];
  DDRB = 0xFF;  // 6812 only
}
```

Program 8.25
Helper functions used to control the stepper motor.

The function in Program 8.26 will step the motor moving to the desired position. It will choose to go clockwise or counterclockwise, depending on which way is closer.

In the routine of Program 8.26 the computer will step the motor to the desired New position. We can use the current position, Pos, to determine if it would be faster to go clockwise or counterclockwise. CWsteps is calculated as the number of steps from Pos to New if the motor were to spin clockwise. If it is greater than 100, then it would be faster to get there by going counterclockwise.

**8.6.2
Basic Operation**   Figure 8.82 shows a simplified stepper motor. The permanent magnet stepper has a rotor and a stator. The rotor is manufactured from a spool-shaped permanent magnet. This rotor has five teeth in both the north and south ends of the magnet. The teeth on the two magnetic poles are offset by half the tooth pitch. A stepper motor with 200 steps per revolution has

```
;6811 or 6812
;Reg B=desired 0<=RegB<=199
desired equ  0    ;desired state
Seek  pshb        ;Save as local
      tsy
      subb Pos    ;Go CW or CCW?
      beq  done   ;Skip if equal
      bhi  high   ;Desired>Pos?
;Desired<Pos
      negb        ;(POS-Desired)
      cmpb #100
      blo  goCCW ;Go CCW if
;Desired<Pos and Pos-Desired<100
goCW  bsr  CW     ;RegA current
      cmpa desired,Y
      bne  goCW   ;Pos=Desired?
      bra  done
high  cmpb #100   ;(Desired-Pos)
      blo  goCW   ;Go CW if
;Desired>Pos and Desired-Pos<100
goCCW bsr  CCW    ;RegA current
      cmpa desired,Y
      bne  goCCW ;Pos=Desired?
done  pulb
      rts
```

```
// 6811 or 6812
void Seek(unsigned char desired){
short CWsteps;
  if((CWsteps=desired-Pos)<0){
    CWsteps+=200;
  } // CW steps is 0 to 199
  if(CWsteps>100){
    while(desired!=Pos){
      CCW();
    }
  }
  else{
    while(desired!=Pos){
      CW();
    }
  }
}
```

Program 8.26
High-level function to control the stepper motor.

Figure 8.82
Simple stepper motor with 20 steps per revolution.

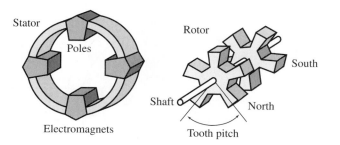

50 teeth on each end of the rotor. The stator consists of multiple iron-core electromagnets whose poles have the same width as the teeth on the rotor but are spaced slightly farther apart. The stator of this simple stepper motor has four electromagnets and four poles. The stator of a stepper motor with 200 steps per revolution has 8 electromagnets each with 5 teeth, making a total of 40 poles.

The operation of the stepper motor is illustrated in the next four figures. In a four-wire (or bipolar) stepper motor, the electromagnets are wired together, creating two phases. The five- and six-wire (or unipolar) stepper motors also have two phases, but each is center-tapped to simplify the drive circuitry.

Since there are five teeth on this simple motor, the shaft will rotate 360°/(4 · 5), or 18° per step. Figure 8.83 shows the initial position of the rotor with respect to the fixed stator. The labels on the permanent magnet rotor are added to clarify the rotation steps. There is no net force on the left and right stator poles because the north and south teeth are equally covered. On the other hand, there is a strong attractive force on both the top and bottom, holding the rotor at this position. In fact, stepper motors are rated according to their holding torque. Typical holding torques range from 10 to 300 oz·in.

Figure 8.83
Stable state 1.

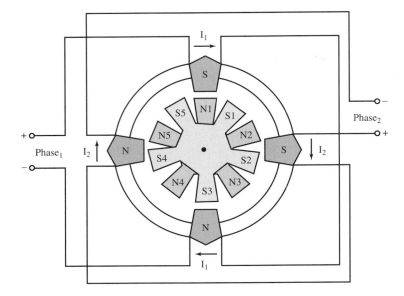

When the polarity of phase$_1$ is reversed, the rotor is in an unstable state, as shown in Figure 8.84. The rotor will move because there are strong repulsive forces on both the top and bottom. By observing the left and right poles, the closest stable state occurs if the rotor rotates clockwise, resulting in the stable state illustrated in Figure 8.85.

Figure 8.84
Unstable state as rotor goes from state 1 to state 2.

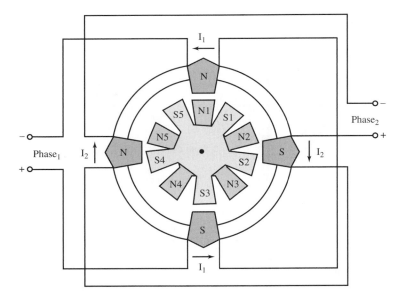

State 2 is exactly 18° clockwise from state 1. Now there is no net force on the top and bottom stator poles because the north and south teeth are equally covered. On the other hand, there are strong attractive forces on both the left and right, holding the rotor at this new position.

Next the polarity of phase$_2$ is reversed. Once again, the rotor is in an unstable state because there are strong repulsive forces on both the left and right. The closest stable state occurs if the rotor rotates clockwise, resulting in the stable state illustrated in Figure 8.86. State 3 is exactly 18° clockwise from state 2, with no net force on the left and right stator poles and strong attractive forces on both the top and bottom.

Figure 8.85
Stable state 2.

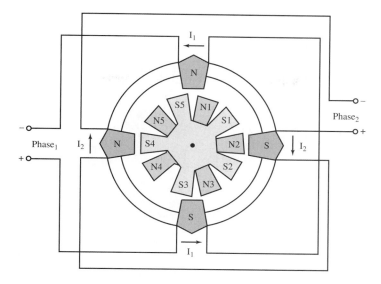

Figure 8.86
Stable state 3.

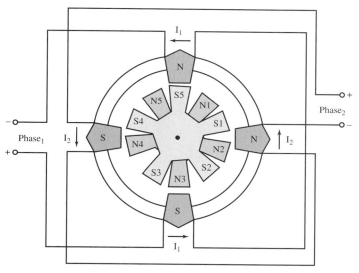

Once again the polarity of phase$_1$ is reversed, and the rotor rotates clockwise by 18°, resulting in the stable state shown in Figure 8.87. State 4 has no net force on the top and bottom stator poles and strong attractive forces on both the left and right.

When the polarity of phase$_2$ is reversed again, and the rotor rotates clockwise by 18°, it results in the stable state similar to Figure 8.83. These four states are cycled, causing the motor to spin clockwise. The rotor will spin in a counterclockwise direction if the sequence is reversed.

The time between states determines the rotational speed of the motor. Let ΔT be the time between steps and θ the step angle; then the rotational velocity v is $\theta/\Delta T$. As long as the load on the shaft is below the holding torque of the motor, the position and speed can be reliably maintained with an open-loop software control algorithm. To prevent skips (digital commands that produce no rotor motion) it is important to limit the change in acceleration, or *jerk*. Let $\Delta T(n-2)$, $\Delta T(n-1)$, $\Delta T(n)$ be the discrete sequence of times between steps.

$$v(n) = \theta/\Delta T$$

The acceleration is given by

$$a(n) = (v(n) - v(n-1))/\Delta T(n) = (\theta/\Delta T(n) - \theta/\Delta T(n-1))/\Delta T(n) = \theta/\Delta T(n)^2 - \theta/(\Delta T(n-1)\Delta T(n))$$

Figure 8.87
Stable state 4.

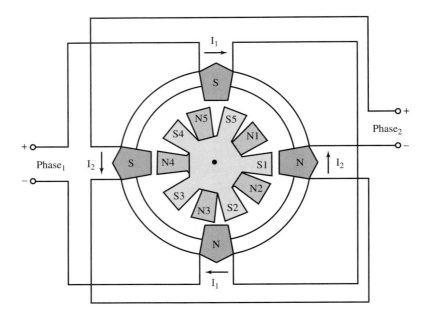

The change in acceleration, or jerk, is given by

$$b(n) = (a(n) - a(n-1))/\Delta T(n)$$

For example, if the time between steps is to be increased from 1000 to 2000 μs, an ineffective approach (as shown in Table 8.14) would simply be to go directly from 1000 to 2000. This produces a very large jerk that may cause the motor to skip.

Table 8.14
An ineffective approach to changing motor speed.

n	ΔT (μs)	$v(n)$ (°/s)	$a(n)$ (°/s^2)	$b(n)$ (°/s^3)
1		1000	1800	
2	1000	1800	0.00E+00	
3	2000	900	−2.50E+05	0.00E+00
4	2000	900	0.00E+00	−2.25E+08
5	2000	900	0.00E+00	2.25E+08
6	2000	900	0.00E+00	0.00E+00

Table 8.15 shows that a more gradual change from 1000 to 2000 produces a ten times smaller jerk, reducing the possibility of skips. The optimal solution (the one with the smallest jerk) occurs when $v(t)$ has a quadratic shape. This will make $a(t)$ linear, and $b(t)$ a constant. Limiting the jerk is particularly important when starting to move a stopped motor.

8.6.3
Stepper Motor
Hardware
Interfaces

The bipolar stepper motor can be controlled with two H-bridge drivers. We could design two H-bridge drivers using individual transistors, as previously shown in Figures 8.72 and 8.73. Unipolar stepper motors can be controlled with four current switches, as previously shown in Figure 8.80. During the design phase of a project it is appropriate to evaluate alternative solutions. In this section, we present stepper motor interfaces that employ integrated circuits to control the stepper motor. In addition to the devices presented in this section, Allegro, Texas Instruments, and ST Microelectronics offer a variety of stepper motor controllers (e.g., A3972, A3980, and UCN5804B). Table 8.16 lists some low cost, low torque stepper motors.

Table 8.15
An effective approach to changing motor speed.

n	ΔT (μs)	$v(n)$ (°/s)	$a(n)$ (°/s^2)	$b(n)$ (°/s^3)
1		1000	1800	
2	1000	1800	0.00E+00	
3	1000	1800	0.00E+00	0.00E+00
4	1008	1786	−1.39E+04	−1.38E+07
5	1032	1744	−4.11E+04	−2.64E+07
6	1077	1671	−6.77E+04	−2.47E+07
7	1152	1563	−9.37E+04	−2.26E+07
8	1275	1411	−1.19E+05	−1.96E+07
9	1500	1200	−1.41E+05	−1.48E+07
10	1725	1044	−9.06E+04	2.91E+07
11	1848	974	−3.78E+04	2.86E+07
12	1923	936	−1.96E+04	9.44E+06
13	1968	915	−1.09E+04	4.43E+06
14	1992	904	−5.65E+03	2.63E+06
15	2000	900	−1.77E+03	1.94E+06
16	2000	900	0.00E+00	8.85E+05
17	2000	900	0.00E+00	0.00E+00

Table 8.16
Specifications of typical stepper motors.

Manufacturer	Phase voltage	Phase current	Step angle	Holding torque	Type	No. of leads	Used cost
Eastern Air	2.7 V	1.9 A	1.8°	53 oz-in	unipolar	8	$8.95
Astro-Syn	5 V	0.45 A	1.8°	10 oz-in	bipolar	4	$4.95
Shinano Kenshi	6 V	1.2 A	1.8°	80 oz-in	unipolar	5	$12.95
AirPax	12 V	0.33 A	7.5°	10.5 oz-in	unipolar	6	$3.95
Oriental Vexta	12 V	0.68 A	1.8°	125 oz-in	unipolar	6	$24.50
Sigma	24 V	0.66 A	1.8°	325 oz-in	unipolar	5	$34.50

The L293 is a popular IC for interfacing stepper motors. It uses Darlington transistors in a double H-bridge configuration (as shown in Figure 8.88), which can handle up to 1A per channel and voltages from 4 to 36V. The 1N4003 snubber diodes protect the electronics from the back EMF generated when currents are switched on and off. The L293D has internal snubber diodes, but can handle only 600 mA. Figure 8.88 shows four digital outputs from the microcontroller connected to the **1A**, **2A**, **3A**, **4A** inputs. The software rotates the stepper motor using either the standard full-step (5-6-10-9. . .) or half-step (5-4-6-2-10-8-9-1. . .) sequence. For example, Programs 8.24 and 8.25 can be used to control this motor.

There are integrated circuits, such as the MC3479 in Figure 8.89, that perform even more of the stepper motor logic in hardware. **Vm** is the motor voltage, and the IC derives an internal +5V to power its digital logic. The circuit is protected from back EMF with internal snubber diodes, and the external zener diode placed between **Vm** and **Vd**. The high current outputs on **L1**, **L2**, **L3**, and **L4** can source or sink 350 mA. The maximum sink current can be limited by the **RB** from pin 6 to ground. The stepper motor is moved on each low-to-high transition of the **Clock** input, therefore the speed of the stepper is controlled by the frequency of the **Clock**. This interface can be controlled using the squarewave generation software in Program 6.10 or 6.11. Either full or half stepping can be selected (pin 9), and the direction is controlled by the **CW/CCW** signal. The **OIC** signal controls the voltage on the coils during the half-stepping off states. For example, the half-step voltages on (**L1,L2**) on a 12V interface can be either

$$\text{OIC}=0 \ (12,0) \ -> (\text{hiZ,hiZ}) \ -> (0,12) \ -> (\text{hiZ,hiZ}) \ -> \ldots \text{ or}$$
$$\text{OIC}=1 \ (12,0) \ -> \ (12,12) \ -> (0,12) \ -> (12,12) \ -> \ldots$$

Figure 8.88
Bipolar stepper motor interface using an L293 driver.

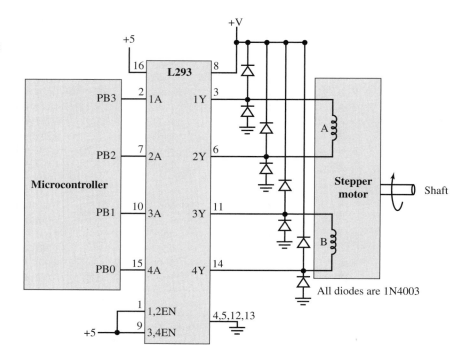

Figure 8.89
The another bipolar stepper motor interface.

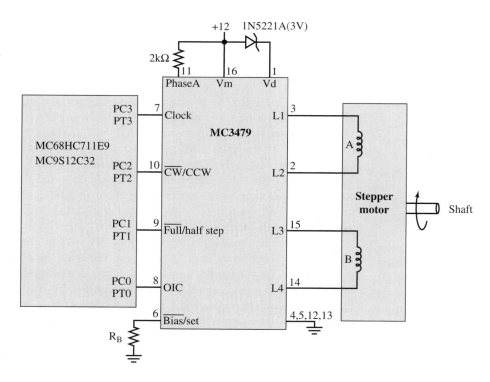

The unipolar stepper architecture provides for bidirectional currents by using a center tap on each phase. The center tap is connected to the +V power source, and the four ends of the phases are controlled with open collector drivers, as shown in Figure 8.90. Only half of the electromagnets are energized at one time. The L293 provides up to 1A current.

Figure 8.90
Unipolar stepper motor
interface.

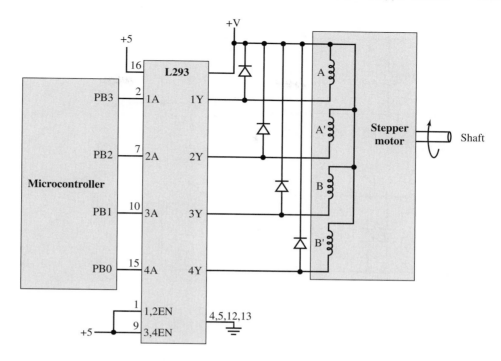

Checkpoint 8.9: What changes could you make to a stepper motor system to increase torque, increasing the probability that a step command actually rotates the shaft?

Checkpoint 8.10: Do you need a sensor feedback to measure the shaft position when using a stepper motor?

8.6.4 Stepper Motor Shaft Encoder

Stepper motors can also be used in a passive mode as shaft encoders, giving both speed and position. When the shaft of a stepper motor is rotated, a series of electric potentials is generated across its coils. The circuit shown in Figure 8.91 converts the AC voltage signals into two digital square waves. The frequency of the waves gives the rotational speed of the shaft, and the phase between the two waves gives the direction (clockwise or counterclockwise).

The 1N914 diodes produce signals that are ± 1 V from the 2.5-V center. Four steps of the motor will produce one large cycle and one small cycle on V_1 and V_2. The distance traveled to produce these two cycles is the distance between two magnetic detents. The positive feedback hysteresis is adjusted so that the output signals (Out_1, Out_2) are sensitive to the large cycle but miss the small cycle. The shaft rotation speed, **r** in rps, is given by

$$r = \frac{4f}{N}$$

where **f** is the frequency of Out_1 in hertz and **N** is the number of steps in the motor. For example, a 200-step motor will yield 50 pulses per rotation. The microcomputer can measure the frequency of Out_1 to obtain shaft speed, or count pulses modulo 50 to establish position.

The direction of the rotation can be determined by the phase between Out_1 and Out_2. For a clockwise rotation, the value of Out_2 at the time of the fall of Out_1 is high (Figure 8.92).

For a counterclockwise rotation, the value of Out_2 at the time of the fall of Out_1 is low (Figure 8.93).

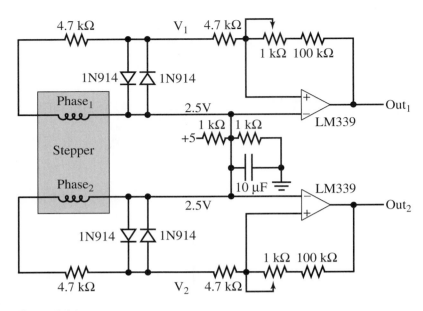

Figure 8.91
Stepper motors can also be used as shaft position sensors.

Figure 8.92
Clockwise timing of
a stepper motor used
as shaft position sensor.

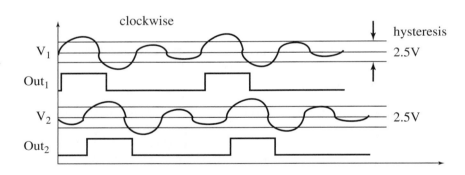

Figure 8.93
Counterclockwise timing
of a stepper motor used
as shaft position sensor.

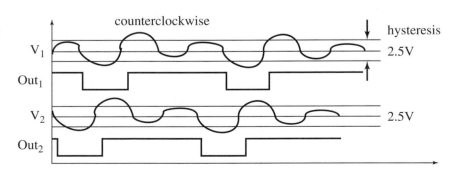

A unipolar stepper with six wires can be used exactly like a bipolar stepper by not connecting the common wires. A unipolar stepper with five wires is connected with the 2.5-V center at the common signal, then only half of each phase is used. This configuration will yield signals with half the amplitude.

8.7 Exercises

8.1 Design a three-digit LED display interfaced only to the 6811/6812 PORTB (Figure 8.94). The software will utilize a 12-bit packed BCD global variable to contain the current value to be displayed. For example, if the value "456" is to be displayed, the main program will set the 16-bit global to $0456. Your interrupt software (part c) will read this global and output the appropriate signals to PORTB.

Figure 8.94
Three common-anode
LED digits.

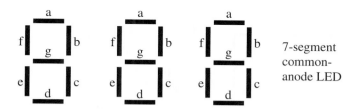

7-segment
common-
anode LED

Available hardware devices include resistors, three common-anode seven-segment LED displays (Figure 8.95), any TTL chip (TTL, LSTTL, or HC) between 7400 and 7499, ULN2074s, 75491s, or 75492s.

a) Show the hardware interface between the display and the 6811 PORTB. Assume the desired LED operating point is 10 mA, 2 V.

b) Show the ritual that initializes the display software/hardware. *Full credit requires the appropriate usage of a circular statically allocated* linked list data structure.

c) Show the interrupt software that maintains the three-digit display. You do not write the `main()` program that sets the global variable. The main program is free to perform other unrelated tasks.

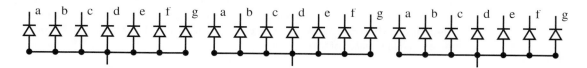

Figure 8.95
Connections for three common-anode LED digits.

8.2 Design the interface between a computer output port, shown as B0 in Figure 8.96, and the EM relay. To activate the alarm the relay coil requires a voltage between +3.5 and 6.0 V and a current of 20 mA.

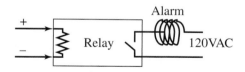

Figure 8.96
Relay interface.

8.3 Design a security code system. Assume there are four keys that have 10 ms of bounce. The keys are configured in a five-wire common line manner (Figure 8.97). Assume there is a global structure stored in EEPROM that contains the valid secret code

```
unsigned char secret[3];
```

Don't worry about how this structure is initialized.

a) Show interface between the keypad and the microcomputer. Any interrupting mechanism can be used (input capture, key wake up, output compare, RTI, etc.).

b) Show the initialization software. Assume there is a main program (which you do not write that will call this ritual, then perform other unrelated tasks).

c) Show the interrupt handler(s) that performs the security functions in the background. Call the function `AlarmOff();` if the proper three-code sequence is typed. You will not write this `AlarmOff();` function.

Figure 8.97
Switch configuration.

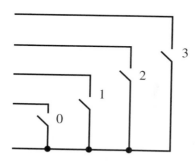

8.4 The objective of this problem is to create a synchronized stepper motor network using single-chip 6811s (Figure 8.98). Each 6811 has an E clock of 2 MHz, 256 bytes of RAM, and 8 kbytes of ROM. Each 6811 is battery-operated and is separated by 100 m. Each system has a serial input, a serial output, and a 4-bit parallel output. Implement an 8-bit, no parity, one stop, 2400-baud serial communication. Upon receipt of a serial input, the 8-bit data are written to the stepper motor interface on Port B. After a 10-ms delay the 8-bit data are then transmitted in serial fashion to the next computer along the chain. At 2400 baud, the smallest interval between receive frames is about 4 ms (10 bits/2400 bits/s). This means it is possible for a second (and a third) input frame to arrive while one is waiting the 10 ms required before transmitting the first data. You will not design the master controller. Your software must be interrupt-driven. There is no BUFFALO. You will specify all the software contained in each 6811 system. The +5-V stepper motor has a coil resistance of 500Ω.

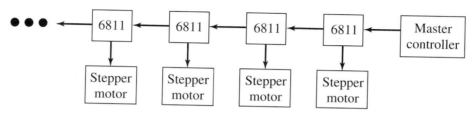

Figure 8.98
Stepper motor distributed network.

a) Show the hardware interface for one 6811 system. Please label chip numbers but not pin numbers. Choose the appropriate interface mechanism and be careful how the systems are grounded to each other. Clearly label all wires in the two 100-m cables (from the previous and to the next). The Stepper motor has +5-V power and the standard A, A', B, B' control.

b) Show all the information that will be located in the 256 byte RAM (0000 to $00FF). Use the assembly language directive `rmb`. Since the RAM contents are unspecified on power up, you cannot use directives `fcb`, `fdb`, or `fcc` in RAM. In the following example notice that the parallel output occurs right away, but the serial output is delayed by 10 ms. Also notice that the serial input may not (but could) occur at the full 2400 bits/s.

Time	Serial input	Parallel output	Serial output
0	6	6	None
4	9	9	None
9	11	11	None
10	None	None	6
13	2	2	None
14	None	None	9
19	None	None	11
23	None	None	2

c) Give *all* the software for each system. Use the `org` directive to place the programs in the ROM located from $E000 to $FFFF. The reset handler (the main program called on power up) will initialize all data structures, perform all the necessary 6811 I/O rituals, and then execute a *do-nothing* infinite loop. The interrupt processes will perform all the I/O operations. Give careful thought as to when and how to interrupt. Poll for 0s and 1s (jump to restart on error). *Good software has* no *backward jumps.* Use the `org` directive to set up the reset and interrupt vectors for your single-chip 6811. Use comments to explain your approach.

8.5 Design the hardware interface that allows the computer to control an EM solenoid (Figure 8.99). The computer controls the solenoid by driving current (activating the solenoid) or no current (deactivating it) through the coil, which has a resistance of 200 Ω. The solenoid activates when the coil voltage is above 4 V (i.e., a coil current above 20 mA). Limit the voltage across the coil to less than 6 V. There will be both positive and negative back EMF voltages, so protect the electronics.

Figure 8.99
Electromagnetic solenoid.

8.6 This question considers the concept of back EMF that can occur on a DC motor, stepper motor coil, relay coil, or solenoid coil when switched with a digital gate. A snubber diode is used to eliminate the back EMF. The choice of which diode to use depends on the magnitude of the back EMF voltage. The circuit configuration depends on the presence of either positive and/or negative amplitude back EMFs.

a) Aside from empirical (experimental) techniques, develop a procedure to select the proper diode in the circuit of Figure 8.100. We are looking for a different answer than the obvious method of measuring it in the lab with a scope (i.e., develop theoretical equations). Be as detailed as possible.

Figure 8.100
Relay interface with one diode.

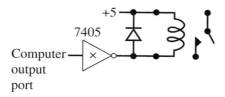

b) Why don't we usually add the second diode (Figure 8.101)? That is, what usually prevents negative-amplitude back EMF?

Figure 8.101
Relay interface with two
diodes.

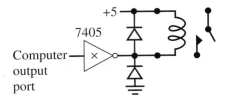

8.8   Lab Assignments

Lab 8.1. The overall objective of this lab is to design a four-function fixed-point calculator. The first task is to interface a 4-by-4 matrix keyboard using interrupt synchronization. You must debounce the keyboard and handle two-key rollover. Rollover is defined as touching the next key before releasing the last key. For example, when some people type "1,2,3," they push "1," push "2," release "1," push "3," release "2," then release "3." Design a low-level device driver for the keyboard interface. Second, you need to interface an LCD display, such as one controlled by the HD44780. The third task is to design a four-function 16-bit signed fixed-point calculator. All numbers will be stored in signed 16-bit fixed-point format with a constant of 0.001. The full-scale range is from -32.767 to $+32.767$. You should include routines to facilitate fixed-point input from the keyboard and output to the LCD. The matrix keyboard will include the numbers '0' – '9', and the letters '+', '−', '*', '/', '=' and '.'. If you want to extend the number of keys, you can define one of the 16 keys as the 'shift' key, creating 30 possibilities. The HD44780 LCD display will show both a 16-bit global accumulator, and a 16-bit temporary register. You are free to design the calculator functionality in any way you wish, but you must be able to (1) clear the accumulator and temporary; (2) type numbers in using the matrix keyboard; (3) add, subtract, multiply, and divide; (4) display the results on the HD44780 LCD display. No SCI input/output is allowed in the calculator program.

Lab 8.2. In this lab, you will control a stepper motor using a finite state machine. The finite state machine must be implemented as a linked data structure. Two switches will allow the operator to control the motor. A background periodic interrupt (either OC or RTI) thread will perform inputs from the switches and outputs to the stepper motor coils. You will use a motor that has fixed-number full-steps to move the shaft one rotation. Two switches determine the motor operation. If both buttons are released, the motor should stop. If switch 1 is pressed, the motor should spin slowly clockwise. If switch 2 is pressed, the motor should spin quickly counterclockwise. The motor should stop when it reaches either the beginning or the end of one rotation even if the operator continues to press the switch. If both switches are pressed, the motor should continuously spin as fast as possible back and forth across the full range of one rotation. If the system is performing one operation and another command is issued, the first operation is terminated and the second command is performed. With an ohmmeter, measure the resistance of one coil. Apply the power across the coil and simultaneously measure, using both a voltmeter and a current-meter, the voltage and current required to activate one coil of your stepper motor. Be sure the interface circuit you select can supply enough current to activate the coil. Measure the inductane of one coil, and determine the turnoff time of your interface. Use these two measurements to estimate the back EMF.

Lab 8.3. The objective of this lab is to design and test a digital alarm clock. You can connect individual push-button switches to input pins of the microcontroller, which the user can use to set the current time and the alarm time. The LCD display will be used to display the current time. You are free to implement whatever features you wish, but there must be a way to set the time. The system maintains three global variables—hour, minute, second. No SCI input is allowed, and the time parameters must be maintained using interrupts. In particular, the interrupt service routine should increment second once a second, increment minute once a minute, and increment hour once an hour. The foreground (main) will output to the LCD display, and interact will the operator via the switch inputs. If the correct sequence of switches is pushed, the main program can initialize the values of hour, minute, second. Design an interface between a speaker and an output port, such that when you toggle the output, an alarm sound is made. The interrupt service routine maintains time, and the main program outputs to the LCD. You can input from the switches either in the foreground or the background. You must be careful not to let the LCD show an intermediate time of 1:00:00 as the time rolls over from 1:59:59 to 2:00:00. You must also be careful not to disable interrupts too long (more than one interrupt period), because a time error will result if any interrupts are skipped. If you use RTI interrupts, you will have to do some 32-bit math to maintain the exact time. For example, if the RTI interrupts at 30.517Hz, then interrupts are requested at exactly 32768 μs. One method is to add 32768 to a 32-bit counter. When the `counter` exceeds 1 million, increment `second` and subtract 1 million from the `counter`.

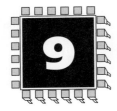

9 Memory Interfacing

Chapter 9 objectives are to:

❑ Present the basic engineering design steps for interfacing memory to the computer
❑ Review the basic building blocks of computer architecture
❑ Develop a descriptive language for defining time-dependent behavior of digital circuits
❑ Use timing diagrams to specify behavior of individual components as well as the entire system
❑ Compare and contrast synchronous, partially asynchronous, and fully asynchronous buses
❑ Discuss address translation as it occurs in a paged-memory system.

For most applications, the embedded system uses a single-chip microcomputer with built-in memory devices. Therefore there is no need to attach additional memory. In other words, we select an available microcontroller with adequate RAM and ROM space to satisfy our constraints. Consequently most embedded system designers can skip this chapter. On the other hand, there are a few situations that might lead one to need to attach external devices to the address/data bus of the microcontroller. Memory interfacing and bus timing are general topics of computer architecture, and you may wish to study this chapter, not because you need to know how to attach memory for the embedded system but because you plan to apply this knowledge toward the design of general-purpose computers. In a similar fashion, understanding how to attach interface memory is an important step in the design of the single-chip microcontroller itself. The other situation that requires memory interfacing occurs when you need more memory than is available on the single-chip microcomputer. In this situation you can attach external memory to satisfy the constraint of the problem. For example, if the embedded system needed 1 Mbyte of RAM, then the best solution would be to interface an external memory.

9.1 Introduction

In this chapter, we will approach memory interfacing using both general concepts and specific examples. The processor is connected to the memory and I/O devices via the bus. The bus contains timing, command, address, and data signals. The timing signals, like the 6811/6812 E clock, determine when strategic events will occur during the transfer. The command signals, like the 6811/6812 R/W line, specify the bus cycle type. During a *read cycle* data flow from the memory or input device to the processor, where the address bus specifies the memory or input device location and the data bus contains the information. During a *write cycle* data are sent from the processor to the memory or output device, where the address bus specifies the memory or output device location and the data bus contains the information. During each particular cycle on most computers there is exactly one device sending the data and exactly one device receiving the data. Some buses like the IEEE488 and SCSI allow for broadcasting, where data is sent from one source to multiple destinations. In a computer system with memory-mapped I/O all slaves share the same bus (Figure 9.1).

Figure 9.1
Architecture of a computer with memory-mapped I/O.

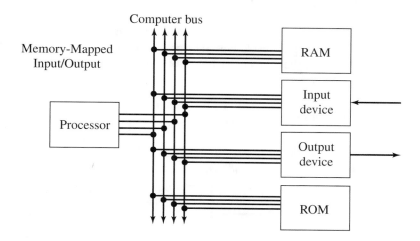

In an Intel x86 computer system with isolated I/O, the slaves share the address and data buses but have separate control signals (Figure 9.2). From a programming perspective, processors with isolated I/O access their I/O devices and memory with separate instructions. On the Intel x86, there are four basic bus cycles: memory read, memory write, I/O read, and I/O write. The memory read/write cycles are similar to the memory-mapped cycles described above. During an *I/O read cycle* data flow from the input device to the processor, where the address bus specifies the input device location and the data bus contains the information. During an *I/O write cycle* data are sent from the processor to the output device, where the address bus specifies the output device location and the data bus contains the information.

Figure 9.2
Architecture of a computer with isolated I/O.

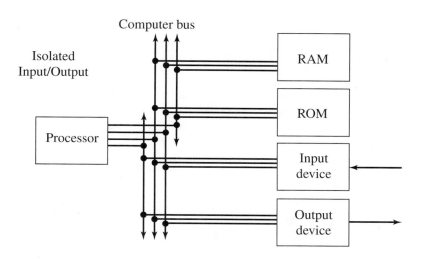

The interface to the slave devices, such as the RAM, ROM, and I/O devices, will consist of two parts. For each cycle, one packet of data is transmitted on the bus. We use the *command* part of the design to specify which slave will be active. The command portion of our slave design will specify whether or not the current cycle is meant for our slave. If the slave can participate in both the read and write cycles, then the command part will also distinguish between the read and write functions. In a system without DMA, the processor will always be the bus master. The bus master will determine the address and bus cycle type for each cycle. The 16 address lines and the R/W signal of the 6811/6812 specify the type of the current cycle (Figure 9.3). *There is no timing information in the address and R/W lines.* On computer systems that contain coprocessors or DMA controllers, there can be more than one bus master. For any given cycle there is only one bus master that will specify the slave address and bus cycle type. The processor, coprocessors, and DMA controllers vie for the use of the shared memory bus. More about DMA can be found in Chapter 10.

Figure 9.3
Both the 6811 and 6812 support external memory interfaces (expanded mode).

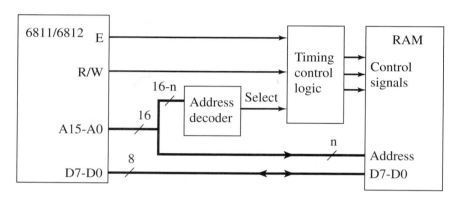

The first step in interfacing a device to the microcomputer is to construct an address decoder. Let SELECT be the positive logic output of the address decoder circuit. The command portion of our interface can be summarized by Table 9.1.

Table 9.1
The address decoder and R/W determine the type of bus activity for each cycle.

SELECT	R/W	Function	Rationale
0	0	Off	Because the address is incorrect
0	1	Off	Because the address is incorrect
1	0	Write	Data flow from 6811/6812 to our device
1	1	Read	Data flow from our device to the 6811/6812

The second part of the interface will be the *timing*. The 6811/6812 uses a *synchronous* bus, thus all timing signals will be synchronized to the E clock. It is important to differentiate timing signals from command signals within the interface. A timing signal is one that has rising and falling edges at guaranteed times during the memory access cycle. In contrast, command signals are generally valid during most of the cycle but do not have guaranteed times for their rising and falling edges. Table 9.2 divides the bus signals into the command or timing category.

Table 9.2
Available timing signals and command signals from the microcomputer.

Microcomputer	Timing signals	Command signals
MC68HC11A8	E AS	R/W A15-A8 AD7-AD0
MC68HC812A4	E CSP0 CSP1 CSD CS3 CS2 CS1 CS0	R/W A15-A0 D15-D0 LSTRB
MC9S12C32	E DE	R/W AD15-AD0 LSTRB

The second objective of this section is to present a shorthand for drawing *timing diagrams*. The 6811 and 6812 bus timing will be presented. Various RAM and PROM memories will be interfaced.

Common error: If control signals in our interface are derived only from command (type and address) signals, then we will have no guarantee when the rising and falling edges will occur.

Observation: We use the command signals to specify which cycles activate our slave, and we use the timing (clock) signals to specify when during the cycle to activate.

9.2   Address Decoding

The address decoder is usually located in each slave interface. The purpose of the address decoder is to monitor the 16 address lines from the bus master (the processor) and determine whether or not the slave has been selected to communicate in the current cycle. Each slave has its own address decoder, uniquely designed to select the addresses intended for that device. Care must be taken to avoid having two devices driving the data bus at the same time. We wish to select exactly one slave device during every cycle. The 6811 and MC9S12C32 use a multiplexed address/data bus. In the 6811, the same eight wires, AD7–AD0, are used for low address (A7–A0) in the first half of each cycle and for the data (D7–D0) in the second half of each cycle. The 6811 address strobe output, AS, is usually used to capture the low address into an external latch (74HC573), as shown in Figure 9.4. In this way, the entire 16-bit address is available during the second half of the cycle.

Figure 9.4
The 6811 multiplexes the low address on the same pins as the data.

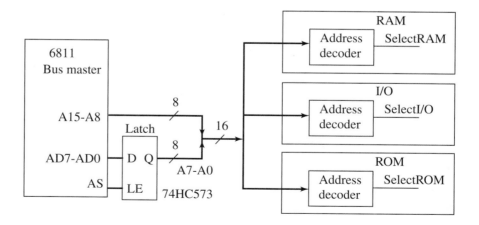

The MC68HC812A4 runs in nonmultiplexed mode, where the address pins are separate from the data pins. In this way, no external address latch is required (Figure 9.5). In the MC9S12C32, the same 16 wires, AD15–AD0, are used for the address (A15–A0) in the first half of each cycle and for the data (D15–D0) in the second half of each cycle. The rising edge of the E clock is used to capture the address into an external latch (two 74FCT374s), as shown in Figure 9.6. In this way, the entire 16-bit address is available during the second half of the cycle.

Figure 9.5
The MC68HC812A4 supports a mode where the lines are dedicated to addresses (nonmultiplexed).

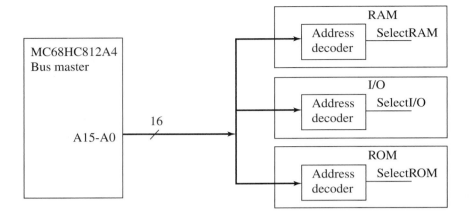

Figure 9.6
The MC9S12C32 multiplexes the 16-bit address on the same pins as the 16-bit data.

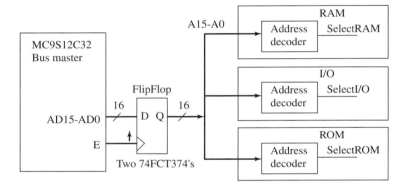

The address decoders should not select two devices simultaneously. In other words, no two select lines should be active at the same time. In this case,

$$SelectRAM \cdot SelectI/O = 0$$
$$SelectRAM \cdot SelectROM = 0$$
$$SelectI/O \quad \cdot SelectROM = 0$$

Common error: If two devices have address decoders with overlapping addresses, then a write cycle will store data at both devices, and during a read cycle the data from the two devices will collide, possibly causing damage to one or both devices.

9.2.1 Full-Address Decoding

Full-address decoding is where the slave is selected if and only if the slave's address appears on the bus. In positive logic,

$$Select \begin{matrix} = 1 \\ = 0 \end{matrix} \quad \begin{matrix} \text{if the slave address appears on the address bus} \\ \text{if the slave address does not appear on the address bus} \end{matrix}$$

Example. Design a fully decoded positive logic Select signal for a 1K RAM at $4000–$43FF.

Step 1. Write specified address using **0,1,X**, using the following rules:

a. There are 16 symbols, one for each of the address bits A15, A14 . . . , A0
b. **0** means the address bit must be 0 for this device

c. **1** means the address bit must be 1 for this device
d. **X** means the address bit can be 0 or 1 for this device
e. All the **X**s (if any) are located on the right-hand side of the expression
f. Let **n** be the number of **X**s. The size of the memory in bytes is 2^n
g. Let **I** be the unsigned binary integer formed from the $16-n$ **0**s and **1**s, then
h. We get the beginning address if all the **X**s are set to 0
i. We get the ending address if all the **X**s are set to 1

$$I = \frac{\text{beginning address}}{\text{memory size}}$$

In this example:

address = **0100,00XX,XXXX,XXXX**
beginning address = $4000
n = 10   size = 1024 bytes = $0400
$$I = 010000_2 = 16_{10} = \frac{\$4000}{\$0400}$$

Step 2. Write the equation using all **0**s and **1**s. A **0** translates into the complement of the address bit, and a **1** translates directly into the address bit. For example,

$$\text{Select} \quad = \overline{A15} \cdot A14 \cdot \overline{A13} \cdot \overline{A12} \cdot \overline{A11} \cdot \overline{A10}$$

Step 3. Build circuit using real TTL gates (Figure 9.7).

Figure 9.7
An address decoder identifies on which cycles to activate.

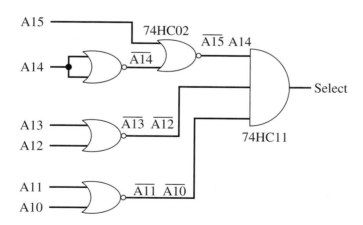

Common error: If all the Xs are not in a group on the right-hand side, then the address range will become discontinuous.

Example. Design a fully decoded select signal for an I/O device at $5500 in negative logic.

Step 1. Address = **0101,0101,0000,0000**

Step 2. The negative logic output can be created simply by inverting the output of a positive logic design. Recall that the address bus of the 6811/6812 is always in positive logic.

$$\text{Select}^* = \overline{\text{A15}} \cdot \text{A14} \cdot \overline{\text{A13}} \cdot \text{A12} \cdot \overline{\text{A11}} \cdot \text{A10} \cdot \overline{\text{A9}} \cdot \text{A8} \cdot \overline{\text{A7}} \cdot \overline{\text{A6}} \cdot \overline{\text{A5}} \cdot \overline{\text{A4}} \cdot \overline{\text{A3}} \cdot \overline{\text{A2}} \cdot \overline{\text{A1}} \cdot \overline{\text{A0}}$$

Step 3. Figure 9.8

Figure 9.8
An address decoder like this could be implemented with programmable logic (PAL).

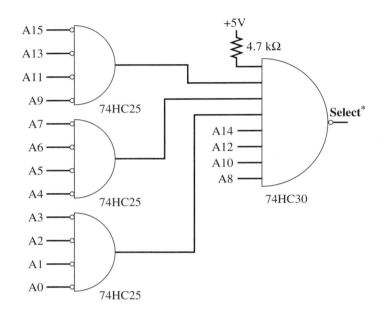

Maintenance tip: Use full-address decoding on systems where future expansion is likely.

Observation: We will use symbols ending in * to signify negative logic. We will also use a line over the symbol to signify negative logic.

Checkpoint 9.1: Write the 0,1,X pattern for the address range $2000 to $3FFF.

Checkpoint 9.2: What address range corresponds to 1101,1XXX,XXXX,XXXX?

**9.2.2
Minimal-Cost
Address Decoding**

Minimal-cost address decoding is a mechanism to simplify the decoding logic by introducing don't care states for unspecified addresses. It does not guarantee lowest cost but often produces simpler solutions than fully decoded. Most embedded microcomputer systems do not use the entire 65,536 available addresses. An unspecified address is one at which there is no device. A reduction of logic may be realized if we introduce *don't care* outputs in the Karnaugh maps when the address equals one of the unspecified addresses. This shortcut should work because the software should never access an unspecified address. If a valid address is accessed, then one and only one device will be selected as expected. But, the address decoders may select zero, one, or more devices if an unspecified address were to be accessed. On most computers, the address bus is in positive logic, but we may construct the **Select** signals in either positive or negative logic, depending on which is more convenient. The formal definition of minimal cost is shown in Table 9.3.

Table 9.3
Minimal-cost decoding optimizes by taking advantage of the don't care states.

Address	Select
Matches our device	True
Matches other specified device	False
Unspecified address	Don't care

Example. Design four minimal-cost select signals in positive logic:

4K RAM	$0000 to $0FFF
Input	$5000
Output	$5001
16K ROM	$C000 to $FFFF

Step 1. Write out the addresses in binary for all devices. Include all specified addresses in the computer, not just the devices that we are designing.

RAM	0000,XXXX,XXXX,XXXX
Input	0101,0000,0000,0000
Output	0101,0000,0000,0001
ROM	11XX,XXXX,XXXX,XXXX

Step 2. Choose as many address lines that are required to differentiate between the devices. Consider the different devices in a pairwise fashion. If the decoder for only one device is being built, then only the addresses that differentiate that device from the others are required. In this example we choose A15, A14, A0. The address bits A15, A12, and A0 also could have been used to differentiate the devices.

Step 3. Draw a Karnaugh map for each device.

a. Put a true for addresses specified by that device
b. Put a false for other devices
c. Put an "X" for unspecified addresses

Positive logic	true = 1 and false = 0
Negative logic	true = 0 and false = 1

Step 4. Minimize using Karnaugh maps and determine equations (Figure 9.9).

$$\text{RAMSelect} = \overline{\text{A14}} \qquad\qquad \text{ROMSelect} = \text{A15}$$
$$\text{InputSelect} = \overline{\text{A15}} \cdot \text{A14} \cdot \overline{\text{A0}} \qquad \text{OutputSelect} = \overline{\text{A15}} \cdot \text{A14} \cdot \text{A0}$$

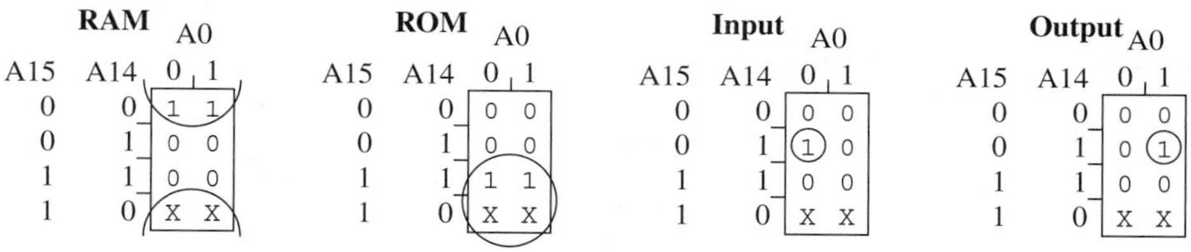

Figure 9.9
Four Karnaugh maps.

Step 5. Build circuit using real TTL gates (Figure 9.10).

Figure 9.10
Four address decoders like these could also be implemented with PAL.

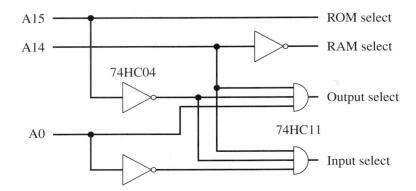

To implement **negative logic** either we can put true = 0, false = 1 into the Karnaugh map or we can build the decoders in positive logic and invert the outputs.

Performance tip: Use minimal-cost decoding on systems where cost and speed are more important than future expansion.

Observation: If there are no unspecified addresses, then minimal-cost decoders will be the same as the full-address decoders.

Performance tip: Address decoders for the entire computer system can sometimes be implemented with a single demultiplexer like the 74HC138 or 74HC139.

Common error: If one does not select enough address lines to differentiate between the devices, then some addresses may incorrectly select more than one device.

Observation: If one selects too many address lines, then the Karnaugh map will be harder to draw, but the resulting solution should be the same.

If all the slave select signals require the same address lines, then all the Karnaugh maps can be combined into a single map using slave names instead of 1s. We can use different colors for each slave. We then make the biggest circles that include each slave one by one. Added protection can be achieved by making none of the slave circles overlap. In this way, if the software inadvertently accesses the unspecified address, at most one slave will be activated. The "Cheaper" example in Figure 9.11 is the same solution as the one above. In this example, if the software reads location is $8000, both the RAM and ROM would drive the data bus. If added protection is desired, we would eliminate the overlapping ROM and RAM circles. The "Safer" solution would activate only the ROM, if the unspecified address is accessed. If the ability to expand in the future is desired, we could leave holes as shown in the "Expandable" solution.

Figure 9.11
A shorthand method for drawing multiple Karnaugh maps.

Cheaper

A15 A14		A0 0	A0 1
0	0	RAM	RAM
0	1	In	Out
1	1	ROM	ROM
1	0	X	X

Safer

A15 A14		A0 0	A0 1
0	0	RAM	RAM
0	1	In	Out
1	1	ROM	ROM
1	0	X	X

Expandable

A15 A14		A0 0	A0 1
0	0	RAM	RAM
0	1	In	Out
1	1	ROM	ROM
1	0	X	X

9.2.3
Special Cases
When Address
Decoding

If the size of the memory is not a power of 2 or if the memory size does not evenly divide the starting address, then the **0,1,X** address specification will require more than one line. To generate the multiline address specification, begin with the starting address and add **X**s to the right side as long as the address range does not exceed the ending address. Start over at the next address until the entire range is covered. The address decoder is the **OR** of the separate lines. Do not add **X**s to the left of a **0** or **1**, because doing so will make a discontinuous address range.

We define a *regular address range* as one with a size that is a power of 2, and one where the beginning address divided by the memory size is an integer. A regular address range can be expressed as a single line of 0s, 1s, and Xs. Two special cases with irregular address ranges are illustrated by the following examples. In the first case, the size of the memory is not a power of 2. In the second case, the beginning address divided by the memory size is not an integer. It requires more than one line of 0s, 1s, and Xs to specify the irregular address range. For irregular address ranges, we break the address range into multiple regular address ranges, solve each part separately, then combine the parts to form the decoder for the whole.

Example. Build a fully decoded positive logic address decoder for a 20K RAM with an address range from $0000 to $4FFF.

Start with

`0000,0000,0000,0000`	Range $0000 to $0000

add Xs while the range is still within $0000 to $4FFF

`0000,0000,0000,000X`	Range $0000 to $0001
`0000,0000,0000,00XX`	Range $0000 to $0003
`0000,0000,0000,0XXX`	Range $0000 to $0007
`0000,0000,0000,XXXX`	Range $0000 to $000F

stop at

`00XX,XXXX,XXXX,XXXX`	*Range $0000 to $3FFF*

Start over

`0100,0000,0000,0000`	Range $4000 to $4000

add Xs while the range is still within $4000 to $4FFF

`0100,0000,0000,000X`	Range $4000 to $4001
`0100,0000,0000,00XX`	Range $4000 to $4003
`0100,0000,0000,0XXX`	Range $4000 to $4007
`0100,0000,0000,XXXX`	Range $4000 to $400F

stop at

`0100,XXXX,XXXX,XXXX`	*Range $4000 to $4FFF*

Combine the two parts to get:

$$\text{RAM SELECT} = \overline{A15} \cdot \overline{A14} + \overline{A15} \cdot A14 \cdot \overline{A13} \cdot \overline{A12}$$

Example. Build a fully decoded negative logic address decoder for a 32K RAM with an address range of $2000 to $9FFF.

Even though the memory size is a power of 2, the size 32,768 does not evenly divide the starting address 8192. We break the $2000 to $9FFF irregular address range into three regular address ranges.

`001X,XXXX,XXXX,XXXX`	Range $2000 to $3FFF
`01XX,XXXX,XXXX,XXXX`	Range $4000 to $7FFF
`100X,XXXX,XXXX,XXXX`	Range $8000 to $9FFF

Combine the two parts to get:

$$\text{SELECT*} = \overline{\overline{A15} \cdot \overline{A14} \cdot A13 + \overline{A15} \cdot A14 + A15 \cdot \overline{A14} \cdot \overline{A13}}$$

Observation: If the beginning address divided by the memory size is not an integer, then the address range cannot be specified by a single row of 0s, 1s, and Xs.

Observation: If the memory size is not a power of 2, then the address range cannot be specified by a single row of 0s, 1s, and Xs.

9.2.4
Flexible Full-Address Decoder

We can design a flexible address decoder (Figure 9.12) with the following features:

It can decode any number of address bits by cascading multiple stages together; there is one stage for each address bit (e.g., the 6811/6812 would use 16 stages to decode all its address bits A15–A0)

It can be configured with DIP switches (there are two switches for each stage/address bit)

Figure 9.12
An approach for designing a programmable address decoder.

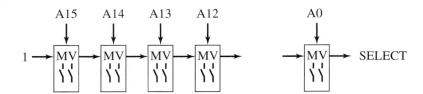

The M switch masks whether or not this address bit is to be considered—an open switch means consider, and a closed switch means ignore

If M is open, the V switch specifies whether this bit should be 1 or 0—an open switch means 1, and a closed switch means 0

Each stage has an address bit input, a cascade control input, and a cascade control output

The cascade control input, In, to the first stage is 1

The cascade control output, Out, of the last stage is the positive logic select

For example, assume the address range is $4000 to $5FFF:

	15	14	13	12	11	10	9	8	7	6	5	4	3	2	1	0
M switch	1	1	1	0	0	0	0	0	0	0	0	0	0	0	0	0
V switch	0	1	0													

For example, assume the address range is $2010 to $2011:

	15	14	13	12	11	10	9	8	7	6	5	4	3	2	1	0
M switch	1	1	1	1	1	1	1	1	1	1	1	1	1	1	1	0
V switch	0	0	1	0	0	0	0	0	0	0	0	1	0	0	0	

In both examples, 0 means closed, 1 means open, blank means either open or closed.

We will show the digital logic for one stage (Table 9.4 and Figure 9.13). We do not have to debounce the switches because they will not be toggled with the power on. There are four inputs and one output for each stage.

Table 9.4
Input/output function of one stage of a programmable address decoder.

In	M_n	A_n	V_n	Out	Meaning
0	X	X	X	0	There was an address mismatch in a previous stage
1	0	X	X	1	Ignore A_n, V_n and pass In to Out
1	1	0	0	1	OK because A_n matches V_n
1	1	0	1	0	Trouble because A_n does not match V_n
1	1	1	0	0	Trouble because A_n does not match V_n
1	1	1	1	1	OK because A_n matches V_n

Figure 9.13
Implementation of a programmable address decoder.

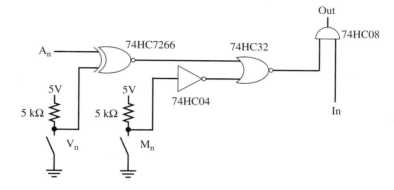

9.2.5 Integrated Address Decoder on the MC68HC812A4

Some microcomputers like the MC68HC812A4 and the MC68HC340 have chip selects integrated into the microcomputer chip itself. In particular, the MC68HC812A4 has seven built-in address decoders that can be used as chip selects for external memory and I/O devices. The chip selects on the MC68HC340 behave similar to the example in the previous section, whereas the MC68HC812A4 negative chip selects have a simple dedicated use. The 5-bit **mmmmm** and **rrrrr** are the values in the INITRM and INITRG configurations, respectively. Notice that the CS3s, CSD, and CSP1 chip selects can be configured at different addresses. Each of these chip selects can be individually enabled or disabled by writing the CSCTL0 register.

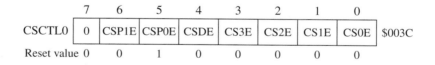

	7	6	5	4	3	2	1	0	
CSCTL0	0	CSP1E	CSP0E	CSDE	CS3E	CS2E	CS1E	CS0E	$003C
Reset value	0	0	1	0	0	0	0	0	

In Table 9.5, the start and end addresses are calculated assuming the default values INITRG=0 and INITRM=$08.

Table 9.5
Built-in address decoder of the MC68HC812A4.

Port	Name	Range	Start	End	Size
PF3	CS3	000000xxxxxxxxxx	$0000	$03FF	1K CS3EP=1,EWDIR=1
PF3	CS3	000001xxxxxxxxxx	$0400	$07FF	1K CS3EP=1,EWDIR=0
PF3	CS3	rrrrr0101xxxxxxx	$0280	$02FF	128 (CS3EP=0)
PF2	CS2	rrrrr0111xxxxxxx	$0380	$03FF	128
PF1	CS1	rrrrr011xxxxxxxx	$0300	$03FF	256
PF0	CS0	rrrrr01xxxxxxxxx	$0200	$03FF	512
PF4	CSD	0111xxxxxxxxxxxx	$7000	$7FFF	4K (CSDHF=0)
PF4	CSD	0xxxxxxxxxxxxxxx	$0000	$7FFF	32K (CSDHF=1)
PF5	CSP0	1xxxxxxxxxxxxxxx	$8000	$FFFF	32K
PF6	CSP1	1xxxxxxxxxxxxxxx	$8000	$FFFF	32K (CSP1FL=0)
PF6	CSP1	xxxxxxxxxxxxxxxx	$0000	$FFFF	64K (CSP1FL=1)

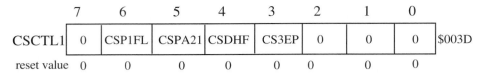

	7	6	5	4	3	2	1	0	
CSCTL1	0	CSP1FL	CSPA21	CSDHF	CS3EP	0	0	0	$003D
reset value	0	0	0	0	0	0	0	0	

Observation: Even though some of the built-in address decoders have overlapping ranges, at most one chip select will activate for each cycle.

When an overlapping address is accessed, the 6812 uses a priority system to determine which chip select is made active. Table 9.6 illustrates the procedure used by the 6812 to resolve overlapping addresses.

Table 9.6
Size and range of CS0 as a function of whether or not CS1, CS2, CS3 are enabled.

CS1	CS2	CS3	CS0 size	CS0 range
Off	Off	Off	512	$0200 to $03FF
Off	Off	Enabled	384	$0200 to $027F and $0300 to $03FF
Off	Enabled	Off	384	$0200 to $037F
Off	Enabled	Enabled	256	$0200 to $027F and $0300 to $037F
Enabled	Off	Off	256	$0200 to $02FF
Enabled	Off	Enabled	128	$0200 to $027F
Enabled	Enabled	Off	256	$0200 to $027F
Enabled	Enabled	Enabled	128	$0200 to $027F

Highest   On-chip registers, 512 bytes at **rrrrr00xxxxxxxxx**, see the INITRG register
    BDM space, 256 bytes at $FFxx when active
    On-chip RAM, 1024 bytes at **mmmmm0xxxxxxxxxx**, see the INITRM register
    On-chip EEPROM if enabled, 4 kbytes, see the INITEE register
    External I/O devices, 128 bytes if CS3 is active
    External I/O devices, 128 bytes if CS2 is active
    External I/O devices, 256 bytes if CS1 is active
    External I/O devices, 512 bytes if CS0 is active
    External Program, if CSP0 is active
    External Data, 16 kbytes if CSD is active
    External Program, if CSP1 is active
Lowest   Remaining external

Observation: The size of the CS0 chip select is affected by whether or not CS1, CS2, or CS3 is active.

This feature greatly simplifies the design of an expanded-mode computer system, because in many applications external memory and I/O devices can be connected directly to the MC68HC812A4 without additional interface logic (Figure 9.14).

Figure 9.14
The MC68HC812A4 has built-in address decoders.

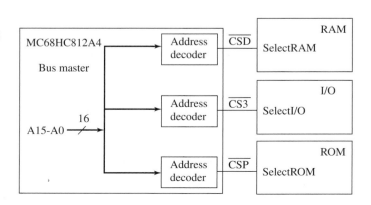

9.3 Timing Syntax

9.3.1
Available and
Required Time
Intervals

To specify an interval of time, we give its start and stop times. These times may be relative numbers in nanoseconds referred to an edge of the system clock, signifying the start of the bus cycle. In the 6811 and 6812, we can refer times to the fall of the E clock because it marks the beginning of the current memory access cycle. For example,

$$(400, 520)$$

means the time interval begins at 400 ns after the fall of the E clock and ends at 520 ns after that same fall, 120 ns later. Sometimes we are not quite sure exactly when it starts or stops, but we can give upper and lower bounds. For example, assuming we know the interval starts somewhere between 400 and 430 ns after the E and stops somewhere between 520 and 530 ns, we would then write

$$([400, 430], \ [520, 530])$$

Sometimes we specify time intervals in terms of the rise and fall of control signals. For example, consider the simple circuit in Figure 9.15. We specify the time that the control signal **A** rises and falls as $\uparrow$**A** and $\downarrow$**A**, respectively. If we assume the propagation delay through the 74LS04 is exactly 10 ns, then the time interval when **Y** is high can be written

$$(\uparrow \mathbf{Y}, \downarrow \mathbf{Y}) = (\downarrow \mathbf{A} + 10, \uparrow \mathbf{A} + 10) = (\downarrow \mathbf{A}, \uparrow \mathbf{A}) + 10$$

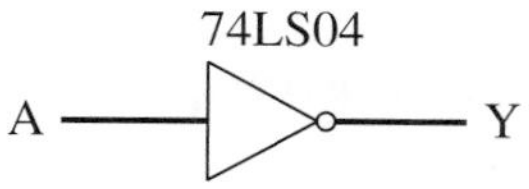

Figure 9.15
A NOT gate.

If we assume the propagation delay through the 74LS04 can range from 0 to 20 ns, then the time interval when **Y** is high can be written

$$(\uparrow \mathbf{Y}, \downarrow \mathbf{Y}) = ([\downarrow \mathbf{A}, \downarrow \mathbf{A} + 20], [\uparrow \mathbf{A}, \uparrow \mathbf{A} + 20]) = (\downarrow \mathbf{A} + [0,20], \uparrow \mathbf{A} + [0,20])$$

When data are transferred from one location (the source) and stored into another (the destination), there are two time intervals that will determine if the transfer is successful. The *data available* interval specifies when the data driven by the source are valid. The *data required* interval specifies when the data to be stored into the destination must be valid. For a successful transfer the data available interval must overlap (start before and end after) the data required interval (Figure 9.16). Let a, b, c, d be times relative to the same time reference, let the data available interval be (a, d), and let the data required interval be (b, c). The data will be successfully transferred if a is less than or equal to b and c is less than or equal to d.

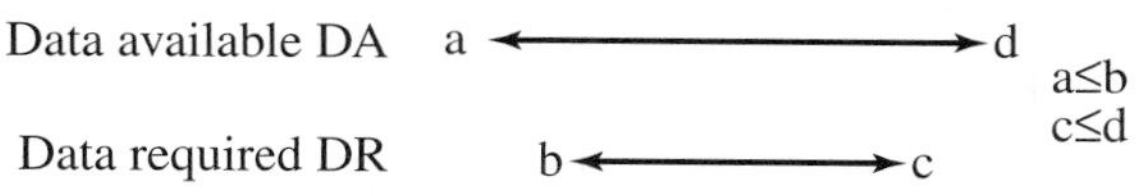

Figure 9.16
The timing of data available should overlap data required.

The following example illustrates the fundamental concepts occurring in a typical memory bus transfer. The objective is to transfer the data from the input **In** to the output **Out**. First, we assume the signal at the **A** input of the 74LS125 is always valid. When the tristate control **G*** is low, then the **A** is copied to the output **Y**. The output **Y** typifies a data bus signal. On the rising edge of **Clk**, the 74LS74 D flip-flop will copy its input **D** to its output **Q** (Figure 9.17).

Figure 9.17
Simple circuit to illustrate that the data available interval should overlap the data required interval.

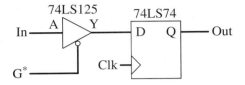

The data available interval is determined by the timing of the source (e.g., the 74LS125). The output of the 74LS125, **Y**, is valid between 10 and 20 ns after the fall of **G***. It will remain valid until 0 to 15 ns after the rise of **G***. The data available interval is

$$DA = (\downarrow G^* + [10,20], \uparrow G^* + [0,15])$$

The data required interval is determined by the timing of the destination (e.g., the 74LS74). The 74LS74 input **D** must be valid from 30 ns before the rise of **Clk** (setup time) and remain valid until 0 after that same rise of **Clk** (hold time). The data required interval is

$$DR = (\uparrow Clk - 30, \uparrow Clk + 0)$$

Since the objective is to make the data available interval overlap the data required window, the worst case situation will be the shortest data available and the longest data required intervals. Without loss of information, we can write:

$$DA = (\downarrow G^* + 20, \uparrow G^*)$$

Thus the data will be properly transferred if $\downarrow G^* + 20$ is less than or equal to $\uparrow Clk - 30$ and if $\uparrow Clk$ is less than or equal to $\uparrow G^*$.

Checkpoint 9.3: Simplify ([2,3] + [4,5],[6,7] + 8)

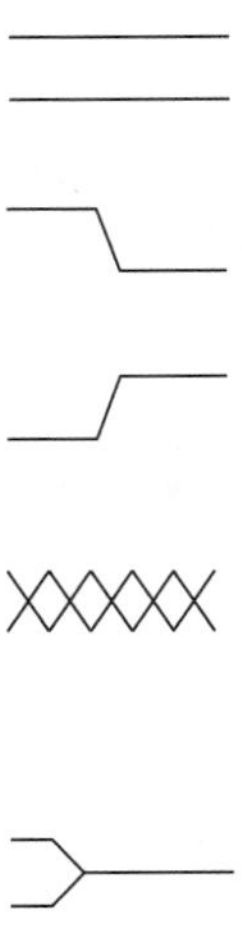

Figure 9.18
Nomenclature for drawing timing diagrams.

Symbol	Input	Output
─────	The input must be valid	The output will be valid
＼	If the input were to fall	Then the output will fall
／	If the input were to rise	Then the output will rise
XXXXX	Don't care, it will work regardless	Don't know, the output value is indeterminate
⟩───	Nonsense	High impedance, tristate, HiZ, Not driven, floating

9.3.2
Timing Diagrams

We will use the timing symbols shown in Figure 9.18 to describe the causal relations in our interface. It is important to have it clear in our minds whether we are drawing an input or an output signal. To illustrate the graphic relationship of dynamic digital signals, we will present some simple examples of timing diagrams. The typical delay time for low-power Schottky logic is 10 ns. The subscript HL refers to the output changing from High to Low (Figure 9.19).

Figure 9.19
Timing of a NOT gate.

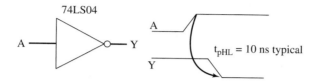

The shape of the tristate buffer timing is similar to the shape of the read cycle timing of the ROM and the RAM. The speed of a 74LS244 is typical of CMOS memories (Figure 9.20).

Figure 9.20
Timing of a tristate buffer gate.

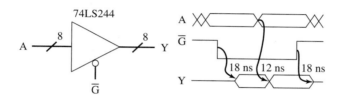

The shape of the timing for clocking the D flip-flop is similar to the shape of the write cycle timing of the RAM. Again, the speed of a 74LS374 is typical of CMOS memories (Figure 9.21). The rising-edge of the Clk stores the D input, with a setup time of 20 ns and a hold time of 0 ns.

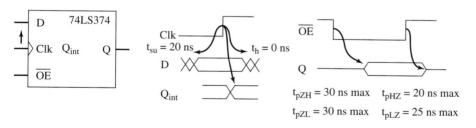

Figure 9.21
Timing of a D flip-flop with tristate outputs.

9.4 General Memory Bus Timing

There are two types of memory access cycles. During a *read cycle* data are passed from a slave device (memory or I/O) into the processor. During a *write cycle* data are passed from the processor to a slave device (memory or I/O). To participate in a read cycle, the slave must have tristate logic so that it can drive data onto the data bus during read cycles involving

Figure 9.22
Simplified diagram
showing circuits used
during a read cycle.

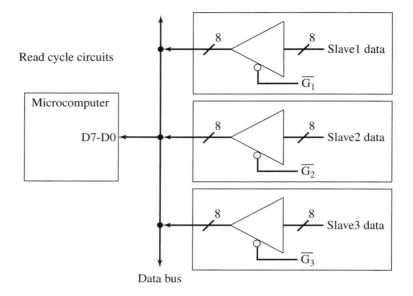

Read cycle circuits

Figure 9.23
Simplified diagram
showing circuits used
during a write cycle.

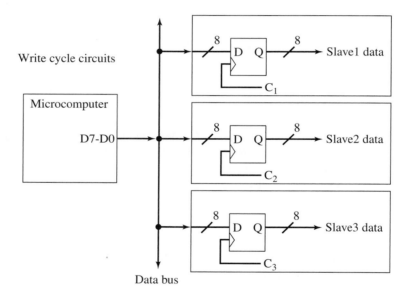

Write cycle circuits

that slave. In the first simplified diagram (Figure 9.22), we show the circuits to drive data on the bus during a read cycle for three slaves. In the second simplified diagram (Figure 9.23), we show the circuits to accept data from the bus during a write cycle for three slaves. Recall that the **command** portion of our slave design specified yes or no as to whether the current cycle is meant for our slave. In this general memory bus timing section, we will discuss three design approaches for generating the bus control signals (like $\overline{G1}$, $\overline{G2}$, $\overline{G3}$, C1, C2, C3). For the TIMING portion of our slave design we will specify when during a read cycle the data are driven onto the bus (e.g., $\overline{G1}$, $\overline{G2}$, $\overline{G3}$) and when during a write cycle data are clocked off the bus and into the slave (e.g., C1, C2, C3).

9.4.1
Synchronous Bus
Timing

The MC68HC11A8, MC68HC812A4, and MC9S12C32 implement synchronous bus transfers. In a *synchronous bus* scheme the bus master (usually the processor) generates a clock (e.g., E clock), and all data transfer timings occur relative, or synchronous, to this

Figure 9.24
Synchronized bus timing.

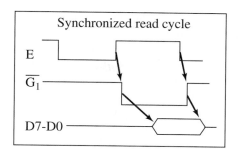 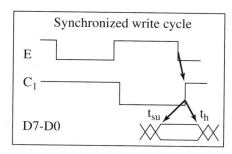

clock. In our simplified circuit example above, a synchronous system has the control signals synchronized to the bus clock. In particular the read and write cycle timing will be as shown in Figure 9.24.

> *Observation:* In a synchronized bus interface the rising and falling edges of the control signals are synchronized (occur immediately) after an edge of the bus clock.

> *Observation:* In a synchronized bus interface there is no feedback from the slave about whether or not the transfer was performed.

> *Observation:* In a synchronized bus interface each bus cycle is exactly the same length, regardless of how fast the slave is.

> *Common error:* If the slave cannot respond fast enough or if no slave exists at the address, data are not properly read or written, but no bus error signal is generated.

> *Common error:* If one slave is replaced with an equivalent but faster device in a synchronized bus interface, the computer will not execute any faster.

9.4.2
Partially Asynchronous Bus Timing

As mentioned above, one of the limitations of a synchronous bus is that there is no feedback from the slave to the master. The two limiting consequences of having no feedback are inefficiency (requires all devices to operate at the speed of the slowest pair) and a lack of flexibility (all devices must be considered together when designed or redesigned.) We define a *partially asynchronous bus* as one that allows slaves to affect bus timing. For example, some computers (like the Freescale 6809 and 680x0 and the Intel x86) allow an external device to slow down the system clock, thus extending the data access time. This is called *cycle stretching.* In a 6809, the E and Q clocks operate normally when MRDY is high. The 6809 E, Q clocks will stop (or stretch) for an integer multiple of quarter bus cycles whenever MRDY is low. The 6809 MRDY must fall within 200 ns of when the end of the cycle would have been for the cycle to be stretched. In Figure 9.25, the cycle is stretched by half

Figure 9.25
Partially asynchronous bus timing of a 6809.

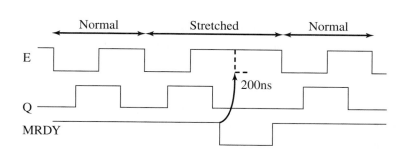

a bus cycle. If a slave can respond in the normal time for the cycle, it will not pull MRDY low. In this way, a simple interface (one without the ability to drive MRDY low) can be designed using the regular rules of a synchronous interface.

Typically a slow memory or I/O device pulls MRDY low until its read or write function is complete. It is much more efficient for the system to slow down only when accessing the slow devices, rather than changing the crystal frequency (which would slow down the system for every cycle). We will use this mechanism to interface dynamic RAMs and slow I/O chips. The 6811 does not allow cycle stretching.

The MC68HC812A4 implements cycle stretching on its built-in chip selects. This mechanism is different from other partially asynchronous interfaces in that there is no feedback from the slave. In a MC68HC812A4 system, the software must have prior knowledge of how fast each slave can respond. The amount of cycle stretching is predetermined by the initialization software using the CSSTR0 and CSSTR1 control registers:

	7	6	5	4	3	2	1	0	
CSSTR0	0	0	SRP1A	SRP1B	SRP0A	SRP0B	STRDA	STRDB	$003E

	7	6	5	4	3	2	1	0	
CSSTR1	STR3A	STR3B	STR2A	STR2B	STR1A	STR1B	STR0A	STR0B	$003F

where the amount of stretch is fixed according to Table 9.7.

Table 9.7
MC68HC812A4 cycle stretching.

STRxxA	STRxxB	Number of E clocks stretched
0	0	0
0	1	1
1	0	2
1	1	3 (default)

Observation: In a partially asynchronous bus, the bus cycle length can vary.

Common error: If the software uses a cycle-counting scheme to implement timing delays, then errors will occur if the programmer does not consider that memory accessed to slow devices will be stretched, thereby causing those cycles to be longer.

Observation: If a slow slave is replaced with an equivalent but faster device in a partially asynchronous bus interface, cycle stretching may be reduced and the computer will execute faster.

Common error: If the slave cannot respond fast enough or if no slave exists at the address, data are not properly read or written, but no bus error signal is generated.

9.4.3
Fully
Asynchronous Bus
Timing

The above two approaches cannot generate a hardware bus error signal when the slave does not or cannot respond. Another limitation of the previous methods is the difficulty in upgrading. To make a synchronous system run faster, each module (master and all slaves) must be redesigned. To upgrade a partially asynchronous system, the new cycle stretching amount must be determined and programmed into the software. In a *fully asynchronous bus* interface, there are control and acknowledge handshake signals that are generated for each bus cycle. Each phase in the following example is signified by the rise or fall of a control signal. The particular example is similar to the protocol used in the SCSI. The key to a fully asynchronous transfer (also called *handshaked* or *interlocked*) is that each phase can vary in length and will wait for the previous phase to occur.

Read Cycle Data Transferred from Slave to Master (Figure 9.26).

Phase 1. Master specifies the address and says, "Please give me data"

I/O=0 and fall of REQ (SCSI)

Phase 2. Slave puts data on the bus and sends an acknowledge saying, "Here is the data"

data are driven and fall of ACK (SCSI)

Phase 3. Master accepts the acknowledge and says "Thank you"

rise of REQ (SCSI)

Phase 4. Slave makes its data HiZ and says "You're welcome"

data are driven and rise of ACK (SCSI)

Figure 9.26
Fully asynchronous (interlocked or handshaked) read cycle bus timing.

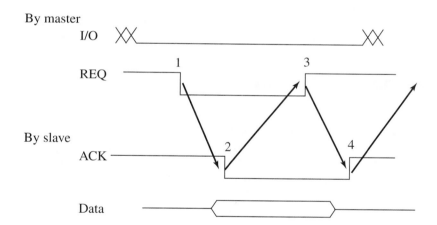

Write Cycle Data Transferred from Master to Slave (Figure 9.27).

Phase 1. Master puts data on the bus and says, "Please save these data"

I/O=1 data driven by master and fall of REQ (SCSI)

Phase 2. Slave accepts the data and sends an acknowledge saying, "I saved the data"

fall of ACK (SCSI)

Phase 3. Master makes its data HiZ and says "Thank you"

rise of REQ (SCSI)

Phase 4. Slave says "You're welcome"

rise of ACK (SCSI)

Figure 9.27
Fully asynchronous
(interlocked or
handshaked) write cycle
bus timing.

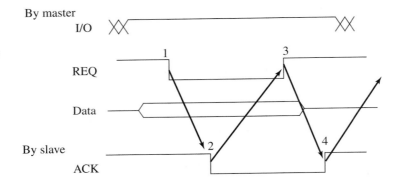

To pass each data correctly and to prevent the same data from being passed twice, there are four transitions in a fully interlocked protocol.

Observation: If you place a time-out mechanism on the time the master waits for the slave response, then a hardware bus error can be generated on a broken or missing module.

Observation: If you replace any module in a fully asynchronous bus system, then the data transfer automatically occurs at the fastest possible rate without changing the software configuration.

Most computer systems do not implement fully asynchronous bus transfers with their memory because the complexity makes the system too expensive and too slow. On the other hand, fully asynchronous transfers are used in many I/O bus protocols such as IEEE-488 and SCSI.

9.5 External Bus Timing

9.5.1 Synchronized Versus Unsynchronized Signals

When designing the control signals for our memory, we have two choices to make. The first choice is relatively easy: Do we use positive or negative logic? A positive logic control signal goes high when the memory is accessed, and a negative logic signal goes low during the cycle of interest. The other choice is whether or not to synchronize to the E clock. These choices are presented in Table 9.8, where **command** refers to a signal derived directly from A15–A0 or R/W. In this discussion, **command** = 1 means activate this cycle. We will also use this table to design the digital logic to create the control signal **CS**.

The four possible shapes for the control signals are shown in Figure 9.28, where the "x" parameter refers to the gate delays through the digital logic required to implement the control signal. If we have a negative logic $\overline{command}$ signal, then the polarity of the command is reversed (Table 9.9).

E	Command	Unsynchronized positive	Unsynchronized negative	Synchronized positive	Synchronized negative
0	0	0	1	0	1
1	0	0	1	0	1
0	1	1	0	0	1
1	1	1	0	1	0
		CS=command	CS=not(command)	CS=E·command	CS=not(E·command)

Table 9.8
A timing signal (like the E clock) can be combined with a positive logic command signal four ways.

Figure 9.28
The timing of four types of control signals.

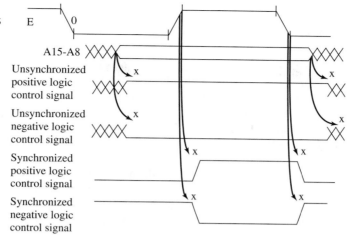

E	$\overline{command}$	Unsynchronized positive	Unsynchronized negative	Synchronized positive	Synchronized negative
0	1	0	1	0	1
1	1	0	1	0	1
0	0	1	0	0	1
1	0	1	0	1	0
		$CS=$ $not(\overline{command})$	$CS=$ $\overline{command}$	$CS=$ $E\cdot not(\overline{command})$	$CS=$ $not(E\cdot not(\overline{command}))$

Table 9.9
A timing signal (like the E clock) can be combined with a negative logic command signal four ways.

9.5.2 Freescale MC68HC11A8 External Bus Timing

9.5.2.1 MC68HC11A8 Execution Modes

The 6811 can run in one of four modes. In *normal single chip* mode, Port B, Port C STRA, and STRB are used for I/O. In *expanded* mode, Port B contains the high address, Port C contains the low address/data, STRA is the address strobe AS, and STRB is the R/W line. *Special bootstrap* mode is a single-chip mode used to load data into the 256-byte RAM, and *special test* mode is used by Freescale for testing. The initial mode is determined by the values of the MODB, MODA pins at the time of the rise of RESET. Table 9.10 also shows the initial values for 4 bits of the HPRIO register.

Table 9.10
There are four execution modes for the MC68HC11A8.

MODB	MODA	Mode description	RBOOT	SMOD	MDA	IRV
1	0	Normal single chip	0	0	0	0
1	1	Normal expanded	0	0	1	0
0	0	Special bootstrap	1	1	0	1
0	1	Special test	0	1	1	1

Observation: To interface external memory devices to the 6811, we will utilize expanded mode.

If the computer is running in one of the special modes, the software can affect the mode by changing the SMOD and MDA bits in the HPRIO register.

	7	6	5	4	3	2	1	0	
HPRIO	RBOOT	SMOD	MDA	IRV	PSEL3	PSEL2	PSEL1	PSEL0	$103C

RBOOT: Read bootstrap ROM
  Can be written only while SMOD equals 1
  1 = Bootstrap ROM enabled at $BF40–$BFFF
  0 = Bootstrap ROM disabled and not present in memory map
  The RBOOT control bit enables or disables the special bootstrap control ROM. This
    192-byte, mask-programmed ROM contains the firmware required to load a user's
    program through the SCI into the internal RAM and jump to the loaded program.
    In all modes other than the special bootstrap mode, this ROM is disabled and does
    not occupy any space in the 64-kbyte memory map. Although it is 0 when the
    MCU comes out of reset in test mode, the RBOOT bit may be written to 1 while
    in special test mode.

SMOD: Special mode
  May be written to 0 but not back to 1
  1 = Special mode variation in effect
  0 = Normal mode variation in effect

MDA: Mode A select
  Can be written only while SMOD equals 1
  1 = Normal expanded or special test mode in effect
  0 = Normal single-chip or special bootstrap mode in effect

IRV: Internal read visibility
  Can be written only while SMOD equals 1; forced to 0 if SMOD equals 0
  1 = Data driven onto external bus during internal reads
  0 = Data from internal reads not visible on expansion bus (levels on bus ignored)

9.5.2.2
MC68HC11A8 Timing

The E clock, AS strobe are timing signal outputs of the 6811. The clock frequency is determined by a crystal placed between the EXTAL and XTAL inputs (Figure 9.29). The 6811
system needs an external crystal that is four times the E clock period (t_1 in the timing
diagram of Figure 9.31). The 2-MHz period is created using an 8-MHz crystal. The 25 pF
includes all stray capacitances. To reduce noise, Freescale suggests placing a ground plane
under the crystal (Figure 9.30).

Figure 9.29
6811 clock circuit.

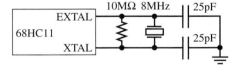

Figure 9.30
Layout for the 6811
clock crystal.

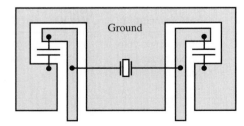

In Figure 9.31 the variables like t_1, t_2, t_3, refer to the timing parameters in Table 9.11.
There are two equivalent ways to define the E clock period:

$$t_1 = t_2 + t_4 + t_3 + t_4$$

Figure 9.31
Simplified bus timing for the 6811 in expanded mode.

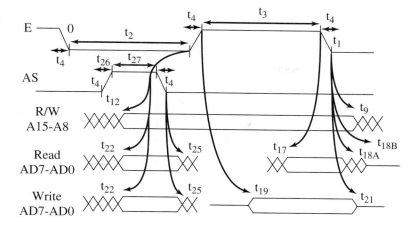

Table 9.11
Timing parameters for the MC68HC11A8.

No.		Characteristic frequency at		
		1.0 MHz	**2.0 MHz**	**2.1 MHz**
t_1	Cycle time (ns)	1000	500	476
t_2	Pulse width E low (ns)	480	230	218
t_3	Pulse width E high (ns)	480	230	218
t_4	Rise/fall time (ns)	20	20	20
t_9	Address hold time (ns)	95.5 min	33 min	30 min
t_{12}	A15–A8, R/W valid time (ns)	281.5 min	94 min	85 min
t_{17}	Read data setup time (ns)	30 min	30 min	30 min
t_{18A}	Read data hold time (ns)	10 min	10 min	10 min
t_{18B}	Read data goes HiZ (ns)	145.5 max	83 max	80 max
t_{19}	Write data delay time (ns)	190.5 max	128 max	125 max
t_{21}	Write data hold time (ns)	95.5 min	33 min	30 min
t_{22}	A7–A0 valid time (ns)	271.5 min	84 min	75 min
t_{25}	A7–A0 hold time (ns)	95.5 min	33 min	30 min
t_{26}	E to AS rise time (ns)	115.5	53	50
t_{27}	AS pulse width (ns)	221	96	90

In some situations, the equivalent expressions might yield different results. This can lead to ambiguity when calculating data available and data required intervals. Another ambiguity results from the fact that a range of parameters may be given. In these situations it is best to design the interface considering the worst case.

Performance tip: It is proper design practice to consider the worse case—that is, the shortest data available interval and the longest data required interval.

For a 2-MHz 6811 the four potential control signals are plotted in Figure 9.32, where the "x" parameter refers to the gate delays through the digital logic required to implement the control signal.

There are three important timing intervals we must determine from the microcomputer timing diagram. The first is the *address available, AA,* interval. This interval defines the time during the cycle when the address is valid.

$$AA = (AdV, AdN) = (later \ (t_2 - t_{12}, t_2 - t_{22} + 35 \ ns), earlier \ (t_1 + t_{26}, t_1 + t_9)) = (181, 533)$$

Figure 9.32
Timing of four types of
control signals with the
6811 in expanded
mode.

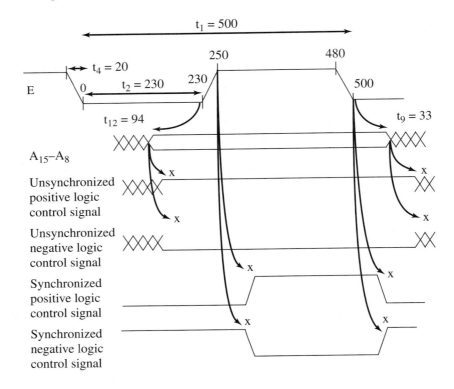

where AdV is the time when the address lines A15–A0 are valid and AdN is the time when
the address lines are no longer valid. The two terms of the later function represent A15–A8
and A7–A0, respectively. The 35 ns comes from the gate delay of the 74HC573 transparent
latch. It will also be useful to know the address available for just A15–A8:

$$AA_{15-8} = (AdV_{15-8}, AdN_{15-8}) = (230 - 94, 500 + 33) = (136, 533)$$

The second important timing interval is *read data required*, RDR. During a read cycle the
data are required by the 6811. Thus to determine the data required interval, we look in the
6811 data sheet. For data required, the worst case is the longest interval:

$$RDR = \text{read data required} = (t_1 - t_4 - t_{17}, t_1 + t_{18}) = (500 - 20 - 30, 500 + 10) = (450, 510)$$

The last important timing interval we get from the microcomputer is *write data available*,
WDA. During a write cycle, the data are supplied by the 6811. Thus, to determine the data
available interval, we look in the 6811 data sheet. We will specify the worst-case timing.
For write data available, the worst case is the shortest interval:

$$WDA = \text{write data available} = (t_2 + t_4 + t_{19}, t_1 + t_{21}) = (230 + 20 + 128, 500 + 33) = (378, 533)$$

9.5.3 Freescale MC68HC812A4 External Bus Timing

9.5.3.1 MC68HC812A4 Execution Modes

The 68HC812 can run in one of eight modes (Table 9.12). Special modes are for testing
and system development, while normal modes are intended for embedded products. In
particular, special modes allow access to many mode control registers, like the MODE
register shown below, that are inaccessible while in normal mode. In the two *single-chip*
modes, Ports A–E are available for I/O. In all four expanded modes, Port E contains the
bus control signals. In the two *expanded narrow* modes, Ports A, B implement the 16-bit
address bus, and Port C implements the 8-bit data bus. In the two *expanded wide* modes,
Ports A, B implement the 16-bit address bus, and Ports C, D implement the 16-bit data bus.

BKGD	MODB	MODA	Mode description	PortA	PortB	PortC	PortD
0	0	0	Special single chip	In/Out	In/Out	In/Out	In/Out
0	0	1	Special expanded narrow	A15–A8	A7–A0	D7–D0	In/Out
0	1	0	Special peripheral	A15–A8	A7–A0	D15–D8	D7–D0
0	1	1	Special expanded wide	A15–A8	A7–A0	D15–D8	D7–D0
1	0	0	Normal single chip	In/Out	In/Out	In/Out	In/Out
1	0	1	Normal expanded narrow	A15–A8	A7–A0	D7–D0	In/Out
1	1	0	Reserved	—	—	—	—
1	1	1	Normal expanded wide	A15–A8	A7–A0	D15–D8	D7–D0

Table 9.12
There are eight execution modes for the MC68HC812A4.

Special peripheral mode is used by Freescale for testing. The initial mode is determined by the values of the BKGD, MODB, MODA pins at the time of the rise of RESET.

Observation: For a simple low-cost external memory interface, we will utilize expanded narrow mode.

Observation: For a high-speed external memory interface, we will utilize expanded wide mode.

When a 16-bit access is required in expanded narrow mode, it is performed by two sequential 8-bit accesses. If the computer is running in one of the special modes, the software can affect the mode by changing the SMODN, MODB, and MODA bits in the MODE register.

Observation: In most situations a program running in expanded memory implemented in narrow mode runs twice as slow as the same program running in wide mode.

	7	6	5	4	3	2	1	0	
MODE	SMODN	MODE	MODA	ESTR	IVIS	0	EMD	EME	$000B

SMODN, MODB, MODA: Mode select special, B and A
These bits show the current operating mode and reflect the status of the BKGD, MODB, and MODA input pins at the rising edge of reset. Read anytime. SMODN may be written only if SMODN = 0 (in special modes), but the first write is ignored; MODB, MODA may be written once if SMODN = 1; anytime if SMODN = 0, except that special peripheral and reserved modes cannot be selected.

ESTR: E clock stretch enable
Determines if the E clock behaves as a simple free-running clock or as a bus control signal that is active only for external bus cycles. Normal modes: write once; special modes: write anytime, read anytime
   0 = E never stretches (always free running)
   1 = E stretches high during external access cycles and low during nonvisible internal accesses

IVIS: Internal visibility
This bit determines whether internal ADDR, DATA, R/W, and LSTRB signals can be seen on the external bus during accesses to internal locations. In special narrow mode if this bit is set and EMD = 1 when an internal access occurs, the data appear wide on Port C and Port D. This allows for emulation. Visibility is not available when the part is operating in a single-chip mode. Normal modes: write

once; special modes: write anytime EXCEPT the first time. Read anytime

0 = No visibility of internal bus operations on external bus

1 = Internal bus operations are visible on external bus

EMD: Emulate Port D

This bit has meaning only in special expanded narrow mode. In expanded wide modes and special peripheral mode, PORTD, DDRD, KWIED, and KWIFD are removed from the memory map regardless of the state of this bit. In single-chip modes and normal expanded narrow mode, PORTD, DDRD, KWIED, and KWIFD are in the memory map regardless of the state of this bit. Removing the registers from the map allows the user to emulate the function of these registers externally. Normal modes: write once; special modes: write anytime EXCEPT the first time. Read anytime

0 = PORTD, DDRD, KWIED, and KWIFD are in the memory map

1 = If in special expanded narrow mode, PORTD, DDRD, KWIED, and KWIFD are removed

EME: Emulate Port E

In single-chip mode PORTE and DDRE are always in the map regardless of the state of this bit. Removing the registers from the map allows the user to emulate the function of these registers externally. Normal modes: write once; special modes: write anytime *except* the first time. Read anytime

0 = PORTE and DDRE are in the memory map

1 = If in an expanded mode, PORTE and DDRE are removed from the internal memory map

Another register that affects expanded mode interfacing is MISC:

	7	6	5	4	3	2	1	0	
MISC	EWDIR	NDRC	0	0	0	0	0	0	$0013

EWDIR: Extra window positioned in direct space

This bit is only valid in expanded modes. If the EWEN bit in the WINDEF register is cleared, then this bit has no meaning or effect

0 = If EWEN is set, then a 0 in this bit places the EPAGE at $0400–$07FF

1 = If EWEN is set, then a 1 in this bit places the EPAGE at $0000–$03FF

NDRC: Narrow data bus for Register Chip Select Space

This function requires at least one of the chip selects CS[3:0] to be enabled. It effects the (external) 512 byte memory space. If the narrow (8-bit) mode is being utilized, this bit has no effect

0 = Makes the register-following chip select active space act as a full 16-bit data bus

1 = Makes the register-following chip selects (2, 1, 0, and sometimes 3) active space [512-byte block] act the same as an 8-bit only external data bus (data goes through only port C externally). This allows 8-bit and 16-bit external memory devices to be mixed in a system.

9.5.3.2
MC68HC812A4 Timing

The E clock, and the seven chip selects are timing signal outputs of the MC68HC812A4. The clock frequency is determined by a crystal placed between the EXTAL and XTAL inputs (Figure 9.33). The MC68HC812A4 system needs an external crystal that is two times the E clock period (t_1 in the timing diagram of Figure 9.34). The 8 MHz period is created using a 16 MHz crystal. The 22 pF includes all stray capacitances. The PCB layout of the crystal clock is shown in Figure 9.30.

Figure 9.33
MC68HC812A4 clock circuit.

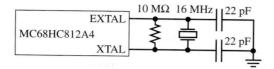

Figure 9.34
Simplified bus timing for the MC68HC812A4 in expanded mode.

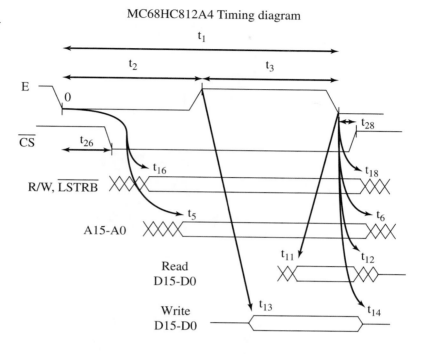

In Figure 9.34 the symbols like t_1, t_2, t_3 refer to the timing parameters in Table 9.13. During a read cycle the MC68HC812A4 clocks the data in on the fall of E; t_{11} is the setup time, and t_{12} is the hold time for this transfer. During a write cycle the MC68HC812A4 drives the data bus.

No.		Characteristic frequency at 8 MHz for			
		0 stretch	**1 stretch**	**2 stretch**	**3 stretch**
t_1	Cycle time (ns)	125	250	375	500
t_2	Pulse width E low (ns)	60 min	60 min	60 min	60 min
t_3	Pulse width E high (ns)	60 min	185 min	310 min	435 min
t_5	A15–A0, R/W delay time (ns)	60 max	60 max	60 max	60 max
t_6	Address hold time (ns)	20 min	20 min	20 min	20 min
t_{11}	Read data setup time (ns)	30 min	30 min	30 min	30 min
t_{12}	Read data hold time (ns)	0 min	0 min	0 min	0 min
t_{13}	Write data delay time (ns)	46 max	46 max	46 max	46 max
t_{14}	Write data hold time (ns)	20 min	20 min	20 min	20 min
t_{16}	R/W delay time (ns)	49 max	49 max	49 max	49 max
t_{18}	R/W hold time (ns)	20 min	20 min	20 min	20 min
t_{26}	CS delay time (ns)	60 max	60 max	60 max	60 max
t_{28}	CS hold time	10 max	10 max	10 max	10 max

Table 9.13
Timing parameters for the MC68HC812A4 with an E clock of 8 MHz.

In expanded narrow mode, the data bus is only 8 bits. The $\overline{CS}$ signal shown in Figure 9.35 is one of the seven built-in chip selects available on Port F. With no cycle stretching, the external E clock follows the internal E clock. With one-cycle stretching, the access cycle is 250 ns long (Figure 9.36). With two-cycle stretching, the access cycle is 375 ns long (Figure 9.37). With three-cycle stretching, the access cycle is 500 ns long (Figure 9.38).

Figure 9.35
CS timing with no cycle stretching.

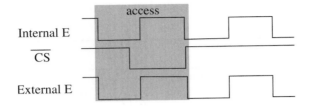

Figure 9.36
CS timing with one-cycle stretching.

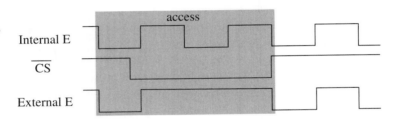

Figure 9.37
CS timing with two-cycle stretching.

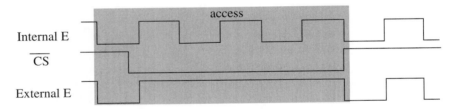

Figure 9.38
CS timing with three-cycle stretching.

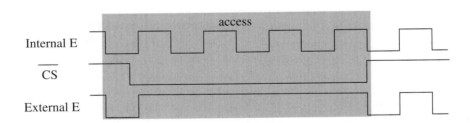

There are three important timing intervals we must determine from the microcomputer timing diagram. The first is the *address available,* AA, interval. This interval defines when during the cycle the address is valid:

$$AA = (AdV, AdN) = (t_5, t_1 + t_6)$$

where AdV is the time when the address lines A15–A0 are valid and AdN is the time when the address lines are no longer valid. For an 8-MHz MC68HC812A4 system, with no cycle stretching,

$$AA = (AdV, AdN) = (60, 145)$$

The second important timing interval is *read data required,* RDR. During a read cycle the data are required by the MC68HC812A4. Thus to determine the data required interval, we look in the MC68HC812A4 data sheet. For data required, the worst case is the longest interval:

$$RDR = \text{read data required} = (t_1 - t_{11}, t_1 + t_{12}) = (125 - 30, 125 + 0) = (95, 125)$$

The last important timing interval we get from the microcomputer is *write data available,* WDA. During a write cycle, the data are supplied by the MC68HC812A4. Thus, to determine the data available interval, we look in the MC68HC812A4 data sheet. We will specify the worst-case timing. For write data available, the worst case is the shortest interval:

$$WDA = \text{write data available} = (t_2 + t_{13}, t_1 + t_{14}) = (60 + 46, 125 + 20) = (106, 145)$$

9.5.4. Freescale MC9S12C32 External Bus Timing

The MC9S12C32 can run in one of eight modes, as listed in Table 9.14. The initial mode is determined by the values of the MODC MODB MODA pins at the time of the rise of RESET. Special and emulation modes are for testing and system development, whereas normal modes are intended for embedded products. In particular, special mode allows you to access many test registers that are inaccessible in normal mode. In the *Single Chip* modes, Ports A, B, E, and K are available for general-purpose input/output. In the *Expanded* modes, Port E contains the bus control signals, and Port K contains the XAB19-14 address lines used in paging. In the *Expanded Narrow* mode, Port A implements the time-multiplexed A15-A8 address D7-D0 data bus, and Port B implements A7-A0, the low 8 bits of the address bus. In the *Expanded Wide* mode, Ports A and B implement the time-multiplexed 16-bit address 16-bit data bus. When paging is used, a 20-bit address allows access to external memory up to 1 Megabyte. *Peripheral* mode is used by Freescale for testing.

MODC	MODB	MODA	Mode description	Port A	Port B	MODx write ability
0	0	0	Special Single Chip	In/Out	In/Out	Write anytime, but not to peripheral
0	0	1	Emulation Expanded Narrow	A15-A8/ D7-D0	A7-A0	Cannot change mode
0	1	0	Special Test	A15-A8/ D15-D8	A7-A0/ D7-D0	Write anytime, but not to peripheral
0	1	1	Emulation Expanded Wide	A15-A8/ D15-D8	A7-A0/ D7-D0	Cannot change mode
1	0	0	Normal Single Chip	In/Out	In/Out	Write once to Normal Expanded Narrow or Wide
1	0	1	Normal Expanded Narrow	A15-A8/ D7-D0	A7-A0	Cannot change mode
1	1	0	Peripheral	—	—	Cannot change mode
1	1	1	Normal Expanded Wide	A15-A8/ D15-D8	A7-A0/ D7-D0	Cannot change mode

Table 9.14
There are eight execution modes for the MC9S12C32.

When a 16-bit access is required in expanded narrow mode, it is performed by two sequential 8-bit accesses. In most situations, running in expanded narrow mode is twice as slow as running in expanded wide mode. Table 9.15 shows the registers we use to configure the MC9S12C32 expanded mode external bus.

Address	Bit 7	6	5	4	3	2	1	Bit 0	Name
$0000	PA7	PA6	PA5	PA4	PA3	PA2	PA1	PA0	PORTA
$0001	PB7	PB6	PB5	PB4	PB3	PB2	PB1	PB0	PORTB
$0002	DDRA7	DDRA6	DDRA5	DDRA4	DDRA3	DDRA2	DDRA1	DDRA0	DDRA
$0003	DDRB7	DDRB6	DDRB5	DDRB4	DDRB3	DDRB2	DDRB1	DDRB0	DDRB
$0008	PE7	PE6	PE5	PE4	PE3	PE2	PE1	PE0	PORTE
$0009	DDRE7	DDRE6	DDRE5	DDRE4	DDRE3	DDRE2	0	0	DDRE
$000A	NOACCE	0	PIPOE	NECLK	LSTRE	RDWE	0	0	PEAR
$000B	MODC	MODB	MODA	0	IVIS	0	EMK	EME	MODE
$000C	PUPKE	0	0	PUPEE	0	0	PUPBE	PUPAE	PUCR
$000D	RDPK	0	0	RDPE	0	0	RDPB	RDPA	RDRIV
$000E	0	0	0	0	0	0	0	ESTR	EBICTL
$0013	0	0	0	0	EXSTR1	EXSTR0	ROMHM	ROMON	MISC
$001E	IRQE	IRQEN	0	0	0	0	0	0	IRQCR
$0032	PK7	PK6	PK5	PK4	PK3	PK2	PK1	PK0	PORTK
$0033	DDRK7	DDRK6	DDRK5	DDRK4	DDRK3	DDRK2	DDRK1	DDRK0	DDRK
$0034	0	0	SYN5	SYN4	SYN3	SYN2	SYN1	SYN0	SYNR
$0033	0	0	0	0	REFDV3	REFDV2	REFDV1	REFDV0	REFDV
$0037	RTIF	PROF	0	LOCKIF	LOCK	TRACK	SCMIF	SCM	CRGFLG
$0039	PLLSEL	PSTP	SYSWAI	ROAWAI	PLLWAI	CWAI	RTIWAI	COPWAI	CLKSEL
$003A	CME	PLLON	AUTO	ACQ	0	PRE	PCE	SCME	PLLCTL

Table 9.15
MC9S12C32 registers used for external memory interfacing.

As the name implies, the **MODE** register specifies the operating mode, (i.e., **MODC MODB MODA**), but there is a complex set of rules about changing the mode, as listed in the last column of Table 9.14. If MODA = 1, then MODC, MODB, and MODA cannot be changed. If MODC = MODA = 0, then MODC, MODB, and MODA are writable with the exception that you cannot change to special peripheral mode. If MODC = 1, MODB = 0, and MODA = 0, then MODC can not be changed. In this case, MODB and MODA are write once, except that you cannot change to special peripheral mode. From normal single-chip, only normal expanded narrow and normal expanded wide modes are available. **IVIS** is the Internal Visibility bit. This bit determines whether internal bus signals can be seen on the external bus during accesses to internal locations. **EMK** and **EME** are the Emulate Port K and E bits. When these bits are set to one while in an expanded mode, PORTK/PORTE and DDRK/DDRE are removed from the internal memory map. In single-chip mode PORTK and PORTE are always in the map regardless of the state of these bits. Removing the registers from the map allows us to emulate the function of these registers externally.

Another register that affects expanded mode interfacing is **PEAR**. The NOACCE, PIPEO, LSTRB, and RDWE bits can be written once in normal mode. This means the initialization software can write to the PEAR register once, then the software will not be able to change any of these bits. In special mode, the bits of PEAR can be written anytime, but in emulation mode, they can not be written. The NOACCE, PIPEO, LSTRB, and RDWE bits have no effect in single-chip or special peripheral modes. **NOACCE** is the CPU No Access Output Enable bit. If NOACCE = 1, PE7 is an output and indicates whether the cycle is a CPU free cycle. If NOACCE = 0, then PE7 is general-purpose I/O. Recall from Chapter 1 that when an instruction is executed, there are four phases: fetch instruction, read memory data, operate, and write memory data. Free cycles occur when the processor is busy executing an instruction and does not require the memory bus. In this situation, the bus interface unit will issue a memory read cycle but ignore the data.

PIPOE is the Pipe Status Signal Output Enable bit. If PIPOE = 1, then PE6:PE5 are outputs and indicate the state of the instruction queue. If PIPOE = 0, then PE6:PE5 are general-purpose I/O. **NECLK** is the No External E Clock bit. As is not the case with the other bits in the PEAR register, we can write to this bit anytime while in Normal and Special modes. If NECLK = 1, PE4 is a general-purpose I/O pin. If NECLK = 0, then PE4 is the external E clock pin. The PE4 external E clock is free-running if ESTR = 0. **LSTRE** is the Low Strobe (LSTRB) Enable bit. The signal LSTRB is used to implement 8-bit writes when running in expanded wide mode. LSTRE = 1 means PE3 is configured as the LSTRB bus control output, otherwise PE3 is a general-purpose I/O pin. After reset in normal expanded mode, LSTRB is disabled to provide an extra I/O pin. If LSTRB is needed, it should be enabled when using expanded narrow mode. External reads do not normally need LSTRB because all 16 data bits can be driven even if the system only needs 8 bits of data. **RDWE** is the Read/Write Enable bit. If RDWE = 1, PE2 is configured as the R/W pin. The R/W signal is 1 during read cycles and 0 during write cycles. If RDWE = 0, then PE2 a general-purpose I/O pin. After reset in normal expanded mode, R/W is disabled to provide an extra I/O pin. If R/W is needed, it should be enabled before any external writes.

> *Observation:* LSTRB is necessary in expanded wide modes but is not needed for expanded narrow interfaces.

The **EBICTL** register contains one bit called **ESTR**, which determines whether the E clock behaves as a simple free-running clock or as a bus control signal that is active only for external bus cycles. We will set ESTR to 1 when interfacing external memory so that E stretches high during stretched external accesses and remains low during non-visible internal accesses. If ESTR is 0, then E never stretches (always free running). This bit has no effect in single-chip modes.

In order to allow fast internal bus cycles to coexist in a system with slower external memories, the 9S12C32 supports the concept of stretched bus cycles. The **EXSTR1 EXSTR0** bits in the **MISC** register specify the amount of stretch for all external memory cycles, as described in Table 9.16 and illustrated in Figures 9.35 through 9.38. While stretching, the CPU state machines are all held in their current state. At this point in the CPU bus cycle, write data would already be driven onto the data bus so that the length of time write data is valid is extended. Read data would not be captured by the system until the E clock falling edge. In the case of a stretched bus cycle, read data is not required until the specified setup time before the falling edge of the stretched E clock. The LSTRB and R/W signals remain valid during the period of stretching (throughout the stretched E high time). **ROMHM** is the FLASH EEPROM Only in Second Half of Memory Map bit. When ROMHM = 1, it disables direct access to the FLASH EEPROM or ROM in the lower half of the memory map. These physical locations of the FLASH EEPROM or ROM can still be accessed through the Program Page window. When ROMHM = 0, the fixed page(s) of FLASH EEPROM or ROM in the lower half of the memory map can be accessed. **ROMON** is the Enable Flash EEPROM bit. If the internal RAM, registers, EEPROM, or BDM ROM (if active) are mapped to the same space as the Flash EEPROM, they will have priority over the Flash EEPROM. ROMON = 0 disables the Flash EEPROM, and ROMON = 1 enables the Flash EEPROM.

Table 9.16
E clock stretching for the external access space.

EXSTR1	EXSTR0	E Clock Stretch
0	0	None (default)
0	1	1
1	0	2
1	1	3

Observation: The address bus portion of the memory cycle is not stretched, just the data portion.

Observation: TCNT, Pulse Accumulator, PWM, RTI, SCI, and SPI are not affected by bus stretching.

Accesses to the PORTA, PORTB, PORTE, and PORTK registers should be made only when the corresponding pins are configured as general purpose I/O. These port and direction registers have no function when a pin is configured as an external memory bus signal. Similarly, we use the Pull-Up Control Register (**PUCR**) to enable pull-up resistors on the corresponding ports when used as a general purpose inputs. We use the Reduced Drive Register (**RDRIV**) to save power when using the corresponding ports as general-purpose outputs.

The alternate function of PE1 is IRQ, and the alternate function of PE0 is XIRQ. In particular, these pins, which are always input, can be used to request interrupts. **IRQEN** is the External IRQ Enable bit. We set IRQEN to one to use PE1 as an external IRQ interrupt request signal. We make IRQE = IRQEN = 1 to configure the 9S12C32 to request IRQ interrupts on falling edges of PE1/IRQ. We set IRQEN = 1 and IRQE = 0 to configure PE1/IRQ as low-level interrupt request signal. We clear IRQEN to use PE1 as a general-purpose input. To use PE0 as a general-purpose input, we simply never enable XIRQ interrupts (i.e., we leave the X-bit in the CCR equal to 1)

Observation: To use low-level IRQ interrupts, the software needs to be able to acknowledge the interrupt making IRQ = 1 again before returning from the interrupt service routine.

Common error: When using PE1 as a regular input, the software must make IRQEN = 0, because the default value of this bit is enable.

The E clock is a timing signal of the MC9S12C32, meaning the rising and falling edges occur at predictable and precise times. The clock frequency is determined by a crystal placed between the EXTAL and XTAL inputs, as shown in Figure 9.39. The MC9S12C32 system requires an external crystal that is twice the E clock frequency. For example, a 4 MHz E clock period is created using an 8 MHz crystal. The values of R_B, C_1, and C_2 are suggested by the crystal manufacturer. The MC9S12C32 allows the software to modify the E clock frequency by programming parameters to the phase-lock loop (PLL). The external components connected to XFC help the PLL circuitry create E clock frequencies faster than the crystal frequency. The values of R_s, C_s, and C_p are calculated from the ratio of the crystal frequency to PLL frequency, and the design equations can be found in the 9S12C data sheet.

Figure 9.39
MC9S12C32 clock circuit, values taken from the Technological Arts Nanocore12.

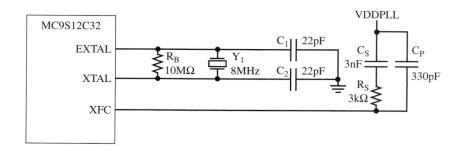

Observation: Microcontroller data sheets suggest a PCB layout pattern for interfacing the crystal.

Figure 9.40 draws a simplified timing diagram, showing both the read and write timing. The bus cycle time is t_1. For example, the 8 MHz crystal (period=125ns) in Figure 9.39

Figure 9.40
Simplified bus timing for
the MC9S12C32 in
expanded mode.

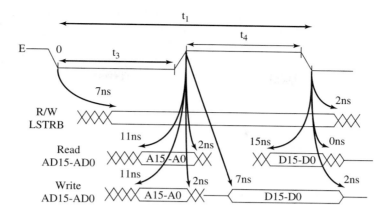

will create a t_1 of 250ns without cycle stretching. Let OSCCLK be the frequency of the crystal, and let PLLCLK be the frequency of the PLL. The **SYNR** and **REFDV** registers determine the PLL frequency,

$$PLLCLK = 2 * OSCCLK * \frac{(SYNR + 1)}{(REFDV + 1)}$$

The **CLKSEL** register contains the PLL Select Bit, **PLLSEL**. When PLLSEL is 1, the system clocks are derived from PLLCLK (bus clock frequency is PLLCLK/2). When PLLSEL is 0, the system clocks are derived from the crystal oscillator (bus clock frequency is OSCCLK/2). Let t_{cyc} be the period of the bus clock.

$$t_{cyc} = \frac{2}{OSCCLK}(if\ PLLSEL = 0) \quad or \quad t_{cyc} = \frac{2}{PLLCLK}(if\ PLLSEL = 1)$$

We can execute software that activates the PLL, creating a bus cycle time as short as 40ns (25MHz). Program 9.1 shows the sequence of steps required to engage the PLL. First, it sets the SYNR and REFDV registers to specify the PLL frequency. Next, it sets the **PLLCTL** register. The Clock Monitor Enable Bit (**CME**) is set to enable the clock monitor, so that slow or stopped clocks will cause a clock monitor reset sequence. The Phase Lock Loop On Bit (**PLLON**) is set to turn on the PLL. The Acquisition Bit (**ACQ**) is set to select the high bandwidth filter. The Self Clock Mode Enable Bit (**SCME**) is set so the detection of crystal clock failure forces the MCU in Self Clock Mode.

Checkpoint 9.4: Modify program 9.1 to run at 24MHZ

When stretching the E clock, confusion sometimes arises. Changing the bus cycle time with the PLL will modify both halves of the bus cycle, t_3 and t_4. On the other hand, when a bus cycle is stretched, times t_1, t_4 are increased, whereas t_3 is fixed. The rising edge of the E clock occurs at the same time regardless of stretching, which will be at $1/2t_{cyc}$. Notice in Figures 9.35 through 9.38 that there is a stretched E clock and an unstretched E clock. The clocks used by SPI, SCI, PulseAcc, PWM, and TCNT are derived from the unstretched E clock. External memory accesses use the stretched E clock. In other words, the only

Program 9.1
MC9S12C32 code to
change the E clock from
8 to 25 MHz.

```
void PLL_Init(void){
    CLKSEL = 0x00;                      // make sure PLL is deselected
    SYNR = 24;  REFDV = 7;              // PLLCLK=2*OSCCLK*(SYNR+1)/(REFDV+1)
    PLLCTL = 0xD1;                      // Turn on PLL
    while((CRGFLG&0x08) == 0){}         // Wait for PLLCLK to stabilize.
    CLKSEL |= 0x80;                     // Switch to PLL clock
```

activity affected by cycle stretching is the time for external memory accesses and hence the execution of the instructions causing those accesses. Internal memory accesses, the timer, and I/O functions are not affected by cycle stretching. In Figure 9.40, the time delays with numerical values (e.g., 7ns, 11ns, etc.) are parameters of the MC9S12C32 and not affected by the PLL or cycle stretching. Let **n** be the number of stretches (0, 1, 2, or 3) as specified by the **EXSTR1 EXSTR0** bits. When interfacing external memory, we need to know when the E clock rises and when it falls.

$$\uparrow E = {}^{1/2}t_{cyc}$$

$$\downarrow E = t_1 = (n + 1) * t_{cyc}$$

> **Observation:** Most of the activities (such as executing instructions, memory access, TCNT, SCI, SPI) are affected by the PLL. Conversely, RTI interrupts always operate using the oscillator crystal.

In expanded narrow mode, the data bus is only 8 bits (on pins PA7–PA0), and the low address is available throughout the cycle (on pins PB7–PB0). In narrow mode, we need only one octal latch to capture the A15–A8 address, as shown in Figure 9.41. In expanded wide mode, the data bus is 16 bits, and we need two octal latches to capture the A15–A0 address from PA7–PA0 on the rising edge of E. The 74FCT374 octal flip flop is chosen because of its speed and the fact that it captures on the rising edge of the clock. The setup and hold times for the 74FCT374 are 2ns and 1.5 ns respectively. In Figure 9.40, we see the 6812 maintains the address 11ns before and 2 ns after the rise of E, so the setup and hold times of the octal D flip-flop are satisfied. The propagation delay from clock to output valid is [2,6.5] ns.

Figure 9.41
Address latch for the MC9S12C32 in expanded narrow and wide modes.

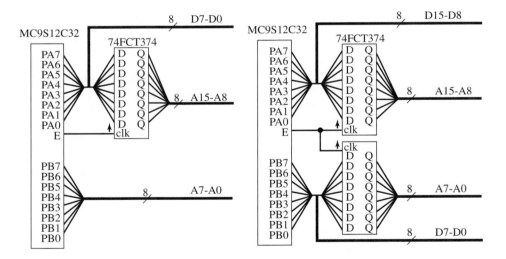

There are three important timing intervals we must determine from the microcomputer timing diagram. The first is the address available (AA) interval. This interval defines when during the cycle the address is valid. In particular, AdV is the time when the address lines A15–A0 are valid, and AdN is the time when the address lines are no longer valid. In order to calculate when the address is available, we first consider the 74FCT374 octal D flips used to capture the address. In expanded wide mode, all 16 bits of the address are clocked into the 74FCT374 on the rising edge of E, so the address is available during the second half of the cycle and continues to be valid until the next rising edge of E. The +[2,6.5] occurs because of the delay in the 74FCT374's.

$$AA = (AdV, AdN) = ({}^{1/2}t_{cyc} + [2,6.5], \; t_1 + {}^{1/2}t_{cyc})$$

In expanded narrow mode the least significant address lines are not latched; therefore, AdV is determined by the latched address, and AdN is determined by the unlatched address, which has the same timing as R/W in Figure 9.40.

$$AA = (AdV, AdN) = (\tfrac{1}{2}\mathbf{t_{cyc}} + [2, 6.5], \ \mathbf{t_1} + 2)$$

The second important timing interval is read data required (RDR). During a read cycle the data is required by the MC9S12C32. Thus, to determine the data required interval, we look in the MC9S12C32 data sheet. For data required, the worst case is the longest interval.

$$RDR = \text{Read Data Required} = (\ \mathbf{t_1} - 15, \mathbf{t_1})$$

The last important timing interval we get from the microcomputer is write data available (WDA). During a write cycle, the data are supplied by the MC9S12C32. Thus, to determine the data available interval we look in the MC9S12C32 data sheet. We will specify the worst-case timing. For write data available, the worst-case is the shortest interval.

$$WDA = \text{Write Data Available} = (\tfrac{1}{2}\mathbf{t_{cyc}} + 7, \mathbf{t_1} + 2)$$

Observation: The speed of CMOS logic is strongly dependent on capacitive load. The 9S12C32 and 74FCT374 timings are specified for capacitive loads of 50pF. If the actual load is larger than 50pF, the timing will be significantly slower.

Table 9.17 presents these three intervals calculated for a 250-ns cycle time with 0,1,2,3 stretches. The address times are given for expanded narrow mode with the shortest interval.

Table 9.17
Timing intervals for the MC9S12C32 with a 4-MHz clock.

n	0	1	2	3
↑ E = $\tfrac{1}{2}t_{cyc}$	125	125	125	125
↓ E = t_1	250	500	750	1000
AA	(131.5,252)	(131.5,502)	(131.5,752)	(131.5,1002)
RDR	(235,250)	(485,500)	(735,750)	(985,1000)
WDA	(132,252)	(132,502)	(132,752)	(132,1002)

Checkpoint 9.5: What are AA, RDR, and WDA with no stretches, if $\mathbf{t_{cyc}}$ is 100ns?

Checkpoint 9.6: What are AA, RDR, and WDA with 2 stretches, if $\mathbf{t_{cyc}}$ is 40ns?

9.6   General Approach to Interfacing

9.6.1
Interfacing to a 6811

The MC68HC11A1 contains the 6811 I/O ports and internal RAM, but it has no internal ROM. We connect MODA, MODB so that the computer executes in expanded mode. The general approach to interfacing a memory to the 6811 is shown in Figure 9.42.

The 74HC573 *Latch* will capture the low address, A7–A0, from the AD7–AD0 signals on the fall of the AS so that the entire 16-bit 6811 address, A15–A0, is available for the duration of the cycle. The faster 74AHC573 can also be used. If the RAM contains 2^n bytes, then the low n bits of the address go directly to the RAM and specify which cell to access. The top $16 - n$ are used by the *Address Decoder* circuit to determine whether or not this RAM should activate during this cycle. If the **CS** signal is true, the *Timing Control Logic* will generate appropriate control signals for the memory. The control signals will activate

Figure 9.42 General approach to memory interfacing on a 6811 in expanded mode.

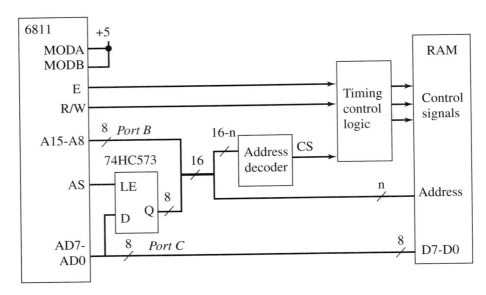

a read operation (drive data out of memory onto the bus) if the **CS** is true and R/W is 1. Similarly, the control signals will activate a write operation (store data from the bus into the memory) if the **CS** is true and R/W is 0.

9.6.2
Interfacing to a
6812 in Expanded
Narrow Mode

In expanded narrow mode, all external memory accesses utilize only 8 bits of data. We connect MODA, MODB so that the 6812 executes in expanded narrow mode. The 6812 will automatically divide 16-bit reads and writes into two separate 8-bit accesses. In many cases, we can interface external devices to the MC68HC812A4 in narrow mode without any additional digital logic. The MC68HC812A4 will use one of its built-in address decoders, shown as **CS** in Figure 9.43. The MC9S12C32 interface will require an address decoder to generate the chip select, **CS** (Figure 9.44). A common reason for using narrow mode is that many I/O devices only support 8-bit accesses.

If the RAM contains 2^n bytes, then the low n bits of the address go directly to the RAM and specify which cell to access. One of the seven built-in negative logic chip selects will be used with the MC68HC812A4. If the **CS** signal is true, the *Timing Control Logic* will generate appropriate control signals for the memory. The control signals will activate a read

Figure 9.43
General approach to memory interfacing on a MC68HC812A4 in narrow expanded mode.

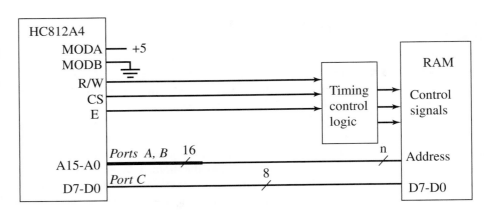

Figure 9.44
General approach to memory interfacing on an MC9S12C32 in narrow expanded mode.

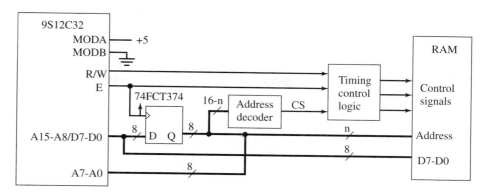

operation (drive data out of memory onto the bus) if the **CS** is true and **R/W** is 1. Similarly, the control signals will activate a write operation (store data from the bus into the memory) if the **CS** is true and **R/W** is 0.

On both 6812 systems, the software must configure the cycle stretching, and enable the E clock and R/W signals as needed in the PEAR register. MC68HC812A4 initialization software should also activate the chip selects.

9.6.3
Interfacing to a 6812 in Expanded Wide Mode

In expanded wide mode, external memory accesses utilize 16 bits of data. We connect MODA, MODB so that the computer executes in expanded wide mode: 16-bit reads and writes to even addresses (aligned) will be performed in a single access; 16-bit reads and writes to odd addresses (misaligned) will be performed in two separate 8-bit accesses. The LSTRB (=0 when low byte data is valid) is used to implement 8-bit memory writes. The advantage of wide mode is execution speed, because opcode fetches will always occur as aligned 16-bit reads (Figures 9.45 and 9.46).

Figure 9.45
General approach to memory interfacing on a MC68HC812A4 in wide expanded mode.

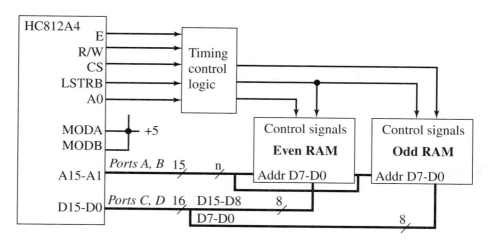

The typical approach to creating a 16-bit memory is to use two 8-bit memories called *even* and *odd*. Assume the size of the memory is 2^n words (16 bits each). The two memory chips share address lines (An–$A1$) and most control signals. One of the seven built-in negative logic chip selects will be used with the MC68HC812A4. The MC9S12C32 will require an address decoder that accepts the most significant $15 - n$ address lines and produces the chip select, **CS**. The 16-bit data bus is divided so that 8 bits go to each chip. The

Figure 9.46
General approach to
memory interfacing on
an MC9S12C32 in wide
expanded mode.

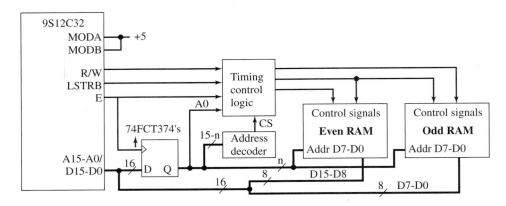

"big endian" format places the most significant data, D15–D8, into the even address and the least significant data, D7–D0, into the odd address. When the 6812 performs an 8-bit read, there is usually no problem if the memory system responds with 16 bits of data (the processor simply takes the data it wants). On the other hand, if the 6812 performs an 8-bit write, it would be a mistake to save all 16 bits D15–D0 into the memory system.

The external signals LSTRB, R/W, and A0 can be used to determine the type of bus access. Accesses to the internal RAM module are the only situation that requires the even and odd bytes to be swapped, LSTRB = A0 = 1 (Table 9.18). The internal RAM is specifically designed to allow misaligned 16-bit accesses in a single cycle. In a misaligned 16-bit access, the data for the address that was accessed are on the low half of the data bus and the data for address +1 are on the high half of the data bus. We will use A0 and LSTRB to handle 8-bit data writes while running in 16-bit data mode.

LSTRB	A0	R/W	Type of Access	Even RAM	Odd RAM
1	0	1	8-bit read of an even address	Activate	No action
0	1	1	8-bit read of an odd address	No action	Activate
1	0	0	8-bit write of an even address	Activate	No action
0	1	0	8-bit write of an odd address	No action	Activate
0	0	1	16-bit read of an even address	Activate	Activate
1	1	1	16-bit read of an odd address*	Not applicable	Not applicable
0	0	0	16-bit write to an even address	Activate	Activate
1	1	0	16-bit write to an odd address*	Not applicable	Not applicable

*Low/high data swapped.

Table 9.18
LSTRB, A0, R/W specify when to activate even and odd RAM modules.

Notice that we should activate the *even RAM* when $\overline{CS}$ = 0 and A0 = 0, and we should activate the *odd RAM* when $\overline{CS}$ = 0 and LSTRB = 0. These digital logic functions can be implemented with one 74AHC32 package. AHC or FCT logic is used because the propagation delay is less than 10 ns (Figure 9.47).

On both 6812 systems, the software must configure the cycle stretching and enable the E clock, LSTRB, and R/W signals as needed in the PEAR register. The NDRC bit in the MISC register can be set to configure the 512 bytes following the register space (typically $0200–$03FF) for narrow mode. When NDRC is set, all addresses except $0200–$03FF can utilize the 16-bit data bus. MC68HC812A4 initialization software should also activate the chip selects.

Figure 9.47
Circuit needed to handle even and odd 8-bit accesses.

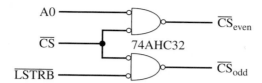

9.7 **Memory Interface Examples**

9.7.1 32K PROM Interface

The objective of this example is to interface a 32K by 8 bit PROM to the microcomputer. We will place the 27256-25 PROM at locations $8000 to $FFFF. The **command portion** determines which CPU cycles will activate the PROM. The 27256-25 PROM will drive the data bus whenever both **CE*** and **OE*** are 0. The other three values of **CE*** and **OE*** will disable the PROM. The PROM chip requires less power (I_{cc} is the + 5 V supply current) whenever **CE*** is high (Table 9.19).

Table 9.19
Function table of a typical PROM.

CE*	OE*	Function	I_{cc} (mA)
1	1	Off	40
1	0	Off	40
0	1	Off	100
0	0	Drive data bus	100

Using full decoding the address **SELECT** line in positive logic would be **SELECT** = A15. Because this chip is read only, the control lines will be activated only during read cycles (R/W=1) from this PROM (A15 = 1). To conserve power (lower I_{cc}), then the **CE*** = 1 state is used to deselect the PROM (Table 9.20). The **timing portion** of the interface determines the time during an active cycle when a control signal should rise and fall. The objective is to design the interface such that the data available interval overlaps the data required interval.

Table 9.20
Command table for the 27256 PROM interface.

SELECT	R/W	Function	CE*	OE*
0	0	Off	1	X
0	1	Off	1	X
1	0	Off	1	X
1	1	Drive data bus	0	0

During a read cycle, the data are supplied by the 27256-25. Thus, to determine the data available interval we look at the 27256-25 data sheet. We will specify the worst-case timing. For data available, the worst case is the shortest interval (Figure 9.48).

$$\text{Read data available} = (\text{later } (\text{AdV} + t_{ACC}, \downarrow CE^* + t_{CE}, \downarrow OE^* + t_{OE}),$$
$$\text{earlier (AdN} + t_{OH}, \uparrow CE^* + t_{OH}, \uparrow OE^* + t_{OH}))$$

To conserve power (minimize I_{cc}), we let CE* be 0 only during read cycles to $8000 to $FFFF. If we simply grounded CE*, the interface would be faster but would require more power. From the 27256-25 data sheet, $t_{ACC} = 250$, $t_{CE} = 250$ $t_{OE} = 100$, $t_{OH} = 0$.

Figure 9.48
Read timing for the
27256-25 32K PROM
chip.

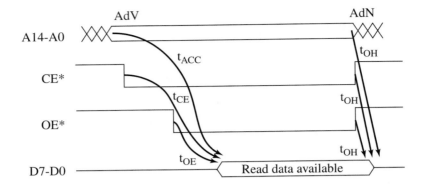

9.7.1.1
32K PROM Interface to a
MC68HC11A1

Entering the AdV, AdN times of a 2-MHz 6811, we calculate

$$\text{Read data available} = (\text{later} (181 + 250, \downarrow CE^* + 250, \downarrow OE^* + 100),$$
$$\text{earlier} (533 + 0, \uparrow CE^* + 0, \uparrow OE^* + 0))$$

During a read cycle the data are required by the 6811.

$$\text{Read data required} = (t_1 - t_4 - t_{17}, t_1 + t_{18}) = (450, 510)$$

We must choose the timing of the fall and rise of both CE* and OE* to make the data available interval overlap the data required window. Note that since CE* and OE* are active low, they will fall, then rise, during the active cycles. Thus,

$181 + 250$	≤ 450	and	533	≥ 510
$\downarrow CE^* + 250$	≤ 450	and	$\uparrow CE^*$	≥ 510
$\downarrow OE^* + 100$	≤ 450	and	$\uparrow OE^*$	≥ 510

or

$\downarrow CE^*$	≤ 200	and	$\uparrow CE^*$	≥ 510
$\downarrow OE^*$	≤ 350	and	$\uparrow OE^*$	≥ 510

The following signal is constructed inside the PROM chip:

$$\text{ON} = \overline{CE^*} \cdot \overline{OE^*}$$

The data bus is driven when ON equals 1. Since the 6811 is a synchronous computer, the operation of the ON signal must be synchronized to the 6811. There are three ways to synchronize the ON signal:

1. Synchronize only OE*
2. Synchronize only CE*
3. Synchronize both CE* and OE*

Performance tip: In general, we will synchronize the control signal with the shortest access time.

It will be less expensive to synchronize just one of the signals. Because $t_{OE} < t_{CE}$, it will be faster to synchronize OE*. This means that CE* will be controlled by bus signals A15 and R/W. The fall of CE* will be AdV plus the maximum LS gate delay (+10 ns). Similarly, the rise of CE* will be AdN plus the minimum LS gate delay (0). Assuming one 10-ns gate delay, we see that the CE* timing is satisfied,

$$\downarrow CE^* = AdV_{15-8} + 10 = 146 \leq 200 \quad \text{and} \quad \uparrow CE^* = AdN_{15-8} \geq 510$$

Since the address bus is indeterminate during the first 181 ns of every cycle, we can never design a control signal that changes with the beginning of the cycle (at 0 ns). Thus, there is only one choice for the high-to-low and low-to-high transitions of OE*: OE* must fall at 260 and rise at 510 (Table 9.21). The beginning of RDA is determined by AdV (bits 7–0) and the end by $\uparrow$OE*:

$$\text{Read data available} = (\text{AdV} + 250, \uparrow\text{OE*} + 0) = (431, 510)$$

Table 9.21
Read timing table for the 27256 PROM interface.

E	OE*
0	1
1	0

Next we find the Boolean expressions that will generate CE* and OE* from E, AS, R/W, and SELECT. The fully decoded positive logic select signal is SELECT = A15. The command table is expanded to include the timing information (Table 9.22). Karnaugh maps are used to determine the logic equations. 74LSxx low-power Schottky logic has a good balance of speed, input loading current (I_{IL}), and power supply current. 74AHCxx advanced high-speed CMOS logic should be used for battery-operated systems because of its low power requirements. 74FCTxx logic is used when speed is important.

Table 9.22
Combined timing table for the 27256 PROM interface.

E	R/W	A15	ON	CE*	OE*	
0	0	0	0	1	X	Address not on the chip
1	0	0	0	1	X	
0	1	0	0	1	X	Address not on the chip
1	1	0	0	1	X	
0	0	1	0	1	X	Correct address
1	0	1	0	1	X	But it is a write
0	1	1	0	0	1	
1	1	1	1	0	0	Enable ROM

Next, we build the interface (Figure 9.49). Finally, we draw the *Combined read cycle timing diagram* and use it to verify that read data available overlaps read data required (Figure 9.50).

9.7.1.2
32K PROM Interface to a
MC68HC812A4

Two of the issues when interfacing to a 6812 are to determine data bus width (narrow or wide) and the number of cycles to stretch. For simplicity we start with an 8-bit narrow example. To stretch the E clock we will have to use one of the seven built-in chip selects (e.g., either CSP0 or CSP1). For now, we will develop the timing equations, and then we will use them to determine the cycle stretching requirements. The AdV time is 60 ns, independent of stretching, while AdN will be $t_1 + 20$ ns. We therefore calculate

Read data available $= (\text{later } (60 + t_{ACC}, \downarrow\text{CE*} + t_{CE}, \downarrow\text{OE*} + t_{OE}), \text{earlier } (\text{AdN} + t_{OH}, \uparrow\text{CE*} + t_{OH}, \uparrow\text{OE*} + t_{OH}))$
$= (\text{later } (310, \downarrow\text{CE*} + 250, \downarrow\text{OE*} + 100), \text{earlier } (t_1 + 20 + 0, \uparrow\text{CE*} + 0, \uparrow\text{OE*} + 0))$

During a read cycle the data are required by the 6812.

$$\text{RDR} = \text{read data required} = (t_1 - t_{11}, t_1 + t_{12}) = (t_1 - 30, t_1)$$

Figure 9.49 Interface between a 6811 and a 27256-25 32K PROM chip.

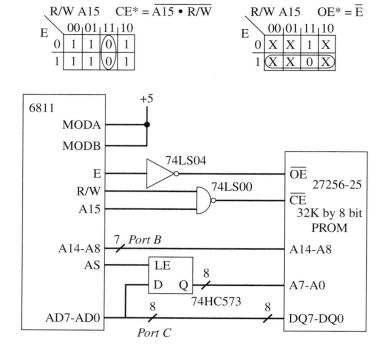

Figure 9.50 Timing of a 6811 and a 27256-25 32K PROM chip.

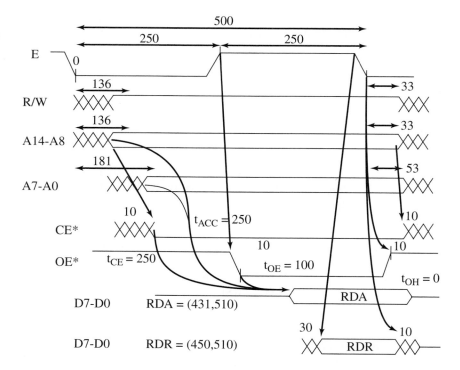

We must choose the timing of the fall and rise of both CE* and OE* to make the data available interval overlap the data required window. Note that since CE* and OE* are active low, they will fall, then rise, during the active cycles. Thus,

310	$\leq t_1 - 30$	and	$t_1 + 20$	$\geq t_1$
$\downarrow$CE* $+ 250$	$\leq t_1 - 30$	and	$\uparrow$CE*	$\geq t_1$
$\downarrow$OE* $+ 100$	$\leq t_1 - 30$	and	$\uparrow$OE*	$\geq t_1$

First line ($340 \leq t_1$) tells us that two extra cycles are needed to stretch the access time to 375 ns. The following signal is constructed inside the PROM chip:

$$ON = \overline{CE^*} \cdot \overline{OE^*}$$

The data bus is driven when ON equals 1. Since the 6812 is a synchronous computer, the operation of the ON signal must be synchronized to the 6812. There are three ways to synchronize the ON signal:

1. Synchronize only OE*
2. Synchronize only CE*
3. Synchronize both CE* and OE*

It will be less expensive to synchronize just one of the signals. Because $t_{OE} < t_{CE}$, it will be faster to synchronize OE*. On the other hand, it will be cheaper to synchronize CE* by connecting it directly to CSP0. This means CE* will be controlled by timing signal CSP0. The fall of CE* will be at 60 ns and, the rise of CE* will be at $t_1 + t_{28}$, where t_{28} is [0,10]. The CE access time must be satisfied:

$$\downarrow CE^* + 250 \leq t_1 - 30 \qquad \text{and} \qquad \uparrow CE^* \geq t_1$$

or

$$60 + 250 \leq t_1 - 30 \qquad \text{and} \qquad t_1 + [0,10] \geq t_1$$

Again we need two extra cycles to stretch the access time to 375 ns. In this particular interface we will ground OE*, so the timing delays with OE* need not be calculated. Now that we have decided to use two-cycle stretching, we can calculate the actual RDA and RDR intervals.

Read data available = (310, 375)

Read data required = (345, 375)

Next we find the Boolean expressions that will generate CE* and OE* from E, CSP0, and R/W. In this simple case, we connected CE* to CSP0 and OE* to ground (Figure 9.51). The MC68HC812A4 will come out of reset in expanded mode, with CSP0 enabled having three-cycle stretching. The reset software should enable any additional chip selects and

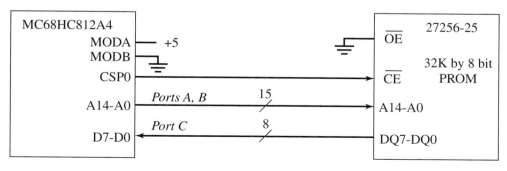

Figure 9.51
Interface between a MC68HC812A4 and a 27256-25 32K PROM chip.

change the cycle stretching on CSP0 to two cycles. Last, we draw the *Combined read cycle timing diagram* and use it to verify that read data available overlaps read data required (Figure 9.52).

Figure 9.52
Timing of a
MC68HC812A4 and a
27256-25 32K PROM
chip.

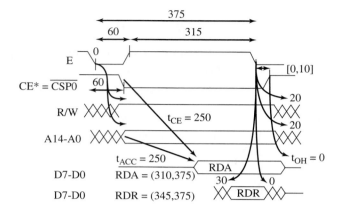

The software to activate this interface is shown in Program 9.2.

Program 9.2
Software to configure the
chip select for the external
PROM.

```
void PROM_Init(void){
    CSSTR0=(CSSTR0&0xF3)|0x08;} // 2 cycle stretch on CSP0
```

9.7.2
8K RAM Interface

In this example we interface an 8K by 8-bit static RAM to the microcomputer using the Motorola MCM60L64. The memory will be placed at \$6000 to \$7FFF. The **command portion** determines which CPU cycles will activate the RAM. Using full decoding the address **Select*** line in negative logic is

$$\text{Select*} = \overline{A15} \cdot A14 \cdot A13$$

A 74HC138 3 input to 8 output decoder could be used to generate the chip selects for multiple 8K devices in a single system. The Y3 output of the decoder implements the above address decoder. This RAM chip has both a positive logic (**E2**) and a negative logic ($\overline{E1}$) chip select. Negative logic is used because the output of the 74HC138 is in negative logic. With E2 tied to + 5 V, the RAM has the functions shown in Table 9.23.

Table 9.23
Function table for the
MCM60L64 RAM.

$\overline{E1}$	$\overline{W}$	$\overline{G}$	Function
1	X	X	Disabled, low I_{cc} = 30 μA
0	1	1	Disabled, high I_{cc} = 3 mA
0	0	X	Write data into RAM
0	1	0	Read data out of RAM

To reduce power, we will let $\overline{E1}$ = 0 only during accesses to this RAM. We will synchronize the read and write functions by synchronizing either $\overline{E1}$ or both $\overline{W}$ and $\overline{G}$. A read operation will occur when both $\overline{E1}$ and $\overline{G}$ are 0 and $\overline{W}$ is 1.

$$RD = \overline{\overline{E1} \cdot \overline{W} \cdot \overline{G}}$$

Similarly, a write operation will occur when both $\overline{E1}$ and $\overline{W}$ are zero.

$$WR = \overline{\overline{E1} \cdot \overline{W}}$$

Table 9.24

Command table for the MCM60L64 RAM interface.

Select	R/W	RD	WR	$\overline{\text{E1}}$	$\overline{\text{W}}$	$\overline{\text{G}}$	Function
0	0	0	0	1	X	X	Disable because address not on the chip
0	1	0	0	1	X	X	Disable because address not on the chip
1	0	0	1	0	0	X	Write cycle to this RAM
1	1	1	0	0	1	0	Read cycle from this RAM

The signals RD and WR are generated internal to the RAM. The command part of the design will have RD = 1 only during read cycles from this RAM, and WR = 1 only during write cycles to this RAM (Table 9.24).

The **timing portion** of the interface determines the timing of the rise and fall of the control signals. The objective is to design the interface such that the data available interval overlaps the data required. Since this is a RAM, we consider both the read and write cycles. Data are read from the memory when $\overline{\text{E1}} = 0$, $\overline{\text{G}} = 0$, and $\overline{\text{W}} = 1$. During a read cycle, the data are supplied by the 60L64. Thus, to determine the read data available interval we look in the Motorola MCM60L64 data sheet. We will specify the worst-case timing. For read data available, the worst case is the shortest interval (Figure 9.53).

$$\text{Read data available} = (\text{later (AdV} + t_{\text{AVQV}}, \downarrow \overline{\text{E1}} + t_{\text{E1LQV}}, \downarrow \overline{\text{G}} + t_{\text{GLQV}}),$$
$$\text{earlier (AdN} + t_{\text{AXQX}}, \uparrow \overline{\text{E1}} + t_{\text{E1HQZ}}, \uparrow \overline{\text{G}} + t_{\text{GHQZ}}))$$

Figure 9.53

Read timing for the 60L64 8K RAM chip.

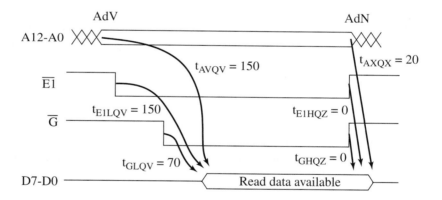

where AdV is the time when the address lines A12–A0 are valid and AdN is the time when the address lines are no longer valid. From the 60L64 data sheet $t_{\text{AVQV}} = 150$, $t_{\text{E1LQV}} = 150$, $t_{\text{GLQV}} = 70$, $t_{\text{AXQX}} = 20$, $t_{\text{E1HQZ}} = 0$, and $t_{\text{GHQZ}} = 0$.

During a write cycle the data are required by the 60L64. Thus, to determine the write data required interval, we look in the 60L64 data sheet. Since the write operation occurs on the overlap of $\overline{\text{E1}}$ and $\overline{\text{W}}$, the first of these two signals to rise will cause data to be written into the memory. There are two possible timing diagrams for the write cycle. If $\overline{\text{E1}}$ is unsynchronized negative logic and $\overline{\text{W}}$ is synchronized negative logic, then it is the rise of $\overline{\text{W}}$ that stores data into the RAM (Figure 9.54).

$$\text{Write data required} = (\uparrow \overline{\text{W}} - t_{\text{DVWH}}, \uparrow \overline{\text{W}} + t_{\text{WHDX}}) = (\uparrow \overline{\text{W}} - 60, \uparrow \overline{\text{W}})$$

For write data required, the worst case is the longest interval. From the 60L64 data sheet, the setup time $t_{\text{DVWH}} = 60$, and the hold time $t_{\text{WHDX}} = 0$. Note also that the address must be valid whenever $\overline{\text{E1}} = 0$. If $\overline{\text{E1}}$ is synchronized negative logic and $\overline{\text{W}}$ is unsynchronized negative logic, then it is the rise of $\overline{\text{E1}}$ that stores data into the RAM (Figure 9.55).

$$\text{Write data required} = (\uparrow \overline{\text{E1}} - t_{\text{DVWH}}, \uparrow \overline{\text{E1}} + t_{\text{WHDX}}) = (\uparrow \overline{\text{E1}} - 60, \uparrow \overline{\text{E1}})$$

Figure 9.54
Write timing controlled
by W for the 60L64 8K
RAM chip.

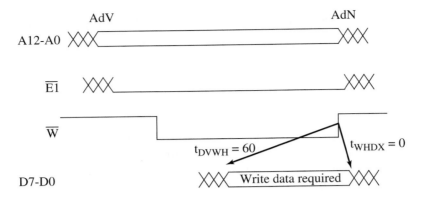

Figure 9.55
Write timing controlled
by E1 for the 60L64 8K
RAM chip.

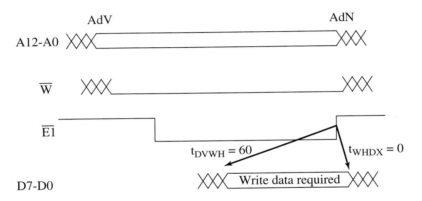

9.7.2.1
8K RAM Interface to a
MC68HC11A1

To simplify the RAM interface to the 6811, we can let $\overline{\text{E1}}$ = **Select*** and make $\overline{\text{G}}$ and $\overline{\text{W}}$ synchronize the read and write cycles, respectively. Recall that for a 6811 running at 2 MHz:

$$AA = (AdV, AdN) = (181, 533)$$

$$
\begin{aligned}
\text{Read data available} &= (\text{later} (AdV + t_{AVQV}, \downarrow \overline{\text{E1}} + t_{E1LQV}, \downarrow \overline{\text{G}} + t_{GLQV}), \\
&\quad \text{earlier} (AdN + t_{AXQX,} \uparrow \overline{\text{E1}} + t_{E1HQZ}, \uparrow \overline{\text{G}} + t_{GHQZ})) \\
&= (\text{later} (181 + 150, \downarrow \overline{\text{E1}} + 150, \downarrow \overline{\text{G}} + 70), \\
&\quad \text{earlier} (533 + 20, \uparrow \overline{\text{E1}}, \uparrow \overline{\text{G}}))
\end{aligned}
$$

During a read cycle the data are required by the 6811. Thus to determine the data required interval, we look in the 6811 data sheet. For read data required, the worst case is the longest interval. Recall that

$$\text{Read data required} = (t_1 - t_4 - t_{17}, t_1 + t_{18}) = (450, 510)$$

We must choose the timing of the fall and rise of $\overline{\text{E1}}$ and $\overline{\text{G}}$ to make the read data available interval overlap the read data required interval. We must make $\overline{\text{W}}$ =1 during a read cycle. Note that since $\overline{\text{G}}$ is active low, it will fall, then rise, during the active cycles. During the inactive cycles, $\overline{\text{G}}$ will be high (with no transitions). Thus,

$181 + 150$	≤ 450	and	$533 + 20$	≥ 510
$\downarrow \overline{\text{E1}} + 150$	≤ 450	and	$\uparrow \overline{\text{E1}}$	≥ 510
$\downarrow \overline{\text{G}} + 70$	≤ 450	and	$\uparrow \overline{\text{G}}$	≥ 510

or

$$\begin{array}{llll} \downarrow \overline{E1} & \leq 300 & \text{and} & \uparrow \overline{E1} & \geq 510 \\ \downarrow \overline{G} & \leq 380 & \text{and} & \uparrow \overline{G} & \geq 510 \end{array}$$

In our design, $\overline{E1}$ is derived only from the address signals A15, A14, and A13. Thus

$$\downarrow \overline{E1} = 230 - 94 + 36 = 172 \qquad \text{and} \qquad \uparrow \overline{E1} = 500 + 33 + 0 = 533$$

The 36 ns represents the maximum gate delay in the 74HC138 decoder, while the 0 ns represents the minimum gate delay in the 74HC138 decoder. Luckily, the actual timing of $\overline{E1}$ satisfies the constraints. If the actual $\downarrow \overline{E1}$ were greater than 300, then one would have to:

- Use a 74AHC138 advanced high-speed CMOS decoder
- Slow down the 6811 by increasing the E period
- Decrease t_{E1LQV} by spending more money on a faster RAM chip

Since the 6811 is a synchronous computer, the $\overline{G}$ control must be synchronized to the 6811. Since the 6811 has only the E clock, only two choices exist for the high-to-low and low-to-high transitions of $\overline{G}$. Assuming a gate delay between E and $\overline{G}$ of 10 ns, the choices are 260 and 510. $\overline{G}$ must fall at 260 and rise at 510. The beginning of RDA is determined by AdV (bits 7–0) and the end by $\uparrow \overline{G}$:

$$\text{Read data available} = (\text{AdV} + 150, \uparrow \overline{G} + 0) = (331, 510)$$

Observation: Synchronizing the memory read signals to the E clock on a 6811 interface prevents the memory from driving the data bus during the first half of the cycle, when the 6811 is outputting the low address on these same lines.

Common error: If you do not synchronize driving the data bus on a multiplexed address/data computer like the 6811 and MC9S12C32, then memory data might be driven onto the bus during the first half of the cycle and conflict with the address being driven at that time.

Data are written into the memory when $\overline{E1} = 0$, $\overline{G} = 1$, and $\overline{W} = 0$. During a write cycle, the data are supplied by the 6811. For write data available, the worst case is the shortest interval. Recall that

$$\text{Write data available} = (t_2 + t_4 + t_{19}, t_1 + t_{21}) = (378, 533)$$

We must choose the timing of the fall and rise of $\overline{W}$ to make the write data available interval overlap the write data required window. We will make $\overline{G} = 1$ during a write cycle (it will still write even if it were 0). Note that since $\overline{W}$ is active low, it will fall, then rise, during the active cycles. During the inactive cycles, $\overline{W}$ will be high (with no transitions).

$$378 \leq \uparrow \overline{W} - 60 \qquad \text{and} \qquad \uparrow \overline{W} \leq 533$$

Thus,

$$438 \leq \uparrow \overline{W} \leq 533$$

Since the 6811 is a synchronous computer, the $\overline{W}$ control line must be synchronized to the E clock. If we did not synchronize $\overline{W}$, then we could not guarantee that write data available will overlap write data required. In some situations a rising edge might not even occur at all. For example, consider the case where $\overline{W}$ is unsynchronized and equal to R/W. The STD $8000 instruction will produce successive write cycles to $8000 and $8001. In this case there may be only one rise of $\overline{W}$ and even that one rise may be too late to properly store data into the RAM (Figure 9.56).

Common error: If you do not synchronize clocking data into a memory during a write cycle, then data might not be stored properly.

Figure 9.56
Unsynchronized signals do not have guaranteed times of their rise and fall.

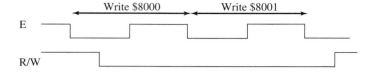

There is only one choice that satisfies the design. The fall of $\overline{W}$ must be 260, and the rise of $\overline{W}$ must be 510. Since the rise of $\overline{W}$ is 510, the WDR becomes (450, 510). We combine the command, and timing aspects of the design to find the Boolean expression that will generate the control signals: $\overline{E1}$, $\overline{G}$, and $\overline{W}$ (Table 9.25).

$$\overline{E1} = \textbf{Select*}$$
$$\overline{G} = \text{R/W} \cdot \text{E}$$
$$\overline{W} = \overline{\text{R/W}} \cdot \text{E}$$

Table 9.25
Combined timing table for the MCM60L64 RAM interface.

E	R/W	Select* = $\overline{E1}$	RD	WR	$\overline{G}$	$\overline{W}$	
0	0	1	0	0	X	X	Address not on the chip
1	0	1	0	0	X	X	
0	1	1	0	0	X	X	Address not on the chip
1	1	1	0	0	X	X	
0	0	0	0	0	1	1	Write cycle
1	0	0	0	1	1	0	
0	1	0	0	0	1	1	Read cycle
1	1	0	1	0	0	1	

The next step is to build the interface (Figure 9.57). To verify proper read cycle timing we draw the *Combined read cycle timing diagram* (Figure 9.58). Notice that read data available overlaps read data required.

Figure 9.57
Interface between the 6811 and the 60L64 8K RAM chip.

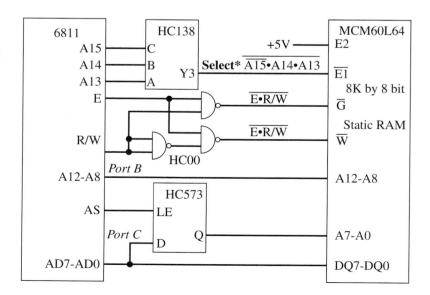

Figure 9.58
Read timing diagram of
the 6811 and the 60L64
8K RAM chip.

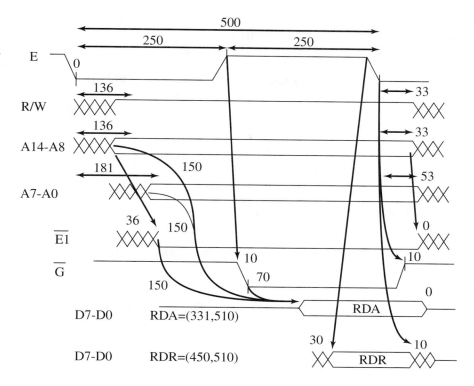

To verify proper write cycle timing we draw the *Combined write cycle timing diagram* (Figure 9.59). Notice that write data available overlaps write data required.

Figure 9.59
Write timing diagram of
the 6811 and the 60L64
8K RAM chip.

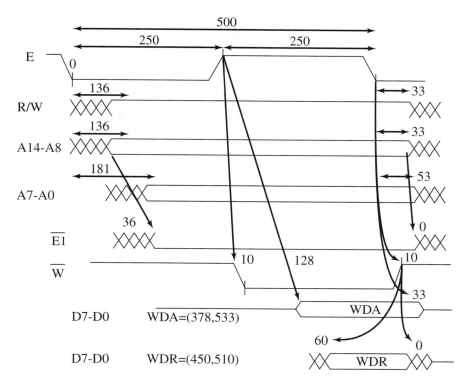

To simplify the RAM interface to the MC68HC812A4, we will operate in expanded narrow mode and connect $\overline{E1}$ to the built-in CSD. We will have to configure CSD for the 32K bytes space $0000 to $7FFF. This will greatly simplify the interface, and we will still be able to add other external devices using other chip selects. Another simplification we will make is to ground $\overline{G}$. Because $\overline{E1}$ is synchronized to CSD, both the read and write operations will be synchronized. Just like the 32K PROM interface to the 6812 earlier, we must select the proper amount of cycle stretching. Recall that for a MC68HC812A4 running at 8 MHz,

$$AA = (AdV, AdN) = (t_5, t_1 + t_6) = (60, t_1 + 20)$$

Note that since $\overline{G}$ is grounded, we will neglect its timing constraints. Since $\overline{E1}$ is connected to CSD, $\downarrow\overline{E1}$ will occur at 60 ns, and $\uparrow\overline{E1}$ will occur at $t_1 + t_{28}$, where t_{28} is [0,10]. Entering this information into the memory timing,

$$\text{Read data available} = (\text{later } (AdV + t_{AVQV}, \downarrow\overline{E1} + t_{E1LQV}), \text{earlier } (AdN + t_{AXQX}, \uparrow\overline{E1} + t_{E1HQZ}))$$
$$= (\text{later } (60 + 150, 60 + 150), \text{earlier } (t_1 + 20 + 20, t_1 + 10))$$

During a read cycle the data are required by the 6812. Thus to determine the read data required interval, we look in the 6812 data sheet. For read data required, the worst case is the longest interval. Recall that

$$RDR = \text{read data required} = (t_1 - t_{11}, t_1 + t_{12}) = (t_1 - 30, t_1 + 0)$$

We must choose the number of cycle stretches to make the read data available interval overlap the read data required interval. We must make $\overline{W} = 1$ during a read cycle. Thus,

Address	$60 + 150 \leq t_1 - 30$	and	$t_1 + 40 \geq t_1$
E1	$60 + 150 \leq t_1 - 30$	and	$t_1 + [0,10] \geq t_1$

First column ($240 \leq t_1$) tells us that one extra cycle is needed to stretch the access time to 250 ns. If the read data available did not overlap the read data required, then one would have to:

- Increase the number of cycle stretches
- Slow down the 6812 by increasing the E period
- Decrease t_{E1LQV} by spending more money on a faster RAM chip

The beginning of RDA is determined by $\downarrow\overline{E1}$ and the end by $\uparrow\overline{E1}$:

$$\text{Read data available} = (\downarrow\overline{E1} + 150, \uparrow\overline{E1}) = (210, 260)$$

Observation: You do not have to synchronize driving the data bus on a nonmultiplexed address/data computer like the MC68HC812A4, because the data bus is dedicated at all times.

Data are written into the memory when $\overline{E1} = 0$, and $\overline{W} = 0$. During a write cycle, the data are supplied by the 6812. For write data available, the worst case is the shortest interval. Recall that

$$WDA = \text{write data available} = (t_2 + t_{13}, t_1 + t_{14}) = (106, t_1 + 20)$$

Since we have chosen to synchronize $\overline{E1}$,

$$\text{Write data required} = (\uparrow\overline{E1} - 60, \uparrow\overline{E1})$$

where $\uparrow\overline{E1}$ is $t_1 + 10$. The number of cycle stretches will also affect whether or not the write data available interval overlaps the write data required interval.

$$106 \leq t_1 + 10 - 60 \qquad \text{and} \qquad t_1 + 10 \leq t_1 + 20$$

Thus,

$$156 \leq t_1$$

Therefore the one-cycle stretch needed for the read cycle also is sufficient for the write cycle.

Observation: In most situations the read cycle timing is more critical than the write cycle timing.

Common error: If you do not synchronize clocking data into a memory during a write cycle, then data might not be properly stored.

We combine the command and timing aspects of the design to find the Boolean expression that will generate the control signals: $\overline{E1}$, $\overline{G}$, and $\overline{W}$ (Table 9.26).

$$\overline{E1} = CSD$$

$$\overline{G} = 0$$

$$\overline{W} = R/W$$

Table 9.26
Combined timing table for the MCM60L64 RAM interface.

E	R/W	CSD	RD	WR	$\overline{E1}$	$\overline{G}$	$\overline{W}$	
0	0	1	0	0	1	X	X	Address not on the chip
1	0	1	0	0	1	X	X	
0	1	1	0	0	1	X	X	Address not on the chip
1	1	1	0	0	1	X	X	
0	0	1	0	0	1	X	X	Write cycle
1	0	0	0	1	0	X	0	
0	1	1	0	0	1	X	1	Read cycle
1	1	0	1	0	0	0	1	

The next step is to build the interface (Figure 9.60). To verify proper read cycle timing we draw the *Combined read cycle timing diagram* (Figure 9.61). Notice that read data available overlaps read data required, although it is kind of close, and you may elect to be more conservative and increase the cycle stretching to two cycles.

Figure 9.60
Interface between the MC68HC812A4 and the 60L64 8K RAM chip.

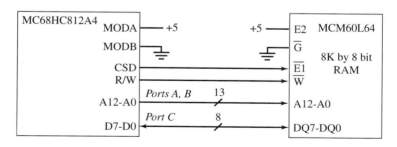

To verify proper write cycle timing we draw the *Combined write cycle timing diagram* (Figure 9.62). Notice that write data available overlaps write data required. One of the difficulties in interfacing additional RAM to a MC68HC812A4 is how to configure the system out of reset. If the above circuit is constructed, the system will come out of reset in expanded narrow mode (MODA = 1, MODB = 0). Unfortunately, the built-in 4K EEPROM is mapped to $1000 to $1FFF in expanded narrow mode. Therefore no memory exists at $FFFE, and the software cannot start. There are two solutions to this situation:

1. Include external PROM (Section 9.7.1.2) and come out of reset into expanded narrow mode and let the software:

   Enable CSD, R/W outputs
   Select one-cycle stretch on CSD

Figure 9.61
Read timing of the
MC68HC812A4 and the
60L64 8K RAM chip.

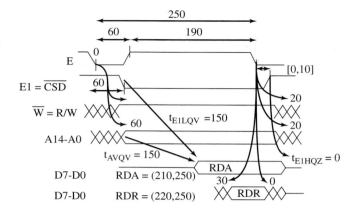

Figure 9.62
Write timing of the
MC68HC812A4 and the
60L64 8K RAM chip.

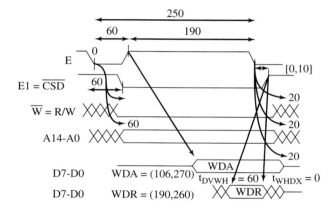

2. Come out of reset in special single-chip mode (BKGD = MODA = MODB = 0) and let the software:

Switch over the special expanded narrow mode
Enable CSD, R/W outputs
Select one-cycle stretch on CSD

The software to activate this interface is shown in Program 9.3.

Program 9.3
Software to configure
the chip select for the
external RAM.

```
void RAM_Init(void){
        MODE=0x3B                         // special expanded narrow mode
        PEAR=0x2C;                        // enable E, R/W, LSTRB(not needed)
        WINDEF=WINDEF&0x7F;               // disable DPAGE
        CSCTL0=CSCTL0|0x10;               // enable CSD
        CSCTL1=CSCTL1|0x10;               // CSD $0000 to $7FFF
        CSSTR0=(CSSTR0&0xFC)|0x01;}       // 1 cycle stretch on CSD
```

Checkpoint 9.7: How fast would the RAM read access time need to be to run the MC68HC814A4/MCM60L64 interface in Figure 9.60 with no cycle stretches?

9.7.2.3. 8K RAM
Interface to a
MC9S12C32

The design goal of this section is to interface an 8K external RAM to the MC9S12C32, minimizing cost. The least expensive way to interface external RAM to the MC9S12C32 is to use expanded narrow mode. We will place the 8K RAM at $8000-$9FFF, because there are no internal devices at these addresses. The address map of our system will be

$0000-$03FF	I/O ports
$3800-$3FFF	Internal RAM
$4000-$7FFF	Internal EEPROM
$8000-$9FFF	External RAM
$C000-$FFFF	Internal EEPROM

First, we design a minimal cost address decoder for $8000 to $9FFF using a 74FCT139. We need addresses A15 and A14 to differentiate our external RAM from the other three devices. The positive logic chip select will be A15 • $\overline{A14}$. Just; with the MC68HC812A4 interfaces earlier, we must select the proper amount of cycle stretching. Recall that for a MC9S12C32 running at 4 MHz ↑**E** will be 125ns, and ↓**E** will be (**n** + 1)* 250ns where **n** is the number of stretches. The propagation delay from clock to output change through the 74FCT374 has a minimum of 2ns and a maximum of 6.5ns. In expanded narrow mode, the address available interval begins with the high address being clocked into the 74FCT374 and ends with the low address end time, which is 2ns after ↓**E**. Thus,

$$AA = (AdV, AdN) = (125 + [2,6.5], \downarrow E + 2) = ([127,131.5], \downarrow E + 2)$$

The high address, which is used by the address decoder, is available all the way through until the next rise of **E**.

$$AA_{15-8} = ([127, 131.5], \downarrow E + 125)$$

To guarantee proper timing on the read and write cycles, we have three options. We must synchronize E2, $\overline{E1}$, or both $\overline{G}$ and $\overline{W}$. The RAM has both a positive logic (E2) and a negative logic ($\overline{E1}$) chip select. Because we are going to use the positive logic chip select for the timing, we will use the negative logic chip select for the address decoder. We choose to synchronize E2 because it is fastest. We need to connect directly to E2 so that the end of WDA overlaps WDR. In summary, we will have

E2	positive logic synchronized to E
$\overline{E1}$	negative logic unsynchronized address decoder
$\overline{G}$	negative logic unsynchronized
$\overline{W}$	negative logic unsynchronized

With time-multiplexed signals, we have to be careful to avoid address/data collisions during a read cycle. All read and write timing will be controlled by the E clock (without any gate delays) by connecting the E directly to E2. In this way, the memory data output will not collide with the microcomputer address output during a read cycle when E = 0. Assuming 9 ns 74FCT139 gate delay max and 1.5 ns delay minimum, $\overline{E1}$ falls at [127,131.5] + [1.5,9] ns and rises at ↓**E** + 125 + [1.5,9] ns. The worst case timing is the latest (maximum = 140.5) time for ↓$\overline{E1}$ and the earliest (minimum = ↓**E** + 126.5) time for ↑$\overline{E1}$. $\overline{G}$ will be grounded, so it is removed from the timing equation. Entering this information into the memory timing

$$RDA = (later (AdV + t_{AVQV}, \downarrow \overline{E1} + t_{E1LQV}, \uparrow E2 + t_{E2HQV}, \downarrow \overline{G} + t_{GLQV}),$$
$$earlier (AdN + t_{AXQX}, \uparrow \overline{E1} + t_{E1HQZ}, \downarrow E2 + t_{E2LQZ}, \uparrow \overline{G} + t_{GHQZ}))$$

$$= (later (131.5 + 150, 140.5 + 150, 125 + 150),$$
$$earlier (\downarrow E + 2 + 20, \downarrow E + 126.5, \downarrow E))$$

During a read cycle the data is required by the 6812. Thus, to determine the read data required interval, we look in the 6812 data sheet. For read data required, the worst case is the longest interval. Recall that

$$RDR = Read\ Data\ Required = (\downarrow E - 15, \downarrow E)$$

We must choose the number of cycle stretches to make the read data available interval overlap the read data required interval. We must make = 1 during a read cycle. Thus,

Address	$131.5 + 150 \leq\ \downarrow\mathbf{E} - 15$	and	$\downarrow\mathbf{E} + 2 + 20 \geq\ \downarrow\mathbf{E}$
$\overline{\mathrm{E1}}$	$140.5 + 150 \leq\ \downarrow\mathbf{E} - 15$	and	$\downarrow\mathbf{E} + 126.5 \geq\ \downarrow\mathbf{E}$
E2	$125 + 150 \leq\ \downarrow\mathbf{E} - 15$	and	$\downarrow\mathbf{E} \geq\ \downarrow\mathbf{E}$

The $\overline{\mathrm{E1}}$ timing ($305.5 \leq\ \downarrow\mathbf{E}$) tells us that 1 extra cycle is needed to stretch the access time to 500ns. If the read data available did not overlap the read data required, then we would have to:

- increase the number of cycle stretches; or
- slow down the 6812 by increasing the E period; or
- decrease t_{E1LQV} by spending more money on a faster RAM chip.

The beginning of RDA is determined by $\downarrow\overline{\mathrm{E1}}$ and the end by $\downarrow\mathrm{E2}$

$$\text{Read Data Available} = (\downarrow\overline{\mathrm{E1}} + 150, \downarrow\mathrm{E2}) = (290.5, 500).$$

Common error: If you do not synchronize driving the data bus on a multiplexed address/data computer such as the 6811 and MC9S12C32, memory data might be driven onto the bus during the first half of the cycle, and it might conflict with the address being driven at that time.

Checkpoint 9.8: How fast would the RAM read access time need to be to run the interface in Figure 9.63 with no cycle stretches?

Data is written into the memory when E2 = 1, $\overline{\mathrm{E1}}$ and $\overline{\mathrm{W}}$ = 0. During a write cycle, the data are supplied by the 6812. For write data available, the worst case is the shortest interval. Recall that

$$\text{WDA} = \text{Write Data Available} = (\tfrac{1}{2}\mathbf{t_{cyc}} + 7, \downarrow\mathbf{E} + 2) = (132, \downarrow\mathbf{E} + 2)$$

Figure 9.63
Interface between the MC9S12C32 and the 60L64 8K RAM chip.

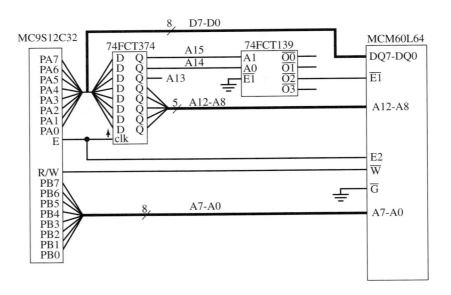

Since we have chosen to synchronize E2 to the E clock,

$$\text{Write Data Required} = (\downarrow\mathbf{E} - 60, \downarrow\mathbf{E})$$

The number of cycle stretches will also affect whether or not the write data available interval overlaps the write data required interval. The worst case delay selects the largest WDR interval

$$132 \le \downarrow E - 60 \quad \text{and} \quad \downarrow E \le \downarrow E + 2$$

Thus,

$$192 \le \downarrow E$$

Therefore, the 1-cycle stretch needed for the read cycle also is also sufficient for the write cycle. Although, the write cycle could have operated with no stretches, we have to choose 1-stretch so that both read and write timing is satisfied.

Common error: If you do not synchronize clocking data into a memory during a write cycle, data might not be properly stored.

Checkpoint 9.9: The write-cycle timing does not work for this 9S12C32/MCM60L64 interface if it is synchronized to E1 or W instead of connecting E2 directly to E. Why?

We combine the command, and timing aspects of the design to find the Boolean expression that will generate the control signals: E2, $\overline{E1}$, $\overline{G}$ and $\overline{W}$.

Common error: A capacitive load occurs both by the physical layout of the PCB as well as with each input pin the output must drive. The 9S12C32 and 74FCT374 timings are specified for capacitive loads of 50 pF. If the actual load is larger than 50 pF, the timing will be significantly slower.

Table 9.27
Combined timing table for the MCM60L64 RAM interface.

E	R/W	A15, A14	RD	WR	E2	$\overline{E1}$	$\overline{G}$	$\overline{W}$	
0	0	00 01 or 11	0	0	X	1	X	X	Address not on the chip
1	0	00 01 or 11	0	0	X	1	X	X	
0	1	00 01 or 11	0	0	X	1	X	X	
1	1	00 01 or 11	0	0	X	1	X	X	
0	0	10	0	0	0	0	X	0	Write cycle
1	0	10	0	1	1	0	X	0	
0	1	10	0	0	0	0	X	1	Read cycle
1	1	10	1	0	1	0	0	1	

$$E2 = E$$
$$\overline{E1} = \overline{A15 \cdot \overline{A14}}$$
$$\overline{G} = 0$$
$$\overline{W} = R/W$$

The next step is to build the interface. MODC, MODB, MODA are set to start the system in Special Single-Chip mode.

To verify proper read cycle timing we draw the *Combined Read Cycle Timing Diagram* as shown in Figure 9.64. We need to verify that read data available overlaps read data required before we build the design.

Similarly, to verify proper write cycle timing we draw the *Combined Write Cycle Timing Diagram,* as shown in Figure 9.65. Notice that write data available overlaps write data required. After reset, the MC9S12C32 will be running in Special Single-Chip mode. The compiler generates initialization code to place the RAM at $3800 to $3FFF. Program 9.4 will change the mode from Special Single-Chip to Normal Expanded Narrow. Internal visibility is turned off to simplify debugging (IVIS = 0). The PEAR register is set to enable the R/W and E clock.

Figure 9.64
Read timing of the
MC9S12C32 and the
60L64 8K RAM chip.

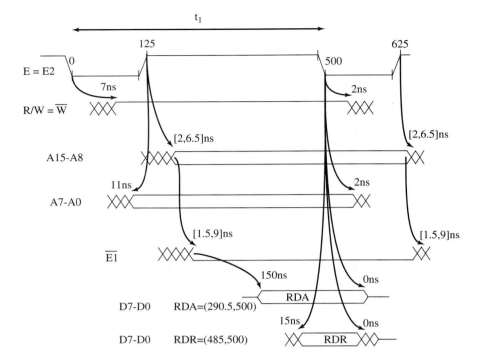

Figure 9.65
Write timing of the
MC9S12C32 and the
60L64 8K RAM chip.

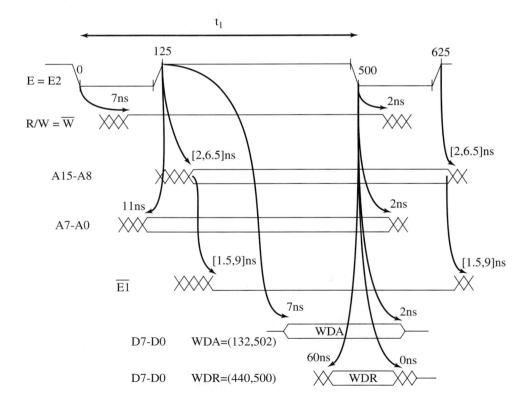

```
void RAM_Init(void){
      MODE = 0xA0;                       // normal expanded narrow mode
      MISC = (MISC&0xF3)|0x04;           // 1-cycle stretch on external
      PEAR = 0x0C;                       // enable E, R/W, LSTRB(not needed)
```

Program 9.4
Software to configure the MC9S12C32 for the external RAM.

> **Observation:** If we come out of reset in normal single-chip mode, the 6812 will allow one write attempt to the MODE register. Sometimes the compiler or debugger uses up this one change during initialization, before our code runs.

> **Checkpoint 9.10:** How many cycle stretches would be needed for the 9S12C32/MCM60L64 interface in Figure 9.63 if the t_{cyc} were reduced from 250 to 125 ns?

9.7.3 32K by 16-bit PROM Interface to a MC68HC812A4

The objective of this example is to interface two 16K by 8-bit PROMs to the 6812 in expanded wide mode. We will place the two 27C128-90 PROMs at locations $8000 to $FFFF. The **command portion** determines which CPU cycles will activate the PROMs. The two 27C128-90 PROMs will drive the 16-bit data bus whenever both $\overline{CE}$ and $\overline{OE}$ are 0. The other three values of $\overline{CE}$ and $\overline{OE}$ will disable the PROM. The PROM chip requires less power (I_{cc} is the +5 V supply current) whenever $\overline{CE}$ is high (Table 9.28).

Using full decoding the address **SELECT** line in positive logic would be **SELECT** = A15. To simplify the interface, we will neglect the R/W and assume the software will not

Table 9.28
Function table for the 27C128 PROM.

$\overline{CE}$	$\overline{OE}$	function	I_{cc}(mA)
1	1	Off	1
1	0	Off	1
0	1	Off	35
0	0	Drive data bus	35

attempt to write into this ROM. To conserve power (lower I_{cc}), the $\overline{CE}$ =1 state is used to deselect the PROM (Table 9.29). The **timing portion** of the interface determines the time during an active cycle when a control signal should rise and fall. The objective is to design the interface such that the data available interval overlaps the data required interval (Figure 9.66).

Table 9.29
Command table for the 27C128 PROM interface.

SELECT	R/W	Function	$\overline{CE}$	$\overline{OE}$
0	0	Off	1	X
0	1	Off	1	X
1	0	Don't care	X	X
1	1	Drive data bus	0	0

During a read cycle, the data are supplied by the 27C128-90. Thus, to determine the data available interval we look at the 27C128-90 data sheet. We will specify the worst-case timing. For data available, the worst case is the shortest interval.

$$\text{Read data available} = (\text{later } (AdV + t_{ACC}, \downarrow\overline{CE} + t_{CE}, \downarrow\overline{OE} + t_{OE}),$$
$$\text{earlier } (AdN + t_{OH}, \uparrow\overline{CE} + t_{OH}, \uparrow\overline{OE} + t_{OH}))$$

To conserve power (minimize I_{cc}), we let $\overline{CE}$ be 0 only during read cycles to $8000 to $FFFF. If we simply grounded $\overline{CE}$, the interface would be faster but would require more power. From the Fairchild 27C128-90 data sheet $t_{ACC} = 90$, $t_{CE} = 90$ $t_{OE} = 50$, $t_{OH} = 0$.

Figure 9.66
Read timing for the
27C128-90 16K PROM
chip.

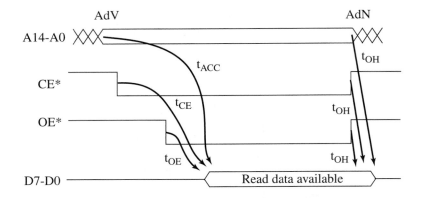

Two of the issues when interfacing to a 6812 are to determine data bus width (narrow or wide) and the number of cycles to stretch. For increased speed, we will utilize 16-bit wide mode. To stretch the E clock we will have to use one of the seven built-in chip selects (e.g., either CSP0 or CSP1). For now, we will develop the timing equations, and then we will use them to determine the cycle stretching requirements. The AdV time is 60 ns, independent of stretching, while AdN will be $t_1 + 20$ ns. We therefore calculate

$$\text{Read data available} = (\text{later} (60 + t_{ACC}, \downarrow\overline{CE} + t_{CE}, \downarrow\overline{OE} + t_{OE}),$$
$$\text{earlier} (\text{AdN} + t_{OH}, \uparrow\overline{CE} + t_{OH}, \uparrow\overline{OE} + t_{OH}))$$

$$= (\text{later} (150, \downarrow\overline{CE} + 90, \downarrow\overline{OE} + 50),$$
$$\text{earlier} (t_1 + 20 + 0, \uparrow\overline{CE} + 0, \uparrow\overline{OE} + 0))$$

During a read cycle the data are required by the 6812.

$$\text{RDR} = \text{read data required} = (t_1 - t_{11}, t_1 + t_{12}) = (t_1 - 30, t_1)$$

We must choose the timing of the fall and rise of both $\overline{CE}$ and $\overline{OE}$ to make the data available interval overlap the data required window. Note that since $\overline{CE}$ and $\overline{OE}$ are active low, they will fall, then rise, during the active cycles. Thus,

150	$\leq t_1 - 30$	and	$t_1 + 20$	$\geq t_1$
$\downarrow \overline{CE} + 90$	$\leq t_1 - 30$	and	$\uparrow \overline{CE}$	$\geq t_1$
$\downarrow \overline{OE} + 50$	$\leq t_1 - 30$	and	$\uparrow \overline{OE}$	$\geq t_1$

First line ($180 \leq t_1$) tells us that one extra cycle is needed to stretch the access time to 250 ns. The following signal is constructed inside the PROM chip. It will be cheaper to synchronize $\overline{CE}$ by connecting it directly to CSP0. This means $\overline{CE}$ will be controlled by timing signal CSP0. The fall of $\overline{CE}$ will be at 60 ns and, the rise of $\overline{CE}$ will be $t_1 + [0,10]$. The CE access time must be satisfied:

$$\downarrow \overline{CE} + 90 \leq t_1 - 30 \quad \text{and} \quad \uparrow \overline{CE} \geq t_1$$

or

$$60 + 90 \leq t_1 - 30 \quad \text{and} \quad t_1 + [0,10] \geq t_1$$

Again one extra cycle is needed to stretch the access time to 250 ns. In this interface we will ground $\overline{OE}$, so the timing delays with $\overline{OE}$ need not be calculated. Now that we have decided to use one-cycle stretching, we can calculate the actual RDA and RDR intervals.

$$\text{Read data available} = (150, 250)$$
$$\text{Read data required} = (220, 250)$$

Next we find the Boolean expressions that will generate $\overline{CE}$ and $\overline{OE}$ from E, CSP0, and R/W. In this simple case, we connected $\overline{CE}$ to CSP0 and $\overline{OE}$ to ground (Figure 9.67). The MC68HC812A4 will come out of reset in expanded mode, with CSP0 enabled having three-cycle stretching. The reset software should enable any additional chip selects and change the cycle stretching on CSP0 to one cycle. Last, we draw the *Combined read cycle timing diagram* and use it to verify that read data available overlaps read data required (Figure 9.68). The software to activate this interface is shown in Program 9.5.

Figure 9.67
Interface between the MC68HC812A4 and the 27C128-90 16K PROM chip.

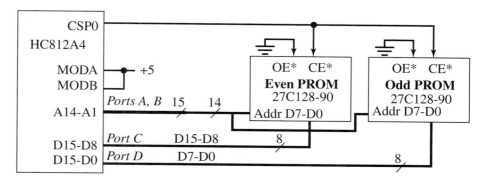

Figure 9.68
Read timing of the MC68HC812A4 and the 27C128-90 16K PROM chip.

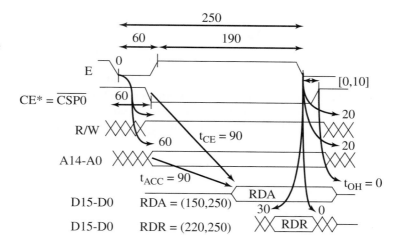

Program 9.5
Software to configure the chip select for the external PROM.

```
void PROMinit(void){
    CSSTR0=(CSSTR0&0xF3)|0x04;} // 1 cycle stretch on CSP0
```

9.7.4
8K by 16-bit RAM Interface

In this example we interface an 8K by 16-bit static RAM to the 6812 using two MCM60L64s. To enhance the execution speed, we will operate in expanded wide mode. The memory will be placed at $4000 to $7FFF.

9.7.4.1
8K by 16-bit RAM Interface to the MC68HC812A4

The MC68HC812A4 interface will utilize the built-in address decoder CSD (0000 to $7FFF). The operation and timing of the MCM60L64 was presented earlier in Section 9.7.2. The solution uses 13 address lines (A13–A1) from the 6812. The 74AHC32 gates create chip selects that handle the situation where 8-bit data are written to this 16-bit RAM (Figure 9.69).

Figure 9.69
Interface between the
MC68HC812A4 and the
MCM60L64 RAM.

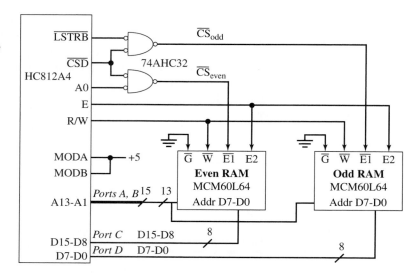

The timing considerations are almost identical to the 8-bit interface. For write data available, the worst case is the shortest interval. Recall that

$$WDA = \text{write data available} = (t_2 + t_{13}, t_1 + t_{14}) = (106, 270)$$

In the 8-bit interface, we connected $\overline{E1}$ directly to CSD, where the $\uparrow$CSD occurred at $t_1 + [0,10] = 250$ ns, making

$$\text{Write data required} = (\uparrow \overline{E1} - 60, \uparrow \overline{E1})$$

In this 16-bit interface, the 74AHC32 will delay the rising edge of $\overline{E1}$. If the gate delay is less than 10 ns, the write data available will overlap the write data required. Just to be safe, the E clock is added to the E2 RAM control signal, so the write data required interval (now triggered on the fall of E2) is not a function of the gate delay through the 74AHC32.

$$\text{Write data required} = (\downarrow E2 - 60, \downarrow E2)$$

The addition of the E2 connection to the 6812 E clock does not affect the read cycle timing.

$$\text{Read data available} = (\text{later } (AdV + t_{AVQV}, \downarrow \overline{E1} + t_{E1LQV}, \uparrow E2 + t_{E2HQV}),$$
$$\text{earlier } (AdN + t_{AXQX}, \uparrow \overline{E1} + t_{E1HQZ}, \downarrow E2 + t_{E2LQZ}))$$

$$= (\text{later } (60 + 150, 60 + 150, 60 + 150),$$
$$\text{earlier } (270 + 20, 260 + 0, 250 + 0) = (210, 250))$$

With one-cycle stretching, recall that

$$RDR = \text{read data required} = (t_1 - t_{11}, t_1 + t_{12}) = (220, 250)$$

The initialization software must enable E, LSTRB, CSD, R/W outputs and select one-cycle stretch on CSD (Program 9.6).

Program 9.6
Software to configure
the chip select for the
external RAM.

```
void RAM_Init(void){
        MODE=0x7B                       // special expanded wide mode
        PEAR=0x2C;                       // enable E, R/W, LSTRB
        WINDEF=WINDEF&0x7F;              // disable DPAGE
        CSCTL0=CSCTL0|0x10;              // enable CSD
        CSCTL1=CSCTL1|0x10;              // CSD $0000 to $7FFF
        CSSTR0=(CSSTR0&0xFC)|0x01;}      // 1 cycle stretch on CSD
```

9.7.4.2. 8K byte 16-bit RAM Interface to the MC9S12C32

We will use full decoding, placing the RAM at addresses $8000 to $BFFF. The MC9S12C32 interface will require this external address decoder, and the positive logic chip select will be A15 • $\overline{A14}$ see Figure 9.70. The operation and timing of the MCM60L64 were presented in Section 9.7.2. The 74FCT139 gates create chip selects that handle the situation where an 8-bit data is written to this 16-bit RAM, as described in Table 9.18. The timing considerations are identical to the 8-bit interface. If the software performs an 8-bit write to an even address, just the even RAM is activated. If the software performs an 8-bit write to an odd address, just the odd RAM is activated. If the software performs a 16-bit write to an even address, both RAMs are activated. If the software performs a 16-bit write to an odd address, the 6812 divides the request into two sequential 8-bit writes.

Figure 9.70
Interface between the MC9S12C32 and the MCM60L64 RAM.

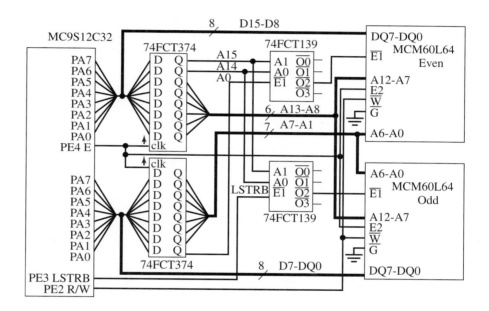

Program 9.7 enables E, LSTRB, R/W outputs and selects 1-cycle stretch on external devices.

Program 9.7
Software to configure the mode for the external RAM.

```
void RAM_Init(void){
    MODE = 0xE0;                    // normal expanded wide mode
    MISC = (MISC&0xF3)|0x04;        // 1-cycle stretch on external
    PEAR = 0x0C;                    // enable E, R/W, LSTRB
```

9.7.5 Extended Address Data Page Interface to the MC68HC812A4

One of the unique features of the MC68HC812A4 is its ability to interface large RAM and ROM using the data page and program page memory, respectively. Up to 1 Mbytes can be configured using the data page system. In this example, two 128K by 8-bit RAM chips are interfaced to the 6812 using 18 address pins (A17–A0). The 628128 is a 128K by 8-bit static RAM similar to the 60L64 static RAM presented earlier. When the paged memory is enabled, Port G will contain address lines A21–A16 (although the data page system can only use up to A19). The built-in address decoder CSD must be used with the data page system. The hardware circuit and timing equations are quite similar to the 8K by 16 bit RAM interfaced earlier. The only hardware difference is that 18 address lines (A17–A0) are needed instead of the 14 (A13–A0) (Figure 9.71).

Figure 9.71
Interface between the
MC68HC812A4 and the
628128 256K RAM.

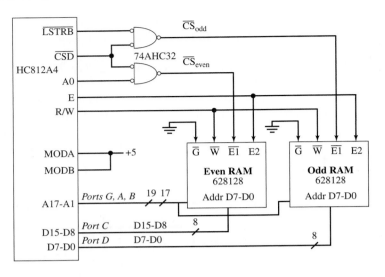

Program 9.8
Software to configure
the mode for the
extended RAM.

```
void RAM_Init(void){
        MODE=0x7B                          // special expanded wide mode
        PEAR=0x2C;                         // enable E, R/W, LSTRB
        WINDEF=WINDEF|0x80;               // enable DPAGE
        MXAR=0x03;                         // enable A17, A16 on Port G
        CSCTL0=CSCTL0|0x10;               // enable CSD
        CSCTL1=CSCTL1&0xEF;               // CSD $7000 to $7FFF
        CSSTR0=(CSSTR0&0xFC)|0x01;}       // 1 cycle stretch on CSD
```

```
struct addr20
{   unsigned char msb;     // bits 19-12, only 17-12 used in this interface
    unsigned short lsw;      // bits 11-0
};
typedef struct addr20 addr20Type;
char Ram_Read(addr20Type addr){ char *pt;
        DPAGE=addr.msb;                    // set address bits 19-12, only 17-12 used
        pt=(char *)(0x7000+addr.lsw);      // set address bits 11-0
        return *pt;}                       // read access
void Ram_Write(addr20Type addr, char data){ char *pt;
        DPAGE=addr.msb;                    // set address bits 19-12, only 17-12 used
        pt=(char *)(0x7000+addr.lsw);      // set address bits 11-0
        *pt=data;}                         // write access
```

Program 9.9
Method for accessing extended RAM.

We divide the software into initialization (Program 9.8) and access (Program 9.9). The initialization software performs the usual steps:

Enable E, LSTRB, CSD, R/W outputs
Clear bit 4 in the CSCTL1 to set up CSD for the $7000–$7FFF range
Select one-cycle stretch on CSD

To enable the data page system we must also:

Set bit 7 in WINDEF to enable the Data Page Window
Set bits 1, 0 in MXAR to enable memory expansion pins A17–A16

Let A17–A0 be the desired 256K RAM location. To access that location requires two steps:

> Set the most significant addresses A17–A12[1] into DPAGE register
> Access $7000 to $7FFF, with the least significant addresses A11–A0

When MXAR is active, the MC68HC812A4 will convert all 16 addresses to the extended 22-bit addresses. When an access is outside the range of any active page window (EPAGE, DPAGE, or PPAGE), the upper 6 bits are 1. In Table 9.30 it is assumed that only the DPAGE

Internal	A21	A20	A19	A18	A17	A16	A15	A14	A13	A12	A11–A0
$0xxx	1	1	1	1	1	1	0	0	0	0	xxx
$1xxx	1	1	1	1	1	1	0	0	0	1	xxx
$2xxx	1	1	1	1	1	1	0	0	1	0	xxx
$3xxx	1	1	1	1	1	1	0	0	1	1	xxx
$4xxx	1	1	1	1	1	1	0	1	0	0	xxx
$5xxx	1	1	1	1	1	1	0	1	0	1	xxx
$6xxx	1	1	1	1	1	1	0	1	1	0	xxx
$7xxx	1	1	DP7	DP6	DP5	DP4	DP3	DP2	DP1	DP0	xxx

Table 9.30
Extended addressing using DPAGE.

is active, the DPAGE register contains DP7–DP0, and MXAR is 0x3F (activating all 22 address bits). From this table, we can see another trick when using paged memory. If we were to switch the initialization so that CSD activated on $0000–$7FFF, then seven 4K pages would overlap the regular address range $0000 to $6FFF. In this particular system, there are 64 data pages, numbered $00 to $3F. Notice that page numbers $30 through $36 overlap with regular data space $0000 to $6FFF. If we avoid using data pages $30–$36, we would have 28K of regular RAM from $0000 to $6FFF plus 57 4K windows of data paged memory space.

9.7.6 Extended Address Program Page Interface to the MC68HC812A4

Up to 4 Mbytes can be configured using the program page system. In this example, two 1 Mbyte by 8-bit PROM chips are interfaced to the 6812 using 21 address pins (A20–A0). The 27C080-12 is a 1 Mbyte by 8-bit PROM similar to the 27C128-90 PROM presented earlier. Each 27C080-12 has 20 address lines, eight data lines, and a 120-ns access time. When the paged memory is enabled, Port G will contain address lines A21–A16 (although this program page system will only use up to A20). The built-in address decoders CSP0 or CSP1 must be used with the program page system. We will use CSP0 so that part of the PROM is available immediately after reset. The hardware circuit and timing equations are quite similar to the 32K by 16 RAM interfaced earlier. The only hardware difference is that 21 address lines (A20–A0) are needed instead of the 15 (A14–A0) (Figure 9.72).

After a reset in expanded wide mode:

> Ports A, B contain the address lines A15–A0
> CSP0 is enabled for the $8000–$FFFF range (CSCTL0 = $20)
> There are three-cycle stretches on CSP0
> The extended memory address lines are inactive (MXAR = 0)
> Port G is an input port (DDRG = 0) but has pull-ups, so A21–A16 will be high

---

[1]With a 1 Mbyte RAM, you would set A19–A12 into DPAGE.

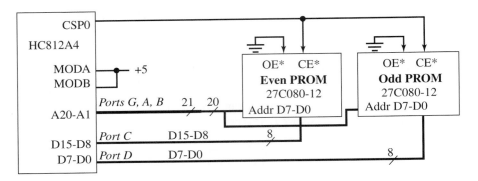

Figure 9.72
Interface between the MC68HC812A4 and the 27C080 1M PROM.

To enable the program page system we must also (Program 9.10):

Set bit 6 in WINDEF to enable the Program Page Window
Set bits 4–0 in MXAR to enable memory expansion pins A20–A16
Reduce cycle stretch to one because of the 120-ns access time

Program 9.10
Software to configure the mode for the extended PROM.

```
void PROM_Init(void){
        MODE=0x7B               // special expanded wide mode
        PEAR=0x2C;              // enable E, R/W, LSTRB (none needed)
        WINDEF=WINDEF|0x40;    // enable PPAGE
        MXAR=0x01F;            // enable A20-A16 on Port G
        CSSTR0=(CSSTR0&0xF3)|0x04;}  // 1 cycle stretch on CSP0
```

Let A20–A0 be the desired 2-Mbyte PROM location. To access that location requires two steps (Program 9.11):

Set the most significant addresses A20–A14[2] into PPAGE register
Access $8000 to $BFFF, with the least significant addresses A13–A0

```
struct addr22
{   unsigned char msb;    // bits 21-14, only 20-14 used in this interface
    unsigned short lsw;   // bits 13-0
};
typedef struct addr22 addr22Type;
char PROM_Read(addr22Type addr){ char *pt;
        PPAGE=addr.msb;                    // set address bits 21-14
        pt=(char *)(0x8000+addr.lsw);  // set address bits 13-0
        return *pt;}                       // read access
```

Program 9.11
Method for accessing extended PROM.

When MXAR is active, the MC68HC812A4 will convert all 16 addresses to the extended 22-bit addresses. When an access is outside the range of any active page window (EPAGE, DPAGE, or PPAGE), the upper 6 bits are 1. In Table 9.31 it is assumed that only the PPAGE is active, the PPAGE register contains PP7–PP0, and MXAR is 0x3F (activating all 22 address bits).

---
[2]With a 4 Mbyte PROM, you would set A21–A14 into PPAGE.

Internal	A21	A20	A19	A18	A17	A16	A15	A14	A13	A12	A11–A0
$8xxx	PP7	PP6	PP5	PP4	PP3	PP2	PP1	PP0	0	0	xxx
$9xxx	PP7	PP6	PP5	PP4	PP3	PP2	PP1	PP0	0	1	xxx
$Axxx	PP7	PP6	PP5	PP4	PP3	PP2	PP1	PP0	1	0	xxx
$Bxxx	PP7	PP6	PP5	PP4	PP3	PP2	PP1	PP0	1	1	xxx
$Cxxx	1	1	1	1	1	1	1	1	0	0	xxx
$Dxxx	1	1	1	1	1	1	1	1	0	1	xxx
$Exxx	1	1	1	1	1	1	1	1	1	0	xxx
$Fxxx	1	1	1	1	1	1	1	1	1	1	xxx

Table 9.31
Extended addressing using PPAGE.

From Table 9.31, we can see another trick when using paged memory. Notice that CSP0 activates on $8000–$FFFF. For the range $8000 to $BFFF, the program page system is activated, but for the range $C000-$FFFF, the CSP0 activates without utilizing the program page system. In this particular system, there are 128 program pages, numbered $00 to $7F. Notice that page number $7F overlaps with regular space $C000 to $FFFF. If we avoid using program page $7F, we would have 16K of regular PROM from $C000 to $FFFF plus 127 16K windows of paged program memory space. Also notice that this regular PROM is available immediately after reset and should contain the reset vectors and reset handler.

This extended memory system can be used to hold programs or fixed data. If it were constructed with a battery-backed nonvolatile static RAM (SRAM) like the following from Dallas Semiconductor,

DS1220Y-150	150 ns 16K
DS1225AD-150	150 ns 64K
DS1230Y-85	85 ns 256K

then the extended memory system could be used for programs and long-term (10-year) storage. When the software wishes to call a function that exists within the program paged system, it uses the `call` op code. This instruction pushes the old PPAGE and PC on the stack and loads a new PPAGE and PC as needed. The `rtc` instruction pulls the old PC and PPAGE from the stack. The effective address of the call instruction includes the most significant byte (8 bits), least significant word (14 bits) components similar to the above C programs. One of the most flexible ways to implement the program division is to use a jump table scheme and write relocatable code (Program 9.12). Notice that the software can

```
*****regular******
    org $C000
main lds #$C00
* call func2 in page1
    call fun2,1
* call func3 in page2
    call fun3,2
* call func1 in page1
    call fun1,1


global

    rts
    org $FFFE
    dc.w main
```

```
******page 1******
    lbra func1
    lbra func2
    lbra func3
func1
* call func2 in page 2
    call fun3,2
    rtc
func2
* call global function
    jsr  global
    rtc
func3
* call func2 same page
    call fun2,1
    rtc
```

```
*****page 2******
    lbra func1
    lbra func2
    lbra func3
func1
    rtc
func2
* call a local function
    bsr  local
    rtc
func3
* call func2 same page
    call fun2,2
    rtc
local
    rts
```

Program 9.12
Method for calling functions in extended PROM.

use `call` to execute a function within the same page. Each of the following columns can be assembled separately, using the common `equ` definitions. For example,

```
fun1 equ 0
fun2 equ 3
fun3 equ 6
```

> ***Observation:*** It is important to always use `call` to execute a function that ends with `rtc`.

> ***Observation:*** It is important to always use `bsr` or `jsr` to execute a function that ends with `rts`.

9.8 Dynamic RAM (DRAM)

Because of their high density and low cost DRAMs are used for situations requiring a large amount of storage. Consequently embedded systems will almost exclusively employ SRAMs for their RAM needs. As the complexity of embedded system software increases, the need for RAM space will grow as well. At some point, DRAMs may become cost-effective for embedded systems. The purpose of this section is to introduce the basic principles involved in DRAMs. The information in a DRAM cell is saved as a charge on a capacitor. The information in a SRAM cell is saved as the state of a set-reset flip-flop. This reduction is size comes at the expense of needing to refresh the capacitor charges. This refresh activity adds a fixed cost. Table 9.32 compares the two memory types.

Table 9.32
Comparison between DRAM and RAM.

DRAMs	SRAMs
High density	Low density
One transistor, one capacitor/bit	Three–four transistors/bit
Slower	Faster
High fixed cost (refresh)	Low fixed cost (address decoder)
Low incremental cost	Higher incremental cost
Address multiplexing	Direct addressing

The higher fixed cost and lower incremental cost of DRAM versus SRAM means for small memory applications (like embedded systems) the SRAMs will be cost-effective. At some point as the memory size gets large enough, the DRAMs will become cost-effective.

To interface a DRAM certain support functions are required. Because the DRAM devices have so many address lines, they employ a technique called address multiplexing. This means the computer first gives half the address and toggles a row address strobe, RAS. Then the computer gives the second half of the address and issues a column address strobe, CAS. A refresh cycle occurs when a RAS is issued without a following CAS. This will refresh the entire row. The interface logic must handle the refresh process controlling RAS and CAS. In addition, the interface must arbitrate between normal read/write and refresh cycles. This arbitration is usually handled using the partially asynchronous bus protocol shown in Figure 9.25. Because the 6811 and 6812 do not support dynamic bus cycle stretching, DRAM interfacing to these computers is difficult.

9.9 Exercises

9.1 The objective is to interface an 8 K by 8-bit MCM60L64 RAM to a new (fictional) microprocessor. This new microprocessor is similar to a 6811 (16-bit address, 8-bit data, synchronous bus). In other ways this microprocessor is different from a 6811 (two control signals **R** and **W** instead of the **R/W** line and **E** clock, the address bus is not time-multiplexed). There are two types of bus cycles, *Read* and *Write*. One byte is transferred from the memory to the microprocessor during a read cycle, when **R** = 1 and **W** = 0. See part d for the specific timing of Read Data Required. One byte is transferred from the microprocessor to the memory during a write cycle, when **R** = 0 and **W** = 1. See part e for the specific timing of Write Data Available.

a) Show the minimal cost address decoder for your 8K RAM at $4000 to $5FFF. Other devices are at addresses $2000 to $27FF and $FF00 to $FFFF. Choose to build either a positive logic or negative logic decoder, depending on which would be most appropriate.

b) Create a combined table that has **Select, R, W** as inputs and $\overline{E1}, \overline{G}$, and $\overline{W}$ as outputs. Put a bar over the top of **Select**, if you decided to build a negative logic decoder in part a. Give the equations for $\overline{E1}, \overline{G}$, and $\overline{W}$ as a function of **Select, R, W**.

c) Show the digital interface between the microprocessor (R, W, A15-A0, D7-D0) and the 8K RAM. Label all chip numbers, but not pin numbers.

d) Finish the **Read Cycle Timing diagram** (Figure 9.73); use arrows to show causal relationships. Show $\overline{E1}, \overline{G}, \overline{W}$, and Read Data Available. Clearly label Read Data Available, showing that it does overlap Read Data Required.

Figure 9.73
Read timing.

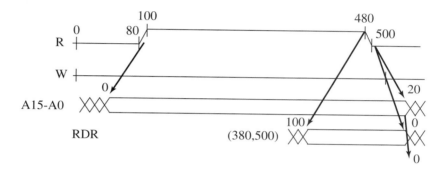

e) Finish the **Write Cycle Timing diagram**, (Figure 9.74); use arrows to show causal relationships. Show **E1, G, W**, and Write Data Required. Clearly label Write Data Required, showing that Write Data Available does overlap Write Data Required.

Figure 9.74
Write timing.

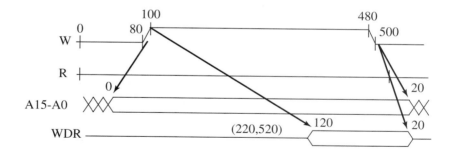

9.2 Interface a 16K by 8-bit RAM to a 6811/6812 running at 2 MHz. Design FULL address decoder, placing the RAM from $4000–$7FFF. The address must be valid whenever the chip is enabled (**CS1** = 1, $\overline{\text{CS0}}$ = 0). Assume a 10-ns gate delay in each TTL chip and a 500-ns **E** clock period. The RAM has three control lines. A read (data out of the RAM) occurs on the overlap of **CS1** = 1 $\overline{\text{CS0}}$ = 0 and **WE** = 0. The RAM read timing is shown in Figure 9.75. A write (data into the RAM) occurs on the overlap of **CS1** =1, $\overline{\text{CS0}}$ = 0 and **WE** = 1. The earlier edge ↓**CS1** or ↑$\overline{\text{CS0}}$ determines the RAM write timing. The RAM write timing is shown in Figure 9.76.

a) Either **CS1** = 0 or $\overline{\text{CS0}}$ = 1 will turn off the RAM. Create a status table with **Select** and **R/W** as inputs and CS1, $\overline{\text{CS0}}$, and WE as outputs.

b) What is the read data available interval? Express your answer as an equation using only the terms like AdV, AdN, ↓ $\overline{\text{CS0}}$, and ↑**WE**. Don't calculate (yet) the actual interval in nanoseconds.

Figure 9.75
Read timing.

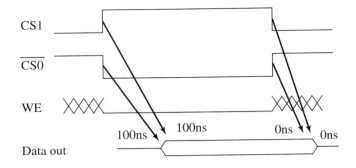

Figure 9.76
Write timing.

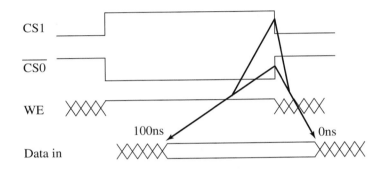

c) What is the write data required interval? Express your answer as an equation using only the terms like AdV, AdN, ↓$\overline{\text{CS0}}$, and ↑**WE**. Don't calculate (yet) the actual interval in nanoseconds.

d) Show the combined table with **Select, R/W,** and **E** as inputs and **CS1**, $\overline{\text{CS0}}$, and **WE** as outputs, and include digital equations for **CS1**, $\overline{\text{CS0}}$, and **WE**.

e) Show digital circuit for the interface between the 6811/6812 and the RAM. Include the address latch if and only if it is necessary for this interface. Please label TTL chip numbers but not pin numbers.

f) Show the combined **write cycle** timing diagram. All signals are outputs except Data Required. Use arrows to signify causal relations.

9.3 Interface an AD557 8-bit DAC directly to the 6811/6812 bus (not to an output port). You will choose the fastest 6811 clock speed that allows this interface to operate properly. For the 6812, assume an 8 MHz E clock, and choose the proper number of cycle stretches. Assume a 2-ns gate delay through each digital gate. The MINIMAL COST address decoder for the 6811 is given, placing the DAC at $7000 (SELECT = $\overline{\text{A15}}$ · A14 · A13 · A12). You *do not* design the address decoder. When both $\overline{\text{CE}}$ and $\overline{\text{CS}}$ are 0, the D/A output follows the 8-bit digital input (transparent mode). If $\overline{\text{CE}}$ rises while **CS** equals zero (or if $\overline{\text{CS}}$ rises while $\overline{\text{CE}}$ equals zero), the 8-bit digital inputs are latched into the chip. The setup time is 300 ns and the hold time is 0 ns. When either $\overline{\text{CE}}$ or $\overline{\text{CS}}$ is 1, the DAC ignores the input digital data and the analog output remains constant, depending on the digital input data at the time of the last rising edge (Figure 9.77).

Figure 9.77
Write timing.

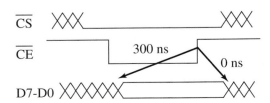

a) Fill in the command table with Select and R/W as inputs and $\overline{\text{CE}}$ and $\overline{\text{CS}}$ as outputs.

b) What is the write data available interval? Express your answer as a function of the E clock period. Let t_{cyc} be the E clock period.

c) What is the write data required interval? Express your answer as an equation using only the terms like $\downarrow\overline{\text{CS}}$ and $\uparrow\overline{\text{CE}}$. Don't calculate (yet) the actual interval in nanoseconds.

d) Show the combined table with Select, R/W, and E as inputs and $\overline{\text{CE}}$ and $\overline{\text{CS}}$ as outputs. Include digital equations for $\overline{\text{CS}}$, and $\overline{\text{CE}}$.

e) Show a digital circuit for the interface between the expanded mode 6811/6812 and the AD557 8-bit DAC. Include the address latch if and only if it is necessary for this interface. Please label TTL chip numbers but not pin numbers.

f) For the 6811, what is the smallest possible E clock period for this interface? For the 6812, what is the smallest number of cycle stretches? *Show your work.*

9.4 Interface a 4K by 8 bit RAM to a 6811/6812 running at 2 MHz. The FULL address decoder is given, placing the RAM from $7000–$7FFF (SELECT= $\overline{\text{A15}} \cdot \text{A14} \cdot \text{A13} \cdot \text{A12}$). The RAM address must be valid whenever the chip is enabled $\overline{\text{MEMR}} = 0$ or $\overline{\text{MEMW}} = 0$. Assume a 2-ns gate delay in each TTL chip and a 500-ns **E** clock period. The RAM has two control lines. A read (data out of the RAM) occurs when $\overline{\text{MEMR}} = 0$. The RAM read timing is shown in Figure 9.78. A write (data into the RAM) occurs when $\overline{\text{MEMW}} = 0$. The RAM write timing is shown in Figure 9.79.

a) $\overline{\text{MEMR}} = 1$ and $\overline{\text{MEMW}} = 1$ turn off the RAM. Fill in the command table that has **Select** and **R/W** as inputs and $\overline{\text{MEMR}}$ and $\overline{\text{MEMW}}$ as outputs.

b) What is the read data available interval? Express your answer as an equation using only terms like $\downarrow\overline{\text{MEMR}}$ and $\uparrow\overline{\text{MEMW}}$. Don't calculate (yet) the actual interval in nanoseconds.

c) What is the write data required interval? Express your answer as an equation using only terms like $\downarrow\overline{\text{MEMR}}$ and $\uparrow\overline{\text{MEMW}}$. Don't calculate (yet) the actual interval in nanoseconds.

d) Show the combined table that has **Select, R/W**, and **E** as inputs and $\overline{\text{MEMR}}$ and $\overline{\text{MEMW}}$ as outputs. Include digital equations for $\overline{\text{MEMR}}$ and $\overline{\text{MEMW}}$.

e) Show a digital circuit for the interface between the 6811/6812 and the RAM. Include an address latch if and only if it is necessary for this interface. Please label TTL chip numbers but not pin numbers.

f) What is the allowable range (minimum and maximum) of values for t_{acc}?

g) What is the allowable range (minimum and maximum) of values for t_{su}?

Figure 9.78
Read timing.

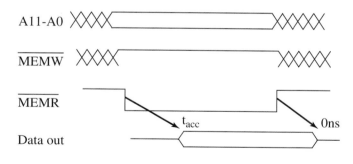

Figure 9.79
Write timing.

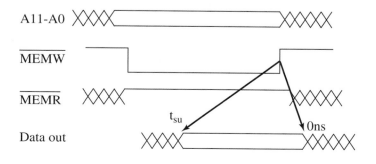

9.5 The overall objective is to select the cheapest PROM that will operate with a **2.1 MHz** 6811. When PROM input **CS** is 1, the PROM **Data** outputs are driven with the 8-bit value stored at the specified **Address**. The address must be valid whenever **CS** is high and the **CS** access time is t_{CS} (Figure 9.80).

Figure 9.80
Read timing.

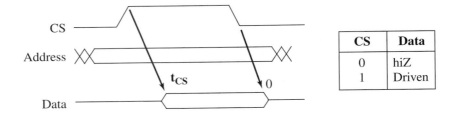

CS	Data
0	hiZ
1	Driven

a) The **CS** signal is positive logic, meaning it is normally low and goes high only during an access to this PROM. *Circle* the best shape for the PROM **CS** chip select control line:

Positive logic synchronized to the E clock
Positive logic synchronized to the AS 6811 signal
Positive logic unsynchronized

b) Why did you choose the above shape? That is, list the bad thing(s) that would happen with the other two choices.

c) What is the **Read Data Required** Interval for this 6811 running at **2.1 MHz**? Give your answer with exact explicit numbers.

d) Develop an equation for the **Read Data Available** Interval that is a function only of t_{CS}. All the other terms in the interval should be exact explicit numbers. Assume a 10-ns gate delay.

e) What is the maximum allowable value for t_{CS}?

9.6 Interface the following *Output Device* (Figure 9.81) to the address/data bus of an expanded mode 6812 running at 8 MHz. (*Hint: Think of this device as a memory that responds only to 6812 write cycles. That is, this device ignores all read cycles.*) The chip has two control inputs **C1, C2**, one address input **A0**, eight data inputs **D7–D0**, and two port outputs **Port0, Port1**. The minimal-cost positive logic address decoder is already completed, locating it at $4000 to $4001. Assume a 2 ns gate delay.

Figure 9.81
Output device.

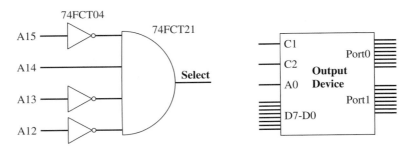

There are two different, but equally effective, timing situations. A write operation occurs on the overlap of **C1**=1 and **C2**=0. You will not be connecting to Port0, Port1 in this problem. The function table is:

C1	C2	A0	Function
0	X	X	Off low I_{cc}=10 μA
1	1	X	Off high I_{cc}=1 mA
1	0	0	Write data into Port0
1	0	1	Write data into Port1

The write timing when controlled by **C1** is shown on the left in Figure 9.82; the write timing when controlled by **C2** is shown on the right in Figure 9.82.

Figure 9.82
Write timing.

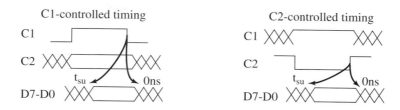

a) What is the **Write Data Available** Interval for this 6812 running at 8 MHz?
b) Show the combined table with **Select, R/W, E** as inputs and **C1** and **C2** as outputs. Include digital equations for **C1** and **C2**.
c) Show the digital circuit interface between the 6812 and this Output Device. Clearly label all the connections to the 6812. Include chip numbers but not pin numbers.
d) What is the allowable range (minimum and maximum) of values for t_{su}?
e) Show the combined **write cycle** timing diagram. Show **E, R/W, A15–A8, A7–A0, Select, C1, C2, Data Available,** and **Data Required**. All signals are outputs except Data Required. Use arrows to signify causal relations, with the timing delays as numbers in nanoseconds.

9.7 Interface a 16K by 8 bit PROM to a **6812** running at 8 MHz. Use FULL address decoding to place the PROM from $4000–$7FFF. The address must be valid whenever the chip is enabled (both **CS1** = 1 and $\overline{CS0}$ = 0). Assume a 2-ns gate delay in each digital chip and a 125-ns **E** clock period. The PROM has two control lines. A read (data out of the PROM) occurs on the overlap of **CS1**=1 and $\overline{CS0}$ = 0. Notice that the **CS1** access time is larger than the $\overline{CS0}$ access time. The read timing is shown in Figure 9.83.

Figure 9.83
Read timing.

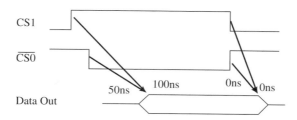

a) Either **CS1**=0 or $\overline{CS0}$ = 1 will turn off the PROM. Fill in the command table that has **Select, R/W** as inputs and **CS1,** $\overline{CS0}$ as outputs.
b) What is the read data available interval? Express your answer as an equation using only the terms like AdV, AdN, $\downarrow\overline{CS0}$, and $\uparrow$CS1 . Don't calculate (yet) the actual interval in nanoseconds.
c) Show the combined has **Select, R/W, E** as inputs and **CS1,** $\overline{CS0}$ as outputs. Include digital equations for **CS1** and $\overline{CS0}$.
d) Show digital circuit for the interface between the 6812 and the PROM. The chip has two control inputs **CS1,** $\overline{CS0}$, 14 address inputs **A13–A0**, eight data outputs **D7–D0**. Include an address latch if and only if it is necessary for this interface. Please label chip numbers but not pin numbers.
e) Show the combined **read cycle** timing diagram. All signals are outputs except Data Required. Show **E, R/W, A15–A8, A7–A0, Select, CS1,** $\overline{CS0}$, **Data Available,** and **Data Required**. Use arrows to signify causal relations.

9.8 Interface a Programmable Keyboard/Display Interface chip (Intel 8279-5) to a **6811**. Use minimal cost address decoding to place the I/O chip from $2018 to $2019. Assume a 10-ns TTL gate delay and a 640-ns **E** clock period. Your 6811 has the timing shown in Figure 9.84.

Figure 9.84
6811 timing.

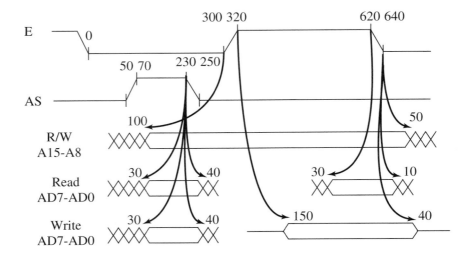

The Intel 8279-5 has the following function table:

$\overline{\text{CS}}$	$\overline{\text{RD}}$	$\overline{\text{WR}}$	A0	Function
0	0	1	0	Read data register
0	0	1	1	Read status register
0	1	0	0	Write data register
0	1	0	1	Write control register
0	0	0	X	Illegal (don't do this)
0	1	1	X	Off
1	X	X	X	Off

a) Design the minimal cost address decoder (your choice as to whether to make it positive or negative logic). Show digital equation and the TTL circuit, including chip numbers. If you choose positive logic, then label the output of the decoder **Select**. If you choose negative logic, then label the output of the decoder **$\overline{\text{Select}}$**. *Do not build all the decoders, just the one for this Intel 8279!*

b) Fill in the command table.

c) What is the read data available interval? Express your answer as an equation using only the terms like AdV, AdN, $\downarrow\overline{\text{RD}}$, and $\uparrow\overline{\text{WR}}$. The address access time is 250 ns, the $\overline{\text{CS}}$ access time is 250 ns, and the $\overline{\text{RD}}$ access time is 150 ns. Don't calculate (yet) the actual interval in nanoseconds.

d) What is the read data required interval? Express your answer in nanoseconds according to the time convention shown later in part i. Remember, this is a 6811 running with an E period of 640 ns. Give the actual numbers [e.g., RDR = (450, 510)].

e) What is the write data available interval? Express your answer in nanoseconds according to the time convention shown later in part i. Remember, this is a 6811 running with an E period of 640 ns. Give the actual numbers [e.g., WDA = (375, 533)].

f) What is the write data required interval? Express your answer as an equation using only the terms like $\downarrow\overline{\text{WR}}$ and $\uparrow\overline{\text{WR}}$. Don't calculate (yet) the actual interval in nanoseconds. The write occurs on the fall of $\overline{\text{WR}}$. The setup time is 150 ns, and the hold time is zero.

g) Show the combined table, and include digital equations for $\overline{\text{CS}}$, $\overline{\text{RD}}$, and $\overline{\text{WR}}$. You will build the circuit later in part h. The read and write functions must be synchronized to the **E** clock.

h) Show digital circuit for the interface between the 6811 and the I/O device. Please label TTL chip numbers but not pin numbers. You need not copy the address decoder circuit if shown earlier in part a, just use the **Select** label.

i) Show the combined **read cycle** timing diagram. All signals are outputs except Data Required. The A0 is the signal connected to the I/O device (not necessarily the AD7–AD0 of the 6811). Use arrows to signify causal relations.

9.9 The ultimate goal of this problem is to design a simple 14-pin LSI chip that supports XIRQ interrupts on an expanded mode 6811/6812. That is, you will be connecting up to the 6811/6812 address/data bus, not to the parallel ports. You will design the device using 74HC or 74 FCT digital logic. We will use two 74HC74 D flip-flops to implement READY and ARM, and a 74HC125 tristate buffer to perform the read status function. Assume a 20-ns gate delay in each 74HC chip and assume a 2-ns gate delay in each 74 FCT chip. Synchronize all read/write/clear operations to the **E** clock period. The device should have the following command table (others are no operation):

$\overline{CS3}$	$\overline{CS2}$	CS1	CS0	R/W	RS	Function
0	0	1	1	1	0	Read status (READY = bit 7)
0	0	1	1	0	0	Write status (READY = 0, regardless of the data bus)
0	0	1	1	1	1	No operation (ARM is write-only)
0	0	1	1	0	1	Write mode (ARM = bit 7)

The READY signal is set on the rise of INPUT and cleared (acknowledged) on a write status. The software can observe the READY signal by performing a read status operation. The hardware can also observe READY because it is a direct output of your interface. The software can arm or disarm the interface by performing a write mode operation. An XIRQ interrupt is requested (open-collector) if READY = 1 and ARM = 1. The device is connected to an expanded mode 6811 like the one shown in Figure 9.85.

Figure 9.85
External device.

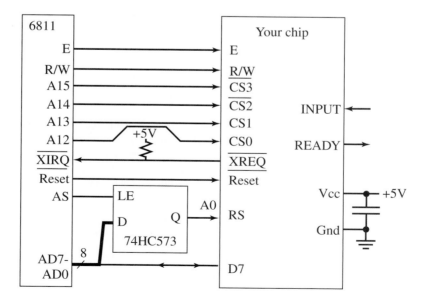

a) Complete the hardware design of your 14-pin chip. Do not change the interface to the 6811/6812 as shown in Figure 9.85. Specify 74HC or 74 FCT chip numbers but not pin numbers. Do not worry about simultaneous events: a rising edge of INPUT and a write status.

b) Show the ritual subroutine that arms and enables the XIRQ device. Use assembly language or C, but be consistent. Initialize an 8-bit unsigned global variable, CNT, to 0.

c) Show the XIRQ software that counts the number of rising edges of input (CNT = CNT + 1). Execute SWI if an XIRQ interrupt occurs that is not from your chip.

9.10 Interface the following 32-kbyte RAM to a 2-MHz 6811 or an 8 MHz 6812 using full-address decoding. The RAM will be located at $4000 to $BFFF. The RAM function table is:

CE	OE	Function
0	0	Off
0	1	Off
1	0	Write data into RAM
1	1	Read data out of RAM

The read timing is (with OE = 1) shown on the left in Figure 9.86; the write timing is (with OE = 0) shown on the right.

Figure 9.86
Read and write timing.

a) Show the interface between the 6811/6812 and this RAM. Clearly label all the connections, especially A14.

b) Show the combined **write cycle** timing diagram. Show **E**, **R/W**, **A15–A8**, **A7–A0**, **Select**, **CE**, **OE**, **WDA**, and **WDR**. All signals are outputs except Data Required. Use arrows to signify causal relations. Show the timing delays as arrows, with numbers in nanoseconds. Calculate the actual WDA and WDR intervals.

9.11 Interface a 27C128-90 16K by 8-bit PROM to the 6812 address/data bus. The address will be $C000 to $FFFF. No other external devices exist. Assume the 6812 is running at 8 MHz in expanded narrow mode.

a) Show a digital circuit for the interface between the 6812 and the PROM. Please label TTL chip numbers but not pin numbers.

b) Develop equations for read data available and read data required, and use them to determine the amount of cycle stretching required to interface this 90-ns memory.

c) Show the software ritual required to initialize this memory interface.

d) Show the combined **read cycle** timing diagram. Assume a 10-ns gate delay through any digital gate. All signals are outputs except Data Required. Use arrows to signify causal relations.

9.12 Consider the 8K by 8-bit static RAM interface between the MC68HC812A4 and MCM60L64. Assume the MC68HC812A4 is running at 8 MHz in expanded narrow mode. Assume the same hardware presented in the MC68HC812A4/MCM60L64 example earlier in this chapter. The objective of this problem is to develop the specifications that allow this RAM interface to operate without cycle stretching. You may use timing equations or timing diagrams to justify your answers.

a) Develop the equations that specify the maximum allowable address access time t_{AVQV} and chip enable access time t_{E1LQV}. In other words, how fast would the RAM chip have to be to operate without cycle stretching?

b) Develop the equations that specify the maximum allowable t_{DVWH}.

9.13 Briefly explain the difference between single-chip, expanded narrow, and expanded wide mode on the MC6812C812A4 as they apply to memory interfacing. Give one sentence for each mode. In particular, explain the effect on Ports A, B, C, D, E.

Lab Assignments

Lab 9.1. The overall objective of this lab is to interface a RAM and design a memory tester. The first step is to design, build, and test an external RAM module. During the testing phase, using a multichannel scope or a logic analyzer, you should draw the actual read and write timing diagrams, and compare them to the theoretical timing diagrams derived from the data sheets. The second part of the lab is to design a memory test program that detects and classifies the two types of errors: address bit not connected and data bit not connected. You can evaluate your software using a DIP socket with some of the pins removed. Insert the faulty DIP socket between the microcontroller circuit and the RAM. The output of your program should be very specific, e.g., "Pins A4, A2, D7, D5, D1 are not connected." Do not attempt to simulate errors such as pins stuck together, pins stuck high, or pins stuck low, because it could cause damage to your system.

Lab 9.2. The overall goal is to develop a solid-state disk. Your system will be able to create files, append data to the end of a file, printout the entire contents of a file, and delete files. In addition, your system will be able to list the names and sizes of the available files. Basically, you will interface a large RAM, then write a series of software functions that make it appear as a disk. The operation of the solid-state disk was described previously in Lab 7.3. On the 6812, you should use the paging mechanism. If you use paging, you should select a disk block size different from the memory page size. You can use a 3V battery and two low-voltage drop diodes to create a nonvolatile RAM.

Lab 9.3. The overall objective of this lab is to interface an external EEPROM (like the 28C64) to the microcontroller and develop mechanisms to program the device. The first step is to design, build, and test an external EEPROM module. During the testing phase, using a multichannel scope or a logic analyzer, you should draw the actual read timing diagram, and compare it to the theoretical timing diagram derived from the data sheets. The second part of the lab is to design a software system that allows you to program the EEPROM directly from the microcontroller.

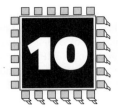

10 High-Speed I/O Interfacing

Chapter 10 objectives are to:

- ❑ Define the terms bandwidth, latency, and priority
- ❑ Introduce the concept of DMA synchronization
- ❑ Discuss the alternatives of hardware FIFOs, dual port memory, and bank-switch memory
- ❑ Use DMA to transfer memory blocks
- ❑ Present high-bandwidth/low-latency applications

Embedded system designers will not need DMA to solve most of their problems. But like the last chapter, there are two motivations for this chapter. The first motivation is basic knowledge, and the second is solving specific high-performance applications. DMA is an important yet complicated interfacing process. One of the advantages of learning DMA on a simple device like graphics controller is that it maintains all the fundamental concepts without most of the complexities found in larger computer systems. As the performance requirements of our embedded system grow, there comes a point when the simple methods of I/O interfacing are not adequate. This chapter introduces a number of techniques that produce high bandwidth and low latency.

10.1 The Need for Speed

Bandwidth, latency, and priority are quantitative parameters we use to evaluate the performance of an I/O interface. The basic function of an input interface is to transfer information about the external environment into the computer. In a similar way, the basic function of an output interface is to transfer information from the computer to the external environment. The bandwidth is the number of information bytes transferred per second. The bandwidth can be expressed as a maximum or peak that involves short bursts of I/O communication. On the other hand, the overall performance can be represented as the average bandwidth. The latency of the hardware/software is the response time of the interface. It is measured in different ways, depending on the situation. For an input device, the *interface latency* is the time from when new input is available to when the data are transferred into memory. We can also define *device latency* as the response time of the external I/O device. For example, if we request that a certain sector be read from a disk, then the device latency is the time it takes to find the correct track and spin the disk (seek) so that the proper sector is positioned under the read head. For an output device, the interface latency is the time from when the output device is idle to when the interface writes new data. A *real-time* system is one that can guarantee a worst-case interface latency. Table 10.1 illustrates specific ways to calculate latency. In each case, however, latency is the time from when the need arises to when the need is satisfied.

Table 10.1
Interface latency is a measure of the response time of the computer to a hardware event.

The time a need arises	The time the need is satisfied
New input is available	Input data are read
New input is available	Input data are processed
Output device is idle	New output data are written
Sample time occurs	ADC is triggered, input data
Periodic time occurs	Output data, DAC is triggered
Control point occurs	Control system executed

If we consider the busy/done I/O states introduced in Chapters 3 and 4, the interface latency is the time from the *busy to done* state transition to the time of the *done to busy* state transition. Sometimes we are interested in the worst-case (maximum) latency and sometimes in the average. If we can put an upper bound on the latency, then we define the system as real time. A number of applications involve performing I/O functions on a fixed-interval basis. In a data acquisition system, the ADC is triggered (a new sample is requested) at the desired sampling rate.

Checkpoint 10.1: What is the difference between bandwidth and latency?

10.2   High-Speed I/O Applications

Before introducing the various solutions to a high-speed I/O interface, we will begin by presenting some typical applications.

10.2.1 Mass Storage

The first application is mass storage, including floppy disk, hard disk, tape, and CD-ROM. Writing data to disk with these systems involves:

1. Establishing the physical location to write (position the record head at the proper block, sector, track, cylinder, etc.)
2. Specifying the block size
3. Waiting for the physical location to arrive under the record head
4. Transmitting the data

Reading data from disk with these systems is similar and involves:

1. Establishing the physical location to read (position the read head at the proper block, sector, track, cylinder, etc.)
2. Specifying the block size
3. Waiting for the physical location to arrive under the read head
4. Receiving the data

Under most situations the size of the data block transferred is fixed. The bandwidth depends on the rotation speed of the disk and the information density on the medium (Figure 10.1). A typical hard drive can sustain about 10 to 40 Mbytes/s. The time to locate the physical location is called the *seek time.* Although seek time has a significant impact on the disk performance, it does not affect the latency or bandwidth parameters. A 32X CD-ROM has a peak bandwidth of 4.8 Mbytes/s. There is a wide range of disk speeds, but it is important to note that for most situations, the disk bandwidth will be

Figure 10.1
Data to/from the
read/write head of a
hard drive has a high
bandwidth.

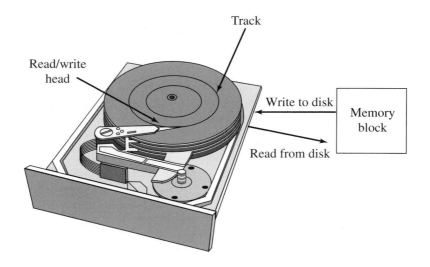

less than the computer bus bandwidth but greater than the maximum bandwidth that a software-controlled interface can achieve. If the disk interface is not buffered, then the interface must respond to each data byte at a rate as fast as the peak disk bandwidth. For example, in a disk read, once the data become available, the interface must capture and store them in memory, before the next data become available. If we do not meet the response time requirement in the disk interface, the rotation speed will have to be reduced. Notice that because of the seek time (time for the physical location to arrive under the head), the average and peak bandwidth will be quite different. Also notice that without buffering, the maximum interface latency will be inversely related to the peak bandwidth.

> *Checkpoint 10.2:* What happens if we are reading data off a hard drive and it doesn't satisfy the latency requirement? In other words, the read data is ready, but we do not capture it in time.

**10.2.2
High-Speed Data
Acquisition**

Examples of high-speed data acquisition are CD-quality sound recording (16-bit, two-channel, 44 kHz), real-time digital image recording, and digital scopes (8-bit 200 MHz). Sound recording actually has two high-speed data channels: one for recording into memory and a second for storing the memory data on hard disk or CD (Figure 10.2). A spectrum analyzer combines the high-speed data acquisition of a digital scope with the Discrete Fourier Transform (DFT) to visualize the collected data in the frequency domain. In the context of this chapter, we will define a high-speed data acquisition as one that samples faster than a software-controlled interface would allow.

Figure 10.2
Sound data from the
microphone are stored
in memory, processed,
then saved on CD.

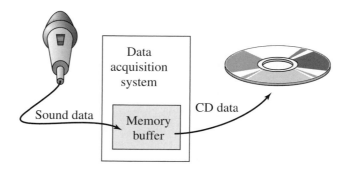

Checkpoint 10.3: What happens to the sound recording if data are missed?

10.2.3
Video Displays

Real-time generation of TV or video images requires an enormous data bandwidth. Consider the information bandwidth required to maintain an image on a graphics display (Figure 10.3). A video graphics array image is 256 colors (8-bit), 480 rows, 640 columns, and is refreshed at about 60 Hz. Calculating the bandwidth in bytes per second, we get $1 \cdot 480 \cdot 640 \cdot 60$, which is 18,432,000 bytes/s. Luckily, we don't have to communicate each pixel for each image but rather just transmit the changes from the previous image. To achieve the necessary bandwidth, video interface hardware will use a combination of DMA and dual port memories.

Figure 10.3
Image data from memory are displayed on the graphics screen.

10.2.4
High-Speed Signal
Generation

Examples of high-speed signal generation are CD-quality sound playback (16-bit, two-channel, 44 kHz), and real-time waveform generation. Sound playback also has two high-speed data channels: one for loading sound data into memory from CD and a second for playing the memory data out to the speakers (Figure 10.4). In the context of this chapter, we will define a high-speed interface as one that samples faster than a software-controlled interface would allow.

Figure 10.4
Sound data from the CD are loaded in memory, processed, then output to the speakers.

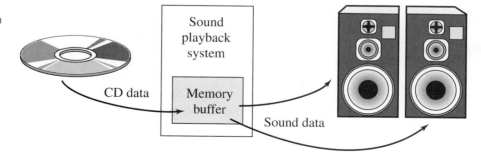

10.2.5
Network
Communications

For many networks the communication bandwidth of the physical channel will exceed the ability of the software to accept or transmit messages. For these high-speed applications, we will look for ways to decouple the software that creates outgoing messages and processes incoming messages from the hardware that is involved in the transmission

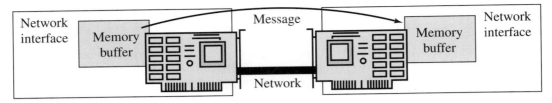

Figure 10.5
Message data are transmitted from the buffer of one computer to the buffer of another.

and reception of individual bits (Figure 10.5). Because the network load will vary, the average bandwidth (determined by how fast the transmission software can create outgoing messages and the reception software can process incoming messages) will be slower than the peak/maximum bandwidth that is achieved by the network hardware during transmission. This mismatch allows one network to be shared among multiple potential nodes.

Checkpoint 10.4: What happens in a communication system when packets are lost?

10.3 General Approaches to High-Speed Interfaces

10.3.1
Hardware FIFO

If the software-controlled interface can handle the average bandwidth but fails to satisfy the latency requirements, then a hardware FIFO can be placed between the I/O device and the computer. A common application of the hardware FIFO is the serial interface. Assume in this situation that the average serial bandwidth is low enough for the software to read the data from the serial port and write them to memory. We saw in Chapter 7 that the latency requirement of a SCI input port like the 6811/6812 is the time it takes to transmit one data frame. To reduce this interface latency requirement (without changing the average bandwidth requirement) we can add a hardware FIFO between the receive shift register and the receive data register, as illustrated in Figure 10.6.

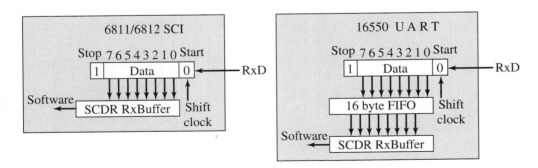

Figure 10.6
The 16550 UART employs a 16-byte FIFO to reduce the latency requirement of the interface.

Observation: With a serial port that has a shift register, a FIFO of size n, and one data register, the latency requirement of the input interface is the time it takes to transmit n+1 data frames.

A hardware FIFO, placed between the SCI data register and the transmit shift register, allows the software to write multiple bytes of data to the interface and then perform other tasks while the frames are being sent.

10.3.2 Dual Port Memory

One approach that allows a large amount of data to be transmitted from the software to the hardware is the dual port memory. A dual port memory allows shared access to the same memory between the software and hardware (Figure 10.7). For example, the software can create a graphics image in the dual port memory using standard memory write operations. At the same time the video graphics hardware can fetch information out of the same memory and display it on the computer monitor. In this way, the data need not be explicitly transmitted from the computer to the graphics display hardware. To implement a dual port memory, there must be a way to arbitrate the condition when both the software and hardware wish to access the device simultaneously. One mechanism to arbitrate simultaneous requests is to halt the processor using a MRDY signal so that the software temporarily waits while the video hardware fetches what it needs. Once the video hardware is done, the MRDY signal is released and the software resumes. None of the 6811/6812 expanded modes supports this sort of hardware-initiated cycle stretching. This hardware-controlled cycle stretching was defined in Chapter 9 as a partially asynchronous memory bus. Notice that except for the access conflict, both the software and graphics hardware can operate simultaneously at full speed.

Figure 10.7
A dual port memory can be accessed by two different modules.

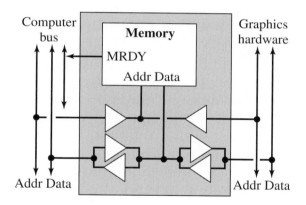

Checkpoint 10.5: Explain how the bidirectional tristate buffers connected to the memory data lines in Figure 10.7 work.

10.3.3 Bank-Switched Memory

Another approach similar to the dual port memory is the bank-switched memory. A bank-switched memory also allows shared access to the same memory between the software and hardware (Figure 10.8). The difference between bank-switched and dual port memory is that the bank-switched memory has two modes. In one mode (M=1), the computer has access to memory bank A, and the I/O hardware has access to memory bank B. In the other mode (M=0), the computer has access to memory bank B, and the I/O hardware has access to memory bank A. Because access is restricted in this way, there are no conflicts to resolve. The master controller determines the value of M.

Observation: With a bank-switched memory, the latency requirement of the software is the time it takes the hardware to fill (or empty) one memory bank.

Figure 10.8
A bank-switched memory can be accessed by two different modules, one at a time.

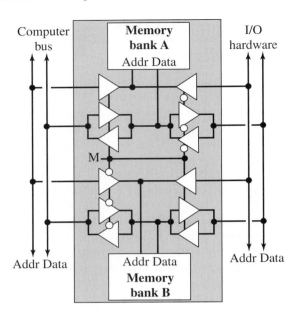

Many high-speed data processing systems employ this approach. The ADC hardware can write into one bank, while the computer software processes previously collected data in the other. When the ADC sampling hardware fills a bank, the memory mode is switched (by changing the M signal), and the software and hardware swap access rights to the memory banks. In a similar way, a real-time waveform generator or sound playback system can use the bank-switched approach. The software creates the data and stores them into one bank, while the hardware reads data from the other bank that was previously filled. Again, when the hardware is finished, the memory bank mode is switched by changing the M signal.

Checkpoint 10.6: How would you redesign the bank-switched memory in Figure 10.8 if the communication channel were simplex (data flows left to right only)?

10.4   Fundamental Approach to DMA

With a software-controlled interface (gadfly or interrupts) if we wish to transfer data from an input device into RAM, we must first transfer them from the input to the processor, then from the processor into RAM. To improve performance, we will transfer data directly from input to RAM or RAM to output using DMA. Because DMA bandwidth can be as high as the bus bandwidth, we will use this method to interface high-bandwidth devices like disks and networks. Similarly, because the latency of this type of interface depends only on hardware and is usually just a couple of bus cycles, we will use DMA for situations that require a very fast response. On the other hand, software-controlled interfaces have the potential to perform more complex operations than simply transferring the data to/from memory. For example, the software could perform error checking, convert from one format to another, implement compression/decompression, and detect events. These more complex I/O operations may preclude the usage of DMA.

10.4.1
DMA Cycles

During a *DMA read cycle,* the processor is halted and data are transferred from memory to output device (Figure 10.9). During a *DMA write cycle,* the processor is halted and data are transferred from input device to memory (Figure 10.10). In some DMA interfaces,

Figure 10.9
A DMA read cycle copies data from RAM or ROM to an output device.

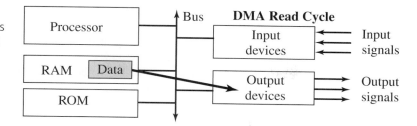

Figure 10.10
A DMA write cycle copies data from the input device into RAM.

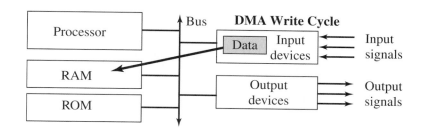

two DMA cycles are required to transfer the data. The first DMA cycle brings data from the source into the DMA module, and the second DMA cycle sends the data to their destination.

10.4.2
DMA Initiation

We can classify DMA operations according to the event that initiates the transfer. A *software-initiated* transfer begins with the program setting up and starting the DMA operation. Using DMA to transfer data from one memory block to another greatly speeds up the function. The efficiency of memory block transfers is very important in larger computer systems. Benchmarks on the Motorola 6808 show that even for block sizes as small as 4 bytes it is faster to initialize a DMA channel and perform the transfer in hardware. As the block size increases, the performance advantage of DMA hardware over traditional software becomes more dramatic.

Most DMA applications, however, involve a *hardware-initiated* DMA transfer. For an input device, the DMA is triggered on new data available. For an output device, the DMA is triggered on output device idle. For periodic events, like data acquisition and signal generation, the DMA is triggered by a periodic timer. These are the exact triggers involved in gadfly and interrupt synchronization discussed in Chapters 3 and 4. The difference with DMA is that the servicing of the I/O need will be performed by the DMA controller hardware without software explicitly having to transfer each byte.

10.4.3
Burst Versus Cycle
Steal DMA

When the desired I/O bandwidth matches the computer bus bandwidth, then the computer can be completely halted, while the block of data is transferred all at once (Figure 10.11). Once an input block is ready, a *burst mode DMA* is requested, the computer is halted, and the block is transferred into memory.

Figure 10.11
An input block is transferred all at once during burst mode DMA.

Figure 10.11 describes an input interface, but the same timing occurs on an output interface using burst mode DMA. For an output interface, the DMA is requested when the interface needs another block of data. During the burst mode DMA, the computer is halted, and an entire block is transferred from memory to the output device.

If the I/O bandwidth is less than the computer bus bandwidth, then the DMA hardware will steal cycles and transfer the data one DMA cycle at a time. In *cycle steal mode,* the software continues to run, although a little bit slower. In either case the processor is halted during the DMA cycles. Figure 10.12 describes an input interface, but the same timing occurs on an output interface using cycle steal mode DMA. For an output interface, the DMA is requested when the interface needs another byte of data. During the cycle steal DMA, one byte is transferred from memory to the output device.

Figure 10.12
Each time an input byte is ready, it is transferred to memory using cycle steal DMA.

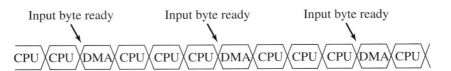

Observation: Some computers must finish the instruction before allowing a burst DMA. In this situation, the latency will be higher than cycle steal DMA, which does not need to finish the current instruction.

Since most I/O bandwidths are indeed less than the memory bandwidth, one technique to enhance speed is I/O buffering. In this approach a dedicated I/O memory buffer exists in the I/O interface hardware. This buffer is like the bank-switched memory discussed earlier. For example, on a hard disk read block operation, raw data comes off the disk and into the buffer. During this time the processor is not halted. When the buffer is full, burst DMA is used to transfer the data into the system memory. Similarly on a hard disk write block operation, the software initiates a burst DMA to transfer data from system memory into the I/O buffer. Once full, the I/O interface can write the data onto the disk.

Checkpoint 10.7: What is the maximum latency in a cycle steal DMA system?

**10.4.4
Single-Address
versus Dual-
Address DMA**

Some computer systems allow the transfer of data between the memory and I/O interface to occur in one bus cycle, while others need two bus cycles to complete the transfer. In a *single-address DMA cycle,* the address and R/W line dictate the memory function to be performed, and the I/O interface is sophisticated enough to know it should participate in the transfer.

In this single-address example, the disk interface is reading bytes from a floppy disk (Figure 10.13). Cycle steal mode will be used because the floppy disk bandwidth is slower than the bus. When a new byte is available, **Request** will be asserted, and this will request a DMA cycle from the DMA controller. The DMA controller will temporarily suspend the processor and drive the address bus with the memory address and the R/W to write. During this cycle the DMA controller will respond to the floppy interface by asserting the **Ack.** The floppy uses the **Ack** (ignoring the address bus and R/W) to know when to drive its data on the bus (Figure 10.14).

Observation: The MC68HC708XL36 does not support single-address DMA.

In a *dual-address DMA cycle,* two bus cycles are required to achieve the transfer (Figure 10.15). In the first cycle the data are read from the source address and copied into the DMA controller. During the first cycle the address bus contains the source address, and R/W signifies read. The information from the data bus is saved in the **Temp** register within

Figure 10.13
Block diagram showing the modules involved in a floppy disk read.

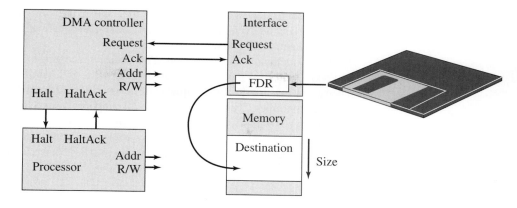

Figure 10.14
Timing diagram of a single-address DMA-controlled floppy disk read.

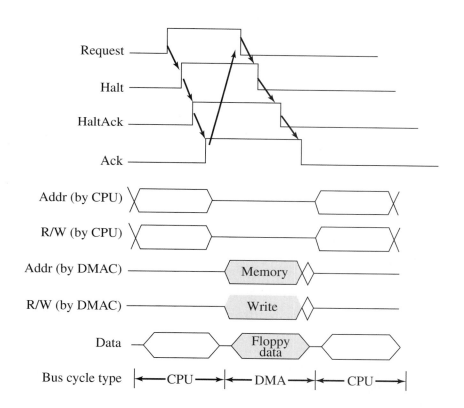

Figure 10.15
Block diagram showing the modules involved in a SPI read.

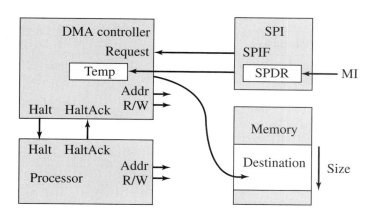

the DMA controller. In the second cycle, the data are transferred to the destination address. During the second cycle, the address bus contains the destination address, the data bus has the **Temp** data, and R/W signifies write.

In this dual-address example, the SPI interface is receiving bytes from a synchronous serial network. Cycle steal mode will be used because the SPI bandwidth is slower than the bus. When a new byte is available, **Request** will be asserted, and this will request a DMA cycle from the DMA controller. The DMA controller will temporarily suspend the processor and first drive the address bus with the SPI data register address (R/W=read), then in the second cycle the DMA controller will drive the address bus with the memory address (R/W=write). The SPI knows it has been serviced, because its data register has been read (Figure 10.16).

Figure 10.16
Timing diagram of a dual-address DMA-controlled SPI read.

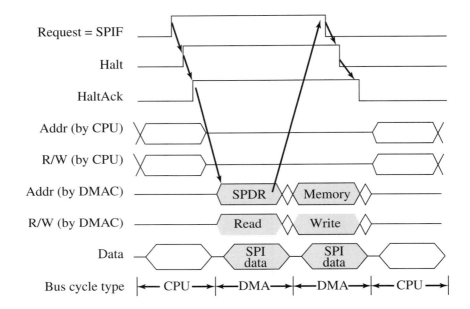

Observation: Single-address DMA is twice as fast as dual-address DMA.

Observation: The dual-address DMA can be used with I/O devices not configured to support DMA.

10.4.5 DMA Programming

Although DMA programming varies considerably from one system to another, there are a few initialization steps that most require. Table 10.2 lists the mode parameters that must be set to utilize DMA. There are two categories of DMA programming: initialization and completion. During initialization, the software sets the DMA parameters so that the DMA will begin.

Parameter	Possible choices
What initiates the DMA	Software trigger, input device, output device, periodic timer
Type	Burst versus cycle steal
Autoinitialization mode	Single event or continuous transfer
Precision	8-bit byte or 16-bit word
Mode	Single address or dual address
Priority	Should software completely halt? Should interrupts be serviced during DMA?
Synchronization	Set gadfly flag, or interrupt on block transfer complete

Table 10.2
DMA initialization usually involves specifying these parameters.

At the end of a block transfer, a done flag is set, and a number of additional actions may occur. If the system is armed, an interrupt can be generated. At the end of a block transfer in a continuous-transfer DMA, the parameters are autoinitialized so that the DMA process continues indefinitely. Table 10.3 lists additional parameters we will need to initialize.

Parameter	Definition
Source address	Address of the module (memory or input) that generates the data
Increment/decrement/static	Automatically $+1/-1/+0*$ the source address after each transfer
Destination address	Address of the module (memory or output) that accepts the data
Increment/decrement/static	Automatically $+1/-1/+0$ the destination address after each transfer
Block size	Fixed number of bytes to be transferred

*Obviously the DMA controller will $+2/-2/+0$ when the precision is 16 bits.

Table 10.3
More DMA initialization parameters.

10.5 LCD Graphics

10.5.1 LCD Graphics Controller

The graphics controller implements a **double buffer** scheme. A double buffer consists of two buffers of fixed size, and this size is typically many bytes. In our case, the size of each of the two buffers will match the size required to store one image of the LCD display. A data flow graph is drawn in Figure 10.17. In this system, the software writes to one buffer, while the hardware simultaneously reads data from the other buffer. The **front buffer** contains the image that we see on the display, whereas the **back buffer** is used to create the image we will see next. The graphics controller is configured to refresh the display over and over from the data contained in the front buffer. Independent from the refresh operation, the software writes data to the back buffer. The differences between a FIFO queue and a double buffer are data size and queue length. The data size of a FIFO is typically one or two bytes. This means that one puts and gets single bytes into and out of the FIFO queue. The data size of the double buffer is typically large (e.g., 80, 256, 1024 bytes.) This means that one always saves and removes big blocks into and out of the double buffer. The FIFO queue length is large (typically ranging from 16 to 60,000 bytes.) On the other hand, the double buffer has exactly two buffers. In Figure 10.17, Buf1 is shown as the back buffer and Buf2 is the front buffer. When a complete image is ready in the back buffer, the software triggers the double buffer to switch, making Buf2 the back buffer and Buf1 the front buffer.

Figure 10.17
A double buffer allows you to store data into one buffer while retrieving data from the other buffer.

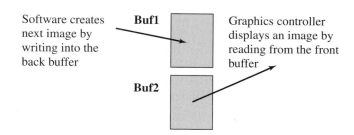

Software creates next image by writing into the back buffer

Buf1

Buf2

Graphics controller displays an image by reading from the front buffer

In general, we cannot interface an LCD graphics display directly to a microcontroller, because microcontrollers don't usually have enough RAM to support graphics and the processing speeds of microcontrollers are often not fast enough to refresh an LCD graphics display and perform the necessary functions of the embedded system in real time. Furthermore, even with systems with more memory and processing power, it is appropriate to employ a separate graphics processor to perform the low-level functions of the graphics display. In this section, we will design a simple graphics controller for a 320 by 240 monochrome LCD display. The specific device we will interface is the Oprex F-51477, but its features are typical of other displays of this size. Although designing a graphics controller with discrete digital logic is not usually practical, the design does illustrate the concepts of a graphics controller and high bandwidth interfacing. However, in the next section we will design a more practical system using integrated components. Figure 10.18 shows a block diagram of the controller. Because the LCD crystal does not have persistence, the image must be redrawn over and over. As long as the refresh rate is faster than 60 times per second, the image will look continuous to our eyes. The 320 by 240 monochrome display has 76800 pixels, and if each pixel is controlled by one binary bit, then 9600 bytes of storage are required to define one image. One refresh cycle involves sending the entire 78,800-bit pattern to the LCD display.

Figure 10.18
A simple graphics controller built with a bank-switched memory.

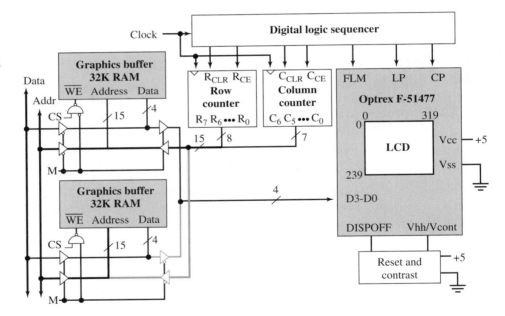

Figure 10.19 shows the basic timing required to send data to the LCD. Four bits are clocked into the device on each falling edge of **CP**. Therefore, our controller will change the data pins on the rising edge of **CP**. In particular, our address counters will be incremented on the rising edge of **CP**. Since there are 320 pixels in each row, it takes 80 **CP** pulses to enter data for one row. The falling edge of **LP** specifies the start of a new row, and there are 240 LP pulses required to update the entire display. The **FLM** signal will be high during the sending of the entire first row. If we wish to refresh at 60 Hz or faster, the period of the **CP** clock period must be smaller than 4s/(60*78,800), which is 846 ns. The column address specifies the x-coordinate (0 to 319), and the row address specifies the y-coordinate (0 to 239). Since the LCD display accepts 4 bits at a time, the **Column Counter** goes from 0 to 79, whereas the **Row Counter** steps from 0 to 239. The row and column counters can be implemented with a pair of 8-bit binary counters, such as 74LS590. To simplify addressing,

Figure 10.19
Basic timing used to refresh a 240 by 320 monochrome LCD display.

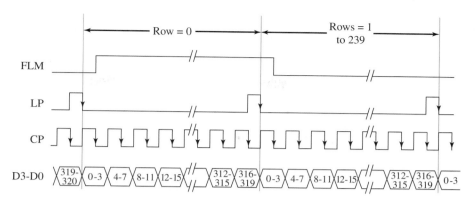

a 128-column by 256-row scheme will be employed. Although we allocate memory for 128 column steps (each step has 4 bits), we will use only 80 of them. Similarly, we allocate memory space for 256 rows, but use only 240 of them. The Optrex display implements a 4-bit data bus, but it will be easier to use an 8-bit RAM and just neglect four data pins. This means we will use a 32,768-byte static RAM (128*256) to store the 9600 bytes contained in one graph image.

Checkpoint 10.8: What would happen if you tried to write 6811/6812 software to output four bits directly to the LCD faster than every 842 ns?

Checkpoint 10.9: What would the display look like if the refresh rate were only 10 Hz?

The other part of the controller is its interface with the microcontroller. In a computer system with DMA, we could use that high speed channel to load the back buffer, but in a simple microcontroller like the 6811/6812 we can interface the back buffer to the address/data bus. It is also possible to connect the back buffer to regular output ports of a 6811/6812 running in single-chip mode. The signal **M** will be interfaced to a standard output pin of the microcontroller, allowing the software to determine which RAM will be the front buffer and which will be the back buffer. When the signal **M** is equal to 1, the top RAM contains the front buffer and the bottom RAM contains the back buffer. In this mode, the microcontroller performs a write cycle to the interface and the 4-bit data is stored in the bottom RAM. Similarly in this mode, the counter address controls the top RAM, and the data from the top RAM is fed to the LCD. When **M** is equal to 0, the microcontroller write cycle stores into the top RAM, and the data from the bottom RAM goes to the LCD. If the address range of the interface is 0 to \$7FFF, then the positive logic **CS** signal combines the E clock, R/W, and address decoder.

$$CS = E \cdot \overline{R/W} \cdot \overline{A15}$$

The last part of the design will be the digital logic sequencer. The input to the sequencer will be a 2 MHz clock, which will be connected to the clock input of both counters. This clock will also be the CP signal. Notice that the LCD captures data on the falling edge, and the 74HC590 will increment on the rising edge. This is fast enough to refresh the display, but slow enough to simplify the memory interfacing. Whether or not the counter will increment is determined by its negative logic counter enable, R_{CE} and C_{CE}. Since the column address is to be incremented each time, it is always enabled.

$$C_{CE} = 0$$

The binary counters have negative logic clears. In order to make the Column Counter step from 0 to 79, we will set its clear condition to occur when the 7-bit counter value equals

80 (1010000). Let C_{CLR} be the negative logic clear signal, and C_0–C_6 be the output of the row counter. Because the counter never reaches values 81–255, we can simplify the clear logic to

$$C_{CLR} = \overline{C_6 \bullet C_4}$$

We will enable the Row Counter to increment when the Column Counter equals 79 (1001111). The negative logic counter enable for the Row Counter will be

$$R_{CE} = \overline{C_6 \bullet \overline{C_5} \bullet \overline{C_4} \bullet C_3 \bullet C_2 \bullet C_1 \bullet C_0}$$

The LP signal will be high when CP=0 and the Column Counter equals 79 (1001111).

$$LP = \overline{CP} \bullet \overline{R_{CE}}$$

In order to make the Row Counter step from 0 to 239, we will set the clear condition to occur when the counter value equals 240 (11110000). Let R_{CLR} be the negative logic clear signal and R_0–R_7 the output of the row counter. Because the counter never reaches values 241–255, we can simplify the clear logic to

$$R_{CLR} = \overline{R_7 \bullet R_6 \bullet R_5 \bullet R_4}$$

Finally, the FLM signal is high during the output of the first row,

$$FLM = \overline{R_7} \bullet \overline{R_6} \bullet \overline{R_5} \bullet \overline{R_4} \bullet \overline{R_3} \bullet \overline{R_2} \bullet \overline{R_1} \bullet \overline{R_0}$$

Observation: A single-buffer implementation of this LCD interface to a 9S12C32 was built and tested, and its details can be found on the website **www.ece.utexas. edu/~valvano/metroworks.**

10.5.2 Practical LCD Graphics Interface

Although HD44780-controlled LCD displays are simple and inexpensive, many embedded applications require the display of both graphics and text. When faced with a project requiring LCD graphics, the most efficient solution is to use an integrated system that includes the graphics controller. Some of the parameters to consider when choosing a graphics system are physical size, the number of x-y pixels, colors, contrast, brightness, and power requirements. Some of the display systems include a touch panel, like those used in a personal digital assistant (PDA). The simplest approach to designing a graphics system is to purchase a graphics controller board like the one shown in Figure 10.20. The interface board includes a graphical processing unit (GPU), such as the Amulet AGB64LV01-QC, which handles the graphical functions in the background. A GPU is similar to a regular CPU, executing instructions, except that it performs many functions specific to graphics. The RAM contains one, two, or three buffers, where each buffer contains a mixture of graphics and text elements to draw on the display. The flash memory can hold GPU programs, graphical images, or fonts. The GPU coordinates the transfer of data from the front

Figure 10.20
A block diagram of an LCD graphics system using an Amulet CB-GT570 interface board.

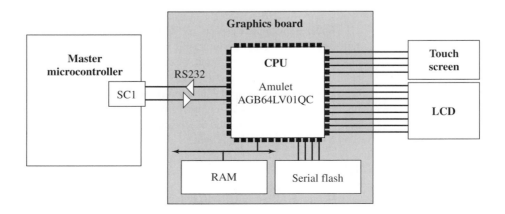

buffer to the display simultaneously with interacting with the master controller while it draws images in the back buffer(s). In this configuration, the master microcontroller communicates with the graphics system using a standard interface such as an RS232 serial. Manufacturers that produce LCD displays (e.g., Amulet, Hantronix, Lumex, Microtips, Optrex, Seiko, and Varitronix) often produce integrated graphics controllers as well.

10.6 Exercises

10.1 The objective of this exercise is to interface various devices to the computer using DMA synchronization. You may assume the bus bandwidth is at least 8 million bytes/s. For each device you are asked to select the most appropriate DMA mode. Assume the devices support single-address DMA. The 16-bit address of the memory buffer used in each case is 0×1234.

> **Write tape drive** Each tape block is 256 bytes. When a tape head is ready, the controller will signal that it is ready to accept all 256 bytes. At this time, the tape interface chip is ready to transfer as fast as possible all 256 bytes from the memory buffer at 0×1234 to the tape.
>
> **Sound input** The sound waveform buffer is located in memory at 0×1234. Your interface will read the 8-bit ADC 1024 times at 22 kHz and store the data in the buffer. Your software will be smart enough to create two 512-byte buffers out of the 1024 bytes (double buffer) so that it can process one buffer, while the A/D data are being stored automatically under DMA control into the other buffer. That is, when the 1024-byte wave buffer has been filled, the DMA system should repeat and fill it up again.
>
> **Read hard drive** There is a 256-byte buffer at 0×1234 that your DMA system will fill with data from the hard disk. When a hard drive read head is ready, the controller will signal that it has the next byte from the disk. It takes 10 ms for the read head to be ready, then the 256 bytes of data can be transferred from the disk to memory at 2 million bytes/s.

Fill in the following table that specifies the most appropriate mode for each device.

	Tape	Sound	Disk
Cycle steal or block transfer			
Read or write transfer			
Autoinitialization (yes or no)			
Address increment or decrement			
DMA address register value			
DMA count register value			

10.2 When a 256-byte block is written to a floppy disk, there are 256 separate single-address DMA cycles in cycle steal mode. This question deals with just one of these DMA transfers. There are 14 events listed below. First you will eliminate the events that do not occur during the DMA cycle that saves one byte on the disk. In particular, list the events that will not occur. Second, you will list the events that do occur in the proper sequence.

a) An interrupt is requested.

b) Registers are pulled from the stack.

c) Registers are pushed on the stack.

d) The DMAC asks the processor to halt by activating its **Halt** signal.

e) The DMAC deactivates its **Halt** request to the processor.

f) The DMAC tells the FDC interface that a FMA cycle is occurring by activating its **Ack** signal; the DMA Controller drives the address bus with the FDC address; the DMAC drives the control bus to signify a write cycle (e.g., R/W=0); the memory drives the data bus; the FDC accepts the data.

g) The DMAC tells the FDC interface that a DMA cycle is occurring by activating its **Ack** signal; the DMAC drives the address bus with the memory address; the DMAC drives the control bus to signify a memory read cycle (e.g., R/W = 1); the memory drives the data bus; the FDC accepts the data.

h) The FDC deactivates its DMA **Request** signal to the DMAC.

i) The FDC requests a DMA cycle to the DMAC by activating its **Request** signal.

j) The interrupt service routine is executed.

k) The write head is properly positioned over the place on the disk.

l) The processor address and control lines float; the processor responds to the DMAC that it is halted by activating its **HaltAck** signal.

m) The processor resumes software execution.

n) Wait until the current instruction is finished executing.

10.7   Lab Assignments

Lab 10.1. The overall objective of this lab is to build an LCD graphics display like the one described in Section 10.5.1. This is a complicated design that requires organization and testing. The first step is to choose digital logic devices fast enough for both the computer interface and the LCD timing. At 2 MHz 74HC logic will suffice, but at 25 MHz careful design and LCD layout will be required. One option for interfacing the system to a microcontroller running in single-chip mode is to interface a 19-bit shift register (15 address lines and 4 data lines) to the SPI. To write data into the back buffer, the software outputs three bytes to the SPI (address and data) then pulses it into the back buffer with a simple digital output line. Another option is to design a single buffered solution. In this system the counters and 19-bit shift registers have tristate outputs. To write to the buffer, the software activates the outputs of the shift registers, disables the outputs of the Row and Column counters, and stops the display clock. After the memory write cycle is complete, the software disables the outputs of the shift registers, enables the outputs of the Row and Column counters, and restarts the display clock.

The second part of the lab is the design of a low-level graphics device driver, which allows for drawing lines, drawing dots, displaying text, and erasing portions of the screen.

Lab 10.2. The overall goal is to design a dual port memory between two microcontrollers, as described in Section 10.3.2. There are a few ways to handle simultaneous requests that do not require dynamic bus cycle stretching with a MRDY signal. The first possibility is to design a hardware semaphore that the two microcontrollers can use to request access to the memory. In this scheme, only one microcontroller can access the memory at a time. A second approach is to give priority to the master computer, and if there is a conflict, let the master have access and ignore the slave. In this scheme, you'll have to add software checking to the slave to verify that a read/write access occurred. The third scheme requires design of both microcontrollers. The microcontrollers can be designed to run off external clocks. In this scheme, you generate a bus clock separately and feed both microcontrollers with the same clock, but out of phase. Then you design the memory interface to occur in the second half of the cycle (when E = 1). In this way, the bus conflict is avoided.

The second part of the lab is the design of low-level device driver software to implement bidirectional communication. Run main programs in each microcontroller that determine the maximum bandwidth of this channel. Will it be faster than a simple SPI channel?

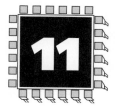

11 Analog Interfacing

Chapter 11 objectives are to:

❑ Design analog amplifiers and filters
❑ Study building blocks for data acquisition, including sample and hold, multiplexers, DAC, and ADC
❑ Discuss the functionality of the built-in ADCs

Most embedded systems include components that measure and/or control real-world parameters. These real-world parameters (like position, speed, temperature, and voltage) usually exist in a continuous (or analog) form. Therefore, the design of an embedded system involving these parameters rarely uses only binary (or digital) logic. Rather, we often will need to amplify, filter, and eventually convert these signals to digital form. In this chapter we will develop the analog circuit building blocks used in the design of data acquisition systems and control systems.

11.1 Resistors and Capacitors

11.1.1 Resistors

As engineers, we use resistors and capacitors for many purposes. The resistor or capacitor type is defined by the manufacturing process, the materials used, and the testing performed. The performance and cost of these devices varies significantly. For example, a 5% 0.25-W carbon resistor costs about 1 cent, while a 0.01% wire-wound resistor may cost $20. It is important to understand both our circuit requirements and the resistor parameters so that we match the correct resistor type to each application, yielding an acceptable cost/performance ratio. We must specify in our technical drawings the device type and tolerance (e.g., 1% metal film) so that our prototype can be effectively manufactured. The characteristics of various resistor types are shown in Table 11.1.

Table 11.1
General specification of various types of resistor components.

Type	Range	Tolerance	Temperature coef	Max power
Carbon composition	$1\ \Omega$ to $22\ M\Omega$	5 to 20%	0.1%/°C	2 W
Wire-wound	$1\ \Omega$ to $100\ k\Omega$	>0.0005%	0.0005%/°C	200 W
Metal film	$0.1\ \Omega$ to $10^{10}\ \Omega$	>0.005%	0.0001%/°C	1 W
Carbon film	$10\ \Omega$ to $100\ M\Omega$	>0.5%	0.05%/°C	2 W

Source: Wolf and Smith, *Student Reference Manual*, Prentice-Hall, p. 272, 1990.

The most common type of resistor is carbon composition. It is manufactured with hot-pressed carbon granules. Various amounts of filler are added to achieve a wide range of resistance values. We add them to digital circuits as +5-V pull-ups. To improve the accuracy and stability of our precision analog circuits, we will use resistors with a lower tolerance and better temperature coefficient. For most applications 1% metal film resistors will be sufficient to build our analog amplifier circuits. For very high precision analog circuits we could use wire-wound resistors. Wire-wound resistors are manufactured by twisting a very long, very thin wire like a spring. The wire is coiled up and down a shaft in such a way to try and cancel the inductance. Since some inductance remains, wire-wound resistors should not be used for high frequency (above 1 MHz) applications.

> **Observation:** All resistors produce white (thermal) noise.

> **Observation:** Metal film and wire-wound resistors do not generate 1/f noise, whereas carbon resistors do.

> **Common error:** If you design an electronic circuit and neglect to specify explicitly the resistor types, then an incorrect substitution may occur in the layout/manufacturing stage of the project.

11.1.2 Capacitors

Similarly, capacitors come in a wide variety of sizes and tolerances. Polarized capacitors operate best when only positive voltages are applied. Nonpolarized or bipolar capacitors operate for both positive and negative voltages. We select a capacitor based on the following parameters: capacitance value, polarized/nonpolarized, voltage level, tolerance, leakage current (resistance), temperature coefficient, useful frequency response, and temperature range. Another parameter of capacitors is the maximum voltage rating. This parameter is important for high-voltage circuits but is of lesser importance for embedded microcomputer systems. Table 11.2 compares various capacitor types we could use in our circuit.

Type	Range	Tolerance	Temp coef	Leakage	Frequencies
Polystyrene	10 pF to 2.7 µF	±0.5%	Excellent	10 GΩ	0 to 10^{10} Hz
Polypropylene	100 pF to 50 µF	Excellent	Good	Excellent	
Teflon	1000 pF to 2 µF	Excellent	Best	Best	
Mica	1 pF to 0.1 µF	±1 to ±20%		1000 MΩ	10^3 to 10^{10} Hz
Ceramic	1 pF to 0.01 µF	±5 to ±20%	Poor	1000 MΩ	10^3 to 10^{10} Hz
Paper (oil-soaked)	1000 pF to 50 µF	±10 to ±20%		100 MΩ	100 to 10^8 Hz
Mylar (polyester)	5000 pF to 10 µF	±20%	Poor	10 GΩ	10^3 to 10^{10} Hz
Tantalum	0.1 to 220 µF	±10%	Poor		
Electrolytic	0.47 µF to 0.01 F	±20%	Ghastly	1 MΩ	10 to 10^4 Hz

Source: Modified from Wolf and Smith, *Student Reference Manual,* Prentice-Hall, p. 302, 1990; and Horowitz and Hill, *The Art of Electronics,* Cambridge University Press, p. 22, 1989.

Table 11.2
General specification of various types of capacitor components.

We will use capacitors for two purposes in our microcomputer-based embedded systems. First, we will place them on the DC power lines to filter the supply voltage to our circuits (Figure 11.1). A voltage supply typically will include ripple, which is added noise on top of the DC voltage level. There are two physical locations to place the supply filters. The first location is at the entry point of the supply voltage onto the circuit board. There are two approaches to this filter. If the supply noise is mostly voltage ripple, then two capacitors in parallel can be used. The large-amplitude polarized capacitor (e.g., 1 to 47 µF electrolytic, tantalum) will remove low-frequency large-amplitude voltage noise, and the nonpolarized capacitor (e.g., 0.01 to 0.1 µF ceramic) will remove high-frequency noise. Two different types of capacitors are used because they are effective (i.e., behave like a capacitor) at different frequencies. The Π filter (CLC) is effective in situations where the supply is very noisy. The Π filter is good for separating boards that have large current spikes from stealing power from each other. The magnitudes of the parameters depend on the amplitude of the supply ripple.

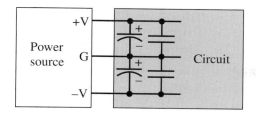 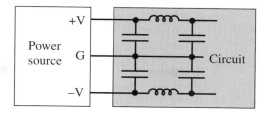

Figure 11.1
DC supply filters.

In addition to the supply filter, we will add bypass capacitors at the supply pins of each chip. It will be important to place these capacitors as close to the pin as physically possible (Figure 11.2). A nonpolarized capacitor (e.g., 0.01 to 0.1 μF ceramic or polyester) will smooth the supply voltage as seen by the chip. Placing the capacitor close to the chip prevents current surges from one chip from affecting the voltage supply of another.

Figure 11.2
Printed circuit board layout positioning the bypass capacitors close to the chip.

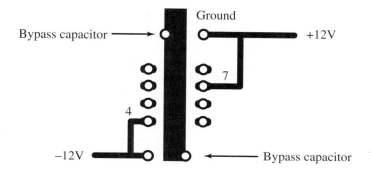

The second application of capacitors in our embedded systems will be in the linear analog circuits of the low-pass filter, the high-pass filter, the derivative circuit, and the integrator circuit. For these applications we will select a nonpolarized capacitor even if the signal amplitude is always positive. In addition, we usually want a low-tolerance, low-leakage capacitor to improve the accuracy of the linear analog circuit.

> *Performance tip:* Teflon, polystyrene, and polypropylene capacitors are excellent choices when designing precision analog filters, and Mylar or ceramic capacitors may be as well.

> *Observation:* The maximum voltage rating of a capacitor is limited by its size.

> *Common error:* Polarized capacitors often have a nonlinear capacitance versus frequency response, therefore using them in an analog filter will cause distortion.

> *Observation:* A computer engineer interested in the field of embedded systems will find more job opportunities if he or she can develop microcontroller skills along with some analog circuit design skills.

11.2 Operational Amplifiers (Op Amps)

11.2.1 Op Amp Parameters

Although the design of analog electronics is not an explicit objective of this book, we will include a brief discussion of analog circuit design issues often related to embedded systems. For example, small size, the ability to operate on a single voltage supply, and low supply current are three parameters typical of embedded systems. Other factors to consider are package size, cost, availability, and temperature range. There are hundreds of op amps currently

Single Op Amp Double Op Amp Quad Op Amp	OPA227 OPA2227 OPA4227	OPA132 OPA2132 OPA4132	TLC2272 TLC2274	MAX495 MAX492 MAX494
Description	High Precision	High-Speed FET	Rail-to-Rail	Rail-to-Rail
K, Open loop gain	160 dB	130 dB	104 dB	108 dB
R_{cm}, Input impedance	1 GΩ ‖ 3pF	10^{13} Ω ‖ 6pF	10^{12} Ω ‖ 8pF	
R_{diff}, Input impedance	10 MΩ ‖ 12pF	10^{13} Ω ‖ 2pF	10^{12} Ω ‖ 8pF	2 MΩ
V_{os}, Offset voltage	0.075 mV	0.5 mV	3 mV	0.5 mV
I_{os}, Offset current	10 nA	50 pA	100 pA	6 nA
I_b, Bias current	10 nA	50 pA	100 pA	60 nA
e_n, Noise density	3 nV/√Hz	23 nV/√Hz	50 nV/√Hz	25 nV/√Hz
f_1, Gain*bandwidth product	8 MHz	8 MHz	2.18 MHz	500 kHz
dV/dt, Slew rate	2.3 V/μs	20 V/μs	3.6 V/μs	0.2 V/μs
±V_s, Voltage supply	±5 to ±15 V	±2.5 to ±18 V	0 to 5 or ±5 V	2.7 to 6 V
±I_s, Supply current	±3.8 mA	±4.8 mA	3 mA	170 μA

Table 11.3
Parameters for various op amps used in this chapter.

available, making the choice of which devices to use confusing. The manufacturers of op amps publish a selection guide of their products to assist in finding an appropriate part. Table 11.3 lists performance parameters of four op amps, and these particular devices were chosen because they typify the kinds of op amps used in embedded system, but the choice is not meant as a recommendation. These devices are all available in small packages with 1, 2, or 4 op amps per package. We use rail-to-rail op amps, such as the TLC2274 and MAX494 to design analog circuits that run on a single +5 V supply. The OPA4227 and OPA4132 are traditional op amps to be used in high-precision, low-noise applications. The MAX494 is a low-current device. We will begin our discussion of op amps with the ideal op amp. For simple analog circuits using high-quality devices, the ideal model will be sufficient. With an ideal op amp, the output voltage, V_{out}, is linearly related to the difference between the input voltages, $V_{out} = K(V_y - V_x)$, where the gain, K, is a very large number, as shown in Figure 11.3.

Figure 11.3
Ideal op amp.

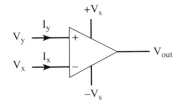

Voltage ranges of the inputs and outputs are bounded by the supply voltages, $+V_s$ and $-V_s$. Op amp circuits found in many traditional analog design textbooks are powered with ±12−V supplies. On the other hand, we will find it convenient to power our embedded systems with a single voltage. More specifically, we will set $+V_s$ to +5 V and $-V_s$ to ground. Either way, we assume the input and output voltages will be between $+V_s$ and $-V_s$. A "rail-to-rail" op amp operates in its linear mode for output voltages all the way from $-V_s$ to $+V_s$. We will see later how to create an effective ±2.5 V analog supply from the single +5 V digital supply. The input currents into the op amp are very small. Because the input impedance of the op amp is large, we will assume I_x and I_y are zero.

If a feedback resistor is placed between the output and the negative terminal of the op amp, then this feedback will select an output such that V_x is very close to V_y. In the ideal model, we let $V_x = V_y$. One way to justify this behavior is to recall that $V_{out} = K \cdot (V_y - V_x)$. Since K is very large, the only way for V_{out} to be between $-V_s$ and $+V_s$ is for V_x to be very close to V_y.

Positive feedback or no feedback drives V_{out} to equal $-V_s$ or $+V_s$. If a feedback resistor is placed between the output and the positive terminal of the op amp, then this feedback will saturate the output to either the positive or negative supply. With no feedback, the output will also saturate. In both cases the output will saturate to the positive supply if $V_y > V_x$, and to the negative supply if $V_x > V_y$. We will see later, that positive feedback can be used to create hysteresis.

Checkpoint 11.1: What is the open loop gain of a MAX494 in units of V/V?

Although the input impedance of an op amp is large, it is not infinite and some current enters the input terminals. Figure 11.4 illustrates the definition of input impedance. We define the *common-mode input impedance,* R_{cm}, as the common-mode voltage divided by the common-mode current. We define the *differential input impedance,* R_{diff}, as the differential voltage divided by the differential current. These parameters vary considerably from op amp to op amp. The CMOS and FET devices have a very large input impedance.

Figure 11.4
Definition of op amp input impedance.

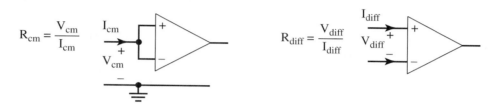

The next realistic parameter we will define is *open-loop output impedance*. When the op amp is used without feedback, the op amp output impedance is defined as the open-circuit voltage divided by the short-circuit current. The output impedance is a measure of how much current the op amp can source or sink. The open-loop output impedance of the TLC2274 is 140 Ω.

Checkpoint 11.2: The MAX494 has an output short-circuit current of 30 mA. Assuming an output of 5 V, what is its output impedance?

Observation: The input and output impedances of the op amp are not necessarily the same as the input and output impedances of the entire analog circuit.

As illustrated in Figure 11.5, the op amp *offset voltage* V_{os} is defined as the voltage difference between V_y and V_x, which yields an output of zero. The *offset current* I_{os} is defined as the current difference between V_y and V_x. There will be an output error equal to the offset voltage times the gain of amplifier. Similarly, the offset current creates an offset voltage through the resistors of the circuit. Some op amps provide the capability to add an external potentiometer to nullify the two offset errors. Alternatively, we can reduce the offset error by using a more expensive lower offset op amp, such as the OPA4227. The op amp bias current I_b is defined as the common current coming out of both V_y and V_x. We can reduce the effect of bias current by selecting resistors in order to equalize the effective impedance to ground from the two input terminals.

Figure 11.5
Definition of op amp offset voltage, offset current, and bias current.

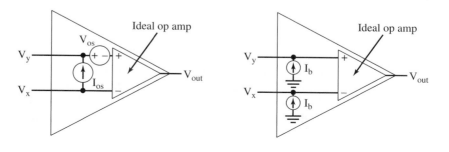

Observation: The use of a null offset pot increases manufacturing costs and incurs a labor cost to adjust it periodically. Therefore, the overall system cost may be reduced by using more expensive op amps that do not require a null offset pot.

Observation: If the gain and offset are small enough not to saturate the output, then the offset error can be corrected in software by adding or subtracting an appropriate constant.

Checkpoint 11.3: Consider the situation where a MAX494 is used to create an analog amplifier with gain 100. What will be the output error due to offset voltage?

Checkpoint 11.4: Why can't a TLC2274 be used to create an analog amplifier with gain 1000?

The *input voltage noise*, V_n, arises from the thermal noise generated in the resistive components within the op amp. Due to the white-noise process, the magnitude of the noise is a function of the bandwidth (BW) of the analog circuit. This parameter varies quite a bit from op amp to op amp. To calculate the RMS amplitude of the voltage noise, we need to calculate, $V_n = e_n \bullet \sqrt{(BW)}$. To reduce the effect of noise, we can limit the bandwidth of the analog system using an analog low pass filter. The *output voltage noise* will be the input voltage noise multiplied by the gain of the circuit.

There are two approaches to defining the transient response of our analog circuits. In the frequency domain we can specify the frequency and phase response. In the time domain, we can specify the step response. For most simple analog circuits designed with op amps, the frequency response depends on the op amp performance and the analog circuit gain. If the unity-gain op amp frequency response is f_1, then the frequency response at gain, G, will be f_1/G. The *bandwidth* (BW) is defined as the frequency at which the gain (V_{out}/V_{in}) drops to 0.707 of the original. The output *slew rate* is the maximum slope that the output can generate. Slew rate is important if the circuit must respond quickly to changes in input (e.g., a sensor detecting discrete events). Alternatively, bandwidth is important if the circuit is responding to a continuously changing input (e.g., audio and video).

Checkpoint 11.5: Consider the situation where a MAX494 is used to create an analog amplifier with a gain of 100. What will be the bandwidth of this circuit? Given this bandwidth, what will be the RMS output voltage noise?

When we consider the performance of a linear amplifier, normally we specify the voltage gain, input impedance, and output impedance. These three parameters can be lumped into a single parameter, A_{db}, called the *power gain*. Let V_{in} R_{in} be the inputs and V_{out} R_{out} be the outputs of our amplifier. The input and output powers are $P_{in} = V^2{}_{in}/R_{in}$ and $P_{out} = V^2{}_{out}/R_{out}$, respectively. Then the power gain in decibels has voltage gain and impedance components.

$$A_{db} = 10 \log_{10}\frac{P_{out}}{P_{in}} = 20 \log_{10}\frac{V_{out}}{V_{in}} + 10 \log_{10}\frac{R_{in}}{R_{out}}$$

11.2.2
Threshold
Detector

Threshold detectors are very important in embedded systems. Recall that all through Chapter 6, a threshold detector was used to convert an external analog signal into a digital signal so that the period, pulse width, or frequency could be measured. An entire class of voltage comparators exist for this purpose. The LM311 is a typical voltage comparator with analog inputs (V_x, V_y), analog voltage supplies ($+V_s$, $-V_s$), digital output (D_{out}), and digital supply (pin 1 ground). The LM311 will operate on a wide range of analog supply voltages ($+V_s$ to $-V_s$) from 0 to +5 V, all the way to −15 to +15 V. When the input V_y is above the input V_x, then the digital output D_{out} becomes high impedance, and the pull-up resistor creates a +5-V output. On the other hand, when V_y is below V_x, then D_{out} saturates to 0.4 V. The LM311 offset voltage of 7.5 mV can be reduced with the null offset pot connected to pins 5, 6. The LM311 offset current is 50 nA, and bias current is 250 nA. The input voltages of the LM311 must be between $-V_s$ and $+V_s$. The LM311 can sink -8 mA output low current. Since the LM311 has an open-collector output, it cannot source any output high current. The LM311 settling time is 200 ns (Figure 11.6).

Figure 11.6
A voltage comparator using a LM311.

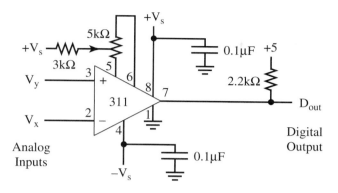

11.2.3
Simple Rules for Linear Op Amp Circuits

A wide range of analog circuits can be designed by following these simple design rules.

1. Choose quality components. It is important to use op amps with good enough parameters. Similarly, we should use low-tolerance resistors and capacitors. On the other hand, once the preliminary prototype has been built and tested, then we could create alternative designs with less expensive components. Because a working prototype exists, we can explore the cost/performance trade-off.

2. Negative feedback is required to create a linear mode circuit. As mentioned earlier, the negative feedback will produce a linear I/O response. In particular, we place a resistor between the negative input terminal and the output.

3. Assume no current flows into the op amp inputs. Since the input impedance of the op amp is large compared to the other resistances in the circuit, we can assume that $I_x = I_y = 0$.

4. Assume negative feedback equalizes the op amp input voltages. If the analog circuit is operating in linear mode with negative feedback, then we can assume $V_x = V_y$.

5. Choose resistor values in the 1 kΩ to 1 MΩ range. To have the resistors in the circuit be much larger than the output impedance of the op amp and much smaller than the input impedance of the op amp, we choose resistors in the 1 kΩ to 1 MΩ range. If we can, it is better to restrict to the 10 kΩ to 100 kΩ range. If we choose resistors below 1 kΩ, then currents will increase. If the currents get too large, the batteries will drain faster and the op amp may not be able to source or sink enough current. As the resistors go above 1 MΩ, the white noise increases, the current errors (I_{os}, I_b, I_n) become more significant. In addition, low-tolerance precision resistors are expensive in sizes above 2 MΩ.

6. The analog circuit BW depends on the gain and the op amp performance. Let the unity gain op amp frequency response be f_1 and the analog circuit gain be G. The frequency response or BW of the analog circuit will be f_1/G.

7. Equalize the effective resistance to ground at the two op amp input terminals. To study the bias currents, consider all other voltage sources as shorts to ground and all other current sources as open circuits. Adjust the resistance values in the circuit so that the impedance from the positive terminal to ground is the same as the impedance from the negative terminal to ground. In this way, the bias currents will create a common-mode voltage, which will not appear at the op amp output because of the common-mode rejection of the op amp (recall the op amp amplifies differential voltage inputs).

8. The input impedance of the analog circuit is the input voltage divided by the input current. If the analog circuit has a single input voltage, then the input impedance Z_{in} is simply the input voltage divided by the input current (Figure 11.7). If the input stage of the analog circuit is a differential amplifier with two input voltages, then we can specify the common-mode input impedance Z_{cm} and the differential mode input impedance Z_{diff} (Figure 11.8).

Figure 11.7
Definition of input impedance for an analog circuit with a single input.

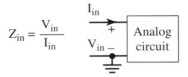

Figure 11.8
Definition of input impedance for an analog circuit with two inputs.

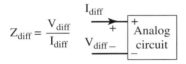

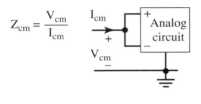

> *Observation:* In most cases, the differential mode input impedance of the analog circuit will be the differential mode input impedance of the op amp.

9. Match input impedances to improve common-mode rejection ratio (CMRR). If the input stage of the analog circuit is a differential amplifier with two input voltages, a very important performance parameter is called CMRR. It is assumed that the signal of interest is the differential voltage, whereas common-mode voltages are considered noise. The CMRR is defined to be the ratio of the differential gain divided by the common-mode gain. In decibels, it is calculated as

$$\text{CMRR} = 20 \bullet \log_{10} \frac{G_{diff}}{G_{cm}}$$

Therefore a differential amplifier with a large CMRR will pass the signal and reject the noise (Figure 11.9).

There are two design sets of impedances we must match to achieve a good CMRR. Each amplifier input has a separate input impedance to ground, shown as Z_{in1} and Z_{in2} in Figure 11.10. To improve CMRR, we make Z_{in1} equal to Z_{in2}. Similarly, signal source has a separate output impedance to ground, Z_{out1} and Z_{out2}. Again, we try to make Z_{out1} equal to Z_{out2}. On the other hand, if Z_{in1} does not equal Z_{in2} or if Z_{out1} does not equal Z_{out2} then a common-mode signal (e.g., added noise in the cable) will appear as a differential signal to the analog circuit and thus be present in the output.

Figure 11.9
Definition of common-
mode rejection ratio
(CMRR).

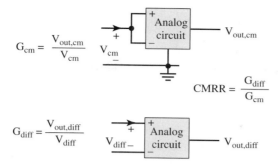

$$G_{cm} = \frac{V_{out,cm}}{V_{cm}}$$

$$CMRR = \frac{G_{diff}}{G_{cm}}$$

$$G_{diff} = \frac{V_{out,diff}}{V_{diff}}$$

Figure 11.10
Circuit model for
improving the CMRR.

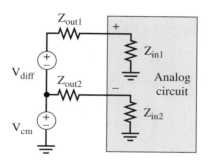

11.2.4
Linear Mode Op
Amp Circuits

We will begin with one of the simplest linear mode analog circuits, which is the inverting amplifier. The gain is the R_2/R_1 ratio. Notice that the gain response is independent of R_3. Thus, we can choose R_3 to be the parallel combination of $R_1 \| R_2$ so that the effect of the bias currents is reduced.

Figure 11.11
Inverting amplifier.

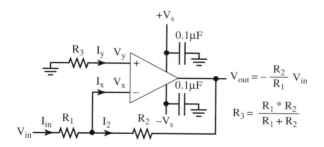

$$V_{out} = -\frac{R_2}{R_1} V_{in}$$

$$R_3 = \frac{R_1 * R_2}{R_1 + R_2}$$

Because I_y is zero, V_y is also zero. Because of negative feedback, V_x equals V_y. Thus, V_x equals zero too. Because V_x is zero, I_{in} is V_{in}/R_1 and I_2 is $-V_{out}/R_2$. Because I_x is zero, I_{in} equals I_2. Setting I_{in} equal to I_2 yields

$$V_{out} = -(R_2/R_1) \cdot V_{in}$$

The input impedance (Z_{in}) of this circuit (defined as V_{in}/I_{in}) is R_1. If the circuit were built with an OPA227 (which has a gain bandwidth product of 8 MHz), the bandwidth of this circuit would be 8 MHz divided by the gain.

Common error: This low input impedance of Z_{in} may cause loading on the previous analog stage.

> ***Observation:*** The inverting amplifier input impedance is independent of the op amp input impedance.

The negative feedback will reduce the output impedance of the amplifier, Z_{out}, to a value much less than the output impedance of the op amp itself, R_{out}. To calculate Z_{out}, we first determine the open circuit voltage

$$V_{open} = -(R_2/R_1) \bullet V_{in}$$

We next determine the short circuit current, I_{short}. This means we consider what would happen if the output were shorted to ground, If the output is shorted, the circuit is no longer in feedback mode, and V_x will not equal V_y. In fact, V_x will be a simple voltage divider from V_{in} through R_1 and R_2 to ground,

$$V_x = V_{in} \bullet R_2/(R_1+R_2)$$

Because of the large open-loop gain, the ideal output will attempt to become

$$V_o = K \bullet (V_y-V_x) = -K \bullet V_{in} \bullet R_2/(R_1+R_2)$$

The short circuit current will be a function of the ideal output voltage, and the output resistance of the op amp,

$$I_{short} = V_o/R_{out}$$

The output impedance of the circuit is defined to be the open circuit voltage divided by the short circuit current, which for this inverting amplifier is

$$Z_{out} = V_{open}/I_{short} = R_{out} \bullet (R_2+R_1)/(K \bullet R_1)$$

> ***Observation:*** The output impedance of analog circuits using op amps with negative feedback is typically in the mΩ's.

For example, if we wished to design an amplifier with $V_{out} = -10V_{in}$, then we would make R_2/R_1 equal to 10. Following the rule to make the resistors in the 1 kΩ to 1 MΩ range, we could choose R_1 equal to 10 kΩ and choose R_2 equal to 100 kΩ.

> ***Checkpoint 11.6:*** If R_1 is equal to 10 kΩ and R_2 is equal to 100 kΩ, what value should you choose for R_3 to remove the error due to the bias currents?

Part	Voltage (V)	$\pm$Accuracy (mV)
AD1580, AD589, REF1004, MAX6120, LT1034, LM385	1.2	1 to 15
ADR420, ADR520, REF191, MAX6191, LT1790, LM4120	2.048	1 to 10
AD580, REF03, REF43, REF1004, MAX6192, MAX6225, LT1389, LM336	2.5	1 to 75
AD1583, ADR530, ADR423, REF193, MAX6163, LT1461, LM4120	3	1.5 to 10
ADR366, REF196, MAX6331, LT1461, LM3411, LM4120	3.3	4 to 10
AD1584, ADR540, ADR292, REF198, MAX6241, LT1790, LM4040	4.096	2 to 8
ADR425, AD586, REF02, REF195, MAX6250, LT1027, LT1236, LM336	5.0	2 to 20
AD581, AD587, AD633, REF01, LT1236, LM4040, LM4050	10.0	5 to 30

Table 11.4
Parameters of various precision reference voltage chips.

> ***Common error:*** Precision reference chips do not provide much output current and should not be used to power other chips.

A *mixed-signal design* includes both analog and digital components. The classic approach to combining analog and digital circuits is to power the analog system with a low noise ±12 V power supply, maintain separate analog and digital grounds, and connect the analog ground to the digital ground only at the ADC. As mentioned earlier, one of the limitations of the ADC built into a microcontroller is that analog signals extending beyond the 0 to +5 V range will permanently damage the microcontroller. One approach to allowing signed analog voltages, while still using a single voltage supply and protecting the microcontroller, is to create an analog ground that is at a different potential from the digital ground. For our 6811/6812 systems, which are powered with a +5V supply, we will create an analog ground that is at 2.5 V relative to the digital ground. Thus, analog signals ranging from −2.5 to +2.5 V are actually 0 to +5 V relative to the 6811/6812 digital ground. The first step to implementing this approach is to use an analog reference chip, like the ones shown in Table 11.4, to create a low-noise +2.50 V signal (the analog ground). The second step is to connect power to the analog circuits with $-V_s$ set to digital ground, and the $+V_s$ set to the +5 V supply. We will use rail-to-rail op amps that operate on 0 to +5 V power, like the TLC2274 or MAX494. The last step is to replace all connections to analog ground with the low-noise +2.50 V signal. Figure 11.12 shows the inverting amplifier from Figure 11.11, redesigned to operate on a single +5 V supply. The signals V_{in} and V_{out} are allowed to vary from 0 to +5 V relative to digital ground, but relative to the analog ground, these signals will vary from −2.5 to +2.5 V.

Figure 11.12
Inverting amplifier with an effective −2.5 V to +2.5 V analog signal range.

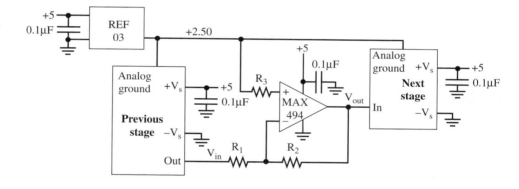

Observation: It is important in this scheme to separate the digital and analog grounds avoiding direct connections between the two grounds.

The second linear mode circuit we will study is the noninverting amplifier, as shown in Figure 11.13. The gain is $1+R_2/R_1$. The noninverting amplifier cannot have a gain less than 1. Just as with the inverting amp, the gain response is independent of R_3. So, we choose R_3 to be the parallel combination of $R_1\|R_2$ so that the effect of the bias currents is reduced.

Figure 11.13
Noninverting amplifier.

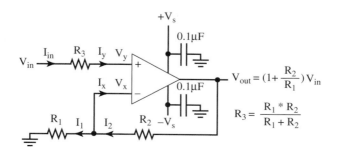

Because I_y is zero, V_y equals V_{in}. Because of negative feedback, V_x equals V_y. Thus, V_x equals V_{in} too. Calculating currents we get I_1 is V_{in}/R_1 and I_2 is $(V_{out}-V_{in})/R_2$. Because I_x is zero, I_1 equals I_2. Setting I_1 equal to I_2 yields

$$V_{out} = (1 + R_2/R_1) \cdot V_{in}$$

Using the simple op amp rules I_y is zero, hence the input impedance (Z_{in}) of this circuit (defined as V_{in}/I_{in}) would be infinite. In this situation we specify the amplifier input impedance to be the op amp input impedance. If the circuit were built with an OPA227 (which has a gain bandwidth product of 8 MHz), the bandwidth of this circuit would be 8 MHz divided by the gain.

> *Checkpoint 11.7:* If the noninverting amplifier in Figure 11.13 were built with an OPA227, what would be the input impedance of the amplifier?

The calculation of the output impedance of this amp follows the same approach as the inverting amp,

$$Z_{out} = R_{out} \cdot (R_2+R_1)/(K \cdot R_1)$$

> *Checkpoint 11.8:* List the design steps you would take to convert the noninverting amplifier, shown in Figure 11.13, to operate on analog signals varying from -2.5 to 2.5 V using a single power supply.

This design example works with any analog circuit in the form of

$$V_{out} = A_1V_1 + A_2V_2 + \ldots + A_nV_n + B$$

Figure 11.14
Boilerplate circuit model for linear circuit design.

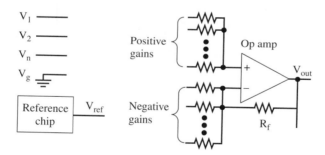

where $A_1\,A_2 \ldots A_n\,B$ are constants and $V_1\,V_2 \ldots V_n$ are input voltages. The circuit will be designed with one op amp, beginning with the boiler plate shown in Figure 11.14.

The **first step** is to choose a reference voltage from available reference voltage chips, like those shown in the Table 11.4. Some of the manufacturers that produce voltage references are Analog Devices, Burr-Brown, Linear Technology, Maxim, and National Semiconductor. The parameters to consider when choosing a voltage reference are voltage, package configuration, accuracy, temperature coefficient, and power. In particular, let V_{ref} be this reference voltage.

> *Common error:* If you use a resistor divider from the power supply to create a voltage constant, then the power supply ripple will be added directly to your analog signal.

The **second step** is to rewrite the design equation in terms of the reference voltage, V_{ref}. In particular, we make $A_{ref} = B/V_{ref}$.

$$V_{out} = A_1V_1 + A_2V_2 + \ldots +A_nV_n + A_{ref}V_{ref}$$

where $A_1\,A_2 \ldots A_n\,A_{ref}$ are constants and $V_1\,V_2 \ldots V_n$ are input voltages.

The **third step** is to add a ground input to the equation. Ground is zero volts ($V_g = 0$), but it is necessary to add this ground so that the sum of all the gains is equal to one.

$$V_{out} = A_1 V_1 + A_2 V_2 + \ldots + A_n V_n + A_{ref} V_{ref} + A_g V_g$$

Choose A_g such that

$$A_1 + A_2 + \ldots + A_n + A_{ref} + A_g = 1$$

In other words, let

$$A_g = 1 - (A_1 + A_2 + \ldots + A_n + A_{ref})$$

The **fourth step** is to choose a feedback resistor, R_f, in the range of 10 kΩ to 1 MΩ. The larger the gains, the larger the value of R_f must be. Then calculate input resistors to create the desired gains. In particular,

$$
\begin{aligned}
|A_1| &= R_f/R_1 & \text{so } R_1 &= R_f/|A_1| \\
|A_2| &= R_f/R_2 & \text{so } R_2 &= R_f/|A_2| \\
|A_n| &= R_f/R_n & \text{so } R_n &= R_f/|A_n| \\
|A_{ref}| &= R_f/R_{ref} & \text{so } R_{ref} &= R_f/|A_{ref}| \\
|A_g| &= R_f/R_g & \text{so } R_g &= R_f/|A_g|
\end{aligned}
$$

Observation: We will get a simple solution if we choose the value of R_f to be a common multiple of the gains A_1, A_2, ... A_n, A_{ref} and A_g.

The **last step** is to build the circuit. If the gain is positive, then the input resistor is connected to the positive terminal of the op amp. Conversely, if the gain is negative, then the input resistor is connected to the negative terminal of the op amp. The feedback resistor, R_f, will always be connected from the negative input to the output.

For example, we will design the following analog circuit

$$V_{out} = 5V_1 - 3V_2 + 2V_3 - 10$$

The **first step** is to choose a reference voltage. The REF02 +5.00 V voltage reference will be used. The **second step** is to rewrite the design equation in terms of the reference voltage.

$$V_{out} = 5V_1 - 3V_2 + 2V_3 - 2V_{ref}$$

The **third step** is to add a ground input to the equation so that the sum of all the gains is equal to one.

$$V_{out} = 5V_1 - 3V_2 + 2V_3 - 2V_{ref} - V_g$$

The **fourth step** is to choose a feedback resistor, $R_f = 150$ kΩ. The value is a common multiple of the gains: 5,3,2,1. Then calculate input resistors to create the desired gains.

$$
\begin{aligned}
R_1 &= R_f/|A_1| &= 150 \text{ k}\Omega/5 &= 30 \text{ k}\Omega \\
R_2 &= R_f/|A_2| &= 150 \text{ k}\Omega/3 &= 50 \text{ k}\Omega \\
R_3 &= R_f/|A_3| &= 150 \text{ k}\Omega/2 &= 75 \text{ k}\Omega \\
R_{ref} &= R_f/|A_{ref}| &= 150 \text{ k}\Omega/2 &= 75 \text{ k}\Omega \\
R_g &= R_f/|A_g| &= 150 \text{ k}\Omega/1 &= 150 \text{ k}\Omega
\end{aligned}
$$

The **last step** is to build the circuit, as shown in Figure 11.15. The positive gain inputs are connected to the plus input of the op amp and the negative gain inputs are connected to the minus input of the op amp input.

11.2.5 Instrumentation Amplifier

The *instrumentation amp* will amplify a differential voltage, shown in Figure 11.17 as $V_2 - V_1$. We use instrumentation amplifiers in applications that require a large gain (above 100), a high input impedance, and a good CMRR. One can be built using three high-quality op amps.

Figure 11.15
A linear op amp circuit.

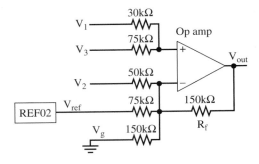

Figure 11.16
Instrumentation
amplifier.

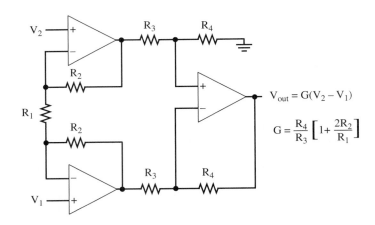

$$V_{out} = G(V_2 - V_1)$$

$$G = \frac{R_4}{R_3}\left[1 + \frac{2R_2}{R_1}\right]$$

Observation: To achieve quality performance with a three-op amp instrumentation amplifier circuit, we must use precision resistors and quality op amps.

Common error: If you use a potentiometer in place of one of the R_3 gain resistors in the Figure 11.16 circuit, then fluctuations in the potentiometer resistance that can occur with temperature, vibration, and time will have a strong effect on the amplifier gain.

Because of the range of applications that require instrumentation amplifiers, chip manufacturers have developed a variety of integrated solutions. In many cases we can achieve higher performance at reduced cost by utilizing one of these ICs. The gain is selected by external jumpers or external resistors. The Analog Devices AD620 is a typical low-cost device ($6) (Figure 11.17). The Max 4460 is a single-supply rail-to-rail amp.

Figure 11.17
Integrated
instrumentation
amplifier.

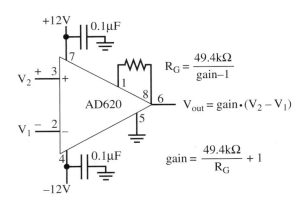

$$R_G = \frac{49.4\text{k}\Omega}{\text{gain} - 1}$$

$$V_{out} = \text{gain} \cdot (V_2 - V_1)$$

$$\text{gain} = \frac{49.4\text{k}\Omega}{R_G} + 1$$

Common error: If you use a potentiometer as the R_G gain resistors in the Figure 11.17 circuit, then fluctuations in the potentiometer resistance that can occur with temperature, vibration, and time will have a strong effect on the amplifier gain.

Checkpoint 11.9: How do you build an instrumentation amp with a gain of 11?

11.2.6
Current-to-Voltage
Circuit

In this circuit, the input is a current, I_{in}. Because no current enters the op amp terminal, this current will cross the feedback resistor R. Because of the negative feedback, the two op amp input terminals will be at approximately the same voltage, in this case zero. Therefore, the output voltage V_{out} will be $-I_{in} \cdot R$ (Figure 11.18).

Figure 11.18
Current-to-voltage
converter.

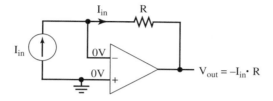

11.2.7
Voltage-to-Current
Circuit

In this circuit, the input is a voltage, V_{in}. The resistance R_1 is fixed and known (e.g., 10 kΩ), but the resistance R_L is external and unknown. The basic idea is that this circuit will deliver a constant current across R_L, independent of the value of R_L. Because of the negative feedback, the two op amp input terminals will be at approximately the same voltage, in this case V_{in}. The current across R_1 will be V_{in}/R_1. Since no current enters the op amp terminal, this same current must also pass through R_L (Figure 11.19).

Observation: A precision reference voltage chip together with this voltage-to-current circuit creates a precision reference current system. In the application notes of the reference chips shown in Table 11.4 are many other circuits that create a precision reference current.

Figure 11.19
Voltage-to-current
converter.

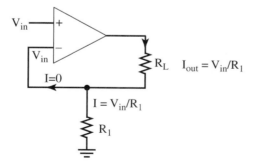

11.2.8
Integrator Circuit

In this circuit, the input is a voltage, V_{in}. The basic idea of this circuit is to create a voltage output that is the integral of the voltage input. Often a computer-controlled switch is added to short the voltage across the capacitor, initializing the output voltage to zero. Because of the negative feedback, the two op amp input terminals will be at approximately the same voltage, in this case zero. The current across the input R will be V_{in}/R. Since no current enters the op amp terminal, this same current must also pass through the capacitor. The second R on the positive terminal was added to subtract the bias current of the op amp (Figure 11.20). A low-offset op amp (e.g., OP227) should be used to reduce the error caused by the offset.

Figure 11.20
Analog integrator circuit.

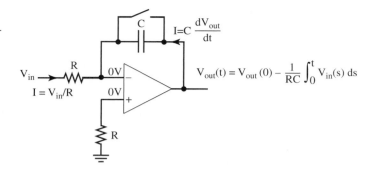

11.2.9
Derivative Circuit

In this circuit, the input is a voltage, V_{in}. The basic idea of this circuit is to create a voltage output that is the derivative of the voltage input. Because of the negative feedback, the two op amp input terminals will be at approximately the same voltage, in this case zero. The current across the input capacitor is related to the derivative of the input. Since no current enters the op amp terminal, this same current must also pass through the feedback resistor. The second R on the positive terminal was added to subtract the bias current of the op amp (Figure 11.21).

Figure 11.21
Analog derivative circuit.

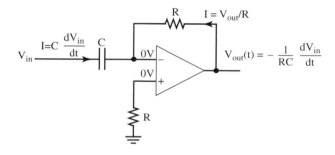

11.2.10
Voltage
Comparators with
Hysteresis

We can use a voltage comparator to detect events in an analog waveform. The range across the input voltage can vary and is usually determined by the analog supply voltages of the comparator. The output takes on two values, shown as V_h and V_l in Figure 11.22. A comparator with hysteresis has two thresholds, V_{t+} and V_{t-}. In both the positive- and negative-logic cases the threshold (V_{t+} or V_{t-}) depends on the present value of the output. Hysteresis prevents small noise spikes from creating a false trigger.

Figure 11.22
Input/output response of voltage converters with hysteresis.

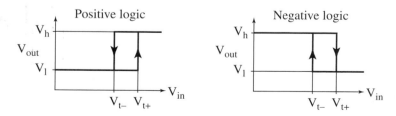

Performance tip: To eliminate false triggering, we select a hysteresis level ($V_{t+} - V_{t-}$) greater than the noise level in the signal.

In this next circuit, a TLC2274 rail-to-rail op amp is used to design a voltage comparator (Figure 11.23). Since the output swings from 0.01 to 4.99 V, it can be connected directly to an input pin of the microcomputer. On the other hand, since +5 and 0 are used to power the

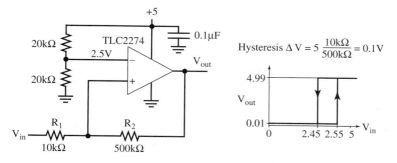

Figure 11.23
A voltage comparator with hysteresis using a rail-to-rail TLC2274.

op amp, the analog input must remain in the 0 to +5V range. The hysteresis level is determined by the R_2/R_1 ratio. If the output is at 0 V, then the input goes above +2.55 before the positive terminal of the op amp reaches 2.5 V. Similarly, if the output is at +5 V, then the input goes below +2.45 before the positive terminal of the op amp falls below 2.5 V. In linear mode circuits we should not use the supply voltage to create voltage references, but in a saturated mode circuit, power supply ripple will have little effect on the response.

There exists a wide range of integrated voltage comparators (Figure 11.24). They vary in price, speed, voltage offset, output configuration, and power consumption. The positive feedback is added to this LM311 circuit to produce hysteresis. If the output is at 0 V, then V_y is 4.9 V and the input goes below +4.9 to make the output rise. Similarly, if the output is at +5 V, then V_y is 5 V and the input goes after 5V to make the output fall.

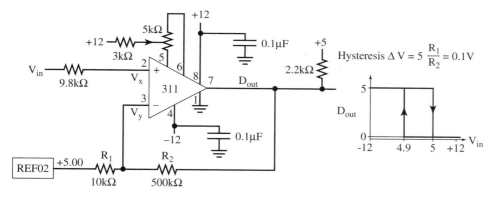

Figure 11.24
A voltage comparator with hysteresis using a LM311.

11.2.11
Analog Isolation

In some medical and industrial applications we need to design analog instrumentation that is isolated from earth ground. In an industrial setting, isolation is one way to reduce noise pickup from large EM fields produced by heavy machinery. In medical applications we need to protect the patient from potentially dangerous microshocks. Thus, the medical instrument must be isolated. There are three approaches to isolation, as shown in Figure 11.25. In the first approach, shown at the top of Figure 11.25, an analog isolation barrier is created between the preamp and the amp. This was the original approach used in analog instruments before the advent of mixed analog-digital systems. It is expensive and bulky and it introduces a very large transfer error. It is not appropriate for embedded applications that use a microcontroller. In the second approach, we use digital isolation. The 6N139 optical isolator, described previously in Chapter 8, is an effective low-cost mechanism to implement digital isolation. This is the most common approach used for new designs when a hard connection between the data acquisition system and building is required. It is fast, small,

Figure 11.25
Analog isolation, digital
isolation, and battery-
powered system all
provide protection from
microshocks.

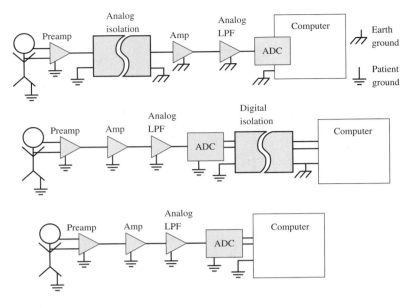

cheap, and will not introduce errors. The third approach runs the entire system with batteries. This is a very attractive approach due to the availability of high-quality low-power LCD displays and wireless networks, such as Bluetooth and 802.11b.

11.3 Analog Filters

11.3.1 Simple Active Filter

We can add capacitors in parallel with the feedback resistor in the inverting amplifier to create a simple *one-pole low-pass filter* (Figure 11.26). The impedance of a resistor R_2 in parallel with the capacitor C is a function of frequency.

$$Z = \frac{R_2}{1 + j\omega R_2 C}$$

Figure 11.26
Complex impedance
model of a resistor in
parallel with a capacitor.

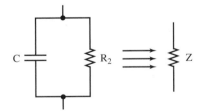

Therefore the gain of the circuit is $-Z/R_1$, which exhibits low-pass behavior. The *cutoff frequency* is defined to be the frequency at which the gain drops to 0.707 of its original value. In this simple low-pass filter, the cutoff frequency f_c is $1/(2\pi R_2 C)$ (Figure 11.27).

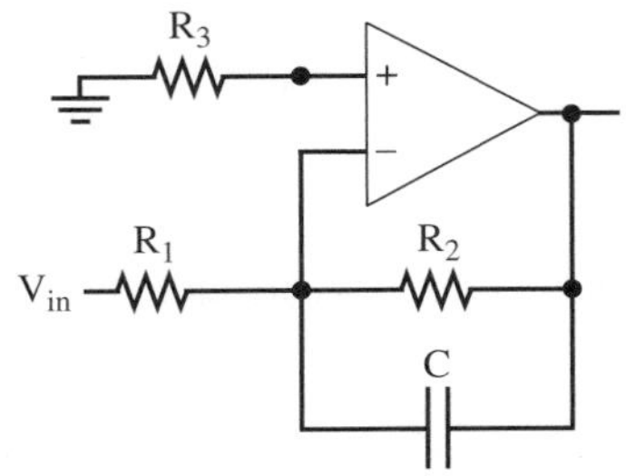

Figure 11.27
One-pole low-pass
analog filter.

$$\frac{V_{out}}{V_{in}} = -\frac{R_2}{R_1}\frac{R_2}{1 + j\omega R_2 C}$$

$$f_c = \frac{1}{2\pi R_2 C}$$

$$R_3 = \frac{R_1 \cdot R_2}{R_1 + R_2}$$

Checkpoint 11.10: What is the value of j?

We classify this low-pass filter as one pole, because the transfer function has only one pair of poles in the s plane. One-pole low-pass filters have a gain versus frequency response as shown in Figure 11.28 (G is a constant).

Figure 11.28
Frequency response of a one-pole low-pass analog filter.

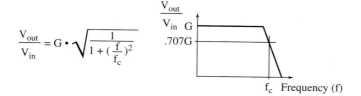

$$\frac{V_{out}}{V_{in}} = G \cdot \sqrt{\frac{1}{1 + (\frac{f}{f_c})^2}}$$

11.3.2 Butterworth Filters

Higher-order analog filters can be designed using multiple capacitors. One of the advantages of the *two-pole Butterworth analog filter* is that as long as the capacitors maintain the 2/1 ratio, the analog circuit will be a Butterworth filter. The design steps for the two-pole Butterworth low-pass filter are as follows (Figure 11.29):

1. Select the cutoff frequency f_c
2. Divide the two capacitors by $2\pi f_c$ (let C_{1A}, C_{2A} be the new capacitor values)

$$C_{1A} = \frac{141.4 \ \mu F}{2\pi f_c} \qquad C_{2A} = \frac{70.7 \ \mu F}{2\pi f_c}$$

3. Locate two standard-value capacitors (with the 2/1 ratio) with the same order of magnitude as the desired values; let C_{1B}, C_{2B} be these standard value capacitors and let x be this convenience factor

$$C_{1B} = \frac{C_{1A}}{x} \qquad C_{2B} = \frac{C_{2A}}{x}$$

4. Adjust the resistors to maintain the cutoff frequency

$$R = 10 \ k\Omega \cdot x$$

Figure 11.29
Two-pole Butterworth low-pass analog filter.

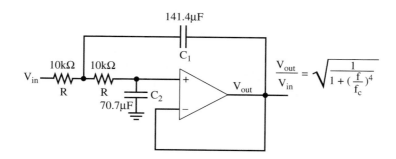

The analog filters in this section all require low-leakage, high-accuracy, and low-temperature coefficient capacitors like Teflon, polystyrene, or polypropylene. Lower-quality (lower cost) Mylar or ceramic capacitors could be used if filter accuracy were less important. This design process is implemented as file LPF.xls on the CD.

Performance tip: If you choose standard-value resistors near the desired values, you will save money and the circuit will still be a Butterworth filter. The only difference is that the cutoff frequency will be slightly off from the original specification.

11.3.3 Bandpass and Band-Reject Filters

Low-pass and high-pass filters can be cascaded to build a bandpass filter (Figure 11.30). Normally we put the high-pass filter first so that the low-pass filter can remove high-frequency noise generated in both stages.

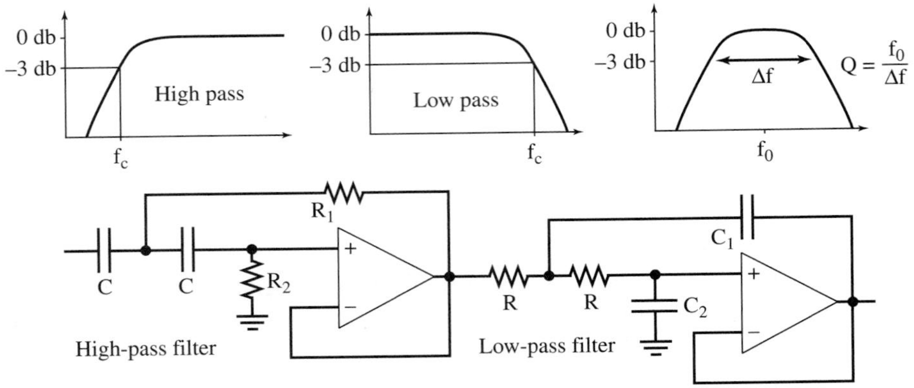

Figure 11.30
Cascade approach to bandpass analog filter design.

This cascade filter should only be used for low-Q applications. Similarly, a band-reject filter can be constructed by running both filters in parallel and summing the output (Figure 11.31). The parallel band-reject filter also should be used only for low-Q applications. One implementation of a high-Q bandpass (frequency select) filter is called the multiple feedback bandpass filter (Figure 11.32). The Q is defined as the center frequency

Figure 11.31
Parallel approach to band-reject analog filter design.

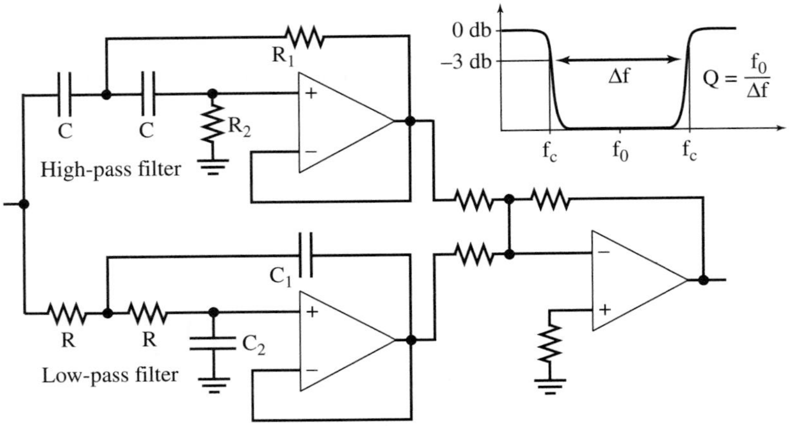

Figure 11.32
Multiple feedback bandpass filter.

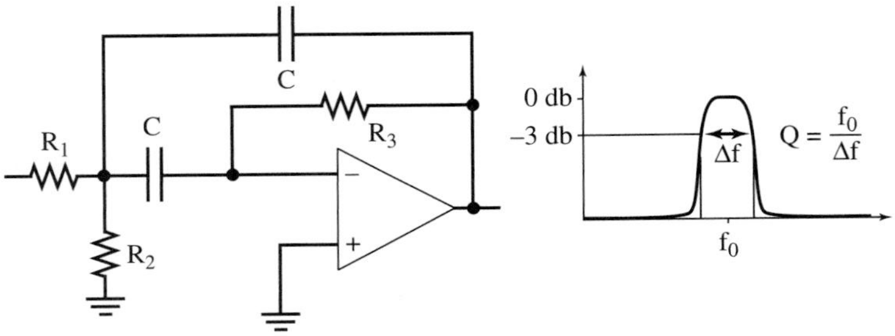

divided by the bandpass range, $f_0/\Delta f$. The two capacitors have the same value and are typically in the range from 0.001 to 10 μF. The design steps are:

1. Select a convenience capacitance value for the two capacitors
2. Calculate the three resistor values for $x = 1/(2\pi f_0 C)$

$$R_1 = Q \cdot x \qquad R_2 = x/(2Q - 1/Q) \qquad R_3 = 2 \cdot Q \cdot x$$

The resistors should be in the range from 5 kΩ to 5 MΩ. If not, then repeat the design steps with a different capacitance value.

11.4 Digital-to-Analog Converters

**11.4.1
DAC Parameters**

A DAC converts digital signals into analog form (Figure 11.33). A microcomputer output can be connected to a DAC. Many DACs can be interfaced to the SPI synchronous serial port. The DAC output can be current or voltage. Additional analog processing may be required to filter, amplify, or modulate the signal. We will also use DACs to design variable-gain or variable-offset analog circuits.

Figure 11.33
DACs provide analog output for our embedded microcomputer systems.

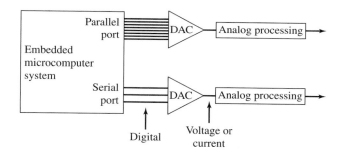

The DAC *precision* is the number of distinguishable DAC outputs (e.g., 256 alternatives, 8 bits). The DAC *range* is the maximum and minimum DAC output (volts, amperes). The DAC *resolution* is the smallest distinguishable change in output. The units of resolution are in volts or amperes depending on whether the output is voltage or current. The resolution is the change that occurs when the digital input changes by 1.

Range (volts) = precision (alternatives) • Resolution (volts)

The DAC *accuracy* is (actual − ideal) / ideal, where ideal is referred to the National Bureau of Standards (NBS). There are two common 8-bit encoding schemes for a DAC:

$$\text{Unsigned:} V_{out} = V_{fs}\left(-\frac{b_7}{2} + \frac{b_6}{4} + \frac{b_5}{8} + \frac{b_4}{16} + \frac{b_3}{32} + \frac{b_2}{64} + \frac{b_1}{128} + \frac{b_0}{256}\right) + V_{os}$$

$$\text{Signed 2s complement:} V_{out} = V_{fs}\left(-\frac{b_7}{2} + \frac{b_6}{4} + \frac{b_5}{8} + \frac{b_4}{16} + \frac{b_3}{32} + \frac{b_2}{64} + \frac{b_1}{128} + \frac{b_0}{256}\right) + V_{os}$$

where b_7, b_6, b_5, b_4, b_3, b_2, b_1, b_0 are the 8-bit digital inputs, V_{fs} is the full-scale voltage (typically +5 or +10), and V_{os} is the offset voltage (hopefully 0).

One can choose the full-scale range of the DAC to simplify the use of fixed-point mathematics. For example, if an 8-bit DAC had a full-scale range of 0 to 2.55 V, then the resolution would be exactly 10 mV. This means that if the DAC digital input were 123_{10}, then the DAC output voltage would be 1.23 V.

The following discussion will focus on a 3-bit DAC with a range of 0 to +7 V. This DAC can be used to generate a variable voltage V_{out}, a variable offset, or a variable gain V_{out}/V_{in} (Figure 11.34). A DAC gain error is a shift in the slope of the V_{out} versus V_{in} static response (Figure 11.35). A DAC offset error is a shift in the V_{out} versus V_{in} static response. The DAC transient response (Figure 11.36) has three components:

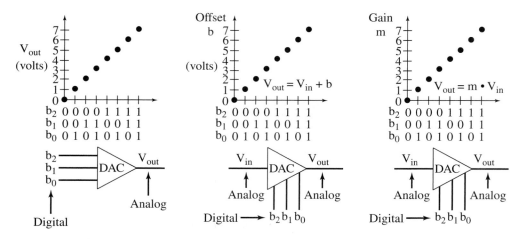

Figure 11.34
Three-bit DAC used to generate a variable output, a variable offset, or a variable gain.

Figure 11.35
Static and dynamic performance measures of DACs.

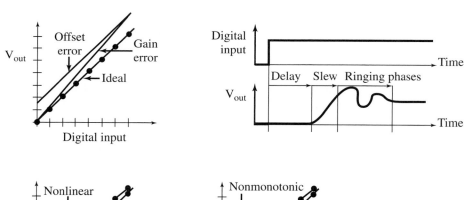

Figure 11.36
Nonlinear and nonmonotonic DACs.

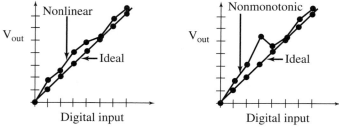

Delay phase
Slewing phase
Ringing phase

For purposes of linearity, let m, n be digital inputs, and let f(n) be the analog output of the DAC. Let Δ be the DAC resolution. The DAC is linear if

$$f(n + 1) - f(n) = f(m + 1) - f(m) = \Delta \qquad \text{for all n, m}$$

The DAC is monotonic Figure 11.64 if

$$\text{sign}[f(n + 1) - f(n)] = \text{sign}[f(m + 1) - f(m)] \qquad \text{for all n, m}$$

Conversely, the DAC is nonlinear if

$$f(n + 1) - f(n) \neq f(m + 1) - f(m) \qquad \text{for some m, n}$$

Practically speaking all DACs are nonlinear, but the worst nonlinearity is nonmonoticity. Table 11.5 lists sources of error. The DAC is nonmonotonic if

$$\text{sign}[f(n + 1) - f(n)] \neq \text{sign}[f(m + 1) - f(m)] \qquad \text{for some n, m}$$

Table 11.5
Sources and solutions to DAC errors.

Errors can be due to	Solutions
Incorrect resistor values	Precision resistors with low tolerances
Drift in resistor values	Precision resistors with good temperature coefficients
White noise	Reduce BW using a low-pass filter, reduce temperature
Op amp errors	Use more expensive devices with low noise and low drift
Interference from external fields	Shielding, ground planes

11.4.2
DAC Using a Summing Amplifier

We will need a -5.00-V reference, which we can create using an inverting op amp and voltage reference (Figure 11.37). We use the summing mode of another op amp to build the DAC. The exponential basis elements of the DAC are created by the resistance values 25 kΩ, 50 kΩ, and 100 kΩ. This method can only be used to build low-precision DACs, because it is difficult to extend the exponential resistance network beyond 5 or 6 bits. In the DAC circuit of Figure 11.38, the SW02 switch is closed (75 Ω) when the digital input (IN) is high, and the SW02 switch is open when the digital input is low. The output is $V_{out} = 4b_2 + 2b_1 + b_0$.

The range of the DAC is 0 to $+7$ V with a resolution of 1 V.

$$V_{out} = \sum_{n=0}^{2} b_n 2^n$$

Figure 11.37
Negative reference that will be used in the DAC design.

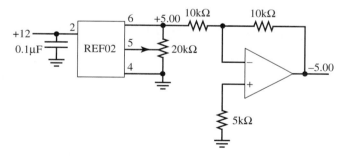

Figure 11.38
Three-bit unsigned summing DAC.

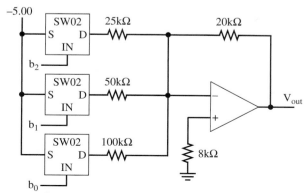

The operation of this 3-bit DAC can be summarized using the basis elements 1, 2, 4. The responses of these basis elements are presented in Table 11.6.

Table 11.6
Responses of the basis elements for an unsigned 3-bit DAC.

b_2	b_1	b_0	V_{out}
0	0	1	+1
0	1	0	+2
1	0	0	+4

11.4.3 Three-Bit DAC with an R-2R Ladder

It is not feasible to construct a high-precision DAC using the summing op amp technique for two reasons. First, if one chooses the resistor values from the practical 10 kΩ to 1 MΩ range, then the maximum precision would be 1 MΩ/10 kΩ = 100, or about 7 bits. The second problem is that it would be difficult to avoid nonmonotonicity because a small percentage change in the small resistor (e.g., the one causing the largest gain) would overwhelm the effects of the large resistor (e.g., the one causing the smallest gain). To address both these limitations, the R-2R ladder is used (Figure 11.39). It is practical to build resistor packages such that all the Rs are equal and all the 2Rs are two times the Rs. Resistance errors will change all resistors equally. This type of error affects the slope V_{fs} but not the linearity or the monotonicity.

Figure 11.39
Three-bit unsigned R-2R DAC.

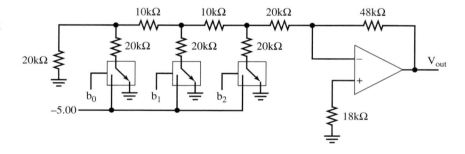

To analyze this circuit we will consider the three basis elements (1, 2, 4). The unsigned basis elements were presented earlier in Table 11.6. If these three cases are demonstrated, then the law of superposition guarantees the other five will work. Notice that these analog switches are "three-pole" and are fundamentally different from the switches in the summing DAC. When one of the digital inputs is true then −5.00 V is connected to the R-2R ladder, and when the digital input is false, then the connection is grounded. In each of the three test cases, the current across the active switch is −1/6 mA. This current is divided by 2 at each branch point. Current injected away from the op amp will be divided more times. Since each stage divides by 2, the exponential behavior is produced. An actual DAC is implemented with a current switch rather than a voltage switch as shown in Figure 11.40. Nevertheless,

Figure 11.40
The DAC output is +1 V, then the digital input is 001.

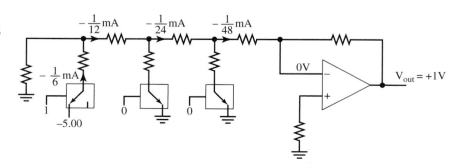

this simple circuit illustrates the operation of the R-2R ladder function. The output voltage is a linear combination of the three digital inputs, $V_{out} = 4b_2 + 2b_1 + b_0$. To increase the precision one simply adds more stages to the R-2R ladder. One can purchase 16-bit DACs that utilize the R-2R ladder technique.

When the input is 001, -5.00 V is presented to the left. The effective impedance to ground is 3R (30 kΩ), so the current injected into the R-2R ladder is $-1/6$ mA. The current is divided three times, and $-1/48$ mA goes across the 48-kΩ resistor to create the $+1$-V output (Figure 11.40).

When the input is 010, -5.00 V is presented in the middle. The effective impedance to ground is again 3R (30 kΩ), so the current injected into the R-2R ladder is $-1/6$ mA. The current is divided twice, and $-1/24$ mA goes across the 48-kΩ resistor to create the $+2$-V output (Figure 11.41).

Figure 11.41
The DAC output is $+2$ V, then the digital input is 010.

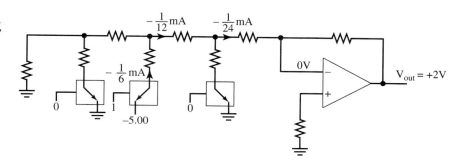

When the input is 100, -5.00 V is presented on the right. The effective impedance to ground is again 3R (30 kΩ), so the current injected into the R-2R ladder is $-1/6$ mA. This time the current is divided only once, and $-1/12$ mA goes across the 48-kΩ resistor to create the $+4$-V output (Figure 11.42).

Figure 11.42
The DAC output is $+4$ V, then the digital input is 100.

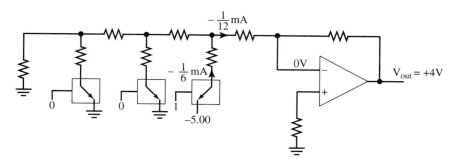

One of the reasons this simple 3-bit DAC is introduced is to illustrate the use of DACs in the design of variable-gain and variable-offset analog amplifiers. In this way, the software has control over the gain and offset of the analog circuit. The variable-offset approach simply connects the DAC output into one of the inputs of an analog adder or analog subtractor, as illustrated in Figure 11.43.

A multiplying DAC is one that allows a voltage input to be connected to the R-2R ladder instead of a voltage reference. If you connect a voltage reference to this pin, the DAC operates in the usual fashion. On the other hand, if this pin is an analog input, then the DAC is a variable-gain amplifier. Using the same 3-bit DAC circuit developed earlier, we can build a variable-gain amplifier. An inverting amplifier is added at the end to create positive-gain amplification. The 48-kΩ resistor is replaced with a 240-kΩ one so that the gain varies from 0 to $+7$ (Figure 11.44).

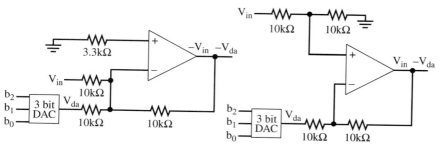

Figure 11.43
Variable-offset analog circuit using a 3-bit DAC.

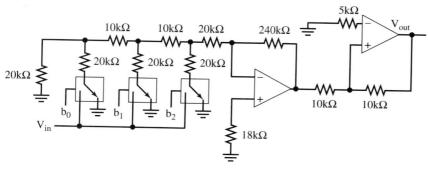

Figure 11.44
Variable-gain analog circuit using a 3-bit multiplying DAC.

11.4.4
Twelve-Bit DAC
with a DAC8043

The synchronous serial interface between the computer and the Analog Devices DAC8043 DAC was introduced previously in Section 7.7.6.1. Here in this section we will focus on the analog aspects of the circuit. If the DAC is used in the regular digital input/analog output mode, then Vin will be a reference voltage (e.g., +5.00 or +10.00 V). If the DAC is used in variable-gain mode (multiplying DAC), then Vin is the analog input (i.e., $-10 \le$ Vin $\le$ +10 V). The optional 100 Ω potentiometer is used to adjust the gain (Figure 11.45).

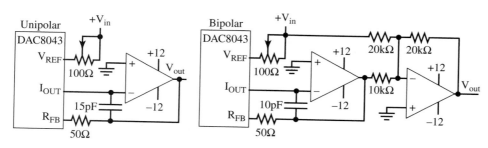

Figure 11.45
Unipolar and bipolar modes of the DAC8043 12-bit DAC.

The Vout in Table 11.7 is calculated for a Vin = +5.00 V reference voltage. The bipolar code is called *offset binary*. To convert to regular 2s complement we simply complement the most significant digital bit. When the unipolar circuit is used in variable-gain mode, it is called two-quadrant because the input (Vin) can be positive or negative, but the gain is always negative. When the bipolar circuit is used in variable-gain mode, it is called four-quadrant because the input (Vin) can be positive or negative, and the gain can be positive or negative.

Digital input	Unipolar Vout (V)	Bipolar Vout (V)	Unipolar gain	Bipolar gain
1111, 1111, 1111	−4.999	4.998	$-\dfrac{4095}{4096}$	$+\dfrac{2047}{2048}$
1000,0000,0001	−2.501	0.002	$-\dfrac{2049}{4096}$	$+\dfrac{1}{2048}$
1000,0000,0000	−2.500	0.000	$-\dfrac{2048}{4096}$	$+\dfrac{0}{2048}$
0111,1111,1111	−2.499	−0.002	$-\dfrac{2047}{4096}$	$-\dfrac{1}{2048}$
0000,0000,0001	−0.001	−4.998	$-\dfrac{1}{4096}$	
0000,0000,0000	0.000	−5.000	$-\dfrac{0}{4096}$	$-\dfrac{2048}{2048}$

Table 11.7
The 12-bit DAC8043 DAC can create an analog output or be used as a variable-gain amplifier.

**11.4.5
DAC Selection**

Many manufacturers, like Analog Devices, Burr Brown, Motorola, Sipex, and Maxim, produce DACs. These DACs have a wide range of performance parameters and come in many configurations. The following paragraphs discuss the various issues to consider when selecting a DAC. Although we assume the DAC is used to generate an analog wave-form, these considerations will generally apply to most DAC applications.

Precision/range/resolution. These three parameters affect the quality of the signal that can be generated by the system. The more bits in the DAC, the finer the control the system has over the waveform it creates. As important as this parameter is, it is one of the more difficult specifications to establish a priori. A simple experimental procedure to address this question is to design a prototype system with a very high precision (e.g., 12, 14, or 16 bits). The software can be modified to use only some of the available precision. For exam-ple, the 12-bit DAC8043 hardware developed in Sections 7.7.6.1 and 11.4.4 can be reduced to 8 or 10 bits using the functions shown in Program 11.1. The bottom bits are set to zero, instead of shifting, so that the rest of the system will operate without change.

Program 11.1
Software used to test how many bits are really needed.

```
void DACout8(unsigned short code){
    DACout(code&0xFFF0);}   // ignore bottom 4 bits
void DACout10(unsigned short code){
    DACout(code&0xFFFC);}   // ignore bottom 2 bits
```

The three versions of the software (e.g., 8-, 10-, 12-bit DAC) are used to see experi-mentally the effect of DAC precision on the overall system performance. Figure 11.46 illustrates how DAC precision affects the quality of the generated waveform.

Figure 11.46
The waveform on the right was created by a DAC with one more bit than the left.

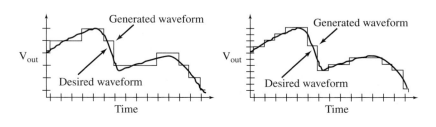

Channels. Even though multiple channels could be implemented using multiple DAC chips, it is usually more efficient to design a multiple-channel system using a multiple-channel DAC. Some advantages of using a DAC with more channels than originally conceived are future expansion, automated calibration, and automated testing.

Configuration. DACs can have voltage or current outputs. Current-output DACs can be used in a wide spectrum of applications (e.g., adding gain and filtering) but do require external components. DACs can have internal or external references. An internal-reference DAC is easier to use for standard digital input/analog output applications, but the external-reference DAC can be used often in variable-gain applications (multiplying DAC). As we saw with the DAC8043, sometimes the DAC generates a unipolar output, while other times the DAC produces bipolar outputs.

Speed. There are a couple of parameters manufacturers use to specify the dynamic behavior of the DAC. The most common is settling time; another is maximum output rate. When operating the DAC in variable-gain mode, we are also interested in the gain/BW product of the analog amplifier. When comparing specifications reported by different manufacturers, it is important to consider the exact situation used to collect the parameter. In other words, one manufacturer may define settling time as the time to reach 0.1% of the final output after a full-scale change in input given a certain load on the output, while another manufacturer may define settling time as the time to reach 1% of the final output after a 1-V change in input under a different load. The speed of the DAC together with the speed of the computer/software will determine the effective frequency components in the generated waveforms. Both the software (rate at which the software outputs new values to the DAC) and the DAC speed must be fast enough for the given application. In other words, if the software outputs new values to the DAC at a rate faster than the DAC can respond, then errors will occur. Figure 11.47 illustrates the effect of DAC output rate on the quality of the generated waveform. We will learn in Chapter 12 that the Nyquist theorem states that to prevent error, the digital data rate must be greater than twice the maximum frequency component of the desired analog waveform.

Figure 11.47
The waveform on the right was created by a system with twice the output rate than the left.

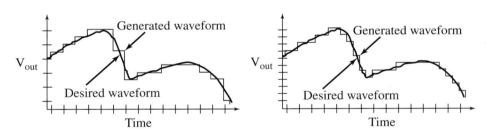

Power. There are three power issues to consider. The first consideration is the type of power required. Some devices require three power voltages (e.g., +5, +12, and −12 V), while many of the newer devices will operate on a single voltage supply (e.g., +2.7, +3.3, +5, or +12 V). If a single supply can be used to power all the digital and analog components, then the overall system costs will be reduced. The second consideration is the amount of power required. Some devices can operate on less than 1 mW and are appropriate for battery-operated systems or for systems where excess heat is a problem. The last consideration is the need for a low-power sleep mode. Some battery-operated systems need the DAC only intermittently. In these applications, we wish to give a shutdown command to the DAC so that it won't draw much current when the DAC is not needed.

Interface. Three approaches exist for interfacing the DAC to the computer (Figure 11.48). In a digital logic interface, the individual data bits are connected to a dedicated computer output port. For example, a 12-bit DAC requires 12-bit output port bits to interface.

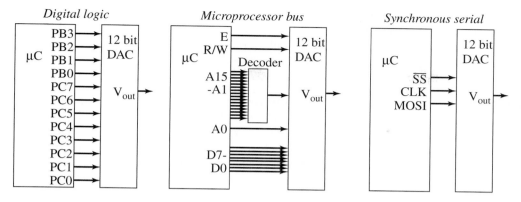

Figure 11.48
Three approaches to interfacing a 12-bit DAC to the microcomputer.

The software simply writes to the parallel port(s) to change the DAC output. The second approach is called µP-bus or microprocessor-compatible. These devices are intended to be interfaced onto the address/data bus of an expanded-mode microcomputer. As we saw in Chapter 9, the MC68HC812A4 has built-in address decoders enabling most µP-bus DACs to be interfaced to the address/data bus without additional external logic. An example of this type of interface can be found in Exercise 9.5. The SPI/DAC8043 interface is an example of the last type of interface. This approach requires the fewest number of I/O pins. Even if the microcomputer does not support the SPI directly, these devices can be interfaced to regular I/O pins via the bit-banging software approach. The DS1620 interface in Section 4.4.9 is an example of attaching a synchronous serial I/O device to regular I/O pins.

Package. The standard DIP is convenient for creating and testing an original prototype. On the other hand, surface-mount packages like the SO and µMax require much less board space. Because surface-mount packages do not require holes in the printed circuit board, circuits with these devices are easier/cheaper to manufacture (Figure 11.49).

Figure 11.49
Integrated circuits come in a variety of packages.

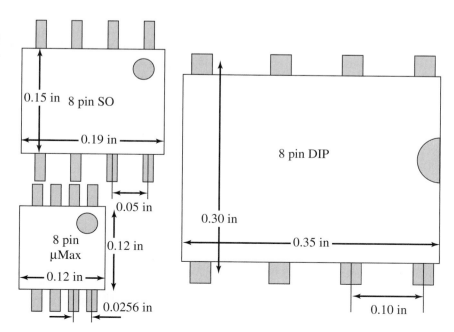

Cost. Cost is always a factor in engineering design. Beside the direct costs of the individual components in the DAC interface, other considerations that affect cost include (1) power supply requirements, (2) manufacturing costs, (3) the labor involved in individual calibration if required, and (4) software development costs.

11.4.6
DAC Waveform
Generation

One application that requires a DAC is waveform generation. In this section, we will discuss various software methods for creating analog waveforms with a DAC. In each case, we will be using the DAC8043 hardware/software interface introduced in Sections 7.7.6.1 and 11.4.4. In addition, we will use an output compare interrupt for the timing so that the waveform generation occurs in the background. The rituals for initializing the periodic interrupt are shown in Section 6.2.3. To get a fair comparison between the various methods, each implementation will generate 32 interrupts per waveform.

In the first approach, we assume there exists a time-to-voltage function, called `wave()`, which we can call to determine the next DAC value to output. For example, the waveform in Figure 11.50 could be generated by Program 11.2. The simplest solution generates an output compare interrupt at a regular rate. The advantage of this approach is that complex waveforms can be encoded with a small amount of data. In this particular example, the entire waveform can be stored as five data points (`2048.0`, `1000.0`, `31.25`, `−500.0`, `125.0`). The disadvantage of this technique is that not all waveforms have a simple function, and this software will run slower as compared to the other techniques (Program 11.3).

In the second approach, we put the waveform information in a large statically allocated global variable. Every interrupt we fetch a new value out of the data structure and output it to the DAC. In this case the output compare interrupt also occurs at a regular rate. Assume the ritual initializes `I=0`. This waveform could be defined as shown in Programs 11.4 and 11.5.

Figure 11.50
Generated waveform using either the function or the table look-up.

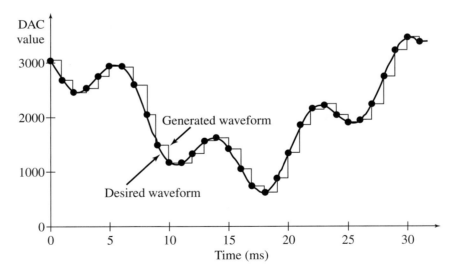

Program 11.2
Periodic interrupt used to create the analog output waveform.

```
unsigned short wave(unsigned short t){
      float result,time;
      time=2*pi*((float)t)/1000.0;
// integer t in msec into floating point time in seconds
      result=2048.0+1000.0*cos(31.25*time)-500.0*sin(125.0*time);
      return (unsigned short) result;}
```

```
// 6811                                          // 6812
#define Rate 2000                                #define Rate 2000
#define OC5  0x08                                #define OC5  0x20
unsigned short Time;  // Inc every 1ms           unsigned short Time;  // Inc every 1ms
#pragma interrupt_handler TOC5handler()          #pragma interrupt_handler TOC5handler()
void TOC5handler(void){                          void TOC5handler(void){
   TFLG1=OC5;        // Ack interrupt               TFLG1=OC5;        // ack C5F
   TOC5=TOC5+Rate; // Executed every 1 ms          TC5=TC5+Rate;     // Executed every 1 ms
   Time++;                                          Time++;
   DACout(wave(Time));}                             DACout(wave(Time));}
```

Program 11.3
Periodic interrupt used to create the analog output waveform.

Program 11.4
Simple data structure for the waveform.

```
unsigned short I;  // incremented every 1ms
const unsigned short wave[32]= {
        3048,2675,2472,2526,2755,2957,2931,2597,
        2048,1499,1165,1139,1341,1570,1624,1421,
        1048,714,624,863,1341,1846,2165,2206,2048,
        1890,1931,2250,2755,3233,3472,3382};
```

```
// 6811                                          // 6812
#define Rate 2000                                #define Rate 2000
#define OC5  0x08                                #define OC5  0x20
#pragma interrupt_handler TOC5handler()          #pragma interrupt_handler TOC5handler()
void TOC5handler(void){                          void TOC5handler(void){
   TFLG1=OC5;          // Ack interrupt             TFLG1=OC5;          // ack C5F
   TOC5=TOC5+Rate; // Executed every 1 ms          TC5=TC5+Rate;     // Executed every 1 ms
   if((++I)==32) I=0;                               if((++I)==32) I=0;
   DACout(wave[I]);}                                DACout(wave[I]);}
```

Program 11.5
Periodic interrupt used to create the analog output waveform.

Since the output rate is equal and fixed, these first two methods have the same performance as illustrated in Figure 11.50. The thick line is the desired waveform, and the thinner line is the actual generated curve. If the size of the table gets large, it is possible to store a smaller table in memory and use linear interpolation to recover the data points between the stored samples. Figure 11.51 shows the generated waveform derived from only 9 of the original 32 data points. To simplify the software, the first data point is repeated as the last data point. For each point we will need to save both the DAC value and time length of the current-line segment. For the 9 saved data points we simply output the data, but for the other points, we must perform a linear interpolation to get the value to output to the DAC.

Assume the ritual initializes I=J=0. Signed 16-bit numbers are used so that the subtractions operate properly. In other words, some of the intermediate calculations can be negative. This waveform could be defined as shown in Programs 11.6 and 11.7.

The software in the previous techniques changes the DAC at a fixed rate. While this is adequate most of the time, there are some waveforms for which an uneven time between outputs seems appropriate. In our test signal, there are phases of the wave where the signal varies slowly while there are phases where the signal changes. Notice the data points in Figure 11.52 are placed at uneven time intervals to match the various phases of

Figure 11.51
Generated waveform using a short table with linear interpolation.

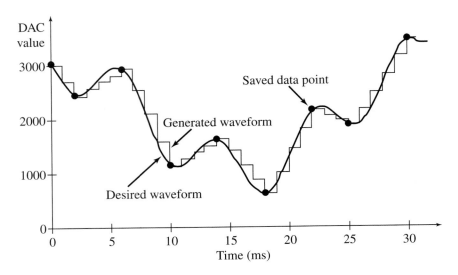

```
short I;   // incremented every 1ms
short J;   // index into these two tables
const short t[10]= {0,2,6,10,14,18,22,25,30,32};   // time in msec
const short wave[10]={3048,2472,2931,1165,1624,624,2165,1890,3472,3048}; //last=first
```

Program 11.6
Data structure with time and value for the waveform.

```
// 6811
#define Rate 2000
#define OC5  0x08
#pragma interrupt_handler TOC5handler()
void TOC5handler(void){
   TFLG1=OC5;        // Ack interrupt
   TOC5=TOC5+Rate; // Executed every 1 ms
   if((++I)==32) {I=0; J=0;}
   if(I==t[J])
      DACout(wave[J]);
   else if (I==t[J+1]){
      J++;
      DACout(wave[J]);}
   else
      DACout(wave[J]+((wave[J+1]-wave[J])
        *(I-t[J]))/(t[J+1]-t[J]));}
```

```
// 6812
#define Rate 2000
#define OC5  0x20
#pragma interrupt_handler TOC5handler()
void TOC5handler(void){
   TFLG1=OC5;        // ack C5F
   TC5=TC5+Rate;     // Executed every 1 ms
   if((++I)==32) {I=0; J=0;}
   if(I==t[J])
      DACout(wave[J]);
   else if (I==t[J+1]){
      J++;
      DACout(wave[J]);}
   else
      DACout(wave[J]+((wave[J+1]-wave[J])
        *(I-t[J]))/(t[J+1]-t[J]));}
```

Program 11.7
Periodic interrupt used to create the analog output waveform.

this signal. This generated waveform is still created with 32 points, but placing the points closer together during phases with large slopes improves the overall accuracy.

The table data structure will encode both the voltage (as a DAC value) and the time. The time parameter is stored as a Δt in cycles to simplify servicing the output compare interrupt. We will assume the TCNT is initialized to count every 500 ns, therefore the maximum Δt that can be generated is 32 ms. Assume the ritual initializes I=0 (Programs 11.8 and 11.9).

Figure 11.52
Generated waveform using the uneven-time table look-up technique.

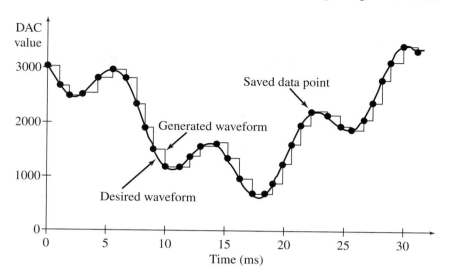

```
unsigned short I;  // incremented every sample
const unsigned short wave[32]= {
   3048,2675,2472,2526,2817,2981,2800,2337,1901,1499,1165,1341,1570,1597,1337, 952,
    662, 654, 863,1210,1605,1950,2202,2141,1955,1876,2057,2366,2755,3129,3442,3382};
const unsigned short dt[32]= { // time increment in 500 ns cycles
  2000,2000,2000,2500,2500,2000,2000,1500,1500,2000,4000,2000,2500,2000,2000,2000,
  2000,1500,1500,1500,1500,2000,2500,2000,2000,2000,1500,1500,1500,2000,2500,2000};
```

Program 11.8
Data structure with delta time and value for the waveform.

```
// 6811
#define OC5   0x08
#pragma interrupt_handler TOC5handler()
void TOC5handler(void){
   TFLG1=OC5;        // Ack interrupt
   if((++I)==32) I=0;
   TOC5=TOC5+dt[I]; // variable rate
   DACout(wave[I]);}
```

```
// 6812
#define OC5   0x20
#pragma interrupt_handler TOC5handler()
void TOC5handler(void){
   TFLG1=OC5;          // ack C5F
   if((++I)==32) I=0;
   TC5=TC5+dt[I];   // variable rate
   DACout(wave[I]);}
```

Program 11.9
Periodic interrupt used to create the analog output waveform.

11.5   Analog-to-Digital Converters

**11.5.1
ADC Parameters**

An ADC converts an analog signal into digital form. The input signal is usually an analog voltage (V_{in}), and the output is a binary number. The ADC *precision* is the number of distinguishable ADC inputs (e.g., 256 alternatives, 8 bits). The ADC *range* is the maximum and minimum ADC input (volts, amperes). The ADC resolution is the smallest distinguishable change in input (volts, amperes). The *resolution* is the change in input that causes the digital output to change by 1.

$$\text{Range (volts)} = \text{precision (alternatives)} \bullet \text{Resolution (volts)}$$

Normally we don't specify accuracy for just the ADC, but rather we give the *accuracy* of the entire instrument (including transducer, analog circuit, ADC, and software). Therefore, accuracy will be described later in Chapter 12 (specifically, Section 12.1.1), part of the systems approach to data acquisition systems. An ADC is *monotonic* if it has no missing codes. This means that if the analog signal is a slow rising voltage, then the digital output will hit all values one at a time. The ADC is *linear* if the resolution is constant through the range. The ADC *speed* is the time to convert, called t_c. The ADC cost is a function of the number and price of internal components. There are four common encoding schemes for an ADC. Tables 11.8 and 11.9 show the four encoding schemes for an 8-bit ADC.

Table 11.8
Unipolar codes for an 8-bit ADC.

Unipolar codes	Straight binary	Complementary binary
+5.00	1111, 1111	0000, 0000
+2.50	1000, 0000	0111, 1111
+0.02	0000, 0001	1111, 1110
+0.00	0000, 0000	1111, 1111

Table 11.9
Bipolar codes for an 8-bit ADC.

Bipolar codes	Offset binary	2s Complement binary
+5.00	1111, 1111	0111, 1111
+2.50	1100, 0000	0100, 0000
+0.04	1000, 0001	0000, 0001
+0.00	1000, 0000	0000, 0000
−2.50	0100, 0000	1100, 0000
−5.00	0000, 0000	1000, 0000

To convert between straight binary and complementary binary, we simply complement (change 0 to 1, change 1 to 0) all the bits. To convert between offset binary and 2s complement, we complement just the most significant bit. The exclusive-OR instruction can be used to complement a single bit.

Just like the DAC, one can choose the full-scale range to simplify the use of fixed-point mathematics. For example, if a 12-bit ADC had a full-scale range of 0 to 4.095 V, then the resolution would be exactly 1 mV. This means that if the ADC input voltage were 1.234 V, then the result would be 1234_{10}.

**11.5.2
Two-Bit Flash ADC**

The flash ADC can be used for high-speed low-precision conversions. Four equal-value resistors are used to create reference voltages 7.50, 5.00, and 2.50. The LM311 voltage comparators perform the ADC conversion. The digital circuit produces the binary code (Z1, Z0). The conversion speed is limited by the comparator delay, which can be as fast as 100 ns (Table 11.10, Figure 11.53).

Some flash converters include two more comparators. The underflow comparator is used to determine if $V_{in} < 0$. The overflow comparator is used to determine if $V_{in} > +10$ V.

A 3-bit flash ADC would place eight equal-value resistors in series to generate the reference voltages 1.25, 2.50, 3.75, 5.00, 6.25, 7.50, and 8.75. Seven comparators would

Table 11.10
Output results for a 2-bit flash ADC.

V_{in}	X3	X2	X1	Z1	Z0
$2.5 > V_{in}$	0	0	0	0	0
$5.0 > V_{in} \geq 2.5$	0	0	1	0	1
$7.5 > V_{in} \geq 5.0$	0	1	1	1	0
$V_{in} \geq 7.5$	1	1	1	1	1

Figure 11.53
Two-bit flash ADC.

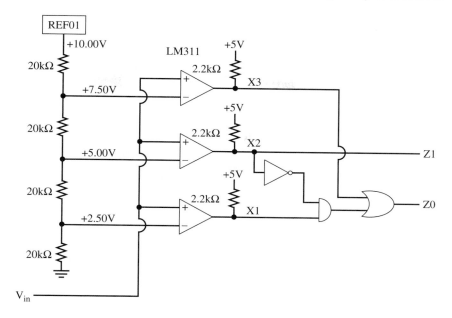

perform the ADC conversion. A 74LS148 8-to-3 encoder could be used to convert the X1,X2,X3,X4,X5,X6,X7 comparator outputs into the 3-bit binary code Z2,Z1,Z0. With a flash ADC there is an exponential relationship between the ADC precision and the number of components. In other words, an 8-bit flash requires 256 resistors, 255 comparators, and a 255 to 8-bit digital encoder.

> *Observation:* The cost of a flash ADC relates linearly with its precision in alternatives.

There are two approaches to produce a signed flash ADC. The direct approach would be to place the equal resistors from a $+10.00$-V reference to a -10.00-V reference. The middle of the series resistors ("zero" reference) should be grounded. The digital circuit would then be modified to produce the desired digital code.

The other method would be to add analog preprocessing to convert the signed input to an unsigned range. This unsigned voltage is then converted with an unsigned ADC. The digital code for this approach would be offset binary. This approach will work to convert any unsigned ADC into one that operates on bipolar inputs.

**11.5.3
Successive
Approximation
ADC**

The most pervasive method for ADC conversion is the successive approximation technique. An ADC with a precision of n is clocked n times. At each clock another bit is determined, starting with the most significant bit. For each clock, the successive approximation hardware issues a new "guess" (V_0) by setting the bit under test to a 1. If V_0 is now higher than the unknown input V_{in}, then the bit under test is cleared. If V_0 is less than V_{in}, then the bit under test remains 1. The C program in Figure 11.54 illustrates the algorithm. In this program, n is the precision in bits, bit is an unsigned integer that specifies the bit under test (e.g., for an 8-bit ADC, bit goes 128, 64, 32, 16, . . ., 1), Dout is the ADC digital output, and z is the binary input that is true if V_0 is greater than V_{in}.

Most successive approximation ADCs use a current-output DAC and a current comparator, instead of the voltage DAC and voltage comparator shown above. In this case, each guess is converted to a current by the DAC. The input voltage is also converted to a current (notice the low-input impedance of this ADC). The current comparator determines if the guess is high or low. If the guess is low, the bit under test is left set to 1. If the guess is high, the bit under test is cleared to 0. This procedure starts with the most significant bit and stops with bit 0. An n-bit ADC requires exactly n cycles to convert, independent of the input voltage level.

Figure 11.54
Successive
approximation ADC.

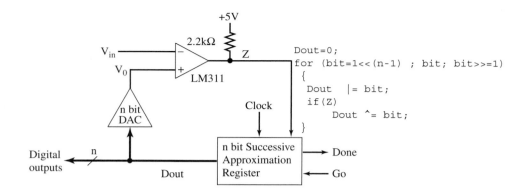

Observation: The speed of a successive approximation ADC relates linearly with its precision in bits.

11.5.4 Sixteen-Bit Dual Slope ADC

The dual slope technique can be used to construct an ADC with precision ranging from 16 to 20 bits (Figure 11.55). This 16-bit ADC will have a range of 0 to +10 V and a conversion time of 130 ms. There are three phases for the conversion. The three digital signals P0, P1, and P2 each control a SPST BiFET analog switch. Only one switch is activated at a time. Let the times t_0, t_1, and t_2 refer to the time at the end of each phase (Figure 11.56). During phase 0, P0 is active, which shorts the capacitor C. This initialization phase is used to make the output V_{out} equal to zero. Thus,

$$V_{out}(t_0) = 0$$

Phase 0 must be long enough to completely discharge the capacitor. During phase 1, P1 is active, which connects the unknown input voltage V_{in} to the integrator. The length of phase 1 is exactly 65,535 μs. We will call this length of phase 1 a time reference,

$$t_{ref} = t_1 - t_0 = 65,535 \text{ μs}$$

Figure 11.55
Hardware required to
implement the dual
slope ADC conversion
technique.

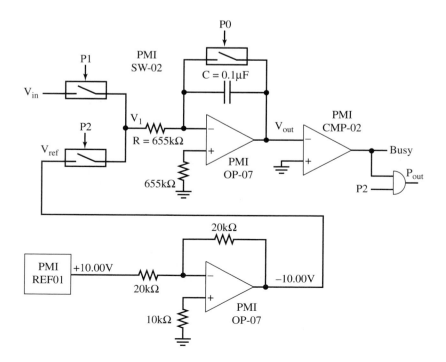

Figure 11.56
Digital and analog waveforms during a dual slope ADC conversion.

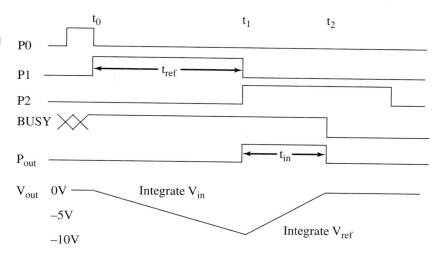

Since the input voltage is a constant and $V_{out}(t_0)$ is known, the integration can be simplified:

$$V_{out}(t_1) = V_{out}(t_0) - \frac{1}{RC} \int_{t_0}^{t_1} V_{in}(s)ds = -\frac{1}{RC} V_{in} t_{ref}$$

The largest negative voltage will occur when V_{in} equals $+10$ V. The values of R,C, and t_{ref} were chosen to prevent the integrating op amp from saturating. Output compare can be used to generate t_{ref}, and input capture can be used to measure t_{in}. During phase 2, P2 is active, which connects the reference voltage V_{ref} to the integrator. The negative voltage reference is created with the REF01 voltage reference and an inverting amplifier. During this phase V_{out} is integrated back to zero. The phase ends when $V_{out} = 0$. Thus,

$$V_{out}(t_2) = 0$$

t_{in} in is the time spent in phase 2, $t_{in} = t_2 - t_1$. The pulse width of P_{out}, t_{in}, is measured with a precision timer. Again, since V_{ref} is constant, the integration can be simplified.

$$V_{out}(t_2) = V_{out}(t_1) - \frac{1}{RC} \int_{t_1}^{t_2} V_{ref}(s)ds = V_{out}(t_1) - \frac{1}{RC} V_{ref} t_{in} = 0$$

Substituting from above the value of $V_{out}(t_1)$

$$\frac{1}{RC} V_{in} t_{ref} - \frac{1}{RC} V_{ref} t_{in} = 0$$

The next algebraic step is critical to explain why this method can be used to produce high-precision ADCs. The RCs cancel!

$$V_{in} t_{ref} - V_{ref} t_{in} = 0$$

We can calculate the unknown voltage V_{in} from the voltage reference V_{ref}, the time reference t_{ref}, and the time measurement t_{in}.

$$V_{in} = \frac{V_{ref}}{t_{ref}} t_{in}$$

The 16-bit precision requires that the time t_{in} be measured with a resolution of 1 μs and a range of 0 to 65,535 μs. The operation can be summarized in Table 11.11.

11.5.5
Sigma Delta ADC

The sigma delta ADC is used in many audio applications (Figure 11.57). It is a cost-effective approach to 16-bit 44-kHz sampling (CD quality). Sigma delta converters have a DAC, a

Table 11.11
Output results of a dual slope ADC.

V_{in} (V)	$V_{out}(t_1)$ (V)	t_{in} (μs)
0.00000	0.00000	0
0.00015	−0.00015	1
0.00031	−0.00031	2
1.00000	−1.00000	6,554
5.00000	−5.00000	32,768
9.99985	−9.99985	65,535

Figure 11.57
Block diagram of the sigma delta ADC conversion technique.

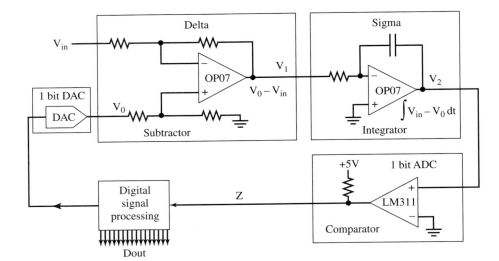

comparator, and digital processing similar to the successive approximation technique. While successive approximation converters have DACs with the same precision as the ADC, sigma delta converters use 1-bit DAC (which is simply a digital signal itself). The digital signal processing will run at a clock frequency faster than the overall ADC system (oversampled). It uses complex signal processing to drive the output voltage V_0 to equal the unknown input V_{in} in a time-averaged sense. The "delta" part of the sigma delta converter is the subtractor, where $V_1 = V_0 - V_{in}$. Next comes the "sigma" part that implements an analog integration. If V_0 to equal the unknown input V_{in} in a time-averaged sense, then V_2 will be zero. The comparator tests the V_2 signal. If V_2 is positive, then V_0 is made smaller. If V_2 is negative, then V_0 is made larger. This DAC-subtractor-integrator-comparator-digital loop is executed at a rate much faster than the eventual digital output rate.

A very simple software algorithm, shown in Program 11.10, is implemented as a periodic interrupt. This algorithm is much too simple to be appropriate in an actual converter, but it does illustrate the sigma delta approach. For an 8-bit conversion, the internal clock rate (interrupt rate) is 256 times the output rate of the 8-bit samples. We assume that the input voltage V_{in} is between 0 and +5 V. In this simple solution, the DAC is set to 1 ($V_0 = +5$) if Z is 0 ($V_2 < 0$). Conversely, the DAC is set to 0 ($V_0 = 0$) if Z is 1 ($V_2 > 0$). Each time the DAC is set to 1, the counter, SUM, is incremented. At the end of 256 passes, the value SUM is recorded as the ADC sample. Since there are 256 passes through the loop, the result will vary from 0 to 255. During each interrupt one pass through the sigma delta algorithm is executed. Similar to the other software ADC examples, Z() is a function that returns the value of the comparator output. Let DACout() be a function that takes a Boolean input parameter and sets the 1-bit DAC. Assume the ritual will clear all three global variables.

**11.5.6
ADC Interface**

Similar to the DAC interface, three approaches exist for interfacing the ADC to the computer (Figure 11.58). In a digital logic interface, the individual data bits are connected to

```
unsigned char DOUT; // 8-bit sample
unsigned char SUM; // number of times Z=0 and V0=1
unsigned char CNT; // 8-bit counter
#pragma interrupt_handler TOC5handler()
void TOC5handler(void){
   TFLG1=OC5;        // ack C5F
   TC5=TC5+rate;     // interrupt 256 times faster than the ADC output rate
   if(Z())           // check input
      DACout(0);     // too high, set DAC output, V0=0
   else {
      DACout(1);     // too low, set DAC output, V0=+5v
      SUM++;
   }
   if(++CNT==0){     // end of 256 loops?
      DOUT=SUM;      // new sample
      SUM=0;         // get ready for the next
   }
}
```

Program 11.10
Soft-ware implementation of a sigma delta ADC.

Figure 11.58
Three approaches to interfacing an ADC to the microcomputer.

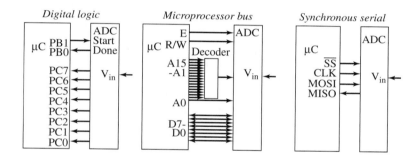

a dedicated computer I/O ports. Usually the software toggles the **Start** ADC signal to begin an ADC conversion. The software can determine when the ADC is finished by reading the **Done** signal. For example, an 8-bit ADC requires one output port bit and 9-bit input port bits to interface. The software toggles the **Start** signal, waits for the **Done** to be true, then reads the ADC result. The second approach is called microprocessor-bus or microprocessor-compatible. These devices are intended to be interfaced onto the address/data bus of an expanded mode microcomputer. A microcomputer-bus ADC can be interfaced to a MC68HC812A4 without additional external logic. Discussion of this type of interface can be found in Chapter 9. The software interfaces to this type of ADC in a manner similar to the internal ADCs (discussed later in Section 11.10). The SPI/ MAX1247 interface (discussed in Section 7.7.6.2) is an example of the last time of interface. This approach requires the fewest number of I/O pins. Even if the microcomputer does not support the SPI directly, these devices can be interfaced to regular I/O pins via the bit-banging software approach.

11.6 Sample and Hold

A sample and hold (S/H) is an analog latch (Figure 11.59). An alternative name for this analog component is track and hold. The purpose of the S/H is to hold the ADC analog input constant during the conversion. The digital input, **control**, determines the S/H mode. The S/H is in *sample mode*, where V_{out} equals V_{in}, when the switch is closed. The S/H is in *hold*

Figure 11.59
The S/H has a digital
input, an analog input,
and an analog output.

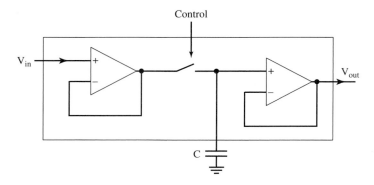

mode, where V_{out} is fixed, when the switch is open. The *acquisition time* is the time for the output to equal the input after the control is switched from hold to sample. This is the time to charge the capacitor C. The *aperture time* is the time for the output to stabilize after the control is switched from sample to hold. This is the time to open the switch that is usually quite fast. The *droop rate* is the output voltage slope (dV_{out}/dt) when **control** equals hold. Normally the gain K should be 1 and the offset V_{off} should be 0. The gain and offset error specify how close the V_{out} is to the desired V_{in} when **control** equals sample,

$$V_{out} = K\,V_{in} + V_{off}$$

where K should be one and V_{off} should be zero.
To choose the external capacitor C:

1. One should use a polystyrene capacitor because of its high insulation resistance and low dielectric absorption
2. A larger value of C will decrease (improve) the droop rate. If the droop current is I_{DR}, then the droop rate will be

$$\frac{dV_{out}}{dt} = \frac{I_{DR}}{C}$$

3. A smaller C will decrease (improve) the acquisition time

11.7   BiFET Analog Multiplexer

The analog multiplexer allows multiple analog signals to be sequentially connected to a single ADC. The operation of this type of computer-controlled switch was discussed previously in Section 8.5.8. Figure 11.60 illustrates the operation of a PMI MUX-08. This chip has four digital inputs, **Enable**, A_2, A_1, and A_0. It also has eight analog inputs (S_1–S_8) and one analog output, **Drain**. When the **Enable** is true, the three address bits determine which analog input is connected to the analog output.

R_{ON} is the on resistance of the switch (200 to 500Ω for the PMI MUX-08.) The "OFF" isolation, ISO_{OFF}, is a measure of how well the output is disconnected from its inputs when the device is shut off (60 dB for the PMI MUX-08.) To measure ISO_{OFF}, we place a 500-kHz +5-V sine wave on channel 8. We make all channels off (Enable = 0) and measure the signal at the output with a R_L=1RΩ, C_L=10pF. ISO_{OFF} is the gain V_{out}/V_{in} when all channels are off. The crosstalk (CT) is a measure of how much the input on one channel spills over into the output of another channel (70 dB for the PMI MUX-08.) To measure CT, we place a 500-kHz +5-V sine wave on channel 8. We activate channel 4 (Enable = 1, $A_2A_1A_0$ = 011) and measure the signal at the output with a R_L=1 MΩ, C_L=10 pF. CT is the gain from V_{in}(channel 8) to V_{out} when channel 4 is on. The settling time t_S is the time it takes a 10-V step to reach within 0.02% of the final output (2.5 μs for the PMI MUX-08.)

Figure 11.60
An analog multiplexer allows the computer to select one of eight analog signals.

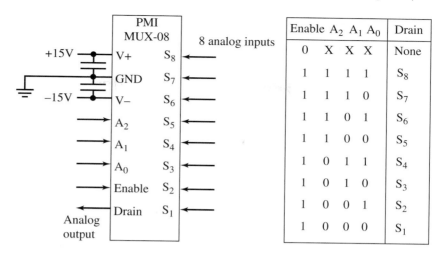

Enable	A_2	A_1	A_0	Drain
0	X	X	X	None
1	1	1	1	S_8
1	1	1	0	S_7
1	1	0	1	S_6
1	1	0	0	S_5
1	0	1	1	S_4
1	0	1	0	S_3
1	0	0	1	S_2
1	0	0	0	S_1

A device similar to the multiplexer is the analog switch. The CD4066B CMOS switch was discussed in Section 8.5.8. The PMI SW02 is a SPST BiFET analog switch with

$$R_{on} = 100\ \Omega\ \text{max}$$
$$ISO_{off} = 58\ \text{dB typical}$$
$$CT = 70\ \text{dB typical}$$
$$t_{on} = 400\ \text{ns max}$$
$$t_{off} = 300\ \text{ns max}$$

Analog switches and/or multiplexers can be used to make variable-gain analog amplifiers. Because of the large (bad) on-resistance (ranges from 85 to 500 Ω), it is important to route the output signal from the switch input into a large input impedance. In this way, there will be a minimal voltage drop across the switch. The two circuits in Figures 11.61 and 11.62, show a good and bad approach to using the bilateral switch to implement a microprocessor-controlled analog amplifier gain. In this first circuit, the gain is either 1+R2B/ R1B or 1+R2A/R1A, independent of the bilateral switch resistance. Because of the large input impedance into the op amp input terminal, the voltage drop across the switch will be zero.

The problem with the circuit in Figure 11.62 is that the resistance of the switch affects the gain of the circuit. In this second circuit, the gain is either 1+(Ron+R2B)/R1 or 1+(Ron+R2A)/R1, where Ron is the bilateral switch resistance. Since Ron can drift with time and temperature, the gain of this circuit will vary as well.

The analog multiplexer in the circuit of Figure 11.63 is used to select the gain from 1, 2, 5, 10, 20, 50, 100, 200.

Figure 11.61
The proper way to use a bilateral switch to make a variable-gain amplifier.

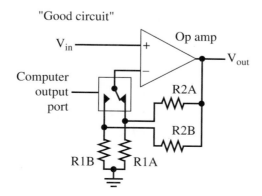

Figure 11.62
The improper way to use a bilateral switch to make a variable-gain amplifier.

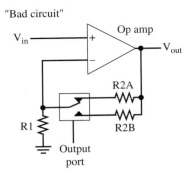

Figure 11.63
The use of an analog multiplexer to make a variable-gain amplifier.

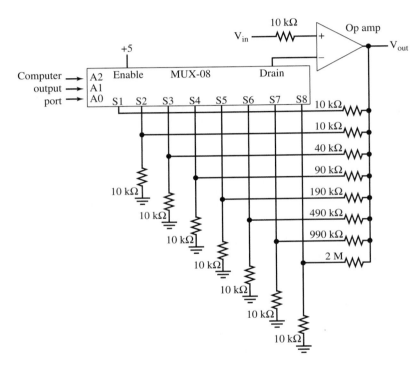

11.8   ADC System

11.8.1 ADC Block Diagram

The block diagram in Figure 11.64 connects the devices introduced in this chapter to build a data acquisition system. The next chapter will discuss the overall design process, but here in this section we discuss the low-level interfacing issues. Recall that the Maxim MAX1147 ADC integrates the multiplexer, S/H, and ADC into a single package.

Figure 11.64
Together, the analog circuit, multiplexer, S/H, and ADC convert external signals to digital form.

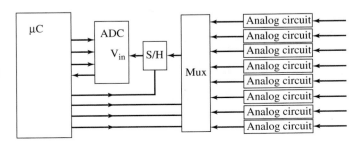

Figure 11.65
Flowcharts for ADC
interrupt software when
no S/H is needed.

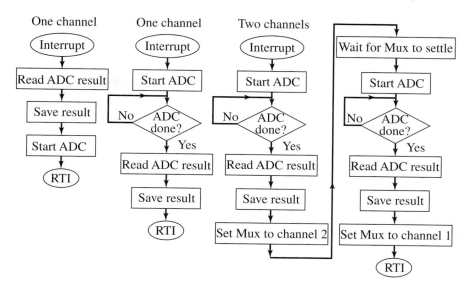

The flowcharts in Figures 11.65 and 11.66 show the tasks performed in the ISRs. It is assumed that the periodic interrupt occurs at the desired sampling rate. In the Figure 11.65, the system does not include a S/H. When only a single channel is being sampled, the multiplexer address is set once in the ritual. There are two approaches to sampling a single channel. If the ADC is started at the end of the interrupt handler, the interrupt software is faster and simpler. Unfortunately, this simple method introduces a delay in the data. In other words, data collected during one interrupt actually represent the signal at the time of the previous interrupt. Multiplexers are usually fast enough that normal software delays from

Figure 11.66
Flowcharts for ADC
interrupt software when
a S/H is needed.

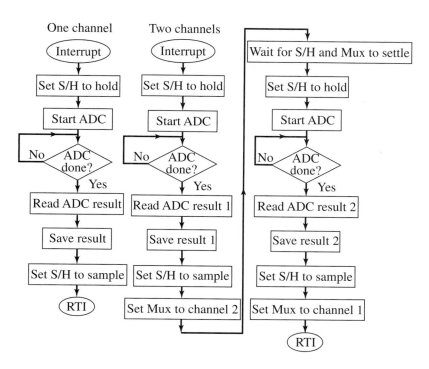

one instruction to the next are sufficient to allow the multiplexer to settle, so no explicit software function need be added for the "Wait for Mux to settle" step.

In Figure 11.66, the system does include a S/H. Just like the last example, if only a single channel is being sampled, the multiplexer address is set once in the ritual. Sample and hold modules are usually so slow that normal software delays from one instruction to the next are not sufficient for the S/H to settle, so some explicit software delay may be needed for the "Wait for S/H and Mux to settle" step.

11.8.2 Power and Grounding for the ADC System

The analog and digital grounds should be connected only at the ADC (Figure 11.67). Many ADC data sheets discuss layout techniques to minimize noise. A "star" pattern is used to decouple the devices. If the ground lines are in series, then a large current passing from an outer device could affect the ground voltage of an inner device. This can be a significant problem for high-speed devices that sink large transient currents during transitions. Because the S/H module requires careful circuit board layout for the capacitor, most manufacturers suggest PC board layout patterns for their S/H modules.

Figure 11.67
The analog and digital grounds are separate, connecting only at the ADC module.

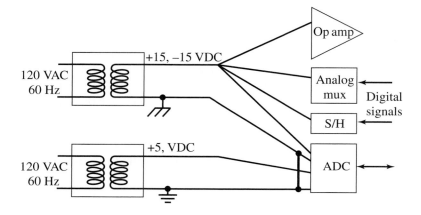

11.8.3 Input Protection for High-Speed CMOS Analog Inputs

It is necessary to protect CMOS inputs from large currents and out-of-range voltages. The microcomputer is particularly vulnerable to negative voltages. The information in this section was derived from the Freescale 6811 data sheets. Although no similar technical information on the 6812 is available, it is safe to assume this information applies to the 6812 as well. The equivalent circuit of the 6811 analog input is shown in Figure 11.68. Even though Port E on the 6811 is an input, part of an output channel provides some input protection

Figure 11.68
Equivalent circuit of an analog input on the 6811.

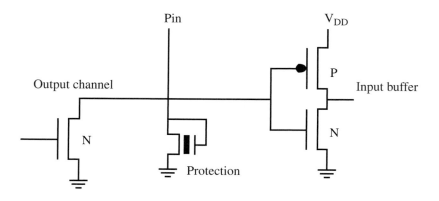

against voltages below zero. The N-channel output transistor and the N-channel thick-field transistor will act like diodes if the input drops below −0.7 V. Similarly, the P-channel transistor on the input buffer will act like a diode if the input rises above 20 V.

Common error: It is a mistake to neglect the possibility of a broken transducer or disconnected cable.

Performance tip: It is good engineering practice to design input protection based Freescale on reasonable expected failure modes. If you develop a collection of broken transducers (open and short) and cables, you could use these devices to design and test your systems.

When designing instrumentation that interfaces to the internal ADC, we must restrict input voltages to be between 0 and +5 V. Freescale advises against using external champing diodes to protect analog inputs from out-of-range voltages. Instead they suggest a 1- to 10-kΩ current-limiting resistor and a low-pass filter. Higher limiting resistor values provide for more protection but may cause conversion errors (Figure 11.69).

Instead of connecting protection diodes at the Port E inputs, it is possible to use protection diodes at an earlier stage of the analog circuit (Figure 11.70). For example, if you are using a unity gain low-pass filter as the last stage, then diodes can be used at the input of the low-pass filter. In this way the Port E input voltage could go out of range only if the low-pass filter were to malfunction.

Figure 11.69
Suggested interface between analog circuits and the analog input.

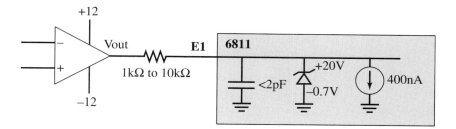

Figure 11.70
One possible approach for protecting the analog input.

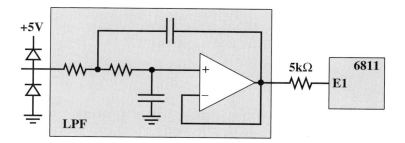

11.9 Multiple-Access Circular Queue

A *multiple-access circular queue* (MACQ) is used for data acquisition and control systems (Figure 11.71). A MACQ is a fixed-length order-preserving data structure. The source process (ADC sampling software) places information into the MACQ. Once initialized, the MACQ is always full. The oldest data are discarded when the newest data are **PUT** into a

Figure 11.71
When data are put into
a MACQ, the oldest
data are lost.

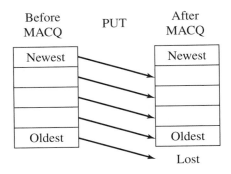

Figure 11.72
Application of the
multiple access circular
queue.

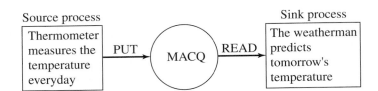

MACQ. The sink process can read any of the data from the MACQ. The **READ** function is nondestructive. This means that the MACQ is not changed by the READ operation. For example, consider the problem of weather forecasting (Figure 11.72). The weatherman measures the temperature every day at 12 noon, and PUTs the temperature into the MACQ. To predict tomorrow's temperature, she looks at the trend over the last 3 days. Let T(0) be today's temperature, T(1) be yesterday's temperature, etc. Tomorrow's predicted temperature might be

$$U = \frac{170 \cdot T(0) + 60 \cdot T(1) + 36 \cdot T(2)}{256}$$

The MACQ is useful for implementing digital filters and linear control systems. One common application of the MACQ is the real-time calculation of derivative. Assume the function `Adin()` returns the current sample in millivolts. Also assume the function `ClkHandler()` is a periodic interrupt executed every 1 ms. x(n) will refer to the current sample, and x(n−1) will be the sample 1 ms ago ($\Delta t = 1ms$). There are a couple of ways to implement the discrete time derivative. The simple approach is

$$d(n) = \frac{x(n) - x(n - 1)}{\Delta t}$$

In practice, this first-order equation is quite susceptible to noise. In most practical control systems, the derivative is calculated using a higher-order equation like

$$d(n) = \frac{x(n) + 3x(n - 1) - 3x(n - 2) - x(n - 3)}{\Delta t}$$

The C implementation of this discrete derivative (Program 11.11) uses a MACQ. Since Δt is 1 ms, we simply consider the derivative to have units millivolts per millisecond and not actually execute the divide by Δt operation.

When the MACQ holds many data points, it can be implemented using a pointer or index to the newest data. In this way, the data need not be shifted each time a new sample is added. The disadvantage of this approach is that address calculation is required during the READ access.

```
// 6811                                          // 6812
#define Rate 2000                                #define Rate 2000
#define OC5  0x08                                 #define OC5  0x20
unsigned short x[4];  // MACQ (mV)               unsigned short x[4];  // MACQ (mV)
unsigned short d;      // derivative (V/s)       unsigned short d;      // derivative (V/s)
#pragma interrupt_handler TOC5handler()          #pragma interrupt_handler TOC5handler()
void TOC5handler(void){                          void TOC5handler(void){
   TFLG1=OC5;         // Ack interrupt              TC5=TC5+Rate;     // Executed every 1 ms
   TOC5=TOC5+Rate; // Executed every 1 ms           TFLG1=0x20;       // ack OC5F
   x[3]=x[2];  // shift MACQ data                   x[3]=x[2];  // shift MACQ data
   x[2]=x[1];  // units of mV                       x[2]=x[1];  // units of mV
   x[1]=x[0];                                       x[1]=x[0];
   x[0]=Adin(); // current data                     x[0]=Adin(); // current data
   d=x[0]+3*x[1]-3*x[2]-x[3];} // mV/ms             d=x[0]+3*x[1]-3*x[2]-x[3];} // mV/ms
```

Program 11.11 Software implementation of first derivative using a MACQ.

11.10 Internal ADCs

The Freescale 6811/6812 microcomputers are available in versions that have built-in ADCs. We will begin by discussing the particular I/O registers used to interface analog signals to the individual microcomputers. The common features include:

- Eight-channel operation
- 8-bit or 10-bit resolution
- Successive approximation conversion technique
- A clock and charge pump is used to create higher voltages needed for the ADC
- Two operation modes: single sequence of conversions, then stop, and continuous conversion
- Two channel selection modes: multiple conversions of a single channel, (e.g., channel 1,1,1,1) and one conversion each on a group of channels (e.g., channels 0,1,2,3)
- External V_{RH}, V_{RL} analog high/low references

11.10.1 6811 ADC System

The 6811 ADC system is enabled by setting ADPU equal to 1 in the OPTION register. When ADPU is 0, Port E operates as a normal digital input port. The CSEL control bit selects the clock for the charge pump. CSEL should be 0 for normal E clock frequencies. For E clock frequencies below 750 kHz, CSEL should be 1.

	7	6	5	4	3	2	1	0	
OPTION	**ADPU**	**CSEL**	IRQE	DLY	CME	0	CR1	CR0	$1039

The control part of the ADCTL register is used to set the operation mode and start the conversion. Each sequence of four conversions requires 128 E clock cycles to complete. For single-sequence mode, we write to the ADCTL with SCAN = 0 to start a new sequence. This register also contains a status bit, CCF, which we use to poll for the ADC conversion

	7	6	5	4	3	2	1	0	
ADCTL	CCF	0	SCAN	MULT	CD	CC	CB	CA	$1030

completion. The CCF flag is cleared by writing data into the ADCTL (i.e., starting a new conversion). To start a continuous conversion, we write to the ADCTL with SCAN=1.

CCF: Conversions complete flag (read only)
    0 = conversions in progress
    1 = conversion complete, ADR1, ADR2, ADR3, ADR4 have valid data
SCAN: Continuous scan control
    0 = single sequence of conversions, then stop
    1 = continuous conversion
MULT: Multiple channel/single channel control
    0 = sequence of conversions on a single channel
    1 = sequence of conversions on multiple channels (0,1,2,3) (4,5,6,7) (8,9,10,11)
        or (12,13,14,15)
        Multiplexer control bits CD and CC specify the group of channel
        Multiplexer control bits CB and CA are ignored
CD, CC, CB, CA: Multiplexer control selects that channel to convert (Table 11.12)

Table 11.12
Multiplexer control for the 6811 ADC.

CD, CC, CB, CA	Channel
0000	PE0 analog channel 0
0001	PE1 analog channel 1
0010	PE2 analog channel 2
0011	PE3 analog channel 3
0100	PE4 analog channel 4
0101	PE5 analog channel 5
0110	PE6 analog channel 6
0111	PE7 analog channel 7
1000	Reserved
1001	Reserved
1010	Reserved
1011	Reserved
1100	V_{RH} analog reference high
1101	V_{RL} analog reference low
1110	$(V_{RH} + V_{RL})/2$
1111	Reserved

The ADC result registers contain the digital outputs of the ADC conversions. In single-sequence mode (MULT = 0), the specified single channel (CD,CC,CB,CA) is converted four times and the results placed in ADR1, ADR2, ADR3, and ADR4. In multiple-sequence mode (MULT = 1), the specified group of four channels (CD,CC) is each converted once and the results placed in ADR1, ADR2, ADR3, and ADR4. For example, if we write $14 into ADCTL, the 6811 will convert analog channels 4, 5, 6, 7 and put the digital results in

	7	6	5	4	3	2	1	0	
ADR1	bit 7	bit 6	bit 5	bit 4	bit 3	bit 2	bit 1	bit 0	$1031
ADR2	bit 7	bit 6	bit 5	bit 4	bit 3	bit 2	bit 1	bit 0	$1032
ADR3	bit 7	bit 6	bit 5	bit 4	bit 3	bit 2	bit 1	bit 0	$1033
ADR4	bit 7	bit 6	bit 5	bit 4	bit 3	bit 2	bit 1	bit 0	$1034

ADR1, ADR2, ADR3, ADR4, respectively. The 8-bit unsigned result will vary from 0 to 255 as the analog input varies from 0 to +5 V.

Performance tip: If we are interested in a single conversion, we could start the ADC, wait 32 cycles, then read the result in ADR1.

11.10.2
6812 ADC System

Most of the 6812 microcontrollers have built-in ADC converters. This section specifically covers the MC9S12C32, but the other 6812 devices operate in a very similar manner. The I/O registers used for the ADC are listed in Table 11.13. The ADC on the MC9S12C32 can be operated in 8-bit mode or 10-bit mode. The 8 pins of Port AD can be individually defined to be analog input, digital output, or digital input. We set the corresponding bit in the ATDDIEN register to be 1 for digital or 0 for analog input. If a pin is digital, then the corresponding bit in the DDRAD register specifies input(0) or output(1). The ADC digital output can be right- or left-justified within the 16-bit result register, and it can be in a signed or unsigned format. When the ADC is triggered, it performs a sequence of conversions, with the sequence length being any number from 1 to 8 conversions. When performing a sequence, it can convert the same channel multiple times or it can convert different channels during the sequence. We can trigger ADC conversions in three ways. The first way is to use an explicit software trigger (write to ATDCTL5); when the conversions are complete, the SCF flag is set. The examples in Programs 11.12 and 11.13 employ the explicit software trigger to start an ADC conversion. The second way to trigger the ADC is continuous mode. In this mode, the ADC sequence is repeated over and over continuously. The third way is to connect an external trigger to the digital input on PAD7. With an external trigger, we can use busy-wait synchronization (gadfly) on the SCF flag, or arm interrupts (ASCIE = 1) on the ASCIF flag. The results of the ADC conversions can be found in the ATDDR0 to ATDDR7 result registers, where the register number refers to the sequence number. In other words, ATDDR0 contains the result of the first conversion in the sequence, ATDDR1 contains the result of the second conversion, . . . and ATDDR7 contains the result of the eighth conversion.

Address	Bit 7	6	5	4	3	2	1	Bit 0	Name
$0082	ADPU	AFFC	AWAI	ETRIGLE	ETRIGP	ETRIG	ASCIE	ASCIF	ATDCTL2
$0083	0	S8C	S4C	S2C	S1C	FIFO	FRZ1	FRZ0	ATDCTL3
$0084	SRES8	SMP1	SMP0	PRS4	PRS3	PRS2	PRS1	PRS0	ATDCTL4
$0085	DJM	DSGN	SCAN	MULT	0	CC	CB	CA	ATDCTL5
$0086	SCF	0	ETORF	FIFOR	0	CC2	CC1	CC0	ATDSTAT0
$008B	CCF7	CCF6	CCF5	CCF4	CCF3	CCF2	CCF1	CCF0	ATDSTAT1
$008D	Bit 7	6	5	4	3	2	1	Bit 0	ATDDIEN
$0270	PTAD7	PTAD6	PTAD5	PTAD4	PTAD3	PTAD2	PTAD1	PTAD0	PTAD
$0272	DDRAD7	DDRAD6	DDRAD5	DDRAD4	DDRAD3	DDRAD2	DDRAD1	DDRAD0	DDRAD

Address	msb															lsb	Name
$0090	15	14	13	12	11	10	9	8	7	6	5	4	3	2	1	0	ATDDR0
$0092	15	14	13	12	11	10	9	8	7	6	5	4	3	2	1	0	ATDDR1
$0094	15	14	13	12	11	10	9	8	7	6	5	4	3	2	1	0	ATDDR2
$0096	15	14	13	12	11	10	9	8	7	6	5	4	3	2	1	0	ATDDR3
$0098	15	14	13	12	11	10	9	8	7	6	5	4	3	2	1	0	ATDDR4
$009A	15	14	13	12	11	10	9	8	7	6	5	4	3	2	1	0	ATDDR5
$009C	15	14	13	12	11	10	9	8	7	6	5	4	3	2	1	0	ATDDR6
$009E	15	14	13	12	11	10	9	8	7	6	5	4	3	2	1	0	ATDDR7

Table 11.13
MC9S12C32 registers used for analog-to-digital conversion.

The ATDCTL2 contains bits that activate the ADC module. The 6812 ADC system is enabled by setting **ADPU** equal to 1. The ADC will request an interrupt on the completion of a conversion sequence if the arm bit **ASCIE** is set. **ASCIF** is the ATD Sequence Complete Interrupt Flag. If ASCIE = 1, the ASCIF flag equals the SCF flag, else ASCIF reads zero. Write operations to ATDCTL2 have no effect on ASCIF. **ETRIGE** is the External Trigger Mode Enable bit. This bit enables an external trigger using the digital input from Port AD bit 7. The external trigger allows us to synchronize sample and ATD conversion processes with external events. If external triggering is enabled, then the type of trigger is defined in the **ETRIGLE** and **ETRIGP** bits as specified in Table 11.14.

Table 11.14
External trigger modes for the MC9S12C32 ADC.

ETRIGLE	ETRIGP	External trigger mode
0	0	Falling edge of PAD7 starts a conversion sequence
0	1	Rising edge of PAD7 starts a conversion sequence
1	0	Perform ADC conversions when PAD7 is low
1	1	Perform ADC conversions when PAD7 is high

The ATDCTL3 and ATDCTL4 contain bits that specify the ADC mode. **S8C, S4C, S2C, S1C** control the number of conversions per sequence. Let **n** be the four-bit number specified by these bits. For values of **n** from 1 to 7, **n** specifies the sequence length. For values of **n** equal to 0 or 8–15, the sequence length is 8. At reset, the default sequence length is 4 (0100), maintaining software continuity to the HC12 family. This book will not discuss FIFO mode or freeze mode. **SRES8** is the ADC Resolution Select bit. This bit selects the resolution of ADC conversion as either 8 (SRES8 = 1) or 10 bits (SRES8 = 0). The ADC converter has an accuracy of 10 bits; however, if low resolution is acceptable, selecting 8-bit resolution will reduce the conversion time.

The time to perform an ADC conversion is determined by the E clock and the ATDCTL4 register. The ATDCTL4 register selects the sample period and the PRS-Clock prescaler. **SMP1, SMP0** are the Sample Time Select bits. These two bits select the length of the second phase of the sample time in units of ATD conversion clock periods, as listed in Table 11.15. The sample time consists of two phases. The first phase is two ATD conversion clock periods long and transfers the sample quickly (via the buffer amplifier) onto the ADC machine's storage node. The second phase attaches the external analog signal directly to the storage node for final charging and high accuracy. Table 11.15 lists the lengths available for the second sample phase. Let **m** be the 5-bit number formed by bits **PRS4-0**. Let f_E be the frequency of the E clock. The ATD conversion clock frequency is calculated as follows:

$$\text{ATD clock frequency} = 1/2\ f_E\ /(m+1)$$

The default (after reset) prescaler value is 5, which results in a default ATD conversion clock frequency that is the E clock divided by 12. For example, if the E clock is 24 MHz, and **m** is 5, then the ATD clock is 2 MHz. The choice of these parameters involves a trade-off between accuracy and speed. Freescale recommends that the ADC clock frequency be restricted to the 500 kHz to 2 MHz range. For analog signals with white noise, we can

Table 11.15
Sampling time for the 6812 ADC.

SMP1	SMP0	First sample phase	Second sample phase	Total sample time
0	0	2 ADC clock periods	2 ADC clock periods	4 ADC clock periods
0	1	2 ADC clock periods	4 ADC clock periods	6 ADC clock periods
1	0	2 ADC clock periods	8 ADC clock periods	10 ADC clock periods
1	1	2 ADC clock periods	16 ADC clock periods	18 ADC clock periods

essentially add an analog low-pass filter by increasing the ADC sample time. To increase conversion speed, we should select a fast clock and short sample period. The last factor to consider is the slewing rate of the input signal. For signals with a high slope, dV/dt, we need to select a faster conversion time (i.e., shorter sample time). More discussion about slewing rate will be made in the next chapter when considering the need for a sample-and-hold analog latch. For a 24 MHz E clock, the possible **m** prescales range from 5 (ADC clock = 2 MHz) to 23 (ADC clock = 500 kHz). Other choices are not recommended. The ADC conversion time is equal to $2 (m + 1) (s + n)/f_E$, where s is the total sample time (Table 11.15) and n is the number of ADC bits (e.g.,10).

Writing to the ATDCTL5 register will start an ADC conversion. To begin continuous conversions, we write to the ATDCTL5 with **SCAN** = 1. On the other hand, if we write to ATDCTL5 with SCAN = 0, only one sequence occurs. **CC, CB, CA** select the analog input channel(s), whose signals are sampled and converted to digital codes. Because the result registers (16 bits) are wider than the ADC digital code (8 or 10 bits), we must choose where in the result register to put the digital code. **DJM** is the Result Register Data Justification bit, where 1 means right-justified and 0 means left-justified data in the result registers. **DSGN** selects between signed and unsigned format. We set DSGN to 1 for signed data representation and we set it to 0 for unsigned data representation. Table 11.16 describes the four possible 10-bit data formations for the MC9S12C32. When **MULT** is 0, the ATD sequence controller samples only from the specified analog input channel for an entire conversion sequence. When MULT is 1, the ATD sequence controller samples across channels. The number of channels sampled is determined by the sequence length value (S8C, S4C, S2C, S1C). The first analog channel examined is determined by channel selection code (CC, CB, CA control bits); subsequent channels sampled in the sequence are determined by incrementing the channel selection code.

The status register contains a status bit, **SCF**, which we use to poll for the ADC conversion completion. The SCF flag is cleared by writing data into the ADCTL5 (i.e., starting a new conversion.) The SCF flag can also be cleared by writing a 1 to it. The CC2,CC1,CC0 bits are the sequence counter as the ADC steps through a conversion sequence. The CCFn bits are individual flags for each of the conversions.

> **Performance tip:** If we are interested in a single conversion, we could start the ADC, wait for CCF0, then read the result in ATDDR0.

11.10.3
ADC Software

Figure 11.73 shows an analog signal that is connected to analog input channel 1. Using one +5 V supply, this system implements a −2.5 to +2.5 V analog range using a reference chip and rail-to-rail op amps. Because the analog ground is 2.5 V above digital ground, the range of voltages on the analog channel will be 0 to +5 V relative to the 6811/6812 ground.

The 6811 subroutine, AD_In, will return an unsigned value from 0 to 255 that represents the analog input. The MC9S12C32 subroutine, AD_In, will return a 10-bit value representing the analog input. Table 11.16 shows the 8-bit format available on the

Figure 11.73
An analog signal is connected to a pin of the internal ADC.

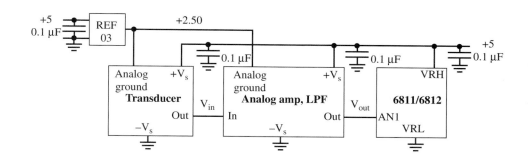

Analog input (V)	8-bit unsigned digital output	10-bit unsigned right-justified	10-bit unsigned left-justified	10-bit signed right-justified	10-bit signed left-justified
0.000	$00 0	$0000 0	$0000 0	$FE00 −512	$8000 −32768
0.005	$00 0	$0001 1	$0040 64	$FE01 −511	$8040 −32704
0.020	$01 1	$0004 4	$0100 256	$FE04 −508	$8100 −32512
2.500	$80 128	$0200 512	$8000 32768	$0000 0	$0000 0
3.750	$C0 192	$0300 768	$C000 49152	$0100 256	$4000 16384
5.000	$FF 255	$03FF 1023	$FFC0 65472	$01FF 511	$7FC0 32704

Table 11.16
Binary formats used by the Freescale internal ADCs.

6811 and the four 10-bit formats available on 6812. When the 6812 inserts 8- or 10-bit data into the 16-bit result register, it will pad the extra bits with 0s, but Table 11.16 is shown with bits 15-10 having been sign-extended from bit 9. The C code to perform this sign extension is

```
Result = ATDDR0;                        // 10-bit signed, right-justified
if(Result&0x200) Result = Result|0xF8;// sign extend if negative
```

For 6811 ADC and the 6812 running with an 8-bit precision, the analog input range is 0 to +5 V, and the analog input resolution is 5V/256, which is about 20 mV. For the 6812 running with a 10-bit precision, the analog input range is 0 to +5 V, the analog input resolution is 5V/1024, which is about 5 mV. In Programs 11.12 and 11.13, the 6811 and 6812 ADC will perform multiple conversions on the same analog signal, but the ADC_In function returns only the first conversion. The format of the input parameter is

```
; MC68HC711E9
; Analog signal to E7-0
ADC_Init ldaa #$80    ;Turn on ADC
         staa OPTION ;ADC power
         rts          ;E clk timing
;In:  RegA has channel Number
;Out: RegA has ADC result
ADC_In   ldx   #ADCTL
         staa  0,X   ;Start ADC
Loop     brclr 0,X,#$80,Loop
         ldaa  ADR1  ;first result
         rts
```

```
; MC9S12C32
; Analog signal connected to PAD7-0
ADC_Init movb  #$80,ATDCTL2 ;power up
         movb  #$05,ATDCTL4 ;10-bit
         rts
;In:  RegA has channel Number
;Out: RegD has ADC result
ADC_In   staa  ATDCTL5 ;Start ADC
Loop     brclr ATDSTAT,$#80,Loop
         ldd   ATDDR0  ;first result
         rts
```

Program 11.12
Assembly software to sample data using the ADC.

```
// MC68HC11A8
void ADC_Init(void){
  OPTION = 0x80;    // Activate ADC
}
unsigned char ADC_In(unsigned char chan){
  ADCTL = chan;     // Start ADC
  while ((ADCTL & 0x80) == 0){};
  return(ADR1);     // 8-bit result
}
```

```
// MC9S12C32
void ADC_Init(void){
  ATDCTL2 = 0x80; // enable ADC
  ATDCTL4 = 0x05; // 10-bit, divide by 12
}
unsigned short ADC_In(unsigned char chan){
  ATDCTL5 = chan;                // start
  while((ATDSTAT1&0x01)==0){}; // CCF0
  return ATDDR0;     // 10-bit result
}
```

Program 11.13
C-language software to sample data using the ADC.

```
bit 7 DJM  1=right , 0=left justified (6812 only)
bit 6 DSGN 1=signed, 0=unsigned (6812 only)
bit 5 SCAN 0=single sequence
bit 4 MULT 0=single channel
bits 2-0   channel number 0 to 7
```

The 6811 version returns an 8-bit result, and the 6812 version returns a 10-bit result.

Checkpoint 11.11: If the input voltage is 1.0 V, what value will the 6811 `ADC_In` return?

Checkpoint 11.12: If the input voltage is 1.0 V, what value, in 10-bit unsigned right-justified mode, will the 6812 `ADC_In` return?

11.11 Exercises

11.1 Consider an ADC system.
  a) Give a definition of the *ADC precision.* State the units.
  b) Give a definition of the *ADC resolution.* State the units.

11.2 A 6811/6812 with *an 8-bit DAC* is used to create a software-controlled analog output. The 8-bit signed DAC (using offset binary format) is connected to the 6811/6812 Port B and has a −5 V to +5V range as shown in the following table. N is the Port B output (DAC input), and V is the DAC output

N=8-bit digital input	V=analog output (mvolts)
0 $00 %00000000	−5000
64 $40 %01000000	−2500
128 $80 %10000000	0
129 $81 %10000001	÷ 39
192 $C0 %11000000	+2500
255 $FF %11111111	+4961

  a) Derive the linear relationship between N and V. Show both the equation that expresses V in terms of N and the equation that expresses N in terms of V.
  b) What is the *DAC precision?* State the units.
  c) What is the *DAC resolution?* State the units.
  d) Write a 6811/6812 assembly language function, DAout, which takes the desired voltage (in millivolts) and outputs the appropriate value to Port B. The 16-bit signed input parameter is passed by reference on the stack. Here are a couple of typical calling sequences:

```
ldx #data pointer to value        ldy #neg2
pshx      by reference            pshy      by reference
jsr DAout                         jsr DAout
pulx      discard parameter       puly
data fdb 2314  means 2.314 volts  neg2 fdb −2000 means −2 volts
```

11.3 What does ADC stand for?

11.4 What does DAC stand for?

11.5 Which port on the HC11A8, HC812A4, and 9S12C32 is used by the ADC?

11.6 If a 10-bit ADC has a range of 0 to +10 V, what is the resolution?

11.7 Give three equivalent ways to specify the precision of a 12-bit ADC.

11.8 If an 8-bit ADC has a range of 5 V, what will be the output for an input voltage of 1 V, assuming straight binary encoding?

11.9 If an 8-bit ADC takes inputs ranging from −2.5 V to +2.5 V, what will the output corresponding to +1 V, assuming offset binary encoding?

11.10 Design an ADC with the transfer function shown in Figure 11.74. Label all chip numbers (but not pin numbers). Specify the +12, −12, and +5 power supply connections, resistor values, and capacitor values. Offset null potentiometers are not required.

Figure 11.74
Digital output versus
analog input for Exercise
11.10.

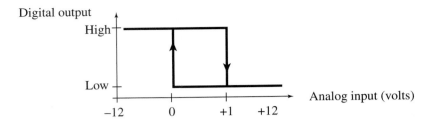

11.11 Design a variable-gain analog amplifier. The analog input is V_{in}, the 3-bit digital inputs (connected to a microcomputer output digital output) are B_2, B_1, B_0, and the analog output is V_{out}. Exactly one of the digital inputs will be 1, and the gain should be

B_2, B_1, B_0	Gain (V_{out}/V_{in})
001	1
010	10
100	100

11.12 In the EKG waveform shown in Figure 11.75, there are long periods of constant value together with short periods of rapidly changing values. Notice that the data points in this figure are placed at uneven time intervals to match the various phases of this signal. Implement software that generates this waveform using uneven time output compare interrupts. In particular, create ROM-based tables that store the nine or ten points and use the DAC8043 12-bit DAC to create the periodic waveform.

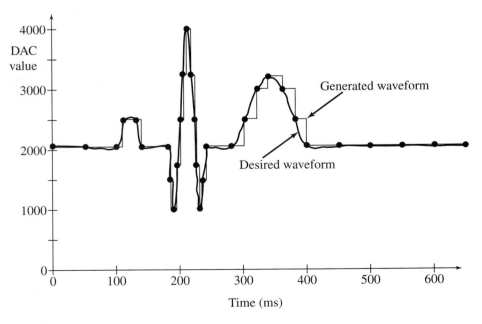

Figure 11.75
EKG output waveform for Exercise 11.12.

11.13 A computer with *an 8-bit DAC* is used to create a software-controlled analog output. The 8-bit signed DAC (using offset binary format) has a -5 V to $+5$ V range as shown in the table. N is the output (DAC input) and V is the DAC output

N=8-bit digital input	V=analog output (mvolts)
0 $00 %00000000	−5000
64 $40 %01000000	−2500
128 $80 %10000000	0
129 $81 %10000001	+39
192 $C0 %11000000	+2500
255 $FF %11111111	+4961

    a) Derive the linear relationship between N and V. Show both the equation that expresses V in terms of N, and the equation that expresses N in terms of V.

    b) What is the **DAC precision**? State the units.

    c) What is the **DAC resolution**? State the units.

11.14 If a 16-bit ADC converter takes inputs ranging from -2.5 volts to $+2.5$ volts, what will the output corresponding to $+1$ volts, assuming 2s complement encoding?

11.15 Design an analog circuit with the following specifications:

Two single-ended inputs (not differential)
Any input impedance is OK
Transfer function $V_{out} = 5 \cdot V_1 - 3 \cdot V_2 + 5$

You are limited to one OPA227 op amp and one reference chip (you choose it). Give chip numbers but not pin numbers. Specify all resistor values. You will use $+12$ and -12 V analog supply voltages.

11.16 Design an instrumentation amp, using an AD620, with the following transfer function:

$$V_{out} = 500*(V_2 - V_1)$$

11.17 Design a two-pole Butterworth analog low-pass filter with a cutoff of 250 Hz.

11.12   Lab Assignments

Lab 11.1. The overall objective of this lab is to design a variable-gain amplifier, where the gain is controlled by a digital output of the microcontroller. The gain settings are 10, 20, 50, and 100. As preparation, design two separate circuits using different approaches. Evaluate the bandwidth, noise, error due to offset voltage, and cost. Build and experimentally measure bandwidth, noise, and offset error.

Lab 11.2. Most digital music devices rely on high-speed DAC converters to create the analog waveforms required to produce high-quality sound. In this lab you will create a very simple sound-generation system that illustrates this application of the DAC. Your goal is to play your favorite song. For the first step, you will interface a 74HC595 serial in/parallel out shift register to the SPI port. Please refer to the 74HC595 data sheets for the synchronous serial protocol. The second step is to create a DAC from the 8-bit digital output of the 74HC595. You are free to design your DAC with a precision anywhere from 5 to 8 bits. You will convert the binary bits (digital) to an analog output current using a simple resistor network. The third step is to convert the DAC analog output to speaker current using an audio amplifier, such as the LM386. It doesn't matter what range the DAC is, as long as there is an approximately linear relationship between the digital data and the speaker current. To do this you will have to run the analog circuit in its linear range. Be careful not to saturate the analog circuit. The performance score is based not on loudness, but on sound quality. On the other hand, sound quality will be a function of the number of DAC bits, the linearity of the analog circuit, and the periodic output rate. If an analog signal is noisy, you can add filter capacitors. It is important to add a 0.1 μF bypass capacitor on the power connection of the 74HC595 to prevent output glitches during serial input transmissions. The fourth step is to design a low-level device driver for the DAC. A single 8-bit SPI frame is all that is required to set the DAC output.

The fifth step is to design a data structure to store the sound waveform. You are free to design your own format, as long as it uses a formal data structure (i.e., `struct`). Compressed data occupies less storage, but requires runtime calculation. On the other hand, a complete list of points will be simpler to process, but requires more storage than is available on the microcontroller. The sixth step is to organize the music software into a device driver. Although you will be playing only one song, the song data itself will be stored in the main program, and the device driver will perform all the I/O and interrupts to make it happen. You will need public functions `Rewind`, `Play`, and `Stop`, which perform operations as does a cassette tape player. The `Play` function has an input parameter that defines the song to play. A background thread implemented with output compare will fetch data from your music structure and send them to the DAC. The last step is to write a main program that inputs from three binary switches and performs the three public functions.

Lab 11.3. The overall objective of this lab is design the music player described in Lab 11.2 with a two-channel DAC chip, and design two audio amplifiers. The system will implement two-channel stereo sound.

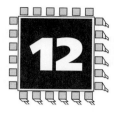

12 Data Acquisition Systems

Chapter 12 objectives are to:

❏ Define performance criteria to evaluate our overall data acquisition system
❏ Introduce specifications necessary to select the proper transducer for our data acquisition system
❏ Describe some typical transducers used in embedded systems
❏ Develop a methodology for designing data acquisition systems
❏ Analyze the sources of noise and suggest methods to reduce their effect
❏ Illustrate concepts of this chapter with case studies

Embedded systems are different from general-purpose computers in the sense that embedded systems have a dedicated purpose. As part of this purpose, many embedded systems are required to collect information about the environment. A system that collects information is called a data acquisition system (DAS). In this chapter, we will use the two terms, DAS and instrument interchangeably. Previous chapters presented the basic building blocks to acquire data into the computer, and in this chapter we will combine these blocks into DASs. Sometimes the acquisition of data is the fundamental purpose of the system, such as with a voltmeter, a thermometer, a tachometer, an accelerometer, an altimeter, a manometer, a barometer, an anemometer, an audio recorder, or a camera. At other times, the acquisition of data is an integral part of a larger system such as a control system or communication system. Control systems and communication systems will be discussed in Chapters 13 and 14.

12.1 Introduction

Figure 12.1 illustrates the integrated approach to instrument design. In this section, we begin with the clear understanding of the problem. We can use the definitions in this section to clarify the design parameters as well as to report the performance specifications. Next, in Section 12.2, we will define the parameters and discuss the selection of a suitable transducer. In Section 12.3, we put together the analog and digital components, introduced in Chapter 11, to build DASs. The use of period/pulse/frequency as a means of collecting information was developed in Chapter 6. The design of digital filters will be developed later in Chapter 15. A closed-loop control system, as we will see in Chapter 13, includes a DAS. Noise can never be eliminated, but we will study techniques in Section 12.4 to reduce its

Figure 12.1
Individual components are integrated into a DAS.

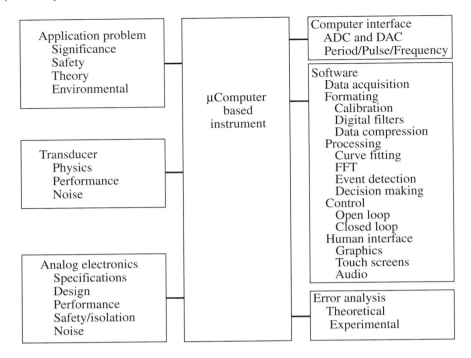

effect on our system. The integrated approach to design will be illustrated using the case studies in Section 12.5.

The *measurand* is the physical quantity, property, or condition that the instrument measures. The measurand can be inherent to the object (like position, size, mass, or color), located on the surface of the object (like the human EKG or surface temperature), located within the object (e.g., fluid pressure or internal temperature), or separated from the object (like emitted radiation). In general, a *transducer* converts one energy type into another. In the context of this book, the transducer converts the measurand into an electric signal that can be processed by the microcomputer-based instrument. Typically, a transducer has a primary sensing element and a variable conversion element. The *primary sensing element* interfaces directly to the object and converts the measurand into a more convenient energy form. The output of the *variable conversion element* is an electric signal that hopefully depends on the measurand. For example, the primary sensing element of a pressure transducer is the diaphragm, which converts pressure into a displacement of a plunger. The variable conversion element is a strain gage that converts the plunger displacement into a change in electric resistance. If the strain gage is placed in a bridge circuit, the voltage output is directly proportional to the pressure. Some transducers perform a direct conversion without having a separate primary sensing element and variable conversion element. The instrumentation contains *signal processing*, which manipulates the transducer signal output to select, enhance, or translate the signal to perform the desired function, usually in the presence of disturbing factors. The signal processing can be divided into stages. The *analog signal processing* consists of instrumentation electronics, isolation amplifiers, amplifiers, analog filters, and analog calculations. The first analog processing involves calibration signals and preamplification. Calibration is necessary to produce accurate results. An example of a calibration signal is the reference junction of a thermocouple. The second stage of the analog signal processing includes filtering and range conversion. The analog signal range should match the ADC analog input range. Examples of analog calculations include RMS calculation, integration, differentiation, peak detection, threshold detection, PLLs, AM and FM modulation/demodulation, and the arithmetic calculations of addition, subtraction, multiplication, division, and square root. When period, pulse width, or frequency measurement is used, we typically use an analog comparator to create a digital logic signal to measure. Whereas Figure 12.1 outlined design components, Figure 12.2 shows the data flow graph for a data acquisition system or control system. The control system uses an actuator to drive a parameter in the real world to a desired value, while the DAS has no actuator because it simply measures the parameter in a noninstrusive manner.

Figure 12.2
Data flow graph of a DAS.

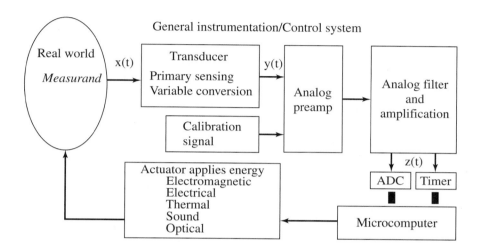

The *data conversion element* performs the conversion between the analog and digital domains. This part of the instrument includes hardware and software computer interfaces, ADC, DAC, S/H, analog multiplexer, and calibration references. The *ADC* converts the analog signal into a digital number. In Chapter 6, we saw that the period, pulse width, and frequency measurement approach provides a low-cost high-precision alternative to the traditional ADC. The *digital signal processing* includes data acquisition (sampling the signal at a fixed rate), data formatting (scaling, calibration), data processing (filtering, curve fitting, Fast Fourier Transform, event detection, decision making, analysis), and control algorithms (open or closed loop). The *human interface* includes the input and output that is available to the human operator. The advantage of computer-based instrumentation is that sophisticated but easy to use and understand devices are possible. The *inputs* to the instrument can be audio (voice), visual (light pens, cameras), or tactile (keyboards, touch screens, buttons, switches, joysticks, roller balls). The *outputs* from the instrument can be numeric displays, cathode ray tube screens, graphs, buzzers, bells, lights, and voice. If the system can deliver energy to the real world, then it is classified as a control system. Control systems will be developed in Chapter 13. In this chapter, we focus on data acquisition. Table 12.1 lists the symbols used in the equations of this chapter.

The rest of this section defines performance criteria we can use to characterize our instrument. Whenever reporting specifications of our instrument, it is important to give the definitions of each parameter, the magnitudes of each parameter, and the experimental conditions under which the parameter was measured, because engineers and scientists apply a wide range of interpretations for these terms.

12.1.1
Accuracy

The *instrument accuracy* is the absolute error referenced to the National Institute of Standards and Technology (NIST) of the entire system including transducer, electronics, and software. Let x_{mi} be the values as measured by the instrument, and let x_{ti} be the true values

Table 12.1

Nomenclature.

Symbol	Signal (units)
a	Thermistor radius (cm)
b	Offset of the analog signal processing (V)
C	Electric capacitance (F)
D	Thermistor dissipation constant (mW/°C)
e_n	Op amp noise voltage (V)
f_s	ADC sampling rate (Hz)
f_c	Analog system cutoff frequency (where gain = 0.707) (Hz)
f_{max}, f_{min}	Range of frequencies of interest (Hz)
G	Gain of the analog signal processing (V/V)
I	Electric current (A)
k	Boltzmann's constant = 1.38×10^{-23} J/K
K	Thermal conductivity [mW/(cm·°C)]
K_m	Thermal conductivity of water [6 mW/(cm· °C)]
m	Number of multiplexer signals
n	Index used in digital signal processing
n	Number of ADC bits = $\log_2 n_z$
n_x	Precision of **x** (alternatives)
n_y	Precision of **y** (alternatives)
n_z	Precision of **z**, the ADC analog input voltage (alternatives)
P	Electric power (W)
r_x	Range of $\mathbf{x} = \mathbf{max_x} - \mathbf{min_x}$ (units of **x**)
r_y	Range of $\mathbf{y} = \mathbf{max_y} - \mathbf{min_y}$ (units of **y**)
r_z	Range of $\mathbf{z} = \mathbf{max_z} - \mathbf{min_z}$ (V)
R	Electric resistance (Ω)
R_0	Thermistor constant (Ω) $\mathbf{R} = \mathbf{R_0}\, e^{\beta/T}$
t_{ap}	S/H aperture time (s)
t_{aq}	S/H acquisition time (s)
t_c	ADC conversion time (s)
t_{mux}	Settling time of the multiplexer (s)
T	Temperature (°C or K)
V	Voltage (V)
V_y	Transducer output (typically V)
V_z	ADC input voltage (V)
V_J	Johnson noise (alternatively thermal or white noise) = $\sqrt{4\,k\,T\,R\,\Delta f}$
x	Real-world signal (cm, mmHg, °C, etc.)
x(n)	ADC output sequence, sampled at $\mathbf{f_s}$
y	Transducer output (typically V)
y(n)	Digital filter output sequence
z	ADC input voltage (V)
τ	Transducer time constant (s)
Z	Electric impedance (Ω)
α	Thermistor thermal diffusivity (cm^2/s)
β	Thermistor constant (K) $\mathbf{R} = \mathbf{R_0}\, e^{\beta/T}$
Δ_x	Resolution of **x** (units of **x**)
Δ_y	Resolution of **y** (units of **y**)
Δ_z	Resolution of **z**, the ADC analog input voltage (V)

from NIST references. In some applications, the signal of interest is a relative quantity (like temperature or distance between objects). For relative signals, accuracy can be appropriately defined many ways:

$$\text{Average accuracy (with units of x)} = \frac{1}{n}\sum_{i=1}^{n}\left|x_{ti} - x_{mi}\right|$$

$$\text{Maximum error (with units of x)} = \max\left|x_{ti} - x_{mi}\right|$$

$$\text{Standard error (with units of x)} = \sqrt{\frac{1}{n-1}\sum_{i=1}^{n}[x_{ti} - x_{mi}]^2}$$

In other applications, the signal of interest is an absolute quantity. For these situations, we can specify errors as a percentage of reading or as a percentage of full scale:

$$\text{Average accuracy of reading (\%)} = \frac{100}{n}\sum_{i=1}^{n}\frac{|x_{ti} - x_{mi}|}{x_{ti}}$$

$$\text{Average accuracy or full scale (\%)} = \frac{100}{n}\sum_{i=1}^{n}\frac{|x_{ti} - x_{mi}|}{x_{tmax}}$$

$$\text{Maximum accuracy of reading (\%)} = 100 \max \frac{|x_{ti} - x_{mi}|}{x_{ti}}$$

$$\text{Maximum accuracy or full scale (\%)} = 100 \max \frac{|x_{ti} - x_{mi}|}{x_{tmax}}$$

> **Observation:** The definitions of the terms accuracy, resolution, and precision vary considerably in the technical literature. It is good practice to include both the definitions of your terms and their values in your technical communication.

Since the Celsius and Fahrenheit temperature scales have arbitrary zeros (e.g., 0°C is the freezing point of water), it is inappropriate to specify temperature error as a percentage of reading or as a percentage of full scale when Celsius and Fahrenheit scales are used. When specifying temperature error, we should use average accuracy, maximum error, or standard error that have units of Celsius or Fahrenheit degrees.

12.1.2 Resolution

The *instrument resolution* is the smallest input signal difference, Δ, that can be detected by the entire system including transducer, electronics, and software. The resolution of the system is sometimes limited by noise processes in the transducer itself (e.g., thermal imaging) and sometimes limited by noise processes in the electronics (e.g., thermistors, resistance temperature devices [RTDs], and thermocouples).

The *spatial resolution* (or *spatial frequency response*) of the transducer is the smallest distance between two independent measurements. The size and mechanical properties of the transducer determine its spatial resolution. When measuring temperature, a metal probe will disturb the existing medium temperature field more than a glass probe. Hence, a glass probe has a smaller spatial resolution than a metal probe of the same size. Noninvasive imaging systems exhibit excellent spatial resolution because the instrument does not disturb the medium that is being measured. The spatial resolution of an imaging system is the medium surface area from which the radiation originates that is eventually focused onto the detector during the imaging of a single pixel, the so-called instantaneous field of view (IFOV). When measuring force, pressure, or flow, the spatial resolution is the effective area over which the measurement is obtained. Another way to illustrate spatial resolution is to attempt to collect a two- or three-dimensional image of the measurand. The spatial resolution is the distance between points in our image.

12.1.3 Precision

Precision is the number of distinguishable alternatives, n_x, from which the given result is selected. Precision can be expressed in alternatives, bits or decimal digits. Consider a thermometer instrument with a temperature range of 0 to 100°C. The system displays the output using three digits (e.g., 12.3°C). In addition, the system can resolve each temperature T from the temperature T + 0.1°C. This system has 1001 distinguishable outputs and hence has a precision of 1001 alternatives, or about 10 bits. For a linear system, there is a simple relationship between range ($\mathbf{r_x}$), resolution ($\mathbf{\Delta_x}$), and precision ($\mathbf{n_x}$). Range is equal to resolution times precision

$$\mathbf{r_x}(100°C) = \mathbf{\Delta_x}(0.1°C) \cdot \mathbf{n_x}(1001 \text{ alternatives})$$

where "range" is the maximum minus minimum temperature and precision is specified in terms of number of alternatives. We will develop later in Section 12.3.3 a more complex relationship for nonlinear systems. Table 12.2 illustrates the relationship between alternatives and decimal digits.

Table 12.2
Definition of decimal digits.

Alternatives	Decimal digits
1000	3
2000	3½
4000	3¾
10000	4

Observation: A good rule of thumb to remember is $2^{10n} \approx 10^{3n}$.

Checkpoint 12.1: How are precision and resolution related?

Checkpoint 12.2: List three major factors that can limit resolution?

12.1.4 Reproducibility or Repeatability

Reproducibility (or repeatability) is a parameter that specifies whether the instrument has equal outputs given identical inputs over some period of time. This parameter can be expressed as the full range or standard deviation of output results given a fixed input, where the number of samples and time interval between samples are specified. One of the largest sources of this type of error comes from transducer drift. *Statistical control* is a similar parameter based on a probabilistic model that also defines the errors due to noise. The parameter includes the noise model (e.g., normal, chi-squared, uniform, salt and pepper)[1] and the parameters of the model (e.g., average, standard deviation).

12.2 Transducers

In this section, we will start with quantitative performance measures for the transducer. Next, specific transducers will be introduced. Rather than give an exhaustive list of all available transducers, the intent in this section is to illustrate the range of possibilities and to provide specific devices to use in the design sections later in the chapter.

12.2.1 Static Transducer Specifications

The input or measurand is **x**. The output is **y**. A transducer converts **x** into **y** (Figure 12.3). In this subsection, we assume that the input parameter **x** is constant or static.

The input **x** and the output **y** can be either absolute or differential. An absolute signal represents a parameter that exists in a single place at a single time. A differential signal is

Figure 12.3
Transducers in this book convert a physical signal into an electric signal.

derived from the difference between two signals that exist at different places or at different times. Voltage is indeed defined as a potential difference, but when the voltage is referred to ground, we consider it an absolute quantity. On the other hand, if the signal is represented by the voltage difference between two points neither of which is ground, then we consider the signal as differential. Table 12.3 illustrates four types of transducers.

---

[1] An example of salt and pepper noise is the white and black spots on a poor Xerox copy.

Type	Input → output	Example
Absolute→absolute	$x \rightarrow y$	Thermistor converts an absolute temperature to a resistance
Relative→absolute	$\Delta x \rightarrow y$	Mass balance converts a mass difference to an angle
Absolute→relative	$x \rightarrow \Delta y$	Strain gauge converts a force to a resistance difference
Relative→relative	$\Delta x \rightarrow \Delta y$	Thermocouple converts a temperature difference to voltage difference

Table 12.3
Four types of transducers.

12.2.1.1
Transducer Linearity

Let x_i, y_i be the I/O signals of the transducer as shown in Figure 12.4. The *linearity* is a measure of the straightness of the static calibration curve. Let $y_i = f(x_i)$ be the transfer function of the transducer. A linear transducer satisfies

$$f(ax_1 + bx_2) = af(x_1) + bf(x_2)$$

for any arbitrary choice of the constants a and b. Let $y_i = mx_i + b$ be the best-fit line through the transducer data. Linearity (or deviation from it) as a figure of merit can be expressed as percentage of reading or percentage of full scale. Let y_{max} be the largest transducer output.

$$\text{Average linearity of reading } (\%) = \frac{100}{n}\sum_{i=1}^{n}\frac{|y_i - mx_i - b|}{y_i}$$

$$\text{Average linearity of full scale } (\%) = \frac{100}{n}\sum_{i=1}^{n}\frac{|y_i - mx_i - b|}{y_{max}}$$

Figure 12.4 illustrates the concept of linearity. In particular, we should not assume this thermocouple to be linear over the 0 to 200°C temperature range because doing so would result in significant temperature errors.

12.2.1.2
Transducer Sensitivity

Two definitions for sensitivity are used for temperature transducers. The *static sensitivity* is

$$m = \frac{\Delta y}{\Delta x}$$

Figure 12.4
Input/output response of a thermocouple transducer.

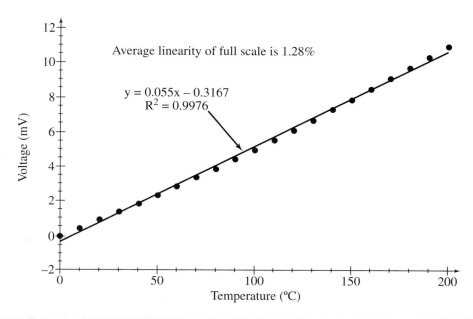

Average linearity of full scale is 1.28%

$y = 0.055x - 0.3167$
$R^2 = 0.9976$

If the transducer is linear, then the static sensitivity is the slope m of the straight line through the static calibration curve that gives the minimum mean squared error. If x_i and y_i represent measured I/O responses of the transducer, then the least squares fit to $y_i = mx_i + b$ is

$$m = \frac{n\sum_{i=1}^{n}x_iy_i - \sum_{i=1}^{n}x_i\sum_{i=1}^{n}y_i}{n\sum_{i=1}^{n}x_i^2 - \left(\sum_{i=1}^{n}x_i\right)^2} \quad \text{and} \quad b = \frac{\sum_{i=1}^{n}y_i\sum_{i=1}^{n}x_i^2 - \sum_{i=1}^{n}x_iy_i\sum_{i=1}^{n}x_i}{n\sum_{i=1}^{n}x_i^2 - \left(\sum_{i=1}^{n}x_i\right)^2}$$

Thermistors can be manufactured to have a resistance value at 25°C ranging from 4 Ω to 20 MΩ. Because the interface electronics can just as easily convert any resistance into a voltage, a 20-MΩ thermistor is not more sensitive than a 30-Ω thermistor. In this situation, it makes more sense to define *fractional sensitivity* as

$$\alpha = \frac{1}{R} \cdot \frac{\partial R}{\partial T} \quad (1/°C) \quad \text{or} \quad \alpha = \frac{1}{y} \cdot \frac{\partial y}{\partial x}$$

Checkpoint 12.3: Why is it better for the transducer to have a higher sensitivity?

12.2.1.3 Specificity	Unfortunately, transducers are often sensitive to factors other than the signal of interest. Environmental issues involve how the transducer interacts with its external surroundings (e.g., temperature, humidity, pressure, motion, acceleration, vibration, shock, radiation fields, electric fields, magnetic fields). *Specificity* is a measure of relative sensitivity of the transducer to the desired signal compared to the sensitivity of the transducers to these other unwanted influences. A transducer with a good specificity will respond only to the signal of interest and be independent of these disturbing factors. On the other hand, a transducer with a poor specificity will respond to the signal of interest as well as to some of these disturbing factors. If all these disturbing factors are grouped together as noise, then the *signal-to-noise (S/N) ratio* is a quantitative measure of the specificity of the transducer.
12.2.1.4 Transducer Impedance	The *input range* is the allowed range of input, **x**. The input impedance is the phasor equivalent of the steady-state sinusoidal effort (voltage, force, chemical concentration, pressure) input variable divided by the phasor equivalent of steady-state flow (current, velocity, molecular flux flow) input variable. The output signal strength of the transducer can be specified by the output resistance $\mathbf{R_{out}}$ and output capacitance $\mathbf{C_{out}}$ (Figure 12.5).

Figure 12.5
Output model of a
transducer.

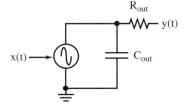

The input impedance of a thermal sensor is a measure of the thermal perturbation that occurs due to the presence of the probe itself in the medium. For example, a thermocouple needle inserted into a laser-irradiated medium will affect the medium temperature because heat will conduct down the stainless steel shaft. A thermocouple has a low input impedance (which is bad) because the transducer itself loads (reduces) the medium temperature. On the other hand, an infrared detector measures surface medium temperature without physical contact. Infrared detectors therefore have a very high input impedance (which is good) because the presence of the transducer has no effect on the temperature to be measured. In the case of temperature sensors, the driving force for heat transfer is the temperature difference ΔT. The resulting heat flow q can be expressed using Fourier's law of thermal conduction:

$$q = KA\frac{\partial T}{\partial x} \approx K4\pi a^2 \frac{\Delta T}{a}$$

where K is the probe thermal conductivity, A is the probe surface area, and a is the radius of a spherical transducer. The steady-state input impedance of a spherical temperature probe can thus be approximated by

$$Z \equiv \frac{\Delta T}{q} \approx \frac{1}{4\pi aK}$$

Again, the approximation in the above equation assumes a spherical transducer. Similar discussions can be constructed for the sinusoidal input impedance. Since most thermal events can be classified as step events rather than as sinusoidally varying events, most researchers prefer the use of a time constant to describe the transient behavior of temperature transducers.

Some transducers are completely passive (e.g., thermocouple, EKG electrode), and others are active, requiring external power (e.g., ultrasonic crystals, strain gage, microphone, thermistors). Electric isolation is a critical factor in medical instrumentation. Some transducers are inherently isolated (e.g., thermistors, thermocouples, microphones), while others are not isolated (e.g., EKG electrodes, pacemakers, blood pressure catheters). Minimization of errors is important for all instruments. The sensitivity to disturbing factors (electric fields, magnetic fields, radiation, vibration, shock, acceleration, temperature, humidity) must be determined before a device can be used.

**12.2.1.5
Transducer Drift**

The *zero drift* is the change in the static sensitivity curve intercept **b** as a function of time or other factor (see Figure 12.6). The *sensitivity drift* is the change in the static sensitivity curve slope **m** as a function of time or some other factor. These drift factors determine how often the transducer must be calibrated. For example, thermistors have a drift much larger than that of RTDs or thermocouples. Transducers may be aged at high temperatures for long periods of time to improve their stability.

Figure 12.6
The two types of transducer drift: sensitivity drift and zero drift.

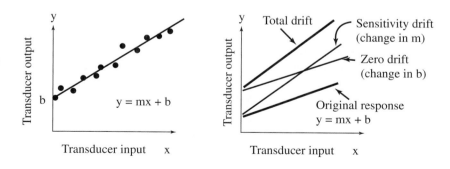

Checkpoint 12.4: How does transducer drift affect accuracy?

**12.2.1.6
Manufacturing Issues**

This subsection discusses issues involved in the manufacturing of the system. The transducer is often a critical device, affecting both the cost and performance of the entire system. A higher quality transducer may produce better signals but at an increased cost. An important manufacturing issue is the availability of components. The availability of a device may be enhanced by having a *second source* (more than one manufacturer that produces the device).

The use of standard cables and connectors will simplify the construction of your system. The power requirements, size, and weight of the device are important in some systems and thus should be considered when selecting a transducer. Some transducers require calibration, which can greatly increase the cost of manufacturing.

12.2.2 Dynamic Transducer Specifications

The *transient response* is the combination of the *delay* phase, the *slewing* phase, and the *ringing* phase (Figure 12.7). The total transient response is the time for the output y(t) to reach 99% of its final value after a step change in input, $x(t) = u_0(t)$.

The transient response of a temperature transducer to a sudden change in signal input can sometimes be approximated by an exponential equation (assuming first-order response):

$$y(t) = y_f + (y_0 - y_f)e^{-t/\tau}$$

Figure 12.7
The step response often has delay, slewing, and ringing phases.

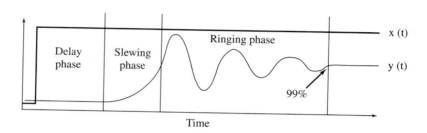

where y_0 and y_f are the initial and final transducer outputs, respectively. The *time constant* τ of a transducer is the time to reach 63.2% of the final output after the input is instantaneously increased. This time is dependent on both the transducer and the experimental setup. Manufacturers often specify the time constant of thermistors and thermocouples in well-stirred oil (fastest) or still air (slowest). In your applications, one must consider the situation. If the transducer is placed in a high-flow liquid like an artery or a water pipe, it may be reasonable to use the stirred-oil time constant. If the transducer is in air or embedded in a solid, then thermal conduction in the medium will determine the time constant almost independently of the transducer.

The *frequency response* is a standard technique to describe the dynamic behavior of linear systems. Let y(t) be the system response to x(t). Let

$$x(t) = A\sin(\omega t) \qquad y(t) = B\sin(\omega t + \phi) \qquad \omega = 2\pi f$$

The magnitude B/A and the phase ϕ responses are both dependent on frequency. Differential equations can be used to model linear transducers. Let x(t) be the time domain input signal, $X(j\omega)$ be the frequency domain input signal, y(t) be the time domain output signal, and $Y(j\omega)$ be the frequency domain output signal (Table 12.4).

Classification	Differential equation	Gain response	Phase response
Zero-order	$y(t) = m\,x(t)$	$Y/X = m$ = static sensitivity	
First-order	$y'(t) + a\,y(t) = b\,x(t)$	$Y/X = \dfrac{b}{\sqrt{a^2 + \omega^2}}$	Phase = arctan$(-\omega/a)$
Second-order	$y''(t) + a\,y'(t) + b\,y(t) = c\,x(t)$		
Time delay	$y(t) = x(t-T)$	$Y/X = \exp(-j\omega T)$	

Table 12.4
Classifications of simple linear systems.

12.2.3 Nonlinear Transducers

Nonlinear characteristics include hysteresis, saturation, bang-bang, breakdown, and dead zone. *Hysteresis* is created when the transducer has memory. We can see in the Figure 12.8 that when the input was previously high it falls along the higher curve, and when the input was previously low it follows along the lower curve. Hysteresis will cause a measurement error, because for any given sensor output, y, there may be two possible measurand inputs. *Saturation* occurs when the input signal exceeds the useful range of the transducer. With saturation, the sensor does not respond to changes in input value when the input is either too high or too low. *Breakdown* describes a second possible result that may occur when in the input exceeds the useful range of the transducer. With breakdown, the sensor output changes rapidly, usually the result of permanent damage to the transducer. Hysteresis, bang bang and dead zone all occur within the useful range of the transducer. *Bang bang* is a sudden large change in the output for a small change in the input. If the bang bang occurs predictably, then it can be corrected for in software. A *dead zone* is a condition where a large change in the input causes little or no change in the output. Dead zones can not be corrected for in software, thus if present will cause measurement errors.

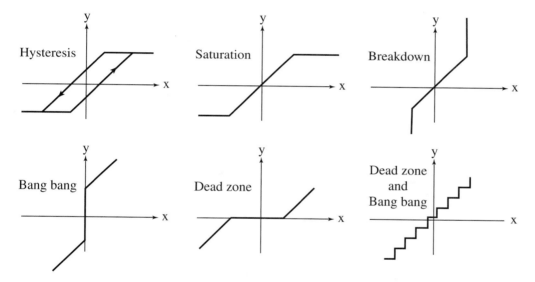

Figure 12.8
Nonlinear transducer responses.

There are many ways to model nonlinear transducers. A nonlinear transducer can be described as a piecewise-linear system. The first step is to divide the range of x into finite subregions, assuming the system is linear in each subregion. The second step is to solve the coupled linear systems so that the solution is continuous. Another method to model a nonlinear system is to use empirically determined nonlinear equations. The first step in this approach is to observe the transducer response experimentally. Given a table of x and y values, the second step is to fit the response to a nonlinear equation.

A third approach to model a nonlinear transducer uses a look-up table located in memory. This method is convenient and flexible. Let x be the measurand and y be the transducer output. The table contains x values, and the measured y value is used to index into the table. Sometimes a small table coupled with linear interpolation achieves equivalent results to a large table. There are two examples of this approach as part of the TExaS application called TBL.RTF and ETBL.RTF.

12.2.4
Position
Transducers

One of the simplest methods to convert position into an electric signal uses a position-sensitive potentiometer. These devices are inexpensive to build and are sensitive to small displacements. The transducer is constructed by wrapping a fine uninsulated wire around an insulating cylinder. The wires are close to one another, but do not touch. The total electric resistance of the transducer is given by

$$R = \rho L / A$$

where ρ is the resistivity of the wire, A is the cross-sectional area of the wire, and L is the length of the wire. A potentiometer is formed by placing a wiper blade that makes electric contact on the wires. The blade is free to move up and down along the axis of the cylinder. If the wire has been uniformly wrapped, then R_{out} will be linearly related to displacement x. The disadvantages of this transducer are its low-frequency response, its high mechanical input impedance, and its degeneration with time. Nevertheless, this type of transducer is adequate for many applications. This transducer will be interfaced two ways later in Section 12.5.5.

Given the fact that this transducer in actuality has multiple discrete outputs (i.e., R_{out} versus x is not linear but has multiple bang-bangs and dead zones), what is the maximum number of ADC bits that can be used? For a discrete wire potentiometer like the one shown in Figure 12.9, the maximum precision is determined by the number of times the wire is wrapped around the post. In other words, the resistance cannot change until the variable arm moves enough to jump to the next wire. For a continuous solid potentiometer there is no fundamental limit.

Figure 12.9
Potentiometer-based
position sensor.

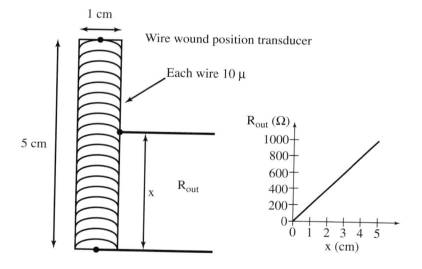

Similar transducers can be constructed using variable capacitors and inductors. The linear variable differential transducer (LVDT) uses a ferrite core to modify the mutual inductance between the single active primary coil and the two passive secondary coils (Figure 12.10). The primary coil is excited with an AC signal. The ferrite core is allowed to move up and down through the middle of the three inductors. If the ferrite core is midway between the two secondaries (x = 0), then the AC voltage across S1 will equal the AC voltage across S2, and the DC output V_{out} will be zero. If the ferrite core is closer to the secondary S1 (x > 0), then the AC voltage across S1 will be larger than the AC voltage across S2, and the DC output V_{out} will be positive. If the ferrite core is closer to the secondary S2 (x < 0), then the AC voltage across S2 will be larger than the AC voltage across S1, and the DC output V_{out} will be negative. An ADC converter is required to generate a digital signal proportional to displacement.

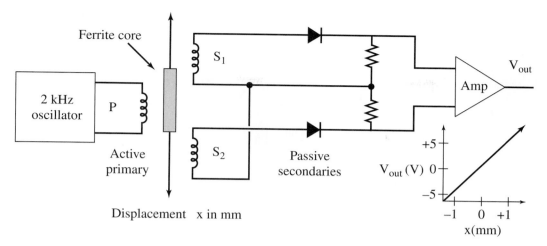

Figure 12.10
A LVDT measures displacement.

A method to measure the distance between two objects is to transmit a sound wave from one object at the other and listen for the reflection (Figure 12.11). The instrument must be able to generate the sound pulse, hear the echo, and measure the time, Δt, between pulse and echo. If the speed of sound, c, is known, then the distance, d, can be calculated. Our microcontrollers also have mechanisms to measure the pulse width Δt. An example of this type of sensor is the **Ping**))) available at www.parallax.com.

$$d = c\Delta t/2$$

Figure 12.11
An ultrasonic pulse-echo transducer measures the distance to an object.

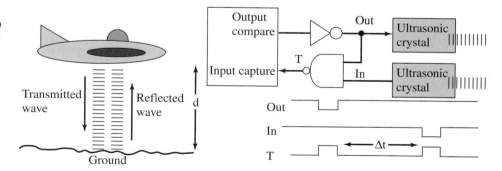

12.2.5
Velocity
Measurements

12.2.5.1
Velocity Transducers

One can use a LED-photosensor pair to measure the angular velocity of a rotating shaft (Figure 12.12). The circular disk is mounted on the rotating axle. The disk has eight equally spaced holes. The LED and sensor are positioned on opposite sides of the disk. The sensor output, Out, will be high when light passes through one of the holes in the disk. Out will be low when no light hits the sensor. If the shaft is rotating at a constant velocity, then Out will be a square wave. The rotations per minute (RPM) of the shaft can be calculated by measuring the frequency of the square wave, f, in hertz. The frequency can also be indirectly calculated from measurement of period or pulse width. Datasheets for this type of sensor can be found on the CD within the "sensor" directory.

$$RPM = 7.5 \cdot f \quad (Hz)$$

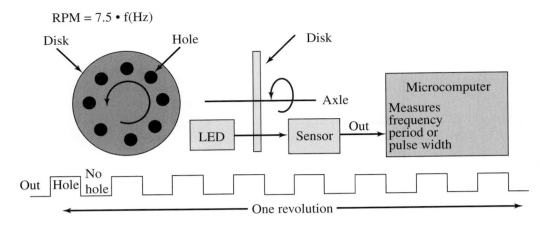

Figure 12.12
A LED-photosensor pair measures shaft rotation.

Checkpoint 12.5: How was the 7.5 calculated? In what way is 7.5 important when designing a system to satisfy a velocity resolution specification?

12.2.5.2
Velocity Calculations

Given a method to measure position, we can use calculus to determine velocity and acceleration.

$$v(t) = \frac{dx(t)}{dt} \qquad a(t) = \frac{dv(t)}{dt}$$

The continuous derivative can be expressed as a limit.

$$v(t) = \lim \frac{x(t + \Delta t) - x(t)}{\Delta t} \qquad \Delta t \to 0$$

The microcomputer can approximate the velocity by calculating the discrete derivative.

$$v(t) = \frac{x(t) - x(t - \Delta t)}{\Delta t}$$

where Δt is the time between velocity samples. This calculation of derivative is very sensitive to errors in the measurement of $x(t)$. A more stable calculation averages two or more derivative terms taken over different time windows. In general we can define such a robust calculation as

$$v(t) = \frac{a}{a + b} \frac{x(t) - x(t - n\,\Delta t)}{n\,\Delta t} + \frac{b}{a + b} \frac{x(t - c\,\Delta t) - x(t - (c + m)\,\Delta t)}{m\,\Delta t}$$

If the integers n, m, and c are all positive, this calculation can be performed in real time. The first term is the derivative over the large time window of $n\Delta t$. The second window term has a smaller size of $m\Delta t$. It normally fits entire inside the first with $c > 0$ and $c - m < n$. The coefficients a, b create the weight for combining the short and long intervals. With $a = b = 1, n = 3, m = 1$, and $c = 1$, we get

$$v(t) = \frac{1}{2} \frac{x(t) - x(t - 3\Delta t)}{3\Delta t} + \frac{1}{2} \frac{x(t - \Delta t) - x(t - 2\Delta t)}{\Delta t} = \frac{x(t) + 3x(t - \Delta t) - 3x(t - 2\Delta t) - x(t - 3\Delta t)}{6\Delta t}$$

The acceleration can also be approximated by a discrete derivative.

$$a(t) = \frac{x(t) - 2x(t - \Delta t) + x(t - 2\Delta t)}{\Delta t^2}$$

To make a more stable calculation of second derivative, you could average two or more second-derivative terms taken over different time windows (Figure 12.13).

Another robust second-derivative calculation concatenates two robust first-derivative calculations (Figure 12.14).

Figure 12.13
Robust calculation of acceleration.

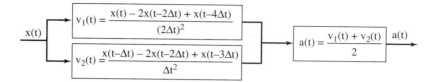

$$v_1(t) = \frac{x(t) - 2x(t-2\Delta t) + x(t-4\Delta t)}{(2\Delta t)^2}$$

$$v_2(t) = \frac{x(t-\Delta t) - 2x(t-2\Delta t) + x(t-3\Delta t)}{\Delta t^2}$$

$$a(t) = \frac{v_1(t) + v_2(t)}{2}$$

$$v(t) = \frac{x(t) + 3x(t-\Delta t) - 3x(t-2\Delta t) - x(t-3\Delta t)}{6\Delta t}$$

$$a(t) = \frac{v(t) + 3v(t-\Delta t) - 3v(t-2\Delta t) - v(t-3\Delta t)}{6\Delta t}$$

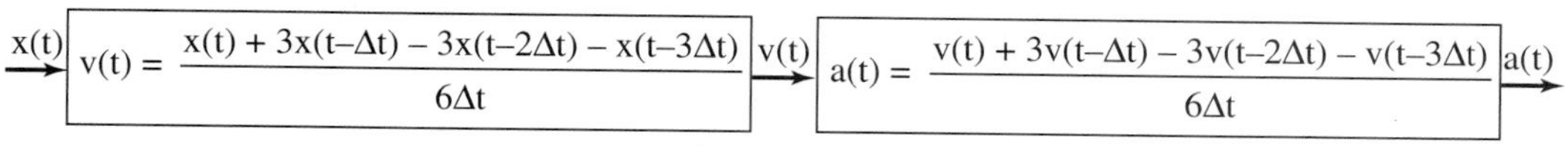

Figure 12.14
Sequential calculation of acceleration.

Observation: In the above calculations of derivative, a single error in one of the x(t) input terms will propagate to only a finite number of the output calculations.

Observation: Although the central difference calculation of v(t) = [x(t + Δt) − x(t − Δt)]/ 2Δt is theoretically valid, we cannot use it for real-time applications, because it requires knowledge about the future, x(t + Δt), which is unavailable at the time v(t) is being calculated.

Similarly, we can perform integration of velocity to determine position.

$$x(t) = \int_0^t v(s)\, ds$$

The microcomputer can perform a discrete integration by summation.

$$x(t) = x(0) + \sum_{n=0}^{t} v(n)\, \Delta t$$

There are two problems with this approach. The first difficulty is determining x(0). The second problem is the accumulation of errors. If one is calculating velocity from position and an error occurs in the measurement of x(t), then that error affects only two calculations of v(t). Unfortunately, if one is calculating position from velocity and an error occurs in the measurement of v(n), then that error will affect all subsequent calculations of x(t).

Observation: In the above calculation of integration, a single error in one of the x(t) input terms will propagate into all remaining output calculations.

The following function, which is quite similar to the derivative, is actually a low-pass digital filter. We will learn more about digital filters in Chapter 15.

$$y(t) = \frac{x(t) + x(t - \Delta t)}{2}$$

12.2.6 Force Transducers

The most common device to measure force is the *strain gage* (Figure 12.15). As a wire is stretched, its length increases and its cross-sectional area decreases. These geometric changes cause an increase in the electric resistance of the wire, R. The transducer is constructed with four sets of wires mounted between a stationary member (frame) and a moving member (armature). As the armature moves relative to the frame, two wires are stretched (increase in R1, R4), and two wires are compressed (decrease in R2, R3). The strain gage is a displacement

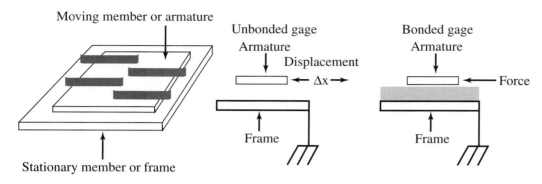

Figure 12.15
Strain gages used for displacement or force measurement.

transducer such that a change in the relative position between the armature and frame, Δx, causes a change in resistance, ΔR. The sensitivity of a strain gage is called its gage factor:

$$G = \frac{\Delta R/R}{\Delta x/x}$$

The gage factor for an Advance strain gage is 2.1. The typical resistance R is 120 Ω. If the gage is bonded onto a material with a spring characteristic

$$F = -kx$$

then the transducer can be used to measure force. The wires each have a significant temperature drift. When the four wires are placed into a bridge configuration, the temperature dependence cancels (Figure 12.16). A high-gain high-input impedance high-CMRR differential amplifier is required.

Figure 12.16
Four strain gauges are placed in a bridge configuration.

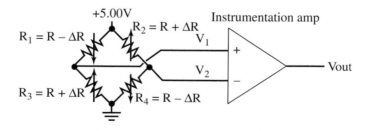

Checkpoint 12.6: How can a force transducer be used to measure pressure?

12.2.7
Temperature Transducers

12.2.7.1
Classification of
Temperature Transducers

An increase in thermal energy causes an increase in the average spacing between molecules. For an ideal gas the kinetic energy $[(\tfrac{1}{2}) mv^2]$ equals the thermal energy $[(\tfrac{3}{2}) kT]$. For liquids and most gases, an increase in kinetic energy is also accompanied by an increase in potential energy. The mercury thermometer and bimetallic strips are common transducers that rely on thermal expansion. A RTD is constructed from long narrow-gage metal wires. Over a moderate range of temperature (e.g., 50°C) the wire resistance is approximately linearly related to its temperature.

$$R = R_0 [1 + \alpha(T - T_0)]$$

The sensitivity α for various metals is given in Table 12.5. R_0 is a function of the resistivity and geometry of the wire. Although many metals such as gold, nichrome (nickel-chromium), nickel, and silver can be used, platinum is typically selected because of its excellent stability.

Table 12.5
Electric properties of various materials.

Material	α, fractional sensitivity (l/°C)	ρ, electric resistivity (Ω−cm)
Gold	+0.004	$2.35 \cdot 10^{-6}$
Nickel	+0.0069	$6.84 \cdot 10^{-6}$
Platinum	+0.003927	10^{-5}
Copper	0.0068	$1.59 \cdot 10^{-6}$
Silver	+0.0041	$1.673 \cdot 10^{-6}$
NTC thermistor	−0.04	10^3
PTC thermistor	+0.1	

The rates of first-order processes are proportional to the Boltzmann factor, $e^{-E/kt}$. This strong temperature dependence can be exploited to make measurements. Rates that are governed by the Boltzmann factor include the evaporation of liquids and the population of charge carriers in a conduction band. The conductance (1/resistance) of a thermistor is proportional to the occupation of charge carriers in the conduction band:

$$G = G_0 e^{-E/kT}$$

The electric resistance is the reciprocal of the conductance:

$$R = R_0 e^{+E/kT} = R_0 e^{+\beta/T}$$

where β is E/k.

A thermocouple is constructed using two wires of different metals welded to form two junctions. One junction is placed at a known reference temperature (e.g., 0°C ice in thermal equilibrium with liquid and gaseous water at 1 atmosphere), and the other junction is used to measure the unknown temperature. The Seebeck effect involves thermal to electric energy conversion. In a closed-loop configuration, the current around the loop is proportional to the temperature difference between the two junctions. In the open-loop configuration, a voltage is generated proportional to the temperature difference:

$$V = a(T_2 - T_1) + b(T_2 - T_1)^2$$

Electromagnetic radiation is emitted by all materials above absolute zero, of which black or gray bodies are a special case. The spectral properties of a surface are a function of the material and its temperature. The total wide-band radiation power W is a function of the fourth power of surface temperature.

$$W = \sigma(T^4 - T_0^4)$$

The sudden transition of state that occurs for a particular substance can be used to measure temperature. In particular, this property is frequently exploited to produce temperature references. An ice bath at 0°C (water's triple point) is often used as the reference junction of a thermocouple. The melting point of gallium is conveniently 29.7714°C. A triple point occurs under pressure when a substance exists simultaneously in all three states: gas, liquid, and solid. The triple point of water is 0.01±0.0005°C at 1 atmosphere, the triple point of rubidium is 39.265°C, and the triple point of succinonitrile is 58.0805°C.

12.2.7.2
Thermistors

Thermistors are a popular temperature transducer made from a ceramic-like semiconductor (Figure 12.17). A negative-temperature coefficient (NTC) thermistor is made from combinations of metal oxides of manganese, nickel, cobalt, copper, iron, and titanium. A mixture of milled semiconductor oxide powders and a binder is shaped into the desired geometry. The mixture is dried and sintered (under pressure) at an elevated temperature.

Figure 12.17
Thermistors come in many shapes and sizes.

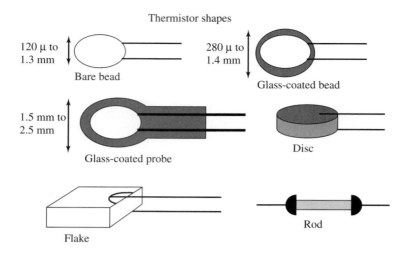

The wire leads are attached and the combination is coated with glass or epoxy. By varying the mixture of oxides, a range of resistance values from 30 Ω to 20 MΩ (at 25°C) is possible. Table 12.6 lists the trade-offs between thermistors and thermocouples. Two thermistor design spreadsheets can be found on the CD called therm.xls and thermID.xls.

Table 12.6
Trade-offs between thermistors and thermocouples.

Thermistors	Thermocouples
More sensitive	More sturdy
Better temperature resolution	Faster response
Less susceptible to noise	Inert, interchangeable V versus T curves
Less thermal perturbation	Requires less frequent calibration
Does not require a reference	More linear

A precision thermometer, an ohmmeter, and a water bath are required to calibrate thermistor probes. The following empirical equation yields an accurate fit over a wide range of temperature:

$$T = \frac{1}{H_0 + H_1 \ln(R) + H_3 [\ln(R)]^3} - 273.15$$

where T is the temperature in °C and R is the thermistor resistance in Ω. The cubic term was added to the above equation to improve accuracy. It is preferable to use the ohmmeter function of the eventual instrument for calibration purposes so that influences of the resistance measurement hardware and software are incorporated into the calibration process.

12.2.7.3
Thermocouples

A *thermocouple* is constructed by spot welding two different metal wires together. Probe transducers include a protective casing that surrounds the thermocouple junction. Probes come in many shapes including round tips, conical needles, and hypodermic needles. Bare thermocouple junctions provide faster response but are more susceptible to damage and noise pickup. Ungrounded probes allow electric isolation but are not as responsive as grounded probes. Commercial thermocouples have been constructed in 16 to 30 gage hypodermic needles—a 30 gage needle has an outside diameter of above 0.03 cm. Bare thermocouples can be made from 30-μm wire, producing a tip with an 80-μm diameter. A spot weld is produced by passing a large current through the metal junction that fuses the two metals together.

If the wires form a loop, and the junctions are at different temperatures, then a current will flow in the loop. This thermal to electric energy conversion is called the Seebeck effect (see Figure 12.18). If the temperature difference is small, then the current I is linearly proportional to the temperature difference $T_1 - T_2$.

Figure 12.18
When the two thermocouple junctions are at different temperatures, current will flow.

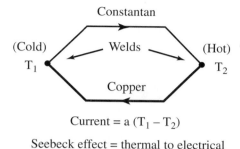

Current = a $(T_1 - T_2)$

Seebeck effect = thermal to electrical

If the loop is broken, and an electric voltage is applied, then heat will be absorbed at one junction and released at the other. This electric to thermal energy conversion is called the Peltier effect (see Figure 12.19). If the voltage is small, then the heat transferred is linearly proportional to the voltage V.

Figure 12.19
When voltage is applied to two thermocouple junctions, heat will flow.

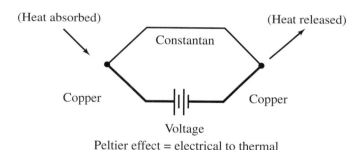

Peltier effect = electrical to thermal

If the loop is broken and the junctions are at different temperatures, then a voltage will develop because of the Seebeck effect. If the temperature difference is small, then the voltage V, is nearly linearly proportional to the temperature difference $T_1 - T_2$. Thermocouples are characterized by (1) low impedance (resistance of the wires), (2) low temperature sensitivity (45 μV/°C for copper/constantan), (3) low power dissipation, (4) fast response (because of the metal), (5) high stability (because of the purity of the metals), and (6) interchangeability (again because of the physics and the purity of the metals).

Figure 12.20 shows a simple approach to measure temperature using a thermocouple. Typically an ice bath ($T_{ref} = 0$°C) is used for the reference junction. A high-gain differential amplifier is used to convert the low-level voltage (e.g., 0 to 0.9 mV) into the range of the ADC (e.g., 0 to 5 V). Table 12.7 gives the sensitivities and temperature ranges of typical thermocouple devices. The amplifier gain is chosen to provide a simple relationship between the A/D output N and the temperature of the medium, T_t. Assuming a linear relationship between T_t and V_1, $T_t = N/10$.

If the temperature range is less than 25°C, then the linear approximation can be used to measure temperature. Let N be the digital sample from the ADC for the unknown medium temperature T_1. A calibration is performed under the conditions of a constant reference temperature: typically, one uses the extremes of the temperature range (T_{min} and T_{max}). A precision thermometer system is used to measure the "truth" temperatures. Let

Figure 12.20
An instrumentation and
low-pass filter are used
to interface a
thermocouple.

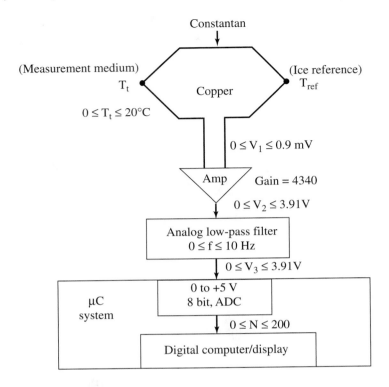

Constantan

(Measurement medium)   (Ice reference)

T_t T_{ref}

Copper

$0 \le T_t \le 20°C$

$0 \le V_1 \le 0.9$ mV

Amp   Gain = 4340

$0 \le V_2 \le 3.91$V

Analog low-pass filter
$0 \le f \le 10$ Hz

$0 \le V_3 \le 3.91$V

μC
system

0 to +5 V
8 bit, ADC

$0 \le N \le 200$

Digital computer/display

Type—Thermocouple	μV/°C at 20°C	Useful range, °C	Comments
T—Copper/constantan	45	−150 to +350	Moist environment
J—Iron/constantan	53	−150 to +1000	Reducing environment
K—Chromel/alumel	40	−200 to +1200	Oxidizing environment
E—Chromel/constantan	80	0 to +500	Most sensitive
R S—Platinum/platinum-rhodium	6.5	0 to +1500	Corrosive environment
C—Tungsten/rhenium	12	0 to +2000	High temperature

Table 12.7
Temperature sensitivity and range of various thermocouples.

N_{min} and N_{max} be the digital samples at T_{min} and T_{max}, respectively. Then the following equation can be used to calculate the unknown medium temperature from the measured digital sample:

$$T_1 = T_{min} + (N - N_{min}) \cdot \frac{T_{max} - T_{min}}{N_{max} - N_{min}}$$

Because the thermocouple response is not exactly linear, the errors in the above linear equation will increase as the temperature range increases. For instruments with a larger temperature range, a quadratic equation can be used,

$$T_1 = H_0 + H_1 \cdot N + H_2 \cdot N^2$$

where H_0, H_1, and H_2 are determined by calibration of the instrument over the range of interest. Linear regression can be used by letting $z = T_1$, $x = N$, and $y = N^2$.

12.3 DAS Design

12.3.1
Introduction and
Definitions

Before designing a DAS we must have a clear understanding of the system goals. We can classify a system as a *quantitative DAS* if the specifications can be defined explicitly in terms of desired range (r_x), resolution (Δx), precision (n_x), and frequencies of interest (f_{min} to f_{max}). If the specifications are more loosely defined, we classify it as a *qualitative DAS*. Examples of qualitative DAS's include systems that mimic the human senses where the specifications are defined using terms like "sounds good," "looks pretty," and "feels right." Other qualitative DAS's involve the detection of events.

We will consider two examples: a burglar detector and an instrument to diagnose cancer. For binary detection systems required to detect the presence or absence of a burglar or the presence or absence of cancer, we define a true positive (TP) when the condition exists (there is a burglar) and the system properly detects it (the alarm rings). We define a false positive (FP) when the condition does not exist (there is no burglar) but the system thinks there is (the alarm rings). A false negative (FN) occurs when the condition exists (there is a burglar) but the system does not think there is (the alarm is silent). A true positive (TN) occurs when the condition does not exist (the patient does not have cancer) and the system properly detects it (the instrument says the patient is normal). **Prevalence** is the probability that the condition exists, sometimes called pre-test probability. In the case of diagnosing the disease, prevalence tells us what percentage of the population has the disease. **Sensitivity** is the fraction of properly detected events (a burglar comes and the alarm rings) over the total number of events (number of robberies). It is a measure of how well our system can detect an event. For the burglar detector, a sensitivity of 1 means when a burglar breaks in the alarm will go off. For the diagnostic instrument, a sensitivity of 1 means every sick patient will get treatment. **Specificity** is the fraction of properly handled nonevents (a patient doesn't have cancer and the instrument claims the patient is normal) over the total number of nonevents (the number of normal patients.) A specificity of 1 means no people will be treated for a cancer they don't have. The **positive predictive value** of a system (PPV) is the probability that the condition exists when restricted to those cases in which the instrument says it exists. It is a measure of how much we believe the system is correct when it says it has detected an event. A PPV of 1 means when the alarm rings, the police will come and arrest a burglar. Similarly, a PPV of 1 means if our instrument says a patient has the disease, then that patient is sick. The **negative predictive value** of a system (NPV) is the probability that the condition doesn't exist when restricted to those cases where the instrument says it doesn't exist. A NPV of 1 means if our instrument says a patient doesn't have cancer, then that patient is not sick. Sometimes the true negative condition doesn't really exist (how many times a day does a burglar not show up at your house?). For these situations, only sensitivity and PPV are relevant.

$$\text{Prevalence} = \frac{\text{TP} + \text{FN}}{\text{TP} + \text{TN} + \text{FP} + \text{FN}}$$

$$\text{Sensitivity} = \frac{\text{TP}}{\text{TP} + \text{FN}} \qquad \text{Specificity} = \frac{\text{TN}}{\text{TN} + \text{FP}}$$

$$\text{PPV} = \frac{\text{TP}}{\text{TP} + \text{FP}} \qquad \text{NPV} = \frac{\text{TN}}{\text{TN} + \text{FN}}$$

Checkpoint 12.7: Explain how a design decision can involve a tradeoff between sensitivity and PPV?

The *transducer* converts the physical signal into an electric signal. The *amplifier* converts the weak transducer electric signal into the range of the ADC (e.g., 0 to +5 V). The *analog filter* removes unwanted frequency components within the signal. The analog filter is required to remove aliasing error caused by the ADC sampling. The *analog multiplexer* is used to select one signal from many sources. The *S/H* is an analog latch used to keep the ADC input voltage constant during the ADC conversion. The *clock* is used to control the sampling process. Inherent in digital signal processing is the requirement that the ADC be sampled on a fixed time basis. The *computer* is used to save and process the digital data. A *digital filter* may be used to amplify or reject certain frequency components of the digitized signal. The MACQ is a convenient data structure to use with the digital filter (Figure 12.21).

Figure 12.21
Block diagram showing how an analog multiplexer is used to sample multiple signals.

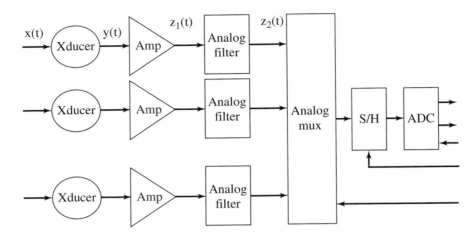

12.3.2 Using Nyquist Theory to Determine Sampling Rate

There are two errors introduced by the sampling process. *Voltage quantizing* is caused by the finite word size of the ADC. The *precision* is determined by the number of bits in the ADC. If the ADC has **n** bits, then the number of distinguishable alternatives is

$$n_z = 2^n$$

Time quantizing is caused by the finite discrete sampling interval. *Nyquist theory* states that if the signal is sampled at f_s, then the digital samples only contain frequency components from 0 to 0.5 f_s. Conversely, if the analog signal does contain frequency components larger than 0.5 f_s, then there will be an *aliasing* error. Aliasing is when the digital signal appears to have a different frequency than the original analog signal.

Simply put, if one samples a sine wave at a sampling rate of f_s,

$$V(t) = A \sin(2\pi f t + \phi)$$

is it possible to determine **A**, **f**, and **ϕ** from the digital samples? Nyquist theory says that if f_s is strictly greater than twice **f**, then one can determine **A**, **f**, and **ϕ** from the digital samples. In other words, the entire analog signal can be reconstructed from the digital samples. But if f_s less than or equal to twice **f**, then one cannot determine **A**, **f**, and **ϕ**. In this case, the apparent frequency, as predicted by analyzing the digital samples, will be shifted to a frequency between 0 and 0.5 f_s.

In this first example the frequency of an input sine wave at 1000 Hz and the sampling rate f_s, is varied

$$V = \sin(2\pi \cdot 1000 \cdot t)$$

where **t** is in seconds. In this case the largest (and only) frequency component f_{max} is 1000 Hz. If the signal is sampled at 1333 Hz ($f_s \leq 2\ f_{max}$), then an aliasing error will occur (Figure 12.22). This error occurs regardless of the phase between the signal and the ADC sampling. Notice that the apparent frequency of the digital samples (333 Hz) is different from the actual frequency of 1000 Hz.

Figure 12.22
Aliasing makes the 1000-Hz signal appear as a 333-Hz signal.

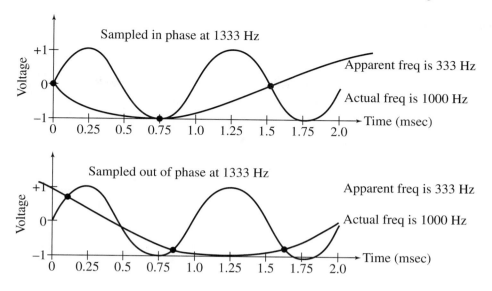

If the 1000-Hz sine wave is sampled in phase at 2000 Hz, then the digital samples appear as a constant (0 Hz). When the sampling frequency is exactly equal to twice the input frequency, the aliasing error is dependent on the phase between the signal and the ADC sampling (Figure 12.23).

Figure 12.23
Right at the Nyquist frequency aliasing may or may not occur.

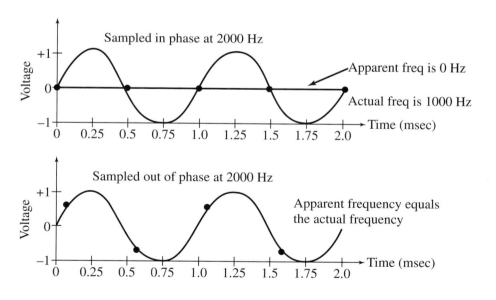

If the 1000-Hz sin wave is sampled above 2000 Hz, then the frequency, magnitude, and phase of the signal can be reconstructed from the digital samples. This reconstruction can be performed regardless of the phase between the signal and the ADC sampling (Figure 12.24).

In this next example the sampling rate f_s will be fixed at 1000 Hz, and the frequency of the input signal will be varied. In the first two cases, the sampling rate is more than twice the input frequency, so the original signal can be properly reconstructed (Figure 12.25).

Figure 12.24
Aliasing does not occur
when the sampling rate
is more than twice the
signal frequency.

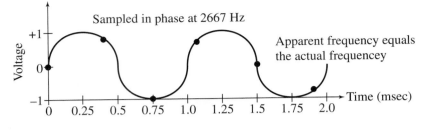

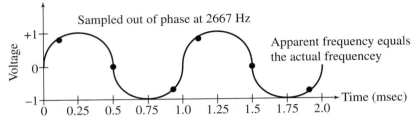

Figure 12.25
Aliasing does not occur
when the sampling rate
is more than twice the
signal frequency.

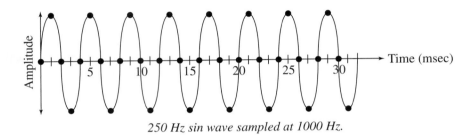

250 Hz sin wave sampled at 1000 Hz.

When sampling rate is exactly twice the input frequency, the original signal may or may not be properly reconstructed. In this specific case, it is frequency shifted (aliased) to (0 Hz) and lost (Figure 12.26).

Figure 12.26
Right at the Nyquist
frequency aliasing may
or may not occur.

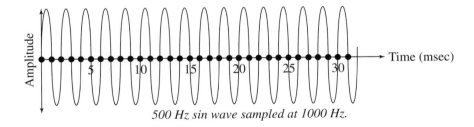

500 Hz sin wave sampled at 1000 Hz.

When sampling rate is slower than twice the input frequency, the original signal cannot be properly reconstructed. It is frequency shifted (aliased) to a frequency between 0 and 0.5 f_s. In this case the 533-Hz wave was aliased to 33 Hz (Figure 12.27).

Figure 12.27
Aliasing makes the
533-Hz signal appear as
a 33-Hz signal.

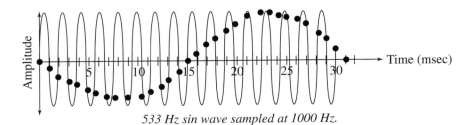

533 Hz sin wave sampled at 1000 Hz.

The choice of *sampling rate* f_s is determined by the maximum useful frequency contained in the signal. One must sample at least twice this maximum useful frequency. Faster sampling rates may be required to implement various digital filters and digital signal processing.

$$f_s > 2f_{max}$$

Even though the largest signal frequency of interest is $\mathbf{f_{max}}$, there may be significant signal magnitudes at frequencies above $\mathbf{f_{max}}$. These signals may arise from the input $\mathbf{x}$, from added noise in the transducer, or from added noise in the analog processing. Once the sampling rate is chosen at $\mathbf{f_s}$, then a low-pass analog filter may be required to remove frequency components above $0.5\ \mathbf{f_s}$. A digital filter cannot be used to remove aliasing.

12.3.3
How Many Bits
Does One Need
for the ADC?

The choice of the *ADC precision* is a compromise of various factors. The overall objective of the DAS will dictate the potential number of useful bits in the signal. If the transducer is nonlinear, then the ADC precision must be larger than the precision specified in the problem statement. For example, let $\mathbf{y}$ be the transducer output and let $\mathbf{x}$ be the real-world signal. Assume for now that the transducer output is connected to the ADC input. Let the range of $\mathbf{x}$ be $\mathbf{r_x}$, let the range of $\mathbf{y}$ be $\mathbf{r_y}$, and let the required precision of $\mathbf{x}$ be $\mathbf{n_x}$. The resolutions of $\mathbf{x}$ and $\mathbf{y}$ are $\mathbf{\Delta x}$ and $\mathbf{\Delta y}$, respectively. Let the following describe the nonlinear transducer.

$$\mathbf{y = f(x)}$$

The required ADC precision $\mathbf{n_y}$ (in alternatives) can be calculated by

$$\mathbf{\Delta_x = \frac{r_x}{n_x}}$$

$$\mathbf{\Delta y = min\ \{f(x + \Delta_x) - f(x)\}\ for\ all\ x\ in\ r_x}$$

$$\mathbf{n_y = \frac{r_y}{\Delta_y}}$$

For example, consider the nonlinear transducer $\mathbf{y = x^2}$. The range of $\mathbf{x}$ is $0 \le \mathbf{x} \le 1$. Thus, the range of $\mathbf{y}$ is also $0 \le \mathbf{y} \le 1$. Let the desired resolution be $\mathbf{\Delta_x} = 0.01$. $\mathbf{n_x = r_x/\Delta_x} = 100$ alternatives, or about 7 bits. From the above equation, $\mathbf{\Delta y} = \min\{(\mathbf{x} + 0.01)^2 - \mathbf{x}^2\} = \min\{0.02\mathbf{x} + 0.0001\} = 0.0001$. Thus, $\mathbf{n_y = r_y/\Delta y} = 10000$ alternatives, or almost 15 bits.

> *Checkpoint 12.8:* How many ADC bits are required, assuming the system is linear, if the range is 0 to 10 cm and the desired resolution is 0.01cm?

12.3.4
Specifications for
the Analog Signal
Processing

If the analog signal processing is linear, then

$$\mathbf{z = Gy + b}$$

where $\mathbf{G}$ is the gain and $\mathbf{b}$ is the offset. The resolution and range of $\mathbf{z}$ can be found from the previous section.

$$\mathbf{\Delta_z = G\ \Delta_y}$$
$$\mathbf{r_z = G\ r_y}$$

Thus, the precision at $\mathbf{z}$ equals the precision at $\mathbf{y}$.

$$\mathbf{n_z = \frac{r_z}{\Delta_z} = \frac{Gr_y}{G\Delta_y} = \frac{r_y}{\Delta_y} = n_y}$$

If the transducer and analog signal processing are both linear, then $\mathbf{n_z = n_y = n_x}$. Another factor to consider in the choice of ADC word size is the electric noise in the signal. For example, if the signal ranges from 0 to $+8$ V and there is 1 mV of noise in the signal, then any ADC bits beyond 13 would be wasteful. Other factors include cost and convenience for digital processing.

An *analog low-pass filter* may be required to remove aliasing. The cutoff of this analog filter should be less than $0.5\ \mathbf{f_s}$. Some transducers automatically remove these unwanted frequency components. For example, a thermistor is inherently a low-pass device. Other types of filters (analog and digital) may be used to solve the DAS objective. One useful filter is a 60-Hz band-reject filter.

Let $\mathbf{X(s)}$ be the Fourier transform of the physical signal, let $\mathbf{Y(s)}$ be the Fourier transform of the transducer output, let $\mathbf{Z_2(s)}$ be the Fourier transform of the ADC input, let $\mathbf{H_1(s)}$ be the Fourier transfer function for the transducer, let $\mathbf{H_2(s)}$ be the Fourier transfer function for the amplifier, let $\mathbf{H_3(s)}$ be the Fourier transfer function for the analog filter, and let $\mathbf{f_{min} \leq f \leq f_{max}}$ be the frequency range in the signal to be processed. The analog system (transducer, amplifier, filter) must pass frequencies $\mathbf{f_{min}}$ to $\mathbf{f_{max}}$. To avoid aliasing, the analog system must reject frequencies above 0.5 $\mathbf{f_s}$ (Figure 12.28).

Figure 12.28
A DAS shown in the frequency domain.

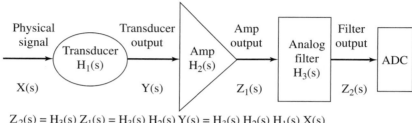

$$Z_2(s) = H_3(s)\, Z_1(s) = H_3(s)\, H_2(s)\, Y(s) = H_3(s)\, H_2(s)\, H_1(s)\, X(s)$$

Let the gain of the analog filter be $\mathbf{G_3 = |\,H_3(s)|}$. Then the system should pass, with no error as seen by the ADC, for signal frequencies between $\mathbf{f_{min}}$ and $\mathbf{f_{max}}$. For example,

$$\frac{2^{n+1} - 1}{2^{n+1}} \leq G_3 \leq \frac{2^{n+1}+1}{2^{n+1}} \qquad \text{for } f_{min} \leq f \leq f_{max}$$

Figure 12.29
Ideal and practical filter responses.

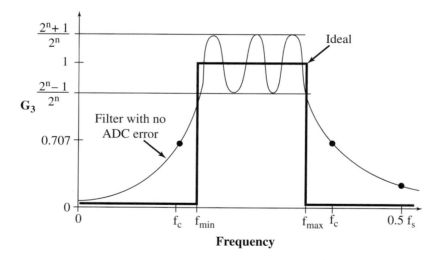

For example, consider a system that uses an 8-bit ADC. If the gain is lower than 255/256 or larger than 257/256, then an error will result in the ADC sample. Conversely, if the gain is between 255/256 and 257/256, then no error should occur in the ADC sample. We will add a safety factor of 1/2 and require that

$$\frac{511}{512} \leq G_3 \leq \frac{513}{512} \qquad \text{for } f_{min} \leq f \leq f_{max}$$

Observation: Most DASs do not need to pass "without ADC error" all frequencies from f_{min} to f_{max}. In this situation we simply place the filter cutoff frequencies at f_{min} and f_{max}.

To prevent aliasing, one must know the frequency spectrum of the ADC input voltage. This information can be measured with a spectrum analyzer. Typically, a spectrum analyzer samples the analog signal at a very high rate (>1 MHz), performs a discrete Fourier transform, and displays the signal magnitude versus frequency. Let $|\mathbf{Z_2}|$ be the magnitude of the ADC input voltage as a function of frequency. For example a bandpass device might have a frequency spectrum like the following. There are three regions in the magnitude versus frequency graph shown in Figure 12.30. We will classify any signal with an amplitude less than the ADC resolution, $\mathbf{\Delta_z}$, to be undetectable. This region is labeled "too small to be detected." Undetectable signals cannot cause aliasing regardless of their frequency. We will classify any signal with an amplitude larger than the ADC resolution at frequencies less than $0.5\ \mathbf{f_s}$ to be properly sampled. This region is labeled "properly represented in digital samples." It is information in this region that is available to the software for digital processing. The last region includes signals with an amplitude above the ADC resolution at frequencies greater than or equal to $0.5\ \mathbf{f_s}$. Signals in this region will be aliased, which means their apparent frequencies will be shifted into the 0 to $0.5\ \mathbf{f_s}$ range.

Figure 12.30
To prevent aliasing there should be no measurable signal above 0.5$\mathbf{f_s}$.

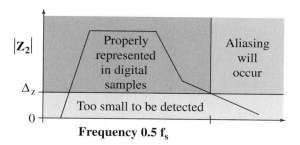

Aliasing will occur if $|\mathbf{Z_2}|$ is larger than the ADC resolution for any frequency larger than or equal to $0.5\ \mathbf{f_s}$. In order to prevent aliasing, $|\mathbf{Z_2}|$ must be less than the ADC resolution. Our design constraint will again include a safety factor of 1/2. Thus, to prevent aliasing we will make

$$|\mathbf{Z_2}| < 0.5\,\Delta_\mathbf{z} \qquad \text{for all frequencies larger than or equal to } 0.5\ \mathbf{f_s}$$

This condition usually can be satisfied by increasing the sampling rate or increasing the number of poles in the analog low-pass filter. We cannot remove aliasing with a digital low-pass filter, because once the high-frequency signals are shifted into the 0 to $0.5\ \mathbf{f_s}$ range, we will be unable to separate the aliased signals from the regular ones.

There are errors caused by *impedance loading* between stages of the system. In general, one tries to maximize the input impedance and minimize the output impedance of the modules. A good rule is to design your system such that the error due to impedance loading causes no error in the ADC sample. The errors in the first stages must be multiplied by the amplifier gain so that they can be compared to the ADC resolution (Figure 12.31).

In this example, assume the $\mathbf{V_y}$ and $\mathbf{V_z}$ are unipolar, and the analog signal processing is linear,

$$0 \le \mathbf{V_y} \le \mathbf{r_y} \qquad 0 \le \mathbf{V_z} \le \mathbf{r_z}$$

$$\mathbf{V_z} = \mathbf{GV_y}$$

Using the simple impedance model, the voltage errors are

$$\mathbf{V_{err1}} = \mathbf{V_y}\frac{\mathbf{R_{out1}}}{\mathbf{R_{out1}} + \mathbf{Z_{in1}}} \approx \mathbf{V_y}\frac{\mathbf{R_{out1}}}{\mathbf{Z_{in1}}}$$

$$\mathbf{V_{err2}} = \mathbf{V_z}\frac{\mathbf{R_{out2}}}{\mathbf{R_{out2}} + \mathbf{R_{on}} + \mathbf{Z_{in2}}} \approx \mathbf{V_z}\frac{\mathbf{R_{out2}}}{\mathbf{Z_{in2}}}$$

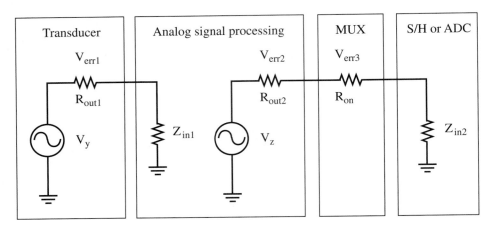

Figure 12.31
Block diagram for considering the problem of impedance loading.

$$V_{err3} = V_z \frac{R_{on}}{R_{out2} + R_{on} + Z_{in2}} \approx V_z \frac{R_{on}}{Z_{in2}}$$

The voltage error at the input increases by the gain, **G**, so the total error referred to the output is

$$G \, V_{err1} + V_{err2} + V_{err3} \approx G \, V_y \frac{R_{out1}}{Z_{in1}} + V_z \frac{R_{out2}}{Z_{in2}} + V_z \frac{R_{on}}{Z_{in2}} = V_z \left(\frac{R_{out1}}{Z_{in1}} + \frac{R_{out2} + R_{on}}{Z_{in2}} \right)$$

One wishes to keep the error below the ADC resolution, $\mathbf{\Delta_z}$. The factor 1/2 is again arbitrarily chosen. The largest error occurs when $\mathbf{V_z}$ equals $\mathbf{r_z}$:

$$r_z \left(\frac{R_{out1}}{Z_{in1}} + \frac{R_{out2} + R_{on}}{Z_{in2}} \right) \approx 0.5 \, \Delta_z$$

or

$$\left(\frac{R_{out1}}{Z_{in1}} + \frac{R_{out2} + R_{on}}{Z_{in2}} \right) \leq \frac{0.5}{n_z}$$

where the units of $\mathbf{n_z}$ are alternatives. In general, the ratio of the input impedance of each stage divided by the output impedance of the previous stage must exceed the precision of the ADC.

The voltage drop across the multiplexer can be a source of impedance-loading error (Figure 12.32). Consider the issue of impedance loading on the following two circuits. The input impedance of the ADC is 10 kΩ. The ADC is configured for −10 to +10 V. The max "On Resistance, at $-25°C \leq T_A \leq 85°C$" of the PMI MUX08EP is 400 Ω. The output impedance of an inverting amplifier is the open-circuit voltage divided by the short-circuit current.

$$Z_{out} \equiv \frac{V_{open}}{I_{short}} = \frac{R_{out}(R_1 + R_2)}{K \, R_1}$$

where R_1 and R_2 are as shown below. K, the op amp open-loop gain, is 100,000. R_{out}, the op amp output impedance, is 60Ω. The output impedance of a unity gain buffer is

$$Z_{out} \equiv \frac{R_{out}}{K}$$

Version 1 (incorrect, see Figure 12.32): $V_{max} = +10$ V

Figure 12.32
A DAS with an impedance-loading program.

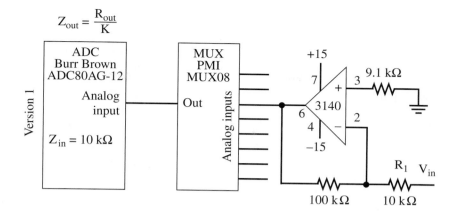

$$Z_{out} = \frac{R_{out}}{K}$$

$$Z_{out}\ (\text{of 3140 inverter}) = \frac{R_{out}(R_1 + R_2)}{K\,R_1}$$

$$= \frac{60(10\ k\Omega + 100\ k\Omega)}{100000 \times 10\ k\Omega} = 0.0066\ \Omega$$

$$R_{on}\ (\text{of MUX08}) = 400\ \Omega$$

$$Z_{in}\ (\text{of ADC}) = 10\ k\Omega$$

$$V_{err} = V_{max}\frac{R_{on} + Z_{out}}{Z_{in} + R_{on} + Z_{out}}$$

$$= 10\ V\frac{400\ \Omega + 0.0066\ \Omega}{10\ k\Omega + 500\ \Omega + 0.0066\ \Omega}$$

$$\approx 10\ V\frac{400\ \Omega}{10400\ \Omega}$$

$$= 0.4\ V\ (80\ \text{times the ADC resolution of } 20/4096 = 0.005\ V)$$

A unity gain amplifier is added between the MUX and the ADC (Figure 12.33). The purpose of this unity gain buffer amplifier in the second design is to prevent a significant voltage drop from occurring across the multiplexer. Ohm's law can be used to calculate the voltage drop across the MUX08 for both circuits.

Version 2 (correct, see Figure 12.33):

Stage 1:

$$Z_{out}\ (\text{of 3140 inverter}) = \frac{R_{out}(R_1 + R_2)}{K\,R_1} = \frac{60\ (10\ k\Omega + 100\ k\Omega)}{100000 \times 10\ k\Omega} = 0.0066\ \Omega$$

$$R_{on}\ (\text{of MUX08}) = 400\ \Omega$$

$$Z_{in}\ (\text{of 3140 buffer}) = 10^{12}\ \Omega$$

$$V_{err} = V_{max}\frac{R_{on} + Z_{out}}{Z_{in} + R_{on} + Z_{out}}$$

$$= 10\ V\frac{400\ \Omega + 0.0066\ \Omega}{10^{12}\ \Omega + 400\ \Omega + 0.0066\ \Omega} \bullet 10\ V\frac{400\ \Omega}{10^{12}\ \Omega}$$

$$= 4\ nV \qquad (\textit{This is well below the ADC resolution} = 20/4096 = 5\ mV)$$

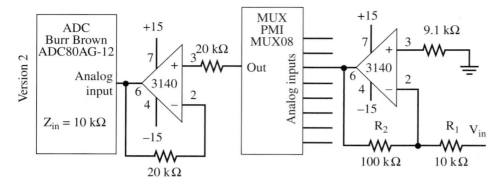

Figure 12.33
The impedance-loading program is solved with a voltage follower before the ADC.

Stage 2:

$$Z_{out} \text{ (of 3140 buffer)} = \frac{R_{out}}{K} = \frac{60}{100000} = 0.0006\ \Omega$$

$$Z_{in} \text{ (of ADC)} = 10\ k\Omega$$

$$V_{err} = V_{max} \frac{Z_{out}}{Z_{in} + Z_{out}} = 10\ V \frac{0.0006\ \Omega}{10\ k\Omega + 0.0006\ \Omega} \approx 10\ V \frac{0.0006\ \Omega}{10\ k\Omega}$$

$$= 0.6\ \mu V \qquad \textit{(This is well below the ADC resolution} = 20/4096 = 5\ mV)$$

12.3.5 How Fast Must the ADC Be?

The *ADC conversion time* must be smaller than the quotient of the sampling interval by the number of multiplexer signals. Let **m** be the number of multiplexer signals that must be sampled at a rate $\mathbf{f_s}$, let $\mathbf{t_{mux}}$ be the settling time of the multiplexer, and let $\mathbf{t_c}$ be the ADC conversion time. Then without a S/H,

$$m \cdot (t_{mux} + t_c) < 1/f_s$$

With a S/H, one must include both the acquisition time, $\mathbf{t_{aq}}$ and the aperture time, $\mathbf{t_{ap}}$:

$$m \cdot (t_{mux} + t_{aq} + t_{ap} + t_c) < 1/f_s$$

12.3.6 Specifications for the S/H

A S/H is required if the analog input changes more than one resolution during the conversion time. Let **dz/dt** be the maximum slope of the ADC input voltage, let $\mathbf{\Delta_z}$ be the ADC resolution, and let $\mathbf{t_c}$ be the ADC conversion time. A S/H is required if

$$\frac{dz}{dt} \cdot t_c > 0.5\ \Delta_z$$

If the transducer and analog signal processing are both linear, the determination of whether or not to use a S/H can be calculated from the input signals. A S/H is required if

$$\frac{dx}{dt} \cdot t_c > 0.5\ \Delta_x$$

Two S/Hs can be used to analyze the timing between two signals. For example, the circuit in Figure 12.34 can be used to measure the phase between two signals with one ADC. The

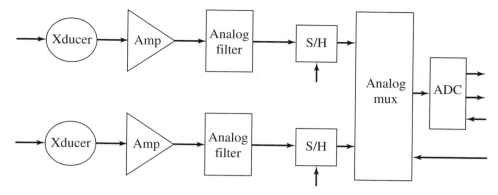

Figure 12.34
Multiple S/Hs can be used to implement synchronized sampling.

two S/Hs are given the hold command simultaneously, then the signals are sequentially converted. The digital samples represent the two signals at the same time.

12.4 Analysis of Noise

The consideration of noise is critical for all instrumentation systems. The success of an instrument does depend on careful transducer design, precision analog electronics, and clever software algorithms. But any system will fail if the signal is overwhelmed by noise. Fundamental noise is defined as an inherent and nonremovable error. It exists because of fundamental physical or statistical uncertainties. We will consider three types of fundamental noise:

> Thermal noise (also called white noise or Johnson noise)
> Shot noise
> 1/f noise

Although fundamental noise cannot be eliminated, there are ways to reduce its effect on the measurement objective. In general, added noise includes the many disturbing external factors that interfere with or are added to the signal. We will consider three types of added noise:

> Galvanic noise
> Motion artifact
> Electromagnetic field induction

12.4.1 Thermal Noise

Thermal fluctuations occur in all materials at temperatures above absolute zero. Brownian motion, the random vibration of particles, is a function of absolute temperature (Figure 12.35).

Figure 12.35
Brownian motion of individual particles.

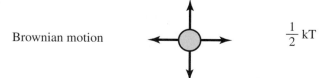

Brownian motion $\frac{1}{2}kT$

As the particles vibrate, there is an uncertainty as to the position and velocity of the particles. This uncertainty is related to the thermal energy:

> Absolute temperature T (K)
> Boltzmann's constant k = $1.67 \ 10^{-23}$ J/K
> Uncertainty in thermal energy $\approx (1/2)$ kT

Because the electric power of a resistor is dissipated as thermal power, the uncertainty in thermal energy produces an uncertainty in electric energy. The electric energy of a resistor depends on:

> Resistance R (Ω)
> Voltage V (V)
> Time (s)
> Electric power = V^2/R (W)
> Electric energy = $V^2 \cdot$ time/R (W $\cdot$ s)

By equating these two energies we can derive an equation for voltage noise similar to the empirical findings of J. B. Johnson. In 1928, he found that the open-circuit RMS voltage noise of a resistor was given by

$$V_J^2 = 4kTR \ \Delta\gamma$$

$$\Delta\gamma = f_{max} - f_{min}$$

where $f_{max} - f_{min}$ is the frequency interval, or bandwidth, over which the measurement was taken. For instance, if the system bandwidth is DC to 1000 Hz, then $\Delta\gamma$ is 1000 cycles/s. Similarly, if the system is a bandpass from 10 to 11 kHz, then $\Delta\gamma$ is also 1000 cycles/s. The term *white noise* comes from the fact that thermal noise contains the superposition of all frequencies and is independent of frequency. It is analogous to optics, where *white light* is the superposition of all wavelengths (Figure 12.36). Table 12.8 illustrates that white noise increases with resistance value and with system bandwidth.

Figure 12.36
White noise exists in all resistors.

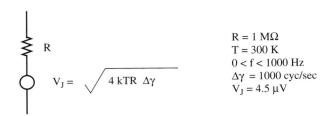

$$V_J = \sqrt{4 \ kTR \ \Delta\gamma}$$

R = 1 MΩ
T = 300 K
$0 < f < 1000$ Hz
$\Delta\gamma$ = 1000 cyc/sec
V_J = 4.5 μV

Table 12.8
White noise for resistors at 300K = 27°C.

	1 Hz	10 Hz	100 Hz	1 kHz	10 kHz	100 kHz	1 MHz
10 kΩ	14 nV	45 nV	142 nV	448 nV	1.4 μV	4.5 μV	14 μV
100 kΩ	45 nV	142 nV	448 nV	1.4 μV	4.5 μV	14 μV	45 μV
1 MΩ	142 nV	448 nV	1.4 μV	4.5 μV	14 μV	45 μV	142 μV

Interestingly, only resistive, not capacitive or inductive, electric devices exhibit thermal noise. Thus a transducer that dissipates electric energy will have thermal noise, and a transducer that simply stores electric energy will not.

Figure 12.37 defines RMS as the square root of the time average of the voltage squared. RMS noise is proportional to noise power. The crest factor is the ratio of peak value divided by RMS. The peak value is half the peak-to-peak amplitude and can be measured easily from recorded data. From Table 12.9, we see that the crest factor is about 4. The crest factor can be defined for other types of noise.

Figure 12.37
RMS is a time average of
the voltage squared.

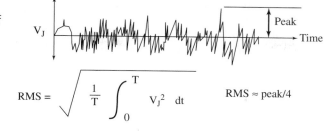

$$RMS = \sqrt{\frac{1}{T} \int_0^T V_J^2 \; dt} \qquad RMS \approx peak/4$$

Table 12.9
Crest factor for thermal
noise.

Percent of the time the peak is exceeded	Crest factor (peak/RMS)
1.0	2.6
0.1	3.3
0.01	3.9
0.001	4.4
0.0001	4.9

Example analysis. The objective of the following DAS is to sample x(t) at 10 Hz and perform a software calculations based on the DC to 5-Hz components of x. The gain of this amplifier is −100. Assume the bandwidth of the amplifier without the capacitor C is 100 kHz. With C = 0.01 μF, the bandwidth of the amplifier is reduced to 100 Hz. This reduction does not affect the DC to 5-Hz components of x (Figure 12.38).

Figure 12.38
A DAS with a gain of
100 and a frequency
range of 100 Hz.

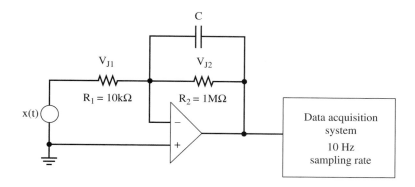

Let V_{J1} be the thermal noise of R_1 and V_{J2} be the thermal noise of R_2. Since the noise is related to the thermal power, the voltage amplitudes are combined at the power level. That is, if two devices generate voltage noises of V_1, and V_2 respectively, then together they will generate

$$V_{total} = \sqrt{V_1^2 + V_2^2}$$

When considering the overall effect of noise, V_{J1} is amplified by the op amp, whereas V_{J2} is not. The presence of the capacitor reduces the noise by 32 without degrading the system performance (Table 12.10). In addition to white noise in the resistor, there is white noise in the op amp.

Table 12.10
Reducing the system
bandwidth reduces the
thermal noise.

	Without C	With C
V_{J1}	4.5 μV	142 nV
V_{J2}	45 μV	1.4 μV
$\sqrt{[(100V_{J1})^2 + (V_{J2})^2]}$	452 μV	14.3 μV

12.4.2
Shot Noise

Shot noise arises from the statistical uncertainty of counting discrete events. Thermal cameras, radioactive detectors, photomultipliers, and O_2 electrodes count individual photons, gamma rays, electrons, and O_2 particles, respectively, as they impinge upon the transducer. Let dn/dt be the count rate of the transducer, and let Δt be the measurement interval or count time. The average count is

$$n = dn/dt \, \Delta t$$

On the other hand, the statistical uncertainty of counting random events is $\sqrt{n}$. Thus the shot noise is

$$\text{Shot noise} = \sqrt{dn/dt \, \Delta t}$$

The S/N ratio is

$$S/N = \frac{n}{\sqrt{n}} = \sqrt{dn/dt \, \Delta t}$$

The solutions are to maximize the count rate (by moving closer or increasing the source) and to increase the count time. There is a clear trade-off between accuracy and measurement time.

12.4.3
1/f, or Pink Noise

Pink noise is somewhat mysterious. The origin of 1/f lacks rigorous theory. It is present in devices that have connections between conductors. 1/f noise results from a fluctuating conductivity. It is of particular interest to low-bandwidth applications because of the 1/f behavior. Wire-wound resistors do not have 1/f noise, but semiconductors and carbon resistors do. One of the confusing aspects of 1/f noise is its behavior as the frequency approaches 0 Hz. The noise at DC is not infinite because although 1/f is infinite at DC, $\Delta\gamma$ is zero. Garrett gives an equation to calculate the 1/f noise of a carbon resistor (Table 12.11):

$$V_c = (10^{-6})\sqrt{1/f} \, R \, I\sqrt{\Delta\gamma}$$

Table 12.11
V_c versus frequency for
$R = 10 \text{ k}\Omega$, $I = 1$ mA,
$\Delta\gamma = 1$ kHz.

f(Hz)	V_c (μV)
1	316
10	100
100	32
1000	10

where
$\quad V_c$ = 1/f voltage noise, V
$\quad f$ = frequency, Hz
$\quad R$ = resistance, Ω
$\quad I$ = average direct current, A
$\quad \Delta\gamma$ = system bandwidth, Hz

12.4.4
Galvanic Noise

The contact between dissimilar metals will induce a voltage, due to the electrochemistry at the metal-metal interface. Voltages will also develop when a conductive liquid contacts a metal. This problem usually arises as a metal surface within a connector oxidizes (corrosion due to moisture). The materials least susceptible to corrosion are silver, graphite, gold, and platinum.

12.4.5
Motion Artifact

Motion can introduce errors in many ways. According to Faraday's law, a conducting wire that moves in a magnetic field will induce an EMF. This voltage error is proportional to the strength of the magnetic field, the length of the wire that is moving, the velocity of the motion, and the angle between the velocity and the field. Another problem occurring with moving cables is that the connector impedance may change or disconnect. Acceleration of the transducer will induce forces inside the device, often affecting its response.

12.4.6
Electromagnetic
Field Induction

Usually, the largest source of noise is electromagnetic field induction. According to Faraday's law, changing magnetic fields can induce a voltage into our circuits. The changing magnetic field must pass through a wire loop, drawn as the shaded area in Figure 12.39. This voltage noise (V_m) is proportional to the strength of the magnetic field, B (wb/m^2), the area of the loop S(m^2), and a geometric factor, K (volts/wb.) The drawing on the left of Figure 12.39 illustrates the physical situation causing magnetic field pickup. A typical situation causing magnetic field noise occurs when AC power is being delivered to a low-impedance load, such as a motor. The voltage V_1 is the 120 VAC 60 Hz power line, and V_s is a signal in our instrument. The alternating current (I_1) will create a magnetic field, B. This magnetic field (B) also alternates as it flows through the loop area (S) formed by the wires in our circuit (such as the lead wires between the transducer and the instrument box). This will induce a current (I_m) along the wire, causing a voltage error (V_m). We can test for the presence of magnetic field pickup by deliberately changing the loop area and observing the magnitude of the noise as a function of the loop area. The drawing on the right of Figure 12.39 illustrates an equivalent circuit we can use to model magnetic field pickup. Basically, we can model magnetic-field–induced noise as a mutual inductance between an AC current flow and our electronics.

Figure 12.39
Magnetic field noise pickup can be modeled as a transformer.

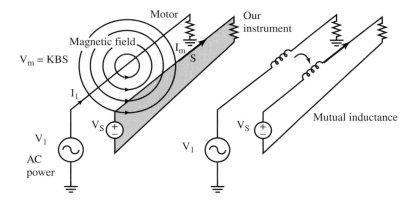

The second way EM fields can couple into our circuits is via the electric field. Changing electric fields will capacitively couple into the lead wires. The drawing on the left of Figure 12.40 illustrates the physical situation causing electric field pickup. The alternating voltage (V_1) will create an electric field. This electric field also traverses near the wires in our circuit (such as the lead wires between the transducer and the instrument box). This will induce a displacement current (I_d) along the wire. We can test for the presence of electric field pickup by placing a shield separating our electronics from the source of the field and observing the magnitude of the noise. The drawing on the right of Figure 12.40 illustrates an equivalent circuit we can use to model electric field pickup. Basically, we can model electric-field–induced noise as a stray capacitance between an AC voltage and our electronics.

> ***Observation:*** Sometimes EM fields originate from inside the instrument box, such as high-frequency digital clocks and switching power supplies.

12.4.7
Techniques to
Measure Noise

There are two objectives when measuring noise. The first objective is to classify the type of noise. In particular, we wish to know whether the noise is broadband (i.e., all frequencies, such as white noise) or does the noise contain specific frequencies (e.g., 60, 120, 180 Hz, . . . such as a 60 Hz EM field pickup). The type of noise is of great importance when determining

Figure 12.40
Electric field noise pickup can be modeled as a stray capacitance.

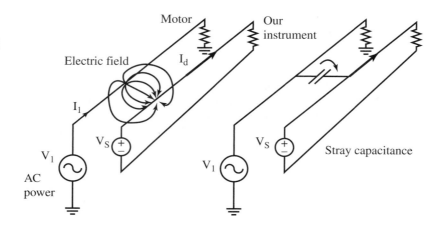

where the noise is coming from. Classifying the noise type is essential in developing a strategy for reducing the effect of the noise. The second objective is to quantify the magnitude of the noise. Quantifying the noise is helpful in determining whether a change to the system design has increased or decreased the noise. The measurement resolution of many data acquisition systems is limited by noise rather than by ADC precision and software algorithms. For these systems, quantitative noise measurements are an important performance parameter of the instrument.

Digital Voltmeter (DVM) in AC Mode.

Root-mean-squared (RMS) is defined the square root of the time-average of the voltage squared (Figure 12.37). If you remove the input signal, the output of the system contains just noise. Because the resistance load is usually constant, squaring the voltage results in a signal proportional to noise power. The averaging calculation gives a measure related to average power, and the square root produces a result with units in volts. RMS noise of a signal can be measured with a DVM using AC mode. Most DVMs in AC mode perform a direct measurement of RMS, hence this method is the most precise. For example, a $3^1/_2$ digit DVM has a precision of about 11 bits. A calibrated voltmeter in AC mode will provide the most accurate quantitative method to measure noise.

Analog Oscilloscope.

The second method is to connect the signal to an oscilloscope and measure the peak-to-peak noise amplitude, as illustrated in Figure 12.41. The crest factor is the ratio of peak value divided by RMS. The peak value is one-half of the peak-to-peak amplitude, and can be estimated from the scope tracing. From Table 12.9, we see for white noise that the crest factor is about 4, so we can approximate the RMS noise amplitude by dividing the peak-to-peak noise by 8. Because the quantitative assessment of noise with a scope requires visual observation, this method can be used only to approximate the quantitative level of noise. One the other hand, oscilloscopes have very high bandwidth, and therefore they are good for classifying high-frequency noise. White noise and 1/f noise look random, like the left graph in Figure 12.41. For white and 1/f noise, the scope trigger will not be able to capture a repeating waveform.

Figure 12.41
Quantifying noise by measuring peak-to-peak amplitude.

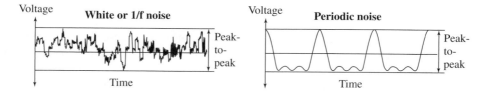

Noise from EM fields on the other hand is repeating and can be triggered by the scope. In fact the **line-trigger** setting on the scope can be used specifically to see whether the noise is correlated to the 60 Hz 120VAC power line. In particular, 60 Hz noise will trigger when using the line-trigger setting of the scope. The shape of the noise varies depending on the relative strengths of the fundamental and harmonic frequencies. The graph on the right is a periodic wave with a fundamental plus a 50% strength first harmonic.

Spectrum Analyzer.

The third method uses a spectrum analyzer, which combines a computer, high-speed ADC sampling, and the Fast-Fourier-Transform (FFT), as illustrated in Figure 12.42. Being able to see the noise in the frequency domain is a particularly useful way to classify the type of noise. Both 1/f and white noise components exist in a typical analog amplifier, but the amplitude is usually small compared to EM field noise. The 1/f and white noise levels occurring in electronics can be reduced by dropping the temperature or spending more money to buy a better device. Typically, we see the fundamental and multiple harmonics for EM field noise. For example, 60 Hz noise also includes components at 120, 180, 240 Hz, and so on.

Figure 12.42
Classifying noise by measuring the amplitude versus frequency with a spectrum analyzer.

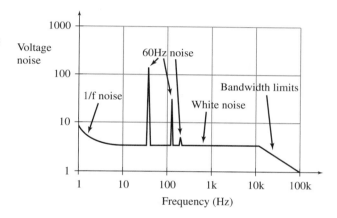

12.4.8
Techniques to Reduce Noise

It is much simpler to reduce noise early in the design process. Conversely, it can be quite expensive to eliminate noise after an instrument has been built. Therefore, it is important to consider noise at every stage of the development cycle. We can divide noise reduction techniques into three categories. The first category involves reducing noise from the source. You can enclose noisy sources in a grounded metal box. If a cable contains a high-frequency noise signal, that signal could be filtered. Magnetic and electric field strength depends on dI/dt and dV/dt. So whenever possible, you should limit the rise and fall times of noisy signals. For example, the square wave on the left will radiate more noise than the smooth signal on the right.

Motors have coils to create electromagnets, and noise can be reduced by limiting the dI/dt in the coil. Cables with noisy signals should be twisted together, so the radiated magnetic fields will cancel. These cables should also be shielded to reduce electric field radiation, and this shield should be grounded on both sides.

Figure 12.43
Limiting rise/fall times can reduce radiated noise.

The second category of noise reduction involves limiting the coupling between the noise source and your instrument. Whenever possible, maximize the distance between the noisy source and the delicate electronics. All transducer cables should use twisted wire, as shown in Figure 12.44. For situations in which a remote sensor is attached to an instrument, the ground shield should be connected only on the instrument side. If the cable is connected between two instruments, then connect the ground shield on both instruments. For high-frequency signals coaxial cable is required. The shield should be electrically insulated to eliminate direct electrical connection to other devices. If noisy signals must exist in the same cable as low-level signals, then separate the two with a ground wire in between. Whenever possible, reduce the length of a cable. Similarly minimize the length of leads that extend beyond the ground shield. Place the delicate electronics in a grounded case. You can use optical or transformer isolation circuits to separate the noisy ground from the ground of the delicate electronics.

Figure 12.44
Proper cabling can reduce noise when connecting a remote transducer or when connecting two instruments.

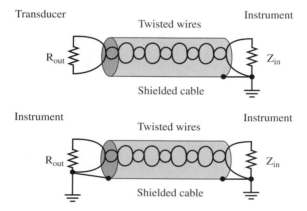

The last category involves techniques that reduce noise at the receiver. The bandwidth of the system should be as small as possible. In particular, you can use an analog low-pass filter to reduce the bandwidth, which will also reduce the noise. You can add frequency-reject digital filters to reduce specific noise frequencies such as 60, 120, 180 Hz. You should use power supply decoupling capacitors on each chip to reduce the noise coupling from the power supply to the electronics. Figure 12.45 illustrates how EM field noise will affect our instrument. If the cable has twisted wires then I_{d1} should equal I_{d2}. The input impedance of the amplifier is usually much larger than the source impedance of the signal. Thus, $V_1 - V_2 = R_{s1} I_{d1} - R_{s2} I_{d2}$.

Some capacitors have a foil wrap surrounding their cylindrical shape. This foil should be grounded. For more information about noise refer to Henry Ott, *Noise Reduction Techniques in Electronic Systems*, Wiley, 1988 or Ralph Morrison, *Grounding and Shielding Techniques*, Wiley, 1998.

Figure 12.45
Capacitively coupled displacement currents.

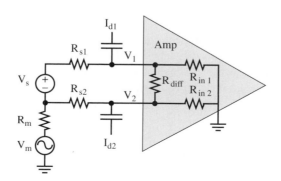

12.5 Data Acquisition Case Studies

We introduced the design process back in Chapter 1. In this section, we will present the designs of three data acquisition systems. The first example is a quantitative measurement of temperature, the second example is a qualitative measurement of electrocardiogram signals, and the third example compares measurements made with the ADC versus timing measurements using input capture. In the **analysis phase**, we determine the requirements and constraints for our proposed system. In the **high-level design** phase, we define our input/output, break the system in modules, and show the interconnection using data flow graphs. In the **engineering design** phase we design the hardware/software subcomponents using techniques such as simulation, mechanical mockups, and call-graphs. We define specific I/O signals, analog circuits, power sources, noise filters, software algorithms, data structures, and testing procedures. During the **implementation** and **testing** phases we build and test the modules. Modularity allows for concurrent development.

12.5.1 Temperature Measurement

The objective of this instrument is to measure temperature within a range of 25 to 50°C, and a resolution is 0.1°C. The frequency range of interest is 0 to 0.1 Hz. The first decision to make is the choice of transducer. The RTD has a linear resistance versus temperature response. RTDs are expensive but a good choice for ease of calibration, interchangeability, and accuracy. Thermocouples are inexpensive and a good choice for large temperature ranges, harsh measurement conditions, fast response time, interchangeability, and large temperature ranges. Interchangeability means we can buy multiple transducers and they will all have similar temperature curves. Thermistors will be used in this design because they are inexpensive and have better sensitivity than RTDs and thermocouples. A thermometer built using a thermistor will be harder to mass-produce because each transducer must be separately calibrated in order to create an accurate measurement. From a first glance, we might expect an 8-bit ADC to generate a temperature resolution of 0.1°C (range/precision = resolution). On the other hand, because the thermistor is nonlinear, we will need to verify that the resolution specification has been met. Figure 12.46 shows the block diagram of the instrument, which also illustrates the data flow in our system.

Figure 12.46
Block diagram of a temperature measurement system using a thermistor.

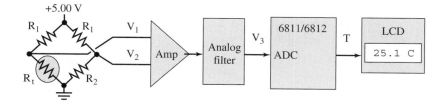

The resistance bridge is a classic means to convert resistance to voltage. Table 12.12 is used during the design phase to show the signal values as they transverse the electronics. The +5.00 reference drives the bridge. The value of resistor R_1 is chosen to eliminate errors due to self-heating of the thermistor (500 kΩ). Since we will be using rail-to-rail electronics, we need to have all voltages within the 0 to +5 V range. We can make the bridge output $(V_1 - V_2)$ positive by selecting the value of resistor R_2 equal to the thermistor resistance at the maximum temperature (200 kΩ). Next, we choose the gain of the amplifier to map the minimum temperature into the +5 V limit of the ADC (4.232). Since the thermistor is nonlinear, we will tabulate explicit values to determine the ADC precision required (Table 12.13). There are two possible approaches to the design of the amplifier. If we use an instrumentation

Table 12.12
Signals as they pass through the temperature data acquisition system.

T (°C)	R_t (kΩ)	V_1 (V)	$V_1 - V_2$ (V)	V_3 (V)	ADC	T (0.1°C)
25	546.0	2.610	1.181	5.000	255	250
30	444.3	2.353	0.924	3.911	199	300
35	364.0	2.106	0.678	2.869	146	350
40	300.1	1.875	0.447	1.890	96	400
45	248.9	1.662	0.233	0.987	50	450
50	207.6	1.467	0.039	0.163	8	500

amp, the input impedance will be large enough not to affect the bridge. If we use a single op amp differential amp, then the amplifier will load the bridge and affect the bridge response. In this system, an instrumentation amp will be used, because the entire amplifier and LPF can be built with a single MAX494 device. The first two columns of Table 12.12 show the resistance temperature calibration of the 500 kΩ thermistor. The third column, V_1, is the voltage across the thermistor. The fourth column is the output of the bridge, $V_1 - V_2$. V_3, the output of the instrumentation amp, is shown in the fifth column. The ADC value gives the digital output of an 8-bit converter, and the last column will be calculated by our software as a decimal fixed-point with resolution 0.1°C. We use this table in two ways. Initially, we use theoretical values to design the electronics and software. During the implementation phase, we substitute resistors with standard values to bring down the cost. During the testing phase, we measure actual values to verify proper operation. Measured values for the last two columns will be stored in software as a calibration table. To measure temperature, the software measures the ADC value, then uses a table look-up and linear interpolation to get the decimal fixed-point temperature (last column). The fixed-point number is output to an LCD display.

Observation: There is an Excel worksheet named Therm.xls on the CD that was used to create this design.

The fastest slew rate for the thermistor is 10°C/sec. Assuming the ADC conversion time is 25 µs, no S/H is needed, because the maximum slope of **T** multiplied by the ADC conversion time is less than the temperature resolution.

$$\textbf{10°C/sec} \bullet \textbf{25 µs} = \textbf{0.00025°C} \ll \textbf{0.25°C}$$

In order to prevent noise in the ADC samples, the noise must be less than the resolution. The resolution of $V_1 - V_2$ is its range (1.181V) divided by its precision (256), which is 4.6 mV. Again a safety factor of 1/2 is included. Thus, in the frequency range 0 to 1 Hz, the maximum allowable noise referred to the input of the differential amp should be

$$\textbf{amplifier noise} \leq \frac{\textbf{resolution}}{2} = \textbf{2.3 mV}$$

A two-pole low pass analog filter is used to pass the temperature signal having frequencies from 0 to 0.1 Hz, to reject noise having frequencies above 0.1 Hz, and to prevent aliasing. We design the filter so the ADC error is less than 1/2 a resolution at 0.1 Hz (i.e. gain = 511/512). The gain of the two-pole low pass filter is

$$G_3 = |H_3| = \frac{1}{\sqrt{1 + (f/f_c)^4}}$$

where f_c must be chosen large enough so the gain, **G**, is at least = $(2^{n+1} - 1)/2^{n+1}$ = 511/512 for frequencies 0 to 0.1 Hz. Thus, f_c must be larger than 1.1 Hz. In order to build the filter with standard components, a 2.25 Hz LPF will be designed (see Figure 12.47). The 0.2 µF capacitor can be created by placing two 0.1 µF capacitors in parallel.

$$f_c = \frac{0.1 \text{Hz}}{\sqrt{(512/511)^4 - 1}} = 1.1 \text{ Hz}$$

Observation: In most situations, we could simply place the LPF cutoff frequency at 0.1 Hz.

Observation: There is an Excel worksheet named LPF.xls on the CD that was used to create this LPF.

In order to prevent aliasing, $\mathbf{Z_2}$ must be less than the ADC resolution for all frequencies larger than or equal to 0.5 $\mathbf{f_s}$. As an extra measure of safety, we make the amplitude less than $\mathbf{0.5\ \Delta_z}$ for frequencies above 0.5 $\mathbf{f_s}$. Thus,

$$|\mathbf{Z_2}| < 0.5\Delta_z = \frac{5\ V}{512} \approx 0.01\ V$$

For discussion, assume the following criteria for frequencies above 1 Hz, which are dominated by the thermistor response and the LPF.

$$|X| = \frac{0.5°C}{\sqrt{1 + (f/1Hz)^2}} \qquad\qquad LPF = \frac{1}{\sqrt{1 + (f/2Hz)^4}}$$

The approximate gain of the entire circuit is 5 V/25°C. Thus

$$|Z_2| = \frac{0.1\ V}{\sqrt{[1 + (f/1Hz)2][1 + (f/2Hz)4]}}$$

We choose the sampling rate, $\mathbf{f_s}$, to prevent aliasing, we need a sampling rate above 6.6 Hz.

$$\frac{0.1V}{\sqrt{[1 + (f_s/2)^2][1 + (f_s/4)^4]}} < 0.01\ V$$

The effective output impedance of the bridge is 500 kΩ. The input impedance of the differential amp must be high enough not to affect the ADC conversion. Although the design equations specified a gain of 4.232, the circuit in Figure 12.47 implements a slightly smaller gain (4.2) in order to use standard resistor values.

Figure 12.47
Amplifier and low-pass filter. $R_1 = 500$ kΩ and $R_2 = 200$ kΩ.

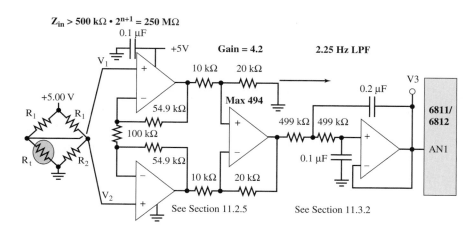

To determine the resolution we work backwards, as illustrated in Table 12.13. The basic approach to verifying the temperature resolution is to work backwards through the circuit,

Table 12.13
Equations calculated in reverse to show that the resolution meets the design specification.

ADC	V_3 (V)	V_1-V_2 (V)	V_1 (V)	R_T (kΩ)	T (°C)	ΔT (°C)
255	5.000	1.181	2.610	546.0	25.000	
254	4.980	1.177	2.605	544.0	25.089	0.089
101	1.980	0.468	1.896	305.5	39.524	
100	1.961	0.463	1.892	304.3	39.628	0.103
1	0.020	0.005	1.433	200.9	50.925	
0	0.000	0.000	1.429	200.0	51.053	0.128

showing that a change in ADC value of 1 corresponds to a temperature change of about 0.1°C.

Checkpoint 12.9: What would be the temperature resolution if the ADC precision were increased from 8 to 10 bits?

There are three possible approaches to converting ADC sample to temperature (the last two columns of Table 12.12). First, we could fit the transfer function to a polynomial equation and save the coefficients of that equation as the calibration file. This approach performs well for simple situations. Second, we could calculate the temperature output for each possible ADC and save it in a 256-entry lookup table. This conversion is fast because we just need to use the ADC data to index into the big table. This method is fast, but requires a lot of memory. The third approach is shown in Programs 12.1 and 12.2, which uses a small table of paired (ADC,T) data. These points are determined from experimental calibration. To find the corresponding temperature for a given ADC value, the program first searches the table for a pair of ADC-values that surround the input. Then, it uses linear interpolation to calculate the temperature, given the four entries in the table and the ADC input. When speed is important, we can consider writing time-critical functions in assembly. Linear interpolation is one of three or four operations on the 6812 for which an assembly solution is significantly faster than a C program (the others are Fuzzy Logic, mixed 16/32-bit math, and the SUM=SUM+X*Y calculation). Program 12.1 is written in Metrowerks assembly syntax. This subroutine is saved as a `LookUp.asm` file and included in the Metrowerks project of the main data acquisition system. The output result is a fixed-point number with a resolution of 0.1°C.

Program 12.1

Assembly language program to convert 8-bit ADC into fixed-point temperature.

```
          xref ADCtable   ; monotonic list of ADC values
          xref Ttable     ; list of corresponding temperature values
          absentry LookUp
;**********Lookup******************
;Inputs: RegD is 0 to 65534 Xdata point, xL
;     RegD input must be greater than or equal to first Xdata point
;     RegD input must be less than last Xdata point
;Output: RegD is 0 to 65535 Ydata point, decimal fixed-point 0.1C
;Registers destroyed: X,Y,B,CCR
LookUp: ldx   #ADCtable   ; first find x1<=xL<x2
        ldy   #Ttable
search  cpd   2,x         ; check xL<x2
        blo   found       ; stops when X points to x1
        leax  2,x
        leay  2,y
        bra   search
found   subd  0,x         ; xL-x1
        pshd
        ldd   2,x         ; x2
        subd  0,x         ; D=x2-x1
        tfr   D,X         ; X=x2-x1
        puld              ; D=(xL-x1)
        fdiv              ; X=(65536*(xL-x1))/(x2-x1)
        tfr   X,D
        tfr   A,B         ; B=(256*(xL-x1))/(x2-x1)
        etbl  0,y         ; Y=>>y1,y2
        rts               ; D=Y1+B*(Y2-Y1)
```

Program 12.2, written in C, performs the real-time data acquisition. The calibration data in **ADCtable** and **Ttable** are stored in EEPROM. Extra entries were added at the beginning and end of the table to guarantee that the search step of **LookUp** function will always be successful. In general, we perform time-critical tasks such as ADC sampling in the background, and noncritical functions such as LCD output in the foreground. The interrupt service routine passes the measured temperature to the foreground through a FIFO queue, and the main program has the responsibility of outputting the result to the LCD.

Program 12.2
Real-time measurement
of temperature.

```
// table of multiple unsigned (x,y), piece-wise linear function
unsigned short const ADCtable[8]= { 0,  8, 50, 96,146,199,255,65535};
unsigned short const Ttable[8]= { 500,500,450,400,350,300,250,  250};
unsigned short LookUp(unsigned short data);
interrupt 11 void TOC3handler(void){ // TCNT is 4us
unsigned short Data;              // raw ADC result, 0 to 255
  Data = ADC_In(0x81);            // 0 to 255
  Fifo_Put(LookUp(Data));         // 250 to 500 (0.1C)
  TC3 = TC3+25000;                // every 0.1s at 4 MHz
  TFLG1 = 0x08;                   // acknowledge OC3
}
```

12.5.2 EKG Data Acquisition System

The objective of this design is to measure electrical activity in the heart. In particular, we wish to measure heart rate. Biopotentials are important measurements in many research and clinical situations. **Biopotentials** are electric voltages produced by individual cells and can be measured on the skin surface. The status of the heart, brain, muscles, and nerves can be studied by measuring biopotentials. Electrodes, which are attached to the skin, interface the machine to the body. Electronic instrumentation amplifies and filters the signal. For example, Figure 12.48 shows a normal Lead II **electrocardiogram**, or **EKG**, which is measured with the positive terminal attached to the left arm, the negative terminal attached to the right arm, and ground connected to the right leg. Each wave represents one heartbeat, and the shape and rhythm of this wave contains a lot of information about the health and status of the heart.

Figure 12.48
Normal II-lead
electrocardiogram.

There are two types of electrodes used to record biopotentials. **Nonpolarizable** electrodes such as silver/silver chloride involve the following chemical reaction in the electrode at the electrode/tissue interface:

$$AgCl + e^- \Leftrightarrow Ag + Cl^-$$

A nonpolarizable electrode has a low electrical impedance, because electrons can freely pass the electrode/tissue interface. In the electrode, current flows by moving electrons, but in the tissue, current flows by physical motion of charged ions (e.g., Na^+, K^+, and Cl^-). A silver/silver chloride electrode does include a half-cell potential of 0.223 V, but since biopotentials are always measured with two electrodes, these half-cell potentials cancel. On the other hand, if you tried to use these electrodes to measure DC voltages, then the foregoing chemical reaction would saturate and fail. Fortunately, biopotentials produced by muscles and nerves are AC only and have no DC component.

Polarizable electrodes, made from metals such as platinum gold or silver, have a high electrical impedance, because electrons cannot freely pass the electrode/tissue interface. Charge can develop at the electrode/tissue interface effectively creating a capacitive barrier. Displacement current can flow across the capacitor, allowing the AC biopotentials to be measured by the electronics. The metallic electrodes also include a half-cell potential, but again, these potentials will cancel.

$$Ag^+ + e^- \Leftrightarrow Ag$$

The graphical display of EKG versus time is an example of a qualitative data acquisition system. The measurement of heart rate is quantitative. The parameters of an EKG amplifier include high input impedance (larger than 1 MΩ), high gain, a 0.05 to 100 Hz bandpass filter, and a good common-mode rejection ratio. The EKG signal is about ± 1 mV, so an overall gain of 2500 will produce a range of 0 to $+5$ V on V3. The data flow graph of this system is similar to Figure 12.46. This EKG amp (Figure 12.49) begins with a preamp stage having a good CMRR, high input impedance, and a gain of 50. Pin 5 of the AD620 is the analog ground, which in this circuit is the 2.50 V reference voltage. The AD620 is not rail-to-rail, but the voltage swing on V1 will be only 2.5 ± 0.05 V. A 0.05 Hz passive high-pass filter is created by R2 and C2. A low-leakage capacitor for C2 is critical for elimination of DC offset drift. A polypropylene or polystyrene would be a good choice for C2, but a low-leakage ceramic is acceptable. The remaining gain is performed with a noninverting amplifier (U2a). The LPF is implemented as a 2-pole Butterworth LPF. The 153 Hz cutoff was chosen because it is greater than 100 Hz and can be implemented with standard components. If the signal V1 saturates, you can reduce the gain on the preamp and increase the gain on the amp.

Figure 12.49
A battery-power EKG instrument.

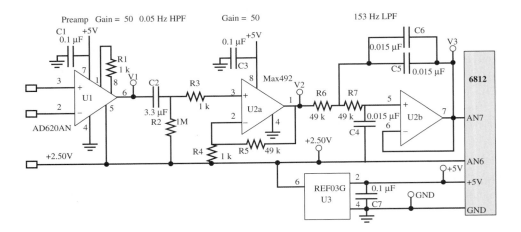

Program 12.3 shows the real-time data acquisition and 60 Hz digital notch filter. The design of digital filters is presented in Chapter 15. The ADC sampling occurs in the background, and the data are passed to the foreground using a FIFO queue. The large pulse in the EKG, originating from the contraction of the ventricles, is called the R-wave, and it occurs once a heartbeat. Program 12.4 shows the foreground process, where there are four calculation steps performed on the EKG data. A low-pass filter followed by a high-pass filter capture a narrow band of information around 8 Hz. The square function calculates power, and the 200 ms wide moving average gives an output very specific for the R-wave. Hysteresis is implemented with two thresholds. A heartbeat is counted (**RCount++**) when the moving average goes below the **LOW** threshold and then above the **HIGH** threshold. This software uses a combined frequency-period method to calculate heart rate. The theory of this method was developed as Exercise 6.3 in Chapter 6. The algorithm to measure heart rate searches for R-waves in a 5-second interval. **Rfirst** is the time (in 1/120 sec units) of the first R-wave, and **Rlast** is the time (also in 1/120 sec units) of the last R-wave. (**RCount-1**) is the number of beat-to-beat intervals between **Rfirst** and **Rlast**. The number 7200 is the conversion between the sample period (1/120 sec) and one minute. For example, at 72 BPM there will be 6 R-waves detected in the 5-second interval, making (**Rcount-1**) equal to 5 and the difference **Rlast-Rfirst** will be 500. For more information on EKG systems, see Webster's book *Medical Instrumentation*, published by Wiley 1997, or Pan and Tompkins, "A Real-Time QRS Detection Algorithm," *IEEE Transactions on Biomedical Engineering*, pp. 230–236, March 1985.

Program 12.3
Real-time sampling of
EKG.

```
interrupt 11 void TOC3handler(void){      // 480Hz
static short sum=0,n=0; short data;
  data = (0x00FF&ADC_In(7))-128;          // real EKG
  sum += data;      // sum=x(n)+x(n-1)+x(n-2)+x(n-3)
  n++;
  if(n==4){
    n = 0;
    Fifo_Put(sum/4);     // 120Hz
    sum = 0;
  }
  TC3 = TC3+16667;       // 2083.375us = 479.99 Hz
  TFLG1 = 0x08;          // acknowledge OC3
}
```

Warning: If you are going to build an EKG, please have a trained engineer verify the
safety of your hardware and software before you attach patients to your machine.

Program 12.4
Measurement of heart
rate.

```
short Data;      // ADC sample, -128 to +127, 8-bit signed ADC sample
short x[50];     // 60Hz notch-filtered EKG, 120Hz
short y[50];     // low pass filter, 120Hz
short z[50];     // high pass filter, 120Hz
short w[50];     // squared result, R-wave power, 120Hz
short Rwav;      // moving average of R-wave power, energy
unsigned short n=25;   // 25,26, ..., 49
unsigned short Trigger;
#define HIGH 100   // trigger when over this
#define LOW 20     // reset when under this
unsigned short Rcount;      // number of R-waves
unsigned short Rfirst;      // time of first R-wave
unsigned short Rlast;       // time of last R-wave
unsigned short HeartRate;  // units bpm
void main(void) { unsigned short time;       // units 1/120sec
short lpfSum=0,hpfSum=0,RwavSum=0;
  LCD_Open(); LCD_OutString("EKG System - Valvano");
  Fifo_Init();
  ADC_Init();      // Activate ADC
  Timer_Init();    // initialize 480Hz OC3
  Trigger =0;      // looking for HIGH
  for(;;) {
    Rcount = 0;
    Rlast = 0;
    for(time=0;time<600;time++){  // 120 Hz, every 5 second
      while(Fifo_Get(&Data)){};    // Get data from background thread
      LCD_Plot(Data);              // draw voltage versus time plot
      n++; if(n==50) n=25;
      x[n] = x[n-25] = Data;         // new data
      lpfSum = lpfSum+x[n]-x[n-4];
      y[n] = y[n-25] = lpfSum/4;       // Low Pass Filter
      hpfSum = hpfSum+y[n]-y[n-10];
      z[n] = z[n-25] = y[n]-hpfSum/10; // High Pass Filter
      w[n] = w[n-25] = (z[n]*z[n])/10; // Power calculation
      RwavSum = RwavSum+w[n]-w[n-24];  // 200ms wide moving average
      Rwav = RwavSum/24;
```

continued on p. 642

Program 12.4
Measurement of heart rate.

continued from p. 641

```
      if(Trigger){
        if(Rpow<LOW){
          Trigger = 0;        // found low
        }
      } else{
        if((Rpow>>HIGH)&&((time-Rlast)>>30)){ // max HR= 240bpm
          Trigger = 1;       // found high
          if(Rcount){
            Rlast = time;   // mark time of last R-wave, units 1/120sec
          } else{
            Rfirst = time;  // mark time of first R-wave
          }
          Rcount++;
        }
      }
    }
    if(Rcount>>=2){
      HeartRate = (7200*(long)(Rcount-1))/(long)(Rlast-Rfirst);
    } else{
      HeartRate = 0;
    }
    LCD_OutUDec(HeartRate);   // display results
  }
}
```

Checkpoint 12.10: What is the theoretical heart rate resolution of this approach when the HR is 60 BPM?

12.5.3 Position Measurement System

In this section we will develop two interfaces for a class of position sensors based on the potentiometer. A potentiometer can be used to convert position into resistance. In this particular interface the full-scale range is 0 to 1000 Ω.

$$R_{out} = 200 \cdot x$$

where the units of R_{out} and x are in Ω and centimeters, respectively. To interface this transducer to the microcomputer we drive the potentiometer with a stable DC voltage using a precision voltage reference (Figure 12.50). If we were to drive the circuit with the +5-V power, then any noise ripple in the power line would couple directly into the measurement. The ADC produces a digital output dependent on its analog input. The resolution is 5cm/256, which is about 0.02cm.

Figure 12.50
Potentiometer interface using an ADC.

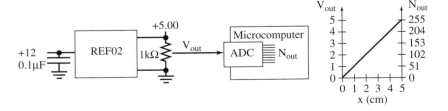

Checkpoint 12.11: What would be the position resolution if the ADC precision were increased from 8 to 10 bits?

Another approach to interface this transducer to the microcomputer would be to use an astable multivibrator. The period of a 555 timer is $0.693 \cdot C_T \cdot (R_A + 2R_B)$. In our circuit, R_A is R_{out}, R_B is 1 kΩ, and C_T is 2.2 μF. Given a fixed R_B, C_T, the period of the square wave, P_{out}, is a linear function of R_{out}. Our microcontrollers have a rich set of mechanisms to measure frequency, pulse width, or period. To change the slope and offset of the conversion between R_{out} and P_{out}, the fixed resistor and capacitor can be adjusted. Even though the period does not include zero, the precision of this measurement is over 3000 alternatives, or more than 11 bits. The precision can be improved by increasing the capacitor C_T, or decreasing the period on the measurement clock (Figure 12.51). The position resolution is 5 cm/3049, which is about 0.002 cm.

Figure 12.51
Potentiometer interfacing using input capture.

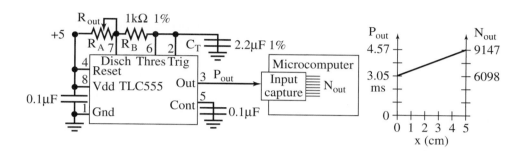

12.6   Exercises

12.1 Design a computer-based DAS that measures pressure. The pressure transducer is built with four resistive strain gages placed in a DC bridge. When the pressure is zero, each gauge has a 120-Ω resistance making the bridge output y zero. When pressure is applied to the transducer, two gages are compressed (which lowers their resistance) and two are expanded (which increases their resistance). At full-scale pressure (p = 100 dynes/cm²), the bridge output is y = 10 mV. The transducer/bridge output impedance is therefore 120 Ω. You may assume the transducer is linear. The desired pressure resolution is 1 dyne/cm². The frequencies of interest are 0 to 100 Hz, and the two-pole Butterworth analog low-pass filter will have a cutoff (gain = 0.707) frequency of 100 Hz (you will design it in part b). In terms of choosing a sampling rate, you may assume the low-pass filter removes all signals above 100 Hz.

a) Show the interface of the ADC to your computer. Justify your ADC precision.

b) Design the analog interface between the transducer/bridge and the your ADC. Use the full-scale ADC range even though it will complicate the conversion software. For example, if the ADC has a range of 0 to +5 V, then a pressure of p = 0 maps to a voltage at the ADC input of 0, and p = 100 dynes/cm² maps to a voltage at the ADC input of +5 V. Include the Butterworth low-pass filter. (Figure 12.52)

Figure 12.52
Transducers are placed in a bridge.

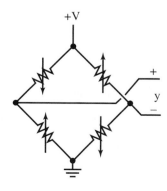

c) Show the ritual that initializes any global variables, the ADC, and an appropriate interrupt.

d) Show the interrupt handler(s) that samples the ADC, calculates pressure in dynes/cm², and stores the value in global variable `Pressure`. The software conversion maps a 0 ADC result into `Pressure=0`, and a 255 ADC result into `Pressure=100`. *Optimize the interrupt handler so that the number of execution cycles is minimized.*

12.2 The objective of this problem is to measure vibrations (displacement versus time) using a strain gage. The displacement range is −100 to +100 μm. The frequencies of interest are 1 to 200 Hz. The four resistors of the strain gage are placed in a bridge: two in compression, two in expansion. The bridge output voltage V_b has a sensitivity of 100 V/m. Each resistance R in the bridge is 100 Ω at a zero displacement. There is a lot of unwanted DC and 60-Hz noise in this system. You will remove the unwanted DC signal using analog signal processing and remove the unwanted 60-Hz signal using digital signal processing.

a) Show the interface between the transducer and the microcomputer. The S/H input impedance is 1 TΩ. The S/H aperture time is 10 μs, and the acquisition time is 100 μs. The ADC input voltage range is −5 to +5 V, the ADC digital output is 12-bit signed 2s complement, and the conversion time is 50 μs. (Figures 12.53 and 12.54)

Figure 12.53
Vibration signals and ADC timing.

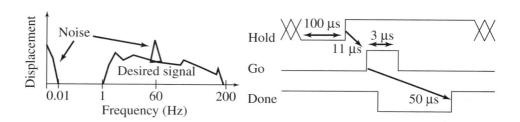

Figure 12.54
Vibration measuring system.

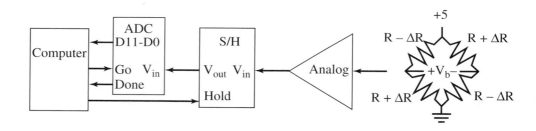

b) What is the maximum allowable droop rate for the S/H? Show your work.

c) What is the displacement resolution? Give units and show your work.

d) What is the slowest sampling frequency that is feasible for this system. Justify your answer. You will implement a sampling rate of 480 Hz and execute the following digital filter to remove the 60 Hz.

$$y(n) = (x(n) + x(n - 4))/2$$

e) Show the ritual subroutine that initializes the necessary microcomputer devices. Initialize all data structures including `DONE`. The output of your software is a global array, `Y`, containing the filtered displacement versus time. The data acquisition and digital filtering will be performed in real time under interrupt control. The main program will call this ritual and perform other unrelated tasks until `DONE` is −1.

```
short Y[1000];  // filtered displacement measurements
// The resolution and units are given in c), fs is 480 Hz
char DONE;      // initially zero, set to -1 when buffer is full
```

f) Show the interrupt software which samples the ADC, implements the digital filter in real time, and saves the filter outputs in the buffer `Y`. Set `DONE` to −1 when the buffer is full.

12.3 Design a remote-sensor temperature acquisition system. The goals are:

Range $1 \leq T \leq 200$ °F
Resolution $\Delta T = 1$°F
Frequencies of interest   DC to 5 Hz

The system has the block diagram shown in Figure 12.55. The timer is Port A on the 6811 and Port T on the 6812.

Figure 12.55
Remote temperature measuring system.

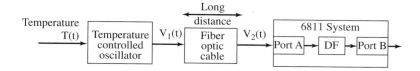

$V_1(t)$ is a low-power Schottky TTL square wave with a period P(t) that is linearly related to the temperature T(t).

$$P(t) = \frac{10 \ \mu s}{°F} T(t)$$

Your interface should connect the fiber-optic cable so that $V_2(t)$ is also a TTL-level square wave with the same period. The 6811/6812 input capture/output compare system should be used to measure T(t) at equal intervals. Let f_s be the fixed sampling rate, and $\Delta t = 1/f_s$. Your system should measure T(t) exactly f_s times per second. Let x(n) be the temperature sampled every Δt with a resolution of 1°F. Your 6811/6812 system will implement the following low-pass Finite Impulse Response (FIR) digital filter:

$$y(n) = \frac{y(n-1) + x(n)}{2}$$

The output of this filter, y(n), should be written to Port B every Δt s.

a) Show the interface between V_1 and the fiber-optic cable. The front end operates like a LED. Make $I_f = 20$ mA (Figure 12.56).

Figure 12.56
Fiber-optic cable transmitter interface.

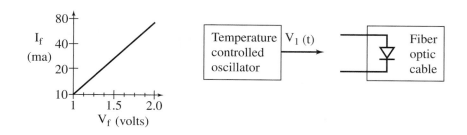

b) Show the interface between the fiber-optic cable and V_2 (Figure 12.57). The output of the fiber-optic cable is a photodiode detector. The detector conducts when light shines on it. It is open when dark. Provide for some noise rejection.

Figure 12.57
Fiber-optic cable receiver interface.

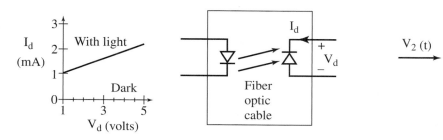

c) Choose the slowest possible sampling rate f_s. Justify your answer.

d) Show connections between the square wave $V_2(t)$ and the 6811/6812.

e) Show global data structure required to implement the digital filter. Full credit will be given to the best dynamic and static efficiency. This is not a question as to whether assembly is more efficient than C, but rather which data structure gets the job done in the simplest manner.

f) Show the ritual and interrupt handler(s). After calling the ritual, the main program (foreground) is free to execute other unrelated tasks. The interrupt handler(s) (background) will implement the fixed rate data acquisition. The output of the digital filter should be output to Port B f_s times per second. You may use either assembly or C.

12.4 The objectives of this problem are (1) to use a single-chip embedded 6811/6812 with its ADC to measure force at various locations on a robot and (2) to transmit the measurements across a master/slave distributed network. The robot system will have multiple slaves, each with its own force-measuring circuitry (Figure 12.58). You will design the slaves and specify/implement the communication channel to the master. The force measurement range is $0 \leq F \leq 200$ dynes. The desired force resolution is 1 dyne. The signals of interest are 0 to 10 Hz. Each slave has $+5$, $+12$ and -12 V power. The force transducers are placed in a resistance bridge powered by the $+5$-V supply (Figure 12.59). The bridge output has a sensitivity of 0.05 mV/dyne. The bridge output is *almost* zero when the force is zero. This transducer offset for each slave is a constant within the range from 0 to 0.2 mV. Each slave will require a no-force calibration that you will decide whether to handle in software or hardware.

Figure 12.58
Distributed force measuring system.

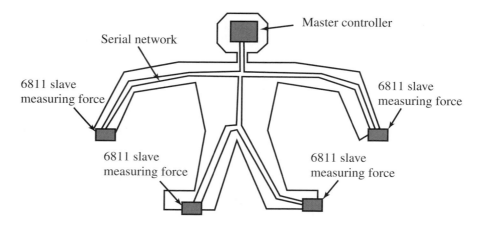

Figure 12.59
The force transducers are placed in a bridge.

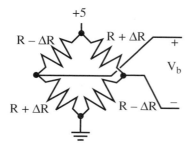

a) Show the analog interface between the bridge and the 6811/6812 ADC channel 1. Choose the appropriate op amp circuit, gain, offset, and analog filter. Include a mechanism to calibrate if you decide to handle the offset adjustment in hardware.

b) Design the distributed network between the slaves and their master. All slaves have identical network connections. To get a measurement from one of the slaves, the master will transmit the slave address that should activate exactly one slave. The slave will respond with the current force with a resolution of 1 dyne. The first priority of your network is to minimize cost (i.e., fewest wires and interface chips). Given the cheapest network, the second priority is to maximize bandwidth. Clearly show whether or not the grounds are connected between the multiple computers.

b1) Show the hardware interface between a single-chip 6811/6812 slave and the network. Remember that the hardware for all slaves will be identical. There is no noise interference.

b2) Show an example communication between the master and a single-chip 6811/6812 slave. Clearly identify the slave address, and force data components. Label the time axis.

c) Include all the software that will exist in each slave. Your solution will be segmented into three parts: RAM, EEPROM, and ROM. Specific software requirements include:

■ Other than a one-time initialization, there will be no foreground (main) program
■ You may use the 6811/6812 ADC continuous scan mode
■ Clearly show where the 6811/6812 is to begin execution on a power on reset
■ Include a mechanism to calibrate if you decide to handle the offset adjustment in software

c1) Show the software that goes in the RAM (uninitialized on power up).
c2) Show the software that goes in the EEPROM (nonvolatile, but can be different for each slave).
c3) Show the software that goes in the ROM (nonvolatile, and must be the same for each slave).

12.5 The objective of this problem is to design a DAS alarm. An alarm should sound when noise is present in the room. Assume the output of the microphone, $x(t)$, is a ± 10-mV differential signal. The signals of interest are 100 to 500 Hz. You will sample the sound and calculate the sound energy once a second (sum of the signal squared). Be careful to subtract off the DC so that a quiet room (microphone = 0) results in an energy calculation of zero. If data is the 8-bit ADC, then $x(n) =$ data $- 128$ represents the current 8-bit signed sample. If you sample at 1000 Hz, sound the alarm if $(x(0) + x(1) + \ldots + x(999))$ greater than 100,000. *32-bit-long integer operations are allowed.* The 6811/6812 system has a 2-MHz E clock. The 6812 system has an 8-MHz E clock. The alarm can be activated by driving 20 mA through a 5-V EM relay. The software should turn on the alarm if the energy level is above a threshold (simply pick any constant). Once on, the alarm should continue until the operator types the code "213" on the keypad. This 6811/6812 will have a secret code of 213, but allow each 6811/6812 to have a different three-number code and a different threshold. The computer is dedicated to this task.

a) Show the analog interface between the differential microphone output and the 6811/6812 ADC channel 2. Choose the appropriate op amp circuit, gain, offset, and analog filter. The Analog Devices AD680 is a three-wire low-cost 2.5-V reference ($+5$ V input, ground $+2.500$ V output). This is a *qualitative* and not a quantitative DAS.

b) Design the interface between the 6811/6812 signal B0 and the EM relay (Figure 12.60). To activate the alarm the relay coil requires a voltage between $+3.5$ and 6.0 V and a current of 20 mA.

Figure 12.60
The alarm is controlled by an EM relay.

c) Show the interface between the keypad and the 6811/6812 (Figure 12.61). Each switch will bounce for about 10 ms, so you must implement a way to debounce. No other 6811/6812 connections may be used for this interface (although you may use the internal features of input capture/output compare). Your solution should minimize cost. *Periodic polling* should be used for this keypad because it is cheaper.

d) Include all the software that will exist in the system. Your solution will be segmented into three parts: RAM, EEPROM, and ROM. Specific software requirements include:

■ Other than a one-time initialization, there will be no foreground (main) program
■ Use the ADC continuous scan mode to simplify the software
■ Clearly show where the 6811/6812 is to begin execution on a power on reset

Figure 12.61
There are four switches
in the keypad.

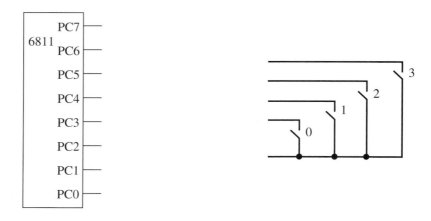

d1) Show the software that goes in the RAM (uninitialized on power up).

d2) Show the software that goes in the EEPROM (nonvolatile, but can be different for each device).

d3) Show the initialization (ritual) software that goes in the ROM (nonvolatile, and must be the same for each device). This is where your system starts executing. Assume the two interrupt vectors are set by the compiler.

d4) Show the 1-kHz DAS interrupt handler that goes in the ROM (nonvolatile, and must be the same for each device). It is here that you will turn on the alarm if the sound energy goes above threshold.

d5) Show the keypad periodic polling interrupt handler that goes in the ROM (nonvolatile, and must be the same for each device). It is here that you will turn off the alarm if the types in the secret code.

12.6 This problem deals with the classification and reduction of noise.

a) Describe a *single experimental procedure* (measurement) that could *identify* (or differentiate) the *type*(s) of *noise* existing on the circuit.

b) For each of these three types of noise (white noise, 60-Hz noise, or 1/f noise), give a typical outcome of the experimental procedure.

c) Give one approach (other than analog or digital filtering) that will reduce white noise.

d) Give one approach (other than analog or digital filtering) that will reduce 60-Hz noise.

e) Give one approach (other than analog or digital filtering) that will reduce 1/f noise.

12.7 A temperature transducer has the relationship

$$R = 200 + 10\,T \qquad \text{where R is in } \Omega \text{ and T is in } °C.$$

The problem specifications are

Range is $30 \le T \le 50°C$
Resolution is $0.01°C$
Frequencies of interest are 0 to 100 Hz
Transducer dissipation constant is 20 mW/°C
ADC range is 0 to +5 V
Sampling rate is 1000 Hz

a) How many ADC bits are required?

b) What is the maximum allowable noise at the amplifier *output*?

c) Design the *analog amplifier/filter*.

12.8 Design an EKG amplifier interface to a 6811/6812 (Figure 12.62). The biopotential is a ± 5-mV signal measured between the left and right arms using silver–silver chloride electrodes. A third electrode is placed on the right leg and is used as a reference. To make the entire system battery-operated, low-power rail-to-rail TLC2274 op amps will be used, each powered with +5 and ground. Notice that the analog system uses the 2.5-V signal as its reference.

Figure 12.62
Block diagram of the
EKG instrumentation.

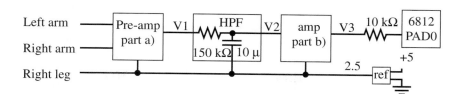

a) Design the preamplifier that has the following characteristics: differential input, gain=10, good CMRR, high Zin, low Zout, bandwidth > 200 Hz. Show resistor values, but not pin numbers.

b) Show the amplifier stage that has the following characteristics: single input, gain=250, high Zin, low Zout, bandwidth > 200 Hz. Show resistor values, but not pin numbers.

c) The signals of interest are 0.1 to 100 Hz. There is unwanted noise at 120, 240, 360, and 480 Hz. Discuss the trade-offs between the following two options:

    c1) Add an active analog low-pass filter with a 100-Hz cutoff, and sample at 200 Hz.

    c2) Sample at 1920 Hz and add a digital 120-, 240-, 360-, 480-Hz reject filter.

Which approach would you take and why? Under what conditions would the other choice be better?

12.9 List eight parameters that we can use to characterize a transducer.

12.10 Design a wind direction measurement instrument using the 8-bit ADC. You are given a transducer with a resistance that is linearly related to the wind direction. As the wind direction varies from 0 to 360 degrees, the transducer resistance varies from 0 to 1000 Ω. The frequencies of interest are 0 to 0.5 Hz, and the sampling rate will be 1 Hz. (See Exercise 6.6.)

a) Show the analog interface between the transducer and the ADC port channel 7. Only the +5-V supply can be used. Show how the analog components are powered. Give chip numbers, but not pin numbers. Specify the type and tolerance of resistors and capacitors.

b) Write the ritual and gadfly function/subroutine that measures the wind direction and returns a 16-bit unsigned result with units of degrees. That is, the value varies from 0 to 359. (You do not have to write software that samples at 1 Hz, simply a function that measures wind direction once.)

12.11 Design a position DAS using a sensor, analog electronics and a Freescale microcomputer with an 8-bit ADC. Let **x** be the position to be measured (Figure 12.63). The input range is −1 to +1 mm, and the signals of interest are 0 to 1 Hz. A LVDT will be used to convert position **x** into voltage **y**. When the input position is −1 mm, the LVDT output **y** is −100 mV. When the input position is zero, the LVDT output, **y**, is zero. When the input position is +1 mm, the LVDT output **y** is +100 mV. In between, the voltage output is linearly related to the position. A REF02 precision reference will provide the constant +5.00 V for your analog circuit and the microcomputer ADC.

Figure 12.63
Block diagram of the
position measuring
system.

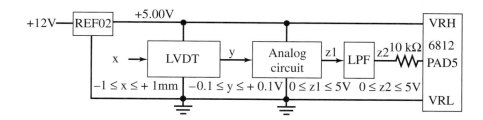

a) What is the transducer sensitivity? Give units.

b) What is the transfer relationship required to convert **y** into **z1**?

c) What is the maximum allowable noise for your analog circuit? Refer the noise to the amplifier input and give units.

    d) Choose an appropriate sampling rate.

    e) Design the preamplifier which has the following characteristics: single input (not differential), gain so that the 0 to +5 V ADC is used, high Zin, low Zout, bandwidth > 200 Hz. Show resistor values, but not pin numbers. You may use any analog op amps, but please include the chip number.

    f) The signals of interest are 0 to 1 Hz. There is unwanted noise at 60 Hz. Add a two-pole low-pass filter with a cutoff of about 10 Hz.

    g) What is the system resolution in millimeters?

12.12 Design an electronic scale using a Freescale microcomputer with an 8-bit ADC (Figure 12.64). Let **x** be in mass to be measured. The input range is 0 to 1 kg and the signals of interest are 0 to 10 Hz. A bonded strain gage bridge will be used to convert mass **x** into voltage, V1 − V2. When the input mass is zero, each arm of the bridge is 100 Ω, and the bridge output (V1 − V2) is zero. At full scale (**x** = 1 kg), two resistors go to 99 Ω and the other two go to 101 Ω. In between 0 and 1 kg, the resistance change is linearly related to the mass:

$$\Delta R = x \qquad \text{where resistance is in ohms and the mass } \mathbf{x} \text{ is in kilograms}$$

Figure 12.64
Block diagram of the mass measuring system.

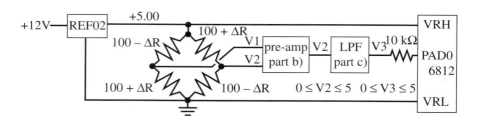

A REF02 precision reference will provide the constant +5.00 V for the bridge and the ADC.

    a) What is the bridge output (V1 − V2) at full scale (**x** = 1 kg)? What gain is required to match the full range of 0 ≤ **x** ≤ 1kg to the 0 to +5 V range of the ADC?

    b) Design the preamplifier that has the following characteristics: differential input, gain so that the 0 to +5 V ADC is used, good CMRR, high Zin, low Zout, bandwidth > 200 Hz. Show resistor values, but not pin numbers.

    c) The signals of interest are 0 to 10 Hz. There is unwanted noise at 60 Hz. Add a two-pole low-pass filter with a cutoff of 20 Hz.

    d) What is the system resolution? Give units.

    e) What sampling rate would you choose? Explain your answer.

12.7 Lab Assignments

Lab 12.1. This experiment will use a thermistor and the ADC converter on the 6811/6812 to construct a digital thermometer. The temperature range should be 20 to 40°C. If the current temperature is above the upper limit in the specified range, a red LED should be turned on. You can test this feature by shorting the thermistor leads together (zero resistance). If it is below the lower limit of the specified range, a yellow LED should be turned on. Similarly, you can test this feature by disconnecting one wire of the thermistor (infinite resistance). Otherwise, a green LED will stay on, indicating that the temperature is within the specified range. The temperature measurements will be displayed as fixed-point numbers on an LCD. Design your system with the best possible resolution. The temperature component is 0 to 1 Hz. Experimentally verify the noise level, time-constant, and accuracy of your system.

Lab 12.2. The objective of this lab is to build a digital sound recorder for human speech. You will first need to interface an external RAM to store the data. Next, you will need to design an analog circuit that interfaces a microphone to the ADC of the 6811/6812. Investigate the frequency components of human speech and design your system to pass these frequencies. You will also need a mechanism to play back the recorded sound, so design an audio amplifier that interfaces a DAC to a speaker. Your human interface should include buttons to trigger sound recording, stop recording, start playback, and stop playback.

Lab 12.3. Design a thermocouple-based thermometer with a range of 0 to 50°C. You can use an ice bucket for the reference, or you could design the thermistor-based thermometer of Lab 12.1 and use it to compensate for the cold junction of the thermocouple. Design your system with the best possible resolution. The temperature measurements will be displayed as fixed-point numbers on an LCD. The temperature component is 0 to 1 Hz. Experimentally verify noise level, time-constant, and accuracy of your system.

Lab 12.4. Design a digital scale using a force transducer. Select the range of the scale to match the linear range of your force transducer. You can build a force transducer using a slide pot and a spring. Design your system with the best possible resolution. The force measurements will be displayed as fixed-point numbers on an LCD. Experimentally verify the noise level, time-constant, and accuracy of your system.

Lab 12.5. Design two digital position measurement systems using a slide potentiometer as the transducer. Select the range of the scale to match the linear range of your transducer. The position measurements will be displayed as fixed-point numbers on an LCD. The first system will use the ADC, and the second system will use an astable multivibrator (LM555) and input capture. Design your systems with the best possible resolution. Experimentally verify the noise level, time-constant, and accuracy of both systems.

Lab 12.6. Design an autoranging voltmeter. The three ranges are 0 to 2V, 0 to 0.2V, and 0 to 0.02V. The hardware/software system automatically adjusts the range providing the best possible measurement resolution. Display both the numerical and graphical results on an LCD display. The HD44780 LCD allows you to define up to 8 new character images (ASCII codes 0 to 7). By dynamically setting these 8 by 5 bit characters, you can create an 8-bit high by 40-bit wide graphical image. Write a graphical device driver for the LCD and use it to graph the time-varying voltage in real time. Experimentally determine the measurement resolution for each range and compare it to the expected theoretical resolution. Analyze the various sources of measurement error in your system.

13 Microcomputer-Based Control Systems

Chapter 13 objectives are to:

❏ Introduce the general approach to digital control systems
❏ Design and implement some simple open-loop control systems
❏ Design and implement some simple closed-loop control systems
❏ Develop a methodology for designing PID control systems
❏ Present the terminology and give examples of fuzzy logic control systems

In the last chapter, we developed systems that collected information concerning the external environment. In this chapter we will use this information to control the external environment. To build this microcomputer-based control system we will need an output device that the computer can use to manipulate the external environment. If we review the list of applications introduced in Section 1.1, we find that many involve control systems. Control systems originally involved just analog electronic circuits and mechanical devices. With the advent of inexpensive yet powerful microcomputers, implementing the control algorithm in software provided a lower cost and a more powerful product. The goal of this chapter is to provide a brief introduction to this important application area. Control theory is a richly developed discipline, and most of the theory is beyond the scope of this book. Consequently, this chapter focuses more on implementing the control system with an embedded computer and less on the design of the control equations.

13.1 Introduction to Digital Control Systems

A *control system* is a collection of mechanical and electric devices connected for the purpose of commanding, directing, or regulating a *physical plant*. The *real state variables* are the properties of the physical plant that are to be controlled. The *sensor* and *state estimator* comprise a DAS, as discussed in Chapter 12. The goal of this DAS is to estimate the state variables. A *closed-loop* control system uses the output of the state estimator in a feedback loop to drive the system to a desired state. The control system compares these *estimated state variables* X′(t) to the *desired state variables* X*(t) to decide appropriate action U(t). The *actuator* is a transducer that converts the control system commands U(t) into driving forces

V(t) that are applied to the physical plant. In general, the goal of the control system is to drive the real state variables to equal the desired state variables. In actuality though, the controller attempts to drive the estimated state variables to equal the desired state variables. It is important to have an accurate state estimator, because any differences between the estimated state variables and the real state variables will translate directly into controller errors. If we define the error as the difference between the desired and estimated state variables,

$$e(t) = X^*(t) - X'(t)$$

then the control system will attempt to drive e(t) to zero. In general control theory, X(t), X'(t), X*(t), U(t), V(t), and e(t) refer to vectors (e.g., x, y, z position or Vx, Vy, Vz velocity), but the examples in this chapter control only a single parameter. The focus of this book is the microcomputer interfacing, and it should be straightforward to apply standard multivariate control theory to more complex problems. We usually evaluate the effectiveness of a control system by determining three properties: steady-state controller error, transient response, and stability. The steady-state controller error is the average value of e(t). The transient response is how long the system takes to reach 99% of the final output after X* is changed. A system is stable if a steady state (smooth constant output) is achieved. An unstable system oscillates (Figure 13.1).

Figure 13.1
Block diagram of a microcomputer-based closed-loop control system.

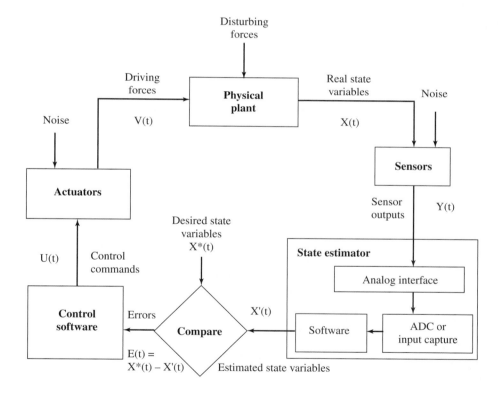

13.2 Open-Loop Control Systems

An open-loop control system does not include a state estimator. It is called open loop because there is no feedback path providing information about the state variable to the controller. It will be difficult to use open loop with the plant that is complex because the disturbing forces will have a significant effect on controller error. On the other hand, if the plant is well-defined and the disturbing forces have little effect, then an open-loop approach may be feasible. (Figure 13.2).

Figure 13.2
Block diagram of a
microcomputer-based
open-loop control
system.

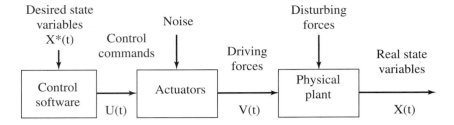

13.2.1
Open-Loop
Control of a
Toaster

This first example is a simple toaster (Figure 13.3). It is an open-loop control system because the control signal (heat applied to the bread) is independent of the state variable (taste of the toast). The controller simply applies heat for a fixed amount of time. If we provide an adjustable heat cycle to our toaster and allow the humans to adjust the heating cycle according to their taste, then the toast/humans combination becomes a closed-loop control system. Because an open-loop control system does not know the current values of the state variables, large errors can occur.

Figure 13.3
A simple open-loop
control system.

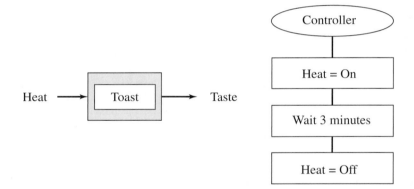

13.2.2
Open-Loop
Robotic Arm

The overall objective is to control the rotational orientation of the tip of a three-axis robotic arm, shown in Figure 13.4. The three parts of rotational orientation are called pitch, roll, and yaw. For this simple robotic arm, there are independent stepper motors controlling each axis. Although a typical robot arm would have more motors, and each

Figure 13.4
A three-axis open-loop
stepper motor control
system.

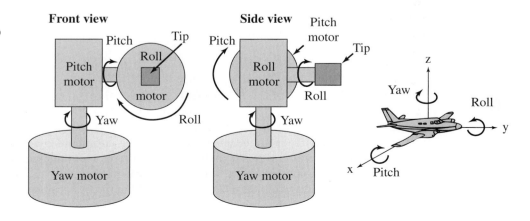

motor would affect both translational and rotational motion, the design process illustrated in this example can be used for a wide range of open-loop controllers. Furthermore, this implementation can easily be extended into a closed-loop system by adding sensors and using the sensor information to adjust the command inputs.

Pitch is rotation around the lateral or transverse axis (x). For an airplane, this axis is parallel to the wings, thus the nose and tail can both pitch up or down. **Roll** is rotation around the longitudinal axis (y). For an airplane, this axis is drawn through the body of the vehicle from tail to nose. **Yaw** is rotation about the axis normal to the ground (z), which is perpendicular to the pitch and roll axes. If an airplane were placed on a flat surface and pivoted about the center of mass (coordinate origin), the motion would be described as yawing.

If the motors are strong enough, the system can be controlled in an open loop fashion. For safety considerations, open-loop systems are not appropriate. An industrial robot needs feedback to prevent the robot from damaging itself or its environment. Safety features can easily be added to this implementation. There are four digital signals the computer uses to control each stepper motor. If the software outputs the sequence 6,5,10,9,6,5,10,9 . . . the motor will spin. For each change in output (6 to 5, 5 to 10, 10 to 9 or 9 to 6), the motor turns 1.8 degrees. Therefore, it takes 200 outputs to turn the shaft one complete rotation. The goal of the system is to simultaneously control the position and speed on three stepper motors. The acceleration and jerk parameters will not be explicitly controlled. Again, without sensors to measure the shaft position, this system is an open-loop controller. The control signals (power to the stepper coils) are independent of the state variables (shaft positions). For low torque applications (when the motor does not skip), the software can keep track of the shaft position without a sensor. For discussions of interfacing and minimizing jerk in the stepper motor section, refer back to Chapter 8.

In the **analysis phase**, we determine the requirements and constraints for our proposed system. For this open-loop system, the three motors must perform the actions specified by the speed and duration, where the acceleration and jerk are not specified. The ten-second pattern will be repeated continuously.

Yaw	Pitch	Roll
1 RPM, 1 sec	0 RPM, 2 sec	0 RPM, 1 sec
−1 RPM, 1 sec	1 RPM, 2 sec	1 RPM, 2 sec
2 RPM, 0.5 sec	−2 RPM, 1 sec	−4 RPM, 0.5 sec
−0.5 RPM, 2 sec	0 RPM, 5 sec	0 RPM, 6.5 sec
0 RPM, 5.5 sec		

Checkpoint 13.1: The 3-axis robot described in Figure 13.4 also generates some translational motion. Describe qualitatively the (x,y,z) position of the tip as a function of the yaw, pitch, and roll angles.

In the **high-level design** phase, we define our input/output, break the system into modules, and show the interconnection using data flow graphs. In this system we will have a central controller, executed using periodic interrupts. A data flow graph is shown in Figure 13.5, where the input is the desired speed of each motor and the outputs are 12 digital outputs, four for each stepper motor. The Yaw and Pitch motors are interfaced to Port T, and the Roll motor is connected to PM3-0. The high-level master controller will perform commands such as rotate at 1 RPM for 1 second, and the three low-level controllers will perform the commands step clockwise and step counterclockwise. There will be a `Time` and `Direction` for each motor. `Time` will be the number of msec in between outputs to the stepper motor. To design the low-level controller, we must consider the time in between outputs for all the desired speeds. For the fastest speed of 4 RPM, there will be

$$1000(\text{ms/s})*60(\text{s/min})/200(\text{steps/rev})/4(\text{rev/min}) = 75 \text{ ms per step.}$$

Similarly, the speeds of 2, 1, 0.5 RPM will require a **Time** of 150, 300, and 600 respectively. A **Time** of zero will be defined as stop. **Direction** specifies the motion as clockwise or counterclockwise.

Figure 13.5
Data flow graph for the control system.

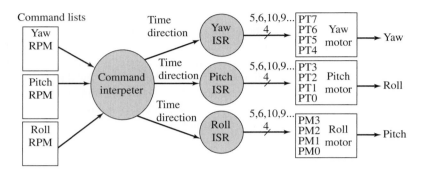

In the **engineering design** phase, we design the hardware/software subcomponents using techniques such as simulation, mechanical mockups, and call-graphs. The weights of the motors and tips together with the lengths of the shafts will determine the required torques of the stepper motors. The required torque should also include friction forces on the shaft. The electrical interface of stepper motors was presented previously in Chapter 8. To design the software controller, we first determine the time-interval between high-level commands, which is 0.5 second in this example, Next, the command specifications are rewritten as a list of nine operations, each defined by a speed, direction, and duration. The other alternative would have been to divide the 10-second pattern into 20 equally-sized 0.5-sec intervals, and specify speed and direction for each interval.

Interval	Yaw			Pitch			Roll		
0.0 to 1.0 sec	1 RPM	300 ms	CW	0 RPM	0 ms	—	0 RPM	0 ms	—
1.0 to 2.0 sec	1 RPM	300 ms	CCW	0 RPM	0 ms	—	1 RPM	300 ms	CW
2.0 to 2.5 sec	2 RPM	150 ms	CW	1 RPM	300 ms	CW	1 RPM	300 ms	CW
2.5 to 3.0 sec	0.5 RPM	600 ms	CCW	1 RPM	300 ms	CW	1 RPM	300 ms	CW
3.0 to 3.5 sec	0.5 RPM	600 ms	CCW	1 RPM	300 ms	CW	4 RPM	75 ms	CCW
3.5 to 4.0 sec	0.5 RPM	600 ms	CCW	1 RPM	300 ms	CW	0 RPM	0 ms	—
4.0 to 4.5 sec	0.5 RPM	600 ms	CCW	2 RPM	150 ms	CCW	0 RPM	0 ms	—
4.5 to 5.0 sec	0 RPM	0 ms	—	2 RPM	150 ms	CCW	0 RPM	0 ms	—
5.0 to 10.0 sec	0 RPM	0 ms	—	0 RPM	0 ms	—	0 RPM	0 ms	—

During the **implementation** and **testing** phases, we build and test the modules. Modularity allows for concurrent development. An important factor when designing data structures is to establish a one-to-one linkage between the foregoing abstract command table and the implementation of the data structure. Program 13.1 shows the implementation of the high-level command table. This one-to-one linkage allows us to separate the testing of the low-level functions (spinning motors) from the high-level control functions.

> ***Observation:*** Open-loop controllers that have simple outputs like this one can also be implemented as a finite state machine with no inputs.

Program 13.1
Table structure used to define high-level commands.

```
#define CW   0
#define CCW 1
#define RPM0    0  // means stop
#define RPM0_5 8  // 8*75ms=600ms => 0.5 RPM
#define RPM1    4  // 4*75ms=300ms => 1 RPM
#define RPM2    2  // 2*75ms=150ms => 2 RPM
#define RPM4    1  // 1*75ms=75ms  => 4 RPM
const struct Command{
  unsigned short Duration;       // time for the command, 0.5s units
  unsigned short YawTime;        // determines Yaw speed, 75ms units
  unsigned short YawDirection;   // 0 for CW, 1 for CCW
  unsigned short PitchTime;      // determines Pitch speed, 75ms units
  unsigned short PitchDirection;
  unsigned short RollTime;       // determines Roll speed, 75ms units
  unsigned short RollDirection;  // 0 for CW, 1 for CCW
  const struct Command *Next;    // circular link
};
typedef const struct Command CommandType;
CommandType Machine[9]={
// Duration Yaw          Pitch       Roll
  {2,        RPM1,   CW, RPM0,   0, RPM0,   0, &Machine[1]},
  {2,        RPM1,  CCW, RPM0,   0, RPM1,  CW, &Machine[2]},
  {1,        RPM2,   CW, RPM1,  CW, RPM1,  CW, &Machine[3]},
  {1,        RPM0_5, CCW, RPM1, CW, RPM1,  CW, &Machine[4]},
  {1,        RPM0_5, CCW, RPM1, CW, RPM4, CCW, &Machine[5]},
  {1,        RPM0_5, CCW, RPM1, CW, RPM0,   0, &Machine[6]},
  {1,        RPM0_5, CCW, RPM2, CCW, RPM0,  0, &Machine[7]},
  {1,        RPM0,    0, RPM2, CCW, RPM0,   0, &Machine[8]},
  {10,       RPM0,    0, RPM0,   0, RPM0,   0, &Machine[0]}};
CommandType *CmdPt;   // Current command
```

Figure 13.6 shows the flowcharts used to implement the robot controller. The high-level controller is implemented with a 0.5-second periodic output compare 0 interrupt. The global High-Count counter implements the high-level delay for each command. The low-level controller is implemented with a 75-ms periodic output compare 1 interrupt. Figure 13.6 just shows the algorithm for the Yaw motor, but the Pitch and Roll functions are identical. The global Yaw-Count PitchCount RollCount counters implement the low-level delays for each motor.

Figure 13.6
Flowchart and linked structure for the control system.

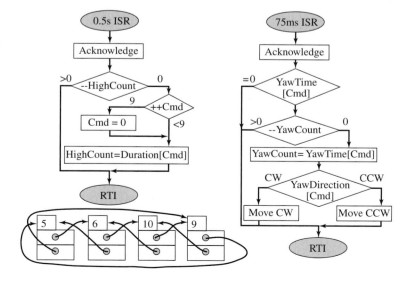

Checkpoint 13.2: Explain how this system could have been designed using just one ISR.

Program 13.2 contains a doubly-circular linked list used to define the stepper motor sequence. The three global pointers **(YawPt PitchPt RollPt)** specify the current motor state for each motor. Both high and low nibbles data are included to speed execution (eliminating a shift operation).

Program 13.2
Linked list structure used to define low-level stepper outputs.

```
const struct StepperState{
  unsigned char OutHigh;  // Output for motors with bits 7-4
  unsigned char OutLow;   // Output for motors with bits 3-0
  const struct StepperState *Next[2]; // CW or CCW
};
typedef const struct StepperState StepperStateType;
StepperStateType Stepper[4]={
  { 0x50,0x05,{&Stepper[1],&Stepper[3]}},
  { 0x60,0x06,{&Stepper[2],&Stepper[0]}},
  { 0xA0,0x0A,{&Stepper[3],&Stepper[1]}},
  { 0x90,0x09,{&Stepper[0],&Stepper[2]}}
};
StepperStateType *YawPt;    // Current Stepper state for Yaw motor
StepperStateType *PitchPt;  // Current Stepper state for Pitch motor
StepperStateType *RollPt;   // Current Stepper state for Roll motor
```

Program 13.3 contains the initialization and the two interrupt service routines used to execute the controller. The global Cmd is an index into the Machine[] table indicating which command is active. It is the primary communication between the high-level and low-level controllers.

Program 13.3
C software to spin three stepper motors.

```
unsigned short HighCount,YawCount,PitchCount,RollCount;
interrupt 8 void TC0handler(void){
  TFLG1 = 0x01;           // acknowledge C0F
  TC0 = TC0+62500U;       // 0.5sec period
  if(--HighCount==0){     // done?
    CmdPt = CmdPt->Next;  // next command
    HighCount = CmdPt->Duration;
  }
}
interrupt 9 void TC1handler(void){
  TFLG1 = 0x02;           // acknowledge C1F
  TC1 = TC1+9375;         // 75ms period
  if(CmdPt->YawTime){
    if(--YawCount == 0){
      YawCount = CmdPt->YawTime;
      YawPt = YawPt->Next[CmdPt->YawDirection];
      PTT = (PTT&0x0F)+YawPt->OutHigh; //set Port T bits 7-4
    }
  }
  if(CmdPt->PitchTime){
    if(--PitchCount == 0){
      PitchCount = CmdPt->PitchTime;
      PitchPt = PitchPt->Next[CmdPt->PitchDirection];
      PTT = (PTT&0xF0)+PitchPt->OutLow; //set Port T bits 3-0
    }
  }
}
```

```
    if(CmdPt->RollTime){
      if(--RollCount == 0){
        RollCount = CmdPt->RollTime;
        RollPt = RollPt->Next[CmdPt->RollDirection];
        PTM = (PTM&0x30)+RollPt->OutLow; //set Port M bits 3-0
      }
    }
  }
}
void Motor_Init(void){ // assumes 4MHz E clock
  PTT = 0x55;         // start at 5, first output moves motor
  PTM = 0x05;
  DDRT = 0xFF;        // PT7-4 is yaw, PT3-0 is pitch
  DDRM |= 0x0F;       // PM3-0 is roll
  TIOS |= 0x03;       // activate TC0, TC1 as output compares
  TSCR1 = 0x80;       // Enable TCNT, 8us
  TSCR2 = 0x05;       // divide by 32 TCNT prescale, TOI disarm
  PACTL = 0;          // timer prescale used for TCNT
  CmdPt = &Machine[8]; // first command will be 0
  YawPt = PitchPt = RollPt = &Stepper[0];
  HighCount = YawCount = PitchCount = RollCount = 1;
  TIE  |= 0x03;       // arm OC1, OC0
  TC0   = TCNT+50;    // first interrupt right away
  TC1   = TC0+50;     // interrupts after TC0
  asm cli
}
```

Checkpoint 13.3: Cmd is a shared global variable with read and write activity. Does this activity cause a critical section?

13.3 Simple Closed-Loop Control Systems

13.3.1 Bang-Bang Temperature Control

This digital control system applies heat to the room to maintain the temperature as close to T* (labeled Tstar in the software) as possible (Figure 13.7). This is a closed-loop control system because the control signals (heat) depend on the state variables (temperature). In this application, the actuator has only two states: *on,* which warms up the room, and *off,* which does not apply heat. For this application to function properly, there must be a passive heat loss that lowers the room temperature when the heater is turned off. A typical digital control algorithm for this type of actuator is *bang-bang* (Figure 13.8). Other names for bang-bang include *two-position, on-off,* or *binary* controller. A bang-bang controller turns on the power if the temperature is too low and turns off the power if the temperature is too high. To implement hysteresis, we need two set point temperatures, T_{HIGH} and T_{LOW}. The controller turns on the power (activates relay) if the temperature goes below T_{LOW} and turns off the power (deactivates relay) if the temperature goes above T_{HIGH}. The difference $T_{HIGH} - T_{LOW}$ is called hysteresis. The hysteresis extends the life of the relay by reducing the number of times the relay opens and closes.

In this implementation, the RTD is made from a thin platinum wire and converts the room temperature into a resistance. The analog circuit matches the full-scale range of the room temperature to the 0 to +5 V ADC input voltage range. The software converts the ADC result into estimated temperature, T′ (labeled T in the software) (Figure 13.9).

Figure 13.7
Flowchart of a bang-bang temperature controller.

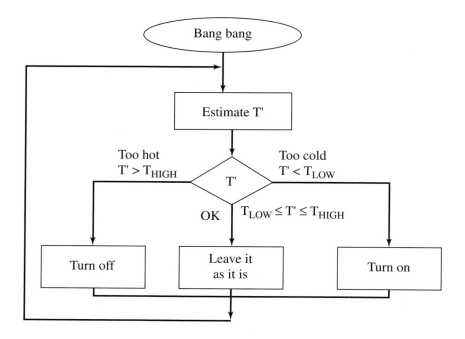

Figure 13.8
Algorithm and response of bang-bang temperature controller.

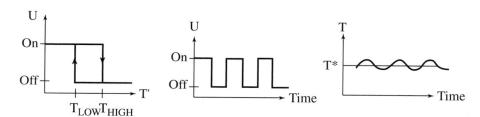

Figure 13.9
Interface of a simple bang-bang temperature controller.

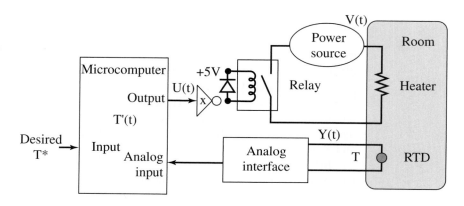

Assume that the function SE() converts the ADC sample into the estimated temperature as a binary fixed-point number with a resolution of 0.5°C. The C code in Program 13.4 uses a periodic interrupt so that the bang-bang controller runs in the background. The interrupt Period is selected to be about 10 times faster than the time constant of the physical plant. The temperature variables Tlow, Thigh, and T could be in any format, as long as the three formats are the same.

Program 13.4
Bang-bang temperature control software.

```
short Tlow,Thigh;      // controller set points, 0.5 C
// PTM bit 0 turns power on/off
interrupt 13 void TC5handler(void){
short T=SE(ADC_In(0)); // estimated temperature, 0.5 C
  if(T < Tlow){
    PTM |= 0x01;       // too cold so on
  }
  else if (T > Thigh){
    PTM &= ~0x01;      // too hot so off
  }                    // leave as is if Tlow<T<Thigh
  TC5 = TC5+Period;    // periodic rate
  TFLG1 = 0x20;        // acknowledge C5F
}
```

Checkpoint 13.4: What happens if Tlow and Thigh are too close together?

Checkpoint 13.5: What happens if Tlow and Thigh are too far apart?

Observation: Bang-bang control works well with a physical plant that has a very slow response.

Observation: The DS1620 interface presented first in Chapter 3 then again in Chapter 7 can be used as the SE function in this example.

13.3.2 Closed-Loop Position Control System Using Incremental Control

The objective of this *incremental control* system is to control the position of the robot arm, X (Figure 13.10). The control signals (power) are dependent on the state variables (position). The LVDT, which was described in Chapter 12, is used to sense the position of the robot arm.

An incremental control algorithm simply adds or subtracts a constant from U depending on the sign of the error. In other words, if X is too small, then U is incremented, and if X is too large, then U is decremented. It is important to choose the proper rate at which the incremental control software is executed. If it is executed too many times per second,

Figure 13.10
Interface of a position controller.

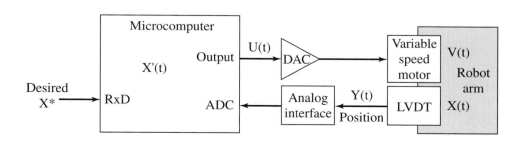

then the actuator will saturate, resulting in a bang-bang system. If it is not executed often enough, then the system will not respond quickly to changes in the physical plant or changes in X*. In this incremental controller we add or subtract "1" from the actuator, but a value larger than "1" would have a faster response at the expense of introducing oscillations (Figure 13.11).

Figure 13.11
Flowchart of a position controller implemented using incremental control.

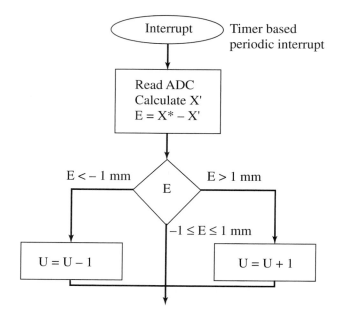

Common error: An error will occur if the software does not check for overflow and underflow after U is changed.

Observation: If the incremental control algorithm is executed too frequently, then the resulting system behaves like a simple bang-bang controller.

Observation: Many control systems operate well when the control equations are executed about ten times faster than the step response time of the physical plant.

Assume that the function SE() converts the ADC sample into position as a signed decimal fixed-point number with a resolution of 0.1 mm. The 6812 C code in Program 13.5 uses a

Program 13.5
Incremental position control software.

```
short X,Xstar,E; // position, fixed-point in 0.1mm
// PTT is connected to the 8-bit DAC
interrupt 13 void TC5handler(void){ short U;
    X = SE(ADC_In(0));    // estimated (0.1 mm)
    E = Xstar-X;          // error (0.1 mm)
    U = (short)PTT;       // promote to 16 bits
    if(E < -10)    U--; // decrease if less than -1mm
    else if(E > 10) U++; // increase if greater than +1mm
                         // leave as is if -1mm<E<1mm
    if(U<0)    U=0;      // underflow
    if(U>255) U=255;     // overflow
    PTT = U;             // output to actuator
    TC5 = TC5+Period;    // periodic rate
    TFLG1 = 0x20;        // acknowledge C5F
}
```

periodic interrupt so that the incremental controller runs in the background. The interrupt `Period` is selected to be about ten times faster than the time constant of the physical plant. Even though the position variables `X` and `Xstar` may be unsigned, the error calculation `E` will be signed.

> *Checkpoint 13.6:* In what ways would the controller behave differently if **-10** and **+10** were to be changed to **0**?

> *Checkpoint 13.7:* What happens if **Period** is too small (i.e., it executes too frequently)?

> *Observation:* It is a good debugging strategy to observe the assembly listing generated by the compiler when performing calculations on variables of mixed types (signed/unsigned, char/short).

> *Observation:* Incremental control will work moderately well (accurate and stable) for an extremely wide range of applications. Its only shortcoming is that the controller response time can be quite slow.

13.4 PID Controllers

13.4.1 General Approach to a PID Controller

The simple controllers presented in the last section are easy to implement, but they will have either large errors or very slow response times. To make a faster and more accurate system, we can use linear control theory to develop the digital controller. There are three components of a *PID controller.*

$$U(t) = Kp\, E(t) + K_I \int_0^t E(\tau)\, d\tau + K_D\, \frac{dE(t)}{dt}$$

The error, E(t), is defined as the present set-point, X*(t), minus the measured value of the controlled variable, X'(t).

$$E(t) = X^*(t) - X'(t)$$

The PID controller calculates its output by summing three terms. The first term is proportional to the error. The second is proportional to the integral of the error over time, and the third is proportional to the rate of change (first derivative) of the error term. The values of K_P, K_I, and K_D are design parameters and must be properly chosen for the control system to operate properly. The proportional term of the PID equation contributes an amount to the control output that is directly proportional to the current process error. The gain term K_P adjusts exactly how much the control output response should change in response to a given error level. The larger the value of K_P, the greater the system reaction to differences between the set-point and the actual temperature. However, if K_P is too large, the response may exhibit an undesirable degree of oscillation or even become unstable. On the other hand, if K_P is too small, the system will be slow or unresponsive. An inherent disadvantage of proportional-only control is its inability to eliminate the steady-state errors (offsets) that occur after a set-point change or a sustained load disturbance.

The integral term converts the first-order proportional controller into a second-order system capable of tracking process disturbances. It adds to the controller output a factor that takes corrective action for any changes in the load level of the system. This integral term is scaled to the sum of all previous process errors in the system. As long as there is a process error, the integral term will add more amplitude to the controller output until the sum of all previous errors is zero. Theoretically, as long as the sign of K_I is correct, any value of K_I will eliminate offset errors. But, for extremely small values of K_I, the controlled variables will return to the setpoint very slowly after a load change or setpoint change occurs. On the other hand, if K_I is too large, it tends to produce oscillatory response of the controlled process and reduces system stability. The undesirable effects of too much integral action can be avoided by proper tuning (adjusting) the controller or by including derivative action which tends to counteract the destabilizing effects.

Checkpoint 13.8: What happens in a PID controller if the sign of K_I is incorrect?

The derivative action of a PID controller adds a term to the controller output scaled to the slope (rate of change) of the error term. The derivative term "anticipates" the error, providing a greater control response when the error term is changing in the wrong direction and a dampening response when the error term is changing in the correct direction. The derivative term tends to improve the dynamic response of the controlled variable by decreasing the process setting time, the time it takes the process to reach steady state. But if the process measurement is noisy, that is, if it contains high-frequency random fluctuations, then the derivative of the measured (controlled) variable will change wildly, and derivative action will amplify the noise unless the measurement is filtered.

We also can use just some of the terms. For example, a proportional/integral controller drops the derivative term. We will analyze the digital control system in the frequency domain. Let $X(s)$ be the Laplace transform of the state variable $x(t)$, let $X^*(s)$ be the Laplace transform of the desired state variable $x^*(t)$, and let $E(s)$ be the Laplace transform of the error. Because the system is linear

$$E(s) = X^*(s) - X(s)$$

Let $G(s)$ be the transfer equation of the PID linear controller. PID controllers are unique in this aspect. In other words, we cannot write a transfer equation for a bang-bang, incremental, or fuzzy logic controller, but for a PID controller we have

$$G(s) = c\left(k_P + k_Ds + \frac{k_I}{s}\right)$$

Let $H(s)$ be the transfer equation of the physical plant. If we assume the physical plant (e.g., a DC motor) has a simple single-pole behavior, then we can specify its response in the frequency domain with two parameters. m is the DC gain and τ is its time constant. The transfer function of a single-pole plant is

$$H(s) = \frac{m}{1 + \tau \cdot s}$$

The overall gain of the control system (Figure 13.12) is

$$\frac{X(s)}{X*(s)} = \frac{G(s)H(s)}{1 + G(s)H(s)}$$

Theoretically we can choose controller constants k_P, k_I, and k_D to create the desired controller response. Unfortunately it can be difficult to estimate c, m, and τ. If the load to the motor varies, then m and τ will change.

Figure 13.12
Block diagram of a linear control system in the frequency domain.

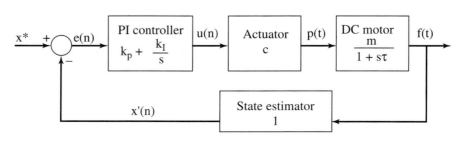

To simplify implementation of the PID controller, we break the controller equation into separate proportional, integral, and derivative terms. That is, let

$$U(t) = P(t) + I(t) + D(t)$$

where $U(t)$ is the actuator output and $P(t)$, $I(t)$, and $D(t)$ are the proportional, integral, and derivative components, respectively. The proportional term makes the actuator output linearly related to the error. Using a proportional term creates a control system that applies more energy to the plant when the error is large.

$$P(t) = K_p \cdot E(t)$$

To implement the proportional term we simply convert the above equation into discrete time.

$$P(n) = K_p \cdot E(n)$$

where the index n refers to the discrete time input of $E(n)$ and output of $P(n)$.

> **Observation:** To develop digital signal-processing equations, it is imperative that the control system be executed on a regular and periodic rate.

> **Common error:** If the sampling rate varies, then controller errors will occur.

The integral term makes the actuator output related to the integral of the error. Using an integral term often will improve the steady-state error of the control system. If a small error accumulates for a long time, this term can get large. Some control systems put upper and lower bounds on this term, called *anti-reset-windup*, to prevent it from dominating the other terms:

$$I(t) = K_I \cdot \int_0^t E(\tau)d\tau$$

The implementation of the integral term requires the use of a discrete integral or sum. If $I(n)$ is the current control output, and $I(n-1)$ is the previous calculation, the integral term is simply

$$I(n) = K_I \cdot \sum_1^n (E(n) \cdot \Delta t) = I(n-1) + K_I \cdot E(n) \cdot \Delta t$$

where Δt is the sampling rate of $E(n)$.

The derivative term makes the actuator output related to the derivative of the error. This term is usually combined with either the proportional and/or integral term to improve the transient response of the control system. The proper value of K_D will provide for a quick response to changes in either the set point or loads on the physical plant. An incorrect value may create an overdamped (very slow response) or an underdamped (unstable oscillations) response.

$$D(t) = K_D \cdot \frac{dE}{dt}$$

There are a couple of ways to implement the discrete time derivative. The simple approach is

$$D(n) = K_D \cdot \frac{E(n) - E(n-1)}{\Delta t}$$

In practice, this first-order equation is quite susceptible to noise. Figure 13.13 shows a sequence of $E(n)$ with some added noise. Notice that huge errors occur when the above equation is used to calculate derivative.

Figure 13.13
Illustration of the effect
noise plays on the
calculation of discrete
derivative.

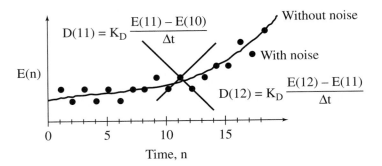

In most practical control systems, the derivative is calculated using the average of two derivatives calculated across different time spans. For example,

$$D(n) = K_D \cdot \left[\frac{1}{2} \frac{E(n) - E(n-3)}{3\Delta t} + \frac{1}{2} \frac{E(n-1) - E(n-2)}{\Delta t} \right]$$

which simplifies to

$$D(n) = K_D \cdot \frac{E(n) + 3E(n-1) - 3E(n-2) - E(n-3)}{6\Delta t}$$

Checkpoint 13.9: How is the continuous integral related to the discrete integral?

Checkpoint 13.10: How is the continuous derivative related to the discrete derivative?

**13.4.2
Design Process
for a PID
Controller**

The first design step is the analysis phase, where we determine specifications such as range, accuracy, stability, and response time for our proposed control system. A data acquisition system will be used to estimate the state variables. Thus, its range, accuracy, and response time must be better than the desired specifications of the control system. We can use time-based techniques from Chapter 6, or develop an ADC-based state estimator using the techniques of Chapters 11 and 12. In addition, we need to design an actuator to manipulate the state variables. It too must have a range and response time better than the controller specifications. The **actuator resolution** is defined as the smallest reliable change in output. For example, a 100 Hz PWM output generated by a 1 μsec clock has 10,000 different outputs. For this actuator, the actuator resolution is MaxPower/10000. We wish to relate the actuator performance to the overall objective of controller accuracy. Thus, we need to map the effect on the state variable caused a change in actuator output equal to 1 resolution. This change in state variable should be less than or equal to the desired controller accuracy.

After the state estimator and actuator are implemented, the controller settings (K_P, K_I, and K_D) must be adjusted so that the system performance is satisfactory. This activity is referred to as **controller tuning** or **field tuning**. If you perform controller tuning by guessing the initial setting and then adjusting them by trial and error, it can be tedious and time consuming. Thus, it is desirable to have good initial estimates of controller settings. A good first setting may be available from experience with similar control loops. Alternatively, initial estimates of controller settings can be derived from the transient response of the physical plant. A simple open-loop method, called the **process reaction curve approach**, was first proposed by Ziegler/Nichols and Cohen/Coon in 1953. In this discussion, the term "process" as defined by Ziegler/Nichols means the same thing as the "physical plant" described earlier in this chapter. This open-loop method requires only that a single step input be imposed on the process. The process reaction method is based on a single experimental test that is made with the controller in the manual mode. A small step change, ΔU, in the actuator output is introduced and the measured process response is recorded, as shown in Figure 13.14. To obtain parameters of the process, a tangent is drawn to the process reaction curve at its point of maximum slope (at the inflection point). This slope is R, which is called the **process reaction rate**. The intersection of this tangent line with the original base line gives an indication of **L**, the process lag. **L** is really a measure of equivalent dead time for the process. If the tangent drawn at the inflection point is extrapolated to a vertical axis drawn at the time when the step was imposed, the amount by which this value is below the horizontal base line will be represented by the product **L • R**. ΔT is the time step for the digital controller. It is recommended to run P and PI controllers with ΔT = 0.1L, and a PID controller at a rate 20 times faster (ΔT = 0.05L.) Using these parameters, Ziegler and Nichol proposed initial controller settings as

Proportional Controller

$$\mathbf{K_P} = \Delta U/(\mathbf{L} \cdot \mathbf{R})$$

Proportional-Integral Controller

$$\mathbf{K_P} = 0.9\ \Delta U/(\mathbf{L} \cdot \mathbf{R})$$

$$\mathbf{K_I} = \mathbf{K_P}/(3.33\mathbf{L})$$

Proportional-Integral-Derivative Controller

$$\mathbf{K_P} = 1.2\ \Delta U/(\mathbf{L} \cdot \mathbf{R})$$

$$\mathbf{K_I} = 0.5\ \mathbf{K_P}\ /\mathbf{L}$$

$$\mathbf{K_D} = 0.5\ \mathbf{K_P}\ \mathbf{L}$$

Checkpoint 13.11: Are the Ziegler/Nichol equations consistent from a dimensional-analysis perspective? In other words, are the units correct?

Figure 13.14
A process reaction curve used to determine controller settings.

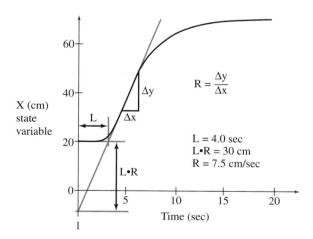

The **response time** is the delay after **X*** is changed for the system to reach a new constant state. **Steady state controller accuracy** is defined as the average difference between **X*** and **X′**. **Overshoot** is defined as the maximum positive error that occurs when **X*** is increased. Similarly, **undershoot** is defined as the maximum negative error that occurs when **X*** is decreased. During the testing phase, it is appropriate to add minimally intrusive debugging software that specifically measures performance parameters, such as response time, accuracy, overshoot, and undershoot. In addition, we can add instruments that allow us to observe the individual P(t), I(t), and D(t), components of the PID equation and their relation to controller error E(t).

Once the initial parameters have been selected, a simple empirical method can be used to fine-tune the controller. This empirical approach starts with proportional term ($\mathbf{K_P}$). As the proportional term is adjusted up or down, evaluate the quickness and smoothness of the controller response to changes in setpoint and to changes in the load. $\mathbf{K_P}$ is too big if the actuator saturates both at the maximum and minimum after $\mathbf{X^*}$ is changed. The next step is to adjust the integral term ($\mathbf{K_I}$) a little at a time to improve the steady-state controller accuracy without adversely affecting the response time. Don't change both $\mathbf{K_P}$ and $\mathbf{K_I}$ at once. Rather, you should vary them one at a time. If the response time, overshoot, undershoot, and accuracy are within acceptable limits, a PI controller is adequate. On the other hand, if accuracy and response are OK but overshoot and undershoot are unacceptable, adjust the derivative term ($\mathbf{K_D}$) to reduce the overshoots and undershoots.

13.4.3
Velocity PID
Controller

The objective of this example is to develop the fixed-point equations that implement a PID velocity controller (Figure 13.15).

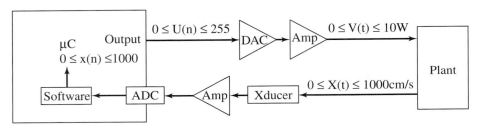

Figure 13.15
Interface of a PID velocity controller.

The maximum output of 255 maps linearly into a maximum applied power of 10 W to the physical plant. Similarly, a maximum speed of 1000 cm/s maps linearly into a maximum ADC result of 255. Also U = 0 converts into V = 0, and X = 0 gives an ADC conversion of 0. Let Xstar be a 16-bit unsigned integer containing the desired speed or set point in centimeters per second. Let current error be defined as

$$e(t) = Xstar - X(t)$$

in centimeters per second. We will implement the following PID control equation

$$V(t) = 0.1e(t) + 0.5\int_0^\tau e(\tau)\, d\tau + 0.005\,\frac{de(t)}{dt}$$

where e(t) is in centimeters per second and V(t) is in watts. To simplify the problem, we will break the controller into separate proportional, integral, and derivative terms. That is, let

$$U(n) = P(n) + I(n) + D(n)$$

where U(n) is the next DAC output and P(n), I(n), and D(n) are the proportional, integral, and derivative components, respectively. Let data be the current 8-bit unsigned ADC sample. x(n) will be the 16-bit estimated current speed in centimeters per second. $x(n)$ ranges from 0 to 1000 cm/s. The fixed-point calculation that converts `data` into the estimated speed $x(n)$ is

$$x(n) = \frac{1000 \cdot \texttt{data}}{256} = \frac{125 \cdot \texttt{data}}{32}$$

We define $e(n)=\texttt{Xstar}-x(n)$ as the 16-bit signed current error in centimeters per second. Let $e(n-1)$ be the previous calculation sampled at 100 Hz. Next, we develop fixed-point equations for $P(n)$, $I(n)$, and $D(n)$ in terms of the current and previous $e(n)$ calculations. The term $I(n-1)$ refers to the previous calculation of the integral.

$$e(n) = \texttt{Xstar}-x(n)$$

For the proportional term we have

$$V(t) = 0.1 \cdot e(t)$$

therefore

$$\frac{10 \cdot P(n)}{256} = 0.1 \cdot e(n)$$

Rearranging we get

$$P(n) = \frac{256 \cdot e(n)}{100}$$

For the integral term we have

$$V(t) = 0.5 \int_0^t e(\tau)d\tau = 0.5\Delta t \sum_{i=0}^{n} e(n) = 0.005 \sum_{i=0}^{n} e(n)$$

therefore

$$\frac{10 \cdot I(n)}{256} = 0.005 \cdot \Sigma e(n)$$

Rearranging we get

$$I(n) = \frac{16 \cdot e(n)}{125} + I(n-1)$$

For the derivative term we have

$$V(t) = 0.005 \frac{de(t)}{dt} = 0.005 \frac{e(n) - e(n-1)}{\Delta t}$$

therefore

$$\frac{10 \cdot D(n)}{256} = 0.5 \cdot [e(n) - e(n-1)]$$

Rearranging, we get

$$D(n) = \frac{64 \cdot [e(n) - e(n-1)]}{5}$$

13.4.4 Proportional-Integral Controller with a PWM actuator

We will design an MC9S12C32 microcomputer-based proportional-integral control system. The overall objective is to control the position of an object with an accuracy of 0.1 cm and a range of 0 to 100 cm, as shown in Figure 13.16. Let **X*** is desired state variable. In this example, **X*** is a decimal fixed-point number and is set by the main program. Let **X′** be the estimated state variable that comes from the **state estimator**, which encodes the current position as the period of a squarewave, interfaced to **PT1**. The period output of the sensor is linearly related to the position **X** with a fixed offset. The accuracy of the

Figure 13.16
PI controller using period measurement and pulse-width modulation.

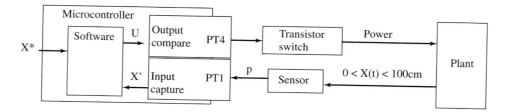

state estimator needs to match the 0.1 cm specification of the controller. If **p** is the measured period in 0.1 ms and **X'** is the estimated position in 0.1 cm, the state estimator measures the period and calculates **X'**.

$$X' = p - 100$$

Let **U** be the actuator control variable ($100 \le U \le 19900$). This system uses **pulse-width modulation** with a 100 Hz squarewave that applies energy to the physical plant, as shown in Figure 13.17. **U** will be the number of E clock cycles (out of 20000) than the **PT4** output is high. There is an external gravitational force pulling down on the object. **PT4** is an output from the computer and an input to the actuator, creating an upward force.

Figure 13.17
Pulse width modulated actuator signals.

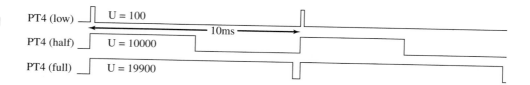

The process reaction curve shown previously in Figure 13.14 was measured for this system after the actuator was changed from 250 to 2000; thus, ΔU is 1750 (units of E clock cycles). From Figure 13.4, the lag **L** is 4.0 sec and the process reaction rate **R** is 7.5cm/sec. The controller rate is selected to be about ten times faster than the lag **L**, so $\Delta T = 0.4$ sec. In this way, the controller runs at a rate faster than the physical plant. We calculate the PI controller settings using the Ziegler/Nichol equations.

$$K_P = 0.9\ \Delta U/(L \cdot R) = 0.9*1750/(4.0*7.5) = 52.5 \text{ cycles/cm}$$

$$K_I = K_P/(3.33L) = 52.5/(3.33*4.0) = 3.94144 \text{ cycles/cm/sec}$$

We will execute the proportional control equation once every 0.4 second. X* and X' are decimal fixed-point numbers with a resolution of 0.1 cm. The constant 52.5 is expressed as 105/2. The extra divide by 10 handles the decimal fixed-point representation of X* and X'.

$$P(n) = K_P \cdot (X* - X')/10 = 105 \cdot (X* - X')/20$$

We will also execute the integral control equation once every 0.4 second. Binary fixed-point is used to approximate 1.57658 as 101/64.

$$I(n) = I(n-1) + K_I \cdot (X* - X') \cdot \Delta T/10 = I(n-1) + 3.94144$$
$$\cdot (X* - X') \cdot 0.4/10 = I(n-1) + 101 \cdot (X* - X')/640$$

Program 13.6 includes the OC5 interrupt service handler, which generates the 10 kHz real time clock. The OC5 handler will establish the current **Time** in 0.1 ms. After 4000 OC5 interrupts (0.4 second), the control algorithm is implemented.

Program 13.6
Integral position control
software.

```
unsigned short Time; // Time in 0.1 msec
short X;              // Estimated position in 0.1 cm, 0 to 1000
short Xstar;          // Desired pos in 0.1 cm, 0 to 1000
short E;              // Position error in 0.1 cm, -1000 to +1000
short U,I,P;          // Actuator duty cycle, 100 to 19900 cycles
unsigned short Cnt;   // once a sec
unsigned short Told;  // used to measure period
void interrupt 13 TC5handler(void){
  TFLG1 = 0x20;       // ack C5F
  TC5 = TC5+200;      // every 0.1 ms
  Time++;             // used to measure period
  if((Cnt++)==4000){  // every 0.4 sec
    Cnt = 0;          // 0<X<100, 0<Xstar<100, 100<U<19900
    E = Xstar-X;
    P = (105*E)/20;
    I = I+(101*E)/640;
    if(I < -500) I=-500;   // anti-reset windup
    if(I > 4000) I=4000;
    U = P+I;               // PI controller has two parts
    if(U < 100) U=100;     // Constrain actuator output
    if(U>19900) U=19900;
  }
}
```

Checkpoint 13.12: What is the output **U** of the controller if the position **X** is much greater than the setpoint **X***? In this situation, what does the object do?

Program 13.7 uses OC4 to generate a 100 Hz real-time clock, creating the 100 Hz variable duty-cycle squarewave connected to the actuator. The OC4 output will be high for **U** cycles and low for 20,000 − **U** cycles. Program 13.8 uses IC1 to interrupt on each rise of the sensor squarewave, measuring the period of the sensor squarewave and estimating the current position. Program 13.9 is the ritual that initializes the global variables and arms the three interrupts. Once initialized, the controller runs in the background. The main program (not shown) is responsible for calling the ritual, and establishing the controller set-point, Xstar with units of 0.1 cm.

Program 13.7
PWM actuator control
software.

```
void interrupt 12 TC4handler(void){
  TFLG1 = 0x10;        // acknowledge C4F
  if(PTT&0x10){
    TC4 = TC4+U;        // PT4 is 1, High for the next U cycles
  }
  else{
    TC4 = TC4+20000-U; // PT4 is 1, Low for the next 20000-U cycles
  }
}
```

Program 13.8
Sensor measurement
software.

```
// Time is incremented every 0.1 ms, by OC5
// This handler is executed on rise of PT1
void interrupt 9 TC1Handler(void){unsigned short p;
  TFLG1 = 0x02;  // Acknowledge C1F
  p = Time-Told; // period in msec
  X = p-100;     // estimated position (0.1 cm)
  Told = Time;
}
```

Program 13.9
Initialization software.

```
void Init(void){ // PT5 output for debugging
  asm sei        // make atomic
  TIOS = 0x30;   // output compare OC5, OC4
  DDRT &= ~0x02; // PT1 is input
  DDRT |= 0x30;  // PT5, PT4 are output
  TSCR1 = 0x80;  // enable
  TSCR2 = 0x01;  // 500 ns clock
  TCTL1 = (TCTL1&0xF0)|0x05; // toggle PT5, PT4
  TCTL4 = 0x08;  // Input capture on rise of IC1
  TIE = 0x32;    // Arm OC5F+OC4F+IC1F
  U = 100;       // Initial U, low power
  Time = Told = 0; Cnt = 0;
  TFLG1 = 0x32;  // clear flags
  TC5 = TCNT+50; // First OC5 in 25us
  asm cli
}
```

Observation: PID control will work extremely well (fast, accurate, and stable) if the physical plant can be described with a set of linear differential equations.

13.5   Fuzzy Logic Control

There are a number of reasons to consider fuzzy logic approach to control. It requires less mathematics than PID systems. It will also require less memory and execute faster. In other words, an 8-bit fuzzy system may perform as well (same steady-state error and response time) as a 16-bit PID system. When complete knowledge about the physical plant is known, then a good PID controller can be developed. Since the fuzzy logic control is more robust (still works even if the parameter constants are not optimal), then the fuzzy logic approach can be used when complete knowledge about the plant is not known or can change dynamically. Choosing the proper PID parameters requires expert knowledge about the plant. The fuzzy logic approach is more intuitive, following more closely the way a human would control the system. It is easy to modify an existing fuzzy control system into a new problem. So if the framework exists, rapid prototyping is possible.

Fuzzy logic was conceived in the mid 1960s by Lotfi Zadeh while at the University of California at Berkeley. However, the first commercial application didn't come until 1987, when Matsushita Industrial Electric used it to control the temperature in a shower head. Named after the nineteenth-century mathematician George Boole, Boolean logic is an algebra in which values are either true or false. This algebra includes the operations of AND OR and NOT. Fuzzy logic is also an algebra, but conditions may exist in the continuum between true and false. While Boolean logic defines two states, 8-bit fuzzy logic consists of 256 states all the way from "not at all" (0) to "definitely true" (255). For example, "128" means half way between true and false. The fuzzy logic algebra also includes the operations of AND, OR, and NOT. A **fuzzy membership set**, a **fuzzy variable**, and a **fuzzy set** all refer to the same entity, which is a software variable describing the level of correctness for a condition within fuzzy logic. If we have a fuzzy membership set for the condition "hungry," then as the value of hungry moves from 0 to 255, the condition "hungry" becomes more and more true.

```
....0.....32.....64.....96.....128.....160.....192.....224.....255
Not at all...a little bit...somewhat...mostly...pretty much...definitely
```

The design process for a fuzzy logic controller solves the following eight components. These components are listed in the order we would draw a data flow graph, starting with the state variables on the left, progressing through the controller, and ending with the actuator output on the right.

- The **Physical plant** has *real state variables*.
- The **Data Acquisition System** monitors these signals, creating the *estimated state variables*.
- The **Preprocessor** may calculate relevant parameters called *crisp inputs*.
- **Fuzzification** will convert crisp inputs into *input fuzzy membership sets*.
- The **Fuzzy Logic** is a set of rules that calculate *output fuzzy membership sets*.
- **Defuzzification** converts output sets into *crisp outputs*.
- The **Postprocessor** modifies crisp outputs into a more convenient format.
- The **Actuator System** affects the physical plant based on these output.

We will work through the concepts of fuzzy logic by considering examples of how we as humans control things like driving a car at a constant speed. During the initial stages of the design, we study the **physical plant** and decide which state variables to consider. For example, if we wish to control speed, then speed is obviously a state variable, but it might be also useful to know other forces acting on the object such as gravity (e.g., going up and down hills), wind speed, and friction (e.g., rain and snow on the roadway). The purpose of the **data acquisition system** is to accurately measure the state variables. It is at this stage that the system converts physical signals into digital numbers to be processed by the software controller. We have seen two basic approaches in this book for this conversion: the measurement of period/frequency using input capture and the analog-to-digital conversion using an ADC. The **preprocessor** calculates **crisp inputs**, which are variables describing the input parameters in our software having units (like miles/hr). For example, if we measured speed, then some crisp inputs we might calculate would include speed error and acceleration. Just as with the PID controller, the accuracy of the data acquisition system must be better than the desired accuracy of the control system as a whole.

The next stage of the design is to consider the **actuator** and postprocessor. It is critical for the actuator to be able to induce forces on the physical plant in a precise and fast manner. The step response of the actuator itself (time from software command to the application of force on the plant) must be faster than the step response of the plant (time from the application of force to the change in state variable). Consider the case where we wish to control the temperature of a pot of water using a kitchen stove. The speed of this actuator is the time between turning the stove on and the time when heat is applied to the pot. The actuator on a gas stove is much faster than the actuator on an electric stove. The resolution of an actuator is the smallest change in output it can reliably generate. Just as with the PID controller, the resolution of the actuator (converted into equivalent units on the input) must be smaller than the desired accuracy of the control system as a whole. A **crisp output** is a software variable describing the output parameters having units (e.g., watts, Newtons, dynes/cm^2). The **postprocessor** converts the crisp output into a form that can be directly output to the actuator. The postprocessor can verify that the output signals are within the valid range of the actuator. One of the advantages of fuzzy logic design is its connection to human intuition. Think carefully about how you control the actuator (gas pedal) when attempting to drive a car at a constant speed. There is no parameter in your brain specifying the exact position of the pedal (e.g., 50% pressed, 65% pressed), unless of course, you are city taxicab driver and your brain allows two actuator states: full gas and full brake. Rather, what your brain creates as actuator commands are statements like "press the pedal little harder" and "press the pedal a lot softer." So, mimicking the way people think, the crisp output of fuzzy logic controller might be change in pedal pressure ΔU, and the postprocessor would calculate $U = U + \Delta U$, then check to make sure U is within an acceptable range.

We continue the design of a fuzzy logic controller by analyzing its crisp inputs. As a design step, we create a list of true/false conditions that together describe the current state of the physical plant. In particular, we define **input fuzzy membership sets**, which are fuzzy logic variables describing conditions related to the state of the physical plant. These fuzzy variables do not need to be orthogonal. In other words, it is acceptable to have variables that are related to each other. When designing a speed controller, we could define multiple fuzzy variables referring to similar conditions, such as **WayTooFast**, **Fast**, and **LittleBitFast**. Given the scenario where we are driving too fast, there should be generous overlap in conditions, such that two or even three fuzzy sets are simultaneously partially true. On the other hand, it is important that the entire list of input membership fuzzy sets, when considered as an ensemble, form a complete description of the status of the physical plant. For example, if we are attempting to drive a car at a constant speed, then we might include input fuzzy variables **SlowingDown**, **GoingSteady**, and **SpeedingUp** describing the car's acceleration. **Fuzzification** is the mathematical step converting the crisp inputs into input fuzzy membership sets. The Freescale 6812 defined a set of assembly language instructions to optimize the implementation of fuzzy logic controllers. The instruction **mem** performs the fuzzification process converting a crisp input into an input fuzzy membership set. When implementing fuzzy logic explicitly with C code, we will have available the full set of AND, OR, NOT fuzzy logic operations. On the other hand, if we are implementing the fuzzy logic controller using the built-in 6812 assembly language instructions, we will have access only to the AND OR fuzzy logic operations. Therefore, if the fuzzy logic controller needs a logic parameter that is the complement of **Fast**, we can create an additional input fuzzy variable, **NotFast**, and calculate it during the fuzzification stage.

The heart of a fuzzy logic controller is the **fuzzy logic** itself, which is set of logic equations that calculate fuzzy outputs as a function of fuzzy inputs. An **output fuzzy membership set** is a fuzzy logic variable describing a condition related to the actuator. **QuickStop**, **SlowDown**, **JustRight**, **MorePower**, and **MaxPower** are examples of output fuzzy variables that might be used to describe the action to perform on the gas pedal. Like input fuzzy variables, output fuzzy variables exist in the continuum from definitely false (0) to definitely true (1). Just as with the input specification, it is also important to create a list of output membership fuzzy sets that, when considered as an ensemble, form a complete characterization of what we wish to be able to do with the actuator. We write fuzzy logic equations using AND/OR functions in a way similar to Boolean logic. The fuzzy logic AND is calculated as the minimum value of the two inputs, and the fuzzy logic OR is calculated as the maximum value of the two inputs. The 6812 instruction **rev** will execute an entire set of fuzzy logic rules converting fuzzy inputs into fuzzy outputs. The design of the rules, like the other aspects of fuzzy control, follows the human intuition. For example, this fuzzy logic equations arises from a human thought.

"We should slow down if we are going too fast or if we are speeding up and going a little bit too fast."

```
SlowDown = WayTooFast + SpeedingUp*LittleBitFast
```

Checkpoint 13.13: If **WayTooFast** is 50, **SpeedingUp** is 40, and **LittleBitFast** is 60, then what would be the calculated value for **SlowDown**?

The **defuzzification** stage of the controller converts the output fuzzy variables into crisp outputs. Although any function could be used, an effective approach is to use a weighted average. The 6812 instructions **wav** and **ediv** together perform the defuzzification step, converting output fuzzy variables into a crisp output. Consider the case where the pedal pressure U varies from 0 to 100; thus, the crisp output ΔU can take on values from -100

to $+100$. We think about what crisp output we want if just **QuickStop** were to be true. In this case, we wish to make ΔU equal to -100. In a similar way, we then define crisp output values for **SlowDown**, **JustRight**, **MorePower**, and **MaxPower** as -10, 0, $+10$, and $+100$ respectfully. We can combine the five factors using a weighted average.

$$\Delta U = \frac{-100*\text{QuicksStop} - 10*\text{SlowDown} + 10*\text{MorePower} + 100*\text{MaxPower}}{\text{QuickStop} + \text{SlowDown} + \text{JustRight} + \text{MorePower} + \text{MaxPower}}$$

Because the fuzzy controller is modular, we begin by testing each of the modules separately. The system-level testing of a fuzzy logic controller follows a procedure similar to the PID controller tuning. Minimally intrusive debugging instruments can be added to record the crisp inputs, fuzzy inputs, fuzzy outputs, and crisp outputs during the real-time operation of the system. Fuzzification parameters are adjusted so that the status of the plant is captured in the set of values contained in the fuzzy input variables. Next, the rules are adjusted so that fuzzy output variables properly describe what we want to do with the actuator. Lastly, the defuzzification parameters are adjusted so the proper crisp outputs are created, optimizing both accuracy and response time.

13.5.1 DAC, ADC Fuzzy Controller

The objective of this example is to design a *fuzzy logic* microcomputer-based DC motor controller (Figure 13.18). The actuator is a DAC and linear power amplifier. The power to the motor is controlled by varying the 8-bit DAC output voltage. An amplifier provides power to the motor that is linearly related to the DAC output. The motor speed is estimated with a tachometer connected to an 8-bit ADC.

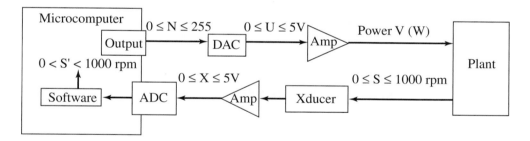

Figure 13.18
Interface of a motor controlled with fuzzy logic.

Our system has:

- Two control inputs
  - S*   desired motor speed, rpm
  - S′   current estimated motor speed, rpm
- One control output
  - N   digital value that we write to the DAC

To utilize 8-bit mathematics, we change the units of speed to $1000/256 = 3.90625$ rpm.

    T* $= (256 \cdot \text{S*})/1000$ desired motor speed, 3.9 rpm
    T′ $= (256 \cdot \text{S′})/1000$ current estimated motor speed, 3.9 rpm

For example, if the desired speed is 500 rpm, then T* will be 128. Notice that the estimated speed T′ is simply the ADC conversion value (i.e., just the ADR1 register without calculations). Inherent in most control systems is the concept of periodic execution. In other words, the control system functions (estimate state variables, control equation calculations, and actuator output) are performed on a regular and periodic basis, every Δt time units. This allows signal-processing techniques to be used. We will let $T'(n)$ refer to the current measurement and $T'(n-1)$ refer to the previous measurement (i.e., the one measured Δt time ago).

In the fuzzy logic approach, we begin by considering how a human would control the motor. Assume your hand were on a joystick (or your foot on a gas pedal) and consider how you would adjust the joystick to maintain a constant speed. We select crisp inputs and outputs to base our control system on. It is logical to look at the error and the change in speed when developing a control system. Our fuzzy logic system will have two crisp inputs

$$E = T^* - T' \qquad \text{error in motor speed, 3.9 rpm}$$
$$D = T'(n) - T'(n-1) \qquad \text{change in motor speed, 3.9 rpm/time}$$

Notice that if we perform the calculations of D on periodic intervals, then D will represent the derivative of T, dT'/dt. T^* and T' are 8-bit unsigned numbers, so the potential range of E varies from -255 to $+255$. Errors beyond ±127 will be adjusted to the extremes $+127$ or -128 without loss of information. Program 13.10 gives the calculations in C.

Program 13.10
Sub-traction with overflow/ underflow checking.

```
char Subtract(unsigned char N, unsigned char M){
/* returns N-M */
unsigned short N16,M16;
short Result16;
    N16=N;          /* Promote N,M */
    M16=M;
    Result16=N16-M16;    /* -255•Result16•+255 */
    if(Result16<-128) Result16 = -128;
    if(Result16>127)  Result16 = 127;
    return(Result16);}
```

Program 13.11 gives the global definitions of the input signals and fuzzy logic crisp input.

Program 13.11
Inputs and crisp inputs.

```
unsigned char Ts;     /* Desired Speed in 3.9 rpm units */
unsigned char T;      /* Current Speed in 3.9 rpm units */
unsigned char Told;   /* Previous Speed in 3.9 rpm units */
char D;               /* Change in Speed in 3.9 rpm/time units */
char E;               /* Error in Speed in 3.9 rpm units */
```

Common error: Neglecting overflow and underflow can cause significant errors.

The need for the special Subtract function can be demonstrated with the following example:

```
E = Ts - T; // if Ts = 200 and T = 50 then E will be -106!!
```

This function can be used to calculate both E and D (Program 13.12). Now, if `Ts` = 200 and `T` = 50, then `E` will be $+127$.

Program 13.12
Calculation of crisp inputs.

```
void CrispInput(void){
    E=Subtract(Ts,T);
    D=Subtract(T,Told);
    Told=T;}     /* Set up Told for next time */
```

To control the actuator, we could simply choose a new DAC value N as the crisp output. Instead, we will select ΔN, which is the change in N, rather than N itself because it better mimics how a human would control it. Again, think about how you control the speed of your car when driving. You do not adjust the gas pedal to a certain position but rather make small or large changes to its position to speed up or slow down. Similarly, when controlling the temperature of the water in the shower, you do not set the hot/cold controls to certain absolute positions. Again you make differential changes to affect the actuator in this control system. Our fuzzy logic system will have one crisp output:

ΔN change in output, $N = N + \Delta N$, in DAC units

Next we introduce fuzzy membership sets that define the current state of the crisp inputs and outputs. Fuzzy membership sets are variables that have true/false values. The value of a fuzzy membership set ranges from definitely true (255) to definitely false (0). For example, if a fuzzy membership set has a value of 128, you are stating the condition is halfway between true and false. For each membership set, it is important to assign a meaning or significance to it. The calculation of the input membership sets is called *fuzzification*. For this simple fuzzy controller, we will define six membership sets for the crisp inputs:

Slow	True if the motor is spinning too slow
OK	True if the motor is spinning at the proper speed
Fast	True if the motor is spinning too fast
Up	True if the motor speed is getting larger
Constant	True if the motor speed is remaining the same
Down	True if the motor speed is getting smaller

We will define three membership sets for the crisp output:

Decrease	True if the motor speed should be decreased
Same	True if the motor speed should remain the same
Increase	True if the motor speed should be increased

The fuzzy membership sets are usually defined graphically, but software must be written to actually calculate each. In this implementation, we will define three adjustable thresholds: TE, TD, and TN. These are software constants and provide some fine-tuning to the control system. We will set each threshold to 20. If you build one of these fuzzy system, try varying one threshold at a time and observe the system behavior (steady-state controller error and transient response). If the error E is −5 (3.9 rpm units), the fuzzy logic will say that *Fast* is 64 (25% true), *OK* is 192 (75% true), and *Slow* is 0 (definitely false). If the error E is +21 (in 3.9 rpm units), the fuzzy logic will say that *Fast* is 0 (definitely false), *OK* is 0 (definitely false), and *Slow* is 255 (definitely true). TE is defined to be the error (e.g., 20 in 3.9 rpm units is 78 rpm) above which we will definitely consider the speed to be too fast. Similarly, if the error is less than −TE, then the speed is definitely too slow (Figure 13.19).

Figure 13.19
Fuzzification of the error input.

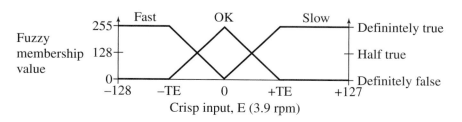

In this fuzzy system, the input membership sets are continuous piecewise-linear functions. Also, for each crisp input value (*Fast, OK, Slow*) sum to 255. In general, it is possible for the fuzzy membership sets to be nonlinear or discontinuous, and the membership values do not have to sum to 255. The other three input fuzzy membership sets depend on the crisp input D. TD is defined to be the change in speed (e.g., 20 in 3.9 rpm/time units is 78 rpm/time) above which we will definitely consider the speed to be going up. Similarly, if the change in speed is less than $-TD$, then the speed is definitely going down (Figure 13.20).

Figure 13.20
Fuzzification of the acceleration input.

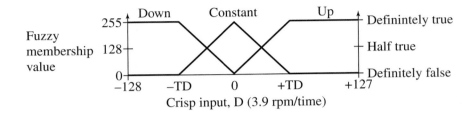

In C, we could define a fuzzy function that takes the crisp inputs and calculates the fuzzy membership set values. Again TE and TD are software constants that will affect the controller error and response time (Program 13.13).

Program 13.13
Calculation of the fuzzy membership variables in C.

```
#define TE 20
unsigned char Fast, OK, Slow, Down, Constant, Up;
#define TD 20
unsigned char Increase,Same,Decrease;
#define TN 20
void InputMembership(void){
        if(E <= -TE) {               /*       E≤-TE */
          Fast=255;
          OK=0;
          Slow=0;}
        else
          if (E < 0) {               /* -TE<E<0     */
            Fast=(255*(-E))/TE;
            OK=255-Fast;
            Slow=0;}
          else
            if (E < TE) {            /*   0<E<TE    */
              Fast=0;
              Slow=(255*E)/TE;
              OK=255-Slow;}
            else {                   /* +TE≤E       */
              Fast=0;
              OK=0;
              Slow=255;}
        if(D <= -TD) {               /*       D≤-TD */
          Down=255;
          Constant=0;
          Up=0;}
        else
        if (D < 0) {                 /* -TD<D<0     */
          Down=(255*(-D))/TD;
          Constant=255-Down;
          Up=0;}
```

```
        else
          if (D < TD) {            /*   0<D<TD     */
            Down=0;
            Up=(255*D)/TD;
            Constant=255-Up;}
          else {                   /*   +TD≤D      */
            Down=0;
            Constant=0;
            Up=255;}}
```

The fuzzy rules specify the relationship between the input fuzzy membership sets and the output fuzzy membership values. It is in these rules that one builds the intuition of the controller. For example, if the error is within reasonable limits and the speed is constant, then the output should not be changed. In fuzzy logic we write:

If *OK* and *Constant* then *Same*

If the error is within reasonable limits and the speed is going up, then the output should be reduced to compensate for the increase in speed. That is,

If *OK* and *Up* then *Decrease*

If the motor is spinning too fast and the speed is constant, then the output should be reduced to compensate for the error. That is,

If *Fast* and *Constant* then *Decrease*

If the motor is spinning too fast and the speed is going up, then the output should be reduced to compensate for both the error and the increase in speed. That is,

If *Fast* and *Up* then *Decrease*

If the error is within reasonable limits and the speed is going down, then the output should be increased to compensate for the drop in speed. That is,

If *OK* and *Down* then *Increase*

If the motor is spinning too slow and the speed is constant, then the output should be increased to compensate for the error. That is,

If *Slow* and *Constant* then *Increase*

These seven rules can be illustrated in a table form (Figure 13.21).

Figure 13.21
Fuzzy logic rules shown in table form.

E \ D	Down	Constant	Up
Slow	Increase	Increase	
OK	Increase	Same	Decrease
Fast		Decrease	Decrease

It is not necessary to provide a rule for all situations. For example, we did not specify what to do for *Fast&Down* or for *Slow&Up,* although we could have added (but did not)

If *Fast* and *Down* then *Same*
If *Slow* and *Up* then *Same*

When more than one rule applies to an output membership set, then we can combine the rules:

Same= (*OK* and *Constant*)
Decrease= (*OK* and *Up*) or (*Fast* and *Constant*) or (*Fast* and *Up*)
Increase= (*OK* and *Down*) or (*Slow* and *Constant*) or (*Slow* and *Down*)

In fuzzy logic, the and operation is performed by taking the minimum and the or operation is the maximum. Thus the C function that calculates the three output fuzzy membership sets is shown in Program 13.14.

Program 13.14
Calculation of the output fuzzy membership variables in C.

```
unsigned char min(unsigned char u1,unsigned char u2){
     if(u1>u2) return(u2);
     else return(u1);}
unsigned char max(unsigned char u1,unsigned char u2){
     if(u1<u2) return(u2);
     else return(u1);}
void OutputMembership(void){
     Same=min(OK,Constant);
     Decrease=min(OK,Up)
     Decrease=max(Decrease,min(Fast,Constant));
     Decrease=max(Decrease,min(Fast,Up));
     Increase=min(OK,Down)
     Increase=max(Increase,min(Slow,Constant));
     Increase=max(Increase,min(Slow,Down));}
```

The calculation of the crisp outputs is called *defuzzification*. The fuzzy membership sets for the output specifies the crisp output, ΔN, as a function of the membership value. For example, if the membership set *Decrease* were true (255) and the other two were false (0), then the change in output should be $-TN$ (where TN is another software constant). If the membership set *Same* were true (255) and the other two were false (0), then the change in output should be 0. If the membership set *Increase* were true (255) and the other two were false (0), then the change in output should be $+TN$ (Figure 13.22).

Figure 13.22
Defuzzification of the ΔN crisp output.

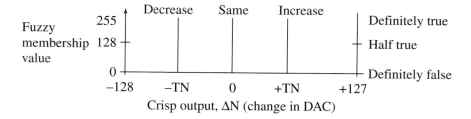

In general, we calculate the crisp output as the weighted average of the fuzzy membership sets:

$$\Delta N = (Decrease \cdot (-TN) + Same \cdot 0 + Increase \cdot TN)/(Decrease + Same + Increase)$$

The C compiler will promote the calculations to 16 bits and perform the calculation using 16-bit signed mathematics that will eliminate overflow on intermediate terms. The output dN will be bounded between $-TN$ and $+TN$. Thus the C function that calculates the crisp output is shown in Program 13.15. When writing in assembly you will

Program 13.15
Calculation of the crisp output in C.

```c
char dN;
void CrispOutput(void){
        dN=(TN*(Increase-Decrease))/(Decrease+Same+Increase);
}
```

need to deal with converting *Increase, Same, Decrease* from 8-bit unsigned to 16-bit signed, and with overflow on intermediate terms. Just like in C, the output, ΔN, will be bounded between $-TN$ and $+TN$. If the calculation is rearranged, you can still use the 8-bit unsigned multiply. The numerator is:

```
TN*Increase−TN*Decrease
```

using 8-bit unsigned multiply and 16-bit subtract. The divide calculation for **dN** needs to be 16-bit signed (Program 13.16).

Program 13.16
Main program for fuzzy logic controller in C.

```c
unsigned int Time;
#define rate 2000   /* Rate must be less than 32767 E clocks */
void Initialize(void){
      OPTION=0x80;    /* Turn on A/D */
      PORTB=0;
      N=0;                /* Initial Actuator */
      Told=0;
      Ts=128; }       /* 500 rpm */
#define CCF 0x80
unsigned char Sample(unsigned char channel){
      ADCTL=channel;           /* Start A/D */
      while((ADCTL&CCF)==0);   /* Wait for CCF */
      return(ADR1);}
void Main(void){ int dT;
      Initialize();                /* Turn on A/D initialize globals */
      Time=TCNT+rate;              /* TCNT value for first calculation */
      while(1){
         while((dT=Time-TCNT)>0){};
         Time=Time+rate;           /* TCNT value for next calculation */
         T=Sample(0);              /* Sample A/D and set T */
         CrispInput();             /* Calculate E,D and new Told */
         InputMembership();        /* Sets Fast,OK,Slow,Down,Constant,Up */
         OutputMembership();       /* Sets Increase,Same,Decrease */
         CrispOutput();            /* Sets dN */
         N=max(0,min(N+dN,255));
         PORTB=N;}}                /* Set Actuator */
```

**13.5.2
PWM Fuzzy
Controller**

The objective of this section is to design a *fuzzy logic* microcomputer-based motor controller (Figure 13.23). The actuator is a PWM digital wave. The power to the motor is controlled by varying the duty cycle of the 200-Hz square wave. A switching power transistor provides current to the motor only when the digital output is high. The diodes protect the electronics from the back EMF (as high as 200 V) that occurs when the current is switched (large dI/dt) to the DC motor coil. The frequency of the square wave is chosen faster than the motor response so that the motor responds only to the duty cycle and not to the individual highs and lows of the square wave.

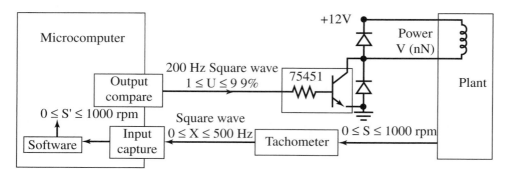

Figure 13.23
Interface of a motor controller with fuzzy logic.

The tachometer output is a frequency (in hertz) that is related to the motor speed (in rpm). The input capture system can be used to measure the period of this signal. Period measurement is faster than frequency measurement, so it is better suited for real-time control. The motor speed is a nonlinear function of the applied power. Because of friction and inertia, there is a range of power below which the motor will not spin. Below 40% duty cycle, the motor has a higher DC gain (Figure 13.24).

Our system has:

- Two control inputs

$S*$	desired motor speed, rpm
S'	current estimated motor speed, rpm

- One control output

U	duty cycle, E clock cycles

- Two crisp inputs

$E = S* - S'$	error in motor speed, rpm
$D = dS'/dt = S'(n) - S'(n-1)$	change in motor speed, Hz/time or rpm/time

- One crisp output

ΔU	change in duty cycle, $U = U + \Delta U$, E clock cycles

Figure 13.24
Steady-state response
of the physical plant.

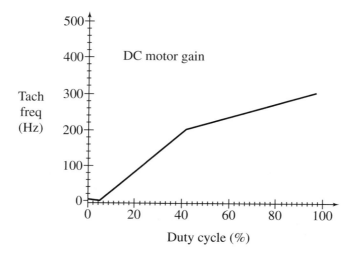

Next we introduce fuzzy membership sets that define the current state of the crisp inputs and outputs. The calculation of the input membership sets is called *fuzzification*. For this simple fuzzy controller, we will define ten membership sets for the crisp inputs:

ENL	True if the motor is spinning much too slow
ENS	True if the motor is spinning a little bit too slow
EZE	True if the motor is spinning at the proper speed
EPS	True if the motor is spinning a little bit too fast
EPL	True if the motor is spinning much too fast
DNL	True if the motor speed is slowing down a lot
DNS	True if the motor speed is slowing down a little
DZE	True if the motor speed is remaining the same
DPS	True if the motor speed is speeding up a little
DPL	True if the motor speed is speeding up a lot

We will define five membership sets for the crisp output:

ONL	True if the motor speed should be decreased a lot
ONS	True if the motor speed should be decreased a little
OZE	True if the motor speed should remain the same
OPS	True if the motor speed should be increased a little
OPL	True if the motor speed should be increased a lot

Notice in the following rules that when the system is operating near the set point (ENS, EZE, EPS) with small changes in speed (DNS, DZE, DPS), then this fuzzy system is similar to the previous implementation. On the other hand, when the system has large errors and/or large acceleration, then it makes large changes in the output (OPL, ONL) in an attempt to reach the desired state faster. The 6811 implementation to this controller would be very similar to the previous example. On the other hand, the 6812 has a rich set of fuzzy logic assembly instructions that will be used to solve this problem. We begin with the fuzzy variables (Program 13.17).

Program 13.17
Global variables for fuzzy controller in 6812 assembly.

```
          org   $3800
; crisp inputs
speed:    ds   1
acceleration: ds 1
; input membership variables
fuzvar:   ds   0     ; inputs
EPL:      ds   1     ; speed way too fast
EPS:      ds   1     ; speed too fast
EZE:      ds   1     ; speed OK
ENS:      ds   1     ; speed too slow
ENL:      ds   1     ; speed way too slow
DPL:      ds   1     ; speed decreasing a lot
DPS:      ds   1     ; speed decreasing
DZE:      ds   1     ; speed constant
DNS:      ds   1     ; speed increasing
DNL:      ds   1     ; speed increasing a lot

fuzout:   ds   0     ; outputs
OPL:      ds   1     ; add a lot of power to system
OPS:      ds   1     ; add some power to system
OZE:      ds   1     ; leave power as is
```

continued on p. 684

Program 13.17
Global variables for fuzzy controller in 6812 assembly.

```
continued from p. 683

ONS:      ds   1      ; subtract some power from system
ONL:      ds   1      ; subtract a lot of power from system
BREAK:    ds   1      ; apply break?
; input membership variables relative offsets
epl:      equ  0      ; speed way too fast
eps:      equ  1      ; speed too fast
eze:      equ  2      ; speed ok
ens:      equ  3      ; speed too slow
enl:      equ  4      ; speed way too slow
dpl:      equ  5      ; speed decreasing a lot
dps:      equ  6      ; speed decreasing
dze:      equ  7      ; speed constant
dns:      equ  8      ; speed increasing
dnl:      equ  9      ; speed increasing a lot
;output membership variables
opl:      equ  10     ; add alot of power to system
ops:      equ  11     ; add some power to system
oze:      equ  12     ; leave power as is
ons:      equ  13     ; subtract some power from system
onl:      equ  14     ; subtract alot of power from system
;crisp outputs
dpower:   ds   1
```

The error fuzzification is plotted in Figure 13.25 and defined in Program 13.18.

Figure 13.25
Fuzzification of the error input.

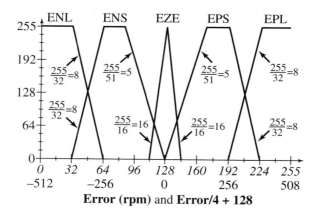

Program 13.18
Definition of the error input function in 6812 assembly.

```
          org  $4000
; format is Point1,Point2,Slope1,Slope2
s_tab:    dc.b  192,255,8,0    ;EPL
          dc.b  128,224,5,8    ;EPS
          dc.b  112,144,16,16  ;EZE
          dc.b  32,128,8,5     ;ENS
          dc.b  0,64,0,8       ;ENL
```

The acceleration fuzzification is plotted in Figure 13.26 and defined in Program 13.19.

Figure 13.26
Fuzzification of the
acceleration input.

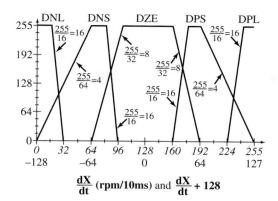

$$\frac{dX}{dt} \text{ (rpm/10ms) and } \frac{dX}{dt} + 128$$

Program 13.19
Definition of the
acceleration input
function in 6812
assembly.

```
a_tab:    dc.b    224,255,16,0      ;DPL
          dc.b    160,255,16,4      ;DPS
          dc.b    64,192,8,8        ;DZE
          dc.b    0,96,4,16         ;DNS
          dc.b    0,32,0,16         ;DNL
```

The fuzzy roles are plotted in Figure 13.27 and defined in Program 13.20.

Figure 13.27
Fuzzy logic rules shown
in table form.

$$D = X(n) - X(n-1)$$

$E = X^* - X$	DNL	DNS	DZE	DPS	DPL
ENL	OPL	OPL	OPL		
ENS	OPL	OPS	OPS		
EZE	OPL	OPS	OZE	ONS	ONL
EPS			ONS	ONS	ONL
EPL			ONL	ONL	ONL

```
rules:    dc.b    enl,dnl,$FE,opl,$FE    ; if ENL and DNL then OPL
          dc.b    ens,dnl,$FE,opl,$FE    ; if ENS and DNL then OPL
          dc.b    eze,dnl,$FE,opl,$FE    ; if EZE and DNL then OPL
          dc.b    enl,dns,$FE,opl,$FE    ; if ENL and DNS then OPL
          dc.b    ens,dns,$FE,ops,$FE    ; if ENS and DNS then OPS
          dc.b    eze,dns,$FE,ops,$FE    ; if EZE and DNS then OPS
          dc.b    enl,dze,$FE,opl,$FE    ; if ENL and DZE then OPL
          dc.b    ens,dze,$FE,ops,$FE    ; if ENS and DZE then OPS
          dc.b    eze,dze,$FE,oze,$FE    ; if EZE and DZE then OZE
          dc.b    eps,dze,$FE,ons,$FE    ; if EPS and DZE then ONS
          dc.b    epl,dze,$FE,onl,$FE    ; if EPL and DZE then ONL
          dc.b    eze,dps,$FE,ons,$FE    ; if EZE and DPS then ONS
          dc.b    eps,dps,$FE,ons,$FE    ; if EPS and DPS then ONS
          dc.b    eps,dps,$FE,onl,$FE    ; if EPL and DPS then ONL
          dc.b    eze,dpl,$FE,onl,$FE    ; if EZE and DPL then ONL
          dc.b    eps,dpl,$FE,onl,$FE    ; if EPS and DPL then ONL
          dc.b    epl,dpl,$FE,onl,$FE    ; if EPL and DPL then ONL
          dc.b    $FF
```

Program 13.20
Definition of the fuzzy rules in 6812 assembly.

The defuzzification is shown in Table 13.1.

Table 13.1
Defuzzification converts
the output memberships
into a crisp output.

Output fuzzy set	Singleton value
ONL	−128
ONS	−10
OZE	0
OPS	10
OPL	127

Program 13.21 initializes the system and Program 13.22 implements the fuzzy logic controller.

Program 13.21
Ritual and main program
for fuzzy controller in
6812 assembly.

```
addsingleton: dc.b 255,138,128,118,0
; 128 subtracted,   +127,10,0,-10,-128

ritual:  sei          ;make atomic
         bset #$20,TIOS  ;OC5
         movb #$80,TSCR1  ;enable, no fast clr
         movb #$02,TSCR2 ;1us clk
         bset #$20,T1E ;Arm OC5
         ldaa #$20        ;clear C5F
         staa TFLG1
         ldd  TCNT        ;current time
         addd #10000      ;first in 10 ms
         std  TC5
         cli              ;enable
         rts
main:    lds  #$4000
         ldd  #100         ; initial duty cycle 1% is off
         std  dutycycle
         jsr  ritual       ; initialize OC interrupt
         bra  *
```

Program 13.22
Interrupt handler for
fuzzy controller in 6812
assembly.

```
timehan: ldaa #$20   ;clear C5F
         staa TFLG1 ;Acknowledge
         ldd  TC5
         addd #10000 ;next in 10 ms
         std  TC5
         jsr  measurespeed ; crisp input speed
;reg A is speed 0 to 255
         ldx  #s_tab
         ldy  #fuzvar
         mem            ; calculate EPL
         mem            ; calculate EPS
         mem            ; calculate EZE
         mem            ; calculate ENS
         mem            ; calculate ENL
         jsr  measureacceleration ; crisp input acceleration
;reg A is acceleration 0 to 255
         ldx  #a_tab
```

```
        mem             ; calculate DPL
        mem             ; calculate DPS
        mem             ; calculate DZE
        mem             ; calculate DNS
        mem             ; calculate DNL
        ldab #6
cloop:  clr  1,y+  ; clear OPL,OPS,OZE,ONS,ONL,BREAK
        dbne b,cloop
        ldx  #rules
        ldy  #fuzvar
        ldaa #$FF
        rev
        ldy  #fuzout
        ldx  #addsingleton
        ldab #5
        wav
        ediv
        tfr  y,d
        subb #128
        stab dpower
change  ldab dpower
        sex  b,d
        addd dutycycle
; 200 Hz squarewave is 10000 cycles
; correct range is 100 to 9600 cycles
        cpd  #100
        bhs  nolow
low:    ldd  #100   ; underflow
        bra  set
notlow: cpd  #9600
        bls  set
high:   ldd  #9600  ; overflow
set:    std  dutycycle
        rti
```

13.5.3 Temperature Controller Using Fuzzy Logic

The objective of this section is to design a *fuzzy logic* microcomputer-based temperature controller. The desired temperature is **T***. The first step in designing a controller is choosing the input sensors and output actuators. There only one sensor input, **Temperature**. There are two actuator outputs. **Heat** is a variable output (0 to 255) that will apply heat to the physical plant. **Fan** is a variable output (0 to 255) that will force air across the physical plant. **Fan+Heat** will warm up a very cold environment, **Heat** alone will slowly warm up the environment, and **Fan** alone will cool down a very hot environment. To conserve power, the use of the fan will be restricted.

The second step is selecting the crisp inputs and crisp outputs. Knowledge of how this physical plant works is critical in this step. What we wish to do depends on both absolute temperature and temperature error, so two crisp inputs will be employed:

T is the temperature (scaled to the range 0 to 255)
E is the temperature error (also scaled to the range 0 to 255)

```
; crisp inputs
T: ds 1 ;temperature (units 0.5 F)
E: ds 1 ;temperature error (128 means no error) (units 0.125 F)
```

The crisp outputs will affect the two actuators:

Heat is the heater actuator control (0 to 255)
Fan is the fan actuator control (0 to 255)

```
;crisp outputs
Heat:   ds   1 ; 0 is off and 255 is maximum heat
Fan:    ds   1 ; 0 is off and 255 is maximum fan
```

The third step is choosing the fuzzy input and output membership sets. Here, we will divide each crisp input and output into three regions. Using five or seven regions would probably create a better controller, but for purposes of illustration, we will use only three. The three fuzzy membership inputs based on temperature are

Cold	means temperature of room is cold
Normal	means the temperature of room is a normal temperature
Hot	means temperature of room is hot

The three fuzzy membership inputs based on temperature error are

TooCold	means temperature of room is below the setpoint
OK	means temperature of room is correct
TooHot	means temperature of room is above the setpoint

The three fuzzy membership outputs based on the heater are

NoHeat	means heater should be off
SomeHeat	means heater should be on a little
MaxHeat	means heater should be on a lot

The three fuzzy membership outputs based on the fan are

NoFan	means fan should be off
SomeFan	means fan should be on a little
MaxFan	means fan should be on full speed

In the fourth step, we choose functions for the conversion of crisp inputs to input membership variables. The particular constants used in these functions will be used as a starting point. Once the system is built and tested, the values can be adjusted as needed (Figure 13.28 and Program 13.23).

Figure 13.28
Fuzzification of the temperature input.

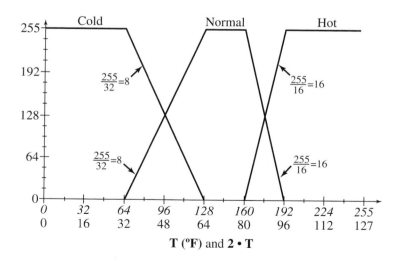

Program 13.23
Fuzzification function for the temperature input in 6812 assembly.

```
; format is Point1,Point2,Slope1,Slope2
T_tab:   dc.b   160,255,16,0      ;Hot
         dc.b   64,192,8,16       ;Normal
         dc.b   0,64,0,8          ;Cold
```

We will define the error as OK if it is within ±2°F of the setpoint (Figure 13.29 and Programs 13.24 and 13.25).

Figure 13.29
Fuzzification of the temperature error input.

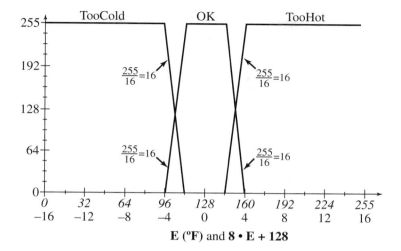

Program 13.24
Fuzzification function for the temperature error input in 6812 assembly.

```
E_tab:    dc.b   144,255,16,0     ;TooHot
          dc.b   64,192,8,16      ;OK
          dc.b   0,96,0,16        ;TooCold
```

Program 13.25
Global variables in 6812 assembly.

```
; input membership variables
fuzvar:  ds   0    ; inputs
Hot:     ds   1
Normal:  ds   1
Cold:    ds   1
TooHot:  ds   1
OK:      ds   1
TooCold: ds   1
; output membership variables
fuzout:  ds   0    ; outputs
NoHeat:  ds   1    ; turn off heater
SomeHeat: ds  1    ; apply some heat
MaxHeat: ds   1    ; turn on heater
NoFan:   ds   1    ; turn off fan
SomeFan: ds   1    ; apply some fan
MaxFan:  ds   1    ; turn on fan
; input membership variables relative offsets
hot:      equ  0
normal:   equ  1
cold:     equ  2
toohot:   equ  3
ok:       equ  4
toocold:  equ  5
;output membership variables
noheat:   equ  6    ; turn off heater
someheat: equ  7    ; apply some heat
maxheat:  equ  8    ; turn on heater
nofan:    equ  9    ; turn off fan
somefan:  equ  10   ; apply some fan
maxfan:   equ  11   ; turn on fan
```

The fifth step is creating the fuzzy rules. Again, we use our intuition as a starting point. During the testing phase of the project, we develop a means to observe the values of the input membership variables and which rules apply in certain situations. The 6812 background debug module is a convenient nonintrusive debugging tool that we can use to observe memory locations while the program is running (Program 13.26).

```
rules:   dc.b hot,toohot,$FE,noheat,maxfan,$FE       ; use fan for max cooling
         dc.b hot,ok,$FE,noheat,somefan,$FE          ; use fan for some cooling
         dc.b normal,toohot,$FE,noheat,nofan,$FE      ; no fan for normal temps
         dc.b normal,ok,$FE,noheat,nofan,$FE          ; perfect
         dc.b normal,toocold,$FE,someheat,nofan,$FE  ; a little heat
         dc.b cold,ok,$FE,noheat,nofan,$FE            ; cold but perfect
         dc.b cold,toocold,$FE,maxheat,maxfan,$FE     ; fast warmup
         dc.b $FF
```

Program 13.26
Fuzzy rules in 6812 assembly.

The last design step is defuzzification. It is here we convert the true/false fuzzy output variables (NoHeat, SomeHeat, MaxHeat, NoFan, SomeFan, MaxFan) to crisp outputs (Heat, Fan) (Program 13.27).

```
Heatsingleton: dc.b 0,50,255
Fansingleton: dc.b 0,128,255
timehan: ldaa #$20   ;clear C5F
         staa TFLG1 ;Acknowledge
         ldd  TC5
         addd #10000 ;next in 10 ms
         std  TC5
         jsr  MeasureTemperatire ; crisp input temperature, T
;reg A is temperature 0 to 255 (units 0.5•F)
         ldx  #T_tab
         ldy  #fuzvar
         mem          ; calculate Hot
         mem          ; calculate Normal
         mem          ; calculate Cold
         jsr  MeasureError ; crisp input error, E
;reg A is error 0 to 255  (128 means no error) (units 0.125•F)
         ldx  #E_tab
         mem          ; calculate TooHot
         mem          ; calculate OK
         mem          ; calculate TooCold
         ldab #6
cloop:   clr  1,y+  ; clear NoHeat, SomeHeat, MaxHeat, NoFan, SomeFan, MaxFan
         dbne b,cloop
         ldx  #rules
         ldy  #fuzvar
         ldaa #$FF
         rev
```

```
ldy  #NoHeat    ; pointer to NoHeat, SomeHeat, MaxHeat
ldx  #Heatsingleton
ldab #3
wav
ediv
tfr  y,d
stab Heat
ldy  #NoFan     ; pointer to NoFan, SomeFan, MaxFan
ldx  #Fansingleton
ldab #3
wav
ediv
tfr  y,d
stab Fan
rti
```

Program 13.27
Fuzzy controller in 6812 assembly.

Output fuzzy set	Singleton value
NoHeat	0
SomeHeat	50
MaxHeat	255

Output fuzzy set	Singleton value
NoFan	0
SomeFan	128
MaxFan	255

Observation: Fuzzy logic control will work extremely well (fast, accurate, and stable) if the designer has expert knowledge (intuition) of how the physical plant behaves.

13.6 Exercises

13.1 The objective of this exercise is to control the rotational speed of a DC motor. The +5 V DC motor has a resistance of 10 Ω. Your control system will apply a variable power to the motor by varying the duty cycle of a 50-Hz signal, as shown in Figure 13.30, for example.

Figure 13.30
PWM output to actuator.

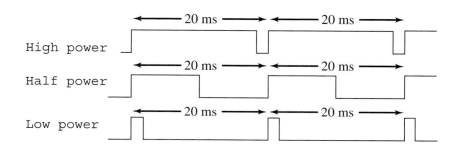

The rotation speed **R** of the motor is measured by a tachometer. You should measure **R** every 20 ms with a resolution of 1 rps. The rotational speed will vary from 0 to 250 rps ($0 \leq \mathbf{R} \leq$ 250 rps). The output of the tachometer is an ugly-looking digital wave with a frequency 100 times the motor speed. Thus, $0 \leq \mathbf{f} \leq 25$ kHz (Figure 13.31).

Figure 13.31
Tachometer frequency is
a function of the motor
rotation speed.

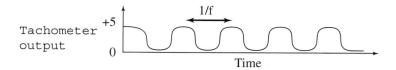

The desired rotational speed (**R*** also in rps) comes from an 8-bit parallel input port. A new
value is available on the rise of Ready. The fall of Done is an acknowledge signal back to the
input device signifying that the microcomputer no longer needs the data. The timing is shown in
Figure 13.32.

Figure 13.32
Sensor timing.

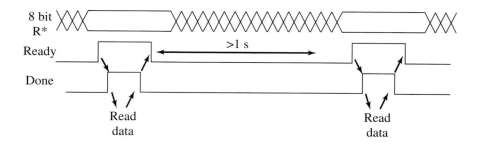

The interrupt-driven control algorithm should be implemented at 50 Hz (**U(t)** is the duty
cycle in percent):

$$U(t) = \int_0^t (R^* - R(t))dt \qquad \text{where } R^* \text{ and } R(t) \text{ are in rps and } t \text{ is in seconds}$$

a) Let **u(n)** equal the cycle count (500 ns each) that controls the duty cycle of the output-
compare variable-duty cycle 50-Hz square wave. If the range of duty cycle is $0 < U(t)$
<100, what is the relationship between **u(n)** and **U(t)**?

b) Let **R(n)** be the sampled sequence of measured rotational speed in rps. Convert the above
integral control equation into discrete form. That is, determine the relationship that calculates **u(n)** from **u(n − 1)**, **u(n − 2)** . . . , **R(n)**, **R(n − 1)**, **R(n − 2)** . . . , and **R***.

c) Show the interface from the input device, the motor, and the tachometer to the microcomputer. Label chip numbers, resistors, and capacitor values (Figure 13.33).

Figure 13.33
Sensor and motor
interface.

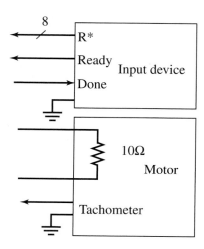

d) Show the ritual software including data structures. The main program executes the ritual, then performs other unrelated tasks (i.e., all processing occurs under interrupt control).

e) Show the interrupt software. You may poll any way you wish.

13.2 The objective of this exercise is to develop the fixed-point equations that implement a PID controller. You are to implement the following control system (Figure 13.34):

```
X(t)            is the state variable (V)
X*              is the desired state (V)
e(t)=X*-X(t)    is the error (V)
V(t)            is the actuator command (V)
```

$$V(t) = K_p e(t) + K_I \int_0^t e(\tau)\, d\tau + K_D \frac{de(t)}{dt}$$

where $K_P = 0.1$ (dimensionless)
$K_I = 10$ (1/s)
$K_D = 0.0001$ (s)

Figure 13.34
Control system for Exercise 13.2.

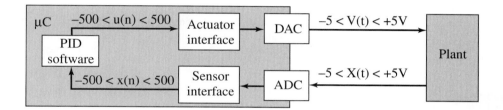

The state estimator and actuator output hardware/software interfaces are given. Your PID software is given a signed 16-bit decimal fixed-point input, x(n), that represents the current state variable. Similarly, your PID software will calculate a signed 16-bit decimal fixed-point output u(n) that will be fed to the actuator interface. For both cases, the fixed-point resolution, Δ, is 0.01 V.

Assuming the digital controller is executed every 1 ms (1000 Hz), show the control equation to be executed in the periodic interrupt handler. In this process you will convert from floating to fixed-point numbers and convert from continuous to discrete time. No software is required; just show the equations. Explain how you would deal with overflow/underflow. You should assume some noise exists on the sensor input.

13.3 Briefly explain why it is important to choose the proper update rate for a fuzzy logic controller. In particular, explain what happens to a fuzzy logic controller if the controller is executed too infrequently. Similarly, explain what happens to a fuzzy logic controller if the controller is executed too frequently.

13.4 Write a 6812 assembly language program for the fuzzy logic controller described in Section 13.5.1 using the special fuzzy logic assembly instructions.

13.5 Assume you have an 8-bit fuzzy logic system like the ones described in this chapter. Write formal descriptions for the complement and exclusive or fuzzy logic operations. Show C code implementations for these two functions.

13.6 The objective of this problem is to use the Ziegler and Nichol approach to develop the PI controller equations that allow an embedded system to control a DC motor. The state variable is speed, which is measured using 16-bit input capture and has a measurement resolution of 1 RPM. The input capture device driver repeatedly updates a global variable, called **Speed**. This 16-bit unsigned variable has units of RPM and a range of 0 to 20000. The microcontroller uses pulse-width modulation to control power to the motor.

The controller software writes to a global variable, called **Duty**, which ranges from 0 (0%) to 10000 (100%). The following plot shows an experimental measurement obtained when **Duty** is changed from 2500 to 5000. The desired speed is stored in the global variable, **Desired**, which has the same units as **Speed**. Design a fixed-point PI controller that takes **Speed** and **Desired** as inputs and calculates **Duty** as an output. From the response graph in Figure 13.35, estimate the **L** and **R** parameters of the Ziegler and Nichol method. How often should the controller be executed? Show just the equations (no software or hardware is required), calculating **Duty** as a function of **Speed** and **Desired**.

Figure 13.35
A process reaction curve for the DC motor.

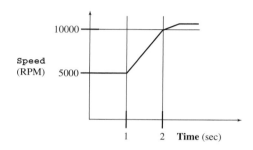

13.7 The objective of this problem is to use the Ziegler and Nichol approach to develop the PID controller equations that allow an embedded system to control the DC motor presented in Question 13.6 (i.e., work through the steps of Question 13.6 for a PID system).

13.7 Lab Assignments

Lab 13.1. The objective of this lab is to design a PID motor controller. The desired speed is received from a user interface (either a keypad or the SCI). You should use a variable-speed DC motor, paying careful attention to the voltage and current specifications of the motor. Attach onto the motor shaft a circular disk, and paint contrast lines in a circular pattern around the disk. Mount a reflectance optical sensor, such as the QRB1113 or QRB1134, pointing towards the disk, and interface it so that there is a digital squarewave with a frequency related to the speed of the shaft. Build the state estimator that measures shaft speed in real time. The first experimental measurement is to determine the current required to spin the motor at full speed. The next step is to design a PWM actuator so the software can control the delivered power to the motor. Measure the inductance of the motor (L) and the turn-off time (Δt) of the PWM switch. Use these parameters to mathematically determine the back EMF ($V = L \cdot \Delta I/\Delta t$) that occurs when the motor is turned off. Make sure the snubber diode can handle this voltage. Use the PWM in open-loop fashion and the state estimator to generate a DC response curve like that shown in Figure 13.24. Furthermore, you should measure a process reaction curve similar to that shown in Figure 13.14. Use the Ziegler and Nichol equations to design the initial PID controller, and implement it using fixed-point math. Run through an experimental fine-tuning, choosing PID parameters that minimize both controller error and response time.

Lab 13.2. The objective of this lab is to design a fuzzy logic motor controller. The desired speed is received from a user interface (either a keypad or the SCI). You should use a variable speed DC motor, playing careful attention to the voltage and current specifications of the motor. Attach onto the motor shaft a circular disk, and paint contrast lines in a circular pattern around the disk. Mount a reflectance optical sensor, such as the QRB1113 or QRB1134, pointing towards the disk, and interface it so that there is a digital squarewave with a frequency related to the speed of the shaft. Build the state estimator that measures both shaft speed and acceleration in real time. Measure the current required to spin the motor at full speed. Design a PWM actuator so the software can control the delivered power to the motor. It is OK if the actuator precision is greater than the 8-bit fuzzy logic numbers. Make sure the snubber diode can handle the back EMF generated when the motor is switched off. Use the PWM in open-loop fashion and the state estimator to generate a DC response curve like that shown in Figure 13.24. Furthermore, you should measure the time-constant of the physical plant (time to 0.69 of final response after a step change in input). Design and implement an initial fuzzy logic controller and add debugging instruments to measure fuzzy variables in real time. Run through an experimental fine-tuning, optimizing the fuzzy parameters in order to minimize both controller error and response time.

Lab 13.3. The objective of this lab is to design a PID temperature controller. The goal is to control temperature of a small object. The desired temperature is received from a user interface (either a keypad or the SCI). You can use a Peltier junction to deliver heat and a thermistor to measure temperature. Use PWM to control power to the Peltier junction, and use the ADC to measure temperature. If you use an H-bridge to control the Peltier junction, you will be able to both heat and cool the object. Be careful that your interface circuit can deliver the current required to activate the Peltier junction. Make sure there is good thermal contact between one side of the Peltier junction, the object, and the thermistor sensor. The next step is to design a PWM actuator so the software can control the delivered power to the junction. Use the PWM in open-loop fashion and the thermistor-based thermometer to generate a DC response curve like that shown in Figure 13.24. Furthermore, you should measure a process reaction curve similar to that shown in Figure 13.14. Use the Ziegler and Nichol equations to design the initial PID controller, and implement it using fixed-point math. Run through an experimental fine-tuning, choosing PID parameters that minimize both controller error and response time.

Lab 13.4. The objective of this lab is to design a fuzzy logic temperature controller. The desired temperature is received from a user interface (either a keypad or the SCI port). The goal is to control temperature of a small object. You can use a Peltier junction to deliver heat and a thermistor to measure temperature. Use PWM to control power to the Peltier junction, and use the ADC to measure temperature. If you use an H-bridge to control the Peltier junction, you will be able to both heat and cool the object. Be careful that your interface circuit can deliver the current required to activate the Peltier junction. Make sure there is good thermal contact between one side of the Peltier junction, the object, and the thermistor sensor. Build the state estimator that measures both temperature and temperature slope in real time. Design a PWM actuator so the software can control the delivered power to the junction. It is OK if the actuator precision is greater than the 8-bit fuzzy logic numbers. Use the PWM in open loop fashion and the state estimator to generate a DC response curve like that shown in Figure 13.24. Furthermore, you should measure the time-constant of the physical plant (time to 0.69 of final response after a step change in input). Design and implement an initial fuzzy logic controller, and add debugging instruments to measure fuzzy variables in real time. Run through an experimental fine-tuning, optimizing the fuzzy parameters in order to minimize both controller error and response time.

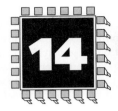

14 Simple Networks

Chapter 14 objectives are to:

❑ Introduce basic concepts of networks
❑ Present master/slave, ring, and multidrop networks based on the SCI interface
❑ Design and implement a controller area network (CAN)
❑ Develop a 9S12-based I^{2}C network
❑ Define some of the terminology and approaches to modem communication
❑ Develop a building-wide control system using the X-10 communication protocol
❑ Introduce the fundamentals of the Universal Serial Bus (USB)

The goal of this chapter is to provide a brief introduction to communication systems. Communication theory is a richly developed discipline, and much of it is beyond the scope of this book. Nevertheless, the trend in embedded systems is to employ multiple intelligent devices, Consequently, their interconnection will be a strategic factor in the performance of the system. Given that various manufacturers are involved in the development of these devices; the interconnection network must be flexible, robust, and reliable. Because the emphasis of this book is on real-time embedded systems, this chapter focuses on implementing communication networks appropriate for embedded systems. The components of an embedded system typically are combined to achieve a common objective. Therefore, the nodes on the communication network must cooperate towards that shared goal. Requirements of an embedded system, in general, involve relatively low to moderate bandwidth, static configuration, and a low probability of corrupted data. In addition, reliability and low latency are important for real-time systems.

14.1 Introduction

In the serial interfacing chapter (Chapter 7), we considered the hardware interfaces between computers. In this chapter, we will build on those ideas and introduce the concepts of networks by investigating a couple of simple networks. A communication network includes both the physical channel (hardware) and the logical procedures (software) that allow users or software processes to communicate with each other. The network provides the transfer of information as well as the mechanisms for process synchronization.

Figure 14.1
A layered approach to
communication systems.

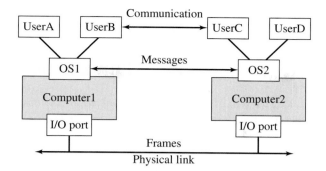

It is convenient to visualize the network in a hierarchical fashion. Figure 14.1 shows a three-layer communication system.

At the lowest level, frames are transferred between I/O ports of the two (or more) computers along the physical link or hardware channel. At the next logical level, the OS of one computer sends *messages* or *packets* to the OS on the other computer. The message protocol will specify the types and formats of these messages. Typically, error detection and correction is handled at this level. Messages typically contain four fields.

1. Address information field
   Physical address specifying the destination/source computers
   Logical address specifying the destination/source processes (e.g., users)
2. Synchronization or handshake field
   Physical synchronization, like shared clock, start, and stop bits
   OS synchronization, like request connection or acknowledge
   Process synchronization, like semaphores
3. Data field
   ASCII text (raw or compressed)
   Binary (raw or compressed)
4. Error detection and correction field
   Vertical and horizontal parity
   Checksum
   Block correction codes (BCC)

Observation: Communication systems often specify bandwidth in total bits per second, but the important parameter is the information transfer rate.

Observation: Often the bandwidth is limited by the software and not the hardware channel.

At the highest level, we consider communication between users or processes. A *process* is a complete software task that has a well-defined goal. For example, when a file is to be printed on a network printer, the OS creates a process that

1. Establishes connection with the remote printer;
2. Reads blocks from the hard disk drive and sends the data to the printer;
   The OS printer driver may have to manipulate graphics/colors for the specific printer;
   The OS network driver will break the data into message packets;
3. Disconnects the printer.

Many embedded systems require the communication of command or data information to other modules at either a near or a remote location. Because the focus of this book is on embedded systems, we will limit our discussion to communication with devices within the same room. A *full-duplex* channel allows data to transfer in both directions at the same time. In a *half-duplex* system, data can transfer in both directions but only in one direction at a time. Half-duplex is popular because it is less expensive (two wires) and allows the addition of more devices on the channel without change to the existing nodes.

14.2 Communication Systems Based on the SCI Serial Port

In this section, we will present three communication networks that utilize the SCI port. If the distances are short, half-duplex can be implemented with simple *open collector* or *open-drain* TTL-level logic. Open collector logic has two output states: low and off. In the off state the output is not driven high or low, it just floats. The 10 kΩ pull-up resistor will passively make the signal high if none of the open collector outputs are low. Both the 6811 and the 6812 can make their **TxD** serial outputs be open collector. This mode allows a half-duplex network to be created without any external logic (although pull-up resistors are often used). Three factors will limit the implementation of this simple half-duplex network: (1) the number nodes on the network, (2) the distance between nodes, and (3) presence of corrupting noise. In these situations, a half-duplex RS485 driver chip like the SP483 made by Sipex or Maxim can be used.

The first communication system uses a **master-slave** configuration, where the master transmit output is connected to all slave receive inputs, as shown in Figure 14.2. This provides for broadcast of commands from the master. All slave transmit outputs are connected together using wire-or open collector logic, allowing for the slaves to respond one at a time. The DWOM bit (SPCR bit 5) in the 6811 slaves should be set to activate open collector mode on PD1. If 6812 machines are used, then the WOMS bit 0 in the slaves should be set to activate open collector mode on PS1. The low-level device driver for this communication system is identical to the SCI driver developed in Chapter 7. When the master performs SCI output, it is broadcast to all the slaves. There can be no conflict when the master transmits, because a single output is connected to multiple inputs. When a slave receives input, it knows it is a command from the master. A potential problem exists in the other direction, because multiple slave transmitters are connected to the same signal. If the slaves transmit only after specifically being triggered by the master, no collisions can occur.

Figure 14.2
A master/slave network implemented with multiple microcomputers.

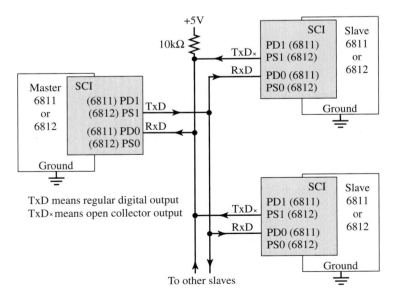

Checkpoint 14.1: What voltage level will the master RxD observe if two slaves simultaneously transmit, one making it a logic high and the other a logic low?

The next communication system is a **ring network**. This is the simplest distributed system to design, because it can be constructed using standard serial ports. In fact, we can build a ring network simply by chaining the transmit and receive lines together in a circle, as shown in Figure 14.3. Building a ring network is a matter as simple as soldering an RS232 cable in a circle with one DB9 connector for each node. Messages will include source address, destination address, and information. If computer A wishes to send information to computer C, it sends the message to B. The software in computer B receives the message, notices it is not for itself, and resends the message to C. The software in computer C receives the message, notices it is for itself, and keeps the message. Although simple to build, this system has slow performance (response time and bandwidth), and it is difficult to add or subtract nodes.

Figure 14.3
A ring network implemented with three microcomputers.

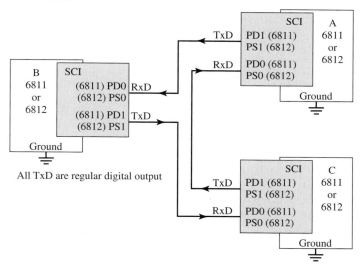

All TxD are regular digital output

Checkpoint 14.2: Assume the ring network has 10 nodes, the baud rate is 100,000 bits/sec, and there are 10 bits per frame. What is the average time it takes to send a 10-byte message from one computer to another?

The third communication system implements a very common approach to distributed embedded systems, called **multi-drop**, as shown in Figure 14.4. To transmit a byte to the other computers, the software activates the SP483 driver and outputs the frame. Since it is half-duplex, the frame is also sent to the receiver of the computer that sent it. This echo can be checked to see whether a collision occurred (two devices simultaneously outputting). If more than two computers exist on the network, we usually send address information first, so that the proper device receives the data. The 6812 SCI has a status bit in the SCISR2 register called RAF that will be true if there is an incoming frame on the RxD line. Many collisions can be avoided by checking this bit before transmitting.

Checkpoint 14.3: How can the transmitter detect whether a collision has corrupted its output?

Checkpoint 14.4: How can the receiver detect whether a collision has corrupted its input?

There are many ways to check for transmission errors. You could use a **longitudinal redundancy check** (LRC) or horizontal even parity. The error check byte is simply the *exclusive-or* of all the message bytes (except the LRC itself). The receiver also performs an *exclusive-or* on the message as well as the error check byte. The result will equal zero if the block has been transmitted successfully. Another popular method is **checksum**, which is simply the $modulo_{256}$ (8-bit) or $modulo_{65536}$ (16-bit) sum of the data packet. In addition, each byte could (but doesn't have to) include even parity.

Figure 14.4
Two multi-drop
networks implemented
with three
microcomputers.

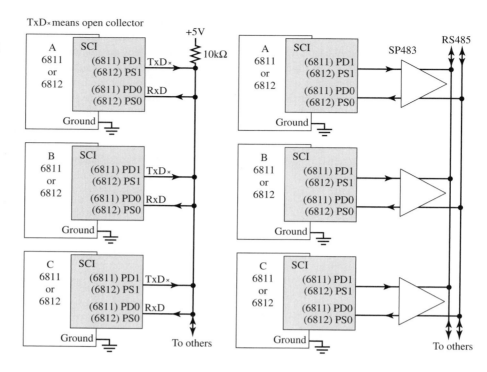

There are two mechanisms that allow the transmission of variable amounts of data. Some protocols use start (STX = $02) and stop (ETX = $03) characters to surround a variable amount of data. The disadvantage of this "termination code" method is that binary data cannot be sent, because a data byte might match the termination character (ETX). Therefore, this protocol is appropriate for sending ASCII characters. Another possibility is to use a byte count to specify the length of a message. Many protocols use a byte count. The S19 records, for example, have a byte count in each line.

14.3 Design and Implementation of a Controller Area Network (CAN)

**14.3.1
The Fundamentals
of CAN**

In this section, we will design and implement a Controller Area Network (CAN). A CAN is a high-integrity serial data communications bus that is used for real-time applications. It can operate at data rates of up to 1 Mbits per second, having excellent error-detection and confinement capabilities. The CAN was originally developed by the Robert Bosch Corporation for use in automobiles, and is now extensively used in industrial automation and control applications. The CAN protocol has been developed into an international standard for serial data communication, specifically the ISO 11989. Figure 14.5 shows the block diagram of a CAN system, which can have up to 112 nodes. There are four components of a CAN system. The first part is the CAN bus, consisting of two wires (CANH, CANL) with 120 Ω termination resistors on each end. The second part is the transceiver, which handles the voltage levels and interfacing the separate receive (**RxD**) and transmit (**TxD**) signals onto the CAN bus. The third part is the CAN controller, which is hardware built into the 9S12C32; it handles message timing, priority, error detection, and retransmission. The last part is software running within the 9S12C32, which handles the high-level functions of generating data to transmit and processing data received from other nodes.

Each node consists of a 9S12C32 microcontroller (with an internal CAN controller) and a transceiver that interfaces the CAN controller to the CAN bus. A **transceiver** is a device capable of transmitting and receiving on the same channel. The CAN is based on the "broadcast communication mechanism," which follows a message-based transmission

Figure 14.5
Block Diagram of a
9S12C32-based CAN
communication system.

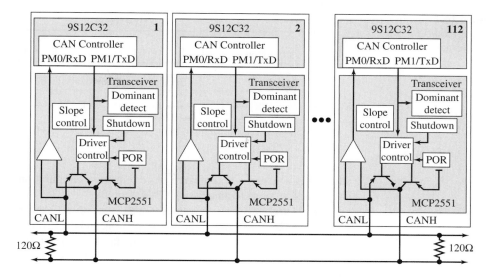

protocol rather than an address-based protocol. The CAN provides two communication services: the sending of a message (data frame transmission) and the requesting of a message (remote transmission request). All other services such as error signaling or automatic retransmission of erroneous frames are user-transparent, which implies that the CAN interface automatically performs these functions. The MC9S12C32 has an integrated CAN interface. The physical channel consists of two wires containing in differential mode one digital logic bit. Because multiple outputs are connected together, there must be a mechanism to resolve simultaneous requests for transmission. In a manner similar to open collector logic, there are **dominant** and **recessive** states on the transmitter, as shown in Figure 14.6. The outputs follow a wired-and-mechanism in such a way that if one or more nodes are sending a dominant state, it will override any nodes attempting to send a recessive state.

Figure 14.6
Voltage specifications
for the recessive and
dominant states.

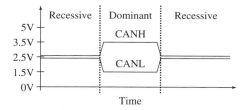

Checkpoint 14.5: What are the dominant and recessive states in open collector logic?

The CAN transceiver is a high-speed, fault-tolerant device that serves as the interface between a CAN protocol controller (located in the 9S12C32) and the physical bus. The transceiver is capable of driving the large current needed for the CAN bus and has electrical protection against defective stations. Typically, each CAN node must have a device to convert the digital signals generated by a CAN controller to signals suitable for transmission over the bus cabling. The transceiver also provides a buffer between the CAN controller and the high-voltage spikes than can be generated on the CAN bus by outside sources. Examples of CAN transceiver chips include the AMIS-30660 high-speed CAN transceiver, the Infineon Technologies TLE6250GV33 transceiver, the ST Microelectronics L9615 transceiver, the Philips Semiconductors AN96116 transceiver, and the Microchip MCP2551 transceiver. These transceivers have similar characteristics and are equally suitable for implementing a CAN system.

In a CAN system, messages are identified by their contents rather than by addresses. Each message sent on the bus has a unique identifier, which defines both the content and the priority of the message. This feature is especially important when several stations compete for bus access, a process called **bus arbitration**. As a result of the content-oriented

Figure 14.7
CAN standard format data frame.

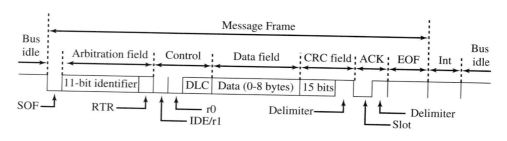

addressing scheme, a high degree of system and configuration flexibility is achieved. It is easy to add stations to an existing CAN network.

Four message types or frames can be sent on a CAN bus. These include the **Data Frame**, the **Remote Frame**, the **Error Frame**, and the **Overload Frame**. This section will focus on the Data Frame, whose parts in standard format are shown in Figure 14.7. The **Arbitration Field** determines the priority of the message when two or more nodes are contending for the bus. For the Standard CAN 2.0A, it consists of an 11-bit identifier. For the extended CAN 2.0B, there is a 29-bit identifier. The identifier defines the type of data. The **Control Field** contains the DLC, which specifies the number of data bytes. The **Data Field** contains zero to eight bytes of data. The **CRC Field** (Cyclic Redundancy Check) contains a 15-bit checksum used for error detection. Any CAN controller that has been able to correctly receive this message sends an Acknowledgement bit at the end of each message. This bit is stored in the Acknowledge slot in the CAN data frame. The transmitter checks for the presence of this bit, and if no acknowledge is received, the message is retransmitted. To transmit a message, the software must set the 11-bit identifier, set the 4-bit DLC, and give the 0 to 8 bytes of data. The receivers can define filters on the identifier field, so that only certain message types will be accepted. When a message is received, the software can read the identifier, length, and data. The **Intermission Frame Space** (IFS) separates one frame from the next. There are two factors that affect the number of bits in a CAN message frame. The ID (11 or 29 bits) and the Data fields (0, 8, 16, 24, 32, 40, 48, 56, or 64 bits) have variable length. The remaining components (36 bits) of the frame have fixed length including SOF (1), RTR (1), IDE/r1 (1), r0 (1), DLC (4), CRC (15), and ACK/EOF/intermission (13). For example, a Standard CAN 2.0A frame with two data bytes has $11 + 16 + 36 = 63$ bits. Similarly, an Extended CAN 2.0B frame with four data bytes has $29 + 32 + 36 = 97$ bits.

If a long sequence of 0's or a long sequence of 1's is being transferred, the data line will be devoid of edges that the receiver needs to synchronize its clock to the transmitter. In this case, measures must be taken to ensure that the maximum permissible interval between two signal edges is not exceeded. **Bit Stuffing** can be utilized by inserting a complementary bit after five bits of equal value. Some CAN systems add stuff bits, where the number of stuff bits depends on the data transmitted. Assuming **n** is the number of data bytes (0 to 8), CAN 2.0A may add $3 + \mathbf{n}$ stuff bits and a CAN 2.0B may add $5 + \mathbf{n}$ stuff bits. Of course, the receiver has to un-stuff these bits to obtain the original data.

The urgency of messages to be transmitted over the CAN network can vary greatly in a real-time system. Typically, there are one or two activities that require high transmission rates or quick responses. Both bandwidth and response time are affected by message priority. Low-priority messages may have to wait for the bus to be idle. There are two priorities occurring as the 9S12C32 CANs transmit messages. The first priority is the 11-bit identifier, which is used by all the CAN controllers wishing to transmit a message on the bus. Message identifiers are specified during system design and cannot be altered dynamically. The 11-bit identifier with the lowest binary number has the highest priority. In order to resolve a bus access conflict, each node in the network observes the bus level bit by bit, a process known as bit-wise arbitration. In accordance with the wired-and-mechanism, the dominant state overwrites the recessive state. All nodes with recessive transmission but dominant observation immediately lose the competition for bus access and become receivers of the message with the higher priority. They do not attempt transmission until the bus is available again. Transmission requests are hence handled according to their importance for the system as a whole. The second priority occurs locally, within each CAN node. When a node has multiple messages ready to be sent, it will send the highest priority messages first.

**14.3.2
Details of the
9S12C32 CAN**

The 9S12C32 CAN receiver has a FIFO queue, which can hold up to five incoming messages, as shown in Figure 14.8. The 9S12C32 CAN transmitter uses a priority queue, which can hold up to three outgoing messages. To transmit a message the software writes the message into addresses $0170 to $017F. The software specifies the priority of the outgoing message (**CANTXTBPR** at $017F). High-priority messages go to the front of the queue and are transmitted next. Low-priority messages go to the back of the queue and are transmitted only when no higher priority messages are ready. Once in the queue, the CAN hardware is responsible for handling priority, timing, transmitting the message, error detection, and retransmission if an error occurs. The 9S12C32 CAN receiver has a FIFO queue, which can hold up to five incoming messages. To retrieve the contents of an incoming message the software reads from addresses $0160 to $016F.

> *Observation:* It is confusing when designing systems that use a sophisticated I/O interface such as a CAN to understand the difference between those activities automatically handled by the CAN hardware module and those your software must perform. The solution is to look at software examples to see exactly the kinds of tasks your software must perform.

Figure 14.8
Data flow through the 9S12C32 CAN controller.

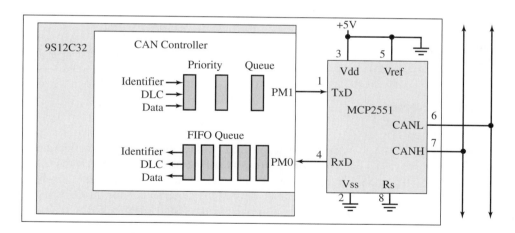

Table 14.1 shows some of the I/O ports used to program the 9S12C32 CAN. The **CANCTL0** and **CANCTL1** registers contain flags and control bits. **RXFRM** is the Received Frame Flag. It is set when a receiver has received a valid message correctly, independently of the filter configuration. Once set, it remains set until cleared by software or reset. Clearing is done by writing a "1" to the bit. **RXACT** is the Receiver Active Status flag. This read-only flag indicates find the CAN is receiving a message. **SYNCH** is the Synchronized Status flag. This read-only flag indicates whether the CAN is synchronized to the CAN bus and, as such, can participate in the communication process. **INITRQ** is the Initialization Mode Request bit. When this bit is set by the CPU, the CAN skips to Initialization Mode. Any ongoing transmission or reception is aborted, and synchronization to the bus is lost. The module indicates entry to Initialization Mode by setting **INITAK** = 1. **SLPRQ** is the Sleep Mode Request bit. This bit requests the CAN to enter Sleep Mode, which is an internal power-saving mode. The Sleep Mode request is serviced when the CAN bus is idle (i.e. the

Address	Bit 7	6	5	4	3	2	1	Bit 0	Name
$0140	RXFRM	RXACT	CSWAI	SYNCH	TIME	WUPE	SLPRQ	INITRQ	CANCTL0
$0141	CANE	CLKSRC	LOOPB	LISTEN	0	WUPM	SLPAK	INITAK	CANCTL1
$0142	SJW1	SJW0	BRP5	BRP4	BRP3	BRP2	BRP1	BRP0	CANBTR0
$0143	SAMP	TSEG22	TSEG21	TSEG20	TSEG13	TSEG12	TSEG11	TSEG10	CANBTR1
$0144	WUPIF	CSCIF	RSTAT1	RSTAT0	TSTAT1	TSTAT0	OVRIF	RXF	CANRFLG
$0145	WUPIE	CSCIE	RSTATE1	RSTATE0	TSTATE1	TSTATE0	OVRIE	RXFIE	CANRIER
$0146	0	0	0	0	0	TXE2	TXE1	TXE0	CANTFLG
$0147	0	0	0	0	0	TXEIE2	TXEIE1	TXEIE0	CANTIER
$014A	0	0	0	0	0	TX2	TX1	TX0	CANTBSEL
$014B	0	0	IDAM1	IDAM0	0	IDHIT2	IDHIT1	IDHIT0	CANIDAC
$0150–$0153	AC7	AC6	AC5	AC4	AC3	AC2	AC1	AC0	CANIDAR0–CANIDAR3
$0154–$0157	AM7	AM6	AM5	AM4	AM3	AM2	AM1	AM0	CANIDMR0–CANIDMR3
$0158–$015B	AC7	AC6	AC5	AC4	AC3	AC2	AC1	AC0	CANIDAR4–CANIDAR7
$015C–$015F	AM7	AM6	AM5	AM4	AM3	AM2	AM1	AM0	CANIDMR4–CANIDMR7
$0160	ID10	ID9	ID8	ID7	ID6	ID5	ID4	ID3	CANRXIDR0
$0161	ID2	ID1	ID0	RTR	IDE=0	0	0	0	CANRXIDR1
$0164–$016B	DB7	DB6	DB5	DB4	DB3	DB2	DB1	DB0	CANRXDSR0–CANRXDSR7
$016C	0	0	0	0	DLC3	DLC2	DLC1	DLC0	CANRXDLR
$0170	ID10	ID9	ID8	ID7	ID6	ID5	ID4	ID3	CANTXIDR0
$0171	ID2	ID1	ID0	RTR	IDE=0	0	0	0	CANTXIDR1
$0174–$017B	DB7	DB6	DB5	DB4	DB3	DB2	DB1	DB0	CANTXDSR0–CANTXDSR7
$017C	0	0	0	0	DLC3	DLC2	DLC1	DLC0	CANTXDLR
$017F	PRIO7	PRIO6	PRIO5	PRIO4	PRIO3	PRIO2	PRIO1	PRIO0	CANTXTBPR

Table 14.1
MC9S12C32 CAN ports.

module is not receiving a message and all transmit buffers are empty). The module indicates entry to Sleep Mode by setting **SLPAK** = 1. **CANE** is the CAN Enable bit, which we set to 1 to enable the CAN module. If it is 0, the module is disabled. **CLKSRC** is the CAN Clock Source bit, which defines the clock source for the CAN module. We set it to 1 to use the Bus Clock, and to 0 to use the Oscillator Clock. The frequency of the Oscillator Clock is equal to the frequency of the external crystal. The Bus Clock is the frequency at which data is accessed on the Bus and is a function of both the crystal and the PLL. We define the time quanta, T_q, as the period of the selected clock. **LISTEN** is the Listen Only Mode bit, which configures the CAN as a bus monitor. When the bit is set, all valid CAN messages with matching IDs are received, but no acknowledgement or error frames are sent out.

The **CANBTR0** and **CANBTR1** registers provide for bus timing control, which can be written only in initialization mode. **SJW1**, **SJW0** are the Synchronization Jump Width bits, which we will set to zero for high-speed communication. **BRP[5–0]** are Baud Rate Prescaler bits, and let **x** be the 6-bit number formed by these bits. The clock period used to create the individual bit timing is $(x + 1)*T_q$. **SAMP** is the Sampling bit, which determines the number of samples of the serial bus to be taken per bit time. If set, three samples per bit are taken the regular one (sample point) and two preceding samples using a majority rule. For higher bit rates, it is recommended that **SAMP** be cleared, which means that only one sample is taken per bit. There are three time segments for each transmitted bit. Segment 0 is exactly one clock period, but the length of the other two periods is programmed using **CANBTR1**. The input bit is sampled at the time in between Segment 1 and Segment 2. **TSEG22-TSEG20** are the three Time Segment 2 bits, and let **y** be the 3-bit number formed by these bits. The length of Segment 2 will be **y** + 1 clock periods. **TSEG13-TSEG10** are the four Time Segment 1 bits, and let **z** be the 4-bit number formed by these bits. The length of Time Segment 1 will be **z** + 2 clock periods. The time for each bit includes all three segments

$$\text{Bit Time} = \mathbf{T_q} *(\mathbf{x} + 1)(3 + \mathbf{y} + \mathbf{z})$$

Checkpoint 14.6: What is the relationship between y and z if we wish to sample the input in the middle of the bit interval?

CANRFLG is the Receiver Flag Register. The **WUPIF CSCIF RSTAT1 RSTAT0 TSTAT1 TSTAT0 OVRIF** and **RXF** flags are cleared by writing a "1" to the corresponding bit position. Every flag has an associated interrupt arm bit in the **CANRIER** register. For low-power applications, we can place the system in Sleep Mode. **WUPIF** is the Wake-Up Interrupt Flag, which is used to detect bus activity while in Sleep Mode. This bit is 1 when it has detected activity on the bus and requested wake-up. **CSCIF** is the CAN Status Change Interrupt Flag. This flag is set when the CAN changes its current bus status as shown in the 4-bit (**RSTAT**[1:0], **TSTAT**[1:0]) status register. The coding for the bits **RSTAT1**, **RSTAT0** is:

```
00 = Rx OK:        0 ≤ Receive Error Counter ≤  96
01 = Rx Warning:  96 < Receive Error Counter ≤ 127
10 = Rx Error:   127 < Receive Error Counter ≤ 255
11 = Bus-Off:    255 < Receive Error Counter
```

The coding for the bits **TSTAT1**, **TSTAT0** is:

```
00 = Tx OK:        0 ≤ Transmit Error Counter ≤  96
01 = Tx Warning:  96 < Transmit Error Counter ≤ 127
10 = Tx Error:   127 < Transmit Error Counter ≤ 255
11 = Bus-Off:    255 < Transmit Error Counter
```

Excessive transmitter errors will turn off both the receiver and the transmitter. **OVRIF** is the Overrun Interrupt Flag, which is set when a data overrun condition occurs. In particular, an overrun occurs when five valid messages are in the receive FIFO and a sixth message is received. **RXF** is the Receive Buffer Full Flag, which is set by the CAN when a new message is shifted in the receiver FIFO. This flag indicates whether the shifted buffer is loaded with a correctly received message (matching identifier, matching Cyclic Redundancy Code (CRC), and no other errors detected). After the CPU has read that message from locations $0160-$016F, the **RXF** flag must be cleared to release the buffer. If armed (**RXFIE**), this bit will request an interrupt.

The software can configure the 9S12C32 CAN to filter incoming messages. Accepted messages will set the **RXF** flag and will be available for processing. Dropped messages will not set the **RXF** flag and will be discarded. **CANIDAR0-7** are the Identifier Acceptance Registers. **CANIDMR0-7** are corresponding the Identifier Mask Registers. These registers can be set only in initialization mode. **CANIDAC** is the Identifier Acceptance Control Register. The two bits **IDAM1 IDAM0** specify the Identifier Acceptance Mode. 00_2 means the eight acceptance registers are configured as two 32-bit filters. 01_2 means the eight acceptance registers are configured as four 16-bit filters. 10_2 means the eight acceptance registers are configured as eight 8-bit filters. 11_2 means the filter is closed, indicating that no message will be accepted and that the foreground buffer is never reloaded. On reception, each message is written into the background receive buffer. The CPU is signaled to read the message only if it passes the criteria in the identifier acceptance and identifier mask registers (accepted); otherwise, the message is overwritten by the next message (dropped). The acceptance registers of the CAN are applied on the IDR0 to IDR3 registers of incoming messages in a bit-by-bit manner. Mask bits **AM7-AM0** are set to 0 to specify that the corresponding bit will be filtered, and a mask bit of 1 means the corresponding bit will match (be acceptable) regardless of ID bit value. **AC7-AC0** comprise a user-defined sequence of bits with which the corresponding bits of the related identifier register (IDRn) of the receive message buffer are compared. The result of this comparison is then masked with the corresponding identifier mask register. The three bits **IDHIT2**, **IDHIT1**, and **IDHIT0** specify which filter is applied to the message currently available in the receive FIFO.

> ***Observation:*** To enable the receiver to accept all messages set the mask registers to 0xFF.

CANTFLG is the Transmitter Flag Register. The flags are cleared by writing a "1" to the corresponding bit position. Every flag has an associated interrupt arm bit in the **CANTIER** register. **TXE2**, **TXE1**, and **TXE0** are the Transmitter Buffer Empty bits, which indicate that the associated transmit message buffer is empty and thus not scheduled for transmission. The CPU must clear the flag after a message is set up in the transmit buffer and is due for transmission. The CAN sets the flag after the message is sent successfully. The flag is also set by the CAN when the transmission request is successfully aborted due to a pending abort request. There are three transmit buffers in the priority, but only one is accessible at addresses $0170-$017F. **CANTBSEL** is the Transmit Buffer Selection register, which buffer will be accessible. In particular, **TX2**, **TX1**, and **TX0** are the Transmit Buffer Select bits. The lowest numbered bit places the respective transmit buffer in the $0170-$017F space (e.g., if **CANTBSEL** is 011_2, transmit buffer 0 is selected). Read and write accesses to the selected transmit buffer will be blocked if the corresponding TXEx bit is cleared and the buffer is scheduled for transmission.

IDE is the ID Extended bit, which indicates whether the extended or standard identifier format is applied in this buffer. In the case of a receive buffer, the flag is set as received and indicates to the CPU how to process the buffer identifier registers. In the case of a transmit buffer, the flag indicates to the CAN what type of identifier to send. **IDE** = 1 means Extended format (29 bit), and **IDE** = 0 means Standard format (11 bit). **RTR** is the Remote Transmission Request bit, which reflects the status of the Remote Transmission Request bit in the CAN frame. In the case of a receive buffer, it indicates the status of the received frame and supports the transmission of an answering frame in software. In the case of a transmit buffer, this flag defines the setting of the RTR bit to be sent. **RTR** = 1 means Remote frame, and **RTR** = 0 means Data frame.

14.3.3 9S12C32 CAN Device Driver

The device driver for the 9S12C32-based CAN network is divided into three components: initialization, transmission, and reception. Although the 9S12C32 can handle standard and extended message formats, this software system will be configured to handle only the standard format. Program 14.1 gives the initialization code for the interface. The high-level software on all nodes of the network will call **CAN_Open()** to initialize the CAN modules. If a node wishes to send 0 to 8 bytes of data to the other nodes, it would pass the information to **CAN_Send()**, which will transmit the message via the CAN bus. This information would then be retrieved by the receiving nodes by calling **CAN_Receive()**. The receiver will generate an interrupt when a new message is ready, and a FIFO queue will be used to pass the message from the background to the foreground. Each entry in the FIFO will be 11 bytes long: two bytes for the 11-bit ID, one byte for the 3-bit length, and eight bytes for the data. The CAN is enabled by setting the **CANE** bit. In order to set the configuration registers, the CAN must be in initialization mode. If the main program calls **CAN_Open** a second time, there may be transmit or receive messages in progress. In order to prevent errors, this ritual will first request a transfer into Sleep Mode. This request will allow incoming and outgoing messages to complete before acknowledging that Sleep Mode has been entered. Once in Sleep Mode, this ritual can safely request the CAN enter Initialization Mode. The initialization

```
void CAN_Open(void){
  asm sei              // make atomic
  CANFifo_Init();      // Initialize FIFO data structure
  CANCTL1 |= 0x80;     // CANE=1, Enable CAN
  CANCTL0 |= 0x02;     // SLPRQ=1, go to sleep first
  while((CANCTL1&0x02)==0){}; // SLPAK signifies Sleep Mode
  CANCTL0 &= ~0x02;    // SLPRQ=0, leave Sleep Mode
  CANCTL0 |= 0x01;     // INITRQ=1, Enter Initialization Mode
  while((CANCTL1&0x01)==0){}; // INITAK signifies Initialization Mode
  CANCTL1 &= ~0x10;    // LISTEN=0, get out of Listen-only mode
  CANCTL1 &= ~0x40;    // CLKSRC=0, use oscillator clock
  CANIDAC = 0x10;      // four 16-bit filters
  CANIDMR0 = 0xFF; CANIDMR1 = 0xFF; CANIDMR2 = 0xFF; CANIDMR3 = 0xFF;
  CANIDMR4 = 0xFF; CANIDMR5 = 0xFF; CANIDMR6 = 0xFF; CANIDMR7 = 0xFF;
  CANBTR0 = 0x03;      // (x+1)=4, assume oscillator is 8 MHz
  CANBTR1 = 0x23;      // (3+y+z)=8, divide by 32 gives 250,000 bits/sec
  CANCTL0 &= ~0x01;    // INITRQ=0, Leave Initialization mode
  while(CANCTL1&0x01){}; // wait for the end of initialization
  CANRIER |= 0x01;     // Arm RxF, interrupt on receive message
  asm cli              // Enable interrupts
}
```

Program 14.1
Initialization of the 9S12C32 CAN network.

sequence turns off Listen Mode, sets the clock, and establishes the acceptance filters. Setting all the acceptance masks to 0xFF means all messages will be accepted.

   Program 14.2 shows the software used to transmit a message. It begins by waiting for an empty transmit buffer. After the first while loop, one or more bits in the CANTFLG register will be set. Each flag bit that is set means its corresponding buffer is free. By writing into **CANTBSEL** the CAN selects which buffer to use. In standard format, the **CANTXIDR0** and **CANTXIDR1** registers contain the 11-bit identifier. **IDE** is set to 0 to create a standard format message with 11-bit identifier. **RTR** is set to 0 to create a data frame. The message length is copied into the **CANTXDLR** register. Zero to eight bytes are copied into the data field of the message. The priority field is set to place this message into the 3-message priority queue maintained by the transmitter. If we write multiple bits into **CANTBSEL** when selecting which buffer to use, reading from it will return which buffer was selected. The last step (writing into **CANTFLG**) causes this message to be flagged as ready to transmit.

```
void CAN_Send(unsigned short id, char length, char *data, char priority) {
  char *pt=(char*)&_CANTXDSR0; // points to transmit message buffer
  while((CANTFLG&0x07)== 0){}; // Wait for transmit buffer available
  CANTBSEL = CANTFLG;          // Request selection of empty transmit buf
  CANTXIDR0 = id>>3;           // Write Identifier into ID registers
  CANTXIDR1 = id<<5;           //    with RTR and IDE=0
  CANTXDLR = length;           // 0 to 8 bytes
  while(length){
    *pt++ = *data++;           // copy data into data registers
    length--;
  }
  CANTXTBPR = priority;        // set priority of this message
  CANTFLG = CANTBSEL;          // flag buffer as ready for transmission
}
```

Program 14.2
Transmitting a message on the 9S12C32 CAN network.

Program 14.3 shows the software used to receive a message. Interrupt synchronization is used so the main program doesn't have to be continuously checking for the presence of an incoming message. The **CANFifo** module implements a FIFO queue that puts and gets 13-byte messages. When a message has been properly received, the **RXF** flag is set and an interrupt is requested. The **CANFifo_Put** function copies 13 bytes (the ID, Length and Data from the CAN buffer) from the CAN receive buffer into the FIFO queue. After the information has been copied, the receive buffer is released by writing a 1 to **RXF**. The high-level program can get the received message by calling **CAN_Receive**.

```
void CAN_Receive(char msg[13]) {
  while (CANFifo_Get(msg) == 0){}; // wait for incoming message
}
interrupt 38 void CANInterruptHandler(void){
  char *msgPtr = (char*)&_CANRXIDR0;
  if(CANRFLG & RXF){
    CANCTL0 |= RXFRM;     // clear Received frame flag
    CANFifo_Put(msgPtr);
    CANRFLG |= RXF;       // clear RXF by writing a 1.
  }
}
```

Program 14.3
Receiving a message from the 9S12C32 CAN network.

Program 14.4 shows an example main program used to send and receive messages. If it receives a message with ID equal to 50, then it will respond with a message ID of 51 and data equal to the sum of the received data.

```
void main(void){    // example foreground program
char msg[13];        // received message
unsigned short id;  // ID of received message
char length;
char i;
short sum;
  CAN_Open();          // activate CAN
  for(;;) {
    CAN_Receive(msg);             // wait for incoming message
    id = (msg[0]<<3)+(msg[1]>>5); // bytes 0,1 are CANTXIDR0-1
    if(id == 50){
      length = msg[12];           // byte 12 is CANTXDLR
      i = 4;
      sum = 0;
      while(length){
        sum += msg[i++];          // bytes 4-11 are data
        length--;
      }
      CAN_Send(51,2,&sum,0);
    }
  }
}
```

Program 14.4
Example main program that sends and receives messages on the 9S12C32 CAN network.

14.4 Inter-Integrated Circuit (I²C) Network

14.4.1
The Fundamentals of the I²C Network

Ever since microcontrollers have been developed, there has been a desire to shrink the size of an embedded system, reduce its power requirements, and increase its performance and functionality. Two mechanisms to make systems smaller are to integrate functionality into the microcontroller and to reduce the number of I/O pins. The inter-integrated circuit (I²C) interface was proposed by Philips in the late 1980s as a means to connect external devices to the microcontroller using just two wires. The SPI interface has been very popular, but it takes three wires for simplex and four wires for full-duplex communication. In 1998, the I²C Version 1 protocol become an industry standard and has been implemented in thousands of devices. The I²C bus is a simple two-wire bidirectional serial communication system that is intended for communication between microcontrollers and their peripherals over short distances. This is typically, but not exclusively, between devices on the same printed circuit board, the limiting factor being bus capacitance. It also provides flexibility, allowing additional devices to be connected to the bus for further expansion and system development. The interface will operate at bit rates of up to 100 kbps with maximum capacitive bus loading. The module can operate up to a baud rate of 400 kbps provided the I²C bus slew rate is less than 100ns. The maximum interconnect length and the number of devices that can be connected to the bus are limited by a maximum bus capacitance of 400pF in all instances. Version 2.0 supports a high-speed mode with a data rate up to 2.4 MHz. This section will focus on Version 1, because the 9S12 does not support Version 2.

Figure 14.9 shows a block diagram of a communication system based on the I²C interface found in many 9S12 microcontrollers. The master/slave network may consist of multiple masters and multiple slaves. The **Serial Clock Line** (SCL) and the **Serial Data line** (SDA) are both bidirectional. Each line is open collector, meaning a device may drive it low or let it float. A logic high occurs if all devices let the output float, and a logic low occurs when at least one device drives it low. The value of the pull-up resistor depends on the speed of the bus. 4.7 kΩ is recommended for data rates below 100 kbps, 2.2 kΩ is recommended for standard mode, and 1 kΩ is recommended for fast mode.

Checkpoint 14.7: Why is the recommended pull-up resistor related to the bus speed?

Figure 14.9
Block diagram of an I²C communication network.

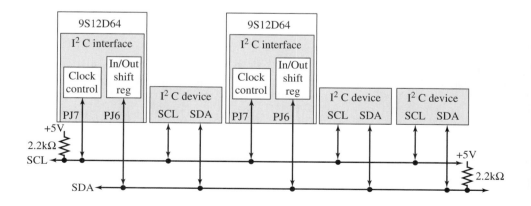

The SCL clock is used in a synchronous fashion to communicate on the bus. Even though data transfer is always initiated by a master device, both the master and the slaves have control over the data rate. The master starts a transmission by driving the clock low, but if a slave wishes to slow down the transfer, it too can drive the clock low (called **clock stretching**). In this way, devices on the bus will wait for all devices to finish. Both address (from master to slaves) and information (bidirectional) are communicated in serial fashion on SDA.

The bus is initially idle where both SCL and SDA are high. This means no device is pulling SCL or SDA low. The communication on the bus, which begins with a START and ends with a STOP, consists of five components:

START (S) is used by the master to initiate a transfer.
DATA is sent in 8-bit blocks and consists of:
   7-bit address and 1-bit direction from the master
   control code for master to slaves
   information from master to slave
   information from slave to master
ACK (A) is used by slave to respond to the master after each 8-bit data transfer.
RESTART (R) is used by the master to initiate additional transfers without releasing the bus.
STOP (P) is used by the master to signal that the transfer is complete and the bus is free.

The basic timings for these components are drawn in Figure 14.10. For now we will discuss basic timing, but we will deal with issues like stretching and arbitration later. A slow slave uses clock stretching to give it more time to react, and masters use arbitration when two or more masters want the bus at the same time. An idle bus has both SCL and SDA high. A transmission begins when the master pulls SDA low and causes a START (S) component. The timing of a RESTART is the same as a START. After a START or a RESTART, the next 8 bits will be an address (7-bit address plus 1-bit direction). There are 128 possible 7-bit addresses; however, 32 of them are reserved as special commands. The address is used to enable a particular slave. All data transfers are 8 bits long, followed by a 1-bit acknowledge. During a data transfer, the SDA data line must be stable (high or low) whenever the SCL clock line is high. There is one clock pulse on SCL for each data bit, the MSB being transferred first. Next, the selected slave will respond with a positive acknowledge (Ack) or a negative acknowledge (Nack). If the direction bit is 0 (write), then subsequent data transmissions contain information sent from master to slave. For a write data transfer, the master drives the RDA data line for 8 bits, then the slave drives the acknowledge condition during the 9th clock pulse. If the direction bit is 1 (read), then subsequent data transmissions contain information sent from slave to master. For a read data transfer, the slave drives the RDA data line for 8 bits, then the master drives the acknowledge condition during the 9th clock pulse. The STOP component is created by the master to signify the end of transfer. A STOP begins with SCL and SDA both low, then it makes the SCL clock high, and it ends by making SDA high. The rising edge of SDA while SCL is high signifies the STOP condition.

Figure 14.10
Timing diagrams of I²C components.

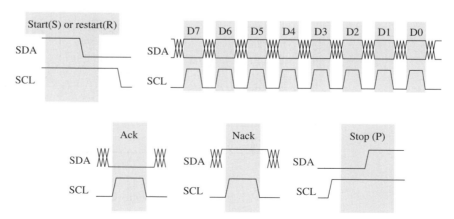

Checkpoint 14.8: What happens if no device sends an acknowledgement?

Figure 14.11 illustrates the case where the master sends 2 bytes of data to a slave. The shaded regions indicate signals driven by the master, and the white areas show those times when the signal is driven by the slave. Regardless of format, all communication begins when the master creates a START component followed by the 7-bit address and 1-bit direction. In this example, the direction is low, signifying a write format. The first through eighth SCL pulses are used to shift the address/direction into all the slaves. In order to acknowledge the master, the slave that matches the address will drive the SDA data line low during the ninth SCL pulse. During the tenth through seventeenth SCL pulses sends the data to the selected slave. The selected slave will acknowledge by driving the SDA data line low during the eighteenth SCL pulse. A second data byte is transferred from master to slave in the same manner. In this particular example, two data bytes were sent, but this format can be used to send any number of bytes, because once the master captures the bus it can transfer as many bytes as it wishes. If the slave receiver does not acknowledge the master, the SDA line will be left high (Nack). The master can then generate a STOP signal to abort the data transfer or a RESTART signal to commence a new transmission. The master signals the end of transmission by sending a STOP condition.

Figure 14.11
I²C transmission of two bytes from master to slave.

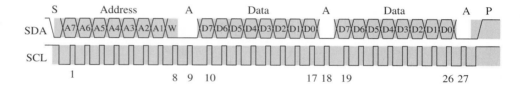

Figure 14.12 illustrates the case where a slave sends 2 bytes of data to the master. Again, the master begins by creating a START component followed by the 7-bit address and 1-bit direction. In this example, the direction is high, signifying a read format. During the tenth through seventeenth SCL pulses the selected slave sends the data to the master. The selected slave can change the data line only while SCL is low and must be held stable while SCL is high. The master will acknowledge by driving the SDA data line low during the eighteenth SCL pulse. Only two data bytes are shown in Figure 14.12, but this format can be used to receive as any many bytes the master wishes. Except for the last byte, all data are transferred from slave to master in the same manner. After the last data byte, the master does not acknowledge the slave (Nack) signifying "end of data" to the slave, so the slave releases the SDA line for the master to generate a STOP or RESTART signal. The master signals the end of transmission by sending a STOP condition.

Figure 14.12
I²C transmission of two bytes from slave to master.

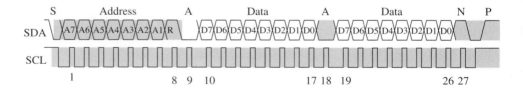

Figure 14.13 illustrates the case where the master uses the RESTART command to communicate with two slaves, reading one byte from one slave and writing one byte to the other. As always, the master begins by creating a START component, followed by the 7-bit address and 1-bit direction. During the first start, the address selects the first slave and the direction is read. During the tenth through seventeenth SCL pulses the first slave sends the data to the selected slave. Because this is the last byte to be read from the first slave, the master will not acknowledge letting the SDA data float high during the eighteenth SCL pulse, so the first slave releases the SDA line. Rather than issuing a STOP at this point, the master issues a repeated start or RESTART. The 7-bit address and 1-bit direction transferred in the twentieth through twenty seventh SCL pulses will select the second slave for writing.

Figure 14.13
I²C transmission of one byte from the first slave and one byte to a second slave.

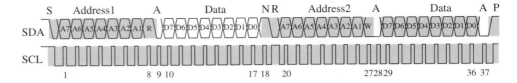

In this example, the direction is low, signifying a write format. The twenty-eighth pulse will be used by the second slave to acknowledge it has been selected. The twenty-ninth through thirty sixth SCL pulses sends the data to the second slave. During the thirty-seventh pulse the second slave to acknowledge the data it received. The master signals the end of transmission by sending a STOP condition.

Table 14.2 lists some addresses that have special meaning. A write to address 0 is a general call address and is used by the master to send commands to all slaves. The 10-bit address mode gives two address bits in the first frame and eight more address bits in the second frame. The direction bit for 10-bit addressing is in the first frame.

Table 14.2
Special addresses used in the I²C network.

Address	R/W	Description
0000 000	0	General call address
0000 000	1	Start byte
0000 001	x	CBUS address
0000 010	x	Reserved for different bus formats
0000 011	0	Reserved
0000 1xx	x	High speed mode
1111 0xx	x	10-bit address
1111 1xx	X	Reserved

14.4.2
I²C
Synchronization

The I²C bus supports multiple masters. If two or more masters try to issue a START command on the bus at the same time, both clock synchronization and arbitration will occur. **Clock synchronization** is a procedure that will make the low period equal to the longest clock low period, and the high is equal to the shortest one among the masters. Figure 14.14 illustrates clock synchronization, where the top set of traces is generated by the first master, and the second set of traces is generated by the second master. Since the outputs are open collector, the actual signals will be the wired-AND of the two outputs. Each master repeats these steps when it generates a clock pulse. It is during step 3 that the faster device will wait for the slower device.

1. Drive its SCL clock low for a fixed amount of time.
2. Let its SCL clock float.
3. Wait for the SCL to be high.
4. Wait for a fixed amount of time, stop waiting if the clock goes low.

Because the outputs are open collector, the signal will be pulled to a logic high by the 2 kΩ resistor only if all devices release the line (output a logic high). Conversely, the signal will be a logic low if any device drives it low. When masters create a START, they first drive SDA low, then drive SCL low. If a group of masters are attempting to create START commands at about the same time, then the wire-AND of their SDA lines has its 1 to 0 transition before the wire-AND of their SCL lines has its 1 to 0 transition. Thus, a valid START command will occur, causing all the slaves to listen to the upcoming address. In the example shown in Figure 14.14, Master #2 is the first to drive its clock low. In general, the SCL clock will be low from the time the first master drives it low (time 1 in this example), until the time the last master releases its clock (time 2 in this example.) Similarly, the SCL clock will be high from the time the last master releases

Figure 14.14
I²C timing illustrating clock synchronization and data arbitration.

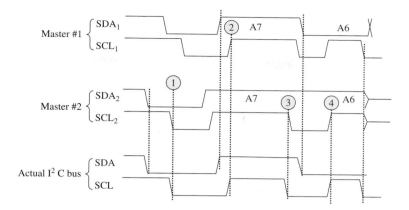

its clock (time 2 in this example), until the time the first master drives its clock low (time 3 in this example.)

The relative priority of the contending masters is determined by a **data arbitration** procedure. A bus master loses arbitration if it transmits logic "1" while another master transmits logic "0." The losing masters immediately switch over to slave receive mode and stop driving the SCL and SDA outputs. In this case, the transition from master to slave mode does not generate a STOP condition. Meanwhile, a status bit is set by hardware to indicate loss of arbitration. In the example shown in Figure 14.14, master #1 is generating an address with A7 = 1 and A6 = 0, while master #2 is generating an address with A7 = 1 and A6 = 1. Between times 2 and 3, both masters are attempting to send A7 = 1, and notice the actual SDA line is high. At time 4, master #2 attempts to make the SDA high (A6 = 1), but notices the actual SDA line is low. In general, the master sending a message to the lowest address will win arbitration.

The third synchronization mechanism occurs between master and slave. If the slave is fast enough to capture data at the maximum rate, the transfer is a simple synchronous serial mechanism. In this case the transfer of each bit from master to slave is illustrated by the following interlocked sequences (See Figure 14.10.)

Master sequence	Slave sequence (no stretch)
1. Drive its SCL clock low	
2. Set the SDA line	
3. Wait for a fixed amount of time	
4. Let its SCL clock float	
5. Wait for the SCL to be high	
6. Wait for a fixed amount of time	**6.** Capture the SDA data
7. Stop waiting if the clock goes low	

If the slave is not fast enough to capture data at the maximum rate, it can perform an operation called **clock stretching**. If the slave is not ready for the rising edge of SCL, it will hold the SCL clock low itself until it is ready. Slaves are not allowed to cause any 1 to 0 transitions on the SCL clock, but rather can delay only the 0 to 1 edge. The transfer of each bit from master to slave with clock stretching is illustrated by the following sequences.

Master sequence	Slave sequence (clock stretching)
1. Drive its SCL clock low	**1.** Wait for the SCL clock to be low
2. Set the SDA line	**2.** Drive SCL clock low
3. Wait for a fixed amount of time	**3.** Wait until it's ready to capture
4. Let its SCL clock float	**4.** Let its SCL float
5. Wait for the SCL clock to be high	**5.** Wait for the SCL clock to be high
6. Wait for a fixed amount of time	**6.** Capture the SDA data
7. Stop waiting if the clock goes low	

Clock stretching can also be used when transferring a bit from slave to master.

Master sequence
1. Drive its SCL clock low
2. Wait for a fixed amount of time

4. Let its SCL clock float
5. Wait for the SCL clock to be high
6. Capture the SDA input
7. Wait for a fixed amount of time,
8. Stop waiting if the clock goes low

Slave sequence (clock stretching)
1. Wait for the SCL clock to be low
2. Drive SCL clock low
3. Wait until next data bit is ready
4. Let its SCL float
5. Wait for the SCL clock to be high

Observation: Clock stretching allows fast and slow devices to exist on the same I²C bus Fast devices will communicate quickly with each other but slow down when communicating with slower devices.

Checkpoint 14.9: Arbitration continues until one master sends a zero while the other sends a one. What happens if two masters attempt to send data to the same address?

14.4.3
9S12 I²C Details

Many 9S12 microcontrollers have an I²C interface, but they implement just a subset of the standard. They support master and slave modes, can generate interrupts on start and stop conditions, and allow I²C networks with multiple masters. On the other hand, the 9S12 microcontrollers do not support general call, 10-bit addressing, or high-speed mode. As shown in Figure 14.9, I/O pins PJ7 and PJ6 can be connected directly to an I²C network. Because I²C networks are intended to connect devices on the same PCB board, no special hardware drivers are required. Stop mode and wait mode are two low-power states. Stop mode occurs when the device is turned off (**IBEN** = 0), and wait mode is a general state issued by the software when it executes a **wai** instruction. Table 14.3 lists the I²C ports on the MC9S12D64.

Table 14.3
MC9S12D64 I²C ports.

Address	Bit 7	6	5	4	3	2	1	Bit 0	Name
$00E0	ADR7	ADR6	ADR5	ADR4	ADR3	ADR2	ADR1	0	IBAD
$00E1	IBC7	IBC6	IBC5	IBC4	IBC3	IBC2	IBC1	IBC0	IBFD
$00E2	IBEN	IBIE	MS/SL	Tx/Rx	TXAK	RSTA	0	IBSWAI	IBCR
$00E3	TCF	IAAS	IBB	IBAL	0	SRW	IBIF	RXAK	IBSR
$00E4	DB7	DB6	DB5	DB4	DB3	DB2	DB1	DB0	IBDR

IBAD is the Bus Address Register. This register contains the address the I²C Bus will respond to when addressed as a slave. Therefore, it is not the address sent on the bus during the address transfer. **IBCR** is the I²C control register, containing many of the bits that configure the I²C interface. **IBEN** is the I²C enable bit, which must be set to activate the interface. **IBIE** is the shared interrupt arm bit for the three flags **IAAS**, **TCF**, and **IBAL**. **MS/SL** is the master/slave bit, where 1 means master and 0 means slave. When this bit is changed from 0 to 1, a START signal is generated. When this bit is changed from 1 to 0, a STOP signal is generated. A STOP signal should be generated only if the **IBIF** flag is set. **MS/SL** is cleared without generating a STOP signal automatically when the master loses arbitration. The **Tx/Rx** bit specifies whether the next data transfer will be an output (equals 1) or an input (equals 0). When operating the interface as a slave and an address match occurs, the **Tx/Rx** bit should be set to match the **SRW** flag received during the address match. When sending an address as a master, **Tx/Rx** should be 1. When a master sends data to a slave, it specifies the R/W bit (bit 0 of the address frame), and sets **Tx/Rx** in the **IBCR**. **TXAK** specifies the value driven onto SDA during data acknowledge cycles for both master and slave receivers. The I²C module will always acknowledge address matches, provided it is enabled, regardless of the value of **TXAK**. **TXAK** is used only when the I²C bus is a receiver, not a transmitter. When receiving data as a master or a selected slave, this bit determines whether an acknowledgement will be sent during the ninth clock bit. 0 means an acknowledgement will be sent (Ack), and 1 means no acknowledgement will be sent (Nack). A repeated start (RESTART) will be sent if the master writes a 1 to the **RSTA** bit, provided this 9S12 is the current bus master. **RSTA** is a write-only bit; reads from this bit always return 0. Attempting a repeated start when the bus is owned by another master will result in loss of arbitration. If **IBSWAI** is 1, then the I²C device will halt during wait mode.

IBFD is the I²C Bus Frequency Divider Register, which determines the baud rate transferred as a master. The bit clock generator is implemented as a prescale divider—IBC7-6, prescaled shift register IBC5-3 select the prescaler divider and IBC2-0 select the shift register tap point. The timing of the 9S12 I²C interface is derived from the bus clock. Table 14.4 presents the three fields of IBFD that define operating speed. Figure 14.15 defines the four timing intervals. t_{start} is the delay from the fall of SDA data to the fall of SCL clock during a START or RESTART. t_{bit} is the time to transfer one bit. t_{hold} is the time after the fall of the clock during which the data will remain valid. t_{stop} is the delay from the rise of SCL clock to the rise of SDA data during a STOP. Table 14.5 gives the ratio of bus frequency to I²C frequency for all possible values of IBFD. For example, if the bus frequency is 8 MHz, and we wish to create an I²C clock frequency of 200 kHz, then we need a divider value of 40. From Table 14.5, we see that IBFD could be chosen as $07, $0B, or $40.

Table 14.4
MC9S12D64 I²C timing components as specified by IBFD.

IBC7-6	MUL		IBC5-3	scl2 start	scl2 stop	scl2 tap	tap2 tap		IBC2-0	SCLTap	SDATap
00	1		000	2	7	4	1		000	5	1
01	2		001	2	7	4	2		001	6	1
10	4		010	2	9	6	4		010	7	2
11	reserved		011	6	9	6	8		011	8	2
			100	14	17	14	16		100	9	3
			101	30	33	30	32		101	10	3
			110	62	65	62	64		110	12	4
			111	126	129	126	128		111	15	4

Figure 14.15
9S12D64 I²C timing intervals.

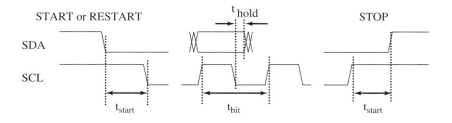

Let t_E be the period of the bus clock, then the four timing intervals are as follows:

$$t_{bit} = t_E \cdot MUL \cdot \{2 \cdot (scl2tap + [(SCLTap - 1) \cdot tap2tap] + 2)\}$$

$$t_{hold} = t_E \cdot MUL \cdot \{scl2tap + [(SDATap - 1) \cdot tap2tap] + 3\}$$

$$t_{start} = t_E \cdot MUL \cdot [scl2start + (SCLTap - 1) \cdot tap2tap]$$

$$t_{stop} = t_E \cdot MUL \cdot [scl2stop + (SCLTap - 1) \cdot tap2tap]$$

IBFD	$0–	$1–	$2–	$3–	$4–	$5–	$6–	$7–	$8–	$9–	$A–	$B–
$–0	20	48	160	640	40	96	320	1280	80	192	640	2560
$–1	22	56	192	768	44	112	384	1536	88	224	768	3072
$–2	24	64	224	896	48	128	448	1792	96	256	896	3584
$–3	26	72	256	1024	52	144	512	2048	104	288	1024	4096
$–4	28	80	288	1152	56	160	576	2304	112	320	1152	4608
$–5	30	88	320	1280	60	176	640	2560	120	352	1280	5120
$–6	34	104	384	1536	68	208	768	3072	136	416	1536	6144
$–7	40	128	480	1920	80	256	960	3840	160	512	1920	7680
$–8	28	80	320	1280	56	160	640	2560	112	320	1280	5120
$–9	32	96	384	1536	64	192	768	3072	128	384	1536	6144
$–A	36	112	448	1792	72	224	896	3584	144	448	1792	7168
$–B	40	128	512	2048	80	256	1024	4096	160	512	2048	8192
$–C	44	144	576	2304	88	288	1152	4608	176	576	2304	9216
$–D	48	160	640	2560	96	320	1280	5120	192	640	2560	10240
$–E	56	192	768	3072	112	384	1536	6144	224	768	3072	12288
$–F	68	240	960	3840	136	480	1920	7680	272	960	3840	15360

Table 14.5
MC9S12D64 I^2C clock divider values as specified by IBFD.

Checkpoint 14.10: Assuming a 24 MHz bus clock, what value can you program into IBFD to create a 100 kHz baud rate?

IBSR is the I^2C Bus Status Register. This status register is read-only, except that **IBIF**, **IAAS**, and **IBAL** can be cleared if the software writes a 1 to the corresponding bit position. **TCF** is the Data Transferring bit. While one byte of data is being transferred, this bit is cleared. It is set by the falling edge of the ninth clock of a byte transfer. Note that this bit is valid only during or immediately following a transfer to the I^2C module or from the I^2C module. **IAAS** is the Addressed as a slave bit, which is set when its own specific address (**IBAD**) is matched with the calling address sent by another master. The **IAAS** flag will request an interrupt if the **IBIE** arm bit is set. After **IAAS** is set, the software should check the **SRW** bit and set its **Tx/Rx** mode accordingly. Writing a 1 to **IAAS** clears this bit. **IBB** is the Bus Busy bit, indicating the status of the bus. When a START signal is detected, the **IBB** is set. If a STOP signal is detected, **IBB** is cleared. A master should wait until **IBB** is 0, signifying the bus is idle before initiating a transfer. **IBAL** is the Arbitration Lost bit, which is set by hardware when the arbitration procedure is lost. Arbitration is lost in the following circumstances:

- SDAs sampled low when the master drives a high during an address or data transmit cycle.
- SDAs sampled low when the master drives a high during the acknowledge bit of a data receive cycle.
- A START cycle is attempted when the bus is busy.
- A RESTART cycle is requested in slave mode.
- A STOP condition is detected when the master did not request it.

This bit must be cleared by software, by writing a 1 to it. **SRW** is the Slave Read/Write bit, which indicates the value of the R/W command bit of the calling address sent from the master. This bit is valid only when the 9S12 is in slave mode, a complete address transfer has occurred with an address match, and no other transfers have been initiated. Checking **SRW**, the CPU can select slave transmit/receive mode according to the command of the master. If **SRW** is 0, it means the master writing data to this 9S12 as a slave. Conversely, if

SRW is 1, master wishes this 9S12 slave to transmit data back to the master. **IBIF** is the I²C Interrupt bit, which is set when one of the following conditions occurs:

- Arbitration lost (**IBAL** bit set)
- Byte transfer complete (**TCF** bit set)
- Addressed as slave (**IAAS** bit set)

These three conditions will request an interrupt if the **IBIE** arm bit is set. **IBIF** must be cleared by software, writing a 1 to it. **RXAK** is the Received Acknowledge bit, which is the value of SDA during the acknowledge bit of a bus cycle. If the **RXAK** is low, it indicates that an acknowledge signal has been received during the ninth clock. If **RXAK** is high, it means no acknowledge signal is detected at the ninth clock.

IBDR is the I²C Bus Data I/O Register. In master transmit mode, when data is written to the **IBDR** a data transfer is initiated. As shown in Figures 14.10 through 14.13, the most significant bit is sent first. In master transmit mode (**MS/SL** = 1 and **Tx/Rx** = 1), writing this register initiates a 9-bit transmission using both SDA data and SCL clock. In master receive mode (**MS/SL** = 1 and **Tx/Rx** = 0), reading this register initiates next byte data receiving, where the 8-bit data from SDA input and the SCL clock is an output. When either a master or a slave is sending, the **TXAK** bit is send during the ninth clock. In slave mode, input/output functions are available only after an address match has occurred. Note that the **Tx/Rx** bit must correctly reflect the desired direction of transfer in master and slave modes for the transmission to begin. Reading the **IBDR** will return the last byte received while the I²C is configured in either master receive or slave receive modes. The **IBDR** does not reflect every byte that is transmitted on the I²C bus, nor can software verify that a byte has been written to the **IBDR** correctly by reading it back. In master transmit mode, the first byte of data written to **IBDR** following assertion of **MS/SL** is used for the address transfer and should consist of the calling address (in position D7-D1) concatenated with the required R/W bit (in position D0).

14.4.4
9S12 I²C Single Master Example

The objective of this example is to present a low-level device driver for an I²C network where this 9S12 is the only master, as shown in Program 14.5. This simple example will employ busy-wait synchronization. **I2C_Open** first enables the I²C interface, starting out in slave mode. Since this is the only master, it does not need a slave address (**IBAD**).

Program 14.5
9S12C32 I²C initialization in single master mode.

```
void I2C_Open(void){
  IBCR = 0x80; // enable, no interrupts, slave mode
  IBFD = 0x1F; // 100kHz, assuming 24 MHz bus clock
  IBSR = 0x02; // clear IBIF
}
```

Program 14.6 contains the function **I2C_Send** that transmits two bytes to a slave, creating a transmission shown in Figure 14.11. In a system with multiple masters it should check to see whether the bus is idle first. Because this system has just one master, the bus should be idle. By setting the **MS/SL** bit, the 9S12 will create a START condition. In a system with multiple masters, it should check to see whether it lost bus arbitration (**IBAL**). The slave address (with bit 0 equal to 0) will be sent. The two data bytes are sent, and then the STOP is issued. If there is a possibility the slave doesn't exist, then this program could have checked **RXAX** after each transfer.

Program 14.7 contains the function **I2C_Recv**, which receives two bytes from a slave, creating a transmission shown in Figure 14.12. By setting the **MS/SL** bit, the 9S12 will create a START condition. During the first transfer, the **Tx/Rx** bit is 1, so the slave address (with bit 0 equal to 1) will be sent. During the second two transfers, the **Tx/Rx** bit is 0, so data flows into the 9S12. To trigger the first data reception, the software performs a dummy read on the **IBDR**. During the first data transfer, TXAK is 0, creating a positive acknowledgement. Conversely, during the second data transfer, TXAK is 1, creating a negative

```
void I2C_Send(char slave, unsigned char data1, unsigned char data2){
  IBCR |= 0x30;              // send START
  IBDR = slave&0xFE;         // send address with D0=0 signifying write
  while((IBSR&0x02)==0){};  // wait for the address to be sent
  IBSR = 0x02;               // clear IBIF
  IBDR = data1;              // send first byte
  while((IBSR&0x02)==0){};  // wait for the data to be sent
  IBSR = 0x02;               // clear IBIF
  IBDR = data2;              // send second byte
  while((IBSR&0x02)==0){};  // wait for the data to be sent
  IBSR = 0x02;               // clear IBIF
  IBCR &=~0x30;              // send STOP
}
```

Program 14.6
9S12C32 I^{2}C transmission in single-master mode.

```
unsigned short I2C_Recv(char slave){ unsigned char data1,data2;
  IBCR |= 0x30;              // send START
  IBDR = slave|0x01;         // send address with D0=1 signifying read
  while((IBSR&0x02)==0){};  // wait for the address to be sent
  IBSR = 0x02;               // clear IBIF
  IBCR &= ~0x18;             // Tx/Rx=0, and TXAK=0
  data1 = IBDR;              // dummy read to initiate receiving
  while((IBSR&0x02)==0){};  // wait for the data to be received
  IBSR = 0x02;               // clear IBIF
  IBCR |= 0x08;              // TXAK=1
  data1 = IBDR;              // capture first byte, initiate second
  while((IBSR&0x02)==0){};  // wait for the address to be sent
  IBSR = 0x02;               // clear IBIF
  IBCR &=~0x38;              // send STOP
  data2 = IBDR;              // capture second byte
  return (data1<<8)+data2;
}
```

Program 14.7
9S12C32 I^{2}C reception in single-master mode.

acknowledgement, signaling to the slave that this is the last data to be transferred. The two data bytes are received, and then the STOP is issued. Notice that the last byte is captured after the STOP is issued, so that a third data transfer is not initiated.

Observation: We can produce the open-collector behavior of any I/O port that has a direction register. We initialize the port by writing a 0 to the data port. On subsequent accesses to the open-collector port, we write the complement to the direction register. That is, if we want the I/O port bit to drive low, we set the direction register bit to 1, and if we want the I/O port bit to float (open collector), we set the direction register bit to 0.

14.5   Modem Communications

14.5.1
FSK Modem

A *modem* is an electronic device that *MO*dulates and *DEM*odulates a communication signal. The transmitter of a FSK modem modulates the digital signals into frequency-encoded sine waves. The receiver demodulates the sine waves back into digital signals. The transmitting computer asynchronous serial port (SCI) converts the digital data into a RS232

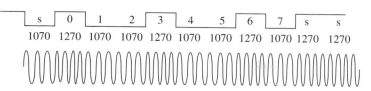

Figure 14.16
300 bits/sec FSK
transmission of the
character 'I' (originate to
answer).

timing serial signal. The Bell103 or V.21 standard achieves a data rate of 300 bits/s. The transmitting modem modulates the digital signal into 3.3-ms bursts of 1070- or 1270-Hz sounds. Figure 14.16 shows the encoded/decoded signals for the transmission of the ASCII character 'I', which is $49.

The transmitting bandpass filter guarantees there are no 2-kHz components in the output signal. The mixer adds the outgoing sound (1070 or 1270 Hz) with the incoming sounds on the telephone. The receiver bandpass filter will pass only the 1070/1270 Hz sounds to the demodulator. The receiver modem will demodulate the waveforms back into digital signals. If all goes well, the input to the receiving SCI, RxD, should be the same as the output from the transmitting SCI, TxD, only delayed in time. The software in the transmitting microcomputer should write data to the transmit serial data register (SCDR) when its TDRE = 1. Eventually, that data will arrive in the receiver's receive serial data register (SCDR), setting the receiver's RDRF (Figure 14.17).

Figure 14.17
Block diagram of a FSK
modem communication
system, showing one
direction transfer.

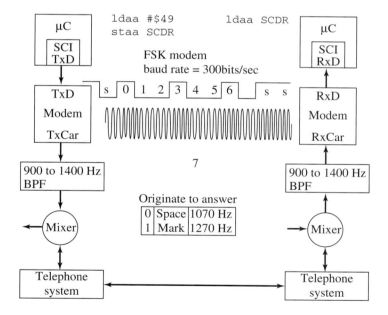

To implement a full-duplex channel, we use a different set of frequencies to simultaneously transmit data in the other direction. We define the *originate* modem as the device that places the telephone call. It dials the telephone, originating the communication. The *answer* modem is the device that answers the phone. The *carrier frequency* is defined as the average or midvalue frequency. In the FSK protocol, the two carrier frequencies are 1170 and 2125 Hz. By convention, the modem that initiates the connection uses the lower carrier frequency. Once a connection has occurred, full-duplex communication can occur. A logical true or Boolean 1 is encoded as a *mark,* while false (0) is defined as a *space*. Table 14.6 defines the full-duplex FSK protocol.

Table 14.6
Frequencies used in a
300 bits/s FSK modem.

Direction	Space, false, /0	Mark, true, 1
Originate to answer Answer to originate	1070 Hz 2025 Hz	1270 Hz 2225 Hz

Figure 14.18 shows the encoded/decoded signals for the answer to originate transmission of a $49.

Figure 14.18
300 bits/s FSK
transmission of the
character 'I' (answer to
originate).

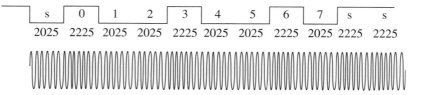

An important function of the bandpass filters is to separate the two pairs of frequencies so that transmission in one direction does not interfere with transmission in the other (Figure 14.19).

Figure 14.19
Block diagram of a FSK
modem communication
system, showing transfer
in both directions.

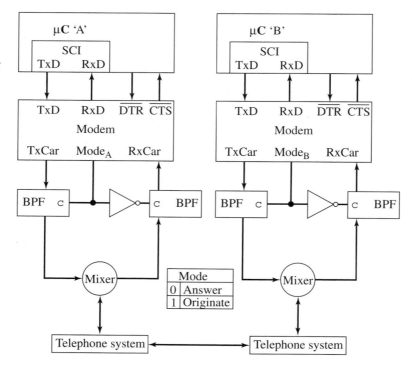

If we develop a modem interface that can both originate and answer telephone calls, then we will need bandpass filters that have digitally controlled center frequencies. We will build a bandpass filter that has the following digital control. c is a digital input from the interface to the bandpass filter (Table 14.7). If computer A originates the call, then $MODE_A = 1$, and $MODE_B = 0$. These values are reversed if computer B originates the communication.

Table 14.7
Filters used in a 300 bits/s
FSK modem.

c	Low frequency	High frequency
0	1900	2400
1	900	1400

The bandwidth of the FSK protocol is limited by the frequency response of the telephone system. At 300 bits/s the four sounds have 3, 4, 6, 7 cycles per bit time. These differences allow for accurate demodulation. If we wished to increase the communication bandwidth by a factor of 10, we would have to increase the carrier frequencies by this factor as well. For example, at 3000 bits/s, we would need carrier frequencies of 11 and 21 kHz to maintain the desirable 3, 4, 6, 7 cycles per bit time property. Unfortunately, the telephone was designed so that humans could comprehend audio speech, and it will not pass 21-kHz sound waves.

14.5.2
Phase-Encoded
Modems

There are three basic components of the wave that we can use to encode the information: frequency, phase, and amplitude. Each of these characteristics can be exploited to encode information. A *phase-shift keying* (PSK) protocol encodes the information as phase changes between the sounds.

A *baud* is defined as a pulse of sound. In the early days of modem communications, the FSK protocol encoded just 1 bit/baud. To increase the communication bandwidth without increasing the carrier frequency, multiple bits are encoded into each change in sound. The *baud rate* is the number of sounds (or sound changes) transmitted per second. With a FSK protocol, the baud rate equals the data rate. On the other hand, with the phase-encoded schemes, the transmission rate will usually be higher than the baud rate. This is a very confusing point because many people improperly interchange the terms baud rate, data rate, and bandwidth. If we encode n bits of data per baud, then

$$\text{data rate} = n \cdot \text{baud rate}$$

For example, the 2400 bits/s International Telegraph and Telephone Consultative Committee (CCITT) protocol uses an 1800-Hz carrier and a 1200 bits/s baud rate. It encodes 2 bits/baud as shown in Table 14.8.

Table 14.8
Phase shifts used in a
1200 bits/s PSK modem.

Data	CCITT phase
00	0°
01	90°
11	180°
10	270°

Encoding is the process of transmitting one or more bits of information with each baud. To improve bandwidth we can also implement data compression. Most data contain repetitive or redundant information. With data compression, the information (the message, digitally encoded sound, or photograph) is reformatted so that it occupies fewer bytes. The compressed data are transmitted across the channel and expanded on the other side. For example, a sampled waveform can be compressed into a sequence of linear line segments (Figure 14.20).

Figure 14.20
Compressed data allow information to be transferred with few data points.

For modem communications, we use Huffman or Lempel-Ziv encoding. We can evaluate a compression/decompression algorithm by three factors. The *compression ratio* is defined as the ratio of the number of original bytes to the number of compressed bytes. If an algorithm achieves a compression ratio of 4:1, then a 1000-byte message can be compressed into 250 bytes. Since only the compressed message is transmitted, the effective bandwidth of the communication system is four times faster than the physical bandwidth of the channel. The second factor to consider is the speed of the compression/decompression algorithm. For most modem applications, we require the compression/decompression functions to proceed in real time so that the operation occurs transparent to the user. The last and most important factor is accuracy. In some compression/decompression applications, such as photography or sound recordings, we can tolerate some imperfections or errors. In most modem applications however, we expect the received message to exactly equal the message transmitted.

Consider the following example of a CCITT 2400 bits/s PSK transmission. To transmit the ASCII 'A' ($41 = %01000001) with one start bit, 8-bit data, even parity, and two stop bits, we transmit the sequence start = 0, data = 10000010, parity = 0, and stop = 11. Remember the data bits are transmitted starting with the least significant bit. The sequence is divided into 2-bit pieces: 01, 00, 00, 01, 00, 11 and encoded as phase shifts: 90°, 0°, 0°, 90°, 0°, 180°. With a carrier frequency of 1800 Hz, we transmit 1.5 cycles and then phase-shift the signal. The baud rate is 1200 because we produce 1200 sound changes per second. The data rate is 2400 bits/s, twice the baud rate, because each sound change encodes 2 bits of data (Figure 14.21).

Figure 14.21
Waveform generated by a phase-encoded modem.

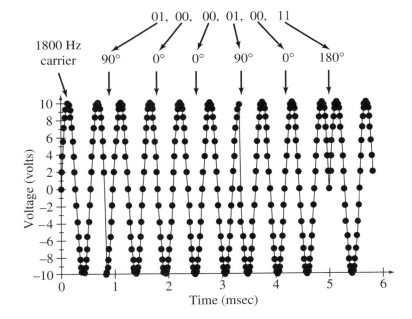

**14.5.3
Quadrature
Amplitude
Modems**

In a quadrature amplitude modem (QAM), we use both the phase and amplitude to encode up to 6 bits onto each baud. For standard QAM, only 4 bits is used to encode data. The *trellis-coded quadrature amplitude modem* (TCQAM or TCM) includes error-correcting signal processes so that all 6 bits can be reliably transmitted (Table 14.9).

Table 14.9
High-speed modems transmit many bits per sound.

Standard	Baud rate	bits/baud	Data rate (bit/s)	Modulation
V.21 Bell103	300	1	300	FSK
V.22 Bell212A	600	2	1,200	PSK
V.26	1200	2	2,400	PSK
V.27	1600	3	4,800	PSK
V.32	2400	4	9,600	QAM
V.33	2400	6	14,400	TCQAM
V.34	3200	10–11	33,600	TCM
V.92	8000	7	56,000	PCM

14.6 X-10 Protocol

X-10 protocol is a simple low-bandwidth communication protocol that can be used to control multiple devices within a single building. The Sears Home Control System and the Radio Shack Plug'n Power System first introduced the code format in 1978. Since then, many home automation and home security systems have been built around this standard. In this section we will develop X-10–based communication systems based on the PL513 and TW523 modules available from X-10 Powerhouse. An X-10 transmitter module sends an 11-bit packet to an X-10 receiver module over the existing power line wires. The key advantage of X-10 is its ability to use existing wires in the building and the wide range of available X-10 interfaces that can be installed. A communication system built with the PL513 allows the microcomputer to send commands that can control remote devices. With a TW523 interface the computer can both send and receive packets. So in addition to the ability to control remote devices, we can also receive information from remote locations. Figure 14.22 depicts a communication system with a single computer controller, but we can add multiple computer interfaces and use the X-10 network to communicate between them.

Figure 14.22
Communication systems built around the PL513 and TW523 modules.

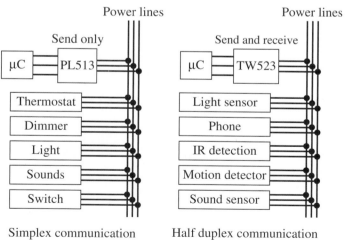

Since the protocol encodes the binary information as 1-ms-long 120-kHz bursts modulated at the zero crossings of the 60-Hz power line, the bandwidth will be less than 60 bits/s. For many applications, however, this is adequate. Figure 14.23 illustrates the transmission of a logic 1, which is signified by the presence of the 120-kHz pulses in the first half cycle

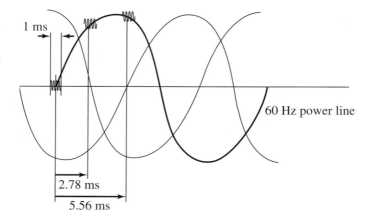

1 ms

60 Hz power line

2.78 ms

5.56 ms

of the 60-Hz power wave. The 120-kHz pulses are repeated at 2.778 and 5.556 ms so that communication can occur between modules that reside on any of the phases in a three-phase distribution system. A binary 0 is represented by the presence of the 120-kHz pulses in the second half cycle of the 60-Hz power wave. A special start code is used to synchronize the transmitter and receiver modules.

Each packet requires 11 power line cycles to transmit. The first two cycles are the Start Code (1110), the next four cycles represent the House Code, and the last five cycles include the Key Code. The House Code is used to address the device to which the packet is being sent. The X-10 protocol uses letters A through P to specify the House Codes (Table 14.10).

Table 14.10
X-10 house codes.

House code	H1 H2 H4 H8
A	0 1 1 0
B	1 1 1 0
C	0 0 1 0
D	1 0 1 0
E	0 0 0 1
F	1 0 0 1
G	0 1 0 1
H	1 1 0 1
I	0 1 1 1
J	1 1 1 1
K	0 0 1 1
L	1 0 1 1
M	0 0 0 0
N	1 0 0 0
O	0 1 0 0
P	1 1 0 0

The X-10 protocol allows for the transmission of a multiple-bit packet over a power grid. Each X-10 packet is defined as 22 bits long. The first 4 bits are a unique "start" code, represented as 1110. The next 8 bits in the X-10 packet are the 4-bit house code, sent with each bit followed by its complement. The last 10 bits in the X-10 packet are the 5-bit function or unit code, also sent with each bit followed by its complement. Thus the X-10 complimentary rule dictates that for both house code and unit/function code, a logic 1 is represented as 10, and a logic 0 is represented as 01. The X-10 protocol also insists that each packet must be transmitted twice, with each transmission separated by 3 (60-Hz) cycles of no-transmission. To activate a specific unit, an X-10 packet must first be sent that contains the house code and the unit code, followed by another X-10 packet that contains the same house code and the function code. Note that each X-10 packet must be sent twice to send house A, unit 1, function on. This complete package consists of a total of 22 + 3 + 22 + 3 + 22 + 3 + 22 = 97 bits. The transmission looks like this:

Start	(4 bits)
House A	(8 bits)
Unit 1	(10 bits)
3-cycle no-transmission	(6 bits)
Start	(4 bits)
House A	(8 bits)
Unit 1	(10 bits)
3-cycle no-transmission	(6 bits)
Start	(4 bits)
House A	(8 bits)
Function on	(10 bits)
3-cycle no-transmission	(6 bits)
Start	(4 bits)
House A	(8 bits)
Function on	(10 bits)

Note that to perform the same function on multiple units in the same house code, the unit codes could be sent next to each other, followed by the function code. For example, to turn on both A1 and A2, send house A, unit 1, house A, unit 2, house A, function on. Also, functions that are addressed to all receivers, such as all units on, do not need the unit code to precede them. Finally, a special note on dim and bright functions: They do not need the 3-cycle delay. The X-10 protocol has 32 Key Codes to specify the data (Number Code) or command (Function Code) information (Table 14.11).

Table 14.11
X-10 number and function codes.

Number code	D1 D2 D4 D8 D16	Function code	D1 D2 D4 D8 D16
1	0 1 1 0 0	All lights off	0 0 0 0 1
2	1 1 1 0 0	All lights on	0 0 0 1 1
3	0 0 1 0 0	On	0 0 1 0 1
4	1 0 1 0 0	Off	0 0 1 1 1
5	0 0 0 1 0	Dim	0 1 0 0 1
6	1 0 0 1 0	Bright	0 1 0 1 1
7	0 1 0 1 0	All lights off	0 1 1 0 1
8	1 1 0 1 0	Extended code	0 1 1 1 1
9	0 1 1 1 0	Hail request	1 0 0 0 1
10	1 1 1 1 0	Hail acknowledge	1 0 0 1 1
11	0 0 1 1 0	Preset dim 0	1 0 1 0 1
12	1 0 1 1 0	Preset dim 1	1 0 1 1 1
13	0 0 0 0 0	Extended data	1 1 0 0 1
14	1 0 0 0 0	Status=on	1 1 0 1 1
15	0 1 0 0 0	Status=off	1 1 1 0 1
16	1 1 0 0 0	Status request	1 1 1 1 1

The hail request is sent to determine if there are any other X-10 transmitters on the system. This allows a transmitter at start-up to assign itself a unique house code. There are 5 bits or 32 levels specified in the preset dim command. In the preset dim command, the D8 bit represents the most significant bit, and H1, H2, H4, H8 are the least significant bits. For example, the H1, H2, H4, H8, D1, D2, D4, D8, D16 code of **011110111** represents the dim level **10111**, or 23/32. Normally there is a 3-cycle gap between packets, except for "dim" and "bright" that can be sent one right after another without gaps.

Except for the start code, the X-10 protocol uses complementary logic in the first and second halves of the 60-Hz cycles. Notice that the logic 1 is encoded by pulses followed by no pulses, and the logic 0 is encoded by no pulses followed by pulses. As an example, if the computer wished to turn on light "P," it would send House Code "P" (1100) with Function Code "on" (00101). Figure 14.24 illustrates this transmission.

Figure 14.24
X-10 transmission used to turn on module "P".

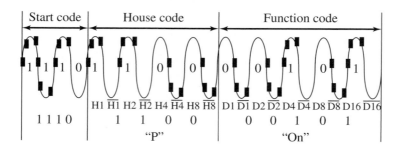

The extended code and extended data formats allow for packets to be longer than 9 bits. There are no standard protocols for these extended formats, except the bit stream must follow the first packet continuously without gaps. For example, we could send the first 8-bit byte as the data count, followed by the raw data. Unfortunately, the PL513 and TW523 interfaces can send extended formats, but neither one can receive extended formats.

When using the PL513 and/or TW523 interfaces, it is necessary to repeat all transmissions. To allow the receiver time to recover and reinitialize, most receivers require a 3 power cycle gap between packets. Figure 14.25 illustrates the waveforms used to first turn on module "P" then turn off module "N".

Figure 14.25
X-10 transmissions to first turn on module "P" then turn off module "N".

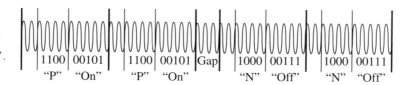

If all we require is the ability to transmit packets, then the PL513 interface can be used (Figure 14.26).

The zero-crossing signal from the PL513 is a square wave that the system uses to synchronize the 120-kHz bursts to the 60-Hz power wave. The microcomputer is responsible for placing a 1-ms (0.95 to + 1.1 ms) positive logic within 50 µs of the rising and/or falling edge of the zero-crossing trigger. Figure 14.27 illustrates the signals that the microcomputer must generate to turn on module "P".

We will connect the zero-crossing input to a microcomputer pin that can generate interrupts on both the rising and falling edges. Either input capture or key wake-up mechanisms can be used. The **Transmit Output** signal can be connected to a simple output pin or to an output compare signal. To transmit a packet, the main program, foreground thread, will

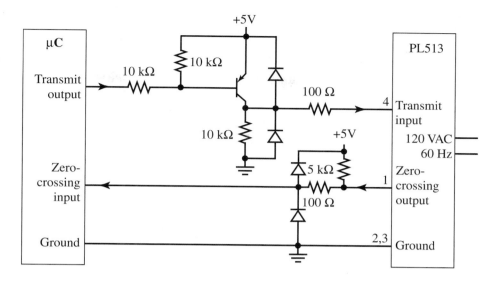

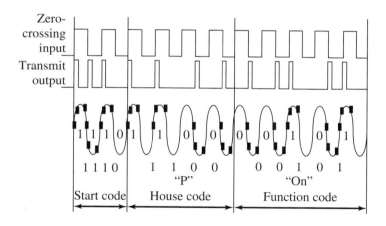

place 50 bits into a BitFifo and arm the zero-crossing interrupt signal so that the first interrupt occurs on the rising edge. At each zero crossing, the interrupt handler will get 1 bit from the BitFifo and issue a 1-ms pulse if the bit is 1. When the BitFifo is empty, the interrupt handler can disarm itself. The C code in Program 14.8 can be used to implement the transmit BitFifo. This is a pointer implementation of the BitFifo. These routines can be used to save (`PutTxBit`) and recall (`GetTxBit`) binary data 1-bit at a time (bit streams). Information is saved/recalled in a FIFO manner. FifoSize is the number of 16-bit words in the FIFO. The FIFO is full when it has 16*FifoSize-1 bits.

To implement both transmission and receiving we will need two BitFifos. Assume we have a second BitFifo with functions `GetRxBit` and `PutRxBit` that manipulate the FIFO for the receive function (Figure 14.28). To both transmit and receive we will build the X-10 interface using the TW523 module (Figure 14.29). Again we will connect the zero-crossing input to a microcomputer pin that can generate interrupts on both the rising and falling edges. After each zero-crossing edge the software will interrogate the receive input to determine if a 120-kHz burst is present. Once the software detects a burst, it can store it in the receive BitFifo, then for the next 21 interrupts it will put a 1 if a burst is present or a 0 if no

```
unsigned int TxFifo[FifoSize]; // storage for Bit Stream
struct BitPointer{
  unsigned int Mask;      // 0x8000, 0x4000,...,2,1
  unsigned int *WPt;};    // Pointer to word containing bit
typedef struct BitPointer BitPointerType;
BitPointerType PutTxPt;    // Pointer of where to put next
BitPointerType GetTxPt;    // Pointer of where to get next
/* BitFIFO is empty if PutTxPt==GetTxPt       */
/* BitFIFO is full  if PutTxPt+1==GeTxtPt after wrap  */
int SameBit(BitPointerType p1, BitPointerType p2){
  if((p1.WPt==p2.WPt)&&(p1.Mask==p2.Mask)) return(1); //yes
  return(0);} // no
//*****************InitTxBit*********************
// initializes the BitFifo to be empty
void InitTxBit(void) {
  PutTxPt.Mask=GetTxPt.Mask=0x8000;
  PutTxPt.WPt=GetTxPt.WPt=&TxFifo[0];}  /* Empty when PutTxPt==GetTxPt */
//*****************PutTxBit**********************
// returns TRUE=1 if successful, FALSE=0 if full and data not saved
// input is boolean FALSE if data==0
int PutTxBit (int data) {  BitPointerType TempPutPt;
  TempPutPt=PutTxPt;
  TempPutPt.Mask=TempPutPt.Mask>>1;
  if(TempPutPt.Mask==0) {
     TempPutPt.Mask=0x8000;
     ++TempPutPt.WPt;    // next word address
     if((TempPutPt.WPt)==&TxFifo[FifoSize]) TempPutPt.WPt=&TxFifo[0];} // wrap
  if (SameBit(TempPutPt,GetTxPt))
     return(0);        /* Failed, fifo was full */
  else {
     if(data)
       (*PutTxPt.WPt) |= PutTxPt.Mask;  // set bit
     else
       (*PutTxPt.WPt)&= ~PutTxPt.Mask;  // clear bit
     PutTxPt=TempPutPt;
     return(1);}}
//*****************GetTxBit**********************
// returns TRUE=1 if successful, FALSE=0 if empty and data not removed
// output is boolean 0 means FALSE, nonzero is true
int GetTxBit (unsigned int *datapt) {
   if (SameBit(PutTxPt,GeTxtPt))
      return(0);        /* Failed, fifo was empty */
   else {
      *datapt=(*GetTxPt.WPt)&GetTxPt.Mask;
      GetTxPt.Mask=GetTxPt.Mask>>1;
      if(GetTxPt.Mask==0) {
        GetTxPt.Mask=0x8000;
        ++GetTxPt.WPt;  // next word
        if((GetTxPt.WPt)==&TxFifo[FifoSize]) GetTxPt.WPt=&TxFifo[0];} // wrap
      return(1); }}
```

Program 14.8
Bit FIFO routines.

burst is detected. The main program can get bits from the receive BitFifo and decodes the transmission (Figure 14.30).

Observation: When a repeated packet is transmitted using the TW523 interface, the second packet is visible on the Receive Input line.

Observation: The receiver signals can be monitored during a TW523 transmission to detect noisy line conditions or the occurrence of a packet collision.

Figure 14.28
Input capture interrupt
occurs on both the
rising/falling edges of
the zero-crossing trigger.

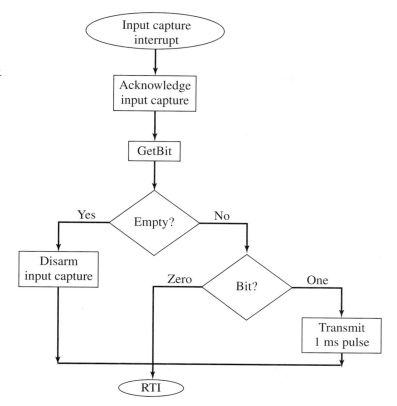

Figure 14.29
Half-duplex interface
between the
microcomputer and the
power system.

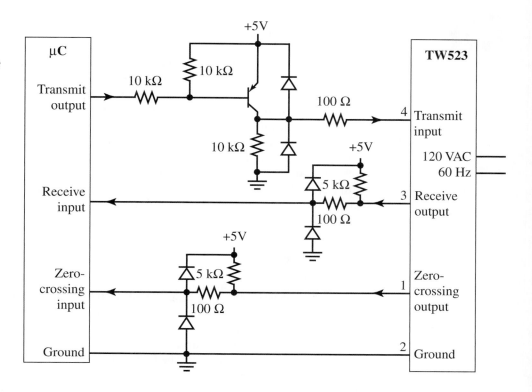

Figure 14.30
Flowchart for receiving
X-10 communication.

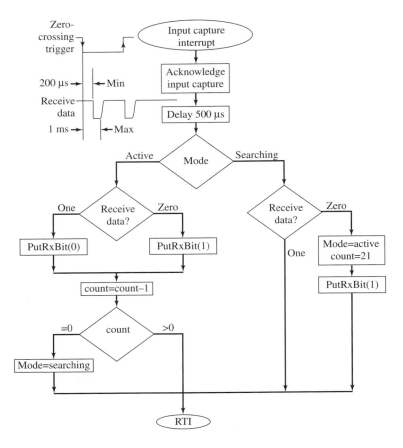

14.7   Universal Serial Bus (USB)

The Universal Serial Bus (USB) is a host-controlled, token-based high-speed serial network that allows communication between many devices operating at different speeds. The objective of this section is not to provide all the details required to design a USB interface; rather, it serves as an introduction to the network. There is 650-page document on the USB standard, which you can download from **http://www.usb.org**. In addition, there are quite a few web sites set up to assist USB designers, such as the one titled "USB in a NutShell" at **http://www.beyondlogic.org/usbnutshell/**. The standard is much more complex than the other networks presented in this chapter. Fortunately, however, there are a number of USB products that facilitate incorporating USB into an embedded system. In addition, the USB controller hardware handles the low-level protocol. USB devices usually exist within the same room and are typically less than 4 meters from each other. USB 2.0 supports three speeds:

> High Speed—480 Mbits/s
> Full Speed—12 Mbits/s
> Low Speed—1.5 Mbits/s

The original USB version 1.1 supported just full-speed mode and a low-speed mode. The Universal Serial Bus is host-controlled, which means the host regulates communication on the bus, and there can be only one host per bus. On the other hand, the On-The-Go specification, added in version 2.0, includes a Host Negotiation Protocol that allows two devices to negotiate for the role of host. The USB host is responsible for undertaking all transactions and scheduling bandwidth. Data can be sent by various transaction methods using a token-based protocol. USB employs a tiered star topology, using a hub to connect additional devices. A hub is at the center of each star. Each wire segment is a point-to-point connection between the host and a hub or function, or a hub connected to another hub or function, as shown in Figure 14.31. Because the hub provides power, the hub can monitor power to each device by switching off a device drawing too much current without disrupting other devices. The hub can filter out high-speed and full-speed transactions so that lower-speed devices do not receive them. Because USB uses a 7-bit address, up to 127 devices can be connected.

Figure 14.31
USB network topology.

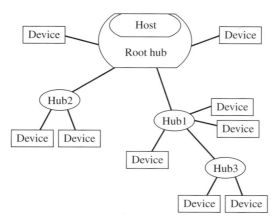

The electrical specification for USB was introduced back in Chapter 7 (Figure 7.18 and Table 7.5), using four shielded wires ($+5$V power, D$+$, D$-$, and ground). The D$+$ and D$-$ are twisted-pair differential data signals. It uses **Non Return to Zero Invert** (**NRZI**) encoding to send data with a sync field to synchronize the host and receiver clocks.

USB drivers will dynamically load and unload. When a device is plugged into the bus, the host will detect this addition, interrogate the device, and load the appropriate driver. Similarly, when the device is unplugged, the host will detect its absence and automatically unload the driver. The USB architecture comprehends four basic types of data transfers:

- **Control Transfers:** Used to configure a device at attach time and can be used for other device-specific purposes, including control of other pipes on the device.
- **Bulk Data Transfers:** Generated or consumed in relatively large quantities and have wide dynamic latitude in transmission constraints.
- **Interrupt Data Transfers:** Used for timely but reliable delivery of data—for example, characters or coordinates with human-perceptible echo or feedback response characteristics.
- **Isochronous Data Transfers:** Occupy a prenegotiated amount of USB bandwidth with a prenegotiated delivery latency. (Also called streaming real time transfers.)

Isochronous transfer allows a device to reserve a defined about of bandwidth with guaranteed latency. This is appropriate for real-time applications such as in audio or video applications. An isochronous pipe is a stream pipe and is, therefore, always unidirectional. An endpoint description identifies whether a given isochronous pipe's communication flow is into or out of the host. If a device requires bidirectional isochronous communication flow, two isochronous pipes must be used, one in each direction.

A USB device indicates its speed by pulling either the D$+$ or D$-$ line to 3.3 V, as shown in Figure 14.32. A pull-up resistor attached to D $+$ specifies full speed, and a pull-up resistor attached to D$-$ means low speed. These device-side resistors are also used by the host or hub to detect the presence of a device connected to its port. Without a pull-up resistor, the host or hub assumes there is nothing connected. A high-speed device begins as a full-speed device (1.5k to 3.3V). Once it has been attached, it will do a high-speed chirp during reset and establish a high-speed connection if the hub supports it. If the device operates in high-speed mode, then the pull-up resistor is removed to balance the line.

Figure 14.32
Pull-up resistors on USB
devices signal specify the
speed.

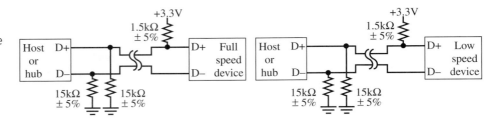

Like most communication systems, USB is made up of several layers of protocols. Like the CAN network presented previously, the USB controllers will be responsible for establishing the low-level communication. Each USB transaction consists of three packets

- **Token Packet** (header)
- **Optional Data Packet** (information)
- **Status Packet** (acknowledge)

The host initiates all communication, beginning with the token packet, which describes the type of transaction, the direction, the device address, and the designated endpoint. The next packet is generally a data packet carrying the information and is followed by a handshaking packet, reporting whether the data or token was received successfully, or whether the endpoint is stalled or not available to accept data. Data is transmitted least significant bit first. Some USB packets are shown in Figure 14.33. All packets must start with a sync field. The **sync** field is 8 bits long at low and full speed or 32 bits long for high speed and is used to synchronize the clock of the receiver with that of the transmitter. **PID** (Packet ID) is used to identify the type of packet that is being sent, as shown in Table 14.12. The **address** field specifies which device the packet is designated for. Being 7 bits in length, it allows for 127 devices to be supported. Address 0 is not valid, as any device that is not yet assigned an address must respond to packets sent to address zero. The **endpoint** field is made up of 4 bits, allowing 16 possible endpoints. Low speed devices, however can have only two additional endpoints on top of the default pipe. **Cyclic Redundancy Checks** are performed on the data within the packet payload. All token packets have a 5-bit CRC, whereas data packets have a 16-bit CRC. EOP stands for **End of packet**. **Start of Frame** Packets (SOF) consist of an 11-bit frame number that is sent by the host every 1ms ± 500ns on a full-speed bus or every 125 μs ± 0.0625 μs on a high-speed bus.

Figure 14.33
USB packet types.

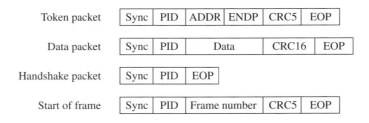

USB **functions** are USB devices that provide a capability or function such as a printer, zip drive, scanner, modem or other peripheral. Most functions will have a series of buffers, typically 8 bytes long. **Endpoints** can be described as sources or sinks of data, shown as **EP0In**, **EP0Out**, and the like in Figure 14.34. As the bus is host-centric, endpoints occur at the end of the communications channel at the USB function. The host-software may send a packet to an endpoint buffer in a peripheral device. If the device wishes to send data to the host, the device cannot simply write to the bus, as the bus is controlled by the host. Therefore,

Table 14.12
USB PID numbers.

Group	PID value	Packet identifier
Token	0001	OUT token, address + endpoint
	1001	IN token, address + endpoint
	0101	SOF token, start-of-frame marker and frame number
	1101	SETUP token, address + endpoint
Data	0011	DATA0
	1011	DATA1
	0111	DATA2 (high speed)
	1111	MDATA (high speed)
Handshake	0010	ACK handshake, receiver accepts error-free data packet
	1010	NAK handshake, device cannot accept data or cannot send data
	1110	STALL handshake, endpoint is halted or pipe request not supported
	0110	NYET (no response yet from receiver)
Special	1100	PREamble, enables downstream bus traffic to low-speed devices
	1100	ERR, split transaction error handshake
	1000	Split, high-speed split transaction token
	0100	Ping, high-speed flow control probe for a bulk/control endpoint

Figure 14.34
USB data flow model shows that the host communicates with endpoints in the device.

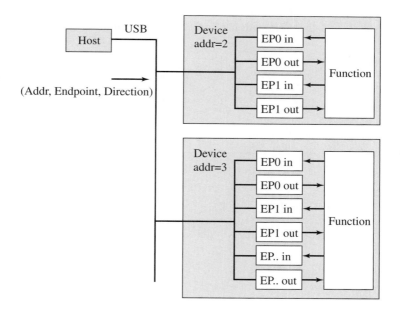

it writes data to the endpoint buffer specified for input, and the data sits in the buffer until such time as the host sends an IN packet to that endpoint requesting the data. Endpoints can also be seen as the interface between the hardware of the function device and the firmware running on the function device.

While the device sends and receives data on a series of endpoints, the client software transfers data through pipes. A **pipe** is a logical connection between host and endpoint(s). Pipes will also have a set of parameters associated with them, such as how much bandwidth is allocated to it, what transfer type (Control, Bulk, Iso or Interrupt) it uses, a direction of data flow, and maximum packet/buffer sizes. **Stream pipes** can be used to send unformatted data. Data flows sequentially and a predefined direction, either in or out. Stream pipes will support bulk, isochronous, and interrupt transfer types. Stream pipes can be controlled by either the host or the device. **Message pipes** have a defined USB format. They are host-controlled and are initiated by a request sent from the host. Data is then transferred in the desired direction, dictated by the request. Therefore, message pipes allow data to flow in both directions but support only control transfers.

14.7.2
Modular USB
Interface

There are two approaches to implementing a USB interface for an embedded system. In the modular approach, we will employ a USB-to-parallel or USB-to-serial converter. The modular approach is appropriate for adding USB functionality to an existing system. For about $30, we can buy a converter cable with a USB interface to connect to the personal computer (PC) and a serial interface to connect to the embedded system, as shown in Figure 14.35. The embedded system hardware and software is standard RS232 serial. These systems come with PC device drivers so that the USB-serial-embedded system looks like a standard serial port (COM) to the PC software. The advantage of this approach is that software development on the PC and embedded system is simple. The disadvantage is that none of the power and flexibility of USB is utilized. In particular, the bandwidth is limited by the RS232 line, and the data stream is unformatted. Similar products are available that convert USB to the parallel port. Companies that make these converters include

IOGear Inc.	http://www.iogear.com
Wyse Technology	http://www.wyse.com
D-Link Corporation	http://www.dlink.com
Computer Peripheral Systems, Inc.	http://www.cpscom.com
Jo-Dan International, Inc.	http://www.jditech.com

Figure 14.35
Modular approach to
USB interfacing.

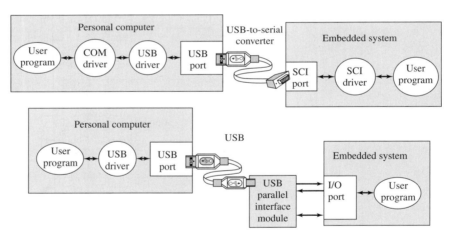

The second modular approach is to purchase a USB parallel interface module. These devices allow you to send and receive data using parallel handshake protocols similar to the input/output examples in Chapter 3. They typically include an USB-enabled microcontroller and receive and transmit FIFO buffers. This approach is more flexible than the serial cable method, because both the microcontroller module and the USB drivers can be tailored and personalized. In particular, some modules allow you to burn PID and VID numbers into EEPROM. The advantages and disadvantages of this approach are similar to the serial cable, in that the data is unformatted and you will not be able to implement high bandwidth bulk transfers or negotiate for real-time bandwidth available with isochronous data transfers. Companies that make these modules include

Future Technology Devices International Ltd.	http://www.ftdichip.com/
ActiveWire, Inc.	http://www.activewireinc.com
DLP Design, Inc.	http://www.dlpdesign.com
Elexol Pty Ltd.	http://www.elexol.com

14.7.3
Integrated USB
Interface

The second approach to implementing a USB interface for an embedded system is to integrate the USB capability into the microcontroller itself. This method affords the greatest flexibility and performance, but requires careful software design on both the microcontroller and the host. During the past ten years, USB has been replacing RS232 serial communication as the preferred method for connecting embedded systems to the personal computer. Manufacturers of microcontrollers have introduced versions of their products with USB capability. Examples include the Microchip PIC18F2455, FTDI FT245BM, and the Freescale 9S12UF32. Figure 14.36 shows a USB-flash disk system designed using the 9S12UF32, which is capable of connecting directly to the USB bus. It has an internal voltage reference needed to create the 3.3 V levels. The flash disk is interfaced to parallel ports of the 9S12UF32. This microcontroller handles the USB 2.0 protocol, including full-speed and high-speed device functions. It has one default control IN/OUT endpoint and five independent configurable physical endpoints, with 64-byte buffers. It can perform high-speed isochronous data toggle communication. The detailed hardware and software for this system, which was developed by Freescale to demonstrate the capabilities of this chip, can be found on the Freescale web site as Reference Design "RDHCS12UF32TD."

Figure 14.36
Thumbdrive system with USB interface built with a 9S12UF32 microcontroller.

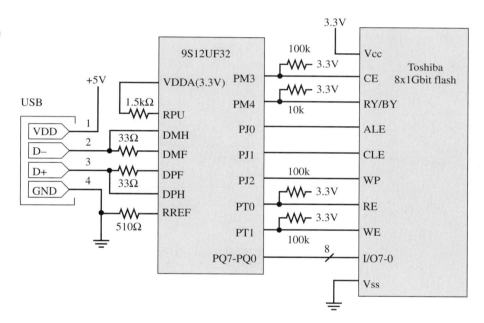

14.8 **Exercises**

14.1 The objective of this exercise is to design a communication network using four single-chip microcomputers. The microcomputers are connected together by fiber-optic cables. The fiber-optic channel is virtually noise-free. The asynchronous simplex serial communication channels link the four computers in a circle. Each 8-bit message consists of a 2-bit source address, a 2-bit destination address, and 4 bits of data. The message data structure has three fields. In memory, messages are stored as 3 bytes, but when transmitted they are compressed into 1 byte.

```
struct message{
   unsigned char source;        // 0,1,2,3 computer that sent the message
   unsigned char destination;   // 0,1,2,3 computer that receives the message
   unsigned char information;}; // 0-15 data part of the message
typedef struct message messageType;
```

Seven steps occur as computer 0 sends "10" to computer 2:

The main program on computer 0 creates a message and passes it to its interrupt software—for example,

```
int SayHi(void){ messagetype Hello;
    Hello.destination=2;
    Hello.information=10;
    return(Send(&Hello)); } // Send knows the source is 0
```

The interrupt software on computer 0 transmits the message to computer 1
The interrupt software on computer 1 receives the message and notices the destination address does not match its computer number (DIP switch)
The interrupt software on computer 1 retransmits the message to computer 2
The interrupt software on computer 2 receives the message and notices the destination address does match its computer number (DIP switch)
The interrupt software on computer 2 passes the message to the main program on computer 2
The main program on computer 2 accepts the message—for example,

```
void WaitFor(void){ messagetype msg;
    while(Accept(&msg)); } // waits for message
    printf("%d %d\n",msg.source,msg.information);}
```

Each microcomputer has a 2-bit DIP switch specifying its computer number. You may assume that each computer has a unique 2-bit address. The electric switch being on (0Ω) signifies a zero address bit, and the electric switch being off (open) signifies a one address bit (e.g., both DIP switches in computer 0 are on (0Ω)) (Figure 14.37).

Figure 14.37
A ring network.

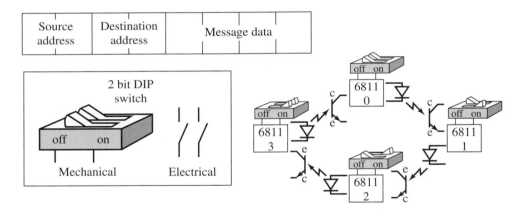

Each of the four fiber-optic cables contains a LED and a light-sensitive transistor. The LED operating point (shines light into the cable) is $V = 2.5$ V, $I = 8$ mA. When light is in the cable, the transistor voltage V_{ce} is less than 0.5 V for any I_{ce} less than 5 mA. If the LED current is zero, then the cable is dark and the transistor is off ($I_{ce} = 0$). The fiber-optic cable can handle baud rates up to 2 Mb/s. Choose the fastest baud rate for this interrupt-driven asynchronous serial communication. You will not write the main program, but you will write subroutines executed by the main program to SEND and ACCEPT messages. Your network software must be interrupt-driven.

a) Show the hardware network interface for one single-chip microcomputer system. Except for the setting of the DIP switch, the network hardware/software is the same for all four computers. Please label chip numbers, but not pin numbers. Be careful how the systems are grounded to each other. If this is computer n, show the outgoing fiber-optic cable to computer n + 1 (diode) and the incoming fiber-optic cable from computer n − 1 (transistor).

b) Show all the information that will be located in the RAM. The globals in RAM are unspecified on power up, so you must initialize them in the ritual in part c. Give careful thought as to the most appropriate data structures.

c) Give the `Ritual`, `Send`, and `Accept` functions along with the interrupt handlers required. You may use FIFO queues by giving the global data structure definitions and the function prototypes without showing the FIFO function definitions. The same network communication software will be placed in each system (you can read the computer address from the DIP switches connected to an input port). The `Ritual` will initialize all data structures and configure the microcomputer. You will write two functions (`Send`, `Accept`) that will be called by the main program (you don't write the main program). The subroutine `Send` will initiate a send message communication. The prototype is shown above. `Send` returns right away, with the result equal to 0 if all is well so far but returns a 1 if the message cannot be sent (e.g., FIFO full). Use comments to explain how and why an error occurred. The subroutine `Accept` will check to see if a message has been received for this computer. The return value (RegD) is 0 if a message is ready and 1 if no incoming message is currently ready. If a message is available, then it returns by reference the source, destination, and information fields as illustrated in the above prototype. Whether or not an incoming message is available, `Accept` returns right away. The interrupt processes will perform the serial I/O operations. You need not poll for 0s and 1s. *Good interrupt software contains no gadfly loops.*

d) What is the bandwidth of your network? Specify units and show the equation that justifies your answer.

14.2 The objective of this exercise is to design a microcomputer-based IEEE488 to RS422 simplex converter (Figure 14.38). Your single-chip microcomputer system will perform handshaked parallel input from a IEEE488 output device, buffer it in an internal FIFO, then transmit the data on an asynchronous RS422 simplex serial channel. The serial protocol is 8-bit data, one stop, no parity, and 300 bits/s baud rate. The input bandwidth can vary from 0 to 1000 bytes/s, with an average of 10 bytes/s. The internal FIFO will allow temporarily the input bandwidth to exceed the output bandwidth. The IEEE488 sequence of events to transmit 1 byte is:

1. Your microcomputer signals it is ready for the next data by making RFD = 1;
2. Eventually the IEEE488 output device will provide the 8-bit data and make $\overline{DAV} = 0$;
3. Your microcomputer makes RFD = 0;
4. Your microcomputer should read the data, put it into the FIFO, then make DAC = 1;
5. Eventually the IEEE488 output device will make $\overline{DAV} = 1$ and remove the data;
6. Your microcomputer makes DAC = 0.

Figure 14.38
An IEEE488 to RS422 interface.

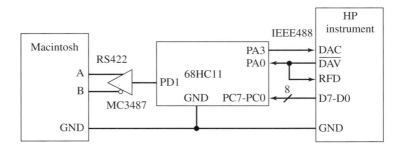

We can simplify this dedicated interface by connecting the $\overline{DAV}$ to RFD, skipping steps 1 and 3:

2. Eventually the IEEE488 output device will provide the 8-bit data and make $\overline{DAV} = 0$;

4. Your microcomputer should read the data, put it into the FIFO, then make DAC = 1;

5. Eventually the IEEE488 output device will make $\overline{\text{DAV}} = 1$ and remove the data;

6. Your microcomputer makes DAC = 0.

  a) What should your microcomputer do when the FIFO is full? Why?

  b) Show all the software including the interrupt handlers, rituals, main program, reset vector, interrupt vector, and ORG statements that place the code into the appropriate addresses on the *single-chip* microcomputer. You may use a FIFO without showing its implementation if you explain where in memory the data, pointers, and subroutines reside.

14.3 The objective of this exercise is to convert data from a parallel SCSI bus to serial RS232 using a single-chip 6811 microcomputer. The baud rate is 2400 bits/s. The RS232 protocol is 8-bit, no parity, and two stop bits. Assume a simple single-initiator/ single-target SCSI system and ignore initialization. The microcomputer acts as the only target device accepting 8-bit data from the Initiator. Your microcomputer will handshake (REQ, ACK) each data byte, buffer it for maximum bandwidth, and output data as fast as possible on the SCI serial output. The 6811 hardware is shown in Figure 14.39. For a 6812, you may use any available ports.

Figure 14.39
A SCSI to RS232 interface.

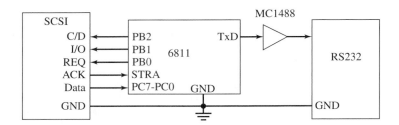

Show all the software, including the main program, interrupt vectors, reset vectors, data structures, interrupt handlers, and I/O. This is a single chip, so the dedicated software must be segmented into RAM (0 to $FF) and ROM ($E000-$FFFF). If you use C, explain the segmentation required to implement single-chip mode. Comments will be graded.

14.4 One of the problems with the X-10 interfaces shown in this chapter is that the digital ground is connected to the AC power line.

  a) Design the interface that uses two 6N139s to isolate the simplex PL513 interface. You will have to use a separate isolated power source different from the supply that powers the microcomputer.

  b) Design the interface that uses three 6N139s to isolate the half-duplex TW523 interface. You will have to use a separate isolated power source different from the supply that powers the microcomputer.

14.5 The objective of this exercise is to develop a message-passing facility that spans across two computers (Figure 14.40). A fully interlocked synchronization method will be implemented. The message is a simple 8-bit byte that is passed from the output port of computer 1 to the input port of computer 2. The following hardware connections are fixed. You may assume both computers initialize together, where computer 1 initializes its PT7 to 0 and computer 2 initializes its PT6 to 0. On the 6811 you may use the timer features of Port A and Port C for the data.

Figure 14.40
A handshaking parallel port interface.

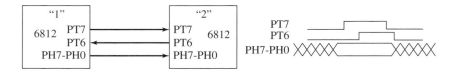

When computer 1 wishes to transmit to computer 2, it puts the data first on its Port H outputs. Next, computer 1 makes a rising edge on PT7, signifying that new data are available. Computer 1 will then wait for a rising edge on PT6, signifying that the data have been accepted. Next, computer 1 makes its PT7 low. Last, it waits for a low signal on its PT6 input line.

   a) Write the software for the transmitting computer 1. The transmission will occur in the background using input capture interrupts. You may use a FIFO queue without writing the three routines: `InitFifo`, `PutFifo`, and `GetFifo`. You may ignore FIFO full errors. There are three routines you will write.

     1. `void Ritual(void)` that clears the FIFO, makes the PT7 output low, and arms/enables the input capture interrupts as appropriate;

     2. `void PutMsg(unsigned char data)` function that is called by the main program (that you do not write); if a current message is in progress, then these data are entered in the FIFO; if there is no current message in progress, then one is started with the new data;

     3. `void IC6Handler(void)` interrupt handler called by the input capture on PT6.

When computer 2 wishes to receive from computer 1, it first waits for a rising edge on its PT7 input. Then it gets the data from its Port H inputs. Next, computer 2 makes a rising edge on its PT6 output, signifying that the data has been accepted. Computer 2 will then wait for a falling edge on its PT7 input. Last, it makes a low signal on its PT6 line.

   b) Write the software for the receiving computer 2. The reception will occur in the background using input capture interrupts. You may use a FIFO queue without writing the three routines: `InitFifo`, `PutFifo`, and `GetFifo`. You may ignore FIFO-full errors. There are three routines you will write

     1. `void Ritual(void)` that clears the FIFO, makes the PT6 output low, and arms/enables the input capture interrupts as appropriate;

     2. `unsigned char data GetMsg(void)` function that is called by the main program (that you do not write); if the FIFO is empty, then this routine will call the `get()` routine over and over until data are available;

     3. `void IC7Handler(void)` interrupt handler called by the input capture on PT7.

14.6 Combine programs 14.5 and 14.6 to create one function that creates the transmission shown in Figure 14.13. In particular, your function should receive 8 bits from addr1 and transmit 8 bits to addr2.

14.7 A CAN system has a baud rate of 100,000 bits/sec, 29-bit ID, and 4 bytes of data per frame. Assuming there is no bit-stuffing, what is the maximum bandwidth of this network, in BYTES/SEC. (Hint: Determine the total number of bits in a frame.)

14.8 A CAN system has a baud rate of 100,000 bits/sec, 11-bit ID, and 8 bytes of data per frame. Assuming there is no bit-stuffing, what is the maximum bandwidth of this network, in BYTES/SEC. (Hint: Determine the total number of bits in a frame.)

14.9 Consider a CAN network that has a variable transmission rate, where the time in between messages can vary from t_{min} to t_{max}. Consider a particular listener on the network that will receive all messages (no filter). If the receive queue in this listener already has five unread messages, then additional CAN transmissions will stall on the network, because this receiver cannot acknowledge any more messages. In order to prevent stalling, this listener requires a real-time interface. The interface latency in this case is defined as the between the Receive CAN Flag (RXF) being set and the execution of the CAN interrupt 38. You may assume executing each CAN ISR will quickly read and remove all messages in its receive queue (clearing RXF). What is the upper bound on the interface latency?

14.9   **Lab Assignments**

Lab 14.1. The overall goal of this lab is to design, implement, and test a peer-to-peer communication system. Peer-to-peer means people on two computers communicate without the people on the other computers seeing the information. The system must use a ring-connected RS232 serial channel (Figure 14.3), must use interrupt-driven I/O, and must have a layered software configuration. The lowest-level software

Figure 14.41
Control and data packets.

Control-code packet

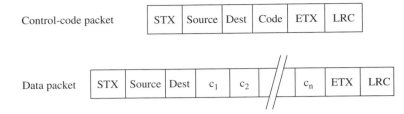

Data packet

performs serial I/O. The middle-level software sends message packets from one computer to another. The highest-level software interfaces with the human operator (keypad/LCD) and provides a mechanism to create a peer-to-peer connection. In a layered system, software in one layer can call routines only within that layer or the layer immediately below it. You need a way to see who is on the network and a way to request/accept or terminate connection between two operators. Local operator input/output will occur via a keypad and LCD. You may assume all nodes on the system are willing to cooperate and will not perform malicious activity. On the other hand, it is possible that another computer on the network may not be plugged in, or the network connection may be broken.

The communication system between two or more microcomputers will be designed in three layers. The first layer, the physical layer is implemented by the SCI hardware and the interrupt-driven device driver. The second layer may consist of a simplified binary synchronous communication protocol (BSC). At this level, message packets will be transmitted between the two machines. Possible formats of the *control-code packet* and the *data packet* are shown in Figure 14.41. This *control-code packet* contains exactly 6 bytes.

> STX—start of text: precedes text block (ASCII $02).
> Source—ASCII code of the source computer (ASCII $31 to $39).
> Dest—ASCII code of the destination computer (ASCII $31 to $39).
> Code—one of the following two control codes:
> > ACK—affirmative acknowledge: last block received correctly (ASCII $06).
> > NAK—negative acknowledge: last block was received in error (ASCII $15).
> > ETX—end of text block: ETX is followed by the error checking byte (ASCII $03).
> > LRC—longitudinal redundancy check, error check byte.

In the data packet, the data $c1, c2, \ldots c_n$ are ASCII characters that constitute the information being sent from source to destination. It is OK to limit message sizes to a maximum of 20 bytes. The destination computer will respond with an ACK control packet if the message was received properly and will respond with a NAK if there are any framing, overrun, noise, or LRC errors. The transmitter will send a message and "stop and wait" for either an ACK or a NAK. If an ACK is received, then it can continue. If a NAK is received or if no response is received after a reasonable delay, then the message is retransmitted. Because there is a ring physical channel, there is no possibility of a collision. You must handle the situations when the destination computer does not exist or when the ring is broken. To solve this fault you will need some time-out mechanism to retransmit the packet if an ACK is not received in some reasonable time. You should choose an upper limit (e.g., three) on the number of times a packet is retransmitted. After three tries an error is reported to the operator.

The highest level will be a keypad interpreter and an LCD display. The LCD should show interactive feedback to the operator creates messages to be sent and should display messages received.

Lab 14.2. The objective of this lab is to design a distributed temperature monitoring system using a CAN network. Each microcontroller node will consist of a temperature data acquisition system, an LCD display, and a CAN network interface. There are three software threads on each node. The producer thread will periodically measure temperature and send its local temperature to the other nodes on the network. The CAN identifier will specify the location (node number) and type (always temperature). The data field will contain the temperature as a decimal fixed-point number. Both data acquisition and network transmission will be performed in the background. The consumer thread, also running in the background, will accept messages from the other nodes and calculate the average and maximum temperature of the room. These data will be transferred to foreground, and the main program will display local, average, and maximum temperature. Choose a CAN bandwidth (e.g., 250,000 bps) and leave it as a constant. Start with a data acquisition and transmission rate so slow (e.g., 1/sec) that collisions will be rare. Then, increase the rate until the maximum bandwidth is reached.

Lab 14.3. The objective of this lab is to design a digital clock using a DS1307 external clock chip. The first step is to interface the clock chip to the microcontroller using an I²C network. The second step is to design low-level drivers to allow the microcontroller to send and receive data from the DS1307. The next software layer includes functions such as **SetTime FormatTime** and **ReadTime**. The highest level is a main program that implements a digital clock using an LED or LCD display. Two, three, or four momentary switches will be used to control the operation of the digital clock.

Lab 14.4. The objective of this lab is to design a digital thermometer using a DS1631A external thermometer chip. The first step is to interface the thermometer chip to the microcontroller using an I²C network. The second step is to design low-level drivers to allow the microcontroller to send and receive data from the DS1631A. The next software layer includes functions such as **SetMode** and **ReadTemperature**. The highest level is a main program that implements a digital thermometer using a LED or LCD display. Two or three momentary switches will be used to control the operation of the digital thermometer.

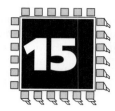

Digital Filters

Chapter 15 objectives are to:

❏ Introduce basic principles involved in digital filtering
❏ Define the Z transform and use it to analyze digital filters
❏ Design simple low-pass, frequency-reject, and high digital filters
❏ Discuss the effect of non-real-time sampling on digital filter error
❏ Develop digital filter implementations

The goal of this chapter is to provide a brief introduction to digital signal processing (DSP). DSP includes a wide range of operations such as digital filtering, event detection, frequency spectrum analysis, and signal compression/decompression. Similar to the goal of analog filtering, a digital filter will be used to improve the S/N ratio in our data. The difference is that a digital filter is performed in software on the digital data sampled by the ADC. The particular problem addressed in several ways in this chapter is removing 60-Hz noise from the signal. Like the control systems and communication systems discussed in the last two chapters, we will provide just a brief discussion to the richly developed discipline of DSP. Again, this chapter focuses mostly on the implementation on the embedded microcomputer. Event detection is the process of identifying the presence or absence of particular patterns in our data. Examples of this type of processing include optical character readers, waveform classification, sonar echo detection, infant apnea monitors, heart arrhythmia detectors, and burglar alarms. Frequency spectrum analysis requires the calculation of the discrete Fourier transform (DFT)[1]. Like the regular Fourier transform, the DFT converts a time-dependent signal into the frequency domain. The difference between a regular Fourier transform and the DFT is that the DFT performs the conversion on a finite number of discrete time digital samples to give a finite number of points at discrete frequencies. Data compression and decompression are important aspects in high-speed communication systems. Although we will not specifically address the problems of event detection, DFT, and compression/decompression in this book, these DSP operations are implemented using similar techniques as the digital filters that are presented in this chapter. Our goal for this chapter is to demonstrate that

---

[1]A fast algorithm to calculate the DFT is called the fast Fourier transform (FFT).

fairly powerful digital signal-processing techniques can be implemented on computers of modest performance like the 6811 and 6812. The limitation of single-chip computers like the 6811 and 6812 is not complexity but rather execution speed. If you have a DAS or control system with unwanted 60-Hz noise, then the techniques described in this chapter provide an effective means for removing this noise.

15.1 Basic Principles

The objective of this section is to introduce simple digital filters. Let $x_c(t)$ be the continuous analog signal to be digitized. $x_c(t)$ is the analog input to the ADC. If f_s is the sample rate, then the computer samples the ADC every T seconds ($T = 1/f_s$). Let $\ldots, x(n) \ldots,$ be the ADC output sequence, where

$$x(n) = x_c(nT) \qquad \text{with } -\infty < n < +\infty. \tag{1}$$

There are two types of approximations associated with the sampling process. Because of the finite precision of the ADC, amplitude errors occur when the continuous signal $x_c(t)$ is sampled to obtain the digital sequence $x(n)$. The second type of error occurs because of the finite sampling frequency. The Nyquist theorem states that the digital sequence $x(n)$ properly represents the DC to ($\frac{1}{2}$) f_s frequency components of the original signal $x_c(t)$. Two important assumptions are necessary to make when using digital signal processing:

1. We assume the signal has been sampled at a fixed and known rate f_s
2. We assume aliasing has not occurred

We can guarantee the first assumption by using a hardware clock to start the ADC at a fixed and known rate. A less expensive but not as reliable method is to implement the sampling routine as a high-priority periodic interrupt process. By establishing a high priority of the interrupt handler, we can place an upper bound on the interrupt latency, guaranteeing that ADC sampling is occurring at an almost fixed and known rate.

To verify aliasing has not occurred, we can observe the ADC input with a spectrum analyzer to prove there are no significant signal components above ($\frac{1}{2}$) f_s. "No significant signal components" is defined as having an ADC input voltage $|Z|$ less than the ADC resolution, Δz,

$$|Z| \le \Delta z \qquad \text{for all } f \ge \frac{1}{2} f_s$$

In a manner similar to the DAS design process developed in Chapter 12, we could impose a more strict constraint of $|Z| \le \frac{1}{2} \Delta z$. The $\frac{1}{2}$ in the expression $\frac{1}{2} \Delta z$ is not the result of fundamental theorems, but rather a safety factor added during the design so that the effect of aliasing will not be present in the digital samples.

A *causal* digital filter calculates $y(n)$ from $y(n-1)$, $y(n-2)$, ... and $x(n)$, $x(n-1)$, $x(n-2)$. ... Simply put, a causal filter cannot have a nonzero output until it is given a nonzero input. The output of a causal filter, $y(n)$, cannot depend on future data [e.g., $y(n+1)$, $x(n+1)$].

A *linear* filter is constructed from a linear equation. A *nonlinear* filter is constructed from a nonlinear equation. One example of a nonlinear filter is the median. To calculate the median of three numbers, one first sorts the numbers according to magnitude, then chooses the middle value.

A *finite-impulse response* (FIR) filter relates $y(n)$ only in terms of $x(n)$, $x(n-1)$, $x(n-2)$. ... In the next section we will determine the gain and phase response of four simple digital filters. These examples are presented in Equations 2 to 5. Equations 2 to 4 describe simple averaging FIR filters:

$$y(n) = \frac{x(n) + x(n-1)}{2} \tag{2}$$

$$y(n) = \frac{x(n) + x(n-1) + x(n-2) + x(n-3) + x(n-4) + x(n-5)}{6} \tag{3}$$

$$y(n) = \frac{x(n) + x(n-3)}{2} \tag{4}$$

An **infinite-impulse response** (IIR) filter relates $y(n)$ in terms of both $x(n)$, $x(n-1)$, ..., and $y(n-1)$, $y(n-2)$. ... Equation 5 describes a simple IIR filter:

$$y(n) = \frac{y(n-1) + x(n)}{2} \tag{5}$$

One way to analyze linear filters is the Z transform. Equation 6 gives the definition of the Z transform:

$$X(z) = Z[x(n)] \equiv \sum_{n=-\infty}^{\infty} x(n)z^{-n} \tag{6}$$

The Z transform is similar to other transforms. In particular, consider the Laplace transform, which converts a continuous time-domain signal $x(t)$ into the frequency domain $X(s)$. In the same manner, the Z transform converts a discrete time sequence $x(n)$ into the frequency domain $X(z)$ (Figure 15.1).

Figure 15.1
A transform is used to study a signal in the frequency domain.

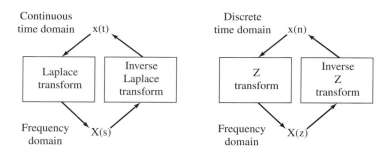

The input to both the Laplace and Z transforms are infinite time signals, having values at times from $-\infty$ to $+\infty$. The frequency parameters **s** and **z** are complex numbers, having a real and imaginary part. In both cases we apply the transform to study linear systems. In particular, we can describe the behavior (gain and phase) of an analog system using its transform, $H(s) = Y(s)/X(s)$. In this same way we will use the $H(z)$ transform of a digital filter to determine its gain and phase response (Figure 15.2).

Figure 15.2
A transform also can be used to study a system in the frequency domain.

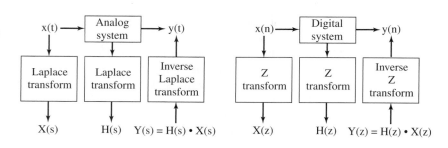

For an analog system we can calculate the gain by taking the magnitude of $\mathbf{H(s)}$ at $\mathbf{s = j2\pi f}$, for all frequencies $\mathbf{f}$ from $-\infty$ to $+\infty$. The phase will be the angle of $\mathbf{H(s)}$ at $\mathbf{s = j2\pi f}$. If we were to plot the $\mathbf{H(s)}$ in the $\mathbf{s}$ plane, the $\mathbf{s = j2\pi f}$ curve is the entire y axis. For a digital system we will calculate the gain and phase by taking the magnitude and angle of $\mathbf{H(z)}$. Because of the finite sampling interval, we will only be able to study frequencies from DC to $(\frac{1}{2})\,\mathbf{f_s}$ in our digital systems. If we were to plot the $\mathbf{H(z)}$ in the $\mathbf{z}$ plane, the $\mathbf{z}$ curve representing the DC to $(\frac{1}{2})\,\mathbf{f_s}$ frequencies will be the unit circle.

We will begin by developing a simple, yet powerful rule that will allow us to derive the $\mathbf{H(z)}$ transforms of most digital filters. Let $\mathbf{m}$ be an integer constant. One can use the definition of the Z transform to prove that

$$\mathbf{Z[x(n-m)]} = \sum_{n=-\infty}^{\infty} \mathbf{z^{-n} x(n-m)}$$

$$= \sum_{p+m=-\infty}^{p+m=+\infty} \mathbf{z^{-p-m} x(p)} \qquad p = n-m,\, n = p+m$$

$$= \sum_{p=-\infty}^{p=+\infty} \mathbf{z^{-p-m} x(p)} \qquad \mathbf{m} \text{ is a constant}$$

$$= \mathbf{z^{-m}} \sum_{p=-\infty}^{p=+\infty} \mathbf{z^{-p} x(p)} \qquad \mathbf{m} \text{ is a constant}$$

$$= \mathbf{z^{-m} Z[x(n)]} \tag{7}$$

For example, if $\mathbf{X(z)}$ is the Z transform of $\mathbf{x(n)}$, then $\mathbf{z^{-2} \cdot X(z)}$ is the Z transform of $\mathbf{x(n-2)}$. To find the Z transform of a digital filter, we take the transform of both sides of the linear equation and solve for

$$\mathbf{H(z) = \frac{Y(z)}{X(z)}} \tag{8}$$

To find the response of the filter, let $\mathbf{z}$ be a complex number on the unit circle

$$\mathbf{z \equiv e^{j2\pi f/f_s}} \qquad \text{for } 0 \leq \mathbf{f} < \tfrac{1}{2}\mathbf{f_s} \tag{9}$$

or

$$\mathbf{z = \cos(2\pi f/f_s) + j\,\sin(2\pi f/f_s)} \tag{10}$$

Let
$$\mathbf{H(f) = a + bj} \qquad \text{where } \mathbf{a} \text{ and } \mathbf{b} \text{ are real numbers} \tag{11}$$

The gain of the filter is the complex magnitude of $\mathbf{H(z)}$ as $\mathbf{f}$ varies from 0 to $(\frac{1}{2})\,\mathbf{f_s}$.

$$\mathbf{Gain \equiv |H(f)| = \sqrt{a^2 + b^2}} \tag{12}$$

The phase response of the filter is the angle of $\mathbf{H(z)}$ as $\mathbf{f}$ varies from 0 to $(\frac{1}{2})\,\mathbf{f_s}$.

$$\mathbf{Phase \equiv angle[H(f)] = \tan^{-1}\frac{b}{a}} \tag{13}$$

15.2 Simple Digital Filter Examples

In this section, we perform the straightforward calculations to determine the filter response (gain and phase) given the filter equation. The gain and phase of these four filters are plotted in Figures 15.3 and 15.4. Even though the four examples are very simple, they can be used in actual applications. The first example, Equation 2, is a simple average of the current sample with the previous sample (Program 15.1).

$$y(n) = \frac{x(n) + x(n - 1)}{2} \tag{2}$$

The first step is to take the Z transform of both sides of Equation 2. The Z transform of $y(n)$ is $Y(z)$, the Z transform of $x(n)$ is $X(z)$, and the Z transform of $x(n–1)$ is $z^{-1}X(z)$. Since the Z transform is a linear operator, we can write

$$Y(z) = \frac{X(z) + z^{-1}X(z)}{2} \tag{14}$$

The next step is to rewrite the equation in the form of $H(z) = Y(z)/X(z)$. We then can use Equations 9 to 13 to determine the gain and phase response of this filter.

$$H(z) \equiv \frac{Y(z)}{X(z)} = \frac{1 + z^{-1}}{2} \tag{15}$$

$$H(f) = \frac{1 + e^{-j2\pi f/fs}}{2} = \frac{1 + \cos(2\pi f/f_s) - j\sin(2\pi f/f_s)}{2} \tag{16}$$

$$\text{Gain} \equiv |H(f)| = 0.5\sqrt{\{1 + \cos(2\pi f/f_s)\}^2 + \{\sin(2\pi f/f_s)\}^2} \tag{17}$$

$$\text{Phase} \equiv \text{angle}[H(f)] = \tan^{-1}\left[\frac{-\sin(2\pi f/f_s)}{1 + \cos(2\pi f/f_s)}\right] \tag{18}$$

Program 15.1
Real time data acquisition with a simple digital filter, Equation 2.

```
// 9S12C32 C implementation, interrupts at a frequency of fs
unsigned short x[2],y;
// x[0] is x(n) current sample
// x[1] is x(n-1) previous sample
void interrupt 13 TC5handler(void){
  TFLG1 = 0x20;        // ack OC5F
  TC5 = TC5+PERIOD;    // Executed every 1/fs
  x[1] = x[0];         // shift MACQ data
  x[0] = ADC_In(CHANNEL);   // new data
  y = (x[0]+x[1])>>1;
}
void DAS_Init(void){ // start sampling
  asm sei          // make atomic
  ADC_Init();      // Program 11.13
  TIOS |= 0x20;    // enable OC5
  TSCR1 = 0x80;    // enable, 4 MHz TCNT
  TIE |= 0x20;     // Arm output compare 5
  TFLG1 = 0x20;    // Initially clear C5F
  TC3 = TCNT+50;   // First one right away
  x[0] = x[1] = 0;
  asm cli
}
```

Observation: Equation 2 is a simple low-pass filter, passing DC and rejecting 0.5 f_s.

Observation: If the ADC is only 8 bits, whether to use char or short for x,y depends on how the compiler implements the mathematical calculations.

Checkpoint 15.1: If TCNT increments at 4 MHz, what value should `PERIOD` be to make the sampling rate 1000 Hz?

Checkpoint 15.2: If the sampling rate is 1000 Hz, what are gains of filter equation 2 at DC, 250 Hz and 500 Hz?

The second simple filter, Equation 3, performs the running average of the last six ADC samples (Program 15.2). It uses a faster method to implement the MACQ.

Program 15.2
Real-time data acquisition with a simple digital filter, Equation 3.

```
// 9S12C32 C implementation, interrupts at 360Hz
unsigned short n,x[12],y;
// x[n]   is x(n) current sample
// x[n-1] is x(n-1) sample 1/360 sec ago
// x[n-2] is x(n-2) sample 2/360 sec ago
// x[n-3] is x(n-3) sample 3/360 sec ago
// x[n-4] is x(n-4) sample 4/360 sec ago
// x[n-5] is x(n-5) sample 5/360 sec ago
void interrupt 13 TC5handler(void){
  TFLG1 = 0x20;      // ack OC5F
  TC5 = TC5+11111;   // Executed at 360 Hz
  n++; if(n==12) n=6;
  x[n] = x[n-6] = ADC_In(CHANNEL);  // new data
  y = (x[n]+x[n-1]+x[n-2]+x[n-3]+x[n-4]+x[n-5])/6;
}
void DAS_Init(void){ // start sampling
  asm sei        // make atomic
  ADC_Init();    // Program 11.13
  n = 6;         // rotates through 6,7,8,9,10,11
  TIOS |= 0x20;  // enable OC5
  TSCR1 = 0x80;  // enable, 4 MHz TCNT
  TIE |= 0x20;   // Arm output compare 5
  TFLG1 = 0x20;  // Initially clear C5F
  TC3 = TCNT+50; // First one right away
  asm cli
}
```

$$y(n) = \frac{x(n) + x(n-1) + x(n-2) + x(n-3) + x(n-4) + x(n-5)}{6} \quad (3)$$

Just like the last example, first we take the Z transform of both sides of Equation 3:

$$Y(z) = \frac{X(z) + z^{-1}X(z) + z^{-2}X(z) + z^{-3}X(z) + z^{-4}X(z) + z^{-5}X(z)}{6} \quad (19)$$

The next step is to rewrite the equation in the form of $H(z) = Y(z)/X(z)$.

$$H(z) \equiv \frac{Y(z)}{X(z)} = \frac{1 + z^{-1} + z^{-2} + z^{-3} + z^{-4} + z^{-5}}{6} \quad (20)$$

We then can use Equations 9 to 13 to determine the gain and phase response of this filter.

$$H(f) = \frac{1 + e^{-j2\pi f/fs} + e^{-j4\pi f/fs} + e^{-j6\pi f/fs} + e^{-j8\pi f/fs} + e^{-j10\pi f/fs}}{6}$$

$$= [1 + \cos(2\pi f/f_s) + \cos(4\pi f/f_s) + \cos(6\pi f/f_s) + \cos(8\pi f/f_s) + \cos(10\pi f/f_s)$$

$$- j\{\sin(2\pi f/f_s) + \sin(4\pi f/f_s) + \sin(6\pi f/f_s) + \sin(8\pi f/f_s) + \sin(10\pi f/f_s)\}]/6 \quad (21)$$

Gain $\equiv |\mathbf{H(f)}|$

$$= 1/6 \cdot [\{1 + \cos(2\pi f/f_s) + \cos(4\pi f/f_s) + \cos(6\pi f/f_s) + \cos(8\pi f/f_s) + \cos(10\pi f/f_s)\}^2$$

$$+ \{\sin(2\pi f/f_s) + \sin(4\pi f/f_s) + \sin(6\pi f/f_s) + \sin(8\pi f/f_s) + \sin(10\pi f/f_s)\}^2] \qquad (22)$$

Phase $\equiv$ **angle[H(f)]**

$$= \frac{\tan^{-1}[-\{\sin(2\pi f/f_s) + \sin(4\pi f/f_s) + \sin(6\pi f/f_s) + \sin(8\pi f/f_s) + \sin(10\pi f/f_s)\}}{\{1 + \cos(2\pi f/f_s) + \cos(4\pi f/f_s) + \cos(6\pi f/f_s) + \cos(8\pi f/f_s) + \cos(10\pi f/f_s)\}]} \qquad (23)$$

Observation: This implementation of the MACQ uses twice as much memory but does not require the data to be shifted for each sample.

Observation: When comparing the execution speed of two potential approaches, we need to measure it directly or observe the assembly listing generated by the compiler.

Checkpoint 15.3: If the sampling rate of Equation 3 is 360 Hz, use the Z transform to prove that the 60 Hz gain is zero.

Observation: Equation 3 is a multiple-notch filter rejecting (⅙) f_s, (⅓) f_s, and (½) f_s. In particular, if f_s is 360 Hz, then the digital filter in Equation 3 rejects 60, 120, and 180 Hz.

The third simple filter, Equation 4, performs the average of the current 8-bit ADC sample with the sample collected three sampling periods ago. Although this filter appears to be simple, we can use it to implement a low-Q 60-Hz notch digital filter (Program 15.3).

$$y(n) = \frac{x(n) + x(n-3)}{2} \qquad (4)$$

Program 15.3
Real-time data acquisition with a simple digital filter, Equation 4.

```
// 9S12C32 C implementation, interrupts at 360Hz
unsigned char x[4],y;   // 8-bit ADC
// x[0] is x(n) current sample
// x[1] is x(n-1) sample 1/360 sec ago
// x[2] is x(n-2) sample 2/360 sec ago
// x[3] is x(n-3) sample 3/360 sec ago
void interrupt 13 TC5handler(void){
  TFLG1 = 0x20;        // ack OC5F
  TC5 = TC5+11111;     // Executed at 360 Hz
  x[3] = x[2];         // shift MACQ data
  x[2] = x[1];
  x[1] = x[0];
  x[0] = ADC_In(CHANNEL);  // new 8-bit data
  y = (x[0]+x[3])>>1;
}
void DAS_Init(void){ // start sampling
  asm sei          // make atomic
  ADC_Init();      // Program 11.13
  TIOS |= 0x20;    // enable OC5
  TSCR1 = 0x80;    // enable, 4 MHz TCNT
  TIE |= 0x20;     // Arm output compare 5
  TFLG1 = 0x20;    // Initially clear C5F
  TC3 = TCNT+50;   // First one right away
  asm cli
}
```

Again we take the Z transform of both sides of Equation 4:

$$Y(z) = \frac{X(z) + z^{-3}X(z)}{2} \tag{24}$$

Next we rewrite the equation in the form of $H(z) = Y(z)/X(z)$.

$$H(z) \equiv \frac{Y(z)}{X(z)} = \frac{1 + z^{-3}}{2} \tag{25}$$

We then can use Equations 9 to 13 to determine the gain and phase response of this filter.

$$H(f) = \frac{1 + e^{-j6\pi f/fs}}{2} = \frac{1 + \cos(6\pi f/f_s) - j\sin(6\pi f/f_s)}{2} \tag{26}$$

$$\text{Gain} \equiv |H(f)| = 0.5\sqrt{\{1 + \cos(6\pi f/f_s)\}^2 + \{\sin(6\pi f/f_s)\}^2} \tag{27}$$

$$\text{Phase} \equiv \text{angle}[H(f)] = \tan^{-1}\left[\frac{-\sin(6\pi f/f_s)}{1 + \cos(6\pi f/f_s)}\right] \tag{28}$$

Checkpoint 15.4: If the sampling rate of Equation 4 is 360 Hz, use the Z transform to prove that the 60 Hz gain is zero.

Observation: Equation 4 is a double-notch filter rejecting ($\frac{1}{6}$) f_s and ($\frac{1}{2}$) f_s. In particular, if f_s is 360 Hz, then the digital filter in Equation 4 rejects 60 and 180 Hz.

When writing digital filters it is important to consider the precision of the both the variables and the calculations. In particular, a serious error will occur if overflow or underflow occurs during the calculations. Consider the calculation x[0]+x[3] performed in Program 15.4. If simple 8-bit addition were to be used, then an error will occur when intermediate calculation x[0]+x[3] goes above 255. Observing the gain versus frequency plot in Figure 15.3, we are confident that the final calculation (x[0]+x[3])>>1 will fit into an 8-bit variable. To verify program correctness, we must look at the output of the compiler. The two listings in Program 15.4 were generated by the ImageCraft ICC11 and Metrowerks CodeWarrior compilers. These are proper implementations because these compilers first promote the 8-bit data into 16 bits, perform the addition and shift using 16-bit operations, then demote the result back to 8 bits when storing into y. The comments were added to clarify the operations.

```
; MC68HC11, Imagecraft ICC11
; y=(x[0]+x[3])>>1;
   ldy  #1
   pshy
   ldab  _x+3   ; 8-bit x[3]
   clra          ; promote into RegD
   pshb          ; save on stack
   psha
   ldab  _x      ; 8 bit x[0]
   clra          ; promote into RegD
   tsy
   addd  0,y     ; 16-bit x[0]+x[3]
   puly
   puly
   jsr  __asrd   ; 16-bit shift
   stab  _y      ; demote to 8 bit
```

```
; 9S12C32, Metrowerks CodeWarrior
; y=(x[0]+x[3])>>1;
   ldab  x:3    ; 8-bit x[3]
   clra          ; promote into RegD
   tfr  D,X      ; save in RegX
   ldab  x       ; 8-bit x[0]
   leax  D,X     ; 16-bit x[0]+x[3]
   asra
   rolb          ; 16-bit shift
   stab  y       ; demote to 8 bit
```

Program 15.4
Compiler-generated assembly listings for the filter implementation.

The fourth simple filter, Equation 5, is an example of an infinite-impulse response filter. It performs the average of the current ADC sample with the last filter output. For this filter we follow the same procedure to determine its gain and frequency response. Notice in Figure 15.3 that the gain is larger than 1 for some frequencies. Therefore it will be important when implementing this filter to use a number system with more bits than the ADC precision. In other words, even if the ADC in Program 15.5 were to have only 8 bits, we will perform the filter using 16-bit precision to avoid overflow.

Program 15.5
Real-time data acquisition with a simple IIR digital filter, Equation 5.

```
// 9S12C32 C implementation, interrupts at a frequency of fs
unsigned short x,y[2];
// y[0] is y(n) current filter output
// y[1] is y(n-1) filter output 1 sample ago
void interrupt 13 TC5handler(void){
  TFLG1 = 0x20;         // ack OC5F
  TC5 = TC5+PERIOD;     // Executed every 1/fs
  y[1] = y[0];          // shift MACQ data
  x = ADC_In(CHANNEL);  // new data
  y[0] = (x+y[1])/2;
}
void DAS_Init(void){ // start sampling
  asm sei          // make atomic
  ADC_Init();      // Program 11.13
  TIOS |= 0x20;    // enable OC5
  TSCR1 = 0x80;    // enable, 4 MHz TCNT
  TIE |= 0x20;     // Arm output compare 5
  TFLG1 = 0x20;    // Initially clear C5F
  TC3 = TCNT+50;   // First one right away
  asm cli
}
```

$$y(n) = \frac{y(n-1) + x(n)}{2} \tag{5}$$

$$Y(z) = \frac{z^{-1}Y(z) + X(z)}{2} \tag{29}$$

$$H(z) \equiv \frac{Y(z)}{X(z)} = \frac{1}{2 - z^{-1}} \tag{30}$$

$$H(f) = \frac{1}{2 - e^{-j2\pi f/fs}} = \frac{1}{2 - \cos(2\pi f/f_s) + j\,\sin(2\pi f/f_s)}$$

$$= \frac{2 - \cos(2\pi f/f_s) - j\,\sin(2\pi f/f_s)}{[2 - \cos(2\pi f/f_s)]^2 - [\sin(2\pi f/f_s)]^2} \tag{31}$$

$$\text{Gain} \equiv |H(f)| = \left| \frac{2 - \cos(2\pi f/f_s) - j\,\sin(2\pi f/f_s)}{[2 - \cos(2\pi f/f_s)]^2 - [\sin(2\pi f/f_s)]^2} \right|$$

$$= \frac{\sqrt{\{2 - \cos(2\pi f/f_s)\}^2 + \{\sin(2\pi f/f_s)\}^2}}{[2 - \cos(2\pi f/f_s)]^2 - [\sin(2\pi f/f_s)]^2} \tag{32}$$

$$\text{Phase} \equiv \text{angle}[H(f)] = \tan^{-1}\left[\frac{-\sin(2\pi f/f_s)}{2 - \cos(2\pi f/f_s)} \right] \tag{33}$$

Checkpoint 15.5: For some frequencies, the filter in Equation 5 has a gain larger than 1. What does that mean?

The response of the four linear digital filters is plotted in Figure 15.3; the phase response of the four linear digital filters is plotted in Figure 15.4.

Figure 15.3
Gain versus frequency response for four simple digital filters.

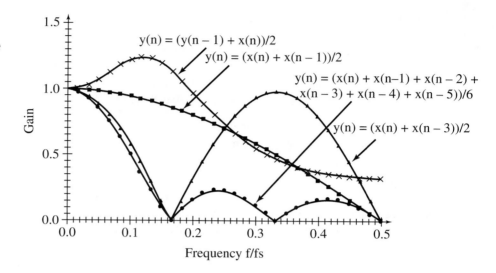

Figure 15.4
Phase versus frequency response for four simple digital filters.

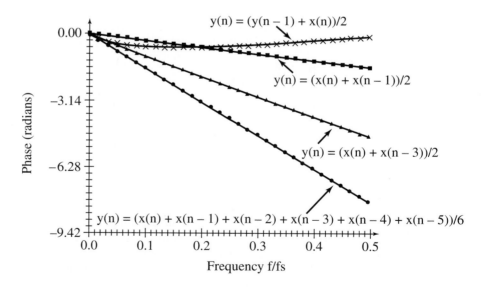

A linear phase versus frequency response is desirable because a linear phase causes minimal waveform distortion. Conversely, a nonlinear phase response will distort the shape or morphology of the signal.

In general, if $\mathbf{f_s}$ is $2 \cdot \mathbf{k} \cdot \mathbf{f_c}$ Hz (where k is any integer $\mathbf{k} \geq 2$), then the following is a $\mathbf{f_c}$ notch filter:

$$y(n) = \frac{x(n) + x(n - k)}{2} \tag{34}$$

Similarly, if f_s is $k \cdot f_c$ Hz (where k is any integer $k \geq 2$), then the following rejects f_c and its harmonics $2f_c$, $3f_c$ . . . :

$$y(n) = \frac{1}{k}\sum_{i=0}^{k-1} x(n-i) \tag{35}$$

15.3   Impulse Response

Another way to analyze digital filters is to consider the filter response to particular input sequences. Two typical test sequences are the step and the impulse.

Step   . . . , 0, 0, 0, 1, 1, 1, 1 . . .
Impulse   . . . , 0, 0, 0, 1, 0, 0, 0 . . .

The step is defined as (Figure 15.5)

$$s(n) \equiv 0 \qquad \text{for } n < 0$$
$$ 1 \qquad \text{for } n \geq 0$$

Figure 15.5
Step input.

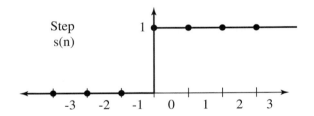

The step signal represents a sharp change (like an edge in a photograph).
    Consider the following three digital filters:

FIR $$y(n) = \frac{x(n) + x(n-1)}{2} \tag{2}$$

IIR $$y(n) = \frac{y(n-1) + x(n)}{2} \tag{5}$$

Median $$y(n) = \text{median}\,[x(n), x(n-1), x(n-2)] \tag{36}$$

The median can be performed on any odd number of data points by sorting the data and selecting the middle value. The median filter can be performed recursively or nonrecursively. For a nonrecursive median filter, the original data points are not modified. That is, the median is calculated each time without regard to the previous filter outputs. For example, a five-wide nonrecursive median filter takes as the filter output the median of $[x(n)$, $x(n-1)$, $x(n-2)$, $x(n-3)$, $x(n-4)]$ On the other hand, a recursive median filter replaces the sample point with the filter output. For example, a five-wide recursive median filter takes as the filter output the median of $[x(n)$, $y(n-1)$, $y(n-2)$, $y(n-3)$, $y(n-4)]$, where $y(n-1)$, $y(n-2)$ . . . are the previous filter outputs. A nonrecursive three-wide median filter is implemented in Program 15.6.

> **Observation:** A median filter can be applied in systems that have impulse or speckle noise. For example, if the noise occasionally causes one sample to be very different from the rest (like a speck on a piece of paper), then the median filter will completely eliminate the noise.

Program 15.6
The median filter is an example of a nonlinear filter.

```
unsigned char median(unsigned char u1,unsigned char u2,unsigned char u3){
unsigned char result;
  if(u1>u2)
    if(u2>u3)       result = u2;   // u1>u2,u2>u3         u1>u2>u3
      else
        if(u1>u3) result = u3;   // u1>u2,u3>u2,u1>u3 u1>u3>u2
        else       result = u1;   // u1>u2,u3>u2,u3>u1 u3>u1>u2
  else
    if(u3>u2)       result = u2;   // u2>u1,u3>u2         u3>u2>u1
      else
        if(u1>u3) result = u1;   // u2>u1,u2>u3,u1>u3 u2>u1>u3
        else       result = u3;   // u2>u1,u2>u3,u3>u1 u2>u3>u1
  return(result);
}
unsigned char x[3],y;
// x[0] is x(n) the current sample
// x[1] is x(n-1) the sample 1/fs ago
// x[2] is x(n-2) the sample 2/fs ago
void sample(void){
  x[2] = x[1];        // shift MACQ data
  x[1] = x[0];
  x[0] = ADC_In(0);  // new data from channel 0
  y = median(x[0],x[1],x[2]);
}
```

The step responses of the three filters are:

FIR	**. . . , 0, 0, 0, 0.5, 1, 1, 1 . . .**
IIR	**. . . , 0, 0, 0, 0.5, 0.75, 0.88, 0.94, 0.97, 0.98, 0.99 . . .**
Median	**. . . , 0, 0, 0, 0, 1, 1, 1, 1, 1 . . .**

Except for the delay, the median filter passes a step without error (Figure 15.6).

The impulse represents a noise spike (like spots on a Xerox copy). The definition of impulse is (Figure 15.7).

$$i(n) \equiv 0 \quad \text{for } n \neq 0$$
$$1 \quad \text{for } n = 0$$

Figure 15.6
Step response of three simple digital filters.

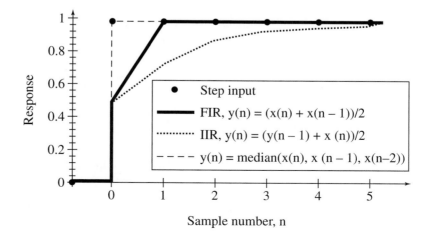

Figure 15.7
Impulse input.

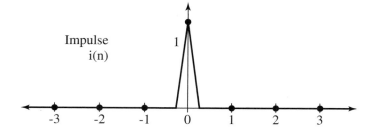

The impulse response of a filter is defined as **h(n)**. The impulse responses of the three filters are:

FIR	. . . , **0, 0, 0, 0.5, 0.5, 0, 0, 0** . . .
IIR	. . . , **0, 0, 0, 0.5, 0.25, 0.13, 0.06, 0.03, 0.02, 0.01** . . .
Median	. . . , **0, 0, 0, 0, 0, 0, 0, 0** . . .

The median filter completely removes the impulse (Figure 15.8). Note that the median filter preserves the sharp edges and removes the spike or impulsive noise. The median filter is **nonlinear**, and hence **H(z)** and **h(n)** are not defined for this particular class of filters.

Figure 15.8
Impulse response of three simple digital filters.

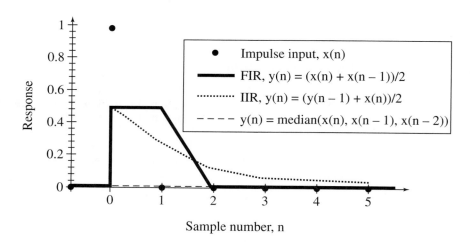

The impulse response **h(n)** can also be used as an alternative to the transfer function **H(z)**. **h(n)** is sometimes called the *direct form*. A causal filter has **h(n)** = 0 for **n** < 0. For a casual filter.

$$H(z) = \sum_{n=0}^{\infty} h(n)\, z^{-n} \tag{37}$$

For a FIR filter, **h(n)** = 0 for **n** ≥ **N** for some **N**. Thus,

$$H(z) = \sum_{n=0}^{N-1} h(n) z^{-n} \tag{38}$$

The output of a filter can be calculated by convolving the input sequence **x(n)** with **h(n)**. For an IIR filter:

$$y(n) = \sum_{k=0}^{\infty} h(k) \, x(n-k) \tag{39}$$

For a FIR filter:

$$y(n) = \sum_{k=0}^{N-1} h(k) \, x(n-k) \tag{40}$$

15.4   High-Q 60-Hz Digital Notch Filter

There are two objectives for this example. The first goal is to show an example of a digital notch filter, and the second goal is to demonstrate the use of fixed-point mathematics. 60-Hz noise is a significant problem in most DASs. The 60-Hz noise reduction can be accomplished by:

1. Reducing the noise source (e.g., shut off large motors)
2. Shielding the transducer, cables, and instrument
3. Implementing a 60-Hz analog notch filter
4. Implementing a 60-Hz digital notch filter

Consider again the analogy between the Laplace and Z transforms. When the H(s) transform is plotted in the s plane, we look for peaks [places where the amplitude H(s) is high] and valleys (places where the amplitude is low). In particular, we usually can identify zeros [H(s) = 0] and poles [H(s) = ∞]. In the same way we can plot the H(z) in the z plane and identify the poles and zeros. The analogies in Table 15.1 apply:

Analog condition	Digital condition	Consequence
Zero near $s=j2\pi f$ line	Zero near $z=e^{j2\pi f/fs}$ circle	Low gain at the f near the zero
Pole near $s=j2\pi f$ line	Pole near $z=e^{j2\pi f/fs}$ circle	High gain at the f near the pole
Zeros in complex conjugate pairs	Zeros in complex conjugate pairs	Output y(t) is real
Poles in complex conjugate pairs	Poles in complex conjugate pairs	Output y(t) is real
Poles in left half plane	Poles inside unit circle	Stable system
Poles in right half plane	Poles outside unit circle	Unstable system
Pole near a zero	Pole near a zero	High-Q response

Table 15.1
Analogies between the analog and digital filter design rules.

It is the 60-Hz digital notch filter that will be implemented in this example. The signal is sampled at $\mathbf{f_s}$ = 244.14 Hz. We wish to place the zeros (gain = 0) at 60 Hz, thus (Figure 15.9):

$$\Theta = \pm 2\pi \cdot \frac{60}{f_s} = \pm 1.54416 \tag{41}$$

The zeros are located on the unit circle at 60 Hz:

$$z_1 = \cos(\theta) + j\sin(\theta) \qquad z_2 = \cos(\theta) - j\sin(\theta) \tag{42}$$

or

$$z_1 = 0.02663 + j\,0.99965 \qquad z_2 = 0.02663 - j\,0.99965 \tag{43}$$

To implement a flat passband away from 60 Hz, the poles are placed next to the zeros, just inside the unit circle. Let α define the closeness of the poles, where $0 < \alpha < 1$.

$$p_1 = \alpha\, z_1 \qquad p_2 = \alpha\, z_2 \tag{44}$$

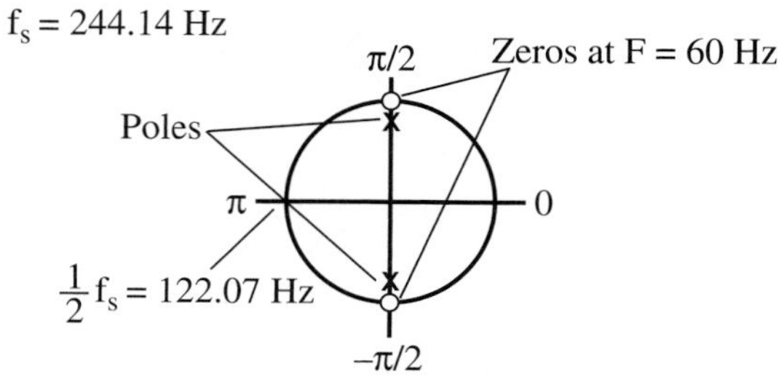

Figure 15.9
Pole-zero plot of a 60-Hz
digital notch filter.

$f_s = 244.14$ Hz

Zeros at F = 60 Hz

for $\alpha = 0.95$

$$p_1 = 0.02530 + j\,0.94966 \qquad p_2 = 0.02530 - j\,0.94966 \tag{45}$$

The transfer function is

$$H(z) = \prod_{i=1}^{k} \frac{z - z_i}{z - p_i} = \frac{(z - z_1)(z - z_2)}{(z - p_1)(z - p_2)} \tag{46}$$

which can be put in standard form (i.e., with terms $1, z^{-1}, z^{-2} \ldots$)

$$H(z) = \frac{1 - 2\cos(\theta)z^{-1} + z^{-2}}{1 - 2\alpha\cos(\theta)z^{-1} + \alpha^2 z^{-2}} \tag{47}$$

or

$$H(z) = \frac{1 - 0.0532672z^{-1} + z^{-2}}{1 - 0.0506038z^{-1} + 0.9025z^{-2}} \tag{48}$$

The digital filter can be derived by taking the inverse Z transform of the H(z) equation

$$y(n) = x(n) - 2\cos(\theta)x(n-1) + x(n-2) + 2\alpha\cos(\theta)y(n-1) - \alpha^2 y(n-2) \tag{49}$$

or

$$y_1(n) = x(n) - 0.0532672x(n-1) + x(n-2) + 0.0506038y_1(n-1) - 0.9025y_1(n-2) \tag{50}$$

To implement this filter in real time without floating-point hardware we rewrite Equation 50 using fixed-point mathematics:

$$y_2(n) = x(n) + x(n-2) + \frac{-14x(n-1) - 13y_2(n-1) - 231y_2(n-2)}{256} \tag{51}$$

It will be important to implement the software such that the intermediate calculations are performed in 16-bit arithmetic. Also, since this filter has a gain larger than 1 for some frequencies, the filter outputs must also be implemented with 16-bit variables.

Table 15.2 shows that the fixed-point calculations do not produce significant filter errors. There are two factors in choosing the fixed value (e.g., 256) in the fixed-point implementation. We choose a value large enough so that the fixed-point value closely approximates the floating-point value (e.g., $13/256 \approx 0.0506038$). The larger fixed-point value we choose, the better the approximation:

$$y_3(n) = x(n) + x(n-2) + \frac{-533x(n-1) + 506y_3(n-1) - 9025y_3(n-2)}{10000} \tag{52}$$

On the other hand, larger fixed-point values will require higher-precision integer mathematics. For an 8-bit microcomputer we would prefer to do 8-bit by 8-bit into 16-bit multiplies, 16-bit addition/subtraction, and 16-bit divided by 8-bit divides. For a 16-bit microcomputer

Table 15.2
Gain versus frequency responses for the floating point and fixed implementations.

f (Hz)	$|Y_1(z)/X(z)|$	$|Y_2(z)/X(z)|$	$|Y_1(z)/X(z)|-|Y_2(z)/X(z)|$
0	1.052	1.049	0.003
10	1.060	1.060	0.000
20	1.048	1.048	0.000
30	1.052	1.052	0.000
40	1.048	1.048	0.000
50	1.040	1.036	0.004
55	0.984	0.980	0.004
56	0.944	0.944	0.000
57	0.888	0.888	0.000
58	0.744	0.740	0.004
59	0.464	0.464	0.000
60	0.012	0.016	−0.004
61	0.396	0.404	−0.008
62	0.752	0.752	0.000
63	0.880	0.880	0.000
64	0.944	0.944	0.000
65	0.980	0.980	0.000
70	1.036	1.036	0.000
80	1.048	1.048	0.000
90	1.052	1.052	0.000
100	1.056	1.060	−0.004
110	1.056	1.056	0.000
120	1.052	1.052	0.000

we would prefer to do 16-bit by 16-bit into 32-bit multiplies, 32-bit addition/subtraction, and 32-bit divided by 16-bit divides. Notice that the multiply by 2^n and divide by 2^n operations are very simple to implement using the arithmetic shift operations.

For $\alpha = 0.90$, the filter becomes

$$y_4(n) = x(n) + x(n-2) + \frac{-14x(n-1) + 12y_4(n-1) - 207y_4(n-2)}{256} \tag{53}$$

For $\alpha = 0.75$, the filter becomes

$$y_5(n) = x(n) + x(n-2) + \frac{-7x(n-1) + 5y_5(n-1) - 72y_5(n-2)}{128} \tag{54}$$

The gain of the three filters (Equations 51, 53, and 54) is plotted in Figure 15.10.

Figure 15.10
Gain versus frequency response of three 60-Hz digital notch filters.

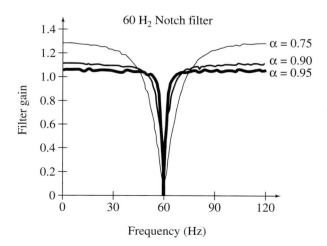

Sometimes we can choose $\mathbf{f_s}$ and/or α to simplify the digital filter equation. For example, if we choose $\mathbf{f_s} = 240$ Hz, then the $\cos(\theta)$ terms in Eq. (48) become zero. If we choose $\alpha = 7/8$ then the fixed-point digital filter becomes

$$y_6(n) = x(n) + x(n-2) - \frac{49 \cdot y_6(n-2)}{64} \tag{55}$$

Another consideration for this type of filter is the fact that the gain in the passbands is greater than 1. The DC gain can be determined two ways. The first method is to use the H(z) transfer equation and set $z = 1$. The H(z) transfer equation for the filter in Eq. (55) is

$$H(z) = \frac{1 + z^{-2}}{1 + (49/64)z^{-2}} \tag{56}$$

At $z = 1$ this reduces to

$$DC\ gain = \frac{2}{1 + 49/64} = \frac{128}{64 + 49} = \frac{128}{113} \tag{57}$$

The second method to calculate DC gain operates on the filter equation directly. In the first step, we set all the $x(n - k)$ terms in the filter to a single variable x and all the $y(n - k)$ terms in the filter to a single variable y.

Next we solve for the DC gain, which is y/x.

$$y = x + x - \frac{49y}{64} \tag{58}$$

This method also calculates the DC gain to be 128/113. We can adjust the digital filter so that the DC gain is exactly 1 by prescaling the input terms [x(n), x(n − 1), x(n − 2) . . .] by 113/128 (Program 15.7).

$$y_7(n) = \frac{113 \cdot x(n) + 113 \cdot x(n - 2) - 98 \cdot y_7(n - 2)}{128} \tag{59}$$

```
; 9S12C32 assembly
        org  RAM
XN      ds   2   ; x(n) current
XN1     ds   2   ; x(n-1) previous
XN2     ds   2   ; x(n-2) 2 samples ago
YN1     ds   2   ; y(n-1) previous
YN2     ds   2   ; y(n-2) 2 samples ago
YN      ds   2   ; y(n) current
acc     ds   4   ; temporary 32-bit
        org  ROM
COEF    dc.w 113,0,113,0,-98
TC5Handler
        movb #$20,TFLG1 ; ack
        ldd  TC5
        addd #16667  ; 240Hz
        std  TC5
        movw YN1,YN2 ; shift MACQ
        movw YN,YN1
        movw XN1,XN2
        movw XN,XN1
        jsr  ADC_In
        std  XN      ; new data
        ldx  #XN     ; data
        ldy  #COEF   ; coef
```

```
// 9S12C32 C implementation
unsigned short x[3],y[3];
// x[0] is x(n) current sample
// x[1] is x(n-1) 1 sample ago
// x[2] is x(n-2) 2 samples ago
// y[0] is y(n) current filter output
// y[1] is y(n-1) filter out 1 ago
// y[2] is y(n-2) filter out 2 ago
void interrupt 13 TC5handler(void){
  TFLG1 = 0x20;          // ack C5F
  TC5 = TC5+16667;       // 240Hz
  y[2]=y[1]; y[1]=y[0];  // shift MACQ
  x[2]=x[1]; x[1]=x[0];
  x[0] = ADC_In(CHANNEL); // new data
  y[0]=(113*(x[0]+x[2])-98*y[2])>>7;
}
```

```
        ldd    #0
        std    acc     ; clear temporary
        std    acc+2
        ldaa   #5      ; number of terms
loop    emacs  acc     ;acc=acc+{X}*{Y}
        leax   2,x
        leay   2,y
        dbne   A,loop
        ldy    acc
        ldd    acc+2
;Y:D=113*(x[0]+x[2])-98*y[2]
        ldx    #128
        edivs
        sty    YN
        rti
```

Program 15.7
Real-time data acquisition with a 60-Hz digital notch filter, Equation 58.

Checkpoint 15.6: For the C language implementation, how large can the addition terms (before the divide by 128) get so that overflow does not occur?

Checkpoint 15.7: For the assembly language implementation, how large can the addition terms (before the divide by 128) get so that overflow does not occur?

Since the gain of this filter is always less than or equal to 1, the filter outputs will fit into the same precision as the filter input. The gain of this filter is shown in Figure 15.11.

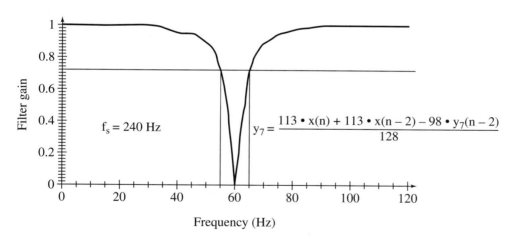

Figure 15.11
Gain versus frequency response of a 60-Hz digital notch filter scaled to have a DC gain of 1.

The Q of a digital notch filter is defined to be

$$Q \equiv \frac{f_c}{\Delta f} \tag{60}$$

where f_c is the center or notch frequency and Δf is the frequency range, where its gain is below 0.707 of the DC gain. For the filter in Equation (59) the gains at 55 and 65 Hz are about 0.707, so its Q is 6.

15.5 Effect of Latency on Digital Filters

The purpose of this analysis is to study the importance of "real time" in a real-time operating system. We define a "hard real-time" system as one that can guarantee that a process will complete a critical task within a certain specified range. One the other hand, we define a "soft real-time" system as one that assigns higher priority to critical functions. We will, in this analysis, derive specific consequences that result when the operating system delays service to a time-critical real-time process. In general, we can define **time-jitter**, δt, as the difference between when a periodic task is supposed to be run and when it is actually run. Let t_n be the time the software task is actually run, and let $n\Delta t$ be the time it was supposed to be run; then the time-jitter at sample n is

$$\delta t_n = t_n - n\Delta t$$

For a real-time system with periodic tasks, we must be able to place an upper bound, k, on the time-jitter.

$$-k \le t_n \le +k \text{ for all } n$$

For a DAS it is often more important to control the time difference between periodic events rather than the absolute time itself. Let Δt_n be the actual time difference between two executions of a software task (e.g., starting the ADC). The desired time difference is $1/f_s$. For a DAS, we define the time-jitter at sample n to be

$$\delta t_n = \Delta t_n - 1/f_s$$

Again, we must be able to place an upper bound, k, on the time-jitter.

$$-k \le \delta t_n \le +k \text{ for all } n$$

If the analog input, $V(t)$, is a slowly varying signal, then time jitter causes little error. In general, the magnitude of the voltage error, ΔV, caused by time jitter, δt, is a function of the slew rate of the input.

$$\Delta V = \frac{dV}{dt} * \delta t$$

Consider a data acquisition system that samples a signal at f_s and performs a 60 Hz digital notch filter to remove noise. To analyze the effect of time jitter we will assume the actual signal is composed of a constant signal, V_s, and a 60 Hz noise signal, V_n.

$$V(t) = V_s + V_n \sin(2\pi60t)$$

The signal-to-noise ratio is Vs/Vn. Assuming the voltage range is between -5 to $+5$ volts, the m bit A/D converter will create an infinite digital sample sequence, $x(1)$ $x(2)$, . . . , $x(n)$, . . . The approximate digital values are

$$x(n) = \frac{2^{m-1}}{5}V\left(\frac{n}{f_s}\right)$$

If the sampling rate, f_s, is 120 • k Hz where k is an integer constant greater than 1, a simple 60 Hz digital notch filter is

$$y(n) = \frac{x(n) + x(n - k)}{2}$$

First we will consider a DAS that samples the ADC and calculates the digital filter and displays the results. In a real-time operating system this process will request service f_s times a second at exact intervals. Unfortunately, time latency occurs between the process request and the process service. What happens to the filter performance if this delay is *always* exactly "**d**"? Assume the delay, **d**, is smaller than 1/**f$_s$**, so that the sample is simply delayed and not skipped altogether. In this situation, we ask how much noise passes the filter?

$$y(n) = \frac{x(n) + x(n-k)}{2} = \frac{\dfrac{2^{m-1}}{5} \cdot V(t_n + d) + \dfrac{2^{m-1}}{5} \cdot V\left(t_n - \dfrac{-k}{f_s} + d\right)}{2}$$

$$= \frac{2^{m-1}}{10}\left(V_s + V_n \sin\left(2\pi60 \cdot \left(\frac{n}{120 \cdot k} + d\right)\right) + V_s + V_n \sin\left(2\pi60 \cdot \left(\frac{n-k}{120 \cdot k} + d\right)\right)\right)$$

$$= \frac{2^{m-1}}{10}\left(2V_s + V_n \sin\left(\frac{\pi n}{k} + 120\pi d\right) + V_n \sin\left(\frac{\pi(n-k)}{k} + 120\pi d\right)\right)$$

$$= \frac{2^{m-1}}{5}V_s \quad \text{independent of d and } V_n$$

For the second consideration, we will look at what happens if *one* ADC sample is delayed but the others occur on time. For example, the request may have come at a time when the OS is performing a higher priority job, and the service is delayed. Let **k** = 2, **f$_s$** = 240, and **t$_n$** be exactly **n/240** for all **n** except at **n** = 50 where **t$_{50}$** = 50/240 + **d**. Assume the delay, **d**, is smaller than 1/240, so that the sample is simply delayed and not skipped altogether. We can derive an expression for the error in y(50) as a function of the delay **d**, the ADC precision **m**, and the 60 Hz noise **V$_n$**.

$$y(n) = \frac{x(50) + x(48)}{2} = \frac{\dfrac{2^{m-1}}{5} \cdot V(t_{50} + d) + \dfrac{2^{m-1}}{5} \cdot V\left(t_{48} - \dfrac{-1}{120}\right)}{2}$$

$$= \frac{2^{m-1}}{10}\left(V_s + V_n \sin\left(\frac{2\pi60 \cdot 50}{240} + 120\pi d\right) + V_s + V_n \sin\left(\frac{2\pi60 \cdot 48}{240}\right)\right)$$

$$= \frac{2^{m-1}}{10}(2V_s + V_n \sin(\pi + 120\pi d))$$

$$= \frac{2^{m-1}}{10}(2V_s - V_n \sin(120\pi d))$$

$$\text{error (50)} = -V_n\frac{2^{m-2}}{5}\sin(120\pi d) \approx -V_n\frac{2^{m-2}}{5}120\pi d \quad \text{for} \quad d \ll \frac{1}{60}$$

We could have derived this same result using the general equation (in this case **d** = δt).

$$\Delta V = \frac{dV}{dt} * \delta t = \frac{2^{m-1}}{5} * \delta t * \frac{d}{dt}(V_s + V_n \sin(2\pi60t))$$

The largest slope occurs when sin() is zero; the magnitude of the error in x(50) is thus bounded by

$$\Delta V \leq \frac{2^{m-1}}{5} * \delta t * 120\pi V_n$$

Because the filter divides x(50) by 2, the error in y(50) is ½ of the preceding equation, yielding the same result.

Checkpoint 15.8: What is the largest source of time-jitter when sampling the ADC using a background interrupt service routine?

15.6   High-Q Digital High-Pass Filters

We will design a *two-pole* high-Q fixed-point digital high-pass filter (Figure 15.12).

Figure 15.12
Gain versus frequency
response of a high-pass
filter.

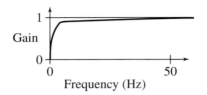

Let the sampling rate f_s be 100 Hz. The pole-zero plot of this type of filter is shown in Figure 15.13.

Figure 15.13
Pole-zero plot of a
high-pass filter.

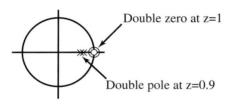

Double zero at z=1

Double pole at z=0.9

Next, we write the transform and convert it to standard form:

$$H(z) = \frac{(z - 1)^2}{(z - 0.9)^2} = \frac{z^2 - 2z + 1}{z^2 - 1.8z + 0.81} = \frac{1 - 2z^{-1} + z^{-2}}{1 - 1.8z^{-1} + 0.81z^{-2}}$$

From the transform, we can write the floating-point filter:

$$y(n) - 1.8y(n-1) + 0.81y(n-2) = x(n) - 2x(n-1) + x(n-2)$$

Next, we arrange the filter in standard form:

$$y(n) = x(n) - 2x(n-1) + x(n-2) + 1.8y(n-1) - 0.81y(n-2)$$

We determine the gain of this filter at 50 Hz by calculating the magnitude of $H(z)$ at $z = -1$:

$$H(-1) = \frac{(-1 - 1)^2}{(-1 - 0.9)^2} = \frac{(-2)^2}{(-1.9)} = \frac{4}{3.61} \qquad 3.61/4 = 0.9025$$

So, to make the gain at 50 Hz equal to 1, we multiply all the **x** terms by 0.9025:

$$y(n) = 0.9025x(n) - 1.805x(n-1) + 0.9025x(n-2) + 1.8y(n-1) - 0.81y(n-2)$$

We can easily convert this filter to fixed-point without introducing any error (the assembly code in Program 15.7 can be used):

$$y(n) = \frac{9025x(n) - 18050x(n-1) + 9025x(n-2) + 18000y(n-1) - 8100y(n-2)}{10000}$$

The following fixed-point approximation is easier to calculate:

$$y(n) = \frac{231x(n) - 462x(n-1) + 231x(n-2) + 461y(n-1) - 207y(n-2)}{256}$$

15.7 Digital Low-Pass Filter

Another class of digital low-pass filters places the zeros equidistant along the unit circle. If k is a positive integer, then the equation

$$1 - z^{-k} = 0$$

has solutions evenly spaced along the unit circle at

$$z = e^{j\,2\pi m/k} \qquad \text{for} \qquad m = 0,1\ldots,k-1$$

If the term $1 - z^{-k}$ appears in the numerator of the filter transform H(z), then zeros will occur at

$$f = \frac{m2\pi f_s}{k} \qquad \text{for} \qquad m = 0,1\ldots,k-1$$

We can create a digital low-pass filter by placing a pole at z = 1 (f = 0) to cancel the zero at z = 1 (Figure 15.14). The resulting transfer function for this class of low-pass filter is

$$H(z) = \frac{1 - z^{-k}}{1 - z^{-1}}$$

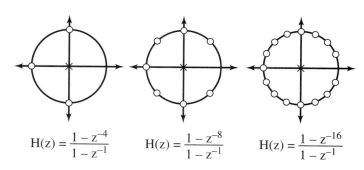

$$H(z) = \frac{1 - z^{-4}}{1 - z^{-1}} \qquad H(z) = \frac{1 - z^{-8}}{1 - z^{-1}} \qquad H(z) = \frac{1 - z^{-16}}{1 - z^{-1}}$$

Figure 15.14
Pole-zero plots of three low-pass filters.

We can use L'Hopital's rule to calculate the DC gain.

$$\lim_{z \to 1} H(z) = \lim_{z \to 1} \frac{kz^{-k-1}}{z^{-2}} = k$$

This class of digital low-pass filters can be implemented with a large MACQ and a simple calculation.

$$y(n) = \frac{x(n) - x(n-k)}{k} + y(n-1)$$

Even though his "looks" like an IIR filter, it is, in fact, exactly equal to the K-sample sliding average. The gain of this class of filter is shown in Figure 15.15 for a sampling rate of 100 Hz.

$$y(n) = \frac{1}{k}\sum_{i=0}^{k-1}x(n-i)$$

Figure 15.15
Gain versus frequency plot of four low-pass filters.

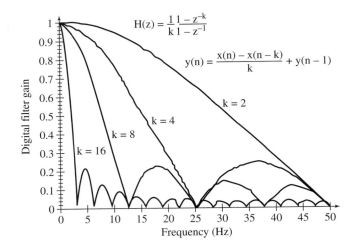

We can decrease the gain in the cutoff region by doubling all the poles and zeros. These double-pole–double-zero digital low-pass filters can also be implemented with a large MACQ (Figure 15.16),

$$H(z) = \left(\frac{1}{k}\frac{1-z^{-k}}{1-z^{-1}}\right)^2 = \frac{1}{k^2}\frac{1-2z^{-k}+z^{-2k}}{1-2z^{-1}+z^{-2}}$$

$$y(n) = \frac{x(n)-2x(n-k)+x(n-2k)}{k^2} + 2y(n-1) - y(n-2)$$

Figure 15.16
Gain versus frequency plot of four low-pass filters, using double poles.

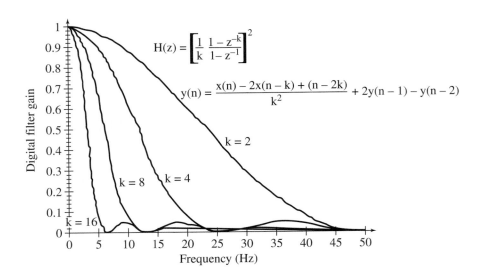

15.8 Direct-Form Implementations

The general form for the transfer function for an IIR filter is

$$H(z) = \frac{Y(z)}{X(z)} = \frac{a_0 + a_1 z^{-1} + a_2 z^{-2} + \ldots + a_M z^{-M}}{1 + b_1 z^{-1} + b_2 z^{-2} + \ldots + b_N z^{-N}}$$

This converts to the standard difference equation

$$y(n) = a_0 x(n) + a_1 x(n-1) + a_2 x(n-2) + \ldots + a_M x(n-M) - b_1 y(n-1) - b_2 y(n-2) - \ldots - b_N y(n-N)$$

The direct-form calculation of this filter requires two MACQs with lengths M and N. There are $(M + N - 1)$ multiplies and $(M + N - 2)$ additions. The data flow picture in Figure 15.17 illustrates the standard implementation. The z^{-1} boxes are time delays (MACQ). The other boxes are multiplications. The + circles are additions.

Figure 15.17
General filter design using a direct-form calculation.

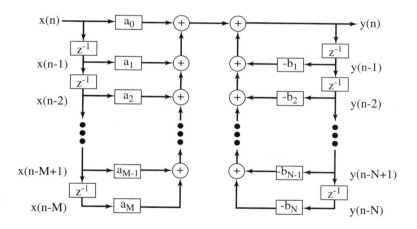

For the next implementation we specify the filter with $N = M$. We can do this without loss of generality by letting some of the coefficients be zero. An alternative implementation, called the *direct-form II realization,* requires only one MACQ of length N. There are still $(2N - 1)$ multiplies and $(2N - 2)$ additions. The data flow picture of Figure 15.18 illustrates the implementation.

Figure 15.18
General filter design using a direct-form II calculation.

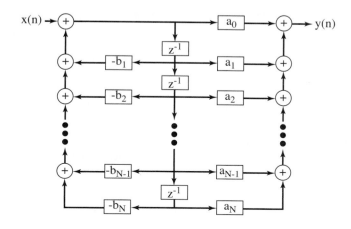

15.9 Exercises

15.1 Consider the following digital filter:

$$y(n) = \frac{x(n) - x(n - 2)}{2}$$

a) Using the Z transform derive general expressions for the gain and phase of the filter.
b) Using the general expressions from part a, calculate the gain and phase of the filter at DC and 60 Hz if the sampling rate is 240 Hz.

15.2 Design a *one-pole* high-Q fixed-point digital high-pass filter. Let $f_s = 100$ Hz, similar to Figure 15.13, but one zero and one pole.
a) Show the pole-zero plot of the filter.
b) Using the pole-zero plot derive the Z transform of the filter.
c) Using the Z transform of the filter, determine the digital filter. The filter gain at $f = 0$ should be 0. The filter gain at $f = 50$ Hz should be 1.
d) Show a fixed-point version of the filter.

15.3 Consider the use of the Z transform in the design and analysis of digital filters.
a) State the definition of the Z transform.
b) Why can't we use the Z transform on a median filter?
c) Use the Z transform to determine the DC gain and phase of the following digital filter:

$$y(n) = x(n) - x(n - 2) + y(n - 1)$$

15.4 Design a 10-Hz digital low-pass filter with a sampling rate of 1000 Hz. Make the gain at DC equal to 1, and the gain at 10 Hz equal to 0.707.
a) Show the pole-zero plot of your filter.
b) Show the H(z) transform.
c) Show the floating-point version of the digital filter.
d) Show the fixed-point version of the digital filter.

15.5 Design a digital filter that rejects both 60 Hz and 120Hz, assuming the sampling rate is 480 Hz. Apply gain scaling so the DC gain is 1. Give the filter in a form that can be implemented with fixed-point math.

15.6 Consider the simple sliding average filter for a general sampling rate of 1000 Hz. This filter is a low-pass filter, as shown in Figure 15.17

$$y(n) = \frac{1}{k} \sum_{i=0}^{i=k-1} x(n - i)$$

What value of **k** should we use to make a gain of about 0.7 at 10 Hz?

15.7 Consider this digital HPF, where **m** and **k** are constants. Figure 15.19 shows the gain versus frequency of this filter for $f_s = 100$ Hz, $m = 2$, and $k = 4, 8$, and 16.

$$y(n) = x(n - m) - \frac{x(n) - x(n - k)}{k} - y(n - 1)$$

a) Is this a FIR or IIR filter? Explain.
b) Derive the H(z) transform of this filter.
c) Let $f_s = 100$ Hz, $m = 8$, and $k = 16$. Use H(z) to plot gain versus frequency from 0 to 50 Hz.

15.8 We defined time-jitter, δt, as the difference between when a periodic task is supposed to be run, and when it is actually run. The goal of a real-time DAS is to start the ADC at a periodic rate, Δt. Let t_n be the nth time the ADC is started. In particular, the goal to make $t_n - t_{n-1} = \Delta t$. The jitter is defined as the constant, δt, such that

$$\Delta t - \delta t < t_i - t_{i-1} < \Delta t + \delta t \quad \text{for all i.}$$

Figure 15.19
Gain versus frequency of
a high-pass filter.

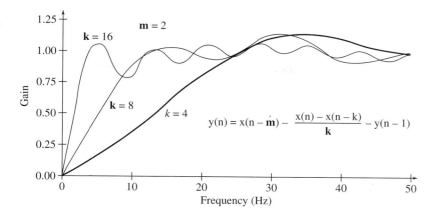

$$y(n) = x(n - m) - \frac{x(n) - x(n - k)}{k} - y(n - 1)$$

Assume the input to the ADC can be described as $V(t) = A + Bsin(2\pi ft)$, where **A, B, f** are constants.

a) Derive an estimate of the maximum voltage error, δV, caused by time-jitter. Basically, solve for the largest possible value of δV as a function of δt, **A**, **B**, and **f**.

b) Consider the situation where this time-jitter is unacceptably large. Which modification to the system will reduce the error the most? Justify your selection.
  A) Run the ADC in condinuous mode
  B) Convert from spinlock semaphores to blocking semaphores
  C) Change from round robin to priority thread scheduling
  D) Reduce the amount of time the system runs with interrupts disabled.
  E) Increase the size of the DataFifo

15.10 Lab Assignments

Lab 15.1. You will design, implement, and test a real-time data acquisition system with a sampling rate of 1000 Hz. This system will include a sliding average filter (data x(n) is sampled at 1000 Hz, and the equation y(n) is also calculated 1000 times/sec).

$$y(n) = (x(n) + x(n - 999))/2$$

The trick to this problem is to implement an MACQ that does not require shifting the data on each sample. You should develop LCD display device driver functions, LCD_Init and LCD_Display, using them to display your results. Using the filtered data, the main program will calculate the RMS, minimum and maximum values over a 1 sec interval. Once a second display these results. Write debugging code to measure the worst-case time-jitter in the sampling process. Use an external sinusoidal source to experimentally measure the gain-versus-frequency response of the digital filter.

Lab 15.2. You will design, implement, and test a real-time data acquisition system with a sampling rate of 1000 Hz. This system will include a sliding average filter (data x(n) is sampled at 1000 Hz, and the equation y(n) is also calculated 1000 times/sec.)

$$y(n) = \frac{1}{1000} \sum_{i=0}^{i=999} \times(n - i)$$

The trick to this problem is to implement an MACQ that does not require shifting the data on each sample and to implement the sum without having to perform 1000 additions for each calculation. You should develop LCD display device driver functions, **LCD_Init** and **LCD_Display**, using them to display your results. Using the filtered data, the main program will calculate the RMS, minimum and maximum values over a 1 sec interval. Once a second display these results. Write debugging code to measure the worst-case time-jitter in the sampling process. Use an external sinusoidal source to experimentally measure the gain-versus-frequency response of the digital filter.

Lab 15.3. The objective in this problem is to design a modular and flexible software implementation of a high-Q digital notch IIR filter. The sampling frequency and notch frequency will be set dynamically (at run time). These parameters will be established when the user calls your **Filter_Init()** function. You may implement any Q you wish, but it should be similar to the high-Q 60 Hz notch IIR filter developed in Section 15.4. Adjust the filter gain so that the gain at DC is 1.0. The data will be passed through the notch filter by calls to your **Filter()** function. You may assume without checking that the cutoff frequency, f_c, is strictly greater than 0 and strictly less than one-half the sampling rate, $0 < f_c < 1/2\ f_s$. Your software handles the establishment of the actual sampling rate. You should use only fixed-point integer calculations. For an example software system that allows for runtime binding of functions to execute, see the OC example on the CD. Use an external sinusoidal source to experimentally measure the gain-versus-frequency response of the digital filter. The following main program illustrates how a user might apply your filter software.

Program 15.8
Example application of the real-time DAS and filter.

```
#include "FILTER.H"    // prototype file for your filter
void Collect(void){ short data;
  data = ADC_In(0);    // sample
  Fifo_Put(data);
}
void main(void){ short x,y;       // filter input/output
  LCD_Init();
  Filter_Init(100,1000,&Collect); // notch 100 Hz, sample at 1KHz
  for(;;){
    while(Fifo_Get(&x)==0){}; // wait for input
    y = Filter(x);           // execute filter
    LCD_Display(y);          // show results
  }
}
```

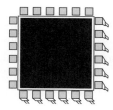

Appendix 1 Glossary

1/f noise A fundamental noise in resistive devices arising from fluctuating conductivity. Same as pink noise.

2's complement (*see* two's complement).

60 Hz noise An added noise from electromagnetic fields caused by either magnetic field induction or capacitive coupling.

accumulator High-speed memory located in the processor, used to perform arithmetic or logical functions. The accumulators on the 6811/6812 are A and B.

accuracy A measure of how close our instrument measures the desired parameter referred to the NIST.

acknowledge Clearing the interrupt flag bit that requested the interrupt.

active thread A thread that is in the ready-to-run circular linked list. It is either running or ready to run.

actuator Electro-mechanical or electro-chemical device that allows computer commands to affect the external world.

ADC Analog-to-digital converter, an electronic device that converts analog signals (e.g., voltage) into digital form (i.e., integers).

address bus A set of digital signals that connect the CPU, memory, and I/O devices, specifying the location to read or write for each bus cycle. *See also* control bus and data bus.

address decoder A digital circuit having the address lines as input and a select line as output (*see* select signal).

aging A technique used in priority schedulers that temporarily increases the priority of low priority threads so they are run occasionally (*see* starvation).

aliasing When digital values sampled at f_s contain frequency components above $\frac{1}{2} f_s$, then the apparent frequency of the data is shifted into the 0 to $\frac{1}{2} f_s$ range. *See* Nyquist theory.

alternatives The total number of possibilities. For example, an 8-bit number scheme can represent 256 different numbers. An 8-bit digital-to-analog converter (DAC) can generate 256 different analog outputs.

anode The positive side of a diode. Current enters the anode side of a diode. Contrast to cathode.

answer modem The device that receives the telephone call.

anti-reset-windup Establishing an upper bound on the magnitude of the integral term in a PID controller, so this term will not dominate, when the errors are large.

arithmetic logic unit (ALU) Component of the processor that performs arithmetic and logic operations.

arm Activate so that interrupts are requested. On the 6811/6812, most flags that can request interrupts will have a corresponding arm bit to allow or disallow that flag to request interrupts. Contrast to enable.

armature The moving structure in a relay; the part that moves when the relay is activated. Contrast to frame.

ASCII American Standard Code for Information Interchange, a code for representing characters, symbols, and synchronization messages as 7-bit, 8-bit or 16-bit binary values.

assembler System software that converts an assembly language program (human-readable format) into object code (machine-readable format).

assembly directive Operations included in the program that are not executed by the computer at run time, but rather are interpreted by the assembler during the assembly process. Same as pseudo-op.

assembly listing Information generated by the assembler in human-readable format, typically showing the object code, the original source code, assembly errors, and the symbol table.

asynchronous bus A communication protocol without a central clock whereby the data is transferred using two or three control lines implementing a handshake interaction between the memory and the computer.

asynchronous communications interface adapter (ACIA) Device to transmit data with asynchronous serial communication protocol. Same as UART and SCI.

asynchronous protocol A protocol whereby the two devices have separate and distinct clocks.

atomic Software execution that can not be divided or interrupted. Once started, an atomic operation will run to its completion without interruption. On most computers the assembly language instructions are atomic.

autoinitialization The process of automatically reloading the address registers and block size counters at the end of a previous block transfer, so that DMA transfer can occur indefinitely without software interaction.

availability The proportion of the total time that the system is working. MTBF is the mean time between failures, MTTR is the mean time to repair, and availability is MTBF/(MTBF + MTTR).

background mode A 6812 mode with the background debug module (BDM) active.

bandwidth In communication systems, the information transfer rate, the amount of data transferred per second. Same as throughput. In analog circuits, the frequency at which the gain drops to 0.707 of the normal value. For a low-pass system, the frequency response ranges from 0 to a maximum value. For a high-pass system, the frequency response ranges from a minimum value to infinity. For a bandpass system, the frequency response ranges from a minimum to a maximum value. Compare to frequency response.

bang-bang A control system where the actuator has only two states, and the system "bangs" all the way in one direction or "bangs" all the way in the other; same as binary controller.

bank-switched memory A memory module with two banks that interfaces to two separate address/data buses. At any given time one memory bank is attached to one address/data bus and the other bank is attached to the other bus. However, this attachment can be switched.

basis Subset from which linear combinations can be used to reconstruct the entire set.

baud rate In general, the baud rate is the total number of bits (information, overhead, and idle) per time that are transmitted. In a modem application it is the total number of sounds per time that are transmitted.

bi-directional Digital signals that can be either input or output.

biendian The ability to process numbers in both big and little endian formats.

big endian Mechanism for storing multiple byte numbers such that the most significant byte exists first (in the smallest memory address). *See also* little endian.

binary A system that has two states, on and off.

binary controller Same as bang-bang.

binary semaphore A semaphore that can have two values. The value = 1 means OK, and the value = 0 means busy. Compare to counting semaphore.

bipolar transistor Either an NPN or a PNP transistor.

bipolar stepper motor A stepper motor in which the current flows in both directions (in/out) along the interface wires; a stepper with four interface wires. Contrast to unipolar stepper motor.

bit Basic unit of digital information taking on the value of either 0 or 1.

bit rate The information transfer rate, given in bits per second. Same as bandwidth and throughput.

bit time The basic unit of time used in serial communication. With serial channel, bit time is 1/baud rate.

blind cycle A software/hardware synchronization method in which the software waits a specified amount of time for the hardware operation to complete. The software has no direct information (blind) about the status of the hardware.

block correction code (BCC) A code (e.g., horizontal parity) attached to the end of a message used to detect and correct transmission errors.

blocked thread A thread that is not scheduled for running because it is waiting on an external event.

blocking semaphore A semaphore whereby the threads will block (so other threads can perform useful functions) when they execute wait on a busy semaphore. Contrast to spinlock semaphore.

borrow During subtraction, if the difference is too small, then we use a borrow to pass the excess information into the next higher place. For example, in decimal subtraction 36-27 requires a borrow from the ones to tens place, because 6-7 is too small to fit into the 0 to 9 range of decimal numbers.

break-before-make In a double-throw relay or double-throw switch, there is one common contact and two separate contacts. Break-before-make means as the common contact moves from one of separate contacts to another, it will break off (finish bouncing and no longer touch) the first contact before it makes (begins to bounce and starts to touch) the other contact. A *form C* relay has a *break-before-make* operation.

break or trap A break or a trap is a debugging instrument that halts the processor. The **TExaS** application will halt both software and hardware simulation when a specific address is encountered. With a resident debugger, the break is created by replacing specific opcode with a software interrupt instruction. When encountered it will stop your program and jump into the debugger. Therefore, a break halts the software. The condition of being in this state is also referred to as a break.

breakdown A transducer that stops functioning when its input goes above a maximum value or below a minimum value. Contrast to dead zone.

breakpoint The place where a break is inserted, the time when a break is encountered, or the time period when a break is active.

buffered I/O A FIFO queue is placed in between the hardware and software in an attempt to increase bandwidth by allowing both hardware and software to run in parallel.

burn The process of programming a ROM, PROM, or EEPROM.

burst DMA An I/O synchronization scheme that transfers an entire block of data all at once directly from an input device into memory, or directly from memory to an output device.

bus A set of digital signals that connect the CPU, memory, and I/O devices, consisting of address signals, data signals, and control signals. *See also* address bus, control bus and data bus.

bus bandwidth The number of bytes transferred per second between the processor and memory.

bus interface unit (BIU) Component of the processor that reads and writes data from the bus. The BIU drives the address and control buses.

busy-waiting A software/hardware synchronization method whereby the software continuously reads the hardware status waiting for the hardware operation to complete. The software usually performs no work while waiting for the hardware. Same as gadfly.

byte Digital information containing eight bits.

carrier frequency The average or midvalue sound frequency in the modem.

carry During addition, if the sum is too large, then we use a carry to pass the excess information into the next higher place. For example, in decimal addition 36 + 27 requires a carry from the ones to tens place because 6 + 7 is too big to fit into the 0 to 9 range of decimal numbers.

cathode The negative side of a diode. Current exits the cathode side of a diode. Contrast to anode.

cathode ray tube (CRT) terminal An I/O device used to input data from a keyboard and output character data to a screen. The electrical interface is usually asynchronous serial.

causal The property whereby the output depends on the present and past inputs, but not on any future inputs.

ceiling Establishing an upper bound on the result of an operation. *See also* floor.

channel The hardware that allows communication to occur.

checksum The simple sum of the data, usually in finite precision (e.g., 8, 16, 24 bits).

closed loop control system A control system that includes sensors to measure the current state variables. These inputs are used to drive the system to the desired state.

CMOS A digital logic system called complementary metal oxide semiconductor. It has properties of low power and small size. Its power is a function of the number of transitions per second. Its speed is often limited by capacitive loading.

command signals The lines that specify general information about the current cycle; signals that specify whether or not to activate during this cycle; the specific times for the rise and fall edges are uncertain. Contrast to timing signals.

common anode LED display A display with multiple LEDs, configured with all of the LED anodes connected together; there are separate connections to the cathodes (current flows in the common anode and out the individual cathodes).

common cathode LED display A display with multiple LEDs, configured with all of the LED cathodes connected together; there are separate connections to the anodes (current flows in the individual anodes and out the common cathode).

common mode For a system with differential inputs, the common mode properties are defined as signals applied to both inputs simultaneously. Contrast to differential mode.

common mode input impedance Common mode input voltage divided by common mode input current.

common mode rejection ratio For a differential amplifier, CMRR is the ratio of the common mode gain divided by the differential mode gain. A perfect CMRR would be zero.

compiler System software that converts a high-level language program (human-readable format) into object code (machine-readable format).

compression ratio The ratio of the number of original bytes to the number of compressed bytes.

condition code register (CCR) Register in the processor that contains the status of the previous ALU operation, as well as some operating mode flags such as the interrupt enable bit.

constraint A condition defining how the system will be developed, generally restricting the range of solutions from which the system will be built.

control bus A set of digital signals that connect the CPU, memory, and I/O devices, specifying when to read or write for each bus cycle. *See also* address bus and data bus.

control unit (CU) Component of the processor that determines the sequence of operations.

cooperative multi-tasking A scheduler that can not suspend execution of a thread without the thread's permission. The thread must cooperate and suspend itself. Same as nonpreemptive scheduler.

counting semaphore A semaphore that can have any signed integer value. The value > 0 means OK, and the value ≤ 0 means busy. Compare to binary semaphore.

CPU bound A situation where the input or output device is faster than the software. In other words it takes less time for the I/O device to process data than for the software to process data. Contrast to I/O bound.

CPU cycle A memory bus cycle where the address and R/W are controlled by the processor. On microcontrollers without DMA, all cycles are CPU cycles. Contrast to DMA cycle.

crisp input An input parameter to the fuzzy logic system, usually with units such as cm, cm/sec, °C, and the like.

crisp output An output parameter from the fuzzy logic system, usually with units such as dynes, watts, and the like.

critical section Locations within a software module at which, if an interrupt were to occur at one of these locations, then an error could occur (e.g., data lost, corrupted data, program crash). Same as vulnerable window.

cross-assembler An assembler that runs on one computer but creates object code for a different computer.

cross-compiler A compiler that runs on one computer but creates object code for a different computer.

cycle steal DMA An I/O synchronization scheme that transfers data one byte at a time directly from an input device into memory, or directly from memory to an output device.

cycle stretch The action whereby some memory cycles are longer, allowing time for communication with slower memories; sometimes the memory itself requests the additional time, and sometimes the computer has a preprogrammed cycle stretch for certain memory addresses.

DAC Digital-to-analog converter, an electronic device that converts digital signals (i.e., integers) to analog form (e.g., voltage).

data acquisition system A system that collects information, same as instrument.

data bus A set of digital signals that connect the CPU, memory, and I/O devices, specifying the value that is being read or written for each bus cycle. *See also* address bus and control bus.

data communication equipment (DCE) A modem or printer connected a serial communication network.

data terminal equipment (DTE) A computer or a terminal connected to a serial communication network.

dead zone A condition of a transducer when a large change in the input causes little or no change in the output. Contrast to breakdown.

deadlock A scenario that occurs when two or more threads are all blocked, each waiting for the other with no hope of recovery.

defuzzification Conversion from the fuzzy logic output variables to the crisp outputs.

desk checking or **dry run** We perform a desk check (or dry run) by determining in advance, either by analytical algorithm or explicit calculations, the expected outputs of strategic intermediate stages and final results for a set of typical inputs. We then run our program to compare the actual outputs with this template of expected results.

device driver A collection of software routines that perform I/O functions.

differential mode For a system with differential inputs, the differential mode properties are defined as signals applied as a difference between the two inputs. Contrast to common mode.

differential mode input impedance Differential mode input voltage divided by differential mode input current.

digital signal processing Processing of data with digital hardware or software after the signal has been sampled by the ADC (e.g., filters, detection and compression/decompression).

direct An addressing mode in which the data or address value for the instruction is located in memory at address $0000 to $00FF.

direct memory access (DMA) The ability to transfer data between two modules on the bus; this transfer is usually initiated by the hardware (device needs service), and the software configures the communication, but the data is transferred without explicit software action for each individual piece of data.

direction register Specifies whether a bidirectional I/O pin is an input or an output. We set a direction register bit to 0 (or 1) to specify the corresponding I/O pin to be input (or output).

disarm Deactivate so that interrupts are not requested, performed by clearing the arm bit.

DMA Direct Memory Access is a software/hardware synchronization method whereby the hardware itself causes a data transfer between the I/O device and memory at the appropriate time when data needs to be transferred. The software usually can perform other work while waiting for the hardware. No software action is required for each individual byte.

DMA cycle A memory bus cycle in which the address and R/W are controlled by the DMA controller. Contrast to CPU cycle.

double byte Two bytes containing 16 bits. Same as word.

double-pole relay Two separate and complete relays, which are activated together. Contrast to single pole.

double-throw relay A relay with three contact connections, one common and two throws. The common will be connected to exactly of one the two throws (*see* single throw).

double-throw switch A switch with three contact connections. The center contact will be connected exactly to one of the other two contacts. Contrast with single-throw.

download The process of transferring object code from the host (e.g., the PC) to the target microcomputer (e.g., the 6812).

dropout An error that occurs after a right shift or a divide, and the consequence is that an intermediate result looses its ability to represent all of the values. For example, I = 100*(N/51) can only result in the values 0, 100, or 200, whereas I = (100*N)/51 properly calculates the desired result.

dual address DMA DMA that requires two bus cycles to transfer data from an input device into memory, or from memory to an output device.

dual-port memory A memory module that interfaces to two separate address/data buses, and allows both systems read/write access the data.

duty cycle For a periodic digital wave, the percentage of time the signal is high. When an LED display is scanned, it is the percentage of time each LED is active. A motor interfaced using pulse width modulation allows the computer to control delivered power by adjusting the duty cycle.

dynamic allocation Data structures such as the TCB that are created at runtime by calling `malloc()` and exist until the software releases the memory block back to the heap by calling `free()`. Contrast to static allocation.

dynamic RAM Volatile read/write storage built from a capacitor and a single transistor having a low cost, but requiring refresh. Contrast with static RAM.

EEPROM Electrically erasable programmable read only memory that is nonvolatile and easy to reprogram. Typically, EEPROM can be erased and reprogrammed 10,000 times.

effective address register (EAR) A register that contains the address for the current memory cycle.

embedded computer system A system that performs a specific dedicated operation where the computer is hidden or embedded inside the machine.

emulator An in-circuit emulator is an expensive debugging hardware tool that mimics the processor pin outs. To debug with a 6811 emulator, you would remove the 6811 processor chip and attach the emulator cable into the 6811 processor socket. The emulator would sense the processor input signals and recreate the processor outputs signals on the socket as if a 6811 chip were actually there, running at 2 MHz. Inside the emulator you have internal read/write access to the registers and processor state. Most emulators allow you to visualize and record strategic information in real time without halting the program execution. You can also remove ROM chips and insert the connector of a ROM-emulator. This type of emulator is less expensive, and it allows you to debug ROM-based software systems.

EPROM Same as PROM. Electrically programmable read only memory that is nonvolatile and requires external devices to erase and reprogram. It is usually erased using UV light.

erase The process of clearing the information in a PROM or EEPROM, using electricity or UV light. The information bits are usually all set to logic 1.

EVB Evaluation Board, a Freescale product used to develop microcomputer software.

even parity A communication protocol whereby the number of ones in the data plus a parity bit is an even number. Contrast with odd parity.

expanded mode The mode in which some of the I/O ports are used to create an external data bus (control, address, data) allowing external memory to be connected.

extended mode More than the usual 16-bit address. The MC9S12C32 allows up to 20 address lines with its paged memory modes.

fan out The number of inputs that a single output can drive if the devices are all in the same logic family.

fast clear A 6812 timer mode in which the associated flag is automatically cleared when the timer register is accessed.

FET Field effect transistor, also JFET.

filter In the debugging context, a filter is a Boolean function or conditional test used to make run-time decisions. For example, if we print information only if two variables x,y are equal, then the conditional (x==y) is a filter. Filters can involve hardware status as well. For example, if we halt when the serial port has an overrun error, then `(SCSR&0x08)` is the filter, and `if(SCSR&0x08)asm("swi");` would be the entire instrument.

finite impulse response filter (FIR) A digital filter in which the output is a function of a finite number of current and past data samples, but not a function of previous filter outputs.

fixed-point A technique whereby calculations involving nonintegers are performed using a sequence of integer operations. For example, 0.123*x is performed in decimal fixed-point as (123*x)/1000 or in binary fixed-point as (126*x) >> 10.

flash EEPROM Electrically erasable programmable read only memory that is nonvolatile and easy to reprogram. Flash EEPROMs are typically larger than regular EEPROM, and have fewer erase-reprogram cycles.

floating A logic state in which the output device does not drive high or pull low. The outputs of open collector and tristate devices can be in the floating state. Same as HiZ.

floor Establishing a lower bound on the result of an operation. *See also* ceiling.

follower An analog circuit with gain equal to 1, large input impedance and small output impedance. Same as voltage follower.

frame A complete and distinct packet of bits occurring in a serial communication channel.

frame The fixed structure in a relay or transducer. Contrast to armature.

framing error An error when the receiver expects a stop bit (1) and the input is 0.

frequency response The frequency at which the gain drops to 0.707 of the normal value. For a low-pass system, the frequency response ranges from 0 to a maximum value. For a high pass system, the frequency response ranges from a minimum value to infinity. For a bandpass system, the frequency response ranges from a minimum to a maximum value. Same as bandwidth.

frequency shift key (FSK) A modem that modulates the digital signals into frequency encoded sine waves.

friendly Friendly software modifies just the bits that need to be modified, leaving the other bits unchanged. Making it easier to combine modules.

full-duplex channel Hardware that allows bits (information, error checking, synchronization or overhead) to transfer simultaneously in both directions. Contrast with simplex and half-duplex channels.

full-duplex communication A system that allows information (data, characters) to transfer simultaneously in both directions.

functional debugging The process of detecting, locating, or correcting functional and logical errors in a program, typically not involving time. The process of instrumenting a program for such purposes is called functional debugging or often simply debugging.

fuzzification Conversion from the crisp inputs to the fuzzy logic input variables.

fuzzy logic Boolean logic (true/false) that can take on a range of values from true (255) to false (0). Fuzzy logic **and** is calculated as the minimum. Fuzzy logic **or** is the maximum.

gadfly A software/hardware synchronization method whereby the software continuously reads the hardware status waiting for the hardware operation to complete. The software usually performs no work while waiting for the hardware. Same as busy-waiting.

gage factor The sensitivity of a strain gage transducer (i.e., slope of the resistance versus displacement response).

half-duplex channel Hardware that allows bits (information, error checking, synchronization, or overhead) to transfer in both directions, but in only one direction at a time. Contrast with simplex and full-duplex channels.

half-duplex communication A system that allows information to transfer in both directions, but in only one direction at a time.

handshake A software/hardware synchronization method whereby control and status signals go both directions between the transmitter and receiver. The communication is interlocked meaning each device will wait for the other.

hard real time A system that can guarantee that a process will complete a critical task within a certain specified range. In data acquisition systems, hard real time means there is an upper bound on the latency between when a sample is supposed to be taken (every $1/fs$) and when the ADC converter is actually started. Hard real time also implies that no ADC samples are missed.

hexadecimal A number system that uses base 16.

hiZ A logic state in which the output device does not drive high or pull low. The outputs of open collector and tristate devices can be in the floating state. Same as floating.

hold time When latching data into a device with a rising or falling edge of a clock, the hold time is the time after the active edge of the clock during which the data must continue to be valid. *See* setup time.

horizontal parity A parity calculated across the entire message on a bit by bit basis (e.g., the horizontal parity bit 0 is the parity calculated on all the bit 0's of the entire message, can be even or odd parity).

hysteresis A condition when the output of a system depends not only on the input, but also on the previous outputs e.g. (a transducer that follows a different response curve when the input is increasing than when the input is decreasing).

I/O bound A situation where the input or output device is slower than the software. In other words, it takes longer for the I/O device to process data than for the software to process data. Contrast to CPU bound.

I/O device Hardware and software components capable of bringing information from the external environment into the computer (input device), or sending data out from the computer to the external environment (output device).

I/O port A hardware device that connects the internal software with external hardware.

IEEE488 A medium-speed handshaking parallel I/O standard used for desktop instruments.

I_{IH} Input current when the signal is high.

I_{IL} Input current when the signal is low.

immediate An addressing mode in which the operand is a fixed data or address value.

impedance loading A condition when the input of stage $n+1$ of an analog system affects the output of stage n, because the input impedance of stage $n+1$ is too small and the output impedance of stage n is too large.

impedance The ratio of the effort (voltage, force, pressure) divided by the motion (current, velocity, flow).

incremental control system A control system where the actuator has many possible states, and the system increments or decrements the actuator value depending on whether an error is positive or negative.

indexed An addressing mode in which the data or address value for the instruction is located in memory pointed to by an index register.

infinite impulse response filter (IIR) A digital filter whose output is a function of an infinite number of past data samples, usually by making the filter output a function of previous filter outputs.

inherent An addressing mode in which there is no operand or the operand is implied (not explicitly stated).

input bias current Difference between currents of the two op amp inputs.

input capture A mechanism to set a flag and capture the current time (TCNT value) on the rising, falling, or rising and falling edge of an external signal. The input capture event can also request an interrupt.

input impedance Input voltage divided by input current.

input noise current Current noise refereed to the op amp inputs.

input noise voltage Voltage noise refereed to the op amp inputs.

input offset current Average current into the two op amp inputs.

input offset voltage Voltage difference between the two op amp inputs that makes the output zero.

instruction register (**IR**) Register in the control unit that contains the op code for the current instruction.

instrument An instrument is the code injected into a program for debugging or profiling. This code is usually extraneous to the normal function of a program and may be temporary or permanent. Instruments injected during interactive sessions are considered to be temporary, because these instruments can be removed simply by terminating a session. Instruments injected in source code are considered to be permanent, because removal requires editing and recompiling the source. An example of a temporary instrument occurs when the debugger replaces a regular op code with the swi instruction. This temporary instrument can be removed dynamically by restoring the original op code. A print statement added to your source code is an example of a permanent instrument, because removal requires editing and recompiling. An embedded system that collects information, same as data acquisition system.

instrumentation The debugging process of injecting or inserting an instrument.

instrumentation amp A differential amplifier analog circuit, which can have large gain, large input impedance, small output impedance, and a good common mode rejection ratio.

interrupt A software/hardware synchronization method whereby the hardware causes a special software program (interrupt handler) to execute when its operation to complete. The software usually can perform other work while waiting for the hardware.

interrupt flag A status bit that is set by the timer hardware to signify that an external event has occurred.

interrupt mask A control bit that, if programmed to 1, will cause an interrupt request when the associated flag is set. Same as **arm**.

interrupt service routine (**ISR**) Program that runs as a result of an interrupt.

interrupt vector Sixteen-bit values at the end of memory specifying where the software should execute after an interrupt request. There is a unique interrupt vector for each type of interrupt including reset.

intrusive (*see* nonintrusive).

I_{OH} Output current when the signal is high.

I_{OL} Output current when the signal is low.

IRQ A interrupt mechanism on the 6811 and the 6812.

isolated I/O A configuration in which the I/O devices are interfaced to the computer in a manner different from the way memories are connected. From an interfacing perspective, I/O devices and memory modules have separate bus signals; from a programmer's point of view, the I/O devices have their own I/O address map separate from the memory map and I/O device access requires the use of special I/O instructions.

jerk The change in acceleration; the derivative of the acceleration.

Johnson noise A fundamental noise in resistive devices arising from the uncertainty about the position and velocity of individual molecules. Same as thermal noise and white noise.

Karnaugh map tabular representation of the input/output relationship for a combinational digital function. The inputs possibilities are placed in the row and column labels, and the output values are placed inside the table.

latch As a noun, it means a register. As a verb, it means to store data into the register.

latched input port An input port at which the signals are latched (saved) on an edge of an associated strobe signal.

latency In this book latency usually refers to the response time of the computer to external events. For example, the time between new input becoming available and the time the input is read by the computer, or the time between an output device becoming idle and the time the input is the computer writes new data to it. There can also be a latency for an I/O device, which is the response time of the external I/O device hardware to a software command.

LCD Liquid crystal display, whereby the computer controls the reflectance or transmittance of the liquid crystal, characterized by its flexible display patterns, low power, and slow speed.

LED Light emitting diode, whereby the computer controls the electrical power to the diode, characterized by its simple display patterns, medium power, and high speed.

light-weight process Same as a thread.

linear filter A filter whose output is a linear combination of its inputs.

linear variable differential transformer (LVDT) A transducer that converts position into electric voltage.

little endian Mechanism for storing multiple byte numbers such that the least significant byte exists first (in the smallest memory address). Contrast with big endian.

loader System software that places the object code into the microcomputer's memory. If the object code is stored in EPROM, the loader is also called an EPROM programmer.

logic analyzer A hardware debugging tool that allows you to visualize many digital logic signals versus time. Real logic analyzers have at least 32 channels and can have up to 200 channels, with sophisticated techniques for triggering, saving, and analyzing the real-time data. In **TExaS**, logic analyzers have only 8 channels and simply plot digital signals versus time.

LSB The least significant bit in a number system is the bit with the smallest significance, usually the rightmost bit. With signed or unsigned integers the significance of the LSB is 1.

maintenance Process of verifying, changing, correcting, enhancing, and extending a system.

make before break In a double-throw relay or double-throw switch, there is one common contact and two separate contacts. Make before break means as the common contact moves from one of separate contacts to another, it will make (finishing bouncing) the second contact before it breaks off (start bouncing) the first contact. A *form D* relay has a *make before break* operation.

mark A digital value of true or logic 1. Contrast with space.

mask As a verb, mask is the operation that selects certain bits out of many bits, using the logical and operation. The bits that are not being selected will be cleared to zero. When used as a noun, mask refers to the specific bits that are being selected.

measurand A signal measured by a data acquisition system.

membership sets Fuzzy logic variables that can take on a range of values from true (255) to false (0).

memory A computer component capable of storing and recalling information.

memory-mapped I/O A configuration in which the I/O devices are interfaced to the computer in a manner identical to the way memories are connected. From an interfacing perspective, I/O devices and memory modules share the same bus signals; from a programmer's point of view, the I/O devices exist as locations in the memory map, and I/O device access can be performed using any of the memory access instructions.

microcomputer A small electronic device capable of performing input/output functions, containing a microprocessor, memory, and I/O devices, where small means you can carry it.

microcontroller A single-chip microcomputer such as the Freescale 6811, Freescale 6812, Intel 8051, PIC16, or the Texas Instruments TMS370.

mnemonic   The symbolic name of an operation code (e.g., like `ldaa psha stx`).

modem   An electronic device that MOdulates and DEModulates a communication signal. Used in serial communication across telephone lines.

monitor or **debugger window**   A monitor is a debugger feature that allows users to passively view strategic software parameters during the real-time execution of a program. An effective monitor is one that has minimal effect on the performance of the system. When debugging software on a windows-based machine, we can often set up a debugger window that displays the current value of certain software variables.

MOSFET   Metal oxide semiconductor field effect transistor.

MSB   The most significant bit in a number system is the bit with the greatest significance, usually the leftmost bit. If the number system is signed, then the MSB signifies positive (0) or negative (1).

multiple access circular queue (MACQ)   A data structure used in data acquisition systems to hold the current sample and a finite number of previous samples.

multi-threaded   A system with multiple threads (e.g., main program and interrupt service routines) that cooperate towards a common overall goal.

mutual exclusion or **mutex**   Thread synchronization whereby at most one thread at a time is allowed to enter.

negative feedback   An analog system with negative gain feedback paths. These systems are often stable.

negative logic   A signal where the true value has a lower voltage than the false value. In digital logic, true is 0 and false is 1; in TTL logic, true is less than 0.7 volts and false is greater than 2 volts; in RS232 protocol, true is -12 volts and false is $+12$ volts. Contrast with positive logic.

negative predictive value (NPV)   is the probability the event is false if our device says it is false (TN/(TN + FN)).

nibble   Four binary bits or one hexadecimal digit.

nonatomic   Software execution that can be divided or interrupted. Most lines of C code require multiple assembly language instructions to execute; therefore, an interrupt may occur in the middle of a line of C code.

noninstrusive   A characteristic whereby the presence of the collection of information itself does not affect the parameters being measured. Nonintrusiveness is the characteristic or quality of a debugger that allows the software/hardware system to operate normally as though the debugger did not exist. Intrusiveness is used as a measure of the degree of perturbation caused in program performance by an instrument. For example, a print statement added to your source code and single-stepping are very intrusive because they significantly affect the real-time interaction of the hardware and software. When a program interacts with real-time events, the performance is significantly altered. On the other hand, an instrument that toggles an LED on and off (requiring just 10 μs to execute) is much less intrusive. A logic analyzer that passively monitors the address and data is completely nonintrusive. An in-circuit emulator is also nonintrusive, because the software input/output relationships will be the same with and without the debugging tool.

nonlinear filter   A filter in which the output is not a linear combination of its inputs. For example, median, minimum, maximum are examples of nonlinear filters. Contrast to linear filter.

nonpreemptive scheduler   A scheduler that cannot suspend execution of a thread without the thread's permission. The thread must cooperate and suspend itself. Same as cooperative multitasking.

nonreentrant   A software module that once started by one thread should not be interrupted and executed by a second thread. Nonreentrant modules usually involve nonatomic accesses to global variables or I/O ports: read modify write, write followed by read or a multistep write.

nonvolatile A condition whereby information is not lost when power is removed. When power is restored, then the information is in the state that occurred when the power was removed.

null cycle A computer bus cycle that fetches data at address $FFFF, but the data is not used.

Nyquist Theorem If an input signal is captured by an ADC at the regular rate of fs samples/ sec, then the digital sequence can accurately represent the 0 to ½ fs frequency components of the original signal.

object code Programs in machine-readable format created by the compiler or assembler. The S19 records contain object code.

odd parity A communication protocol in which the number of ones in the data plus a parity bit is an odd number. Contrast with even parity.

op amp An integrated analog component with two inputs, (V2,V1) and an output (Vout), where Vout = K • (V2-V1). The amp has a very large gain, K. Same as operational amplifier.

op code, **opcode**, or **operation code** A specific instruction executed by the computer. The op code along with the operand completely specifies the function to be performed. In assembly language programming, the op code is represented by its mnemonic (e.g., ldaa). During execution, the op code is stored as a machine code loaded in memory. The ldaa instruction with immediate addressing has a machine code of $86.

open collector A digital logic output that has two states, low and HiZ. Same as wire-or-mode.

open loop control system A control system that does not include sensors to measure the current state variables. An analog system with no feedback paths.

operand The second part of an instruction that specifies either the data or the address for that instruction. An assembly instruction typically has an op code (e.g., ldaa) and an operand (e.g., #55). Instructions that use inherent addressing mode have no operand field.

operating system System software for managing computer resources and facilitating common functions such input/output, memory management, and file system.

originate modem The device that places the telephone call.

oscilloscope A hardware debugging tool that allows you to visualize one or two analog signals versus time. In **TExaS**, oscilloscopes can plot up to eight channels.

output compare A mechanism to cause a flag to be set and an output pin to change when the TCNT matches a preset value. The output compare event can also request an interrupt.

output impedance Open circuit output voltage divided by short circuit output current.

overflow An error that occurs when the result of a calculation exceeds the range of the number system. For example, with 8-bit unsigned integers, 200 + 57 will yield the incorrect result of 1. Also, when TCNT increments from $FFFF back to $0000, setting the TOF flag. This overflow event can also request an interrupt.

overrun error An error that occurs when the receiver gets a new frame but the data register and shift register already have information.

paged memory A memory organization whereby logical addresses (used by software) have multiple and distinct components or fields. The number of bits in the least significant field defines the page size. The physical memory is usually continuous, having sequential addresses. There is a dynamic address translation (logical to physical).

parallel port A port at which all signals are available simultaneously. In this book the parallel ports are 8 bits wide.

partially asynchronous bus A communication protocol that has a central clock, but the memory module can dynamically extend the length of a bus cycle (cycle stretch) if it needs more time.

PC relative An addressing mode in which the effective address is calculated by its position relative to the current value of the program counter.

performance debugging or profiling The process of acquiring or modifying timing characteristics and execution patterns of a program. The process of instrumenting a program for such purposes is called performance debugging or profiling.

periodic polling A software/hardware synchronization method that is a combination of interrupts and busy-waiting. An interrupt occurs at a regular rate (periodic) independent of the hardware status. The interrupt handler checks the hardware device (polls) to determine whether its operation is complete. The software usually can perform other work while waiting for the hardware.

phase shift key (PSK) A protocol that encodes the information as phase changes between the sounds.

photosensor A transducer that converts reflected or transmitted light into electric current.

physical plant The physical device being controlled.

PID controller A control system where the actuator output depends on a linear combination of the current error (P), the integral of the error (I) and the derivative of the error (D).

pink noise A fundamental noise in resistive devices arising from fluctuating conductivity. Same as 1/f noise.

pole A place in the frequency domain where the filter gain is infinite.

poll for zeros and ones An interrupt handler that checks both for the interrupt flag and for the presence of other ones and zeros in the status register.

polling A software function to look and see which of the potential sources requested the interrupt.

port External pins through which the microcomputer can perform input/output. Same as I/O port.

positive feedback An analog system with positive gain feedback paths. These systems will saturate.

positive logic A signal in which the true value has a higher voltage than the false value. In digital logic, true is 1 and false is 0; in TTL logic, true is greater than 2 volts and false is less than 0.7 volts; in RS232 protocol, true is $+12$ volts and false is -12 volts. Contrast with negative logic.

positive predictive value (PPV) is the probabity that an event is true when our device says it is (TP/(TP + FP)).

potentiometer A transducer that converts position into electric resistance.

precision A term specifying the degree of freedom from random errors. For an input signal, it is the number of distinguishable input signals that can be reliably detected by the measurement. For an output signal, it is the number of different output parameters that can be produced by the system. For a number system, precision is the number of distinct or different values of a number system in units of "alternatives." The precision of a number system is also the number of binary digits required to represent all its numbers in units of "bits."

preemptive scheduler A scheduler that has the power to suspend execution of a thread without the thread's permission.

priority When two requests for service are made simultaneously, priority determines the order in which to process them.

private Can be accessed only by software modules in that local group.

private variable A global variable that is used by a single thread, and not shared with other threads.

process The execution of software that does not necessarily cooperate with other processes.

producer-consumer A multithreaded system in which the producers generate new data, and the consumers process or output the data.

profiling (*see* performance debugging).

program counter (PC) A register in the processor that points to the memory containing the instruction to execute next.

PROM Same as EPROM. Programmable read only memory that is nonvolatile and requires external devices to erase and reprogram. It is usually erased using UV light.

promotion Increasing the precision of a number for convenience or to avoid overflow errors during calculations.

pseudo op Operations included in the program that are not executed by the computer at run time, but rather are interpreted by the assembler during the assembly process. Same as assembly directive.

pseudo-code A shorthand for describing a software algorithm. The exact format is not defined, but many programmers use their favorite high-level language syntax (e.g., C) without paying rigorous attention to the punctuation.

public Can be accessed by any software module.

public variable A global variable that is shared by multiple programs or threads.

pulse-width modulation A technique to deliver a variable signal (voltage, power, energy) using an on/off signal with a variable percentage of time the signal is on (duty cycle). Same as **variable duty cycle**.

Q The Q of a bandpass filter (passes f_{min} to f_{max}) is the center pass frequency ($f_o = (f_{max} + f_{min})/2$) divided by the width of the pass region, $Q = f_o/(f_{max} - f_{min})$. The Q of a bandreject filter (rejects f_{min} to f_{max}) is the center reject frequency ($f_o = (f_{max} + f_{min})/2$) divided by the width of the reject region, $Q = f_o/(f_{max} - f_{min})$.

quadrature amplitude modem (QAM) A protocol that uses both the phase and amplitude to encode up to 6 bits onto each baud.

qualitative DAS A DAS that collects information not in the form of numerical values, but rather in the form of the qualitative senses (e.g., sight, hearing, smell, taste, and touch). A qualitative DAS may also detect the presence or absence of conditions.

quantitative DAS A DAS that collects information in the form of numerical values.

R/W signal A bus signal specifying whether it is a read (R/W = 1) or a write (R/W = 0) cycle.

RAM Random Access Memory, a type of memory in which the information can be stored and retrieved easily and quickly. Since it is volatile, the information is lost when power is removed.

range Includes both the smallest possible and the largest possible signal (input or output). The difference between the largest and smallest input that can be measured by the instrument. The units are in the units of the measurand. When precision is in alternatives, range = precision • resolution. Same as span.

read cycle Data flows from the memory or input device to the processor, the address bus specifies the memory or input device location, and the data bus contains the information at that address.

read data available The time interval (start, end) during which the data will be valid during a read cycle, determined by the memory module.

real time A system that can guarantee an upper bound (worst case) on latency.

real-time computer system A system in which time-critical operations occur when needed.

recursion A programming technique whereby a function calls itself.

reentrant A software module that can be started by one thread and interrupted and executed by a second thread. A reentrant module allow both threads to properly execute the desired function. Contrast with nonreentrant.

registers High-speed memory located in the processor. The registers on the 6811/6812 are CCR, A, B, X, Y, SP, and PC.

relay A mechanical switch that can be turned on and off by the computer.

reliability The ability of a system to operate within its specified parameters for a stated period of time. Given in terms of mean time between failures (MTBF).

reproducibility (or **repeatability**) A parameter specifying how consistent over time the measurement is when the input remains fixed.

requirement Detailed performance parameter that the system must satisfy, generally derived from the overall objective of the system.

reset vector The 16-bit value at memory locations $FFFE and $FFFF specifying where the software should start after power is turned on or after a hardware reset.

resistance temperature device (RTD) A linear transducer that converts temperature into electric resistance.

resolution For an input signal, the smallest change in the input parameter that can be reliably detected by the measurement. For an output signal, the smallest change in the output parameter that can be produced by the system; range equals precision times resolution. The units are in the units of the measurand. When precision is in alternatives, range = precision • resolution.

response time Similar to latency, it is the delay between the time an event occurs and the time the software responds to the event.

ritual Software, usually executed once at the beginning of the program, that defines the operational modes of the I/O ports.

ROM Read-Only Memory, a type of memory in which the information is programmed into the device once but can be accessed quickly. It is low cost, must be purchased in high volume, and can be programmed only once. See also EPROM, EEPROM, and flash EEPROM.

round-robin scheduler A scheduler that runs each active thread equally.

roundoff The error that occurs in a fixed-point or floating-point calculation when the least significant bits of an intermediate calculation are discarded so that the result can fit into the finite precision.

RTD Resistance temperature device, a sensor used to measure temperature, usually made from platinum.

sample and hold A circuit used to latch a rapidly changing analog signal, capturing its input value and holding its output constant.

sampling rate The rate at which data is collected in a data acquisition system.

saturation A device that is no longer sensitive to its inputs when its input goes above a maximum value or below a minimum value.

scan or ScanPoint Any instrument used to produce a side effect without causing a break (halt) is a scan. Therefore, a scan may be used to gather data passively or to modify functions of a program. Examples include software added to your source code that simply outputs or modifies a global variable without halting. A ScanPoint is triggered in a manner similar to a breakpoint, but a ScanPoint simply records data at that time without halting execution.

scheduler System software that suspends and launches threads.

Schmitt Trigger A digital interface with hysteresis making it less susceptible to noise.

scope A logic analyzer or an oscilloscope, hardware debugging tool that allows you to visualize multiple digital or analog signals versus time.

SCSI Small Computer Systems Interface, a high-speed handshaking parallel I/O standard.

select signal The output of the address decoder (each module on the bus has a separate address decoder); a Boolean (true/false) signal specifying whether or not the current address of the bus matches the device address.

semaphore A system function with two operations (wait and signal) that provide for thread synchronization and resource sharing.

sensitivity The sensitivity of a transducer is the slope of the output versus input response. The sensitivity of a qualitative DAS that detects events is the percentage of actual events that are properly recognized by the system (TP/(TP + FN)).

serial communication A process whereby information is transmitted one bit at a time.

serial communications interface (SCI) A device to transmit data with asynchronous serial communication protocol (same as UART and ACIA).

serial peripheral interface (SPI) A device to transmit data using the synchronous serial communication protocol.

serial port An I/O port with which the bits are input or output one at a time.

setup time When latching data into a register with a clock, the time before an edge at which the input must be valid. Contrast with hold time.

shot noise A fundamental noise that occurs in devices that count discrete events.

signed two's complement binary A mechanism to represent signed integers where 1 followed by all 0s is the most negative number, all 1s represents the value -1, all 0s represents the value 0, and 0 followed by all 1s is the largest positive number.

sign-magnitude binary A mechanism to represent signed integers whereby the most significant bit is set if the number is negative, and the remaining bits represent the magnitude as an unsigned binary.

simple poll An interrupt handler that simply checks the interrupt flag.

simplex channel Hardware that allows bits (information, error checking, synchronization, or overhead) to transfer only in one direction. Contrast with half-duplex and full-duplex channels.

simplex communication A system that allows information to transfer only in one direction.

simulator A simulator is a software application, such as **TExaS**, that simulates or mimics the operation of a processor or computer system. Most simulators recreate only simple I/O ports and often do not effectively duplicate the real-time interactions of the software/hardware interface. On the other hand, they do provide a simple and interactive mechanism to test software. Simulators are especially useful when learning a new language, because they provide more control and access to the simulated machine, than one normally has with real hardware.

single-address DMA DMA that requires only one bus cycle to transfer data from an input device into memory, or from memory to an output device.

single-pole relay A simple relay with only one copy of the switch mechanism. Contrast with double pole.

single-pole switch A simple switch with only one copy of the switch mechanism. Contrast with double pole.

single-throw switch A switch with two contact connections. The two contacts may be connected or disconnected. Contrast with double-throw.

Small Computer Systems Interface (SCSI) a handshaking parallel data communication channel used to connect high-speed high-volume I/O devices to the computer.

software interrupt vector The 16-bit value at memory locations $FFF6 and $FFF7 specifying where the software should go after executing a software interrupt instruction, `swi`.

software maintenance Process of verifying, changing, correcting, enhancing, and extending software.

solenoid A discrete motion device (on/off) that can be controlled by the computer, usually by activating an electromagnet. For example, electronic door locks on automobiles.

source code Programs in human-readable format created with an editor.

space A digital value of false or logic 0. Contrast with mark.

span Same as range.

spatial resolution The volume over which the DAS collects information about the measurand.

specificity The specificity of a transducer is the relative sensitivity of the device to the signal of interest versus the sensitivity of the device to other unwanted signals. The sensitivity of a qualitative DAS is the fraction of properly handled non-events (TN/(TN + FP)).

spinlock semaphore A semaphore in which the threads will spin (run but perform no useful function) when they execute wait on a busy semaphore. Contrast to blocking semaphore.

stabilize The debugsing process of stabilizing a software system involves specifying all its inputs. When a system is stabilized, the output results are consistently repeatable. Stabilizing a system with multiple real-time events, such as input devices and time-dependent conditions, can be difficult to accomplish. It often involves replacing input hardware with sequential reads from an array or disk file.

stack Last-in-first-out data structure located in RAM and used to temporarily save information.

stack pointer (SP) A register in the processor that points to the RAM location of the stack.

start bit An overhead bit(s) specifying the beginning of the frame, used in serial communication to synchronize the receiver shift register with the transmitter clock. *See also* stop bit, even parity, and odd parity.

starvation A condition that occurs with a priority scheduler whereby low-priority threads are never run (*see* aging).

static allocation Data structures such as an FSM or TCB that are defined at assembly or compile time and exist throughout the life of the software. Contrast to dynamic allocation.

static RAM Volatile read/write storage built from three transistors having fast speed and not requiring refresh. Contrast with dynamic RAM.

stepper motor A motor that moves in discrete steps.

stop bit An overhead bit(s) specifying the end of the frame, used in serial communication to separate one frame from the next. *See also* start bit, even parity, and odd parity.

strain gage A transducer that converts displacement into electric resistance. It can also be used to measure force or pressure.

string A sequence of ASCII characters, usually terminated with a zero.

symbol table A mapping from a symbolic name to its corresponding 16-bit address, generated by the assembler in pass one and displayed in the listing file.

synchronous bus A communication protocol that has a central clock; there is no feedback from the memory to the processor, so every memory cycle takes exactly the same time; data transfers (put data on bus, take data off bus) are synchronized to the central clock.

synchronous protocol A system whereby the two devices share the same clock.

tachometer A sensor that measures the revolutions per second of a rotating shaft.

thermal noise A fundamental noise in resistive devices arising from the uncertainty about the position and velocity of individual molecules. Same as Johnson noise and white noise.

thermistor A nonlinear transducer that converts temperature into electric resistance.

thermocouple A transducer that converts temperature into electric voltage.

thread The execution of software that cooperates with other threads. A thread embodies the action of the software. One concept describes a thread as the sequence of operations including the input and output data.

thread control block (TCB) Information about each thread.

three-pole relay Three separate and complete relays, which are activated together. *See also* single pole.

three-pole switch Three separate and complete switches, which are activated together. *See also* single pole.

throughput The information transfer rate; the amount of data transferred per second. Same as bandwidth.

time constant The time to reach 63.2% of the final output after the input is instantaneously increased.

time profile and execution profile Time profile refers to the timing characteristic of a program, and execution profile refers to the execution pattern of a program.

timing signals The lines used to clock data onto or off of the bus; signals that specify when to activate during this cycle; the specific times for the rise and fall edges are synchronized to the E clock. Contrast to command signals.

tolerance The maximum deviation of a parameter from a specified value.

transducer A device that converts one type of signal into another type.

tristate The state of a tristate logic output when off or not driven.

tristate logic A digital logic device that has three output states: low, high, and off (HiZ).

truncation The act of discarding bits as a number is converted from one format to another.

two's complement A number system used to define signed integers. The MSB defines whether the number is negative (1) or positive (0). To negate a two's complement number, one first complements (flips from 0 to 1 or from 1 to 0) each bit, then add 1 to the number.

two-pole relay Two separate and complete relays, which are activated together; same as double-pole.

two-pole switch Two separate and complete switches, which are activated together; same as double-pole.

ultrasound A sound with a frequency too high to be heard by humans, typically 40 kHz to 100 MHz.

unbuffered I/O The hardware and software are tightly coupled so that both wait for each other during the transmission of data.

unipolar stepper motor A stepper motor in which the current flows in only one direction (on/off) along the interface wires; a stepper with five or six interface wires.

universal asynchronous receiver/transmitter (UART) A device to transmit data with asynchronous serial communication protocol, same as SCI and ACIA.

unsigned binary A mechanism to represent unsigned integers where all 0s represents the value 0, and all 1s represents the largest positive number.

vector An address at the end of memory containing the location of the interrupt service routines. *See also* reset vector and interrupt vector.

vertical parity The normal parity bit calculated on each individual frame; can be even or odd parity.

V_{OH} The smallest possible output voltage when the signal is high.

V_{OL} The largest possible output voltage when the signal is low.

volatile A condition whereby information is lost when power is removed.

voltage follower An analog circuit with gain equal to 1, large input impedance, and small output impedance. Same as follower.

vulnerable window Locations within a software module. If an interrupt were to occur at one of these locations, then an error could occur (e.g., data lost, corrupted data, program crash). Same as critical section.

white noise A fundamental noise in resistive devices arising from the uncertainty about the position and velocity of individual molecules. Same as Johnson noise and thermal noise.

wire-or-mode A digital logic output that has two states: low and HiZ. Same as open collector.

word Two bytes containing 16 bits. Same as double byte.

workstation A powerful general-purpose computer system having a price in the $10K to 50K range and used for handling large amounts of data and performing many calculations.

write cycle Data is sent from the processor to the memory or output device, the address bus specifies the memory or output device location, and the data bus contains the information.

write data available Time interval (start, end) during which the data will be valid during a write cycle, determined by the processor.

write data required Time interval (start, end) during which the data should be valid during a write cycle, determined by the memory module.

XIRQ A high-priority interrupt mechanism available on the 6811 and the 6812.

XON/XOFF A protocol used by printers to feedback the printer status to the computer. XOFF is sent from the printer to the computer in order to stop data transfer, and XON is sent from the printer to the computer in order to resume data transfer.

Z Transform A transform equation converting a digital time-domain sequence into the frequency domain. In both the time and frequency domain it is assumed the signal is band limited to 0 to $\frac{1}{2}$ fs.

zero A place in the frequency domain where the filter gain is zero.

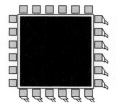

Appendix 2
Solutions to Checkpoints

Checkpoint 1.1: A microcomputer is a small computer that includes a processor memory and input/output. A microprocessor is a small processor that includes registers, ALU, a control unit, and a bus interface unit. A microcontroller is a single-chip microcomputer.

Checkpoint 1.2: An input port is hardware that is part of the computer and the channel through which information enters into the computer. An input interface includes hardware components external to the computer, the input port, and software, which together perform the input function.

Checkpoint 1.3: Typical input devices include the keys on the keyboard, the mouse and its buttons, joystick, CD reader, and microphone. The floppy disk can be used for input and output.

Checkpoint 1.4: Typical output devices include the LEDs on the keyboard, monitor, speaker, printer, CD burner, and speaker. The floppy disk can be used for input and output.

Checkpoint 1.5: CU (control unit) BIU (bus interface unit) and ALU (arithmetic logic unit) are all part of the processor. DMA stands for direct memory access, which is a high-speed mechanism to move data directly from input to memory, or move data directly from memory to output.

Checkpoint 1.6: An embedded system is a microcomputer with mechanical, chemical, and electrical devices attached to it, programmed for a specific dedicated purpose, and packaged as a complete system.

Checkpoint 1.7: The software in the alarm clock must maintain time using a real-time clock, output the current time on the display, respond to button pushes updating parameters as required, and check to see whether the current time matches the alarm time.

Checkpoint 1.8: A requirement is a detailed performance parameter that the system must satisfy, generally derived from the overall objective of the system. A constraint is a condition defining how the system will be developed, generally restricting the range of solutions from which the system will be built.

Checkpoint 1.9: If two modules output to the same port, then the second module will undo the function of the first one. For example, if one module says "go" and the other one says "stop," then the order of execution determines the resulting function. A similar error can occur for input ports.

Checkpoint 1.10: V_{IL} is 1.5 V, and V_{IH} is 3.5. Therefore, 0 and 1 V are considered low, and 4 and 5 V are considered high. 2 and 3 V are indeterminate.

Checkpoint 1.11: Yes, V_{OL}(0.5 V) is less than V_{IL}(0.8 V), V_{OH}(4.4 V) is greater than V_{IH}(2 V), I_{OH}(4 mA) is greater than I_{IH}(20 μA), and I_{OL}(4 mA) is greater than I_{IL}(0.4 mA).

Checkpoint 1.12: No, because $V_{OH}(2.4V)$ is less than $V_{IH}(3.5V)$. To fix this situation we can add a pull-up resistor on the output of the 74LS04 gate. In fact, the 6812 has a feature allowing you to enable internal pull-up resistors on its inputs (see the **PERT** register).

Checkpoint 1.13: 2½ decimal digits is 200 alternatives, which is about 8 bits.

Checkpoint 1.14: The rule of thumb says 2^{60} is about 10^{18}, which is 18 decimal digits. 2^4 is 16, which is about 1½ decimal digits. Together, we have 19½ decimal digits.

Checkpoint 1.15: First, break into nibbles %1110, %1110, %1011, then convert each, $EEB.

Checkpoint 1.16: First, convert each hex digit one at a time 0011 1000 0000 0000, then combine to get %0011100000000000.

Checkpoint 1.17: Each hex digit needs 4 bits, so a total of 20 bits will be required

Checkpoint 1.18: %01101010 equals $64 + 32 + 8 + 2 = 106$.

Checkpoint 1.19: $32 equals $3*16 + 2 = 50$.

Checkpoint 1.20: 35 equals $32 + 2 + 1 = $ %00100011=$23.

Checkpoint 1.21: 200 equals $128 + 64 + 8 = $ %11001000=$C8.

Checkpoint 1.22: They are the same, both equally 53.

Checkpoint 1.23: -35 equals $-128 + 64 + 16 + 8 + 4 + 1 = $ %11011101=$DD.

Checkpoint 1.24: The range of 8-bit signed numbers is -128 to $+127$.

Checkpoint 1.25: The character '0' is represented in ASCII as $30.

Checkpoint 1.26: Converting each character to ASCII yields "48656C6C6F20576F726C6400"

Checkpoint 1.27: π is about 3142, with a resolution of 0.001.

Checkpoint 1.28: π is about 804, with a resolution of 1/256.

Checkpoint 1.29: $y = (1000 \cdot x - 53 \cdot x_1 + 1000 \cdot x_2 + 51 \cdot y_1 - 903 \cdot y_2)/1000$.

Checkpoint 1.30: We set the I bit to 1 to disable interrupts.

Checkpoint 1.31: Registers D, X, Y, SP, and PC can hold 16-bit addresses.

Checkpoint 1.32: The addressing mode specifies where the instruction will read or write data.

Checkpoint 1.33: They have the same machine code and thus perform the same function when executed. The only difference is programming style, or how it looks to the programmer.

Checkpoint 1.34: The instruction ldaa #36 puts the number 36 into Register A, whereas the instruction ldaa 36 fetches the 8-bit data from memory location 36 and puts that data into Register A.

Checkpoint 1.35: The instruction ldx #$0801 puts the number $0801 into Register X, whereas the instruction ldx $0801 fetches the 16-bit data from memory locations $0801, $0802 and puts that data into Register X.

Checkpoint 1.36: The instruction ldaa $12 fetches the 8-bit data from memory location $12 and puts that data into Register A, whereas the instruction ldx $12 fetches the 16-bit data from memory locations $0012, $0013 and puts that data into Register X.

Checkpoint 1.37: The 9 means flash EEPROM, and the 64 means 64 K bytes.

Checkpoint 1.38: Nothing happens if the software writes to an input port.

Checkpoint 1.39: This is the way the 6811 output pins operate. If the software reads an output port, it will retrieve the value on the external pin. On the 6812, when the software reads from an output port, it will get the value that was previously written to it.

Checkpoint 1.40: Since there are 8 bits in a port and 8 bits in the direction register, each bit can be individually programmed as input or output.

Checkpoint 1.41: It will still operate according to specifications, but it may be more expensive to build or it may be harder to order components to build it.

Checkpoint 1.42: It will no longer operate according to specifications.

Checkpoint 1.43: Our eyes can process visual information only at frequencies less than about 10–15 Hz. Information occurring at rates faster than that cutoff is seen as an average value. If you want to see the LEDs with your eyes, you will have to slow it down, placing four delays, one after each output. The value of the delay should be greater than 0.1 sec.

Checkpoint 2.1: The assembler builds the symbol table.

Checkpoint 2.2: The assembler creates the object code and listing file.

Checkpoint 2.3: `leay 10,x`

Checkpoint 2.4: 6811 code

```
xgdx
addd #2000
xgdx
```

6812 code

```
leay 2000,y
```

Checkpoint 2.5: Since the values in Registers A and B range from 0 to 255, their product must range from 0 to 65025. Therefore, all potential results will fit in Register D.

Checkpoint 2.6: dividend = quotient*divisor + remainder.

Checkpoint 2.7:

```
ldaa N
ldab #10
mul       ;D=10*N
ldx  #51
idiv      ;X=(10*N)/51
xgdx
stab M
```

Checkpoint 2.8:

```
ldd  N     ;2.5 is about 65536/26214
ldx  #26214
fdiv       ;RegX=(65536*N)/26214
stx  M
```

Checkpoint 2.9:

```
ldaa PORTB
anda #$EF  ;clear bit 4
staa PORTB
```

Checkpoint 2.10:

```
ldaa PORTB
oraa #$08  ;set bit 3
staa PORTB
```

Checkpoint 2.11:

```
ldaa M
psha
ldaa N
staa M
pula
staa N
```

Checkpoint 2.12: It determines whether to use the **bls** or **ble** instruction. That is,

```;unsigned version ldaa N cmpa #25 bls next ; skip if N<=25 jsr isGreater ; N>25 next```	```;signed version ldaa N cmpa #25 ble next ; skip if N<=25 jsr isGreater ; N>25 next```

Checkpoint 2.13: The assembler must create the symbol table during pass 1, so it must know the size of each assembly line during pass 1. A forward reference would prevent the assembler from knowing how many bytes to allocate during pass 1. It would probably create a phasing error.

Checkpoint 2.14: Public functions have an underline (e.g., SCI_OutString). Private functions have no underline in the name.

Checkpoint 2.15: Local variables begin with a lower-case letter (e.g., myKey). Global variables begin with an upper case letter (e.g., TheKey).

Checkpoint 2.16: In a Moore FSM, the output depends only on the state. In a Mealy FSM, the output depends on the state and the input.

Checkpoint 2.17: The period of the TCNT timer will be the resolution of the delay function. The maximum delay for the assembly version is 32767 times the resolution, and the maximum delay for the C implementation will be 65,000 times the resolution.

Checkpoint 2.18: It will first sit up, then it will stand up.

Checkpoint 2.19: Define the variable within the scope of the function. For example,

```
void MyFunction(void){ short myLocalVariable;
}
```

Checkpoint 2.20: Define the variable outside the scope of the function. For example,

```
short MyGlobalVariable;  // accessible by all programs
void MyFunction(void){
}
```

Checkpoint 2.21: A static local has permanent allocation, which means it maintains its value from one call to the next. It is still local in scope, meaning it is accessible only from within the function. In the following example, count contains the number of times MyFunction is called:

```
void MyFunction(void){ static short count;
count++;
}
```

A static global reduces scope. Regular globals can be accessed from any function in the system, whereas a static global can be accessed only by functions within the same file. Static globals are private. Functions can be static also, meaning they can be called only from other functions in the file. For example,

```
static short myPrivateGlobalVariable;  // accessible by this
void static MyPrivateFunction(void){   // file only
}
```

Checkpoint 2.22: A const global is read-only. It is allocated in the ROM portion of memory. When const modifies a function parameter, it means that parameter can't be changed by the function. Function parameters are always allocated in registers or on the stack.

Checkpoint 2.23: You don't explicitly define the size of the stack. It can be implicitly calculated if the global variables are contiguously allocated starting at the beginning of RAM and the stack pointer is initialized to the end of RAM. Then, the stack size is defined as the total RAM bytes minus the size of the global variables. Some systems have a heap, which is used by `malloc` and `free` and also is defined in RAM.

Checkpoint 2.24: The local variable name can be reused in other subroutines, just as in C.

Checkpoint 2.25:

`; 6811 or 6812`
`  pshx`
`  psha    allocate 3 locals`
`; access locals`
`  ins`
`  ins`
`  ins     deallocate`
	`; 6812 only`
	`     leas -3,sp allocate 3 locals`
	
	`; access locals`
	`     leas 3,sp  deallocate locals`

Checkpoint 2.26:

```
psha          ;save registers
ldaa  PORTA
eora  #$08  ;bit 3
staa  PORTA
pula          ;restore
```

Checkpoint 3.1: PIOC is set to $01 (hand pattern with HNDS = 0, INVB = 1).

Checkpoint 3.2: EGA bit is set to 0 (PIOC bit 1).

Checkpoint 3.3: First we set the direction register bit (DDRC) to 1, then we set CWOM (PIOC bit 5) to 1.

Checkpoint 3.4: In C, we execute **`KWIFH=0x80;`** and in assembly, **`movb #$80,KWIFH`**.

Checkpoint 3.5: On the MC68HC812A4, any of the 24 pins on Ports D, H, or J could be used. On the MC9S12C32, any of the 10 pins on Ports J and P could be used.

Checkpoint 3.6: Port P outputs are working properly if PTP bits equal the PTIP bits.

Checkpoint 3.7: In C, we execute **`PIFJ=0x80;`** and in assembly, **`movb #$80,PIFJ`**.

Checkpoint 3.8: It clears all the bits in the flag register, because of the read-modify-write sequence.

Checkpoint 3.9: The bandwidth equals 9600/10 = 960 bytes/sec.

Checkpoint 3.10: 2,000,000/16 = 125,000 bits/sec.

Checkpoint 3.11: BAUD = $30. 2,000,000/(16∗13) = 9615 bits/sec.

Checkpoint 3.12: 4,000,000/(16∗13) = 19231, which is about 19200 bits/sec.

Checkpoint 3.13: SCIBD = **41.** 25,000,000/(16∗41) = 38110 bits/sec, which is about 38400 bits/sec.

Checkpoint 3.14: The `ldab` instruction will set the N bit with TDRE:

`SCI_OutChar`
`   ldab SCSR    output ready?`
`   bpl  SCI_OutChar wait for TDRE`
`   staa SCDR    write ASCII`
`   rts`
	`SCI_OutChar`
	`   ldab SCISR1  ; output ready?`
	`   bpl  SCI_OutChar ; wait`
	`   staa SCIDRL  ; write ASCII`
	`   rts`

Checkpoint 3.15: Read status register with RDRF set, followed by read SCI data register.

Checkpoint 3.16: Read status register with TDRE set, followed by write SCI data register.

Checkpoint 3.17: Each transmitted frame will be received as two frames, both wrong and the first one might have a framing error.

Checkpoint 3.18: Each transmitted frame will be received as one frame, which will be wrong and might set the noise flag. In fact two transmitted frames might be received as one frame, which will be wrong.

Checkpoint 4.1: I = 0 enables interrupts.

Checkpoint 4.2: If the average rate at which data input into the FIFO is less than the average rate at which data is removed, then increasing the FIFO size will solve a full error. If the average rate at which data input into the FIFO is more than the average rate at which data is removed, the FIFO will become full regardless of its size.

Checkpoint 4.3: The bandwidth of the input is slow compared to the speed of the computer, so it is I/O bound.

Checkpoint 4.4: The bandwidth of the output is fast compared to the speed of the computer, so it is CPU bound.

Checkpoint 4.5: First CCR is pushed (with I = 0), then I = 1.

Checkpoint 4.6: What makes XIRQ interrupts such a high priority is that once enabled (X = 0), they cannot be disabled. This means an XIRQ will interrupt anywhere, even inside an IRQ interrupt service routine. On the other hand, an IRQ interrupt cannot interrupt an XIRQ interrupt service routine.

Checkpoint 4.7: In a typical system, the largest component to latency is the time running with I = 1, such as while inside other ISRs.

Checkpoint 4.8: `PACTL=2;`   `// 61.035Hz`

Checkpoint 4.9: `#define PERIOD 10000`

Checkpoint 4.10: `RTICTL=0x85; // 61.035Hz`

Checkpoint 4.11: `#define PERIOD 10000`

Checkpoint 4.12: `RTICTL=0x71;` or `RTICTL=0x63;` or `RTICTL=0x47;` or `RTICTL=0x4F;`

Checkpoint 4.13: Change `TSCR2=0x07;` to make a 16 μsec TCNT and `#define PERIOD 62500`

Checkpoint 5.1: A program is a list of commands, whereas a thread is the action caused by the execution of software. For example, there might be one copy of a program that searches the card catalog of a library, while separate threads are created for each user that logs into a terminal to perform a search. Similarly, there might be one set of programs that implement the features of a window (open, minimize, maximize, etc.), while there will be a separate thread for each window created.

Checkpoint 5.2: Threads can't communicate with each other using the stack, because they have logically separate stacks. Global variables will be used, because one thread may write to the global, and another can read from it.

Checkpoint 5.3: It is efficient not to schedule a thread when it cannot continue executing at this point (it needs something not currently available). In this way, we schedule only threads that can produce work.

Checkpoint 5.4: A blocked thread is placed back in the ready queue when the output display becomes available.

Checkpoint 5.5: The `RunPt` points to the TCB of the tread currently being executed.

Checkpoint 5.6: The `sts` instruction suspends the current thread, and the `lds` instruction will activate the next thread.

Checkpoint 5.7: It will be suspended (its time slice is over), and the next thread will run. It will essentially sleep until its next turn in the queue. It is a simple way to implement cooperative multitasking.

Checkpoint 5.8: If the semaphore is zero, it will crash because the semaphore will never be incremented (signal cannot be called).

Checkpoint 5.9: There is a read-modify-write critical section. One thread could call wait and read the semaphore. At this point there are two copies of the semaphore, one in memory and the other copy in a register of the first thread. If a thread switch interrupt occurs, then a second thread calls wait, decrements the semaphore, and continues. When the first thread is run again, it does not refetch the semaphore, from memory; rather, it will use the copy in the register. Both threads called wait, but the memory copy of the semaphore got decremented only once.

Checkpoint 5.10: On average, over a long period of time, the number of calls to wait equals the number of calls to signal.

Checkpoint 5.11: In Program 5.17, the stack pointer is the first entry of the TCB, whereas in the previous examples it was the second entry.

Checkpoint 5.12: Since each instruction runs six times faster, the time-jitter will be reduced by 1/6.

Checkpoint 6.1: An input capture event occurs on the selected edge on an input pin.

Checkpoint 6.2: TCNT is copied into the input capture latch, the flag is set, and, if armed, an interrupt is requested.

Checkpoint 6.3: PA2 PA1 PA0 are input only, PA4 PA5 PA6 are output only, and PA7 PA3 are bidirectional.

Checkpoint 6.4: PA0 is input capture channel 3. For the rising edge, make EDG3B, EDG3A equal to 0,1. To arm the channel, set IC3I = 1.

Checkpoint 6.5: PA3 is input capture channel 4. You have to clear the DDRA3 bit and set the I4/O5 bit. For the falling edge, make EDG4B, EDG4A equal to 1, 0. To arm the channel, set IC4I = 1.

Checkpoint 6.6:

`ldaa #$02` `staa TFLG1`	`TFLG1=0x02;`

Checkpoint 6.7: Clear bit 0 of TIOS to specify input capture on channel 0. Clear bit 0 of DDRT to make it an input. Set TEN to enable the timer. For the rising edge, make EDG0B, EDG0A equal to 0, 1. To arm the channel, set C0I = 1.

Checkpoint 6.8: Clear bit 3 of TIOS to specify input capture on channel 3. Clear bit 3 of DDRT to make it an input. Set TEN to enable the timer. For the falling edge, make EDG3B, EDG3A equal to 1, 0. To arm the channel, set C3I = 1.

Checkpoint 6.9:

`movb #$40,TFLG1`	`TFLG1=0x40;`

Checkpoint 6.10: Change the TCNT period to 2 µs.

```
TMSK2=0x01;    // MC68HC711E9 with 2MHz E clock
TSCR2=0x03;    // MC9S12C32 with 4MHz E clock
```

Checkpoint 6.11: Resolution*65535 equals the maximum pulse-width.

Checkpoint 6.12: An output compare event occurs when the TCNT value equals the value in the output compare register.

Checkpoint 6.13: The output can change (set high, clear low, or toggle), the flag is set, and, if armed, it will request an interrupt.

Checkpoint 6.14:

`ldaa #$10` `staa TFLG1`	`TFLG1=0x10;`

Checkpoint 6.15:

movb #$10,TFLG1	TFLG1=0x10;

Checkpoint 6.16: This is a read-modify-write operation. It first reads the register, and any flag that is set will return a 1 in its bit location, the logical or will set bit 5, then when the result is written back, all flags will be cleared. Writing ones to this register clears the flags.

Checkpoint 6.17: Set bit 0 of TIOS to specify output compare on channel 0. Set bit 0 of DDRT to make it an output. Set TEN to enable the timer. For the toggle output, make OM0 = 0 and OL0 = 1. To arm the channel, set C0I = 1.

Checkpoint 6.18: In the existing program, High + Low equals 10000. You can change it so High + Low equals 40000.

Checkpoint 6.19: 1 Hz. If the frequency changes from 100 to 101 Hz, the system can detect the change.

Checkpoint 6.20: The existing system counts for 1 second. In general, the frequency resolution is 1 divided by the wait time. To make a 1 kHz resolution, count pulses for 1ms.

Checkpoint 6.21: On the 6811, the resolution is 32 µs, and on the 6812 it is 16 µs. If the period of the input wave changes by more than a resolution, system can detect the change.

Checkpoint 6.22: On the 6811, the result will be 1234.5/32 = 38. On the 6812, the result will be 1234.5/16 = 77.

Checkpoint 6.23: (A = E/8, SA = A/20) or (A = E/4, SA = A/40) or (A = E/2, SA = A/80) or (A = E, SA = A/160).

Checkpoint 6.24: Change `PWMSCLA=5;` to `PWMSCLA=50;`

Checkpoint 6.25: The precision would be reduced from 16 bits (62501 alternatives) to 10 bits (1001 alternatives).

Checkpoint 6.26: Yes, one uses Clock A and the other uses Clock B.

Checkpoint 7.1: Because every RS232 protocol has one start bit, there are 10 bits in this frame.

Checkpoint 7.2: Full-duplex allows communication in both directions simultaneously, whereas half-duplex allows communication in both directions but only one direction at a time.

Checkpoint 7.3: With synchronous serial (SPI), the transmitter and receiver operate on the same clock, which is included in the cable. With asynchronous serial (SCI), the transmitter and receiver operate with clocks of similar frequencies, and the receiver uses transitions in the data to synchronize with the transmitter.

Checkpoint 7.4: If the baud rates are more than 5% different, there is a good chance the receiver will get framing and noise errors.

Checkpoint 7.5: Read status register with RDRF set, followed by read SCI data register.

Checkpoint 7.6: Read status register with TDRE set, followed by write SCI data register.

Checkpoint 7.7: Because the TxFifo is empty, meaning there is no data to output.

Checkpoint 7.8: Because it is in the ISR, running with interrupts disabled, meaning no other software could run that might empty the RxFifo.

Checkpoint 7.9:

```
BAUD = 0x33;    // 6811 1200 bits/sec
SCIBD = 208;    // 6812 1200 bits/sec
```

Checkpoint 7.10: Read status register with SPIF set, followed by read SPI data register.

Checkpoint 7.11: Connect SS-SS, SCLK-SCLK, MOSI-MOSI and MISO-MISO. Configure one as a master and the other as a slave. It doesn't matter which mode you use for CPOL CPHA as long as they are the same. Similarly, it doesn't matter which baud rate is used, but there is no need not to use the fastest available rate. First the slave writes; then, when the master writes, the two SPDR registers are exchanged.

Checkpoint 8.1:

```
DDRM = 0x00;   // PM5-0 inputs
PPSM = 0x00;   // select pull-up
PERM = 0x3F;   // enable pull-up
```

Checkpoint 8.2: The solution will be a sinusoid with frequency $\sqrt{(K/m)}/(2\pi)$.

Checkpoint 8.3: No, the bounce time is a function of the mass, friction, and spring constant of the switch.

Checkpoint 8.4: Because the Ctrl, Alt, and Shift keys are simultaneously pressed, they need to be interfaced directly. The remaining 100 keys can easily be interfaced with a 10 by 10 scanned matrix.

Checkpoint 8.5: Because you touch just one point at a time, we can use a multiplexed/demultiplexed scanned matrix. A 10 to 1024 multiplexer will drive the rows, and a 1024 to 10 demultiplexer will sense the columns.

Checkpoint 8.6: $R = (4.5 - 2)/0.001 = 2500 \ \Omega$.

Checkpoint 8.7: (1) make the data pins input, (2) set R/W to 1 and RS to 0, (3) set E to 1, (4) Read the data pins (bit 7 will be busy flag), (5) set E to 0, and (6) make the data pins output again.

Checkpoint 8.8: No changes are necessary; just connect $+12$ V to the Vcc input of the TPIC0107B.

Checkpoint 8.9: There are three approaches to increasing torque. You can buy a larger stepper with more torque. You can increase the torque by using a reducing gear box. Your software may be exceeding the maximum jerk, so you can adjust the way you change speeds, as shown in Table 8.15.

Checkpoint 8.10: Feedback is important to know the exact position when the system starts, or whether there is a chance that a step command does not actually move the motor.

Checkpoint 9.1: 001X,XXXX,XXXX,XXXX. Setting all Xs to 0 gives $2000; setting all Xs to 1 gives $3FFF.

Checkpoint 9.2: $D800 to $DFFF.

Checkpoint 9.3: $([2,3] + [4,5],[6,7] + 8) = ([6,8],[14,15])$

Checkpoint 9.4: Change to

```
SYNR = 2;  REFDV = 0;    // PLLCLK=2*OSCCLK*(SYNR+1)/(REFDV+1)
```

Checkpoint 9.5:

$$t_{cyc} = t_1 = 100 \text{ ns}$$
$$RDR = \text{Read Data Required} = (t_1 - 15, t_1) = (85, 100)$$
$$WDA = \text{Write Data Available} = (\tfrac{1}{2}t_{cyc} + 7, t_1 + 2) = (57, 102)$$
$$AA \text{ (narrow)} = (\tfrac{1}{2}t_{cyc} + [2,6.5], t_1 + 2) = ([52,56.5], 102)$$
$$AA \text{ (wide)} = (\tfrac{1}{2}t_{cyc} + [2,6.5], t_1 + \tfrac{1}{2}t_{cyc}) = ([52,56.5], 150)$$

Checkpoint 9.6:

$$t_{cyc} = 40 \text{ ns and } t_1 = 120 \text{ ns}$$
$$RDR = \text{Read Data Required} = (t_1 - 15, t_1) = (105, 120)$$
$$WDA = \text{Write Data Available} = (\tfrac{1}{2}t_{cyc} + 7, t_1 + 2) = (27, 122)$$
$$AA \text{ (narrow)} = (\tfrac{1}{2}t_{cyc} + [2,6.5], t_1 + 2) = ([22,26.5], 122)$$
$$AA \text{ (wide)} = (\tfrac{1}{2}t_{cyc} + [2,6.5], t_1 + \tfrac{1}{2}t_{cyc}) = ([22,26.5], 140)$$

Checkpoint 9.7:

$$t_1 = 125 \text{ ns}, 60 + t_{acc} \leq t_1 - 30, \text{ so } t_{acc} \leq 35 \text{ ns}$$

Checkpoint 9.8:

$$\downarrow E = 250 \text{ ns}, 140.5 + t_{acc} \leq 250 - 15, \text{ so } t_{acc} \leq 94.5 \text{ ns}$$

Checkpoint 9.9: The end of WDA occurs at $\downarrow E + 2$. Let t_p be the propagation delay through external digital logic. To synchronize E1 or W, it would require external logic, so the end of WDR occurs at $\downarrow E + t_p$. If $t_p > 2$ ns, then WDA would not overlap WDR.

Checkpoint 9.10:

$$\tfrac{1}{2}t_{cyc} = 62.5 \text{ ns}, \downarrow E = t_1 = (n + 1)*125, \text{ timing is limited by read cycle } \overline{EI} \text{ access}$$
$$\text{AA (narrow)} = (\tfrac{1}{2}t_{cyc} + [2,6.5], t_1 + 2) = ([64.5,69], t_1 + 2)$$
$$\downarrow \overline{EI} + [64.5,69] + [1.5,9] = [66,78]$$
$$\text{RDR} = \text{Read Data Required} = (\downarrow E - 15, \downarrow E)$$
$$78 + 150 \leq \downarrow E - 15, \text{ so } 243 \leq \downarrow E, \text{ so 1 stretch will be OK}$$

Checkpoint 10.1: Latency is the response time (time delay from device ready to device gets service), while bandwidth is the amount of information transferred per time.

Checkpoint 10.2: If we do not meet the latency requirement, that data is lost. If it happens every time, the system doesn't work. If it happens occasionally, it will run slow because we will have to wait for the disk to spin around one revolution and try it again.

Checkpoint 10.3: A portion of the sound is lost, and it will sound like a skip. We may also hear a click because the waveform is discontinuous.

Checkpoint 10.4: The system runs slow, because the transmitter will time out and try to resend the packets.

Checkpoint 10.5: The bidirectional driver has three possibilities, determined by two control pins. An example of this type of logic is the 74HC245. It can drive data left to right, making the left input and right output. It can drive data right to left, making the right input and left output. The third possibility is that the device can be off, driving neither the left nor the right. This is a noninverting driver, so the output equals the input.

Checkpoint 10.6: Substitute the four bidirectional data bus drivers with four unidirectional tristate drivers. All four data bus drivers operate in the direction of the simplex transfer (left to right). The bank-switched memory looks like a write-only memory to the computer and a read-only memory to the I/O hardware. An example of this simplex channel can be seen in Figure 10.18.

Checkpoint 10.7: The maximum latency for cycle steal DMA is one bus cycle, assume there is only one DMA channel active. If there is more than one DMA channel operating, one DMA request may have to wait for another.

Checkpoint 10.8: 842 ns is not enough time to output one 4-bit packet to the display. If you tried to control it from software it will be much slower and the LCD screen will flicker.

Checkpoint 10.9: 10 Hz is a particularly disturbing frequency for humans. It is too fast to see the individual on/off cycles, but too slow to be seen as continuous.

Checkpoint 11.1: The open loop gain is 108 dB. In V/V, the gain is $10^{108/20} = 10^{5.4} = 251{,}000$.

Checkpoint 11.2: The output impedance is open-circuit voltage divided by short-circuit current, 5V/30 mA, which is 167 Ω.

Checkpoint 11.3: The offset voltage is 0.5 mV. The error will be 0.5 mV times 100, which will be 50 mV.

Checkpoint 11.4: The offset voltage is 3 mV. The error is 3 mV times 1000, which would be 3 V.

Checkpoint 11.5: The gain-bandwidth product of the MAX494 is 500 kHz. The bandwidth of the circuit will be 500 kHz divided by the gain, which will be 5 kHz. The noise density of the MAX494 is 25 nV/√Hz. The noise will be 25 nV*sqrt(5000), which is about 1.8 μV.

Checkpoint 11.6: 10 kΩ in parallel with 100 kΩ is about 9.1 kΩ.

Checkpoint 11.7: The input impedance of the op amp determines the input impedance of a noninverting amplifier, which for the OPA227 is 10 MΩ.

Checkpoint 11.8: Use rail-to-rail op amps, powered by +5 V and ground. Replace all connections to analog ground to a 2.50 V reference.

Checkpoint 11.9: Use an AD620 with R_G of 4.94kΩ.

Checkpoint 11.10: j is the square root of -1, a complex number.

Checkpoint 11.11: 255*1 V/5 V is 51.

Checkpoint 11.12: 1023*1 V/5 V is 205.

Checkpoint 12.1: If precision is given in alternatives, then range = precision*resolution.

Checkpoint 12.2: Noise in the transducer, noise in the electronics, and the ADC resolution.

Checkpoint 12.3: A higher sensitivity means a lower amplifier gain, resulting in less noise at the output.

Checkpoint 12.4: The stability (low drift) and reproducibility of a transducer are major factors affecting accuracy. Accuracy combines resolution and calibration. If you calibrate a device, then the calibration parameters change and the instrument will measure incorrectly.

Checkpoint 12.5: The are 8 slots in the disk, and 60 seconds in a minute. 7.5 = 60/8. The velocity resolution in RPM will be 7.5 times the frequency resolution of the input capture system in Hz.

Checkpoint 12.6: Pressure is force per area. So a force transducer can be directly used to measure pressure if we know the area over which the force is measured.

Checkpoint 12.7: Detection system will have thresholds. Adjusting these thresholds will trade FPs for FNs.

Checkpoint 12.8: 10/0.01 is 1000, so 10 bits are needed.

Checkpoint 12.9: The resolution would decrease by a factor of 4 from about 0.1°C to 0.025°C.

Checkpoint 12.10: At 60 BPM, the **Rcount** will be 4 and the difference **Rlast-Rfirst** will be 480. If the heart rate increases a little, then the difference will reduce to 479, giving a measurement of 60.13. Thus, the inherent measurement resolution is about 0.12 BPM. Because integer math is used, the system will have a resolution of 1 BPM.

Checkpoint 12.11: The resolution would decrease by a factor of 4 from about 0.1°C to 0.025°C.

Checkpoint 13.1: Assume the yaw motor position is fixed. The location of the pitch motor is fixed relative to the yaw motor and has no translational motion. The location of the roll motor is affected by length of the pitch shaft and rotating the yaw motor. The translational motion on the tip caused by rotating the pitch motor depends on the length of the roll shaft, the pitch angle, and the yaw angle.

Checkpoint 13.2: The greatest common factor of 75 ms and 500 ms is 25 ms. A single ISR interrupting every 25 ms could have been used to solve this system. The **YawTime PitchTime RollTime** parameters would be multiplied by 3, and the **Duration** values multiplied by 20.

Checkpoint 13.3: There are no critical sections, because the global is accessed within an ISR. The ISR is atomic, because interrupts are disabled.

Checkpoint 13.4: If they are too close, then the system can turn on-off-on-off- . . . very quickly, causing the electromagnetic relays to prematurely fail.

Checkpoint 13.5: If they are too far apart, then the system will oscillate with large positive and negative errors.

Checkpoint 13.6: At every interrupt, the actuator would be increased or decreased, causing a lot of output changes.

Checkpoint 13.7: If it were too fast, the actuator would be increased to maximum or decreased to minimum, causing it to behave like a bang-bang controller.

Checkpoint 13.8: The output will saturate. The error increases to a very large positive value or decreases down to a very large negative value.

Checkpoint 13.9: The limit of the discrete integral as Δt goes to zero is the continuous integral.

Checkpoint 13.10: The limit of the discrete derivative as Δt goes to zero is the continuous derivative.

Checkpoint 13.11: Yes. Let **watts** be the units of the actuator output and **cm** be the units of the sensor input. The units of the lag **L** is **sec.** The units of the rate **R** is **cm/sec.** The units of ΔU is **watts.**

Proportional	$\mathbf{K_P} = 1.2\ \Delta U/(L*R)$	**watts/(sec*(cm/sec)) = watts/cm**
Integral	$\mathbf{K_I} = 0.5\ \mathbf{K_P}/\mathbf{L}$	**watts/(cm-sec)**
Derivative	$\mathbf{K_D} = 0.5\ \mathbf{K_P}\ \mathbf{L}$	**(watts-sec)/cm**

Checkpoint 13.12: $E = X^* - X$, so the error is very negative, causing the P term to be very negative, making $U = 100$. This removes power, and gravity will force it down.

Checkpoint 13.13:
```
SlowDown=WayTooFast+SpeedingUp*LittleBitFast=50+(40*60)=50
```

Checkpoint 14.1: With open collector outputs, the low will dominate over HiZ. The signal will be low.

Checkpoint 14.2: On average, it will take N/2 transmissions for the message to go from one computer to another. There are 10 bits/frame, so there are 10,000 bytes/sec. Because there are 10 bytes/message, it takes 1ms to transmit a message. Because it has to be sent 5 times, it takes 5ms on average.

Checkpoint 14.3: The frame sent by a transmitter is echoed to its own receiver. If the data does not match, or if there are any framing or noise errors, then a collision occurred.

Checkpoint 14.4: Parity could be used to detect collisions. Also the message could have checksum added. Framing or noise errors can also indicate a collision.

Checkpoint 14.5: Open collector has two output states: low and HiZ. The low state will dominate over HiZ. HiZ is the recessive state.

Checkpoint 14.6: $z + 2$ is equal to $y + 1$. So, make **y** equal to $z + 1$.

Checkpoint 14.7: The rise time of the open collector signal will follow an R-C relationship. As the number of devices increases so does the capacitance. It will rise faster with a smaller R.

Checkpoint 14.8: If no device sends an acknowledgement, the SDA signal will float high, generating a negative acknowledgement.

Checkpoint 14.9: If they both send the same address and the same sequence of data bits, both will finish without getting a lost-arbitration error. If they both send the same address but different data values, an arbitration will occur during the data transfer, and the master with the smaller data value will win arbitration.

Checkpoint 14.10: t_E/t_{bit} is $24000/100 = 240$. The only solution is to make IBFD equal to $1F, making MUL = 1, scl2start = 6, scl2stop = 9, scl2tap = 6, tap2tap = 8, SCLTap = 15

$$MUL \cdot \{2 \cdot (scl2tap + [(SCLTap - 1) \cdot tap2tap] + 2)\}$$
$$= 1 \cdot \{2 \cdot (6 + [(15 - 1) \cdot 8] + 2)\} = 240$$

Checkpoint 15.1: $4{,}000{,}000/1000$ is 4000.

Checkpoint 15.2:

For DC, $z = 1$

$$\mathbf{H(z)} = \frac{1 + z^{-1}}{2} = 1$$

For 250, $z = j$

$$\mathbf{H(z)} = \frac{1-j}{2} = 0.707$$

For 500, $z = -1$

$$\mathbf{H(z)} = \frac{1-1}{2} = 0$$

Checkpoint 15.3: Using Equation 22, at 60 Hz, $\mathbf{f/f_s}$ is 1/6.

$$\mathbf{Gain} = 1/6\sqrt{[\{1 + \cos(2\pi/6) + \cos(4\pi/6) + \cos(6\pi/6) + \cos(8\pi/6) + \cos(10\pi/6)\}^2}$$
$$+ \{\sin(2\pi/6) + \sin(4\pi/6) + \sin(6\pi/6) + \sin(8\pi/6) + \sin(10\pi/6)\}^2]$$

The gain is zero using the fact that $\mathbf{\cos(x + \pi)} = -\mathbf{\cos(x)}$ and $\mathbf{\sin(x + \pi)} = -\mathbf{\sin(x)}$.

Checkpoint 15.4: Using Equation 27, at 60 Hz, f/f_s is 1/6.

$$\mathbf{Gain} = \mathbf{0.5} \ \sqrt{\{1 + \cos(6\pi/6)\}^2 + \{\sin(6\pi/6)\}^2} = \sqrt{\{1 - 1\}^2 + 0^2} = 0$$

Checkpoint 15.5: If the gain is larger than one, amplification occurs. For example if the gain is 1.2, if you put in a sinusoidal wave with amplitude 100, then the output of the filter will be a sinusoidal wave with amplitude 120. This is important because a filtered signal from an 8-bit ADC will not fit into an 8-bit variable.

Checkpoint 15.6: In this case it is unsigned short (16-bit), so the numerator must be less than 65535.

Checkpoint 15.7: In this case the math is signed 32-bit, so the numerator must be less than 2147483647.

Checkpoint 15.8: Some of the minor causes of time-jitter are the variability in which instruction is being executed when the interrupt request is issued, and the variability in software execution paths caused by conditional branching. The largest source of time-jitter occurs when the software runs with interrupts disabled, which can occur either while executing other ISRs or during the foreground when the software temporarily disables interrupts to perform tasks in a critical section.

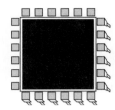

Index

Pin assignments for the RS232 EIA-574 and EIA-561 protocols

DB25 Pin	RS232 Name	DB9 Pin	EIA-574 Name	RJ45 Pin	EIA-561 Name	Signal	Description	True (V)	DTE	DCE
1						FG	Frame Ground/Shield			
2	BA	3	103	6	103	TxD	Transmit Data	−12	out	in
3	BB	2	104	5	104	RxD	Receive Data	−12	in	out
4	CA	7	105/133	8	105/133	TSR	Request to Send	+12	out	in
5	CB	8	106	7	106	CTS	Clear to Send	+12	in	out
6	CC	6	107			DSR	Data Set Ready	+12	in	out
7	AB	5	102	4	102	SG	Signal Ground			
8	CF	1	109	2	109	DCD	Data Carrier Detect	+12	in	out
9							Positive Test Voltage			
10							Negative Test Voltage			
11							Not Assigned			
12						sDCD	Secondary DCD	+12	in	out
13						sCTS	Secondary CTS	+12	in	out
14						sTxD	Secondary TxD	−12	out	in
15	DB					TxC	Transmit Clk (DCE)		in	out
16						sRxD	Secondary RxD	−12	in	out
17	DD					RxC	Receive Clock		in	out
18	LL						Local Loopback			
19						sRTS	Secondary RTS	+12	out	in
20	CD	4	108	3	108	DTR	Data Terminal Rdy	+12	out	in
21	RL					SQ	Signal Quality	+12	in	out
22	CE	9	125	1	125	RI	Ring Indicator	+12	in	out
23						SEL	Speed Selector DTE		in	out
24	DA					TCK	Speed Selector DCE		out	in
25	TM					TM	Test Mode	+12	in	out

General specification of various types of capacitor components[2, 3]

Type	Range	Tolerance	Temperature coef	Leakage	Frequencies
Polystyrene	10 pF to 2.7 μF	±0.5	Excellent	10 GΩ	0 to 10^{10} Hz
Polypropylene	100 pF to 50 μF	Excellent	Good	Excellent	
Teflon	1000 pF to 2 μF	Excellent	Best	Best	
Mica	1 pF to 0.1 μp	±1 to ±20%		1000 MΩ	10^3 to 10^{10} Hz
Ceramic	1 pF to 0.01 μF	±5 to ±20%	Poor	1000 MΩ	10^3 to 10^{10} Hz
Paper (oil-soaked)	1000 pF to 50 μF	±10 to ±20%		100 MΩ	100 to 10^8 Hz
Mylar (polyester)	5000 pF to 10 μF	±20%	Poor	10 GΩ	10^3 to 10^{10} Hz
Tantalum	0.1 μF to 220 μF	±10%	Poor		
Electrolytic	0.47 μF to 0.01 F	±20%	Ghastly	1 MΩ	10 to 10^4 Hz

[2] Modified from Wolf and Smith, *Student Reference Manual,* Prentice Hall, pg. 302, 1990.

[3] From Horowitz and Hill, *The Art of Electronics,* Cambridge University Press, pg. 22, 1989.